Production Optimization
Optimization

Using NODAL™ Analysis

H. Dale Beggs

OGCI and Petroskills Publications
Tulsa, Oklahoma

Production
Optimization
Using NODAL™ Analysis

Contents

3   Flow in Pipes and Restrictions   57

4   Total System Analysis   133

5   Artificial Lift Design   155

Nomenclature   187

Appendix A   191

Appendix B   197

1

Production Systems Analysis

INTRODUCTION

Any production well is drilled and completed to move the oil or gas from its original location in the reservoir to the stock tank or sales line. Movement or transport of these fluids requires energy to overcome friction losses in the system and to lift the products to the surface. The fluids must travel through the reservoir and the piping system and ultimately flow into a separator for gas-liquid separation. The production system can be relatively simple or can include many components in which energy or pressure losses occur. For example, a diagram of a complex production system, which illustrates a number of the components in which pressure losses occur, is shown in Fig. 1-1.

The pressure drop in the total system at any time will be the initial fluid pressure minus the final fluid pressure, $\bar{p}_R - p_{sep}$. This pressure drop is the sum of the pressure drops occurring in all of the components of the system. Since the pressure drop through any component varies with producing rate, the producing rate will be controlled by the components selected. The selection and sizing of the individual components is very important, but because of the interaction among the components, a change in the pressure drop in one may change the pressure drop behavior in all the others. This occurs because the flowing fluid is compressible, and, therefore, the pressure drop in a particular component depends not only on the flow rate through the component, but also on the average pressure that exists in the component.

The final design of a production system cannot be separated into reservoir performance and piping system performance and handled independently. The amount of oil and gas flowing into the well from the reservoir depends on the pressure drop in the piping system, and the pressure drop in the piping system depends on the amount of fluid flowing through it. Therefore, the entire production system must be analyzed as a unit.

The production rate or deliverability of a well can often be severely restricted by the performance of only one component in the system. If the effect of each component on the total system performance can be isolated, the system performance can be optimized in the most economical way. Past experience has shown that large amounts of money have been wasted on stimulating the formation when the well's producing capacity was actually being restricted because the tubing or flowline was too small. Another example of errors in completion design is to install tubing that is too large. This often happens on wells that are expected to produce at high rates. It will be shown that this practice not only wastes money on oversized equipment, but tubing that is too large can actually reduce the rate at which a well will flow. This can cause the well to load up with liquids and die, which necessitates the early installation of artificial lift equipment or compression.

A method for analyzing a well, which will allow determination of the producing capacity for any combination of components, is described in the following section. This method may be used to determine locations of excessive flow resistance or pressure drop in any part of the system. The effect of changing any component on the total well performance can be easily determined.

1

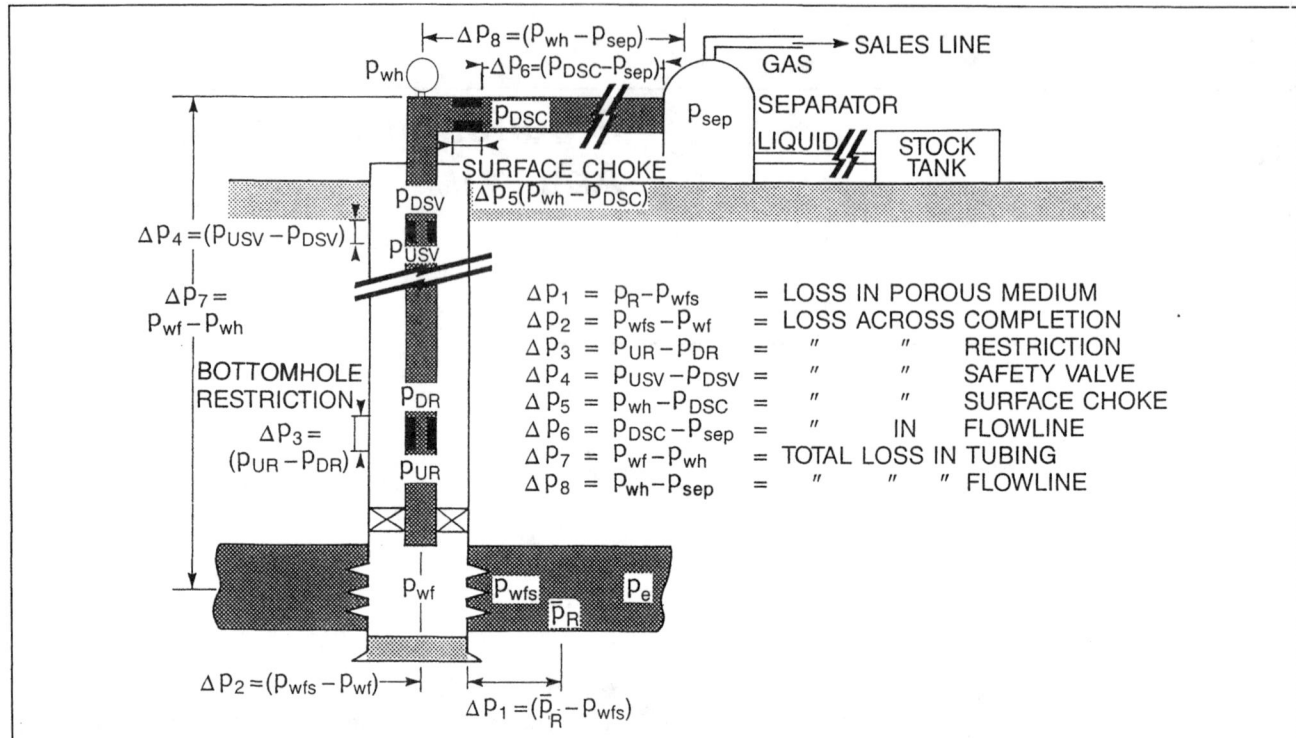

Fig. 1-1. Possible pressure losses in complete system.

SYSTEMS ANALYSIS APPROACH

The systems analysis approach, often called NODAL™ Analysis,* has been applied for many years to analyze the performance of systems composed of interacting components. Electrical circuits, complex pipeline networks and centrifugal pumping systems are all analyzed using this method. Its application to well producing systems was first proposed by Gilbert[1] in 1954 and discussed by Nind[2] in 1964 and Brown[3] in 1978.

The procedure consists of selecting a division point or node in the well and dividing the system at this point. The locations of the most commonly used nodes are shown in Fig. 1-2.

All of the components upstream of the node comprise the inflow section, while the outflow section consists of all of the components downstream of the node. A relationship between flow rate and pressure drop must be available for each component in the system. The flow rate through the system can be determined once the following requirements are satisfied:

1. Flow into the node equals flow out of the node.
2. Only one pressure can exist at a node.

At a particular time in the life of the well, there are always two pressures that remain fixed and are not functions of flow rate. One of these pressures is the average reservoir pressure $\bar{p}_R$, and the other is the system outlet pressure. The outlet pressure is usually the separator pressure p_{sep}, but if the well is controlled by a surface choke the fixed outlet pressure may be the wellhead pressure p_{wh}.

Once the node is selected, the node pressure is calculated from both directions starting at the fixed pressures.

Inflow to the node:

$$\bar{p}_R - \Delta p \text{ (upstream components)} = p_{node}$$

Outflow from the node:

$$p_{sep} + \Delta p \text{ (downstream components)} = p_{node}$$

The pressure drop, Δp, in any component varies with flow rate, q. Therefore, a plot of node pressure versus flow rate will produce two curves, the intersection of which will give the conditions satisfying requirements 1 and 2, given previously. The procedure is illustrated graphically in Fig. 1-3.

The effect of a change in any of the components can be analyzed by recalculating the node pressure versus flow rate using the new characteristics of the component

---

*"NODAL Analysis" is a trademark of Flopetrol Johnston, a division of Schlumberger Technology Corporation, and is protected by U.S. Patent #4,442,710.

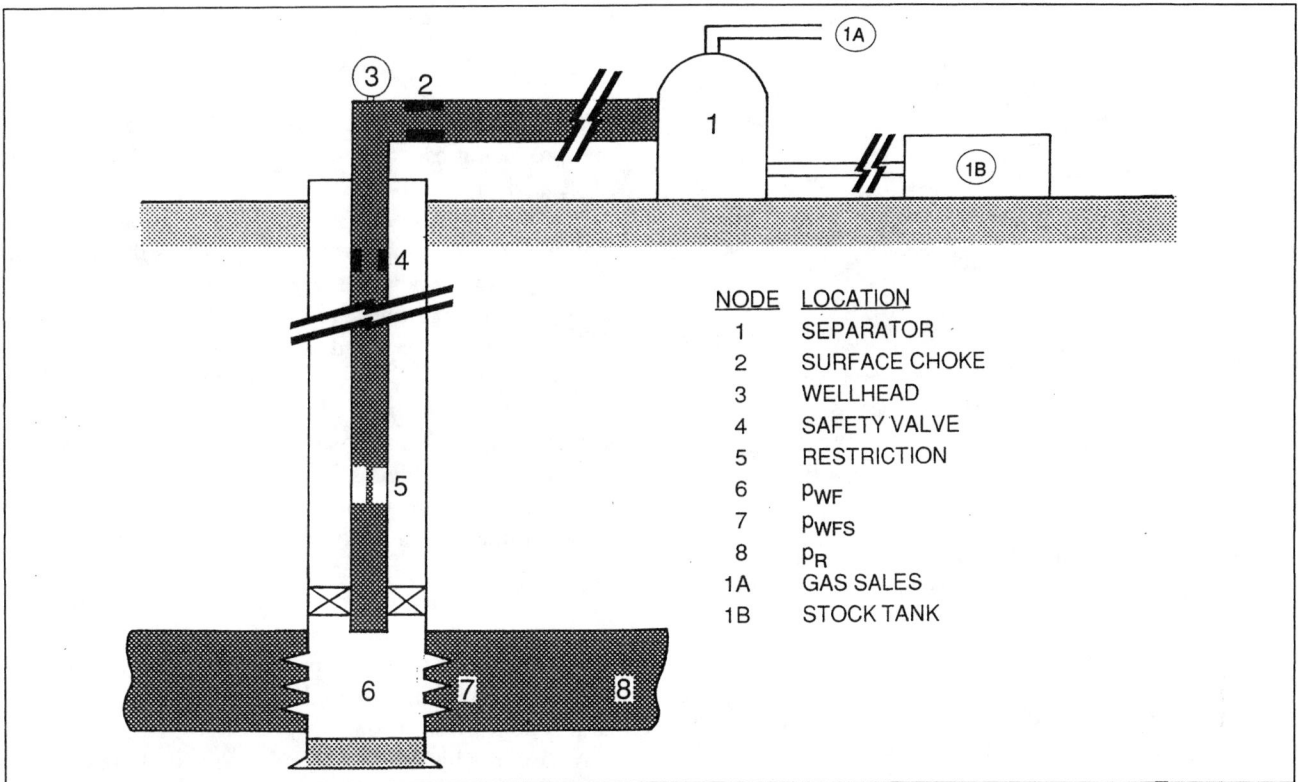

Fig. 1-2. Location of various nodes.

NODE	LOCATION
1	SEPARATOR
2	SURFACE CHOKE
3	WELLHEAD
4	SAFETY VALVE
5	RESTRICTION
6	p_{WF}
7	p_{WFS}
8	p_R
1A	GAS SALES
1B	STOCK TANK

that was changed. If a change was made in an upstream component, the outflow curve will remain unchanged. However, if either curve is changed, the intersection will be shifted, and a new flow capacity and node pressure will exist. The curves will also be shifted if either of the fixed pressures is changed, which may occur with depletion or a change in separation conditions.

The procedure can be further illustrated by considering the simple producing system shown in Fig. 1-4 and selecting the wellhead as the node.

Inflow to node:

$$\bar{p}_R - \Delta p_{res} - \Delta p_{tubing} = p_{wh}$$

Outflow from node:

$$p_{sep} + \Delta p_{flowline} = p_{wh}$$

The effect on the flow capacity of changing the tubing size is illustrated in Fig. 1-5, and the effect of a change in flowline size is shown in Fig. 1-6.

The effect of increasing the tubing size, as long as the

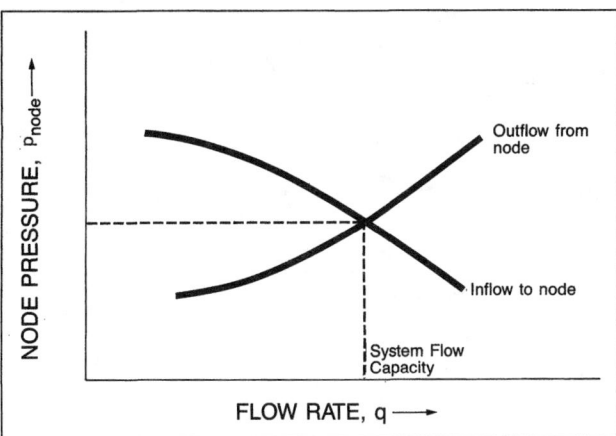

Fig. 1-3. Determination of flow capacity.

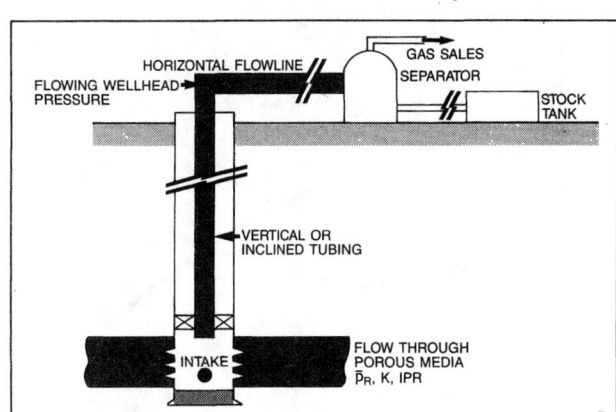

Fig. 1-4. Simple producing system.

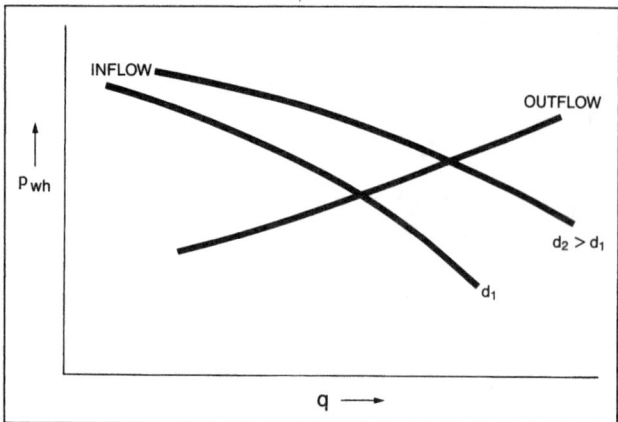

Fig. 1-5. Effect of tubing size.

tubing is not too large, is to give a higher node or well-head pressure for a given flow rate, because the pressure drop in the tubing will be decreased. This shifts the inflow curve upward and the intersection to the right.

A larger flowline will reduce the pressure drop in the flowline, shifting the outflow down and the intersection to the right. The effect of a change in any component in the system can be isolated in this manner. Also, the effect of declining reservoir pressure or changing separator pressure can be determined.

A more frequently used analysis procedure is to select the node between the reservoir and the piping system. This is labeled as point 6 on Fig. 1-2, and the node pressure is p_{wf}. Selecting the node at this point essentially divides the well into a reservoir dominated component and a piping system dominated component. The inflow and outflow expressions for the simple system will then be:

Inflow to node:

$$\bar{p}_R - \Delta p_{res} = p_{wf}$$

Outflow from node:

$$p_{sep} + \Delta p_{flowline} + \Delta p_{tubing} = p_{wf}$$

The effect of a change in tubing size on the total system producing capacity when p_{wf} is the node pressure is illustrated in Fig. 1-7.

A producing system may be optimized by selecting the combination of component characteristics that will give the maximum production rate for the lowest cost. Although the overall pressure drop available for a system, $p_R - p_{sep}$, might be fixed at a particular time, the producing capacity of the system depends on where the pressure drops occur. If too much pressure drop occurs in one component or module, there may be insufficient pressure drop remaining for efficient performance of the other modules. This is illustrated in Fig. 1-8 for a system in which the tubing is too small. Even though the reservoir may be capable of producing a large amount of fluid, if too much pressure drop occurs in the tubing, the well performance suffers. For this type of well completion, it is obvious that improving the reservoir performance by stimulation would be a waste of effort unless larger tubing were installed.

A case in which the well performance is controlled by the inflow is shown in Fig. 1-9. In this case, the excessive pressure drop could be caused by formation damage or inadequate perforations. It is obvious from the plot that improving the performance of the piping system or outflow or placing the well on artificial lift would be fruitless unless the inflow performance were also improved.

An increase in production rate achieved by increasing tubing size is illustrated in Fig. 1-7. However, if tubing is too large, the velocity of the fluid moving up the tubing may be too low to effectively lift the liquids to the surface. This could be caused by either large tubing or low production rates. This phenomenon will be dis-

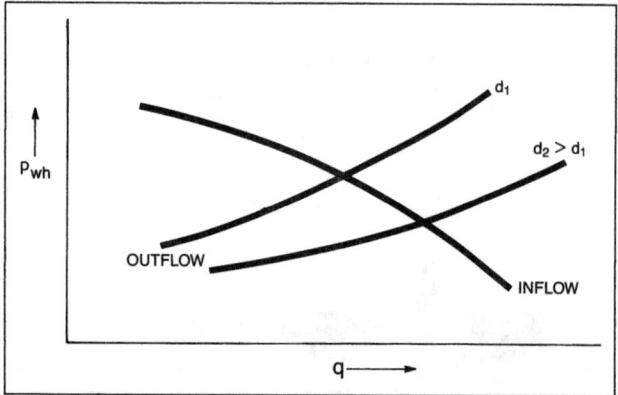

Fig. 1-6. Effect of flowline size.

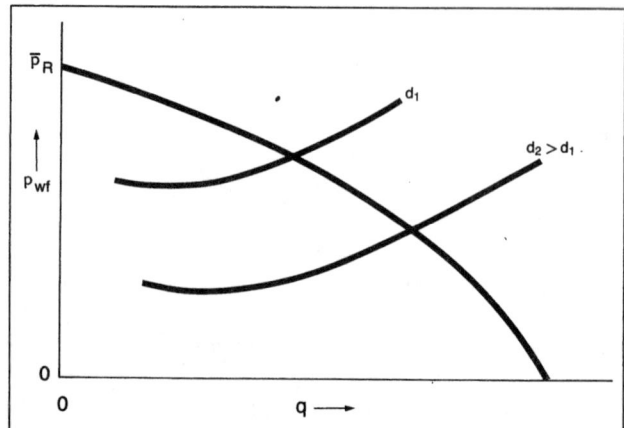

Fig. 1-7. Effect of tubing size.

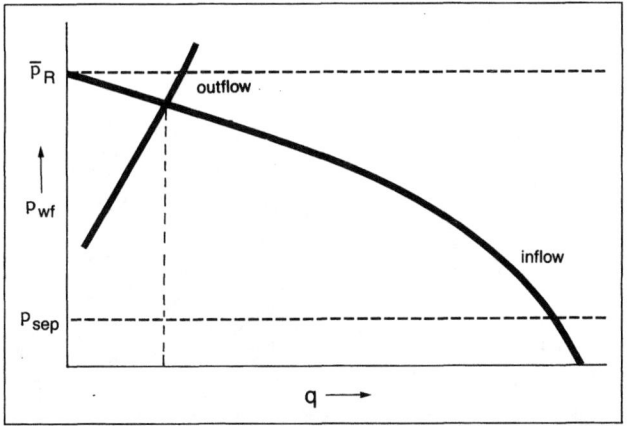

Fig. 1-8. Well restricted by piping system.

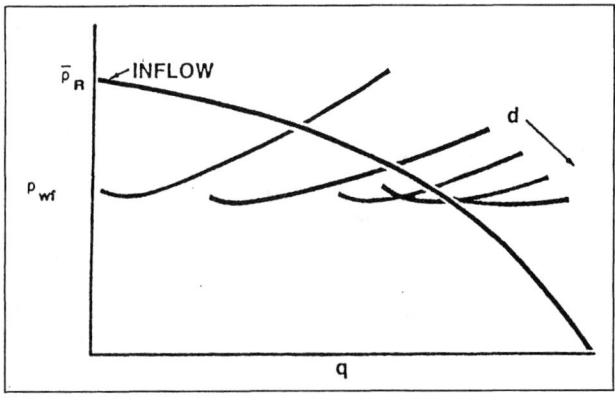

Fig. 1-10 Effect of tubing size.

cussed in detail in Chapter 3. The fluid velocity is the production rate divided by the area of the tubing. A qualitative example of selecting the optimum tubing size for a well that is producing both gas and liquids is shown in Fig. 1-10 and 1-11.

As tubing size is increased, the friction losses decrease, which results in a lower p_{wf} and, therefore, a larger inflow. However, as the tubing size is further increased, the well begins loading with liquid and the flow becomes intermittent or unstable. As the liquid level in the well builds the well will eventually die. Fig. 1-11 illustrates this graphically.

Once a well that is producing liquids along with the gas reaches the stage in which it will no longer flow naturally, it will usually be placed on artificial lift. Application of NODAL™ Analysis to artificial lift wells is discussed extensively in Chapter 5, but an example of determining the optimum gas injection rate for a well on gas lift is illustrated in Fig. 1-12 and 1-13. The purpose of injecting gas into the tubing is to decrease the density of the flowing gas-liquid mixture and, therefore, de-

crease the required flowing bottom hole pressure. However, as the gas rate is increased, the fluid velocity and, therefore, the friction losses also increase. A point will eventually be reached such that the friction losses increase more than the density or hydrostatic losses decrease with an increase in gas rate. This can be determined using NODAL™ Analysis as illustrated in Fig. 1-12.

A plot of liquid production rate versus gas injection rate can be constructed by reading the intersections of the inflow and outflow curves for various injection rates. A plot of this data is shown in Fig. 1-13. This method can also be used to allocate the injection gas available among several wells in a field producing by gas lift.

In recent years, it has been found that an inadequate number of perforations can be very detrimental to the performance of some wells. If the bottom hole flowing pressure is selected as the node pressure, the inflow can be broken down into pressure drop through the rock and pressure drop through the perforations. The inflow and outflow expressions would then consist of:

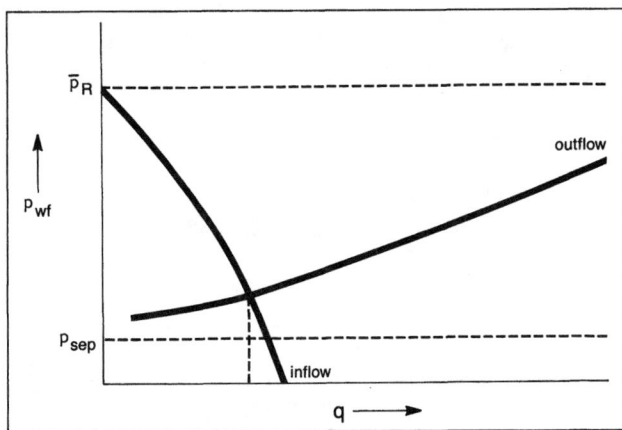

Fig. 1-9. Well restricted by inflow.

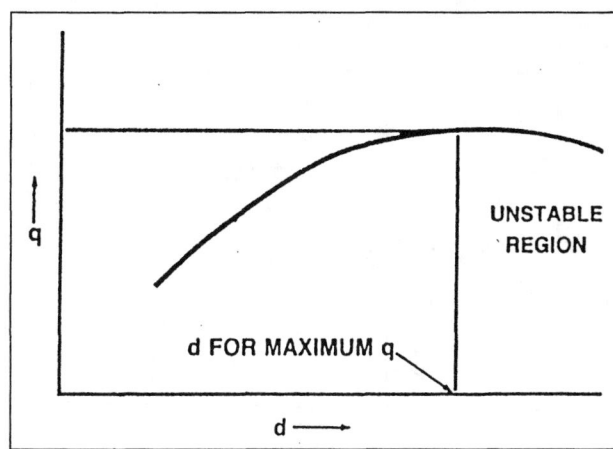

Fig. 1-11. Finding optimum tubing size.

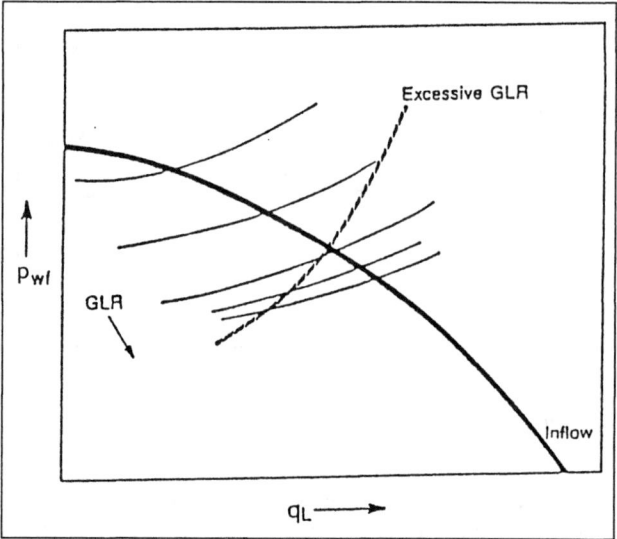

Fig. 1-12. Effect of gas rate on outflow.

Inflow to node:

$$\bar{p}_R - \Delta p_{res} - \Delta p_{perfs} = p_{wf}$$

Outflow from node:

$$p_{sep} + \Delta p_{flowline} + \Delta p_{tubing} = p_{wf}$$

Since the perforation pressure drop is a function of the number of perforations open, as well as production rate, a different inflow curve would exist for each perforating density. This is illustrated qualitatively in Fig. 1-14.

As the number of perforations is increased, a point will eventually be reached such that the perforation pressure drop is negligible, and, therefore, a further increase in perforating density would be useless. A plot of the production rate resulting from various perforating densities, that is, the intersection of the various inflow curves

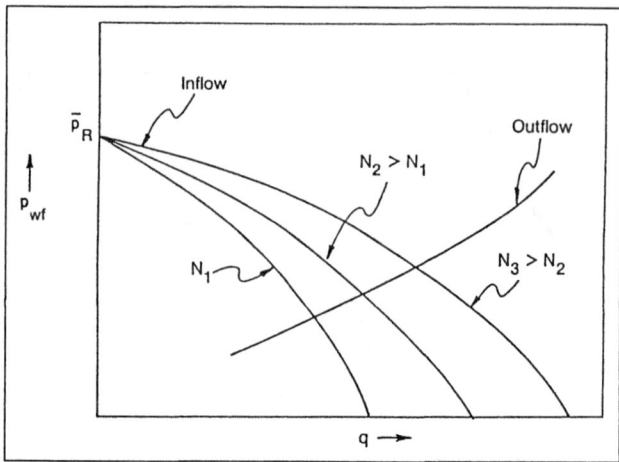

Fig. 1-14. Effect of perforating density on inflow.

with the outflow curve, is shown in Fig. 1-15. Methods for calculating perforation pressure drop are discussed in Chapter 3.

A suggested procedure for applying NODAL™ Analysis is given as follows:

1. Determine which components in the system can be changed. Changes are limited in some cases by previous decisions. For example, once a certain hole size is drilled, the casing size and, therefore, the tubing size is limited.
2. Select one component to be optimized.
3. Select the node location that will best emphasize the effect of the change in the selected component. This is not critical because the same overall result will be predicted regardless of the node location.
4. Develop expressions for the inflow and outflow.
5. Obtain required data to calculate pressure drop versus rate for all the components. This may require more data than is available, which may necessitate per-

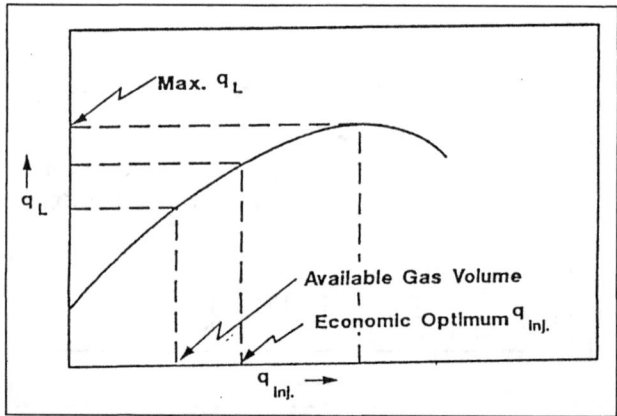

Fig. 1-13. Effect of gas injection rate on liquid rate.

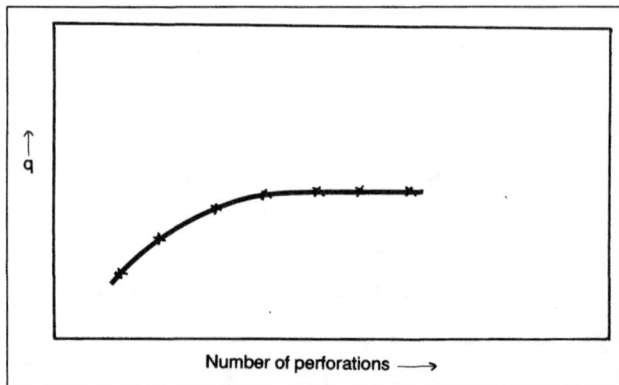

Fig. 1-15. Effect of perforating density on rate.

the analysis over possible ranges of conditions.

6. Determine the effect of changing the characteristics of the selected component by plotting inflow versus outflow and reading the intersection.

7. Repeat the procedure for each component that is to be optimized.

III. APPLICATIONS

The nodal systems analysis approach may be used to analyze many producing oil and gas well problems. The procedure can be applied to both flowing and artificial lift wells, if the effect of the artificial lift method on the pressure can be expressed as a function of flow rate. The procedure can also be applied to the analysis of injection well performance by appropriate modification of the inflow and outflow expressions. A partial list of possible applications is given as follows:

1. Selecting tubing size.

2. Selecting flowline size.

3. Gravel pack design.

4. Surface choke sizing.

5. Subsurface safety valve sizing.

6. Analyzing an existing system for abnormal flow restrictions.

7. Artificial lift design.

8. Well stimulation evaluation.

9. Determining the effect of compression on gas well performance.

10. Analyzing effects of perforating density.

11. Predicting the effect of depletion on producing capacity.

12. Allocating injection gas among gas lift wells.

13. Analyzing a multiwell producing system.

14. Relating field performance to time.

IV. SUMMARY

The nodal systems analysis approach is a very flexible method that can be used to improve the performance of many well systems. To apply the systems analysis procedure to a well, it is necessary to be able to calculate the pressure drop that will occur in all the system components listed in Fig. 1-1. These pressure drops depend not only on flow rate, but on the size and other characteristics of the components. Unless accurate methods can be found to calculate these pressure drops, the systems analysis can produce erroneous results.

The following sections in this book present the latest and most accurate methods for calculating the relationship between flow rate and pressure drop for all the components. This requires a thorough review of reservoir engineering concepts to determine reservoir inflow performance; an understanding of multiphase flow in pipes to calculate tubing and flowline performance; procedures to determine the performance of perforated completions; gravel-pack completions, and damaged or stimulated wells; and an understanding of artificial lift-systems.

Once procedures are presented to analyze each component separately, the systems analysis approach will be applied to many different wells to demonstrate the procedures to optimize well performance.

V. REFERENCES

1. Gilbert, W. E.: "Flowing and Gas-Lift Well Performance," API Drill. Prod. Practice, 1954.
2. Nind, T. E. W.: *Principles of Oil Well Production,* McGraw-Hill, 1964.
3. Brown, K. E. and Beggs, H. D.: *The Technology of Artificial Lift Methods,* Vol. 1, Penn Well Publ. Co., Tulsa, Oklahoma, 1978.

2

Reservoir Performance

I. INTRODUCTION

One of the most important components in the total well system is the reservoir. Unless accurate predictions can be made as to what will flow into the borehole from the reservoir, the performance of the system cannot be analyzed. As discussed in the previous section, one of the fixed pressures, at any time in the life of the reservoir, is the average reservoir pressure $\bar{p}_R$. The flow into the well depends on the drawdown or pressure drop in the reservoir, $\bar{p}_R - p_{wf}$. The relationship between flow rate and pressure drop occurring in the porous medium can be very complex and depends on parameters such as rock properties, fluid properties, flow regime, fluid saturations in the rock, compressibility of the flowing fluids, formation damage or stimulation, turbulence, and drive mechanism. It also depends on the reservoir pressure itself and, depending on the drive mechanism, this may decrease with time or cumulative production.

The reservoir component will always be an upstream component. That is, it will hardly ever be practical to select $\bar{p}_R$ as the node pressure, although the sandface pressure p_{wfs} is sometimes selected. This will isolate the effects of the pressure drop across the perforations or gravel pack.

The flow from the reservoir into the well has been called "inflow performance" by Gilbert[1] and a plot of producing rate versus bottomhole flowing pressure is called an "inflow performance relationship" or IPR. This should not be confused with the inflow to a node as discussed in Chapter 1. The inflow to the node can include the flow through other components, depending on the location of the node selected.

In this chapter the well performance equations will be presented for various reservoir types and drive mechanisms. These equations will permit the calculation of $\Delta p_1 = \bar{p}_R - p_{wfs}$ or, if there is negligible pressure loss across the completion, $\Delta p_1 = \bar{p}_R - p_{wf}$ where p_{wf} is the flowing wellbore pressure. The effects of changing conditions on the accuracy of the equations will be discussed and empirical methods to correct for failure of the theory will be presented. Methods for predicting IPRs for both the present or real time and for future times will be discussed. Both oil and gas reservoir performance will be presented.

Finally, methods for obtaining the necessary rock and fluid properties for use in the equations will be outlined, and the accuracy of the data will be discussed.

II. WELL PERFORMANCE EQUATIONS

To calculate the pressure drop occurring in a reservoir, an equation that expresses the energy or pressure losses due to viscous shear or friction forces as a function of velocity or flow rate is required. Although the form of the equation can be quite different for various types of fluids, the basic equation on which all of the various forms are based is Darcy's law.

A. Darcy's Law

In 1856, while performing experiments for the design of sand filter beds for water purification, Henry Darcy proposed an equation relating apparent fluid velocity to pressure drop across the filter bed. Although the experiments were performed with flow only in the downward vertical direction, the expression is also valid for horizontal flow, which is of most interest in the petroleum industry.

It should also be noted that Darcy's experiments involved only one fluid, water, and that the sand filter was completely saturated with the water. Therefore, no effects of fluid properties or saturation were involved.

Darcy's sand filters were of constant cross-sectional area, so the equation did not account for changes in velocity with location. Written in differential form, Darcy's law is:

$$v = \frac{k}{\mu}\frac{dp}{dx} \tag{2-1}$$

or in terms of volumetric flow rate q

$$q = vA = -\frac{kA}{\mu}\frac{dp}{dx} \tag{2-2}$$

where

k = permeability of the porous medium,
v = apparrent fluid velocity,
q = volumetric flow rate,
A = area open to flow,
μ = fluid viscosity, and
dp/dx = pressure gradient in the direction of flow (negative).

1. Linear Flow

For linear flow, that is for constant area flow, the equation may be integrated to give the pressure drop occurring over some length L:

$$\int_{p_1}^{p_2} \frac{k\,dp}{\mu} = -\frac{q\mu}{kA}\int_o^L dx$$

If it is assumed that k, μ and q are independent of pressure, or that they can be evaluated at the average pressure in the system, the equation becomes

$$\int_{p_1}^{p_2} dp = -\frac{q\mu}{kA}\int_o^L dx \tag{2-3}$$

Integration gives:

$$p_2 - p_1 = \frac{-q\mu}{kA}L \tag{2-4}$$

or

$$q = \frac{CkA(p_1 - p_2)}{\mu L}$$

where C is a unit conversion factor. The correct value for C is 1.0 for Darcy Units and 1.127×10^{-3} for Field Units (See Table 2-1).

TABLE 2-1

Units for Darcy's Law

Variable	Symbol	Units Darcy	Field
Flow rate	q	cc/sec	bbl/day
Permeability	k	darcys	md
Area	A	cm²	ft²
Pressure	p	atm.	psi
Viscosity	μ	cp	cp
Length	L	cm	ft

The geometry of the linear system is illustrated in Fig. 2-1.

It can be observed from Equation 2-3 that a plot on cartesian coordinates of p vs. L will produce a straight line of constant slope, $-q\mu/kA$. That is, the variation of pressure with distance is linear.

If the flowing fluid is compressible, the in-situ flow rate is a function of pressure. Using the fact that the mass flow rate ρq must be constant and expressing the density in terms of pressure, temperature and gas specific gravity, it can be shown that Equation 2-3 becomes:

$$p_1^2 - p_2^2 = \frac{8.93ZT\mu L}{kA}q_{sc} \tag{2-5}$$

where

p = psia,
T = °R,
μ = cp,
L = ft,
k = md,
A = ft²,
q_{sc} = scf / day

For high-velocity flow in which turbulence or non-Darcy flow can exist, Darcy's law must be modified to account for the extra pressure drop caused by the turbulence. Applying the turbulence correction to Equations 2-3 and 2-5 gives:

Oil Flow

$$p_1 - p_2 = \frac{\mu_o B_o L}{1.127 \times 10^{-3} k_o A}q_o \\ + \frac{9.08 \times 10^{-13} B_0^2 \beta \rho_o L}{A^2}q_o^2 \tag{2-6}$$

where

p_1 = upstream pressure, psia,
p_2 = downstream pressure, psia,
μ_o = oil viscosity, cp,
B_o = Oil formation volume factor, bbl/STB,
L = Length of flow path, ft,

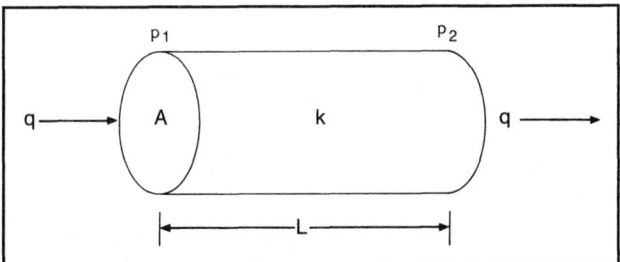

Fig. 2-1. Geometry for linear flow.

k_o = permeability to oil, md,
A = area open to flow, ft²,
ρ_o = oil density, lbm/ft³,
β = velocity coefficient, ft⁻¹, and
q_o = oil flow rate, STB/day.

Gas Flow

$$p_1^2 - p_2^2 = \frac{8.93 Z \mu_g LT}{k_g A} q_{sc}$$

$$+ \frac{1.247 \times 10^{-16} \beta ZTL \gamma_g}{A^2} q_{sc}^2 \qquad (2\text{-}7)$$

where

Z = gas deviation factor evaluated at T, $\bar{p}$,
T = flowing temperature, °R
γ_g = gas gravity (air = 1),
q_{sc} = gas flow rate at 14.7 psia, 60°F, scf/day,
μ_g = gas viscosity at $\bar{T}, \bar{p}$, cp,
k_g = permeability to gas, md, and
A = flow area, ft²

An estimate for the velocity coefficient β can be obtained from:

$$\beta = ak^{-b} \qquad (2\text{-}8)$$

where

β = ft⁻¹,
k = md, and a and b are approximated from:

Formation Type	a	b
Consolidated	2.329 x 10¹⁰	1.2
Unconsolidated	1.47 x 10⁷	0.55

Although linear flow rarely occurs in a reservoir, these equations will be used later to calculate the pressure drop across a gravel pack completion, that is, $\Delta p = p_{wfs} - p_{wf}$.

2. Radial Flow

Darcy's law can be used to calculate the flow into a well where the fluid is converging radially into a relatively small hole. In this case, the area open to flow is not constant and must therefore be included in the integration of Equation 2-2. Referring to the flow geometry illustrated in Figure 2-2, the cross-sectional area open to the flow at any radius is $A = 2\pi rh$.

Also, defining the change in pressure with location to be negative with respect to the direction of flow, dp/dx becomes $-dp/dr$. Making these substitutions in Equation 2-2 gives:

$$q = \frac{2\pi rhkdp}{\mu \ dr} \qquad (2\text{-}9)$$

a. Oil Flow. When applying the Darcy equation to flow of oil in a reservoir, it is assumed that the oil is only slightly compressible. The small change in q with pressure is handled with the oil formation volume factor B_o, so that the flow rate can be expressed in surface or stock tank volumes. For oil flow, Equation 2-9 becomes:

$$q_o B_o = \frac{2\pi rhk_o}{\mu_o} \left(\frac{dp}{dr} \right) \qquad (2\text{-}10)$$

or

$$2\pi h \int_{p_{wf}}^{p_e} \frac{k_o}{\mu_o B_o} dp = q_o \int_{r_w}^{r_e} \frac{dr}{r} \qquad (2\text{-}11)$$

When integrating this equation, it is usually assumed that the pressure function, $f(p) = k_o/\mu_o B_o$, is independent of pressure or that it can be evaluated at average pressure in

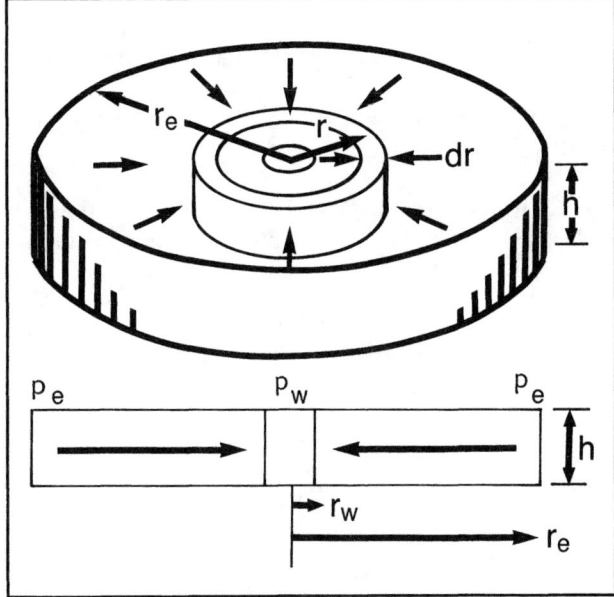

Fig. 2-2. Radial flow system.

the well's drainage volume. This is necessary because no simple analytical equation for this term as a function of pressure can be formulated. Utilizing this assumption and integrating Equation 2-11 over the drainage radius of the well gives:

$$q_o = \frac{2\pi k_o h(p_e - p_{wf})}{\mu_o B_o \ln (r_e / r_w)} \quad (2\text{-}12)$$

For field units, Equation 2-12 becomes:

$$q_o = \frac{0.00708 k_o h(p_e - p_{wf})}{\mu_o B_o \ln (r_e / r_w)} \quad (2\text{-}13)$$

where

q_o = inflow rate, STB / day
k_o = effective oil permeability, md,
h = reservoir thickness, ft,
p_e = pressure at $r = r_e$, psia,
p_{wf} = wellbore flowing pressure at $r = r_w$, psia
r_e = well's drainage radius, ft,
r_w = wellbore radius, ft,
μ_o = oil viscosity, *cp*, and
B_o = oil formation volume factor, bbl/STB.

Equation 2-13 applies for steady-state (p_e = constant), laminar flow of a well in the center of a circular drainage area. It is more useful if expressed in terms of average reservoir pressure $\overline{p}_R$, and for pseudo-steady state or stabilized flow ($\overline{p}_R - p_{wf}$ = constant) as:

$$q_o = \frac{0.00708 k_o h(\overline{p}_R - p_{wf})}{\mu_o B_o \ln (.472 r_e / r_w)} \quad (2\text{-}14)$$

where

$\overline{p}_R$ = average pressure in the drainage volume of the well.

The other terms are the same as those defined for Equation 2-13.

b. Gas Flow. To integrate Equation 2-9 for flow of gases, the fact that ρq is constant is used along with the gas equation of state

$$\rho = \frac{pM}{ZRT} \quad (2\text{-}15)$$

$$\rho q = \rho_{sc} q_{sc}$$

or

$$q_{sc} = \frac{\rho q}{\rho_{sc}} = \frac{p T_{sc}}{p_{sc} TZ} \frac{2\pi r h k_g dp}{\mu_g dr}$$

$$\int_{P_{wf}}^{P_e} p \, dp = \frac{q_{sc} \mu_g T p_{sc} Z}{2\pi h k_g T_{sc}} = \int_{r_w}^{r} \frac{dr}{r}$$

which gives upon integration:

$$p_e^2 - p_{wf}^2 = \frac{q_{sc} \mu_g Z T p_{sc} (\ln r_e / r_w)}{\pi k_g h T_{sc}} \quad (2\text{-}16)$$

Modifying Equation 2-16 for stabilized flow, average reservoir pressure, defining p_{sc} = 14.7 psia and T_{sc} = 520 °R gives an equation for gas inflow rate in field units.

$$q_{sc} = \frac{703 \times 10^{-6} k_g h(\overline{p}_R^2 - p_{wf}^2)}{\mu_g Z T \ln (.472 r_e / r_w)} \quad (2\text{-}17)$$

where

q_{sc} = gas flow rate, Mscfd,
k_g = permeability to gas, md,
h = reservoir thickness, ft,
$\overline{p}_R$ = average reservoir pressure, psia
p_{wf} = wellbore flowing pressure, psia,
μ_g = gas viscosity at T, $\overline{p}$ =.5 ($\overline{p}_R + p_{wf}$), cp
Z = gas compressibility factor at T, $\overline{p}$,
T = reservoir temperature, °R,
r_e = drainage radius, ft, and
r_w = wellbore radius, ft.

c. Reservoir Pressure Profile. The behavior of the pressure in the reservoir as a function of radius can be analyzed by plotting pressure versus radius as predicted by Equation 2-14. Assuming a fixed average reservoir pressure $\overline{p}_R$ at $r = 0.472 r_e$ and solving for pressure, Equation 2-14 gives:

$$p = \overline{p}_R - \frac{141.2 q_o \mu_o B_o}{k_o h} \ln(.472 r_e)$$
$$+ \frac{141.2 q_o \mu_o B_o}{k_o h} \ln r \quad (2\text{-}18)$$

A plot of pressure versus radius for typical well conditions, Figure 2-3, shows the large increase in pressure gradient as the fluid increases in velocity near the wellbore. Approximately one-half of the total pressure drawdown occurs within a 15 ft radius from the well. For gas flow, the pressure drop around the wellbore is even more severe.

Examination of Equation 2-18 reveals that a plot of p versus *ln r* will result in a straight line of constant slope m,

where

$$m = \frac{141.2 q_o \mu_o B_o}{k_o h} \quad (2\text{-}19)$$

This type of plot is illustrated in Figure 2-4. It should be emphasized that the slope remains constant only if all of the terms on the right-hand side of Equation 2-19 remain constant. A different slope and, therefore, a differ-

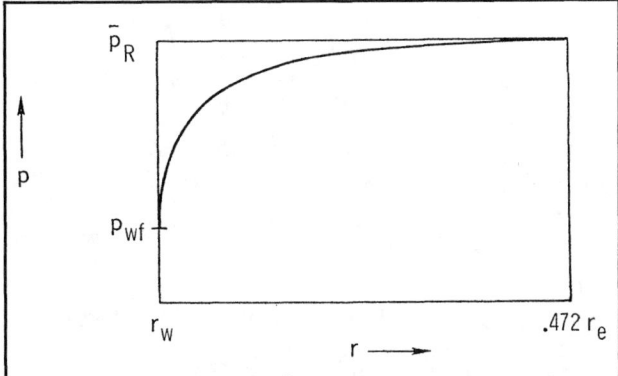

Fig. 2-3. Reservoir pressure profile.

ent value of p_{wf}, would be obtained for each flow rate q_o.

A similar analysis of Equation 2-17 for gas flow reveals that a plot of p^2 versus *ln r* results in a straight line of slope:

$$m = 1422 q_{sc} \mu_g ZT / k_g h$$

3. Productivity Index Concept

The relationship between well inflow rate and pressure drawdown has often been expressed in the form of a *Productivity Index J,*

where

$$J = \frac{0.00708 k_o h}{\mu_o B_o \ln (.472 r_e / r_w)} \qquad (2\text{-}20)$$

The inflow equation for oil flow can then be written as

$$q_o = J(\bar{p}_R - p_{wf}) \qquad (2\text{-}21)$$

or

$$J = \frac{q_o}{\bar{p}_R - p_{wf}} \qquad (2\text{-}22)$$

Solving for p_{wf} in terms of q_o reveals that a plot of p_{wf}

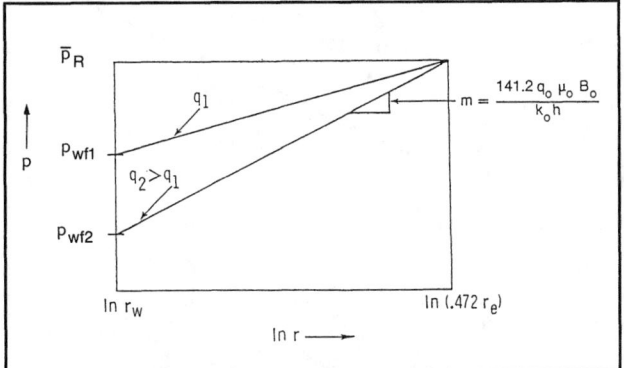

Fig. 2-4. Semi-log plot of pressure vs. radius.

versus q_o on Cartesian coordinates results in a straight line having a slope of $-1/J$ and an intercept of $\bar{p}_R$ at $q_o = 0$.

$$p_{wf} = \bar{p}_R - \frac{q_o}{J} \qquad (2\text{-}33)$$

If conditions are such that J is constant with drawdown, once a value of J is obtained from one production test or calculated using Equation 2-20, it may be used to predict inflow performance for other conditions.

Example 2-1:

A well that is producing from a reservoir having an average pressure of 2085 psig produced at a rate of 282 STB/day when bottomhole flowing pressure was 1765 psig.

Calculate:
1. The productivity index J.
2. The producing rate if p_{wf} is decreased to 1485 psig.
3. The bottomhole pressure necessary to obtain an inflow of 400 STB/day.
4. The inflow rate if p_{wf} is reduced to zero, i.e., Absolute Open Flow potential (AOF) or $q_{o(max)}$.

Solution:

1. $J = \dfrac{q_o}{\bar{p}_R - p_{wf}} = \dfrac{282}{2085 - 1765} = 0.88$ STB/day-psi

2. $q_o = J(\bar{p}_R - p_{wf}) = 0.88(2085 - 1485) = 528$ STB/day

3. $p_{wf} = \bar{p}_R - q_o/J = 2085 - 400/0.88 = 1630$ psig

4. $q_{o(max)} = J(\bar{p}_R - 0) = 0.88(2085) = 1835$ STB/day

The predictions made in Example 2-1 are valid only if J remains constant. This implies that the pressure function $f(p) = k_o/\mu_o B_o$ remains constant, which is seldom the case, as will be discussed further in following sections. The productivity index can also be expressed as:

$$J = \frac{0.00708 h}{(\bar{p}_R - p_{wf}) \ln (.472 r_e / r_w)} \int_{p_{wf}}^{\bar{p}_R} \frac{k_o}{\mu_o B_o} dp \qquad (2\text{-}24)$$

The productivity index concept could also be applied to gas well inflow performance by defining a gas productivity index as

$$J_g = \frac{q_{sc}}{\bar{p}_R^2 - p_{wf}^2} \qquad (2\text{-}25)$$

or

$$J_g = \frac{703 \times 10^{-6} k_g h}{\mu_g ZT \ln (.472 r_e / r_w)} \quad (2\text{-}26)$$

A plot of p_{wf} versus q_{sc} would not be linear on Cartesian coordinates. A more common procedure for gas well analysis will be discussed in following sections.

4. Permeability Alteration and Turbulence

Darcy's law was based on the assumptions that permeability to the flowing fluid was constant in the entire drainage area of the well and that only laminar flow existed. The effective permeability to oil is the product of the relative permeability to oil and the absolute permeability of the reservoir, that is,

$$k_o = k \, k_{ro}$$

The absolute permeability k, can be either increased around the wellbore by well stimulation or decreased by formation damage, such as clay swelling or pore plugging. This would change the slope of the pressure profile out to the radius to which the permeability was altered. This is illustrated in Figure 2-5.

Figure 2-5 illustrates that for a constant flow rate, less pressure drawdown would be required if the well had been stimulated and more drawdown would be required for a damaged well. The bottomhole flowing pressure required for no change in permeability is labeled p'_{wf}.

It is often impossible to determine either the altered radius r_a or the altered permeability k_a. In this case it is assumed that the pressure change due to the altered permeability occurs at the wellbore in the form of a skin effect. The skin effect is defined as a dimensionless quantity and can be included in Equations 2-14 and 2-17 as:

$$q_o = \frac{0.00708 k_o h (\bar{p}_R - p_{wf})}{\mu_o B_o [\ln (.472 r_e / r_w) + S']} \quad (2\text{-}27)$$

and

$$q_{sc} = \frac{703 \times 10^{-6} k_g h (\bar{p}_R^2 - p_{wf}^2)}{\mu_g ZT [\ln (.472 r_e / r_w) + S']} \quad (2\text{-}28)$$

The skin factor S' includes the effects of both turbulence and actual formation damage as:

$$S' = S + Dq \quad (2\text{-}29)$$

where

S = skin factor due to permeability change, and

D = turbulence coefficient

The term S will be positive for damage, negative for improvement, or zero for no change in permeability. The turbulence coefficient, D, will be either positive or zero. The effects of S' on the pressure profile for an oil reservoir are illustrated in Figure 2-6. Although a sudden large pressure drop could occur at the wellbore as indicated for a positive S', if, for example, a small number of perforations are open, it would be physically impossible for a pressure increase to occur as illustrated for a negative S'. The actual situation is illustrated in Figure 2-5.

Equations 2-27 and 2-28 are commonly used to describe pseudosteady state flow in a circular drainage area. If the drainage radius is not circular, then the use of Equations 2-27 and 2-28 may lead to appreciable errors. Odeh[2] developed the following equations to describe pseudosteady state flow in a noncircular drainage area.

$$q_o = \frac{0.00708 k_o h (\bar{p}_R - p_{wf})}{\mu_o B_o (\ln (.472 x) + S')}$$

$$q_{sc} = \frac{703 \times 10^{-6} k_g h (\bar{p}_R^2 - p_{wf}^2)}{\mu_g ZT (\ln (.472 x) + S')}$$

$$J = \frac{q_o}{\bar{p}_R - p_{wf}} = \frac{0.00708 k_o h}{\mu_o B_o (\ln (.472 x) + S')}$$

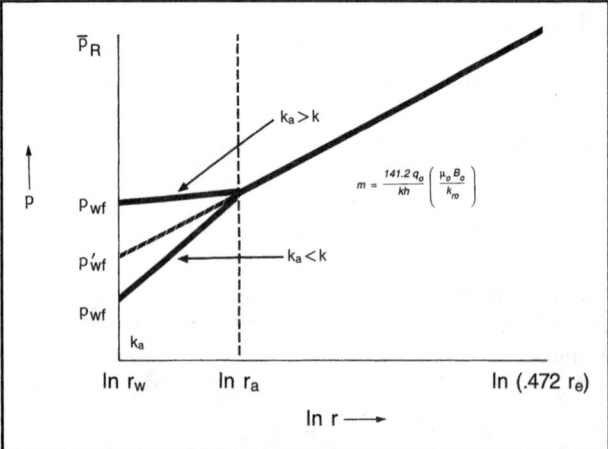

Fig. 2-5. Effects of altered permeability.

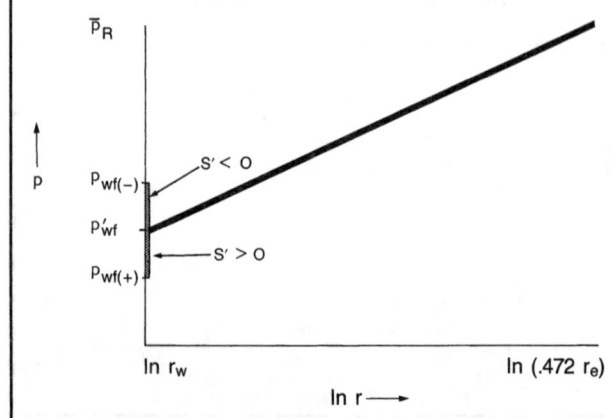

Fig. 2-6. Effects of skin factors.

where x is given in Figure 2-7 for different drainage area shapes and well locations.

The magnitude of the pressure change due to the skin and turbulence, defined as:

$$\Delta p_{skin} = p_{wfs} - p_{wf} \qquad (2\text{-}30)$$

can be calculated from:

$$\Delta p_{skin} = \frac{141.2 q_o \mu_o B_o S'}{k_o h} \qquad (2\text{-}31)$$

A value for S' can be obtained from analysis of various types of pressure transient tests.

B. Factors Affecting Productivity Index

The expression for the productivity index for an oil well, including skin effect, can be written as:

$$J = \frac{0.00708 kh}{(\bar{p}_R - p_{wf})[\ln(.472 r_e / r_w) + S']} \int_{p_{wf}}^{\bar{p}_R} \frac{k_{ro}}{\mu B_o} dp \qquad (2\text{-}32)$$

From this expression, it can be observed that J will not be constant unless the pressure function is independent of pressure. In arriving at Equation 2-14, it was assumed that k_{ro} was constant and that μ_o and B_o could be evaluated at reservoir temperature and $\bar{p} = 0.5 \ (\bar{p}_R + p_{wf})$. Also, the dependence of S' on q_o was neglected. In this section, some of the factors that can cause J to change are discussed. The behavior of μ_o, B_o and k_{ro}, with pressure and fluid saturation change will be presented.

1. Phase Behavior in Reservoirs

A thorough discussion of the phase behavior of oil reservoirs will not be given here, but may be found in books on fluid properties, such as McCain[3] and Amyx, Bass and Whiting.[4]

The concept of bubblepoint pressure and dewpoint pressure will be reviewed because of the importance of gas saturation on the relative permeability to oil. A typical pressure-temperature phase diagram for an oil reservoir is shown in Figure 2-8. The liquid, gas and two-phase regions are shown, and the bubblepoint pressure is indicated as the pressure at which free gas first forms in the reservoir as pressure is reduced at constant reservoir temperature.

The reservoir fluid depicted in Figure 2-8 is above the bubblepoint pressure p_b, at initial reservoir pressure $\bar{p}_{Ri}$ and, therefore, no free gas would exist anywhere in the reservoir. However, if the pressure at any point in the reservoir drops below p_b, free gas will form and k_{ro} will be reduced. Therefore, if a well is produced at a rate that requires that p_{wf} be less than p_b, k_{ro} and, therefore, J will

be decreased around the wellbore. This situation can occur even though $\bar{p}_R$ may be well above p_b.

As pressure depletion in the reservoir occurs, $\bar{p}_R$ will likely drop below p_b and free gas will exist throughout the reservoir.

2. Relative Permeability Behavior

As free gas forms in the pores of a reservoir rock, the ability of the liquid phase to flow is decreased. Even though the gas saturation may not be great enough to allow gas to flow, the space occupied by the gas reduces the effective flow area for the liquids. The behavior of the relative permeability to oil as a function of liquid saturation is shown in Figure 2-9. The relative permeability is defined as the ratio of effective permeability to a particular fluid to the absolute permeability of the rock, $k_{ro} = k_o/k$. The absolute permeability, k, is the permeability to a fluid when the fluid completely saturates the rock and is independent of the fluid as long as the fluid is Newtonian. The relative permeability to gas will be decreased if liquid saturation develops in a gas reservoir, either as a result of retrograde condensation or water formation in the pores.

3. Oil Viscosity Behavior

The viscosity of oil saturated with gas at constant temperature will decrease as pressure is decreased from initial pressure to bubblepoint pressure. Below p_b the viscosity will increase as gas comes out of solution, leaving the heavier molecules in the liquid phase. Figure 2-10 illustrates qualitatively the behavior of μ_o versus pressure at constant temperature. Equations for calculating behavior of viscosity with pressure and temperature change will be presented in Chapter 3.

4. Oil Formation Volume Factor Behavior

As pressure is decreased on a liquid, the liquid will expand. When the bubblepoint pressure of an oil is reached, gas coming out of solution will cause the oil to shrink. The behavior of B_o versus p at constant temperature is shown graphically in Figure 2-11.

The oil formation volume factor is defined as:

$$B_o = \frac{\text{Volume of oil plus its dissolved gas at } p, T}{\text{Volume of oil at stock tank conditions, } p_{sc}, T_{sc}}$$

C. Factors Affecting Inflow Performance

The Inflow Performance Relationship (IPR) for a well is the relationship between flow rate into the wellbore and wellbore flowing pressure p_{wf}. The IPR is illustrated graphically by plotting p_{wf} versus q. If the IPR can be represented by a constant productivity index J, the plot will be linear and the slope of the line will be $-1/J$, with inter-

SYSTEM	X	SYSTEM	X
 ⊙ (circle)	$\dfrac{r_e}{r_w}$	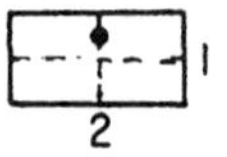 rectangle, well top-center, 2	$\dfrac{0.966\,A^{1/2}}{r_w}$
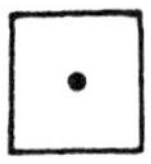 square, centered	$\dfrac{0.571\,A^{1/2}}{r_w}$	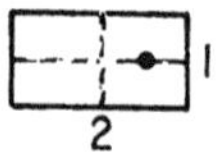 rectangle, 2	$\dfrac{1.44\,A^{1/2}}{r_w}$
 hexagon, centered	$\dfrac{0.565\,A^{1/2}}{r_w}$	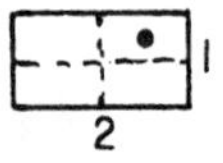 rectangle, 2	$\dfrac{2.206\,A^{1/2}}{r_w}$
 triangle, centered	$\dfrac{0.604\,A^{1/2}}{r_w}$	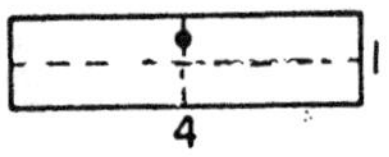 rectangle, 4	$\dfrac{1.925\,A^{1/2}}{r_w}$
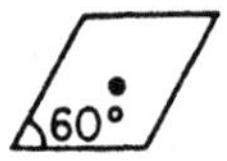 parallelogram, 60°	$\dfrac{0.61\,A^{1/2}}{r_w}$	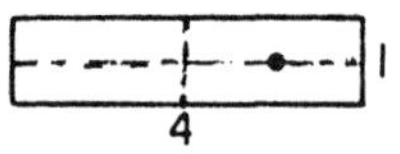 rectangle, 4	$\dfrac{6.59\,A^{1/2}}{r_w}$
 right triangle, $\frac{1}{3}$	$\dfrac{0.678\,A^{1/2}}{r_w}$	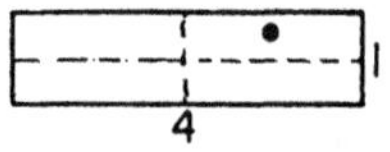 rectangle, 4	$\dfrac{9.36\,A^{1/2}}{r_w}$
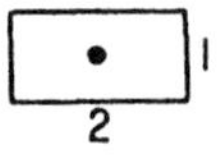 rectangle, 2	$\dfrac{0.668\,A^{1/2}}{r_w}$	 square	$\dfrac{1.724\,A^{1/2}}{r_w}$
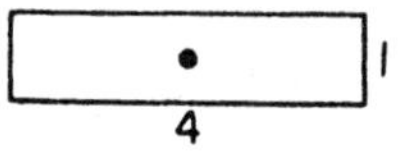 rectangle, 4	$\dfrac{1.368\,A^{1/2}}{r_w}$	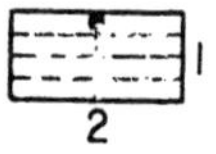 rectangle, 2	$\dfrac{1.794\,A^{1/2}}{r_w}$
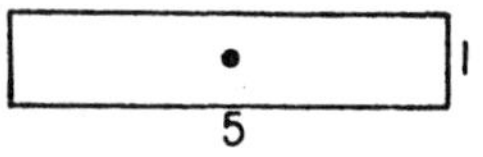 rectangle, 5	$\dfrac{2.066\,A^{1/2}}{r_w}$	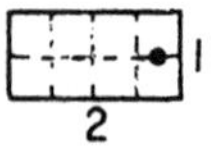 rectangle, 2	$\dfrac{4.072\,A^{1/2}}{r_w}$
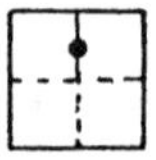 square, 2	$\dfrac{0.884\,A^{1/2}}{r_w}$	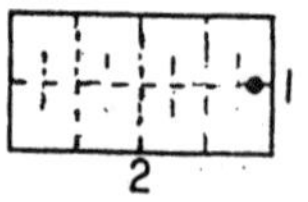 rectangle, 2	$\dfrac{9.523\,A^{1/2}}{r_w}$
 square, 2	$\dfrac{1.485\,A^{1/2}}{r_w}$	 triangle	$\dfrac{10.135\,A^{1/2}}{r_w}$

Fig. 2-7. Factors for different shapes and well positions in a drainage area.[2]

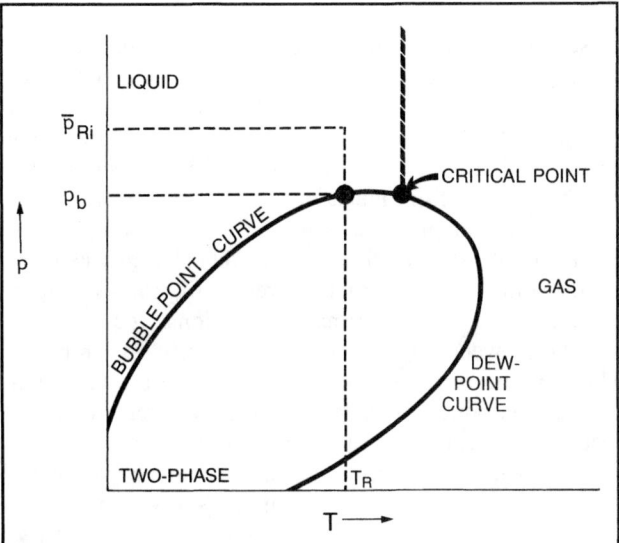

Fig. 2-8. Oil reservoir phase diagram.

cepts of $p_{wf} = \bar{p}_R$ and $q = q_{max}$ at values of $q = o$ and $p_{wf} = o$, respectively.

In the previous section the theoretical expression for J was given in Equation 2-32, and it was pointed out that changes occurring in some of the variables could cause J to change. If the value of J changes, the slope of the IPR plot will change, and a linear relationship between p_{wf} and q will no longer exist. For oil reservoirs, the principal factors affecting the IPR are:

1. A decrease in k_{ro} as gas saturation increases.

2. An increase in oil viscosity as pressure decreases and gas is evolved.

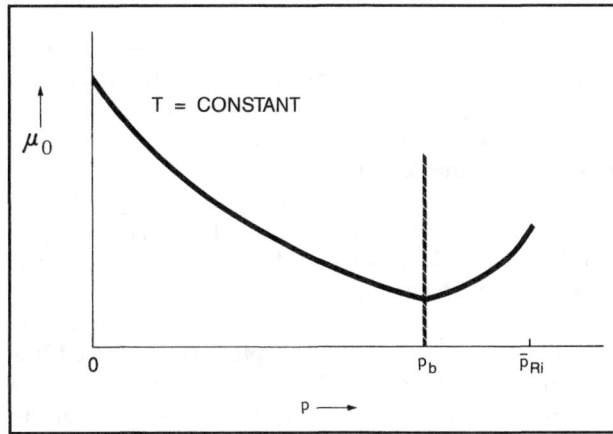

Fig. 2-10. Oil viscosity behavior.

3. Shrinkage of the oil as gas is evolved when pressure on the oil decreases.

4. Formation damage or stimulation around the well-bore ($S \neq 0$) as reflected in the term $S' = S + Dq_o$

5. An increase in the turbulence term Dq_o, as q_o increases.

These factors can change either as a result of draw-down change at a constant value of $\bar{p}_R$ or as $\bar{p}_R$ declines because of depletion. Changes in the skin factor can result from formation damage or stimulation. The effects on inflow performance of different drive mechanisms, draw-down and depletion are discussed qualitatively in this section. Methods to quantitatively predict these ef-fects will be presented subsequently.

1. Drive Mechanisms

The source of pressure energy to cause the oil and gas to flow into the wellbore has a substantial effect on both the performance of the reservoir and the total production

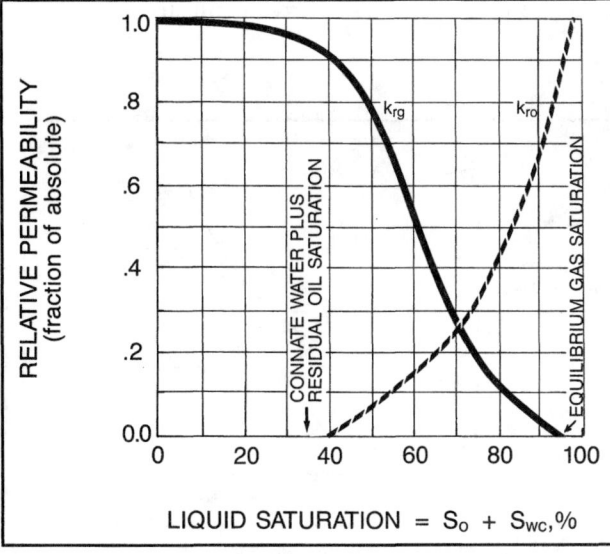

Fig. 2-9. Gas-oil relative permeability data.

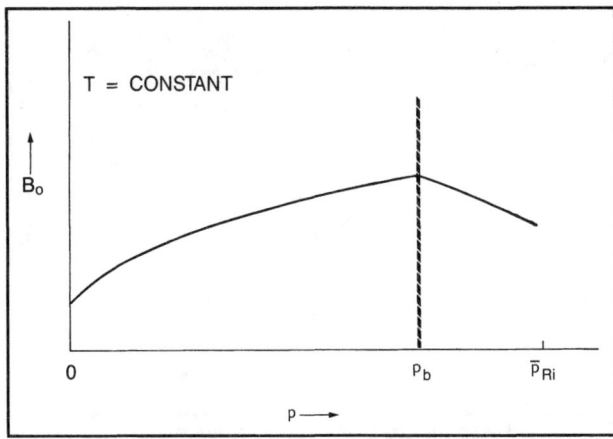

Fig. 2-11. Oil formation volume factor behavior.

system. General descriptions of the three basic types of drive mechanisms are presented. The behavior of reservoir pressure $\bar{p}_R$, the pressure function evaluated at $p = \bar{p}_R$, $f(\bar{p}_R)$, and surface producing gas/oil ratio, R, versus cumulative recovery, N_p, is presented graphically for each drive mechanism.

a. Dissolved Gas Drive. A dissolved-gas-drive reservoir is closed from any outside source of energy, such as water encroachment. Its pressure is initially above bubblepoint pressure, and, therefore, no free gas exists. The only source of material to replace the produced fluids is the expansion of the fluids remaining in the reservoir. Some small but usually negligible expansion of the connate water and rock may also occur.

The reservoir pressure declines rapidly with production until $\bar{p}_R = p_b$, since only the oil is expanding to replace the produced fluids. The producing gas/oil ratio will be constant at $R=R_{si}$ during this period. Also, since no free gas exists in the reservoir, $f(\bar{p}_R)$ will remain fairly constant.

Once $\bar{p}_R$ declines below p_b free gas will be available to expand, and $\bar{p}_R$ will decline less rapidly. However, as soon as the gas saturation exceeds the critical gas saturation, R will increase rapidly, further depleting the reservoir energy. As abandonment conditions are reached, R will begin to decrease because most of the gas has been produced, and, at low reservoir pressures, the reservoir gas volumes are more nearly equal to the standard surface volumes.

Recovery at abandonment conditions will range between 5% and 30% of original oil in place. However, in most cases, some type of pressure maintenance is applied to supplement the reservoir energy and increase recoveries. Typical dissolved-gas drive performance under primary depletion is shown in Figure 2-12.

b. Gas Cap Drive. A gas cap drive reservoir is also closed from any outside source of energy, but the oil is saturated with gas at its initial pressure and, therefore, free gas will exist. As oil is produced the gas cap will expand and help to maintain the reservoir pressure. Also, as the reservoir pressure declines from production, gas will be evolved from the saturated oil.

The reservoir pressure will decline more slowly than for a dissolved-gas drive, but as the free gas cap expands, some of the upstructure wells will produce at high gas/oil ratios. Under primary conditions, the recovery may be between 20% and 40% of the initial oil in place. This may be increased by re-injecting the produced gas into the gas cap. Also, the effects of gravity may in-crease recovery, especially if producing rates are low and the formation has an appreciable dip. Primary performance for a gas cap drive reservoir is shown in Figure 2-13.

c. Water Drive. In a water-drive reservoir, the oil zone is in contact with an aquifer that can supply the material to replace the produced oil and gas.. The water that encroaches may come from expansion of the water only, or the aquifer could be connected to a surface outcrop. The oil will be undersaturated initially, but if the pressure declines below the bubblepoint, free gas will form and the dissolved-gas drive mechanism will also contribute to the energy for production.

The recovery to be expected from a water-drive reservoir may vary from 35% to 75% of the initial oil in place. If the producing rate is low enough to allow water to move in as rapidly as oil and gas are produced or if the water drive is supplemented by water injection, recovery may be even higher. If reservoir pressure remains above bubblepoint, no free gas will form and the pressure function, based on $\bar{p}_R$, will remain fairly constant. The performance of a strong water drive is illustrated in Figure 2-14.

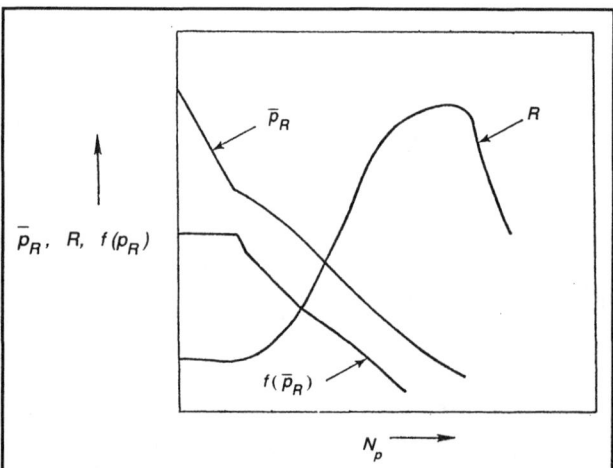

Fig. 2-12. Dissolved gas drive performance.

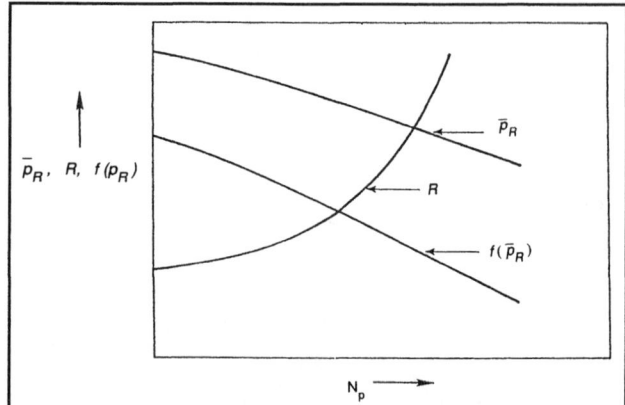

Fig. 2-13. Gas cap drive performance.

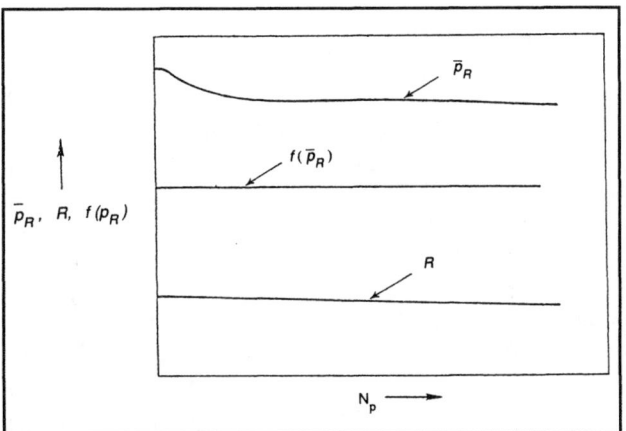

Fig. 2-14. Water drive performance.

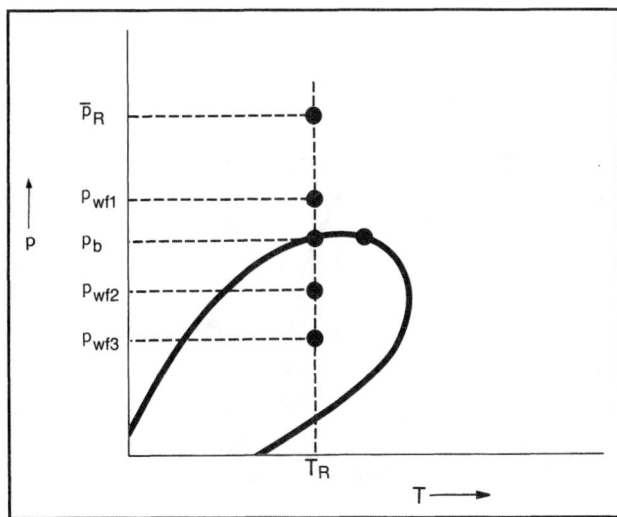

Fig. 2-15. Phase diagram.

d. Combination Drive. In many cases, an oil reservoir will be both saturated and in contact with an aquifer. In this case, all three of the previously described mechanisms may be contributing to the reservoir drive. As oil is produced, both the gas cap and aquifer will expand and the gas/oil contact will drop as the oil/water contact rises, which can cause complex production problems.

It is impossible to generalize on the expected recovery and performance of a combination-drive reservoir because of the wide variation in gas cap and aquifer sizes. The drive mechanisms may be supplemented by both gas and water injection.

2. Drawdown or Producing Rate

It was shown earlier that the principal reason for a change in the productivity index was the change in the pressure function, $f(p) = k_{ro}/\mu_o B_o$. If the pressure anywhere in the reservoir drops below bubblepoint pressure, gas will evolve and the permeability to oil will decrease, causing a decrease in J. Even though the average reservoir pressure may be above p_b, to attain a reasonable inflow rate it may be necessary to reduce p_{wf} below p_b. When this happens, a zone of reduced k_{ro} exists around the wellbore out to the radius at which the pressure in the reservoir equals p_b.

The pressure profile in a reservoir in the drainage area of a well depends on the well's skin factor, as was illustrated in Figures 2-5 and 2-6. The effects of drawdown on inflow performance will be discussed first for a well with zero skin factor. The effects of both positive and negative skin factors will then be discussed.

a. Zero Skin Factor. The effects of drawdown or production rate on inflow performance can best be illustrated graphically. The first case considered will be one in which the reservoir pressure is above bubblepoint pressure, that is, $\bar{p}_R > p_b$. The location of all of the pressures

referred to are shown on a phase diagram, Figure 2-15, a pressure profile, Figure 2-16, and an IPR, Figure 2-17.

If the desired inflow rate can be obtained with $p_{wf} \geq p_b$, the value of $f(p)$ will be fairly constant at all values of radius, and the value of J will be essentially constant.

If a production rate greater than q_{o1} is required, p_{wf} must be decreased further. If p_{wf2} is less than p_b, free-gas saturation will exist out to radius r_2, the value of k_{ro} will be decreased and the slope of the pressure profile will be increased over the distance $r_2 - r_w$. Since J depends on k_{ro}, it will also be decreased as p_{wf} drops below p_b. This is illustrated in Figure 2-17.

Further reduction in p_{wf} to p_{wf3} will extend the zone of reduced k_{ro} out to radius r_3 and will further increase the slope of the inflow performance plot. Methods to quan-

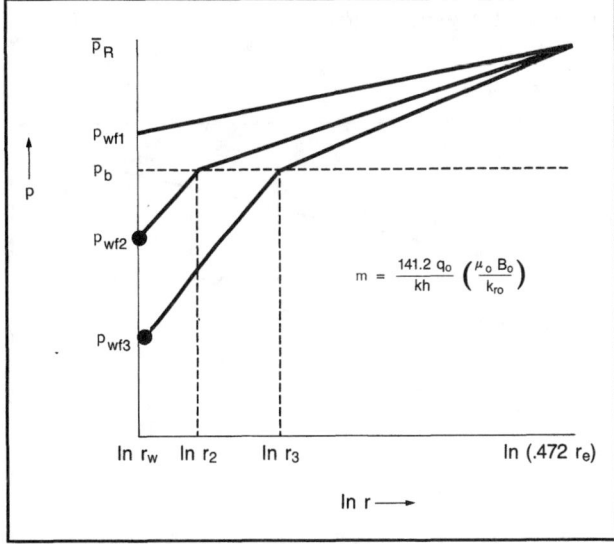

Fig. 2-16. Reservoir pressure profile.

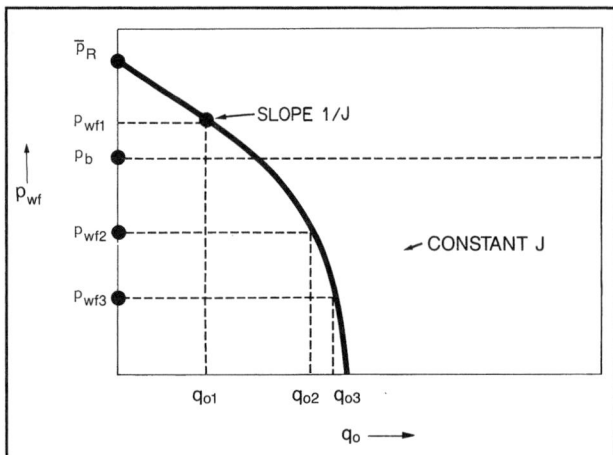

Fig. 2-17. Inflow performance relationship.

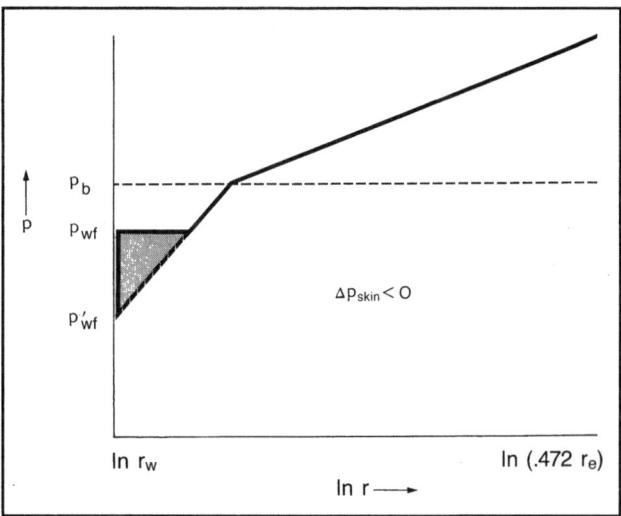

Fig. 2-19. Effect of negative skin.

tify these effects will be presented in a subsequent section.

b. Non-zero Skin Factor. In Figure 2-6, the concept of a zero-radius skin effect was illustrated by assuming that the extra pressure drop caused by damage or the decreased pressure drop caused by stimulation occurred at the wellbore. The construction of an IPR for a well with a non-zero skin factor can be more complex, especially for the case where $\overline{p}_R > p_b$. For the case of a damaged well ($S' > 0$), it is possible that essentially no gas saturation would exist in the reservoir even though $p_{wf} < p_b$. For the case of a stimulated well ($S' < 0$), there may be a negligible pressure drop through a highly stimulated zone out to a significant radius. This will distort the assumed pressure profile. These phenomena, which are illustrated in Figures 2-18 and 2-19, can cause difficulties in constructing an IPR from test data, especially for cases in which $\overline{p}_R > p_b$.

3. Effect of Depletion

In any reservoir in which the average reservoir pressure is not maintained above the bubblepoint pressure, gas saturation will increase in the entire drainage volume of the wells. This will cause a decrease in the pressure function in the form of decreased k_{ro}, which will cause an increase in the slope of the pressure profile and the IPR. Therefore, to maintain a constant inflow rate to a well, it will be necessary to increase the drawdown as $\overline{p}_R$ declines from depletion. These effects are illustrated qualitatively in Figures 2-20 and 2-21.

4. IPR Behavior of Gas Wells

The IPR for a gas well will not be linear because the inflow rate is a function of the square of p_{wf}. For dry-gas and wet-gas reservoirs, in which no liquid condenses in the reservoir, gas saturation and, therefore, permeability to gas will remain constant as $\overline{p}_R$ declines. If turbulent flow exists, the pressure drop due to turbulence will increase with flow rate, causing a deterioration in the inflow performance.

If no liquid forms in the reservoir, the effect of depletion will not cause a decrease in k_{rg}, but turbulence may increase due to the higher actual velocity required to maintain a constant-mass flow rate. Also, the value of the product μZ will change as reservoir pressure changes.

In the case of a retrograde condensate-gas reservoir, that is, where T_R is between the critical temperature and the cricondentherm, if the pressure anywhere in the reservoir drops below the dewpoint pressure p_d, liquid will form and decrease k_{rg}. This can occur from either reducing p_{wf} below p_d or as $\overline{p}_R$ declines below p_d from depletion. Prediction of retrograde-gas reservoir behavior or

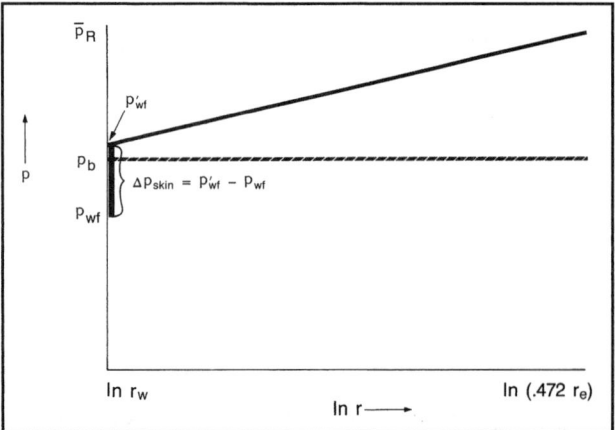

Fig. 2-18. Effect of positive skin.

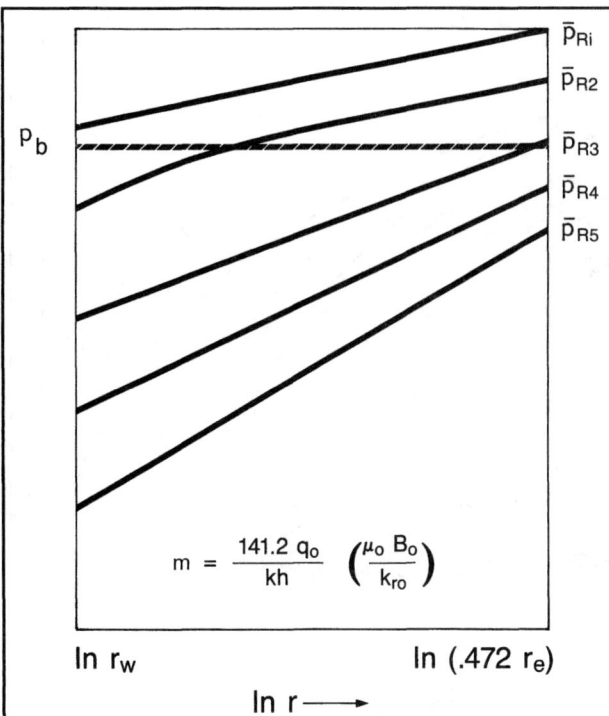

$$m = \frac{141.2\,q_o}{kh}\left(\frac{\mu_o\,B_o}{k_{ro}}\right)$$

$\ln r_w$ $\ln (.472\,r_e)$

$\ln r \longrightarrow$

Fig. 2-20. Effect of depletion on the pressure profile.

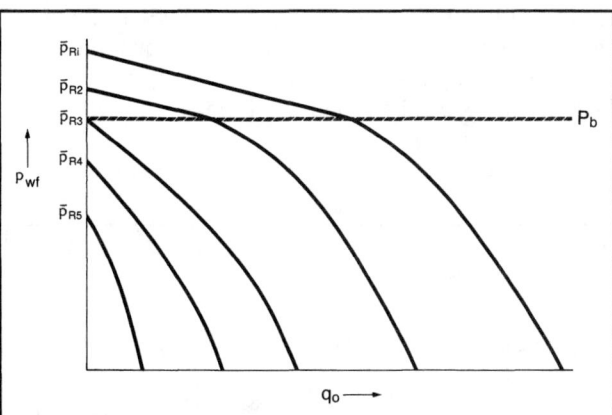

Fig. 2-21. Effect of depletion on the IPR.

water-drive gas reservoir behavior is very complex and, in most cases, requires the use of a reservoir model. Fortunately, since the condensed liquid will occupy the small pore spaces first, the reduction of k_{rg} may be small. This is illustrated qualitatively in Figure 2-9.

III. PREDICTING PRESENT TIME IPR's FOR OIL WELLS

The factors affecting the inflow performance for oil wells were discussed qualitatively in the previous section. If all of the variables in the inflow equations could be calculated, the equations resulting from integration of

Darcy's law could be used to quantify the IPR. Unfortunately, sufficient information rarely exists to accomplish this and, therefore, empirical methods must be used to predict the inflow rate for a well.

Several of the most widely used empirical methods for predicting an IPR for a well are presented in this section. Most of these methods require at least one stabilized test on a well, and some require several tests in which p_{wf} and q_o were measured. A procedure for estimating the IPR when no stabilized tests are available is also outlined.

Methods to account for the effects of drawdown only are first presented, that is, $\bar{p}_R$ is assumed constant. Modification of the methods for depletion will then be discussed.

A. Vogel Method

Vogel[5] reported the results of a study in which he used a mathematical reservoir model to calculate the IPR for oil wells producing from saturated reservoirs. The study dealt with several hypothetical reservoirs including those with widely differing oil characteristics, relative permeability characteristics, well spacing and skin factors. The final equation for Vogel's method was based on calculations made for 21 reservoir conditions.

Although the method was proposed for saturated, dissolved-gas-drive reservoirs only, it has been found to apply for any reservoir in which gas saturation increases as pressure is decreased.

Vogel's original method did not account for the effects of a non-zero skin factor, but a later modification by Standing[6] extended the method for application to damaged or stimulated wells.

The Vogel method was developed by using the reservoir model proposed by Weller[7] to generate IPR's for a wide range of conditions. He then replotted the IPR's as reduced or dimensionless pressure versus dimensionless flow rate. The dimensionless pressure is defined as the flowing wellbore pressure divided by average reservoir pressure, $p_{wf}/\bar{p}_R$. The dimensionless flow rate is defined as the flow rate that would result for the value of p_{wf} being considered, divided by the flow rate that would result from a zero wellbore pressure, that is $q_o/q_{o(max)}$. It was found that the general shape of the dimensionless IPR was similar for all of the conditions studied. Examples of these plots from the original paper are illustrated in Figures 2-22 through 2-25.

After plotting dimensionless IPR curves for all the cases considered, Vogel arrived at the following relationship between dimensionless flow rate and dimensionless pressure:

$$\frac{q_o}{q_{o(max)}} = 1 - 0.2\frac{p_{wf}}{\bar{p}_R} - 0.8\left(\frac{p_{wf}}{\bar{p}_R}\right)^2 \qquad (2\text{-}33)$$

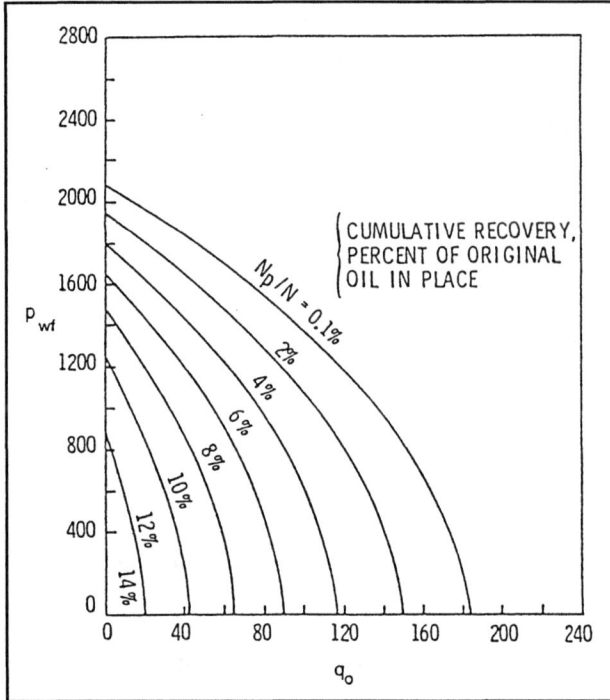

Fig. 2-22. IPR change with depletion.

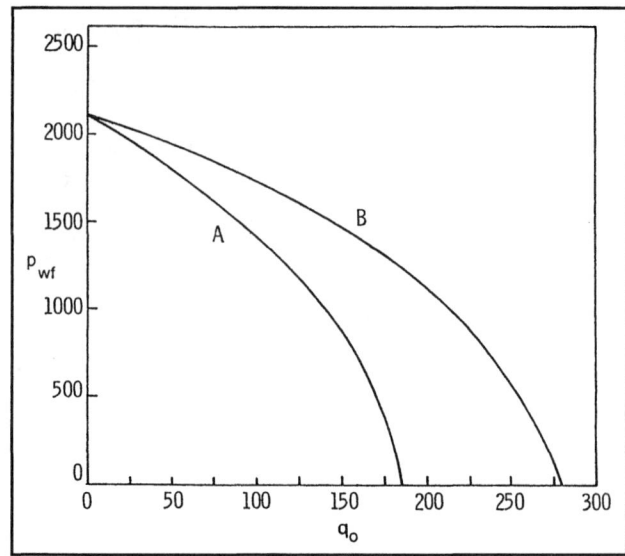

Fig. 2-24. IPR change with reservoir conditions.

$\bar{p}_R =$ average reservoir pressure existing at the time of interest.

where

q_o = inflow rate corresponding to wellbore flowing pressure p_{wf},

$q_{o(max)}$ = inflow rate corresponding to zero wellbore flowing pressure, (AOF), and

The pressures used to calculate the dimensionless pressure ratio should be gage pressures. A plot of the dimensionless IPR represented by Equation 2-33 is shown in figure 2-26, which can be used in lieu of Equation 2-33.

The dimensionless IPR for a well with a constant productivity index can be calculated from

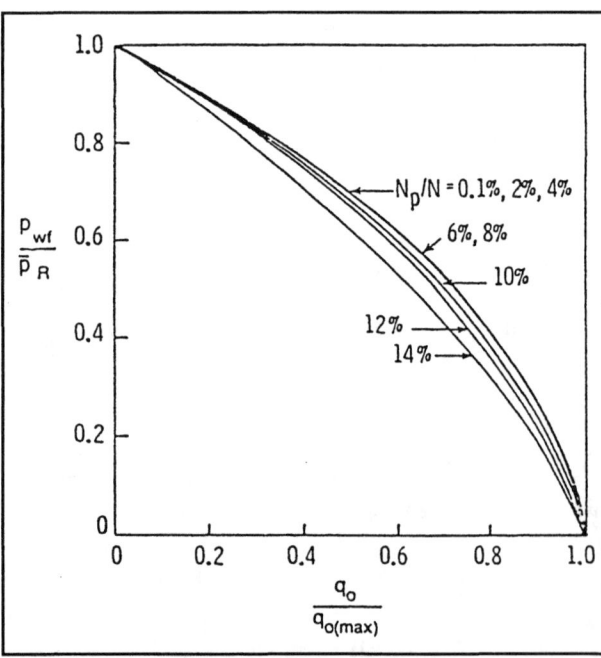

Fig. 2-23. Dimensionless IPR for Fig. 2-22.

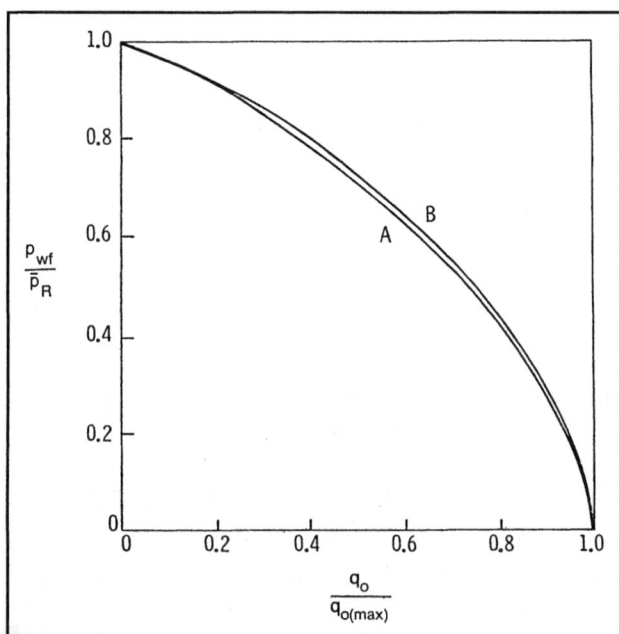

Fig. 2-25. Dimensionless IPR for Fig. 2-24.

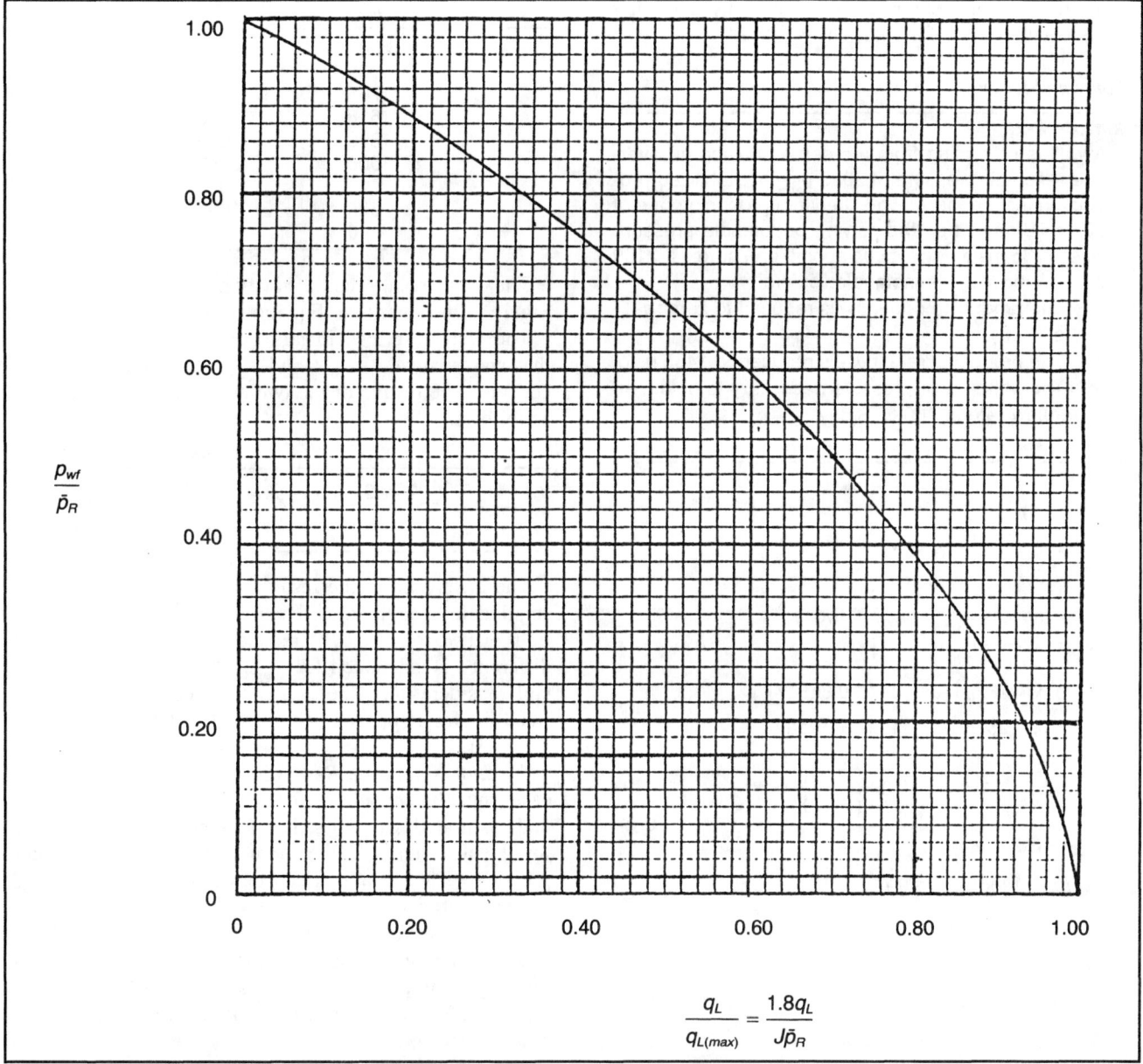

$\dfrac{q_L}{q_{L(\max)}} = \dfrac{1.8 q_L}{J \bar{p}_R}$

Fig. 2-26. Vogel's dimensionless IPR.

$$\frac{q_o}{q_{o(\max)}} = 1 - \frac{p_{wf}}{\bar{p}_R} \qquad (2\text{-}34)$$

Vogel pointed out that in most applications of his method, the error in the predicted inflow rate should be less than 10%, but could increase to 20% during the final stages of depletion. Errors made by assuming a constant J (Equation 2-34) were found to produce errors on the order of 70% to 80% at low values of p_{wf}.

It has also been shown that Vogel's method can be applied to wells producing water along with the oil and gas, since the increased gas saturation will also reduce the permeability to water. Therefore, the ratio $q_o/q_{o(\max)}$ can be replaced by $q_L/q_{L(\max)}$ where $q_L = q_o + q_w$. This has proved to be valid for wells producing at water cuts as high as 97%.

Application of Vogel's method is almost as simple as the constant J method in that only one actual well test is required. The application will be illustrated by examples for conditions in which $\bar{p}_R \leq p_b$ and for conditions in which $\bar{p}_R > p_b$.

1. Application of Vogel Method–Zero Skin Factor

In the original paper by Vogel, only cases in which the reservoir was saturated were considered. The method can be applied to undersaturated reservoirs by applying Vogel's equation only for values of $p_{wf} < p_b$.

a. Saturated Reservoirs.

Example 2-2:

A well is producing from a reservoir having an average reservoir pressure of 2085 psig. A stabilized production test on the well resulted in a producing rate of 282 STB/day when the flowing bottomhole pressure was 1765 psig. The bubblepoint pressure is 2100 psig. Using Vogel's method calculate:

1. The producing rate if p_{wf} is reduced to zero (q_{max} or AOF).
2. The producing rate if p_{wf} is reduced to 1485 psig.
3. The bottomhole pressure necessary to obtain an inflow rate of 400 STB/day.

Solution:

1. $q_o(max) = q_o / \left[1 - 0.2 \dfrac{p_{wf}}{\overline{p}_R} - 0.8 \left(\dfrac{p_{wf}}{\overline{p}_R} \right)^2 \right]$

    From the test, for $q_o = 282$ STB/day,

    $p_{wf}/\overline{p}_R = 1765/2085 = 0.847$

    $q_o(max) = 282 / [1 - 0.2(.847) - 0.8(.847)^2]$

    $q_o(max) = 282/0.257 = 1097$ STB/day

2. $q_o = q_{o(max)} \left[1 - 0.2 \dfrac{p_{wf}}{\overline{p}_R} - 0.8 \left(\dfrac{p_{wf}}{\overline{p}_R} \right)^2 \right]$

    For the new value of p_{wf},

    $p_{wf}/\overline{p}_R = 1485/2085 = 0.712$

    $q_o = 1097 \left[1 - 0.2(.712) - 0.8(712)^2 \right]$

    $\quad = 1097(0.452)$

    $q_o = 496$ STB/day

3. Solving Equation 2-33 for $p_{wf}/\overline{p}_R$:

    $\dfrac{p_{wf}}{\overline{p}_R} = [1.266 - 1.25 q_o/q_{o(max)}]^{0.5} - 0.125$

    $\dfrac{p_{wf}}{\overline{p}_R} = [1.266 - 1.25(400)/1097]^{0.5} - 0.125$

    $\dfrac{p_{wf}}{\overline{p}_R} = 0.901 - 0.125 = 0.776$

    $p_{wf} = 2085(0.776) = 1618$ psig

A complete IPR could be constructed by assuming other values of p_{wf} and calculating the corresponding q_o:

p_{wf}	q_o
2085	0
1800	253
1765	282
1618	400
1300	618
1000	790
700	923
300	1046
0	1097

The same results could have been obtained by using Figure 2-26 rather than Equation 2-33.

The well described in this example is the same well that was analyzed in Example 2-1 using the constant J concept. The results of the two analyses are summarized as follows:

	Constant J	Vogel
q for p_{wf} = 1485 psig	528	496
p_{wf} for q = 400	1645	1618
q(max) or AOF	1835	1097

The difference in the results from the two methods is small as long as the drawdown is close to the test conditions. However, as p_{wf} is reduced to zero, a substantial difference is calculated for $q_{(max)}$. This can be an important consideration if the well is being considered for artificial lift where p_{wf} can be reduced to a low value.

b. Undersaturated Reservoirs ($\overline{p}_R > p_b$). Two test cases must be considered for applying Vogel's method to undersaturated reservoirs. The flowing wellbore pressure for the test can be either above or below bubblepoint pressure. The equations can be derived by considering the productivity index to be constant for $p_{wf} \geqq p_b$ and assuming that Vogel's equation applies for $p_{wf} < p_b$. Also, it is assumed that the complete IPR is continuous, that is the slopes of the two segments are equal at $p_{wf} = p_b$. Figure 2-27 is used to illustrate the IPR for an undersaturated reservoir.

Applying Vogel's equation for any flow rate greater than the rate q_b, corresponding to $p_{wf} = p_b$:

$$\frac{q_o - q_b}{q_{o(max)} - q_b} = 1 - 0.2 \frac{p_{wf}}{p_b} - 0.8 \left(\frac{p_{wf}}{p_b} \right)^2$$

or

$$q_o = q_b + (q_{o(max)} - q_b) \left[1 - 0.2 \frac{p_{wf}}{p_b} - 0.8 \left(\frac{p_{wf}}{p_b} \right)^2 \right] \quad (2\text{-}35)$$

The reciprocal slope is defined as the change in flow rate with respect to the change in p_{wf}, or

$$\frac{dq_o}{dp_{wf}} = (q_{o(max)} - q_b) \left[\frac{-0.2}{p_b} - \frac{1.6 p_{wf}}{p_b^2} \right]$$

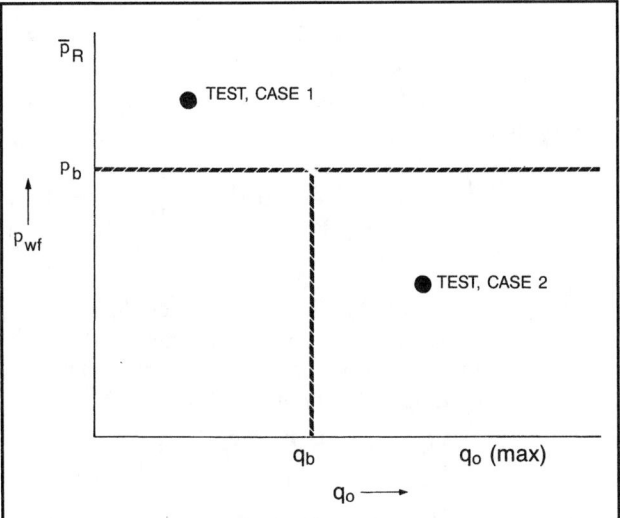

Fig. 2-27. IPR for an undersaturated reservoir.

Evaluating the reciprocal slope at $p_{wf} = p_b$ gives

$$-\frac{dq_o}{dp_{wf}} = \frac{q_{o(max)} - q_b}{p_b}(0.2 + 1.6)$$

$$-\frac{dq_o}{dp_{wf}} = \frac{1.8(q_{o(max)} - q_b)}{p_b} \tag{2-36}$$

The productivity index is defined as the negative of the reciprocal slope, and if J is evaluated at any value of $p_{wf} \geq p_b$, Equation 2-36 becomes:

$$J = \frac{1.8(q_{o(max)} - q_b)}{p_b}$$

or

$$q_{o(max)} - q_b = \frac{Jp_b}{1.8} \tag{2-37}$$

This equation also establishes a relationship between J and $q_{o(max)}$ for saturated reservoirs, that is for $p_b \geq p_R$ and $q_b = 0$. In this case:

$$q_{o(max)} = \frac{J\overline{p}_R}{1.8}$$

Substituting Equation 2-37 into Equation 2-35 gives:

$$q_o = q_b + \frac{Jp_b}{1.8}\left[1 - 0.2\frac{p_{wf}}{p_b} - 0.8\left(\frac{p_{wf}}{p_b}\right)^2\right] \tag{2-38}$$

Once a value of J at $p_{wf} \geq p_b$ is determined, Equation 2-38 can be used to generate an IPR. If the well test is taken with $p_{wf} \geq p_b$, J and q_b can be calculated directly, since:

$$J = \frac{q_o}{\overline{p}_R - p_{wf}} \tag{2-22}$$

and

$$p_b = J(\overline{p}_R - p_b) \tag{2-39}$$

If the test is such that $p_{wf} < p_b$, the calculation for J is more complex since q_b will not be known. This is illustrated as Case 2 in Figure 2-27. An expression for J to use in Equation 2-38 can be obtained by combining Equations 2-38 and 2-39:

$$J = \frac{q_o}{\overline{p}_R - p_b + \frac{p_b}{1.8}\left[1 - 0.2\frac{p_{wf}}{p_b} - 0.8\left(\frac{p_{wf}}{p_b}\right)^2\right]} \tag{2-40}$$

Case 1 Procedure (Test $p_{wf} \geq p_b$)

1. Calculate J using test data in Equation 2-22.

2. Calculate q_b using Equation 2-39.

3. Generate the IPR for values of $p_{wf} < p_b$ using Equation 2-38. The IPR for $p_{wf} \geq p_b$ is linear.

Example 2-3:

The following data pertain to an undersaturated reservoir:
$\overline{p}_R$ = 4000 psig, p_b = 2000 psig, S = 0
Test data: p_{wf} = 3000 psig for q_o = 200 STB/day
Generate an IPR.

Solution:

1. $J = \dfrac{q_o}{\overline{p}_R - p_{wf}} = \dfrac{200}{4000 - 3000} = 0.2\dfrac{STB}{day\text{-}psi}$

2. $q_b = J(\overline{p}_R - p_b) = 0.2(4000 - 2000) =$
   400 STB/day

3. $q_o = q_b + \dfrac{Jp_b}{1.8}\left[1 - 0.2\dfrac{p_{wf}}{p_b} - 0.8\left(\dfrac{p_{wf}}{p_b}\right)^2\right]$

$$q_o = 400 + \frac{0.2(2000)}{1.8}\left[1 - 0.2\frac{p_{wf}}{2000} - \frac{0.8p_{wf}^2}{(2000)^2}\right]$$

p_{wf}	q_o
4000	0
3000	200
2000	400
1500	489
1000	556
500	600
0	622

Case 2 Procedure (Test $p_{wf} < p_b$).

1. Calculate J using test data and Equation 2-40.

2. Calculate q_b using Equation 2-39.

3. Generate the IPR using Equation 2-38 for $p_{wf} < p_b$. The IPR for $p_{wf} \geq p_b$ is linear and can be calculated using $q_o = J(\bar{p}_R - p_{wf})$.

Example 2-4:

The well described in Example 2-3 was retested and the following results obtained:

p_{wf} = 1200 psig for q_o = .532 STB/day

Generate an IPR using this test data.

1. $J = \dfrac{q_o}{\bar{p}_R - p_b + \dfrac{p_b}{1.8}\left[1 - 0.2\dfrac{p_{wf}}{p_b} - 0.8\left(\dfrac{p_{wf}}{p_b}\right)^2\right]}$ (2-40)

$J = \dfrac{532}{4000 - 2000 + \dfrac{2000}{1.8}\left[1 - 0.2\left(\dfrac{1200}{2000}\right) - 0.8\left(\dfrac{1200}{2000}\right)^2\right]}$

$J = \dfrac{532}{2658} = 0.20\ \dfrac{\text{STB}}{\text{day - psi}}$

2. $q_b = J(\bar{p}_R - p_b) = 0.2(4000 - 2000) =$ 400 STB/day

3. $q_o = q_b + \dfrac{Jp_b}{1.8}\left[1 - 0.2\dfrac{p_{wf}}{p_b} - 0.8\left(\dfrac{p_{wf}}{p_b}\right)^2\right]$

$q = 400 + \dfrac{0.2(2000)}{1.8}\left[1 - \dfrac{0.2\,p_{wf}}{2000} - \dfrac{0.8\,p_{wf}^2}{(2000)^2}\right]$

p_{wf}	q_o
4000	0
3000	200
2000	400
1500	489
1200	532
700	585
500	600
0	622

2. Application of Vogel Method–Non-Zero Skin Factor (Standing Modification)

The method for generating an IPR presented by Vogel did not consider an absolute permeability change in the reservoir. Standing[6] proposed a procedure to modify Vogel's method to account for either damage or stimulation around the wellbore. The degree of permeability alteration can be expressed in terms of a Productivity Ratio PR or Flow Efficiency FE, where:

$\text{FE} = \dfrac{\text{ideal drawdown}}{\text{actual drawdown}} = \dfrac{\bar{p}_R - p'_{wf}}{\bar{p}_R - p_{wf}} = \dfrac{q/J'}{q/J} = \dfrac{J}{J'}$ (2-41)

The relationship between p_{wf} and p'_{wf} is shown in Figures 2-5 and 2-6. The flow efficiency can also be expressed in terms of Δp_{skin} and S' as:

$\text{FE} = \dfrac{\bar{p}_R - p'_{wf} - \Delta p_{skin}}{\bar{p}_R - p_{wf}} = \dfrac{\ln(.472 r_e / r_w)}{\ln(.472 r_e / r_w) + S'}$ (2-42)

Using the previous definition for flow efficiency, Vogel's equation becomes:

$\dfrac{q_o}{q_{o(\max)}^{\text{FE}=1}} = 1 - 0.2\dfrac{p'_{wf}}{\bar{p}_R} - 0.8\left(\dfrac{p'_{wf}}{\bar{p}_R}\right)^2$ (2-43)

where
$q_{o(\max)}^{\text{FE}=1}$ = the maximum inflow which could be obtained for the well if FE = 1 or $S' = 0$.

A relationship among p_{wf}, p'_{wf} and FE can be obtained by solving Equation 2-41 for p'_{wf}:

$p'_{wf} = \bar{p}_R - \text{FE}(\bar{p}_R - p_{wf})$ (2-44)

or

$\dfrac{p'_{wf}}{\bar{p}_R} = 1 - \text{FE} + \text{FE}(p_{wf}/\bar{p}_R)$ (2-45)

The following procedure was used by Standing to construct dimensionless IPR curves for flow efficiencies not equal to one:

1. Select a value for FE.

2. Assume a range of values for $p_{wf}/\bar{p}_R$.

3. For each value assumed in Step 2, calculate the corresponding value of $p'_{wf}/\bar{p}_R$ using Equation 2-45.

4. Calculate $q_o/q_{o(\max)}^{\text{FE}=1}$ for each value of $p_{wf}/\bar{p}_R$ assumed in Step 2 using Equation 2-43. Plot $p_{wf}/\bar{p}_R$ versus $q_o/q_{o(\max)}^{\text{FE}=1}$.

5. Select a new FE and go to Step 2.

The dimensionless IPR curves as presented by Standing are shown in Figure 2-28.

The fact that Standing selected the maximum inflow based on a flow efficiency of one as the normalizing flow rate limits the inflow rate that can be calculated by this method to $q_o = q_{o(\max)}^{\text{FE}=1}$. This can be seen by considering the value of $p'_{wf}/\bar{p}_R$ calculated by Equation 2-45. If FE is greater than one, negative values of $p'_{wf}/\bar{p}_R$ could be obtained at large drawdowns or small values of p_{wf}. Vogel's equation would no longer apply, since the square of the negative would become positive. The actual value of $q_{o(\max)}$ for flow efficiencies greater than one can be estimated from an extrapolation of the IPR to $p_{wf} = 0$.

The ratio of Flow Efficiencies after and before stimulation can be expressed in terms of the ratio of productivity indices or Folds of Increase as:

$\dfrac{\text{FE}_f}{\text{FE}_o} = \dfrac{J_f}{J_o} = \text{FOI}$

Standing's graph, Figure 2-28, can be put in equation form by combining Equations 2-43 and 2-45. This gives:

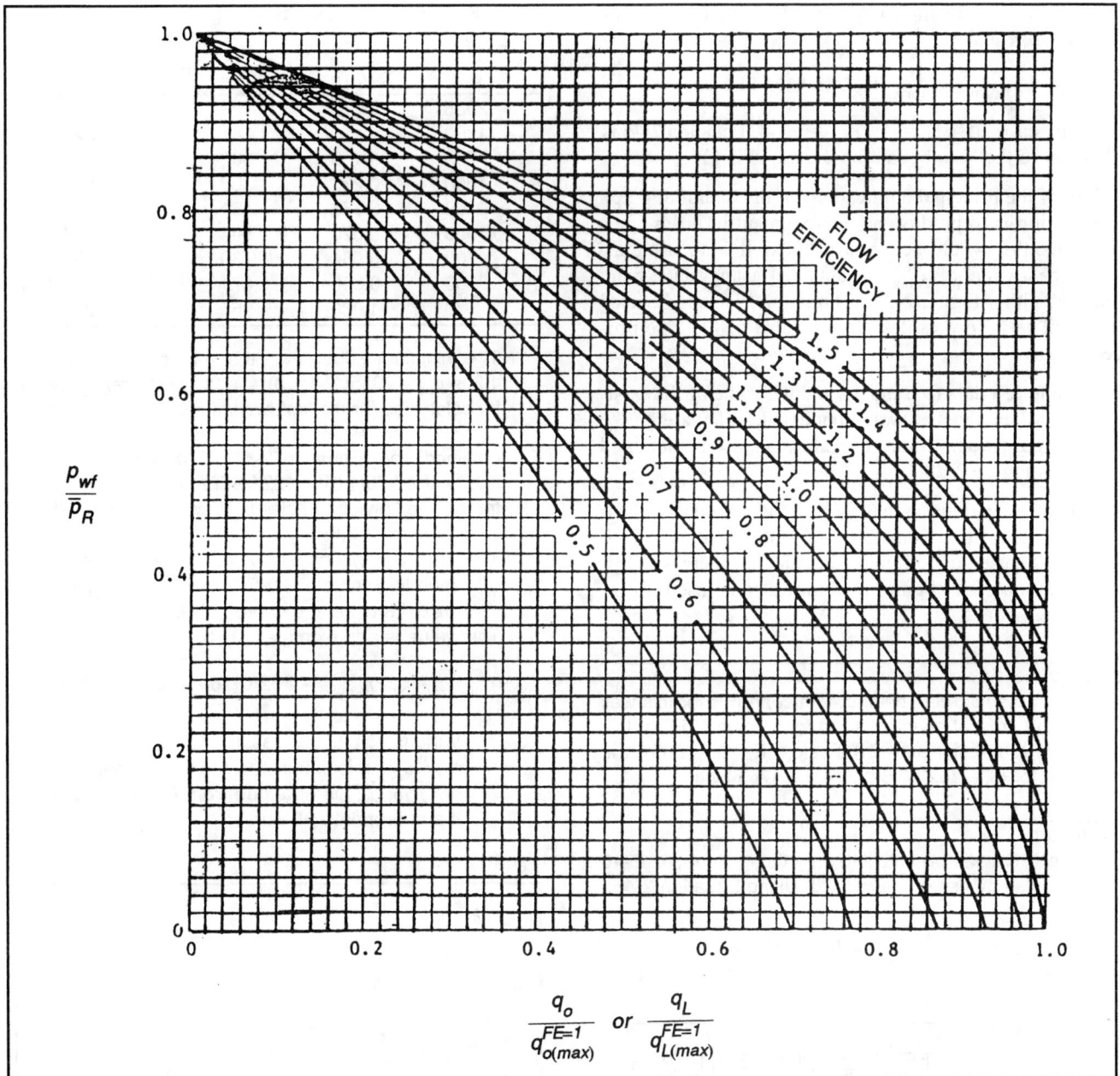

Fig. 2-28. IPR for damaged or stimulated wells.[11]

$$\frac{q_o}{q_{o(\max)}^{\mathrm{FE}=1}} = 1 - 0.2\left[1 - \mathrm{FE} + \mathrm{FE}\left(\frac{p_{wf}}{\overline{p}_R}\right)\right]$$

$$-0.8\left[1 - \mathrm{FE} + \mathrm{FE}\left(\frac{p_{wf}}{\overline{p}_R}\right)\right]^2$$

or

$$\frac{q_o}{q_{o(\max)}^{\mathrm{FE}=1}} = 1.8(\mathrm{FE})\left(1 - \frac{p_{wf}}{\overline{p}_R}\right)$$

$$-0.8(\mathrm{FE})^2\left(1 - \frac{p_{wf}}{\overline{p}_R}\right)^2$$

Because of the restriction that $p'_{wf} \geqq 0$, Equation 2-46 is valid only if

$$q_o \leq q_{o(\max)}^{\mathrm{FE}=1} \text{ or } p_{fw} \geq \overline{p}_R\left(1 - \frac{1}{\mathrm{FE}}\right)$$

This restriction will always be satisfied if FE $\leq$ 1. For values of FE > 1, an approximate relationship between the actual $q_{(max)}$ and $q_{(max)}^{FE=1}$ is

$$q_{(max)} = q_{(max)}^{FE=1}(0.624 + 0.376\,FE) \qquad (2\text{-}47)$$

For the case of FE = 1 ($p_{wf} = p'_{wf}$), Equation 2-46 is identical to the Vogel equation, Equation 2-43.

One of the principal applications of the Standing graph or equation is to predict the improvement in inflow performance that would be attained if a well were stimulated. Once a value of $q_{o(max)}^{FE=1}$ is obtained using data from one test, either Figure 2-28 or Equation 2-46 can be used to calculate inflow values for any value of FE. The procedure is:

1. Using test data (p_{wf} and q_o) and the value of FE existing when the test was conducted, calculate $q_{o(max)}^{FE=1}$ using Equation 2-46. This value can also be obtained from Figure 2-28.

2. Assume various values of p_{wf} and calculate q_o for each p_{wf} from Equation 2-46. Other values of FE may be used to determine the effect of increasing FE by stimulation. Figure 2-28 may also be used if preferred.

Example 2-5A:

Using the following data, construct an IPR for this well for the present conditions and for a value of FE = 1.3. $\bar{p}_R$ = 2085 psig, p_b = 2100 psig, FE = 0.7. From the test, for q_o = 202 STB/day, p_{wf} = 1765 psig.

Solution:

The IPR's will be calculated by using Equation 2-46 and also by using Figure 2-28. Using Equation 2-46:

1. $\left(1 - \dfrac{p_{wf}}{\bar{p}_R}\right) = 1 - \dfrac{1765}{2085} = 0.153$

$$q_{o(max)}^{FE=1} = \dfrac{q_o}{1.8(FE)\left(1 - \dfrac{p_{wf}}{\bar{p}_R}\right) - 0.8(FE)^2\left(1 - \dfrac{p_{wf}}{\bar{p}_R}\right)^2}$$

$$q_{o(max)}^{FE=1} = \dfrac{202}{1.8(.7)(.153) - 0.8(.7)^2(.153)^2}$$

$$= 1100\,STB/day$$

2. $q_o = 1100\left[1.8(FE)\left(1 - \dfrac{p_{wf}}{2085}\right)\right.$

$$\left. - 0.8(FE)^2\left(1 - \dfrac{p_{wf}}{2085}\right)^2\right]$$

Example 2-5A Results Using Equation 2-46

p_{wf}	$1 - \dfrac{p_{wf}}{\bar{p}_R}$	q_o	
		FE = 0.7	FE = 1.3
2085	0	0	0
1800	0.137	181	324
1765	0.153	202	360
1600	0.233	300	518
1300	0.376	461	758
1000	0.520	604	937
700	0.664	730	1054
300	0.856	871	–
0	1.000	955	1224

The minimum value of p_{wf} that may be used for FE = 1.3 is p_{wf} = 2085(1 -1/1.3) = 482. Therefore, q_o cannot be calculated for the last two values of p_{wf} in the table. However, an estimate of the actual $q_{(max)}$ can be obtained from Equation 2-47. That is $q_{(max)}$ = 1100 [0.624 + 0.376(1.3)] = 1224 STB/day. A plot of the IPR for the two flow efficiencies is shown in Figure 2-29.

Using Figure 2-28:

1. Using $p_{wf}/\bar{p}_R$ = 1765/2085 = 0.847, the corresponding value of $q_o/q_{o(max)}^{FE=1}$ obtained from Figure 2-28 using the curve for FE = 0.7.

This value is approximately 0.18.

$$q_{o(max)}^{FE=1} = q_o/q_{o(max)}^{FE=1} = 202/0.18 = 1122\ STB/day$$

2. Various values are assumed for p_{wf}, the ratio of $p_{wf}/\bar{p}_R$ is calculated, and the corresponding ratio of $q_o/q_{o(max)}^{FE=1}$ is obtained from Figure 2-28 from the appropriate FE curve:

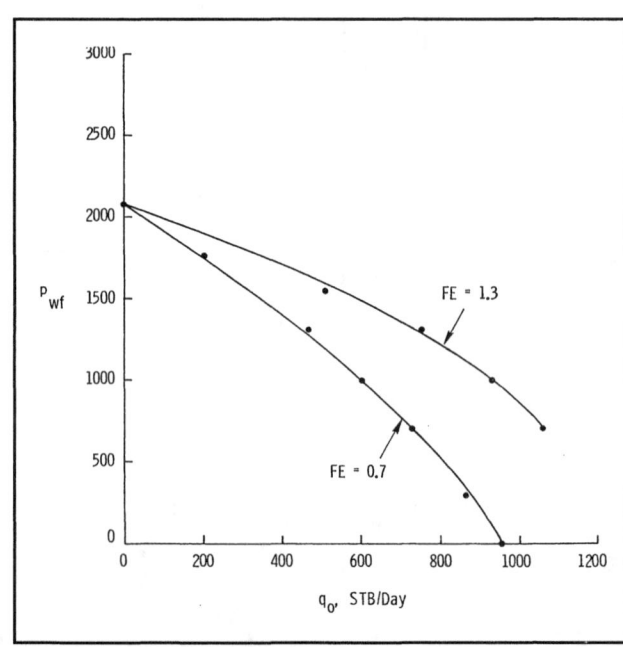

Fig. 2-29. Example 2-5A solution.

Example 2-5A Results Using Fig. 2-28

p_{wf}	$\dfrac{p_{wf}}{\bar{p}_R}$	$q_o/q_{o(max)}^{FE=1}$		q_o	
		FE = 0.7	FE = 1.3	FE = 0.7	FE = 1.3
2085	1	0	0	0	0
1600	0.767	0.267	0.472	300	530
1000	0.480	0.550	0.852	617	956
700	0.336	0.668	0.963	750	1080
300	0.144	0.792	–	890	–
0	0	0.870	–	976	–

There is some difference between the values obtained using the equation and those obtained using the graph. Most of this is caused by the 22 STB/day difference obtained in the value of $q_{o(max)}^{FE=1}$ using the test which was conducted at a low drawdown. The curves are difficult to read at low values of drawdown or high values of $p_{wf}/\bar{p}_R$. Use of the equation is therefore recommended.

a. Undersaturated Reservoirs with FE ≠ 1.0. Standing's modification of Vogel's method to be used when the flow efficiency is not equal to one may also be applied to undersaturated reservoirs.

Equation 2-38 may be modified for a nonzero S' or FE not equal to one to obtain

$$q_o = J(\bar{p}_R - p_b) + \frac{Jp_b}{1.8}\left[1.8\left(1 - \frac{p_{wf}}{p_b}\right)\right.$$
$$\left. -0.8(FE)\left(1 - \frac{p_{wf}}{p_b}\right)^2\right] \qquad (2\text{-}48)$$

The following procedures may be used to generate an IPR for any value of FE, including the case for FE = 1.

Case 1 Procedure (Test $p_{wf} \geq p_b$)

1. Calculate J using test data in Equation 2-22.

2. Generate the IPR for values of $p_{wf} < p_b$ using the known value of FE in Equation 2-48. The IPR for $p_{wf} \geq p_b$ is linear.

3. For values of FE other than that which existed during the test, the value of J is modified by
$$J_2 = J_1(FE)_2/(FE)_1$$

where

$$\begin{aligned}
J_2 &= \text{new value for use in Step 2.}\\
J_1 &= \text{value calculated from test data using } (FE)_1\\
(FE)_1 &= \text{flow efficiency existing during test, and}\\
(FE)_2 &= \text{any other flow efficiency.}
\end{aligned}$$

Case 2 Procedure (Test $p_{wf} < p_b$)

1. Calculate J using test data in equation 2-48.

2. Generate the IPR for values of $p_{wf} \leq p_b$ using

Equation 2-48.

3. For other values of FE, modify J as discussed previously.

Example 2-5B:

Using the following data, calculate the flow rate which will result if p_{wf} is 1500 psig for this well for the present conditions and for a value of FE = 1.4.

$\bar{p}_R$ = 4000 psig, p_b = 2000 psig, FE = 0.7

From a test, q_o =378 STB/day, p_{wf} = 1200 psig

Solution:

Since p_{wf} for the test is below p_b, use Case 2 Procedure.

$$\left(1 - \frac{p_{wf}}{p_b}\right) = 1 - \frac{1200}{2000} = 0.4$$

1. $J = \dfrac{378}{4000 - 2000 + \dfrac{2000}{1.8}\left[1.8.4 - 0.8\,(.7)(.4)^2\right]}$

$J = \dfrac{378}{2000 + 700} = 0.14$

$q_o = 0.14(4000 - 2000) + \dfrac{0.14(2000)}{1.8}$
$$\bullet\left[1.8\left(1 - \frac{p_{wf}}{2000}\right) - 0.8(0.7)\left(1 - \frac{p_{wf}}{2000}\right)^2\right]$$

2. For FE = 0.7 :

For p_{wf} = 1500 psig, $1 - \dfrac{1500}{2000} = 0.25$

$q_o = 280 + 156\,[1.8(.25) - 0.8(0.7)(.25)^2]$

$q_o = 280 + 156(0.415) = 345$ STB/day

For FE = 1.4 :

$J_2 = J_1(FE)_2/(FE)_1 = 0.14(1.4)/0.7 = 0.28$

$q_o = 0.28(2000) + \dfrac{0.28(2000)}{1.8}\left[1.8\left(1 - \dfrac{p_{wf}}{2000}\right)\right.$
$$\left. -0.8(1.4)\left(1 - \frac{p_{wf}}{2000}\right)^2\right]$$

For p_{wf} = 1500 psig , q_o = 560 + 311 [1.8(.25) - 0.8(1.4)(.25)^2] = 678 STB/day

b. Determining FE from Well Tests. A value for the flow efficiency can be calculated if the skin factor is known from a pressure transient test, using Equation 2-42. If values of r_e and r_w are not known exactly, an approximation for FE can be made by assuming that $\ln(.472\, r_e/r_w) \approx 7$. Then:

$$FE \approx \frac{7}{7 + S'} \qquad (2\text{-}49)$$

If two correct stabilized tests are available and $\bar{p}_R$ is accurately known, FE can be calculated directly. Solving Equation 2-46 for $q_{o(max)}^{FE=1}$:

$$q_{o(max)}^{FE=1} = \frac{q_o}{1.8(FE)\left(1-\dfrac{p_{wf}}{p_R}\right)-0.8(FE)^2\left(1-\dfrac{p_{wf}}{p_R}\right)^2}$$

The parameter $q_{o(max)}^{FE=1}$ is a constant and, therefore, must be equal for any two tests. That is:

$$\frac{q_{o1}}{1.8(FE)\left(1-\dfrac{p_{wf1}}{\bar{p}_R}\right)-0.8(FE)^2\left(1-\dfrac{p_{wf1}}{\bar{p}_R}\right)^2}$$
$$=\frac{q_{o2}}{1.8(FE)\left(1-\dfrac{p_{wf2}}{\bar{p}_R}\right)-0.8(FE)^2\left(1-\dfrac{p_{wf2}}{\bar{p}_R}\right)^2}$$

where subscripts 1 and 2 refer to the two tests.
Simplifying and solving for flow efficiency gives:

$$FE = \frac{2.25\left[\left(1-\dfrac{p_{wf1}}{\bar{p}_R}\right)q_{o2}-\left(1-\dfrac{p_{wf2}}{\bar{p}_R}\right)q_{o1}\right]}{\left(1-\dfrac{p_{wf1}}{\bar{p}_R}\right)^2 q_{o2}-\left(1-\dfrac{p_{wf2}}{\bar{p}_R}\right)^2 q_{o1}} \qquad (2\text{-}50)$$

Example 2-6:

Two stabilized tests were conducted on a well that was producing from a reservoir in which the average pressure was 2085 psig. Calculate the flow efficiency for this well.

Test	p_{wf}, psig	q_o, STB/day
1	1605	3000
2	1020	6000

Solution:

$$\left(1-\frac{p_{wf1}}{\bar{p}_R}\right)=1-\frac{1605}{2085}=0.230$$

$$\left(1-\frac{p_{wf2}}{\bar{p}_R}\right)=1-\frac{1020}{2085}=0.511$$

$$FE=\frac{2.25\left[(0.230)(6000)-(0.511)(3000)\right]}{(0.230)^2(6000)-(0.511)^2(3000)}=0.74$$

Equation 2-50 is extremely sensitive to small changes in pressures. It should be emphasized that the value for FE calculated by this method is only approximate and requires accurate test data. Obtaining values of FE using the skin factor from a transient test in Equation 2-42 is much more accurate.

c. Summary of the Vogel-Standing Equations. Equation 2-48 can be used for all of the cases considered previously, that is, for both saturated and undersaturated reservoirs and for wells having formation damage or that have been stimulated. In this section it will be shown that Equation 2-51 degenerates to all of the simpler cases.

$$q_L = J(\bar{p}_R - p_b)+\frac{Jp_b}{1.8}$$
$$\bullet\left[1.8\left(1-\frac{p_{wf}}{p_b}\right)-0.8[FE]\left(1-\frac{p_{wf}}{p_b}\right)^2\right] \qquad (2\text{-}51)$$

For FE = 1 ($S' = 0$):

$$q_L = J(\bar{p}_R - p_b)+\frac{Jp_b}{1.8}$$
$$\bullet\left[1-0.2\left(\frac{p_{wf}}{p_b}\right)-0.8\left(\frac{p_{wf}}{p_b}\right)^2\right] \qquad (2\text{-}52)$$

For FE = 1 and $p_b \geq \bar{p}_R$

$$q_L = \frac{J\bar{p}_R}{1.8}\left[1-0.2\left(\frac{p_{wf}}{\bar{p}_R}\right)-0.8\left(\frac{p_{wf}}{\bar{p}_R}\right)^2\right]$$

$$q_L = q_{L(max)}\left[1-0.2\left(\frac{p_{wf}}{\bar{p}_R}\right)-0.8\left(\frac{p_{wf}}{\bar{p}_R}\right)^2\right] \qquad (2\text{-}53)$$

B. Fetkovich Method

Fetkovich[8] proposed a method for calculating the inflow performance for oil wells using the same type of equation that has been used for analyzing gas wells for many years. The procedure was verified by analyzing isochronal and flow-after-flow tests conducted in reservoirs with permeabilities ranging from 6 md to greater than 1000 md. Pressure conditions in the reservoirs ranged from highly undersaturated to saturated at initial pressure and to a partially depleted field with a gas saturation above the critical.

In all cases, oil-well back-pressure curves were found to follow the same general form as that used to express the inflow relationship for a gas well. That is:

$$q_o = C(\bar{p}_R^2 - p_{wf}^2)^n \qquad (2\text{-}54)$$

where

q_o = producing rate, .
$\bar{p}_R$ = average reservoir pressure,
p_{wf} = flowing wellbore pressure,
C = flow coefficient, and
n = exponent depending on well characteristics.

The value of n ranged from 0.568 to 1.000 for the 40 field tests analyzed by Fetkovich. The applicability of Equation 2-54 to oil well analysis was justified by writing Darcy's equation as:

$$q = \frac{.00708kh}{\ln (.472r_e / r_w) + S'} \int_{p_{wf}}^{\overline{p}_R} f(p)\,dp \qquad (2\text{-}55)$$

where

$$f(p) = \frac{k_{ro}}{\mu_o B_o}$$

For an undersaturated reservoir, the integral is evaluated over two regions as:

$$q_o = C' \int_{p_{wf}}^{p_b} f_1(p)\,dp + C' \int_{p_b}^{\overline{p}_R} f_2(p)\,dp \qquad (2\text{-}56)$$

where

$$C' = \frac{.00708kh}{\ln (.472r_e / r_w) + S'}$$

It was assumed that for $p > p_b$, k_{ro} is equal to one and that μ_o and B_o could be considered constant at $\overline{p} = (\overline{p}_R + p_b)/2$. It was also assumed that for $p < p_b$, $f(p)$ could be expressed as a linear function of pressure, that is:

$$f_1(p) = ap + b \qquad (2\text{-}57)$$

Making these substitutions into Equation 2-55 and integrating gives:

$$q_o = C_1(p_b^2 - p_{wf}^2) + C_2(\overline{p}_R - p_b) \qquad (2\text{-}58)$$

Fetkovich then stated that the composite effect results in an equation of the form:

$$q_o = C(\overline{p}_R^2 - p_{wf}^2)^n \qquad (2\text{-}54)$$

Once values for C and n are determined from test data, Equation 2-54 can be used to generate a complete IPR. As there are two unknowns in Equation 2-54, at least two tests are required to evaluate C and n, assuming $\overline{p}_R$ is known. However, in testing gas wells it has been customary to use at least four flow tests to determine C and n because of the possibility of data errors. This is also recommended for oil well testing.

By taking the log of both sides of Equation 2-54 and solving for $\log (\overline{p}_R^2 - p_{wf}^2)$, the expression can be written as:

$$\log (\overline{p}_R^2 - p_{wf}^2) = \frac{1}{n}\log q_o - \frac{1}{n}\log C$$

A plot of $p_R^2 - p_{wf}^2$ versus q_o on log-log scales will result in a straight line having a slope of $1/n$ and an intercept of $q_o = C$ at $\overline{p}_R^2 - p_{wf}^2 = 1$. The value of C can also be calculated using any point on the linear plot once n has been determined. That is:

$$C = \frac{q_o}{(\overline{p}_R^2 - p_{wf}^2)^n}$$

Three types of tests are commonly used for gas-well testing to determine C and n. These tests can also be used for oil wells and will be described in this section. The type of test to choose depends on the stabilization time of the well, which is a function of the reservoir permeability. If a well stabilizes fairly rapidly, a conventional flow-after-flow test can be conducted. For tight wells, an isochronal test may be preferred. For wells with very long stabilization times, a modified isochronal test may be more practical. The stabilization time for a well in the center of a circular or square drainage area may be estimated from:

$$t_s = \frac{380\phi\mu_o C_t A}{k_o} \qquad (2\text{-}59)$$

where

t_s	=	stabilization time, hrs.,
ϕ	=	porosity,
C_t	=	total fluid compressibility, psi^{-1},
A	=	drainage area, ft^2,
k_o	=	permeability to oil, md,
μ_o	=	oil viscosity, cp

1. Flow-After-Flow testing

A flow-after-flow test begins with the well shut in so that the pressure in the entire drainage area is equal to $\overline{p}_R$. The well is placed on production at a constant rate until the flowing wellbore pressure becomes constant. The flowing pressure should be measured with a bottomhole pressure gage, especially for oil-well tests. Once p_{wf} has stabilized, the production rate is changed, and the procedure is repeated for several rates.

The idealized behavior of production rate and wellbore pressure with time is shown in Figure 2-30. The test may also be conducted using a decreasing rate sequence.

The test is analyzed by plotting $\overline{p}_R^2 - p_{wf}^2$ versus q_o on log-log coordinates and drawing the best straight line through the points. The exponent n is determined from the reciprocal of the slope of the line. That is:

$$n = \frac{\Delta \log q_o}{\Delta \log (\overline{p}_R^2 - p_{wf}^2)} \qquad (2\text{-}60)$$

It is common practice to read the change in q_o over one log cycle of change in $\overline{p}_R^2 - p_{wf}^2$, since the difference in the log value over one cycle is equal to one.

2. Isochronal Testing

If the time required for the well to stabilize on each choke size or producing rate is excessive, an isochronal or equal time test is preferred. The procedure for conducting an isochronal test is:

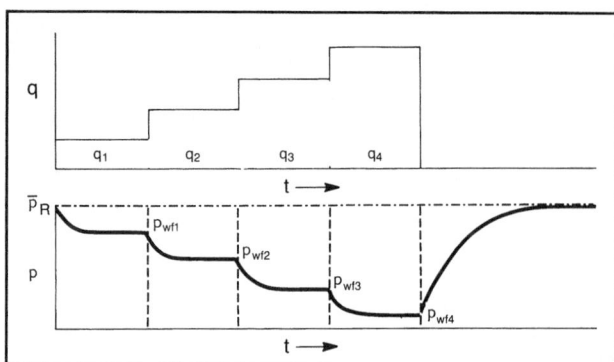

Fig. 2-30. Conventional test—producing rate and pressure diagrams.

a: Starting at a shut-in condition, open the well on a constant production rate and measure p_{wf} at specific time periods. The total production period for each rate may be less than the stabilization time.

b. Shut the well in and allow the pressure to build up to $\bar{p}_R$.

c. Open the well on another producing rate and measure the pressure at the same time intervals.

d. Shut the well in again until $p_{ws} = \bar{p}_R$.

e. Repeat this procedure for several rates.

The values of $\bar{p}_R{}^2 - p_{wf}{}^2$ determined at the specific time periods are plotted versus q_o and n is obtained from the slope of the line. To determine a value for C, one test must be a stabilized test. The idealized behavior of producing rate and pressure as a function of time is shown in figure 2-31.

3. Modified Isochronal Testing

If the shut-in time required for the pressure to build back up to $\bar{p}_R$ between flow periods is excessive, the isochronal test may be modified. The modification consists of shutting the well in between each flow period for

a period of time equal to the producing time. The static well bore pressure p_{ws}, may not reach $\bar{p}_R$, but a plot of $p_{wsi}{}^2 - p_{wfi}{}^2$ versus q_o will usually produce a straight line, from which n may be obtained. A stabilized test is still required to calculate a value for C. The testing procedure is illustrated in Figure 2-32.

Example 2-7A:

A flow-after-flow test was conducted on a well producing from a reservoir in which $\bar{p}_R = 3600$ psia. The test results were:

q_o, STB/day	p_{wf}, psia
263	3170
383	2897
497	2440
640	2150

Construct a complete IPR for this well and determine $q_{o(max)}$.

Solution:

q_o, STB/day	p_{wf}, psia	$(\bar{p}_R^2 - p_{wf}^2) \times 10^{-6}$, psia^2
263	3170	2.911
383	2897	4.567
497	2440	7.006
640	2150	8.338

The data are plotted on Figure 2-33. To calculate n, the producing rates corresponding to a change in Δp^2 over one cycle are used.

$$n = \frac{\Delta \log q_o}{\Delta \log \Delta p^2} = \frac{\log 750 - \log 105}{\log 10^7 - \log 10^6} = 0.854$$

$$C = \frac{q_o}{\left(\bar{p}_R^2 - p_{wf}^2\right)^n} = \frac{750}{\left(10^7\right)^{854}}$$

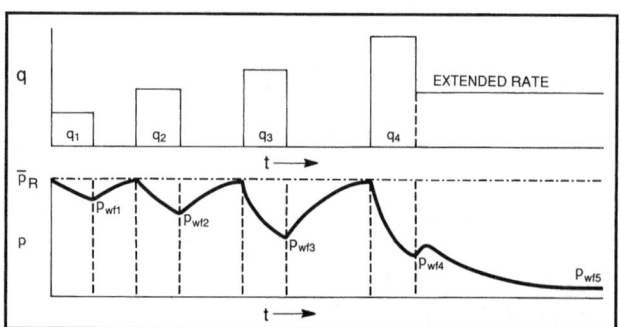

Fig. 2-31. Isochronal test-producing rate and pressure diagrams.

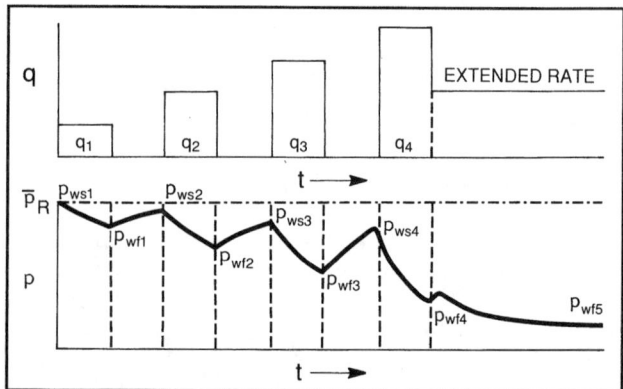

Fig. 2-32. Modified isochronal test—producing rate and pressure diagrams.

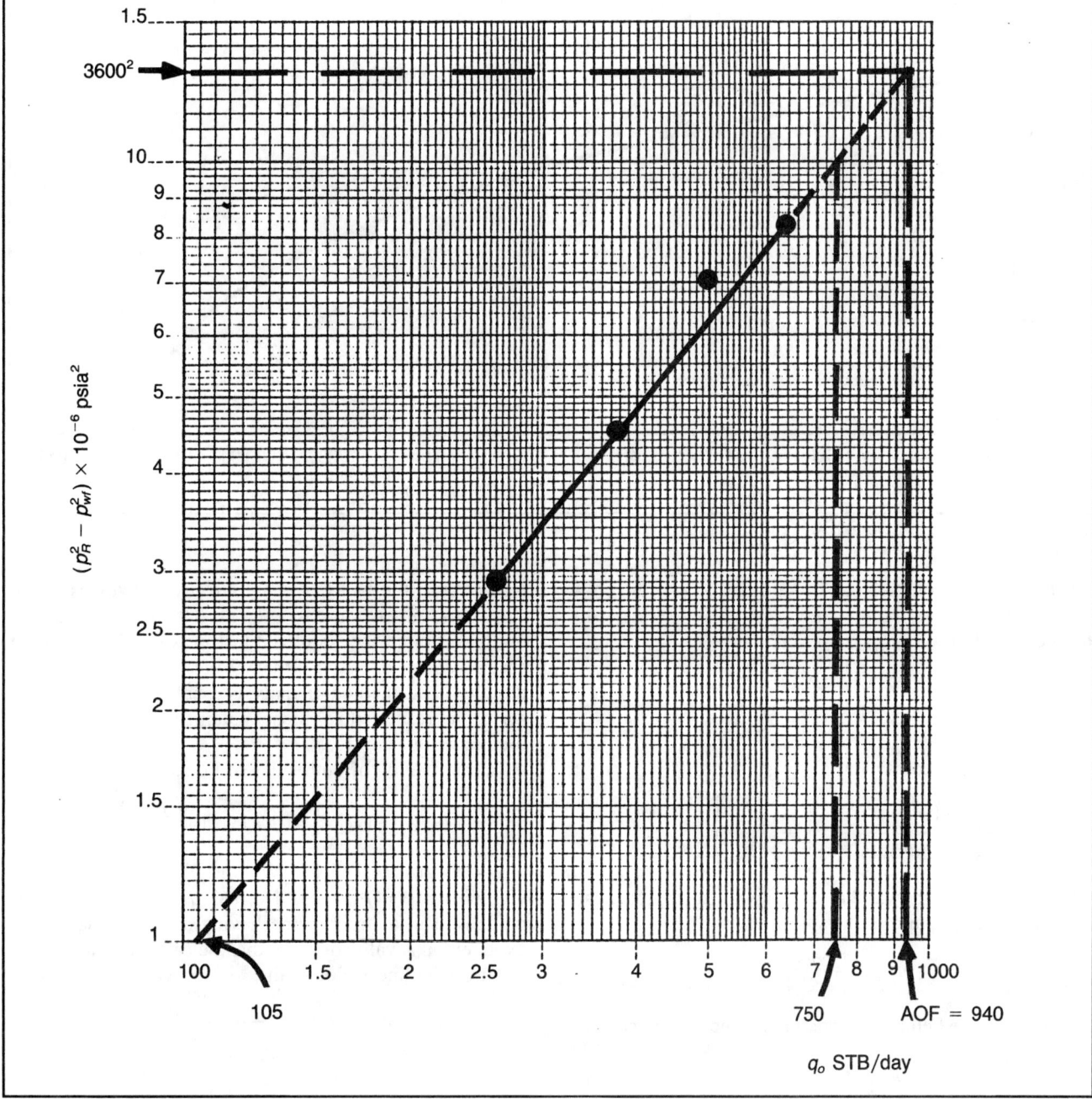

Fig. 2-33. Flow-after-flow results.

$$=0.00079 \frac{\text{STB}}{\text{day-psia}^{1.71}}$$

The inflow equation is therefore:

$$q_o = 0.00079 \, (3600^2 - p_{wf}^2)^{.854}$$
$$q_{o(\text{max})} = 0.00079(3600^2 - 0)^{.854} = 937 \text{ STB/day}$$

To generate the data for an IPR, assume values of p_{wf} and calculate the corresponding q_o:

p_{wf}, psia	q_o, STB/day
3600	0
3000	340
2500	503
2000	684
1500	796
1000	875
500	922
0	937

The IPR is shown in Figure 2-34.

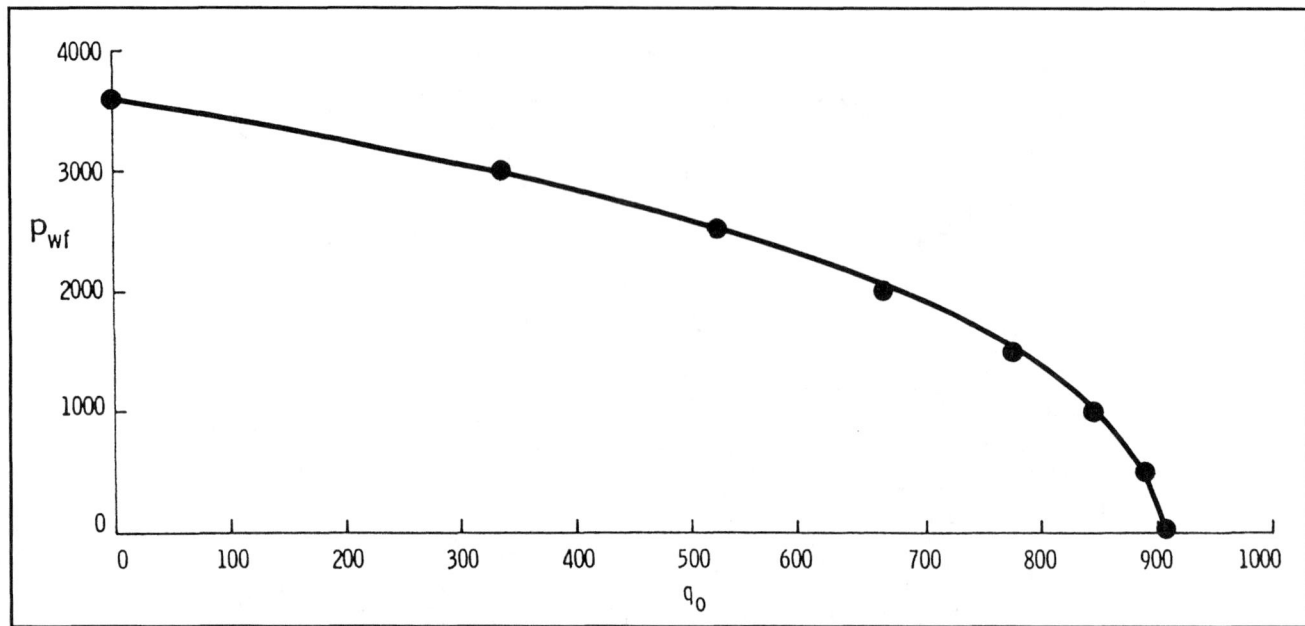

Fig. 2-34. IPR for Example 2-7 A.

The Fetkovich equation can be modified to a form similar to Vogel's equation and stated in terms of Productivity Index J or $q_{L(max)}$ (AOF).

$$q_L = C(\overline{p}_R^2 - p_{wf}^2)^n$$

$$q_{L(max)} = \text{AOF} = C(\overline{p}_R^2 - 0)^n$$

Eliminating the coefficient C gives:

$$\frac{q_L}{q_{L(max)}} = \frac{(\overline{p}_R^2 - p_{wf}^2)^n}{(\overline{p}_R^2)^n} = \left[1 - \left(\frac{p_{wf}}{\overline{p}_R}\right)^2\right]^n$$

It can also be shown that as drawdown approaches zero, that is as p_{wf} approaches $\overline{p}_R$,

$$q_{L(max)} = \frac{J\overline{p}_R}{2}$$

Therefore, the Fetkovich equation can be expressed as:

$$q_L = \frac{J\overline{p}_R}{2}\left[1 - \left(\frac{p_{wf}}{p_R}\right)^2\right]^n \qquad (2\text{-}61)$$

Fetkovich also suggested that the analysis could be further broken down for undersaturated reservoirs as:

$$q_L = J(\overline{p}_R - p_b) + \frac{Jp_b}{2}\left[1 - \left(\frac{p_{wf}}{p_b}\right)^2\right]^n \qquad (2\text{-}62)$$

Application of either Equation 2-54 or Equation 2-61 to analyze a flow-after-flow test requires at least two stabilized production tests. For isochronal testing at least two transient rates and one stabilized rate are required. This

results from the fact that there are two unknowns in the equations, either C and n or J and n. It should be pointed out that if only one stabilized test is available, n is often assume to be one and either C or J can be calculated directly. This method of analysis usually gives more conservative results than those obtained using the Vogel method with FE = 1.

Taking the log of both sides of Equation 2-61 gives:

$$\log(q_L) = \log\left(\frac{J\overline{p}_R}{2}\right) + n\log\left[1 - \left(\frac{p_{wf}}{\overline{p}_R}\right)^2\right]$$

A plot of $[1 - (p_{wf}/\overline{p}_R)^2]$ versus q_L on log-log scales will result in a straight line having a slope equal to the exponent n. A value of J can then be calculated using any point on the linear plot from:

$$J = \frac{2q_L}{\overline{p}_R\left[1 - \left(\frac{p_{wf}}{\overline{p}_R}\right)^2\right]^n}$$

Example 2-7B

The well described in Examples 2-1 and 2-2 is to be analyzed using the Fetkovich equation with the assumption that $n = 1$. One production test on the well resulted in a rate of 282 STB/day for $p_{wf} = 1765$ psig = 1780 psia. The static reservoir pressure is 2085 psig = 2100 psia.

Calculate:
1. Productivity Index J
2. The new producing rate if $p_{wf} = 1500$ psia

3. The value of p_{wf} required for q_L = 400 STB/day
4. $q_{L(max)}$ or AOF.

Solution:

1. $$J = \frac{2q_L}{\overline{p}_R\left[1-\left(\frac{pwf}{pR}\right)^2\right]^{1.0}} = \frac{2(282)}{(2100)\left[1-\left(\frac{1780}{2100}\right)^2\right]}$$

$$J = \frac{564}{591} = 0.95 \text{ STB/day-psi}$$

2. $$q_L = \frac{J\overline{p}_R}{2} = \left[1-\left(\frac{p_{wf}}{\overline{p}_R}\right)^{2n}\right]$$

$$= \frac{0.95(2100)}{2}\left[1-\left(\frac{1500}{2100}\right)^2\right]$$

$$q_L = 489 \text{ STB/day}$$

3. Solving Equation 2-61 for p_{wf} and assuming n = 1:

$$p_{wf} = \overline{p}_R\left(1-\frac{2q_L}{J\overline{p}_R}\right)^{0.5}$$

$$p_{wf} = 2100\left[1-\frac{2(400)}{0.95(2100)}\right]^{0.5} = 1625 \text{ psia}$$

4. $$q_{L(max)} = \text{AOF} = \frac{J\overline{p}_R}{2} = \frac{0.95(2100)}{2}$$
$$= 998 \text{ STB/day}$$

The values for $q_{L(max)}$ obtained from the three methods used to analyze this well test may be compared:

Method	$q_{L(max)}$
Constant J	1835
Vogel	1097
Fetkovich (n = 1)	998

C. Jones, Blount and Glaze Method

In 1976 Jones, Blount and Glaze[9] established a paper discussing the effects of turbulence or non-Darcy flow on well performance. Methods were presented to analyze well completion efficiency and to isolate the rate-dependent component of the total pressure drawdown. Although this paper will be discussed in greater detail in the section on Well Completion Effects, use of their plotting procedure to determine real time IPR's will be presented here.

Equation 2-27 can be written with the turbulence term included as:

$$\overline{p}_R - p_{wf} = Aq_o + Bq_o^2 \qquad (2-63)$$

where

$$A = \frac{141.2\mu_o B_o}{k_o h}\left[\ln(0.472r_e/r_w) + S\right]$$

$$B = \frac{2.3\times10^{-14}\beta B_o^2\rho_o}{h^2 r_w} = \frac{141.2\mu_o B_o}{k_o h}D$$

ρ_o = oil density evaluated at T_R and

$0.5(\overline{p}_R + p_{wf})$, lbm/ft^3, and

β = velocity coefficient, ft^{-1}

The other terms in Equation 2-63 have been defined previously and β can be estimated from Figure 2-35[10] or calculated using Equation 2-8.

$$\beta = \frac{2.329\times10^{10}}{k_o^{1.2}}$$

where k_o is in millidarcies.

The contribution to the pressure drawdown due to laminar or Darcy flow is expressed as Aq_o while the non-Darcy or turbulent contribution is expressed as Bq_o^2. Dividing Equation 2-63 by q_o gives:

$$\frac{\overline{p}_R = p_{wf}}{q_o} = A + Bq_o \qquad (2-64)$$

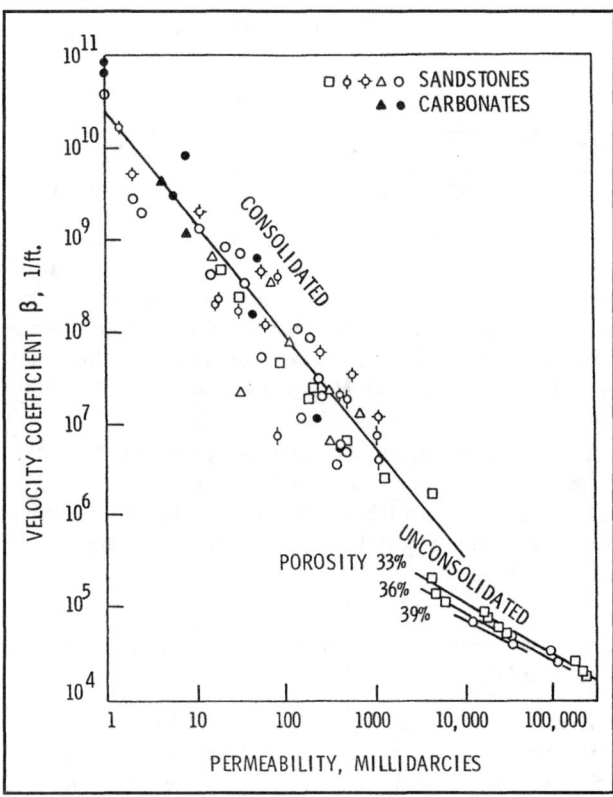

Fig. 2-35. Velocity coefficient correlation.[10]

A plot of $(\bar{p}_R - p_{wf})/q_o$ versus q_o on Cartesian coordinates should yield a straight line of slope B and intercept A as q_o approaches zero. Once A and B are determined, a complete IPR can be constructed using Equation 2-63 or 2-64. At least two stabilized tests are required to evaluate A and B, but usually more tests will be used to smooth out the effects of errors in measurements.

Equation 2-63 can be solved for flow rate to yield:

$$q_o = \frac{-A + \left[A^2 + 4B(\bar{p}_R - p_{wf}) \right]^{0.5}}{2B}$$

Example 2-8A:

The test data presented in Example 2-7A are to be analyzed using the Jones method. Reservoir pressure $\bar{p}_R$ is 3600 psia. Using the data from the four tests, find A, B and $q_{o(max)}$.

Solution:

q_o, STB/day	p_{wf}, psia	$(\bar{p}_R - p_{wf})/q_o$, psia/STB/day
263	3170	1.635
383	2897	1.850
497	2440	2.334
640	2150	2.266

The test data are plotted in Figure 2-36. The slope B is 2.2×10^{-3} psia/(STB/d)^2 and the intercept at $q_o = 0$ gives a value of $A = 1.05$ psia/STB/d. To determine $q_{o(max)}$, set $p_{wf} = 0$:

$$q_{o(max)} = \frac{-1.05 + \left[1.05^2 + 4(2.2 \times 10^{-3})(3600 - 0) \right]^{0.5}}{2(2.2 \times 10^{-3})}$$

$$q_{o(max)} = 1063 \text{ STB/day}$$

The value obtained for $q_{o(max)}$ using the Fetkovich method was 937 STB/day. As can be observed in Figures 2-33 and 2-36, there is considerable scatter in the data, and the choice of which points to use for constructing the lines is questionable. If a least–squares procedure had been used to obtain the slopes, the results would likely have been in closer agreement.

This method may give optimistic results for oil wells because of the assumption that the A term is constant. As the A term includes the reciprocal of the pressure function, reducing p_{wf} below bubblepoint pressure will increase A.

D. Constructing IPR's When No Stabilized Tests Are Available

It is frequently necessary to estimate the inflow performance of a well before the well has been completed, and therefore no stabilized tests would be available. All of the previously described methods for constructing IPR's require at least one stabilized test.

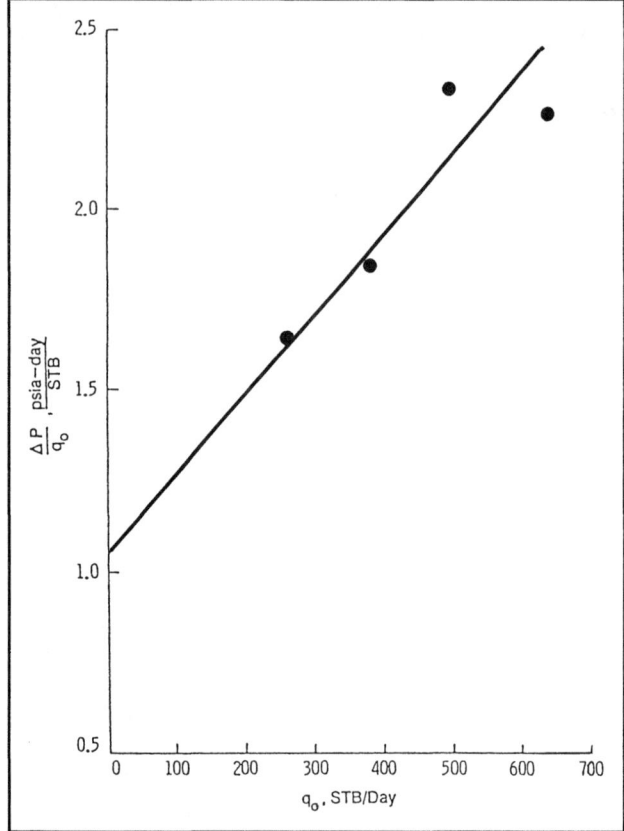

Fig. 2-36. Example 2-8A solution.

The construction of an IPR before completion is required to determine the tubing size, the number of perforations, the need for stimulation, and for sizing of surface equipment. If all the parameters in Equation 2-38 could be determined, and if the bubblepoint pressure p_b, for the reservoir were known, q_b could be calculated, and then the Vogel method could be used to construct an IPR, using $p_{wf} = p_b$ and $q_o = q_b$ as a test point. If a fluid sample is available, then B_o and μ_o can be determined. The wellbore radius r_w, will be known from the bit size and r_e can be estimated, depending on anticipated well spacing. This leaves the permeability to oil k_o, and the skin factor S' to be determined. These can be determined from transient well tests such as a drill stem test. Reservoir pressure, $\bar{p}_R$ can also be obtained from the drill stem test. The procedure is:

1. Using fluid property data and the values of $k_o h$ and S' from the transient test, calculate:

$$J = \frac{0.00708 \, k_o h}{\mu_o B_o [\ln(.472 r_e / r_w) + S']}$$

2. Generate the IPR for values of $p_{wf} < p_b$ using Equation 2-48. If $p_b \geq \bar{p}_R$, use $p_b = \bar{p}_R$ in the equation. The IPR is linear for values of $p_{wf} \geq p_b$. If p_b is

unknown, assume $p_b = \bar{p}_R$. Then use Equation 2-46, on page 27.

E. IPR Construction for Special Cases

All the methods for constructing present-time IPR's discussed previously were based on analysis of wells drilled vertically through a single zone. It was also assumed that the wells had reached pseudo steady state or stabilized flow. Construction of IPR's for some special cases will be discussed in this section. These cases are: horizontal wells, wells in waterflood projects, wells producing from stratified formations and cases in which $\bar{p}_R$ is unknown.

1. Horizontal Wells

It is becoming more common to complete wells by drilling a long horizontal hole into the producing formation. This method of completion can have several advantages when compared to conventional vertical completions. The produced fluid does not have to converge into such a small area, and, therefore, the velocities and friction losses are reduced. It has been found that water and gas coning tendencies are lower for horizontal wells and that a larger volume of the reservoir can be drained by each well.

The actual production mechanism or reservoir flow regimes are more complicated than those for a vertical well, especially if the horizontal section is of considerable length. Some combination of both linear and radial flow actually exists, and the well may behave in a manner similar to that of a well that has been extensively fractured. Generation of data to construct an IPR is best accomplished with a numerical reservoir model, and this has been discussed by Sherrard, et al.[19] They also reported that the shape of measured IPR's for horizontal wells completed in the Prudhoe Bay field was similar to those predicted by the Vogel or Fetkovich methods. That is, the Productivity Index J decreased with increased drawdown. The productivity index for a horizontal well in which permeability difference in the vertical and horizontal directions is small was described by Giger, et al.,[20] as:

$$J = \frac{0.00708 \, k_H}{\mu_o B_o X} \qquad (2\text{-}65)$$

where

$$X = \frac{1}{h} \ln \left[1 + \frac{\sqrt{1 + (L/2r_e)^2}}{L/2r_e} \right] + \frac{B^2}{L}(h/2r_w)$$

$$B = \sqrt{k_H / k_v}$$

k_H = effective permeability to oil in the horizontal direction, md,

k_v = effective permeability in the vertical direction, md, and,

L = length of the horizontal section, ft, and

h = vertical thickness of the formation, ft.

Sherrard, et al.,[19] found that productivity gains of from two to four times those for vertical wells could be obtained for L values of 1500 ft.

The complex flow regime existing around a horizontal wellbore probably precludes using a method as simple as that of Vogel to construct an IPR. However, if at least two stabilized tests can be obtained, the Fetkovich equation, Equation 2-62, could be used. With two tests, values of both J and n could be calculated. In this case, these values would not only account for effects of turbulence and gas saturation around the wellbore, but also for the effects of the nonradial flow regime existing in the reservoir. Bendakhlia and Aziz[24] used a complex reservoir model to generate IPR's for a number of wells and found that the Vogel equation would fit the generated data if expressed as:

$$\frac{q_o}{q_{o(\max)}} = \left(1.0 - V \frac{p_{wf}}{\bar{p}_R} - (1-V)\left(\frac{p_{wf}}{\bar{p}_R} \right)^2 \right)^n$$

In order to apply this equation to well test data, at least three stabilized tests are required to evaluate the three unknowns, $q_{o(\max)}$, V and n.

2. Waterflood Wells

The inflow performance of a well producing from a reservoir that is being waterflooded can be influenced by the facts that the static reservoir pressure, $\bar{p}_R$, usually remains constant and bubblepoint pressure may be fairly low. Also, the water saturation in the reservoir will change with time, causing the liquid productivity index to change. If the reservoir was depleted to a fairly low pressure before water injection was started, solution gas will be low, and the bubblepoint pressure depends on solution gas/oil ratio. A waterflood bubblepoint pressure can be calculated using the solution gas/oil ratio existing at the start of water injection and by assuming that some of the free gas existing at that time will be reabsorbed in the oil as pressure is increased. It can be assumed that the IPR will be linear for values of $p_{wf} \geq p_b$. For $p_{wf} < p_b$, Vogel's equation may be used to account for the effect of gas saturation developing around the wellbore when p_{wf} is below p_b.

The total liquid Productivity Index is:

$$J = \frac{0.00708kh}{[\ln(.472r_e / r_w) + S]} \left[\frac{k_{ro}}{\mu_o B_o} + \frac{k_{rw}}{\mu_w B_w} \right] \qquad (2\text{-}66)$$

As water saturation increases, k_{ro} will decrease and k_{rw}

will increase. The sum of k_{ro} and k_{rw} will first decrease and then increase. Therefore, J will follow the same trend, and even though $\bar{p}_R$ may remain constant, both producing water fraction f_w and J will change with time. Waterflood theory can be used to predict the change in water saturation and, therefore, f_w with time.

3. Stratified Formations

All the previous discussion on constructing Inflow Performance Relationships for producing wells has been concerned with wells that are producing from a single formation. In many cases the produced liquid will contain water, and the water fraction may increase during the life of the well. This is true especially of water-drive reservoirs or reservoirs undergoing pressure maintenance by water injection. Also, some wells are perforated into two or more zones, and the production from all zones is commingled in the wellbore. This can cause both the producing water cut and gas/liquid ratio (GLR) to change with drawdown if the commingled zones have different characteristics. As will be shown in Chapter 3, calculation of the outflow or piping system performance requires accurate values for f_w and GLR.

Analyzing the performance of a commingled well can be illustrated by considering the case where two zones having different values of $\bar{p}_R$, f_w, GLR, and q_{max} or J are producing into a common wellbore. This is illustrated in Figure 2-37.

Consider the case where $\bar{p}_{R2}$ is greater than $\bar{p}_{R1}$. If p_{wf} is greater than $\bar{p}_{R1}$, liquid will flow into Zone 1 from Zone 2. There will be no net production until the wellbore pressure is low enough so that flow from the higher zone is more than the rate that will flow into the lower pressure zone. This value of wellbore pressure at which net production begins must be determined to construct a composite or total IPR. It can be calculated by setting $q_2 = q_1$, and assuming that the productivity index is linear at small drawdowns.

$$q_2 = J_2(\bar{p}_{R2} - p_{wf}^*) = q_1 = J_1(p_{wf}^* - \bar{p}_{R1})$$

or

$$p_{wf}^* = \frac{\bar{p}_{R1} + \bar{p}_{R2}(J_2/J_1)}{1 + J_2/J_1}$$

When p_{wf} is lower than p_{wf}^*, net production will occur. p_{wf}^* will correspond to the condition of zero inflow on the total IPR. Construction of the total IPR requires calculation of the inflow from each zone at various p_{wf} values. The inflow from each zone is added for the total q_L corresponding to each p_{wf}. The characteristics of each zone would have to be known from core and log data or from production logging. The individual and total IPR's are illustrated in Figure 2-38.

Values for total f_w and GLR applying at any value of p_{wf} can be calculated from:

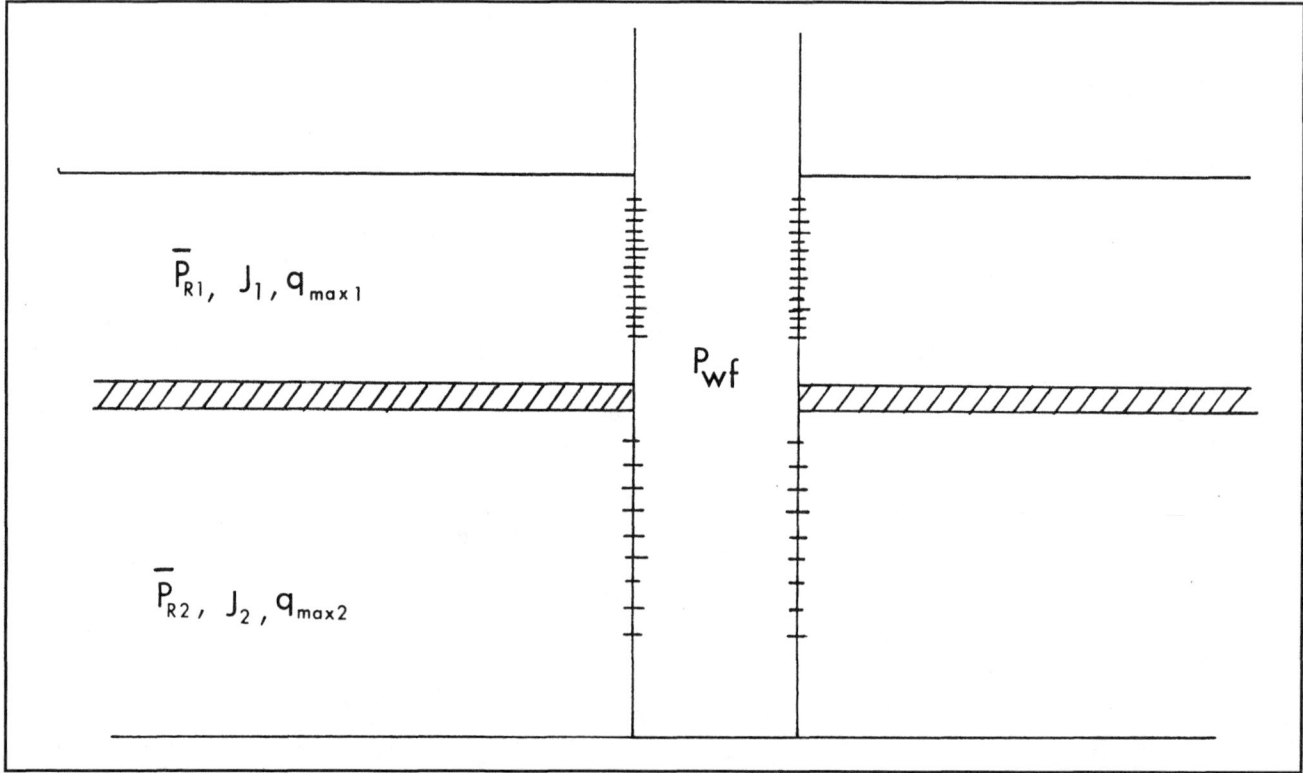

Fig. 2-37. Stratified reservoir.

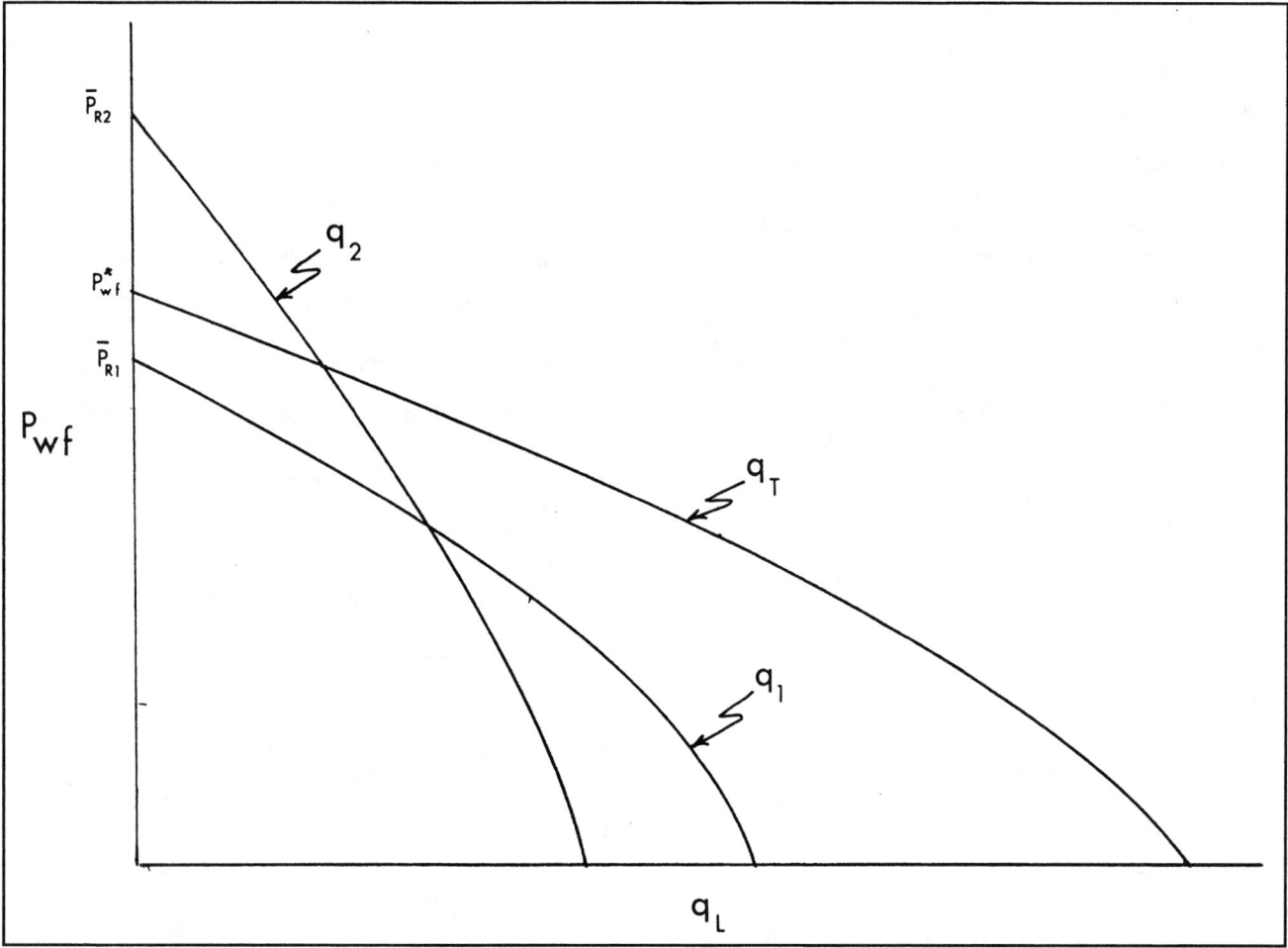

Fig. 2-38. Composite IPR.

$$f_w = \frac{\sum_{i=1}^{N} q_{Li} f_{wi}}{q_T}$$

$$\text{GLR} = \frac{\sum_{i=1}^{N} q_{Li} \text{GLR}_i}{q_T}$$

All of the previously discussed IPR methods for oil wells can be applied to wells that are producing considerable amounts of water along with the oil. This can be accomplished by replacing the oil rate q_o used in the graphs and equations with total liquid rate q_L. This is valid because the change in J is caused mostly from the reduction of relative permeability to liquid as gas saturation forms in the reservoir. The gas will reduce the permeability to water in the same way that it reduces the permeability to oil.

If Vogel's method is used, his graph or equation can be considered to be a relationship between $p_{wf}/\overline{p}_R$ and $q_L/q_{L(\text{max})}$ where $q_L = q_o + q_w$. Field experience has shown that this method of handling water production is valid in wells producing at 97% water cut.

4. Static Reservoir Pressure Unknown

Cases frequently arise in which an analysis must be made on a well at is producing from a reservoir in which $\overline{p}_R$ is not accurately known. $\overline{p}_R$ can be obtained by measuring pressure in a well that has been shut in for a long period of time or from a pressure buildup test. Obtaining an accurate value for $\overline{p}_R$ may be expensive and time consuming. If it is unknown, IPR's may be constructed for a possible range of static pressures, and the maximum and minimum production rates can then be determined through Nodal Analysis™. However, if several stabilized production tests are available, the value of $\overline{p}_R$ required to force one of the IPR equations to reproduce the test data may be calculated.

If the simple form of Vogel's equation, Equation 2-33 is used, there are only two unknowns, that is $q_{(\text{max})}$ and $\overline{p}_R$. Therefore, if two tests are available, both may be calculated. However, as discussed previously, this model assumes that the flow efficiency is one or that $S' = 0$, which is rarely the actual case. If the Fetkovich equation, Equation 2-54, is used, there will be three unknowns: $\overline{p}_R$, C and n. Solving for all of these unknowns would require three stabilized tests. If only two tests are available, n could be assumed to be one and $\overline{p}_R$ and C could be calculated. This would give results similar to those obtained using Vogel's equation, but would be easier to solve for $\overline{p}_R$.

Several methods could be used to solve for the three unknowns, but a simple trial and error or iterative method will be suggested here. Assuming that available test data include three producing rates and three corresponding values of p_{wf}, C can be eliminated from the Fetkovich equation to give:

$$C = \frac{q_1}{\left(\overline{p}_R^2 - p_{wf1}^2\right)^n} = \frac{q_2}{\left(\overline{p}_R^2 - p_{wf2}^2\right)^n} = \frac{q_3}{\left(\overline{p}_R^2 - p_{wf3}^2\right)^n}$$

$\overline{p}_R^2$ can be calculated using tests 1 and 2 and also using tests 1 and 3 to yield:

$$\overline{p}_{R(1-2)}^2 = \frac{p_{wf1}^2(q_2/q_1)^{1/n} - p_{wf2}^2}{(q_2/q_1)^{1/n} - 1}$$

$$\overline{p}_{R(1-3)}^2 = \frac{p_{wf1}^2(q_3/q_1)^{1/n} - p_{wf3}^2}{(q_3/q_1)^{1/n} - 1}$$

Only one value for $\overline{p}_R$ actually exists, and a value for n can be determined that will give $\overline{p}_{R(1-2)} = \overline{p}_{R(1-3)}$. This can be accomplished by trial and error. This procedure will be illustrated by choosing three of the tests reported in Example 2-7A and assuming that $\overline{p}_R$ is unknown:

Example 2-8B:
Test Data:

q_o, STB/day	p_{wf}, psia
263	3170
383	2897
640	2150

Use this data to estimate $\overline{p}_R, n, C$ and $q_{o(max)}$ or AOF.

Solution:

$$\overline{p}_{R(1-2)}^2 = \frac{(3170)^2(383/263)^{1/n} - (2897)^2}{(383/263)^{1/n} - 1}$$

$$\overline{p}_{R(1-3)}^2 = \frac{(3170)^2(640/263)^{1/n} - (2150)^2}{(640/263)^{1/n} - 1}$$

Estimated n	$\overline{p}_{R(1-2)}$	$\overline{p}_{R(1-3)}$
1.00	3699	3719
0.50	3395	3339
0.75	3549	3526
0.85	3609	3604

The two values for $\overline{p}_R$ are approximately equal for $n = 0.85$ and $\overline{p}_R = 3606$ psia. Using the largest flow rate to calculate C gives:

$$C = \frac{q_o}{(\overline{p}_R^2 - p_{wf}^2)^n} = \frac{640}{(3606^2 - 2150^2)^{0.85}}$$

$$C = 0.00083 \text{ STB/day-psia}^{1.7}$$

$$q_{(max)} = \text{AOF} = 0.00083 \, (3606^2 - 0)^{0.85} =$$

$$925 \text{ STB/day}$$

IV. PREDICTING FUTURE IPR's FOR OIL WELLS

As the pressure in an oil reservoir declines from depletion, the ability of the reservoir to transport oil will also decline. This is caused from the decrease in the pressure function as relative permeability to oil is decreased due to increasing gas saturation.

Planning the development of a reservoir with respect to sizing equipment and planning for artificial lift, as well as evaluating the project from an economics standpoint, requires the ability to predict reservoir performance in the future. The effect of depletion was discussed previously, and in this section several methods to quantify this effect will be presented.

A. Standing Method
Standing[11] published a procedure that can be used to predict the decline in the value of $q_{o(max)}$ as gas saturation in the reservoir increases from depletion. Vogel's equation (Equation 2-33) can be rearranged to yield:

$$\frac{q_o}{q_{o(max)}} = \left(1 - \frac{p_{wf}}{\overline{p}_R}\right)\left(1 + 0.8\frac{p_{wf}}{\overline{p}_R}\right) \qquad (2\text{-}67)$$

Substituting the expression for the productivity index (Equation 2-22) into Equation 2-67 and rearranging gives:

$$J = \frac{q_{o(max)}}{\overline{p}_R}\left(1 + 0.8\frac{p_{wf}}{\overline{p}_R}\right) \qquad (2\text{-}68)$$

Standing then defined a "zero drawdown" productivity index as:

$$J^* = \lim_{p_{wf} \to \overline{p}_R} J = \frac{1.8 q_{o(max)}}{\overline{p}_R} \qquad (2\text{-}69)$$

or

$$q_{o(max)} = \frac{J^* \overline{p}_R}{1.8}$$

If the change in J^* with depletion can be predicted, then the change of $q_{o(max)}$ can be calculated. Standing observed that another definition of J^* is:

$$J* = \frac{.00708kh}{\ln\left(\frac{0.472r_e}{r_w}\right)}(f(\bar{p}_R)) \qquad (2\text{-}70)$$

where

$$f(\bar{p}_R) = \frac{k_{ro}}{\mu_o B_o}$$

The pressure function will change with depletion since μ_o and B_o are functions of $\bar{p}_R$, and k_{ro} is a function of oil and gas saturation. The relationship between the present or real time $J*$ and some future time value of $J*$ can be expressed as:

$$\frac{J_F^*}{J_P^*} = \frac{f(\bar{p}_{RF})}{f(\bar{p}_{RP})} \qquad (2\text{-}71)$$

where

J_F* = value of $J*$ when $\bar{p}_{RP}$ has declined to $\bar{p}_{RF}$

J_P* = value of $J*$ at the present reservoir pressure.

This future J may be used directly in Equation 2-48 or if the version requiring a value for $q_{(max)}$, Equation 2-33, is used, combining Equations 2-71 and 2-69 gives a relationship between $q_{o(max)F}$ and $q_{o(max)p}$ as:

$$q_{o(max)F} = q_{o(max)p}\left[\frac{\bar{p}_{RF}f(\bar{p}_{RF})}{\bar{p}_{RP}f(\bar{p}_{RP})}\right] \qquad (2\text{-}72)$$

Once a value of $q_{o(max)p}$ is determined from a well test conducted at the present or real time, future values of $q_{o(max)}$ can be predicted at $\bar{p}_{RF}$. The value of the oil saturation as a function of $\bar{p}_R$ can be estimated using a material balance calculation or other reservoir model, and then k_{ro} can be determined if relative permeability data for the reservoir in question are available. The fluid properties μ_o and B_o can be obtained from a fluid sample analysis or from empirical correlations.

Once the value of $q_{o(max)}$ or J has been adjusted, future IPR's can be generated from

$$q_{o(F)} = q_{o(max)F}\left[1 - 0.2\frac{p_{wf}}{\bar{p}_{RF}} - 0.8\left(\frac{p_{wf}}{\bar{p}_{RF}}\right)^2\right] \qquad (2\text{-}73)$$

or

$$q_{o(F)} = \frac{J_F^* \bar{p}_{RF}}{1.8}\left[1 - 0.2\frac{p_{wf}}{\bar{p}_{RF}} - 0.8\left(\frac{p_{wf}}{\bar{p}_{RF}}\right)^2\right]$$

The procedure for generating a future IPR is:

1. Calculate $q_{o(max)p}$ using present-time well test data and either Equation 2-33 or Figure 2-26.

2. Using fluid property, saturation and relative permeability data, calculate both $f(\bar{p}_{RP})$ and $f(\bar{p}_{RF})$.

3. Calculate $J*_F$ using Equation 2-71 or $q_{o(max)F}$ using Equation 2-72.

4. Generate the future IPR using Equation 2-73 or Figure 2-26.

Example 2-9:

The following example was used by Standing to illustrate the method of generating a future IPR.

	Present Time	*Future Time*
$\bar{p}_R$	2250 psig	1800 psig
μ_o	3.11 cp	3.59 cp
B_o	1.173 bbl/STB	1.150 bbl/STB
S_o	0.768	0.741
k_{ro}	0.815	0.685

Present time test data:
q_o = 400 STB/day,
p_{wf} = 1815 psig
Generate IPR's for both the present and future times.

Solution:

1. $q_{o(max)p} = q_o / \left[1 - 0.2\frac{p_{wf}}{\bar{p}_{RP}} - 0.8\left(\frac{p_{wf}}{\bar{p}_{RP}}\right)^2\right]$

$q_{o(max)p} = 400/\left[1 - 0.2\left(\frac{1815}{2250}\right) - 0.8\left(\frac{1815}{2250}\right)^2\right]$

$q_{o(max)p} = 400/0.318 = 1257$ STB/day

The present or real time IPR can be calculated from:

2. $q_{o(p)} = 1257\left[1 - 0.2\frac{p_{wf}}{2250} - 0.8\frac{p_{wf}^2}{(2250)^2}\right]$

$f(\bar{p}_{RP}) = (k_{ro}/i_o^m B_o)_p = 0.815/(3.11)(1.173)$

$f(\bar{p}_{RP}) = 0.223$

$f(\bar{p}_{RF}) = (k_{ro}/i_o^m B_o)_F = 0.685/(3.59)(1.150)$

$f(\bar{p}_{RF}) = 0.166$

3. $q_{o(max)F} = 1257\left[\frac{1800(.166)}{2250(.223)}\right] = 749$ STB/day

4. The future time IPR can now be calculated from:

$q_{o(F)} = 749\left[1 - 0.2\frac{p_{wf}}{1800} - 0.8\frac{p_{wf}^2}{(1800)^2}\right]$

p_{wf}	$q_{o(P)}$	$q_{o(F)}$
2250	0	–
2000	197	–
1800	378	0
1600	542	142
1400	690	270
500	1148	661
0	1257	749

The real time and future IPR's are plotted in Figure 2-39.

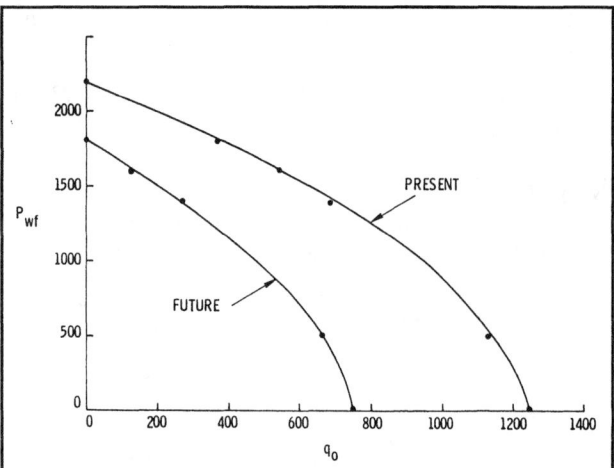

Fig. 2-39. Example 2-9 solution.

Both the present time and future IPR's are plotted in Figure 2-40.

C. Combining Vogel and Fetkovich

The method proposed by Fetkovich for adjusting C can also be used to adjust $q_{o(\max)}$ if a value for the exponent n is assumed. The expressions for $q_{o(\max)p}$ and $q_{o(\max)F}$ can be expressed using the Fetkovich equation as:

$$q_{o(\max)P} = C_P(\bar{p}_{RP}^2)^n \qquad (2\text{-}75)$$

$$q_{o(\max)F} = C_P\left(\frac{\bar{p}_{RF}}{\bar{p}_{RP}}\right)(\bar{p}_{RP}^2)^n \qquad (2\text{-}76)$$

Combining Equations 2-75 and 2-76 and solving for $q_{o(\max)F}$ gives

$$q_{o(\max)F} = q_{o(\max)p}\left(\frac{\bar{p}_{RF}}{\bar{p}_{RP}}\right)^{2n+1} \qquad (2\text{-}77)$$

If a value of n equal to one is assumed, then:

$$q_{o(\max)F} = q_{o(\max)p}\left(\bar{p}_{RF}\ /\ \bar{p}_{RP}\right)^3 \qquad (2\text{-}78)$$

Use of this method is analogous to using the Fetkovich method for both present and future IPR construction if it is assumed that $n = 1$.

Adjustment of $q_{o(\max)}$ for declining $\bar{p}_R$ using this method is illustrated in the following example:

B. Fetkovich Method

The method proposed by Fetkovich[8] to construct future IPR's consists of adjusting the flow coefficient C in Equation 2-54 for changes in $f(\bar{p}_R)$. He assumed that $f(\bar{p}_R)$ was a linear function of $\bar{p}_R$ and, therefore, the value of C can be adjusted as

$$C_F = C_P(\bar{p}_{RF}\ /\ \bar{p}_{RP})$$

A value of C_p is obtained from present time production tests, that is, tests conducted when $\bar{p}_R = \bar{p}_{RP}$. Fetkovich assumed that the value of the exponent n would not change.

Future IPR's can thus be generated from

$$q_{o(F)} = C_P(\bar{p}_{RF}\ /\ \bar{p}_{RP})(\bar{p}_{RF}^2 - p_{wf}^2)^n \qquad (2\text{-}74)$$

The Fetkovich method for generating future IPR's is illustrated in the following example.

Example 2-10:

Using the data from Example 2-7A, construct an IPR for the time when $\bar{p}_R$ has declined to $\bar{p}_{RF} = 2000$ psia. The following data were obtained in Example 2-7A: $\bar{p}_{RP} = 3600$ psia, $n = 0.854$, $C_p = 0.00079$.

Solution:

$q_{o(F)} = 0.00079\ (2000/3600)\ (2000^2 - p_{wf}^2)^{0.854}$
$q_{o(F)} = 0.00044\ (2000^2 - p_{wf}^2)^{0.854}$

p_{wf}	$q_{o(F)}$
2000	0
1500	94
1000	150
500	181
0	191

Example 2-11:

In Example 2-2, it was found that $q_{o(\max)}$ was equal to 1097 STB/day for $\bar{p}_R = 2085$ psig. Using this data and Equation 2-78, calculate:
1. $q_{o(\max)}$ when $\bar{p}_R = 1900$ psig = 1915 psia
2. q_o when $\bar{p}_R = 1900$ psig and $p_{wf} = 1485$ psig

Solution:

1. $q_{o(\max)F} = 1097\ (1915/2100)^3 = 832$ STB/day

2. $q_{o(F)} = q_{o(\max)F}\left[1 - 0.2\left(\dfrac{p_{wf}}{\bar{p}_{RF}}\right) - 0.8\left(\dfrac{p_{wf}}{\bar{p}_{RF}}\right)^2\right]$

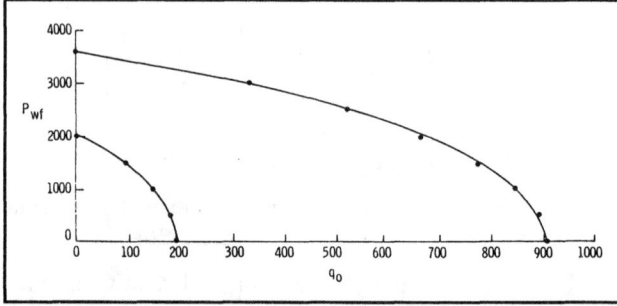

Fig. 2-40. IPR for Example 2-10.

$$q_{o(F)} = 832 \left[1 - \frac{0.2(1485)}{1900} - \frac{0.8(1485)^2}{1900^2} \right]$$

$$q_{o(F)} = 295 \text{ STB/day}$$

Note that the correction factor is calculated using the ratio of *absolute* pressures since these were used by Fetkovich. The pressures used in Vogel's equations should be gage pressures.

V. PREDICTING PRESENT TIME IPR's FOR GAS WELLS

Darcy's equation for radial gas flow including permeability alteration and turbulence was derived previously as Equation 2-28. This equation may be expressed as follows:

$$q_{sc} = \frac{703 \times 10^{-6} k_g h (\overline{p}_R^2 - p_{wf}^2)}{\mu_g Z T[\ln(.472 r_e / r_w) + S']} \qquad (2\text{-}28)$$

Solving for $\overline{p}_R^2 - p_{wf}^2$ and collecting terms yields:

$$\overline{p}_R^2 - p_{wf}^2 = A q_{sc} + B q_{sc}^2 \qquad (2\text{-}79)$$

where

$$A = \frac{1422 \mu_g Z T[\ln(.472 r_e / r_w) + S]}{k_g h}$$

$$B = \frac{3.161 \times 10^{-12} \beta \gamma_g Z T}{h^2 r_w} = \frac{1422 \mu_g Z T}{k_g h} D$$

This definition of B includes the assumption that r_e is much greater than r_w. The effects of turbulence can also be accounted for by including an exponent in the pressure term of Equation 2-28. This results in the familiar back-pressure form of the equation.

$$q_{sc} = C(\overline{p}_R^2 - p_{wf}^2)^n \qquad (2\text{-}80)$$

Observation of Equation 2-80 reveals that for negligible turbulence ($B = 0$), the value of n is 1.0. For a negligible contribution from the laminar or Darcy term ($A = 0$), the value of n is 0.5. The actual value of n usually ranges between 0.5 and 1.0 for gas wells and is an indication of the degree of turbulence or non-Darcy flow taking place. Although the value of n is usually considered to be independent of flow rate, it actually will be rate dependent since it is a measure of the turbulence effects which depend on flow rate. Mattar, et al.[12] have shown that errors in AOF involved in considering n to be constant are usually less than 0.1 percent if the test data do not have to be extrapolated too much to obtain the AOF.

Either Equation 2-79 or 2-80 may be used to generate a present-time IPR for a gas well once the coefficients and exponent are evaluated from test data. Use of both equations will be illustrated in this section.

A. Use of the Back Pressure Equation

Examination of Equation 2-80 reveals that only two flow tests would be required to evaluate C and n when $\overline{p}_R$ is known. However, due to the possibility of errors in measuring values of q_{sc} and p_{wf} it is customary to use at least four flow tests and to determine n by constructing the best straight line through the four tests. Many regulatory agencies require multirate tests to establish allowable production rates. A plot of $\overline{p}_R^2 - p_{wf}^2$ versus q_{sc} on log-log coordinates will result in a straight line having a slope of $1/n$ and an intercept of $C = q_{sc}$ at a value of $\overline{p}_R^2 - p_{wf}^2$ equal to one.

There are essentially three types of multipoint or back-pressure tests that can be used to evaluate C and n. These are the flow-after-flow, the isochronal and the modified isochronal. Each of these tests, including the advantages and disadvantages of each, was described in detail in Section III-B of this Chapter and will therefore not be repeated here. The test requiring the least amount of time is the modified isochronal and will be illustrated for a gas well by means of an example. It should be recalled that at least one fully stabilized test is required to evaluate C even for the modified test. A procedure for conducting a modified isochronal test on a gas well consists of:

1. Start at a shut-in condition ($p_{ws1} = \overline{p}_R$), open the well on a constant flow rate and measure p_{wf} at specific time periods. The total flow period may be less than the stabilization time for the well.

2. Shut the well in for the same period of time at which it was allowed to flow. The bottomhole pressure will not necessarily build back up to $\overline{p}_R$ but the static pressure at the end of the shut-in period (p_{ws2}) is assumed to be the reservoir pressure for the second flow period.

3. Repeat Steps 1 and 2 until the data are obtained for at least two flow rates. A plot of $p_{wsi}^2 - p_{wfi}^2$ versus q_{sc} on log-log coordinates will produce a straight line of slope equal to $1/n$ for each time at which p_{wf} was measured. A value for C can be calculated from the stabilized test, which is usually conducted after the transient tests are run. The procedure is illustrated graphically in Figure 2-32.

Example 2-12:

A modified isochronal test was conducted on a well completed in a reservoir having an average pressure of 1948 psia. The flow and shut-in periods were six hours long, and only the values of p_{wf} measured at the end of each flow period are to be used to determine a value for n. The extended flow test was run for a period of 72 hours at a flow rate of 8 MMscfd, at which time p_{wf} had stabilized at 1233 psia. Using the following data calculate:

1. C and n
2. AOF
3. Producing rate for p_{wf} = 800 psia

Test No.	p_{ws}, psia	p_{wf}, psia	q_{sc}, MMscfd
1	1948	1784	4.50
2	1927	1680	5.60
3	1911	1546	6.85
4	1887	1355	8.25
Extended	1948	1233	8.00

Solution:

Test No.	q_{sc}, MMscfd	$(p_{ws}^2 - p_{wf}^2) \times 10^{-6}$, psia^2
1	4.50	0.612
2	5.60	0.891
3	6.85	1.262
4	8.25	1.725
Extended	8.00	2.274

The test data are plotted in Figure 2-41.

1. To calculate n, read the flow rate change over one log cycle of pressure squared change:

$$n = \frac{\Delta \log q_{sc}}{\Delta \log (p_{wsi}^2 - p_{wfi}^2)} = \frac{\log 24 - \log 6}{\log 10 - \log 1} = 0.60$$

Use the stabilized test to calculate C:

$$C = \frac{q_{sc}}{(\bar{p}_R^2 - p_{wf}^2)^n} = \frac{8.00 \, MMscfd}{(2.274 \times 10^6)^{0.6}}$$

$$= 0.00123 \frac{MMscfd}{psia^{1.2}}$$

2. $\text{AOF} = C (\bar{p}_R - 0)^n$
   $\text{AOF} = 0.00123 \, (1948^2)^{0.6} = 10.9$ MMscfd

3. $q_{sc} = 0.00123 \, (1948^2 - 800^2)^{0.6} = 9.76$ MMscfd

As noted previously, at least one stabilized test must be

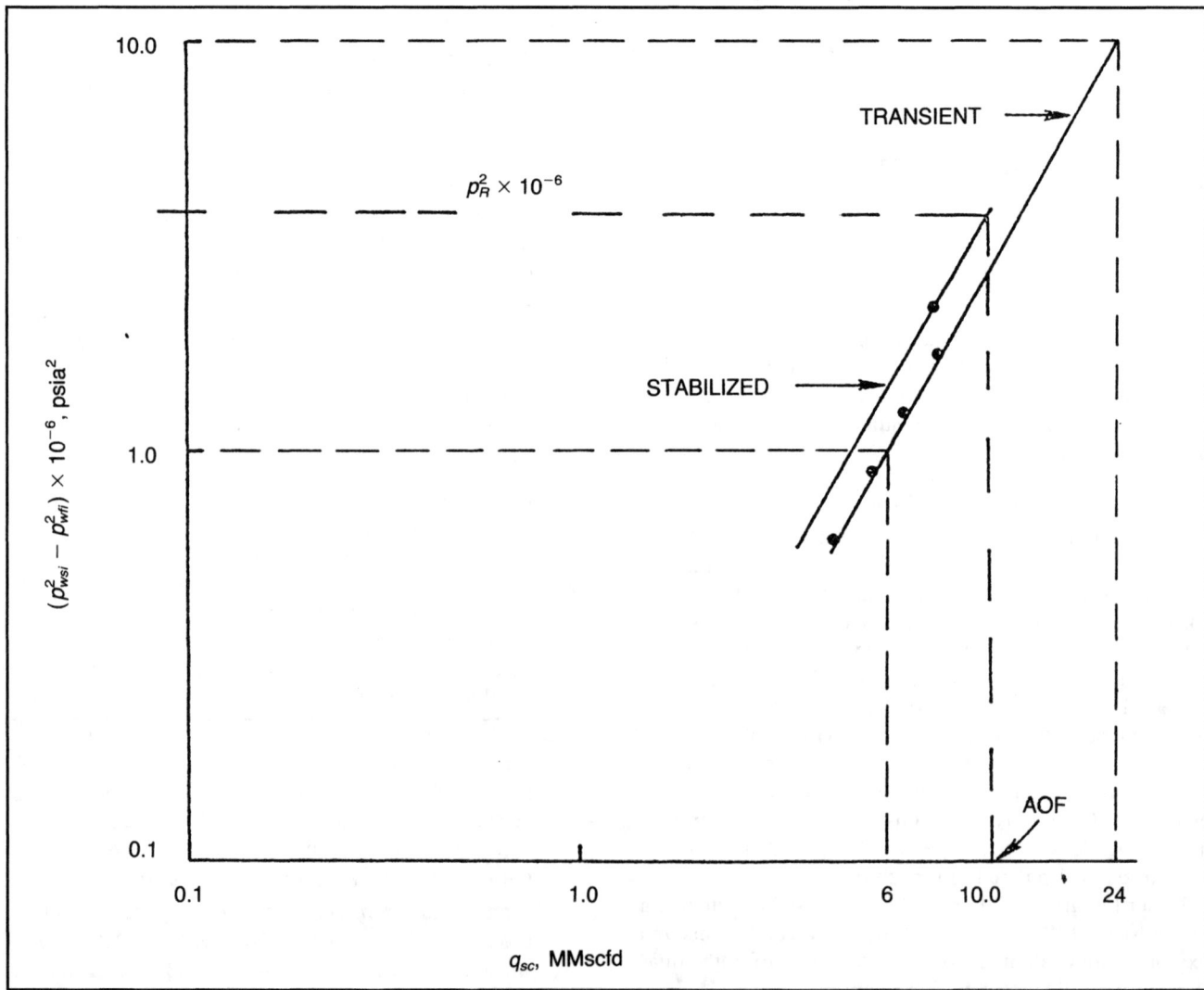

Fig. 2-41. Example 2-12 solution.

run during a modified isochronal test to determine a value for the coefficient C. This sometimes requires an extended testing time and, if the gas is being flared, can waste a considerable amount of energy. Referring to Equation 2-28, if the effects of non-Darcy flow are included in the exponent n, the definition of C is:

$$C = \frac{703 \times 10^{-6} k_g h}{\mu_g ZT[\ln(.472 r_e / r_w) + S]} \qquad (2\text{-}81)$$

If all the terms in Equation 2-81 can be evaluated, a value of C can be calculated and the extended test can be delayed until the well is connected to a sales line. The principal unknowns in Equation 2-81 are permeability to gas k_g, and skin factor S. Both k_g and $S' = S + Dq$ can be obtained from a drawdown or buildup test on the well. The shut-in periods during the modified isochronal test can be estimated. Also, if S' is different for the different flow rates, both S and D can be calculated using two values of S'. Alternatively, values for A and B in Equation 2-79 can be calculated and the inflow performance evaluated using this form of the equation.

B. Jones, Blount and Glaze Method

The method of plotting test data, which was proposed by Jones, et al.[9] can be applied to gas-well testing to determine real or present time inflow performance relationships. The analysis procedure allows determination of turbulence or non-Darcy effects on completion efficiency irrespective of skin effect and laminar flow. The procedure also evaluates the laminar flow coefficient A, and if $k_g h$ is known, an estimate of skin effect can be made. The data required are either two or more stabilized flow tests. At least one stabilized flow test is required to obtain a stabilized value of the laminar coefficient A. No transient tests are required to evaluate the completion efficiency if this method is applied. Jones, et al., also suggested methods to estimate the improvement in inflow performance that would result from reperforating a well to lengthen the completion interval or increase the perforating density and presented guidelines to determine if the turbulent effects were excessive.

Equation 2-79 can be written as:

$$\frac{\bar{p}_R^2 - p_{wf}^2}{q_{sc}} = A + Bq_{sc} \qquad (2\text{-}82)$$

where A and B are the laminar and turbulent coefficients respectively and are defined in Equation 2-79. From Equation 2-82, it is apparent that a plot of $(\bar{p}_R^2 - p_{wf}^2)/q_{sc}$ or $(\Delta p^2/q_{sc})$ versus q_{sc} on Cartesian coordinates will yield a line, which has a slope of B, and an intercept of $A = \Delta p^2/q_{sc}$ as q_{sc} approaches zero. These plots apply to both linear and radial flow, but the definitions of A and B would depend on the type of flow. The definitions of A and B for linear flow are given in Equation 2-7.

To have some qualitative measure of the importance of the turbulent contribution to the total drawdown, Jones et al. suggested comparison of the value of A calculated at the AOF of the well (A') to the stabilized value of A. The value of A' can be calculated from:

$$A' = A + B(\text{AOF}) \qquad (2\text{-}83)$$

where

$$\text{AOF} = \frac{-A + [A^2 + 4B\bar{p}_R^2]^{0.5}}{2B} \qquad (2\text{-}84)$$

Jones, et al. suggested that if the ratio of A' to A was greater than 2 or 3, then it is likely that some restriction in the completion exists. They also suggested that the formation thickness h used in the definition of B could be replaced by the length of the completed zone h_p, since most of the turbulent pressure drop occurs very near the wellbore. The effect of changing completion zone length on B and, therefore, on inflow performance can be estimated from:

$$B_2 = B_1 \left(\frac{h_{p1}}{h_{p2}} \right)^2 \qquad (2\text{-}85)$$

where
B_2 = turbulence coefficient after recompletion,
B_1 = turbulence coefficient before recompletion,
h_{p2} = new completion length, and
h_{p1} = old completion length

Example 2-13:

A four-point test was conducted on a gas well that had a perforated zone of 20 ft. Static reservoir pressure is 5250 psia. Using the Jones et al., method, determine:

1. A and B
2. AOF
3. Ratio of A'/A
4. New AOF if the perforated interval is increased to 30 ft.

Test Data

Test No.	q_{sc}, Mscfd	p_{wf}, psia
1	9300	5130
2	6000	5190
3	5200	5203
4	3300	5225

Solution:

Test No.	q_{sc}, Mcsfd	$(\overline{p}_R^2 - p_{wf}^2)/q_{sc}$, psia^2/Mscfd
1	9300	133.9
2	6000	104.1
3	5200	94.5
4	3300	79.4

1. The data points are plotted in Figure 2-42, from which it is found that:

    $A = 48$ psia^2/Mscfd

    $B = 9.24 \times 10^{-3}$ psia^2/Mscfd^2

2. $\text{AOF} = \dfrac{-48 + [48^2 + 4(9.24 \times 10^{-3})(5250)^2]^{0.5}}{2(9.24 \times 10^{-3})}$

    $\text{AOF} = 52,080$ Mscfd

3. $A' = 48 + 9.24 \times 10^{-3}(52080) = 529$

    $A'/A = 529/48 = 11$

4. $B_2 = B_1(h_{p1}/h_{p2})^2$

$B_2 = 9.24 \times 10^{-3}(20/30)^2 = 4.1 \times 10^{-3}$

$\text{AOF}_2 = \dfrac{-48 + [48^2 + 4(4.1 \times 10^{-3})(5250)^2]^{0.5}}{2(4.1 \times 10^{-3})}$

$\text{AOF}_2 = 76,340$ Mscfd

The value of A'/A calculated in the previous example indicates a large degree of turbulent pressure drop. The effect of increasing the perforated interval on the AOF is substantial. It has also been found that the effect on B of increasing the total number of perforations open can be estimated from $B_2 = B_1(N_1/N_2)^2$ where N represents the number of perforations open. Further implications of the effects of well completion efficiency will be discussed in the section on Well Completion Effects.

C. Predicting Future IPR's for Gas Wells

As reservoir pressure declines from depletion in a gas reservoir, the change in the IPR is not as significant as it is for an oil reservoir. This is due primarily to the fact that effective permeability to gas remains fairly constant since the gas saturation remains constant. This is true for either

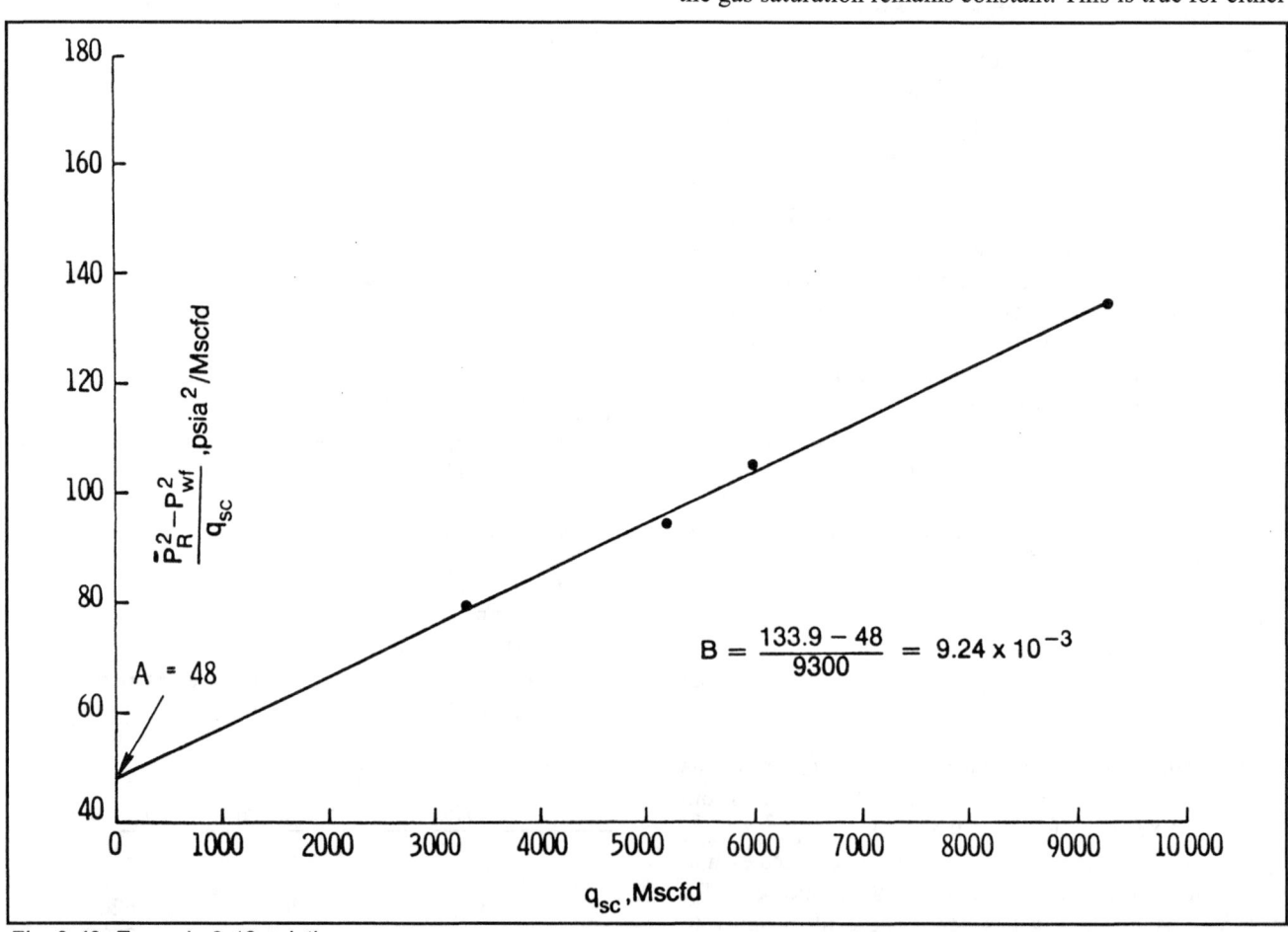

Fig. 2-42. Example 2-13 solution.

a dry or wet gas reservoir, but not for a retrograde condensate reservoir. The factors that will change as pressure changes can be seen by referring to the definition of C, Equation 2-81. The only terms depending on pressure are the gas viscosity μ_g and the gas compressibility factor Z. These same terms appear in the definition of A in Equation 2-79, with only the Z-factor appearing in the definition of B.

If no changes are made in r_e, S or h, the values of C, or A and B can be adjusted for reservoir pressure changes as follows:

$$A_F = A_p (\mu Z)_F / (\mu Z)_p \qquad (2\text{-}86)$$

$$C_F = C_p (\mu Z)_p / (\mu Z)_F \qquad (2\text{-}86A)$$

$$B_F = B_p Z_F / Z_p \qquad (2\text{-}86B)$$

where the subscript p refers to present or real time and the subscript F refers to some future time.

VI. WELL COMPLETION EFFECTS

In many cases, the inflow into a well is controlled more by the completion efficiency than by the actual reservoir characteristics. This was discussed briefly earlier when the inflow performance equations were modified to include a skin factor or flow efficiency.

There are basically three types of completions that may be made on a well depending on the type of well, the well depth, and the type of reservoir or formation. In some cases, the well is completed open hole. That is, the casing is set at the top of the producing formation and the formation is not exposed to cement. Also, no perforations are required. This type of completion is not nearly as common as it was several years ago. Most wells are now completed by cementing the casing through the producing formation.

The most widely used completion method is one in which the pipe is set through the formation, and cement is used to fill the annulus between the casing and the hole. This, of course, requires perforating the well to establish communication with the producing formation. This type of completion permits selection of the zones that are to be opened. The efficiency of the completion is highly dependent on the number of holes or perforations, the depth to which the perforations extend into the formation, the perforation pattern, and whether there is a positive pressure differential existing from the well to the formation or vice versa during the perforating operation. Compaction of the formation immediately around the perforation can reduce the efficiency considerably.

In some reservoirs, the lack of cementing material in the reservoir allows sand to be produced into the well. When completing wells in which the formation is incompetent or unconsolidated, a gravel pack completion scheme is frequently employed. In this type of completion, a perforated or slotted liner or a screen liner is set inside the casing opposite the producing formation. The annulus between the casing and the liner is then filled with a sand that is coarser than the formation sand. The size of the sand or gravel depends on the reservoir sand characteristics and on the type of gravel pack. The gravel-pack sand also fills the perforation tunnels and, in some cases, a zone is washed out behind the pipe, which is also filled with pack sand. Even though the pack sand is loosely packed and has a high permeability, non-Darcy or turbulent flow through the sand-filled perforation tunnels can cause a considerable pressure drop across the gravel pack. This pressure drop not only decreases inflow into the wellbore but also creates high velocities that may destroy the gravel pack if the velocities are too high.

To calculate the extra pressure drop caused by the completion, the general inflow equations can be modified to include the completion efficiency for any type of completion. The equations for both oil and gas flow were derived earlier and are given as

$$q_o = \frac{0.00708 k_o h (\bar{p}_R - p_{wf})}{\mu_o B_o ([\ln(0.472 r_e / r_w) + S']} \qquad (2\text{-}27)$$

$$q_{sc} = \frac{703 \times 10^{-6} k_g h (\bar{p}_R^2 - p_{wf}^2)}{\mu_g \bar{Z} T [\ln(0.472 r_e / r_w) + S']} \qquad (2\text{-}28)$$

where

$$S' = S + Dq \qquad 2\text{-}29$$

The value of S' can be obtained from a single transient test, but obtaining values for S and D requires transient tests conducted at two different rates.

Equations 2-27 and 2-28 may be written in a different form as

$$\bar{p}_R - p_{wf} = A q_o + B q_o^2 \qquad (2\text{-}63)$$

$$\bar{p}_R^2 - p_{wf}^2 = A q_{sc} + B q_{sc}^2 \qquad (2\text{-}79)$$

where A is the laminar coefficient and B is the turbulence coefficient. These coefficients may be written as composites of several terms that depend on the completion characteristics.

$$A = A_R + A_p + A_G \qquad (2\text{-}87)$$

$$B = B_R + B_p + B_G \qquad (2\text{-}88)$$

where

A_R = laminar reservoir component,
A_p = laminar perforation component,

A_G = laminar gravel-pack component,
B_R = turbulent reservoir component,
B_P = turbulent perforation component, and
B_G = turbulent gravel-pack component.

These components have different definitions for oil and gas flow, which will be given as each is discussed. Only values of the overall coefficients A and B can be obtained from production tests on wells that are already completed. Therefore, equations for estimating the value of the components must be available if the effects of each are to be isolated.

A. Open Hole Completions

The only effect of the completion on inflow performance of an openhole completion will be caused by alteration of the reservoir permeability by damage or stimulation. The inflow equations become

$$\bar{p}_R - p_{wf} = A_R q_o + B_R q_o^2 \qquad (2\text{-}63)$$

or

$$\bar{p}_R^2 - p_{wf}^2 = A_R q_{sc} + B_R q_{sc}^2 \qquad (2\text{-}79)$$

The laminar reservoir component includes the effect of Darcy or laminar flow in the reservoir plus any actual formation damage or stimulation. The defining equations are

Oil

$$A_R = \frac{141.2\mu_o B_o}{k_{OR}h}[\ln(0.472\, r_e / r_w) + S_d] \qquad (2\text{-}89)$$

Gas

$$A_R = \frac{1422\mu_g \bar{Z}T}{k_{gR}h}[\ln(0.472\, r_e / r_w) + S_d] \qquad (2\text{-}90)$$

where
k_{OR} = unaltered reservoir permeability to oil,
k_{gR} = unaltered reservoir permeability to gas, and
S_d = skin factor due to permeability alteration around the wellbore.

A value for S_d may sometimes be estimated from the following equation[13]

$$S_d = \left(\frac{k_R}{k_d} - 1\right)\ln(r_d / r_w) \qquad (2\text{-}91)$$

where
k_R = reservoir permeability
k_d = altered zone permeability,
r_w = wellbore radius, and
r_d = altered zone radius.

The actual calculation of an accurate value of S_d is difficult because values of k_d and r_d must be estimated. If a value of S can be obtained from a transient test, this will be equal to S_d for an openhole completion.

The value of B_R may be calculated from

Oil

$$B_R = \frac{2.3\times10^{-14}\beta_R B_o^2 \rho_o}{h^2 r_w} \qquad (2\text{-}92)$$

Gas

$$B_R = \frac{3.161\times10^{-12}\beta_R \gamma_g \bar{Z}T}{h^2 r_w} \qquad (2\text{-}93)$$

Values of the velocity coefficient β may be calculated from

$$\beta_R = \frac{2.33\times10^{10}}{k_R^{1.2}}$$

A value for B_R can be calculated if a value of D is available from a transient test on an openhole completion. The units to be used in all the equations presented in this chapter are the field units described earlier.

B. Perforated Completions

One of the problems involved in designing a perforated completion is estimating the efficiency of the perforations to transmit fluid from the reservoir to the wellbore. The efficiency depends on conditions such as the number of perforations actually open, perforation diameter, penetration depth, degree of damage around the perforation, and phasing.

Figure 2-43 from Bell[21] illustrates the perforating process when shaped-charge perforating is used. A zone of reduced permeability, called the crushed zone or compacted zone, is formed around the perforation. This crushed zone can be the source of considerable pressure drop because of the high fluid velocities caused by the fluid converging into the perforations.

Underbalanced perforating, in which the pressure in the wellbore is less than reservoir pressure, results in an immediate backflow or surge through the perforation, thus minimizing plugging from debris and crushed-zone damage. Essentially three perforating techniques, as described by Bell[21] are available for achieving underbalance while perforating. These are illustrated in Figure 2-44.

Advantages and disadvantages of each method are discussed by Bell[21] but the consensus is that the tubing-conveyed method, even though considerably more expensive, is the best method. Almost any degree of underbal-

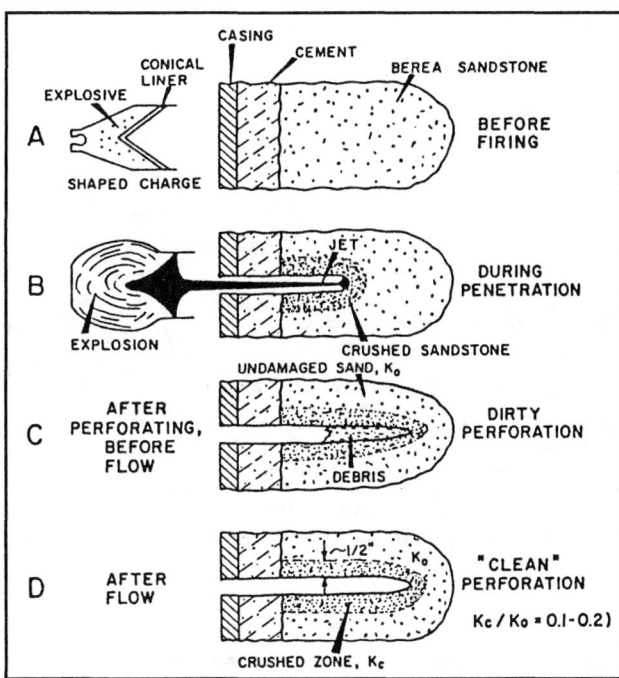

Fig. 2-43. Damage to formation due to jet penetrating process (API-RP 43 Berea sandstone).[21]

ance can be created, perforating density can be high, long zones can be perforated and the well can be produced immediately after perforating. A typical, tubing-conveyed perforating operation is shown in Figure 2-45.

One problem faced in designing an underbalanced perforating job is the optimum degree of underbalance pressure to use. A field study conducted by King, et al.[22] resulted in correlations between reservoir permeability and required underbalance pressure necessary to obtain perforation performance which was not improved by subsequent acidizing. Figures 2-46 and 2-47 can be used to estimate the required underbalance. In these figures, the minimum underbalance pressure for various permeabilities can be obtained from the line dividing improved and unimproved perforations. They observed that a higher degree of underbalance is generally required for gas wells as opposed to oil wells.

Regalbuto and Riggs[23] conducted extensive laboratory tests on cores to determine the optimum degree of underbalance. The tests included pressure differentials from 500 psi overbalance to 1000 psi underbalance. They defined a radial flow ratio (RFR) as the flow rate through the core after perforating divided by the flow rate before perforating, both measured with the same pressure drop across the core. The effects of surging or backflowing fluid through the perforations was also studied. In some

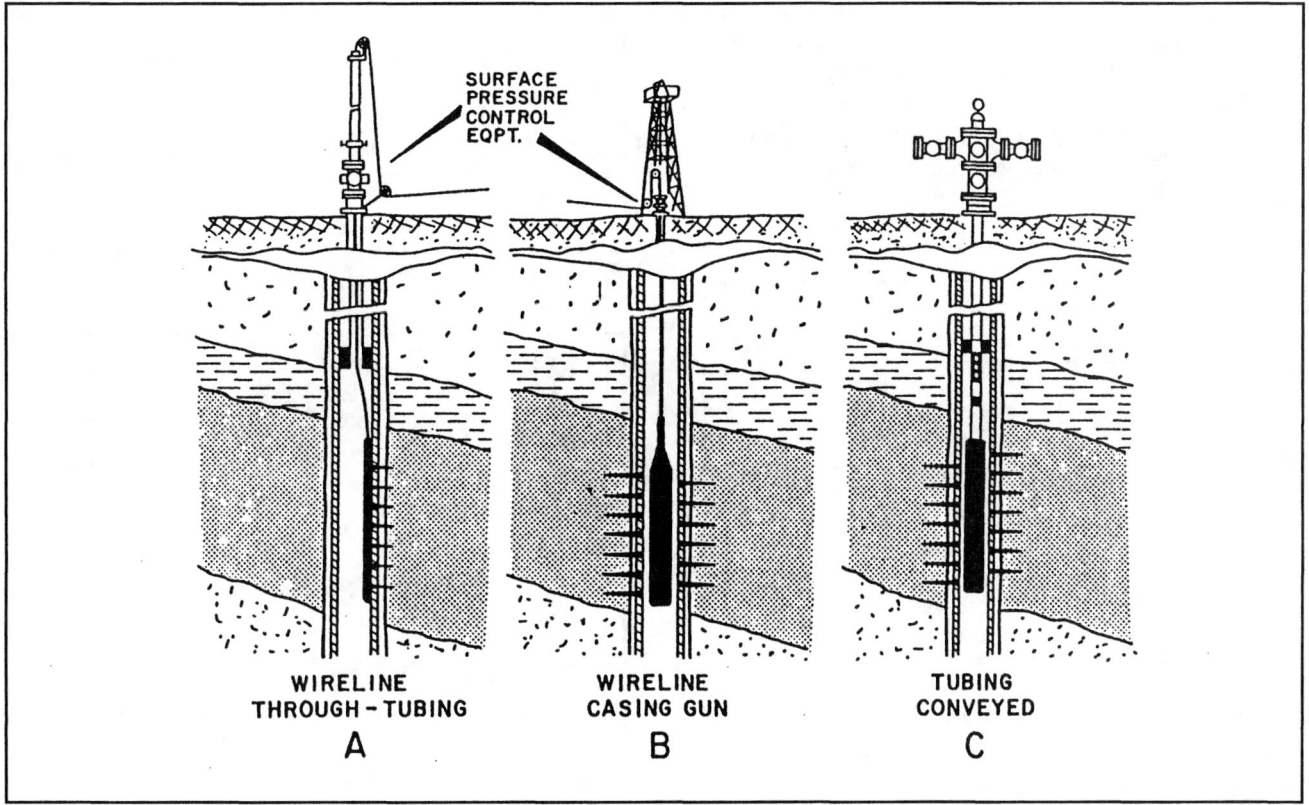

Fig. 2-44. Methods of underbalanced perforating.[21]

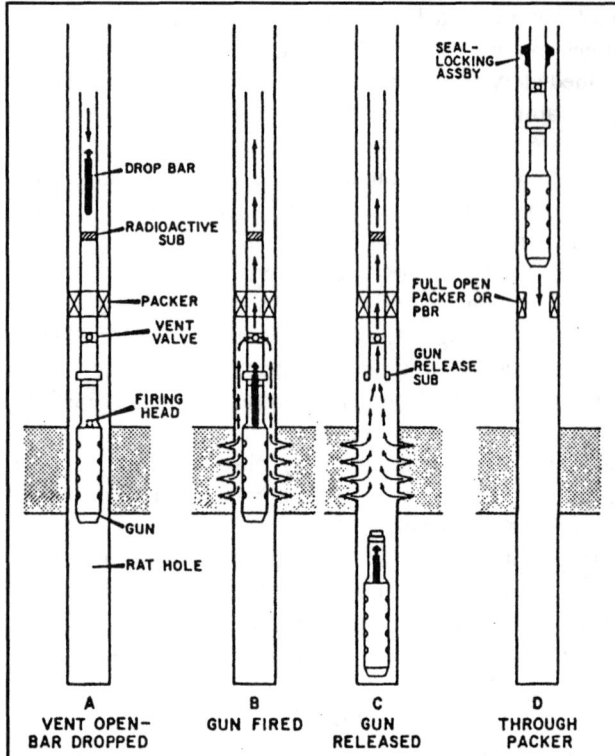

Fig. 2-45. Tubing-conveyed perforating techniques.[21]

cases, the surging was delayed, and, in other cases, the surging was immediate. Their results are summarized in Figure 2-48. It was also observed that a high degree of underbalance followed by surging increased the perforation size, thus removing some of the crushed or compacted zone surrounding the perforation. These results are shown in Figure 2-49.

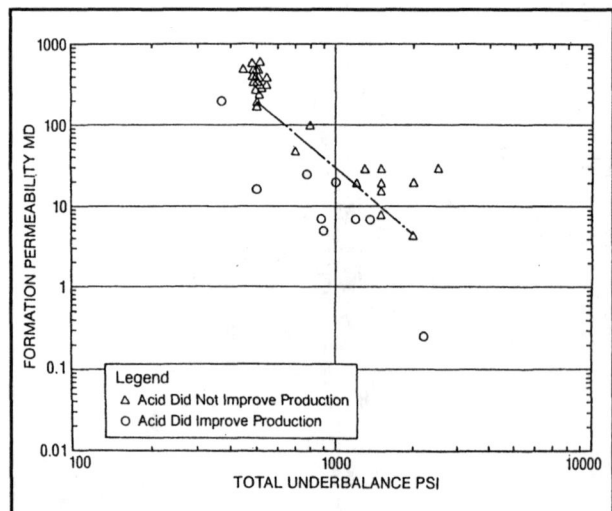

Fig. 2-46. Underbalance used on tubing-conveyed perforating in oil zones in sandstone.

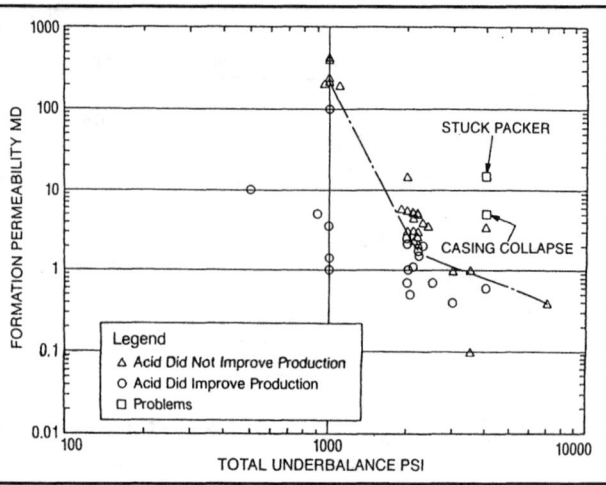

Fig. 2-47. Underbalance pressure used on tubing-conveyed perforating in gas zones in sandstone.[22]

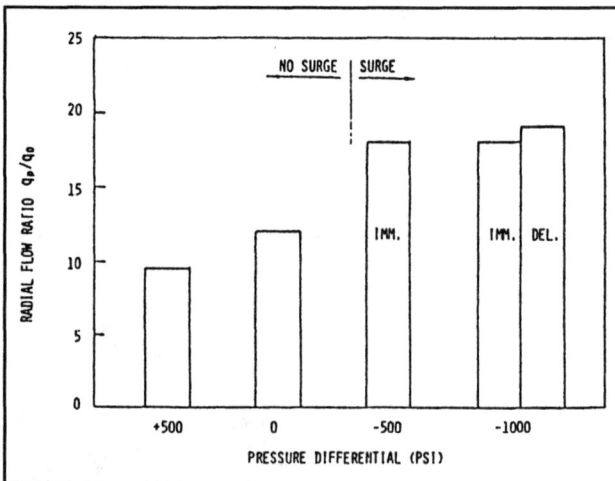

Fig. 2-48. RFR as affected by test conditions.[23]

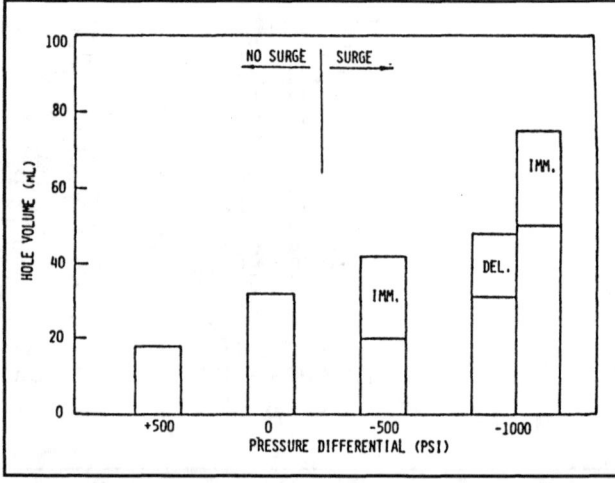

Fig. 2-49. Hole volume as affected by test condition.[23]

A model that can be used to estimate the pressure drop through perforations was presented by McLeod.[14] This model is intended to be used for wells that have not been stimulated through the perforations and requires estimates of several parameters that cannot be measured. McLeod presented equations for gas wells only, but similar equations for oil wells are also presented here. The oil flow equations do not take into account any reduction in permeability that may be caused by two-phase flow below the bubblepoint pressure.

The efficiency of a perforated completion depends on both the reservoir and perforation components in Equations 2-63 and 2-79.

That is

$$\bar{p}_R - p_{wf} = (A_R + A_p)q_o + (B_R + B_p)q_o^2 \qquad (2\text{-}94)$$

and

$$\bar{p}_R^2 - p_{wf}^2 = (A_R + A_p)q_{sc} + (B_R + B_p)q_{sc}^2 \qquad (2\text{-}94A)$$

The laminar perforation component includes the effects of the number and types of perforations, and the effects of compaction around the perforations, These effects were discussed in detail by McLeod[14] and the discussion on perforated completions presented here is based largely on McLeod's work. The equations for oil and gas are

Oil

$$A_p = \frac{141.2\mu_o B_o}{k_{OR}h}(S_p + S_{dp}) \qquad (2\text{-}95)$$

Gas

$$A_p = \frac{1422\mu_g \bar{Z}T}{k_{gR}h}(S_p + S_{dp}) \qquad (2\text{-}96)$$

If sufficient data regarding the perforations are known, values for S_p and S_{dp} may be calculated. S_p is a function of perforating density, perforation length, perforation diameter, phasing, wellbore radius, damaged-zone permeability, ratio of vertical-to-horizontal permeability, and damaged-zone radius.

Values of S_p may be obtained from nomographs published by Hong[15] or Locke.[16] An equation for estimating S_p, that requires a value for vertical permeability, was given by Saidikowski.[17]

$$S_p = \left(\frac{h}{h_p} - 1\right)\left[\ln\left(\frac{h}{r_w}\left(\frac{k_R}{k_v}\right)^{0.5}\right) - 2\right] \qquad (2\text{-}97)$$

where
h = total formation thickness
h_p = perforated-interval length

k_R = reservoir permeability in the horizontal direction, and
k_v = vertical permeability.

The nomograph presented by Locke[16] is shown in Figure 2-50.

Locke presented the following procedure for obtaining a value for S_p from the nomograph:

1. Enter with the perforation length L_p on the upper left stem.

2. Proceed horizontally to the appropriate perforation diameter.

3. Proceed vertically to the appropriate damage or invaded zone thickness. Measure along the damaged zone line horizontally from the vertical axis to the appropriate k_d/k_R line. This is distance *b–c* in the nomograph. Shift the vertical line from the perforation diameter to the right by the distance *b–c* and proceed to Step 4.

4. Go vertically to the appropriate, crushed-zone ratio or relative-permeability line.

5. Proceed horizontally to the shot-density line. Note that for 8 spf two lines exist, one for zero phasing and another for 90° phasing.

6. Proceed vertically to the angular phasing line.

7. Proceed horizontally and read the Productivity Ratio or Flow Efficiency, if required, and the skin factor S_p.

The nomograph is drawn for a borehole size of six in. However, the effects of borehole size are partially compensating. An additional correction for a 12-in. borehole with 160 acre spacing is shown. To apply this correction in Step 7 the horizontal line is shifted up or down before proceeding to the skin factor scale.

McLeod derived an equation for calculating the effect of flow through the compacted zone as

$$S_{dp} = \left(\frac{h}{L_p N}\right)\left(\frac{k_R}{k_{dp}} - \frac{k_R}{k_d}\right)\ln(r_{dp}/r_p) \qquad (2\text{-}98)$$

where
h = total formation thickness
L_p = perforation length,
N = total number of perforations
k_R = unaltered reservoir permeability,
k_{dp} = compacted or crushed zone permeability,
r_p = perforation radius, and
r_{dp} = compacted zone radius.

Figure 2-51 (reference 14) shows a schematic of a perforated completion and the relationship among the various parameters in Equation 2-98.

The largest part of the pressure drop through a perfora-

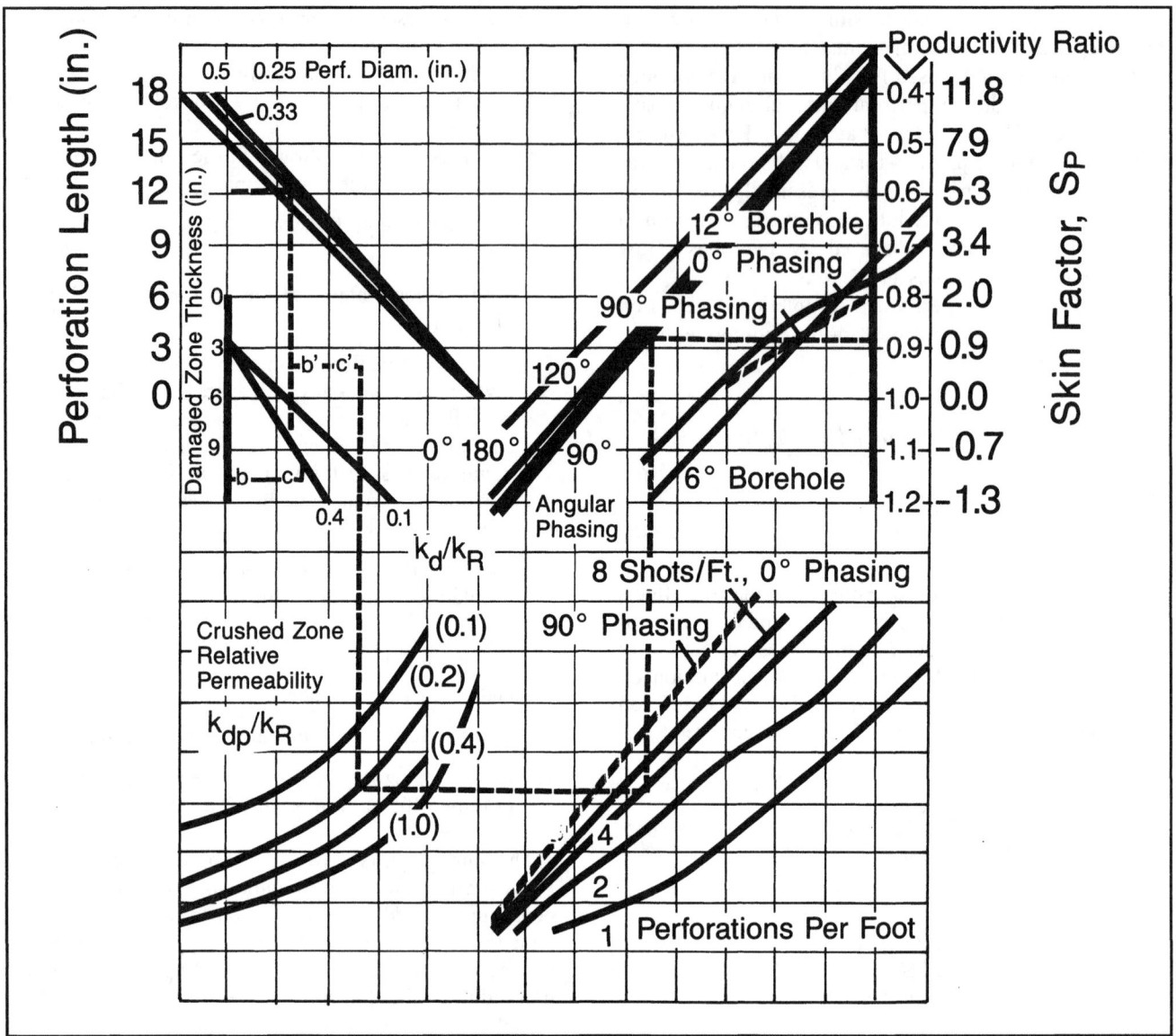

Fig. 2-50. Locke's nomograph.[16]

tion is caused by turbulent or non-Darcy flow through the compacted zone. The equations for calculating this effect are

Oil

$$B_p = \frac{2.3 \times 10^{-14} \beta_{dp} B_o^2 \rho_o}{r_p L_p^2 N^2} \qquad (2\text{-}99)$$

Gas

$$B_p = \frac{3.161 \times 10^{-12} \beta_{dp} \gamma_g \overline{Z} T}{r_p L_p^2 N^2} \qquad (2\text{-}100)$$

The value of the velocity coefficient must be calculated using the compacted-zone permeability. The equation is

$$\beta_{dp} = \frac{2.33 \times 10^{10}}{k_{dp}^{1.2}} \qquad (2\text{-}101)$$

There are several variables in the equations for perforated completions that are difficult to determine. These include the altered-zone permeability, the compacted-zone radius, the perforation length and the altered-zone radius. Some of these parameters can be estimated from API RP-43 test data published by the perforating companies. The following guidelines have been recommended by McLeod:[14]

For wells perforated in mud

$$\frac{k_{dp}}{k_R} = \frac{k_c}{k} \qquad (2\text{-}102)$$

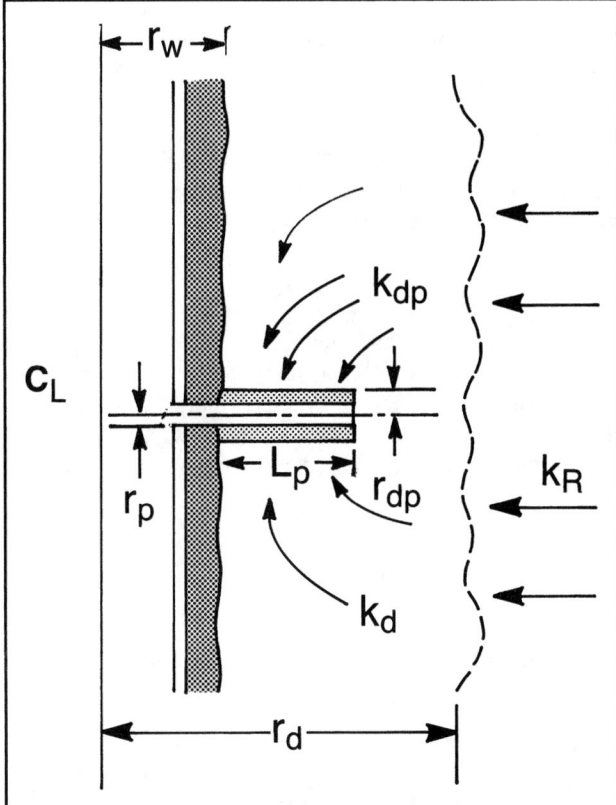

Fig. 2-51. Flow into a performation.[14]

For wells perforated in brine

$$\frac{k_{dp}}{k_d} = \frac{k_c}{k}$$

(2-103)

where k_c/k values are obtained from the API test data. Guidelines for estimating k_c/k when no tests are available were also presented by McLeod[14] in Table 2-2.

TABLE 2-2

Perforating Parameter Guidelines

Fluid in Hole	Pressure Conditions	k_c/k
high solids mud	overbalance	0.01-0.03
low solids mud	overbalance	0.02-0.04
unfiltered brine	overbalance	0.04-0.06
filtered brine	overbalance	0.08-0.16
clean fluid	underbalance	0.30-0.50
ideal fluid	underbalance	1.00

McLeod also suggests that the compacted zone thickness is usually about 0.5 in. That is, $r_{dp} = r_p + 0.5$ if r_p is in inches. If no information is available regarding the altered-zone radius, a value of $r_d = r_w + 1$ may be used, where r_w is given in ft.

C. Perforated, Gravel-Packed Completions

The equations for a gravel-packed completion are

$$\overline{p}_R - p_{wf} = (A_R + A_p + A_G)q_o + (B_R + B_p + B_G)q_o^2 \quad (2\text{-}94)$$

and

$$\overline{p}_R^2 - p_{wf}^2 = (A_R + A_p + A_G)q_{sc} + (B_R + B_p + B_G)q_{sc}^2 \quad (2\text{-}95)$$

For most gravel-packed wells the formation will have a high permeability because of the unconsolidated nature of the sand. This will also result in negligible damage from the compacted zone around the perforations. However, the effect of the linear flow through the perforation tunnel that is filled with pack sand can cause a significant non-Darcy-flow pressure drop. The equations for A_G and B_G are

Oil

$$A_G = \frac{282.4\mu_o B_o L}{k_G N r_p^2} \quad (2\text{-}104)$$

$$B_G = \frac{9.20 \times 10^{-14} \beta_G B_o^2 \rho_o L}{N^2 r_p^4} \quad (2\text{-}105)$$

Gas

$$A_G = \frac{2844\overline{Z}T\mu_g L}{k_G N r_p^2} \quad (2\text{-}106)$$

$$B_G = \frac{1.263 \times 10^{-11} \beta_G \gamma_g \overline{Z}TL}{N^2 r_p^4} \quad (2\text{-}107)$$

where

N = total number of perforations,
k_G = gravel permeability,
L = perforation tunnel length, and

$$\beta_G = \frac{1.47 \times 10^7}{k_G^{0.55}} \quad (2\text{-}108)$$

The following table, from Gurley,[18] may be used to estimate the gravel permeability based on its size.

Sieve Size	k_G, md
10-20	5.00×10^5
16-30	2.50×10^5
20-40	1.20×10^5
40-60	4.0×10^4

A schematic of a gravel-packed completion is illustrated in Figure 2-52.

As illustrated in Figure 2-52, the tunnel length is defined as the radius of the hole minus the outside radius

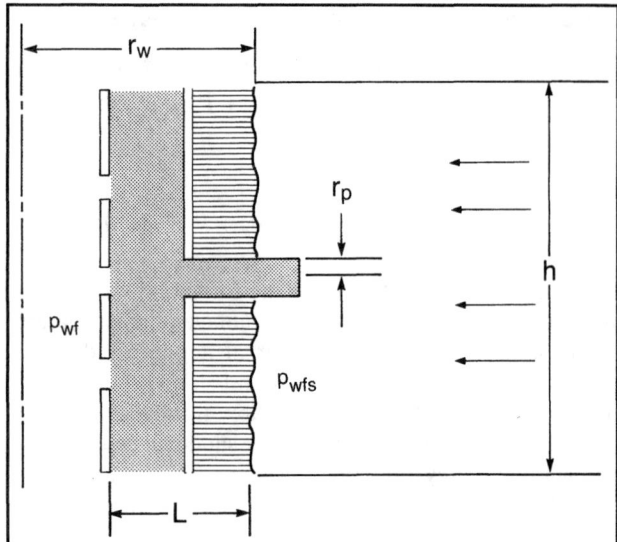

Fig. 2-52. Gravel-pack schematic.

of the screen. In some cases, it is defined as the hole radius minus the inside radius of the casing.

In analyzing perforated completions, it is sometimes convenient to break down the total pressure drawdown into two separate components, that is, the pressure drop in the reservoir and the pressure drop across the gravel pack. This can be expressed as

$$\overline{p}_R - p_{wf} = \overline{p}_R - p_{wfs} + (p_{wfs} - p_{wf}) \qquad (2\text{-}109)$$

where p_{wfs} is the pressure existing at the sand face, as illustrated in Figure 2-52. Most operators agree that the pressure drop across a gravel pack, $p_{wfs} - p_{wf}$, should be less than about 300 psi. The equations for the two pressure drops may be written as

Oil

$$\overline{p}_R - p_{wfs} = A_R q_o + B_R q_o^2 \qquad (2\text{-}110)$$

$$p_{wfs} - p_{wf} = A_G q_o + B_G q_o^2 \qquad (2\text{-}111)$$

Gas

$$\overline{p}_R^2 - p_{wfs}^2 = A_R q_{sc} + B_R q_{sc}^2 \qquad (2\text{-}112)$$

$$p_{wfs}^2 - p_{wf}^2 = A_G q_{sc} + B_G q_{sc}^2 \qquad (2\text{-}113)$$

The effects of completion method on a well's producing capacity will be illustrated in a subsequent chapter.

VII. INFLOW PERFORMANCE SUMMARY

Methods have been presented for constructing for real or present-time conditions for both oil and gas wells. Prediction of the effects of depletion or decreasing reservoir pressure on the inflow performance was discussed,

and methods for quantifying these effects were presented. A summary of these methods will be presented in this section.

A. Oil Wells

An expression for the productivity index for an oil well was presented in Equation 2-32 as

$$J = \frac{0.00708\,kh}{(\overline{p}_R - p_{wf})[\ln(.472 r_e / r_w) + S']} \int_{p_{wf}}^{\overline{p}_R} \frac{k_{ro}}{\mu_o B_o} dp \qquad (2\text{-}32)$$

Any phenomenon that causes a change in any parameter in this equation will cause a change in J and thus affect inflow performance. The main parameters that can change are the pressure function, $f(p) = k_{ro}/\mu_o B_o$ and the skin factor $S' = S + Dq_o$. These parameters can be changed by:

1. Drawdown, which affects k_{ro} around the wellbore and also affects Dq_o,

2. Formation damage or stimulation, which affects S,

3. Depletion, which affects $f(p)$ in the entire drainage volume of the well as $\overline{p}_R$ declines below p_b, and

4. Perforation effects.

Methods were presented for calculating each of these effects. The methods that may be applied are:

1. Drawdown effects:
   a. Vogel[5] (Equation 2-33)
   b. Fetkovich[8] (Equation 2-54)
   c. Jones, et al.[9] (Equation 2-63)

2. Formation damage or stimulation:
   a. Standing modification of Vogel (Flow Efficiency) (Equation 2-46 or 2-48)
   b. Fetkovich (S is included in the coefficient C)
   c. Jones, et al. (S is included in the coefficient A)

3. Depletion:
   a. Standing (adjustment of $q_{o(max)}$ or J) (Equation 2-71 or 2-72)
   b. Fetkovich (adjustment of C) (Equation 2-74)
   c. Vogel-Fetkovich combined (adjustment of $q_{o(max)}$ or J) (Equation 2-78)

4. Perforation Effects:
   a. Locke
   b. McLeod (Equation 2-94)

B. Gas Wells

The inflow-performance equation for gas wells is not as sensitive to pressure as that for oil wells because the gas saturation and therefore the permeability to gas remain fairly constant except for the case of retrograde condensate reservoirs. The following methods were presented for accounting for various effects.

1. Drawdown effects:
   a. Back-pressure equation (Equation 2-80)
   b. Jones, et al. (Equation 2-79)
   c. Darcy radial-flow equation (Equation 2-28)

2. Formation damage or stimulation:
   a. Back-pressure equation (S is included in the coefficient C)
   b. Jones, et al. (S is included in the coefficient A)
   c. Darcy (S is included in the equation)

3. Turbulence effects:
   a. Back-pressure equation (the value of the exponent n is an indication of turbulence)
   b. Jones, et al. (the coefficient B or the value of A'/A indicates the effects of turbulence)
   c. Darcy (the turbulence coefficient D is an indication of turbulence)

4. Depletion—the values of C or A and B can be adjusted for changes in μ_g and Z-factor with pressure change. (Equation 2-86)

5. Perforation Effects:
   a. Locke
   b. McLeod (Equation 2-94A)

VIII. REFERENCES

1. Gilbert, W. E.: "Flowing and Gas-Lift Well Performance," API Drill. Prod. Practice, 1954.
2. Odeh, A. S.: "Pseudo Steady-State Flow Equation and Productivity Index for a Well with Non-circular Drainage Area," Mobil Research and Development Corporation.
3. McCain, A. E.: *The Properties of Petroleum Fluids*, Petroleum Publishing Co., Tulsa, Okla., 1973.
4. Amyx, J. W., Bass, D. M., and Whiting, R. L.: *Petroleum Reservoir Engineering*, McGraw-Hill, New York, 1960.
5. Vogel, J. V.: "Inflow Performance Relationships for Solution Gas Drive Wells," *JPT*, Jan., 1968.
6. Standing, M. B.: "Inflow Performance Relationships for Damaged Wells Producing by Solution Gas Drive," *JPT*, Nov., 1970.
7. Weller, W. T.: "Reservoir Performance During Two-Phase Flow," *JPT*, Feb., 1966.
8. Fetkovich, M. J.: "The Isochronal. TestMg of Oil Wells": Paper 4529, 48th Annual Fall Meeting of SPE, Las Vegas, Nev., 1973.
9. Jones, L. G., Blount, E. M., and Glaze, O. H.: "Use of Short Term Multiple Rate Flow Tests to Predict Performance of Wells Having Turbulence," SPE 6133, presented at SPE 51st Annual Fall Meeting, New Orleans, LA, 1976.
10. Firoozabadi, A. and Katz, D. L.: "An Analysis of High Velocity Gas Flow Through Porous Media," *JPT*, Feb., 1979.
11. Standing, M. B.: "Concerning the Calculation of Inflow Performance of Wells Producing from Solution Gas Drive Reservoirs," *JPT*, Sept., 1971.
12. Mattar, L. and Lin, C.: "Validity of Isochronal and Modified Isochronal Testing of Gas Wells," SPE 10126, presented at 56th Annual Fall Meeting, San Antonio, TX, 1981.
13. Matthews, C. S. and Russell, D. G.: *Pressure Buildup and Flow Test in Wells*, SPE Monograph 1, 1967.
14. McLeod, H. O.: "The Effect of Perforating Conditions on Well Performance," *JPT*, Jan,.1983.
15. Hong, K. C.: "Productivity of Perforated Completions in Formations With and Without Damage," *JPT*, Aug., 1975.
16. Locke, S.: "An Advanced Method for Predicting the Productivity Ratio of a Perforated Well," *JPT*, Dec., 1981.
17. Saidikowski, R. M.: "Numerical Simulations of the Combined Effects of Wellbore Damage and Partial Penetration," SPE 8204, Sept., 1979.
18. Gurley, D. G., Copeland, C. T., and Hendrick, J. O.: "Design Plan and Execution of Gravel-Pack Completions," *JPT*, Oct., 1977.
19. Sherrard, D. W., Brice, W. B., and MacDonald, D. G.: "Application of Horizontal Wells at Prudhoe Bay," *JPT*, Nov., 1987.
20. Giger, F. M., Combe, J., and Reiss, L. H.: "L'interest du forage horizontal pour l'exploitation des guisements d'hydrocarbures," *Revue de l'Inst. Francais de Petrole*, May-June, 1983.
21. Bell, W. T.: "Perforating Underbalanced-Evolving Techniques," *JPT*, Oct., 1984.
22. King, G. E., Anderson, A., and Bingham, M.: "A Field Study of Underbalance Pressures Necessary to Obtain Clean Perforations Using Tubing-Conveyed Perforating," *JPT*, June, 1986.
23. Regalbuto, J. A. and Riggs, R. S.: "Underbalanced Perforation Characteristics as Affected by Differential Pressure," *SPE Prod. Engineering*, Feb., 1988.
24. Bendakhlia, H., and Aziz, K.: "Inflow Performance Relationships for Solution-Gas Drive Horizontal Wells," SPE 19823, 64th Annual Fall Meeting of SPE, San Antonio, Texas, Oct., 1989.

3

Flow in Pipes and Restrictions

I. INTRODUCTION

In Chapter 1, it was pointed out that to determine the performance of any producing well, it is necessary to be able to calculate the pressure losses in all the components in the system. These pressure losses and where they occur in the system are illustrated in Figure 3-1.

Procedures for calculating the pressure loss in the reservoir, $\Delta p_1 = \overline{p}_R - p_{wfs}$, were presented in Chapter 2 for both oil and gas wells. Calculation of pressure loss across the completion, $\Delta p_2 = p_{wfs} - p_{wf}$, was also outlined.

In this Chapter, methods to calculate values for Δp_3 through Δp_8 will be presented. These pressure losses may occur in either the inflow to the node or the outflow from the node. In many cases, the node pressure will be selected as flowing bottomhole pressure p_{wf}. Calculation of the node pressure for the outflow would then take the following form:

$$p_{sep} + \Delta p_{fl} + \Delta p_{choke} + \Delta p_{tubing}$$
$$+ \Delta p_{sssv} + \Delta p_{rst} = p_{wf} \qquad (3\text{-}1)$$

where

$$
\begin{aligned}
p_{sep} &= \text{separator pressure,} \\
\Delta p_{fl} &= \text{pressure drop in the flowline,} \\
\Delta p_{choke} &= \text{pressure drop in the surface choke,} \\
\Delta p_{tubing} &= \text{pressure drop in the tubing,} \\
\Delta p_{sssv} &= \text{pressure drop in the subsurface safety valve,} \\
& \quad \text{and} \\
\Delta p_{rst} &= \text{pressure drop in any other restriction.}
\end{aligned}
$$

As was discussed earlier all these pressure drops are functions of producing rate and the characteristics of the components. In the case of single-phase flow, either liquid or gas, the pressure drops can be calculated easily, as long as component characteristics such as size and roughness are known. Unfortunately, most producing oil or gas wells operate under multiphase conditions. There will usually be some free gas produced along with the oil in an oil well, and most gas wells will produce either water or condensate along with the gas.

The presence of both liquid and gas in the component complicates the pressure loss calculations immensely. As average pressure existing in a component changes, phase changes occur in the fluids. This causes changes in densities, velocities, volumes of each phase, and fluid properties. Also, temperature changes occur for flow in the piping system and restrictions. This was not a problem in calculating the reservoir performance, since reservoir temperature remains constant. Calculation of the pressure change with distance, or pressure gradient, at any point in the system, requires knowledge of the temperature existing at that point. Therefore, procedures to estimate heat or temperature losses must be available.

Design and analysis of a system in which two-phase flow is occurring requires a thorough understanding of the physical phenomena as well as the basic theory and equations. In this chapter, the basic equations and concepts will first be presented in considerable detail. Procedures to estimate the necessary fluid properties as functions of pressure and temperature will then be given. Empirical correlations for calculating pressure losses in both wells and pipeline will be presented and suggestions of which method to use for particular conditions will be made.

The use of prepared pressure traverse curves for making rough estimates of pressure losses in wells and pipelines will be discussed, and the effects of changing conditions in wells or fields will be presented. Finally,

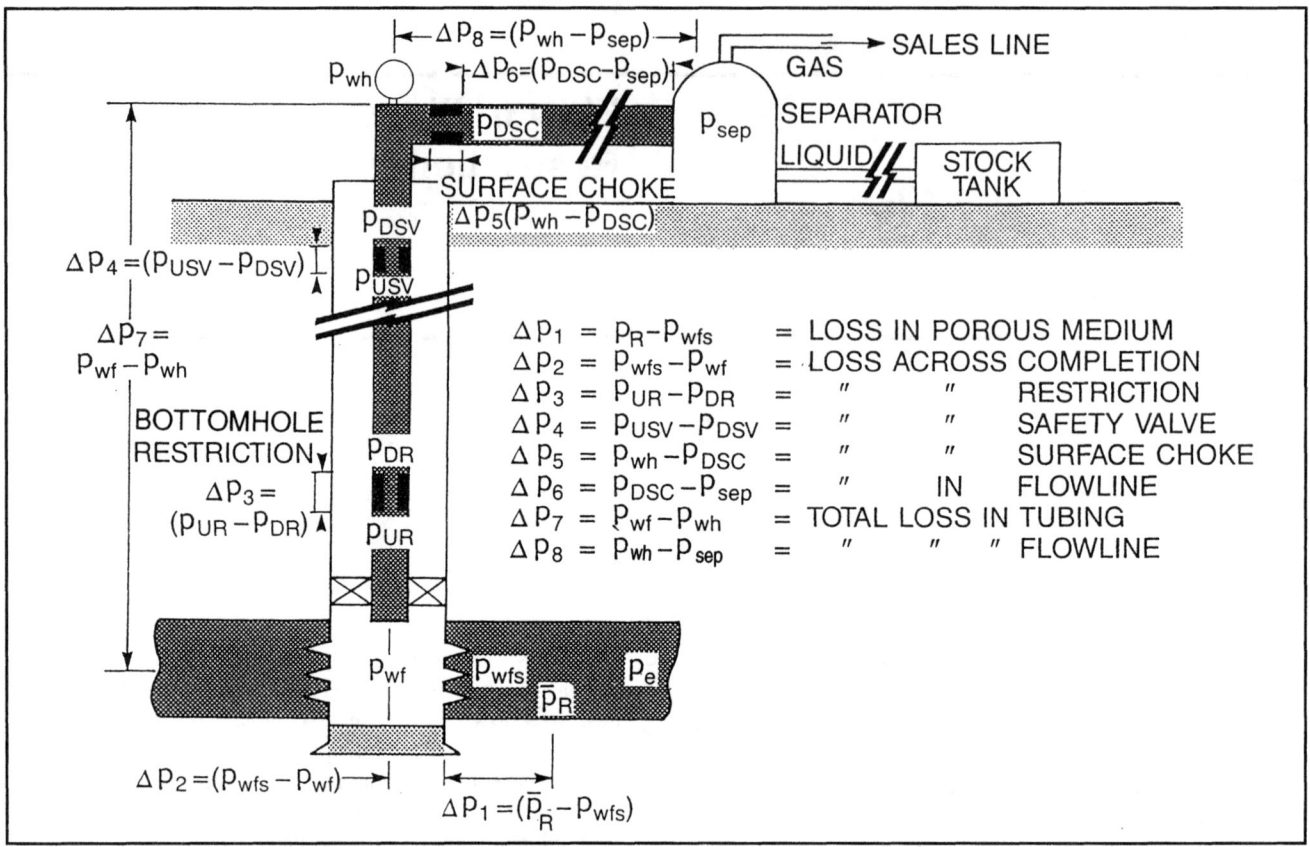

Fig. 3-1. *Possible pressure losses in complete system.*

methods for calculating pressure losses in short restrictions, such as chokes, SSSV's, and pipe fittings, will be given.

II. BASIC EQUATIONS AND CONCEPTS

Pressure gradients occurring during two-phase flow in pipes can be calculated if all the energy changes that take place in the fluids can be predicted. In this section, the basic pressure gradient equation will be derived that will be applicable for flow of any fluid in any piping system. This equation will then be adapted for various piping system conditions and fluid conditions.

A. The General Energy Equation

The theoretical basis for most fluid flow equations is the general energy equation, an expression for the balance or conservation of energy between two points in a system. The energy equation is developed first and, using thermodynamic principles, is modified to a pressure gradient equation form.

The steady state energy balance simply states that the energy of a fluid entering a control volume, plus any shaft work done on or by the fluid, plus any heat energy added to or taken from the fluid, must equal the energy leaving the control volume.

Figure 3-2 may be used to illustrate this principle.

Considering a steady state system, the energy balance may be written as

$$U_1' + p_1V_1 + \frac{mv_1^2}{2g_c} + \frac{mgZ_1}{g_c} + q' + W_s'$$

$$= U_2' + p_2V_2 + \frac{mv_2^2}{2g_c} + \frac{mgZ_2}{g_c} \qquad (3\text{-}2)$$

where

$$
\begin{aligned}
U' &= \text{internal energy,} \\
pV &= \text{energy of expansion or compression,} \\
mv^2/2g_c &= \text{kinetic energy,} \\
mgZ/g_c &= \text{potential energy,} \\
q' &= \text{heat energy added to fluid, and} \\
w_s' &= \text{work done on the fluid by the surroundings}
\end{aligned}
$$

Dividing Equation 3-2 by m to obtain an energy per unit mass balance and writing in differential form gives:

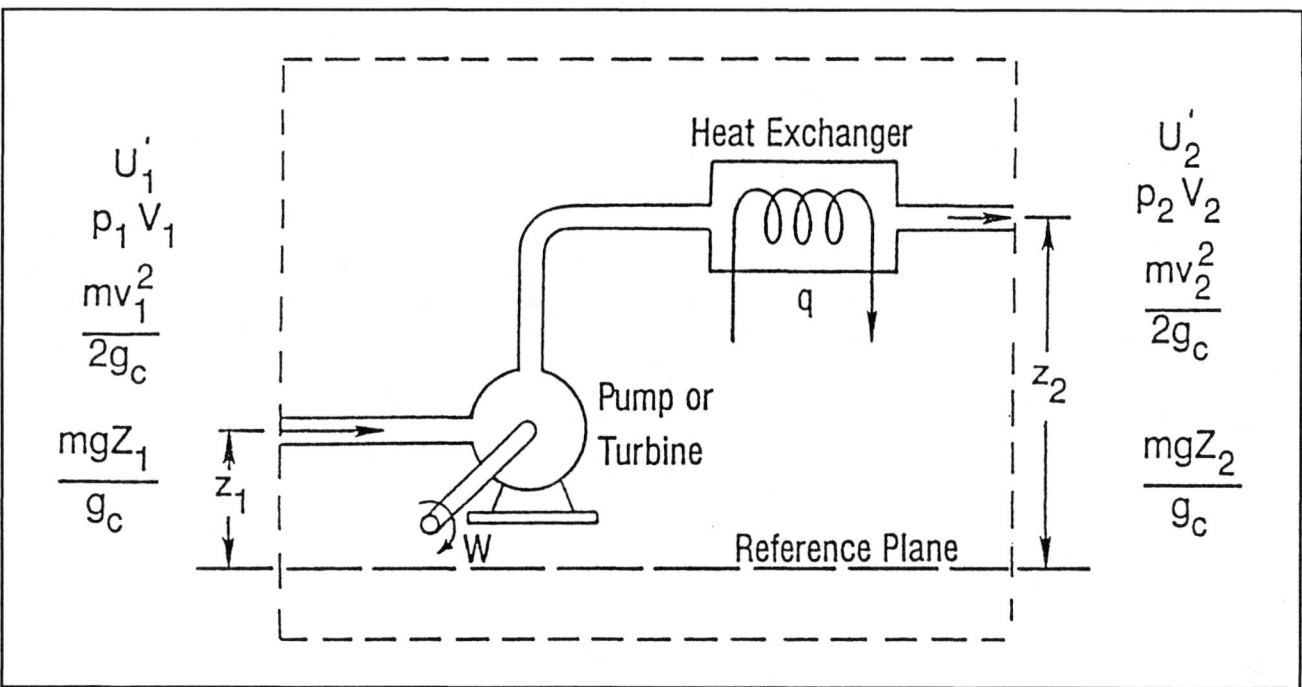

Fig. 3-2. Flow system control volume.

$$dU + d\left(\frac{p}{\rho}\right) + \frac{vdv}{g_c} + \frac{g}{g_c}dZ + dq + dW_s = 0 \qquad (3\text{-}3)$$

This form of the energy balance equation is difficult to apply because of the internal energy term, so it is usually converted to a mechanical energy balance using well-known thermodynamic relations. From thermodynamics:

$$dU = dh - d\left(\frac{p}{\rho}\right) \qquad (3\text{-}3A)$$

and

$$dh = TdS + \frac{dp}{\rho}$$

or

$$dU = TdS + \frac{dp}{\rho} - d\left(\frac{p}{\rho}\right) \qquad (3\text{-}4)$$

where

h = enthalpy,
S = entropy, and
T = temperature.

Substituting Equation 3-4 into Equation 3-3 and simplifying results in:

$$TdS + \frac{dp}{\rho} + \frac{vdv}{g_c} + \frac{g}{g_c}dZ + dq + dW_s = 0 \qquad (3\text{-}5)$$

For an irreversible process, the Clausis inequality states that:

$$dS \geq \frac{-dq}{T},$$

or

$$TdS = -dq + dL_w,$$

where dL_w = losses due to irreversibilities, such as friction. Using this relationship and assuming no work is done on or by the fluid, Equation 3-5 becomes:

$$\frac{dp}{\rho} + \frac{vdv}{g_c} + \frac{g}{g_c}dZ + dL_w = 0 \qquad (3\text{-}6)$$

If we consider a pipe inclined at some angle θ to the horizontal, as in Figure 3-3, since $dZ = dL \sin \theta$:

$$\frac{dp}{\rho} + \frac{vdv}{g_c} + \frac{g}{g_c}dL \sin\theta + dL_w = 0$$

Multiplying the equation by ρ/dL gives:

$$\frac{dp}{dL} + \frac{\rho vdv}{g_c dL} + \frac{g}{g_c}\rho \sin\theta + \rho\frac{dL_w}{dL} = 0 \qquad (3\text{-}7)$$

Equation 3-7 can be solved for pressure gradient, and if a pressure *drop* is considered as being positive in the direction of flow

$$\frac{dp}{dL} = \frac{g}{g_c}\rho \sin\theta + \frac{\rho vdv}{g_c dL} + \left(\frac{dp}{dL}\right)_f \qquad (3\text{-}8)$$

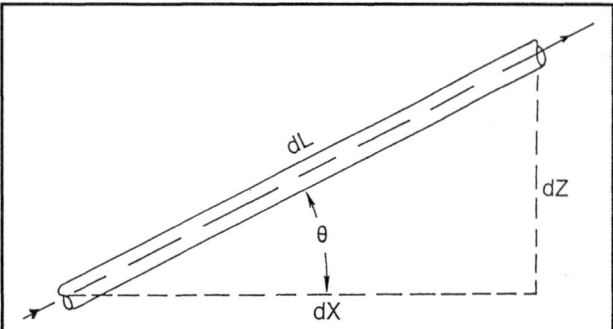

Fig. 3-3. Flow geometry.

where

$$\left(\frac{dp}{dL}\right)_f \equiv \rho \frac{dL_w}{dL}$$

is the pressure gradient due to viscous shear or friction losses.

In horizontal pipe flow, the energy losses or pressure drop are caused by change in kinetic energy and friction losses only. Since most of the viscous shear occurs at the pipe wall, the ratio of wall shear stress (τ_w) to kinetic energy per unit volume ($\rho v^2/2g_c$) reflects the relative importance of wall shear stress to the total losses. This ratio forms a dimensionless group and defines a friction factor.

$$f' = \frac{\tau_w}{\rho v^2 / 2g_c} = \frac{2\tau_w g_c}{\rho v^2} \qquad (3\text{-}9)$$

To evaluate the wall shear stress, a force balance between pressure forces and wall shear stress can be formed. Referring to Figure 3-4:

$$\left[p_1 - \left(p_1 - \frac{dp}{dL} dL\right)\right]\frac{\pi d^2}{4} = \tau_w (\pi d) dL$$

or

$$\tau_w = \frac{d}{4}\left(\frac{dp}{dL}\right)_f \qquad (3\text{-}10)$$

Substituting Equation 3-10 into Equation 3-9 and solving for the pressure gradient due to friction gives:

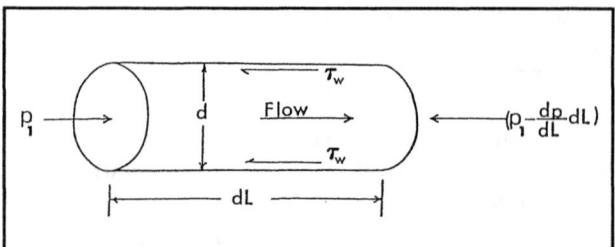

Fig. 3-4. Force balance.

$$\left(\frac{dp}{dL}\right)_f = \frac{2f'\rho v^2}{g_c d}$$

which is the well-known Fanning equation. In terms of a Darcy-Wiesbach or Moody friction factor, $f = 4f'$, and

$$\left(\frac{dp}{dL}\right)_f = \frac{f \rho v^2}{2g_c d} \qquad (3\text{-}11)$$

The friction factor for laminar flow can be determined analytically by combining Equation 3-11 with the Hagen-Poiseuille equation for laminar flow:

$$v = \frac{d^2 g_c}{32\mu}\left(\frac{dp}{dL}\right)_f$$

or:

$$\left(\frac{dp}{dL}\right)_f = \frac{32\mu v}{g_c d^2}$$

Equating the expressions for frictional pressure gradient gives

$$\frac{32\mu v}{g_c d^2} = \frac{f\rho v^2}{2g_c d}$$

or:

$$f = \frac{64\mu}{\rho v d} = \frac{64}{N_{Re}}$$

The dimensionless group, $N_{Re} = (\rho \, v \, d)/\mu$, is the ratio of fluid momentum forces to viscous shear forces and is known as the Reynolds number. It is used as a parameter to distinguish between laminar and turbulent fluid flow. For engineering calculations, the dividing point between laminar and turbulent flow can be assumed to occur at a Reynolds number of 2100 for flow in a circular pipe.

The ability to predict flow behavior under turbulent flow conditions is a direct result of extensive experimental studies of velocity profiles and pressure gradients. These studies have shown that both velocity profile and pressure gradient are very sensitive to characteristics of the pipe wall. A logical approach to defining friction factors is to begin with the simplest case, i.e., the smooth wall pipe, proceed to the partially rough wall and finally to the fully rough wall. Only the most accurate empirical equations available for friction factors are presented here.

For smooth wall pipes, several equations have been developed, each valid over different ranges of Reynolds numbers. The equation that is now used most commonly since it is explicit in f and also covers a wide range of Reynolds numbers, $3000 < N_{Re} < 3 \times 10^6$, was presented by Drew, Koo, and McAdams.[1]

$$f = 0.0056 + 0.5 N_{Re}^{-0.32} \qquad (3\text{-}12)$$

An equation proposed by Blasius may be used for Reynolds numbers up to 100,000 for smooth pipes.

$$f = 0.316 N_{Re}^{-0.25} \qquad (3\text{-}13)$$

The inside wall of a pipe is not normally smooth, and in turbulent flow, the roughness can have a definite effect on the friction factor and thus the pressure gradient. Wall roughness is a function of the pipe material, the method of manufacture, and the environment to which it has been exposed.

From a microscopic sense, wall roughness is not uniform. Individual protrusions, indentations, etc., vary in height, width, length, shape and distribution. The absolute roughness of a pipe, ε, is the mean protruding height of relatively uniformly distributed and sized, tightly packed sand grains that would give the same pressure gradient behavior as the actual pipe.

Dimensional analysis suggests that the effect of roughness is not due to its absolute dimensions, but rather to its dimensions relative to the inside diameter of the pipe, ε/d. In turbulent flow, the effect of wall roughness has been found to be dependent on both the relative roughness and on the Reynolds number. If the laminar sublayer that exists within the boundary layer is thick enough, then the behavior is similar to a smooth pipe. The sublayer thickness is directly related to the Reynolds number.

Nikuradse's[2] famous sand grain experiments formed the basis for obtaining friction factor data for rough pipes. His correlation for fully rough wall pipe is still the best one available. The friction factor may be calculated explicitly from:

$$\frac{1}{\sqrt{f}} = 1.74 - 2 \, \text{Log} \left(\frac{2\varepsilon}{d} \right)$$

The equation that is used as the basis for modern friction factor charts was proposed by Colebrook and White[3] in 1939.

$$\frac{1}{\sqrt{f}} = 1.74 - 2 \, \text{Log} \left(\frac{2\varepsilon}{d} + \frac{18.7}{N_{Re}\sqrt{f}} \right) \qquad (3\text{-}14)$$

The friction factor cannot be extracted readily from the Colebrook equation. By rearranging the equation as follows, a trial and error procedure may be used to solve the equation for friction factor.

$$f_c = \left[\frac{1}{1.74 - 2 \, \text{Log} \left(\dfrac{2\varepsilon}{d} + \dfrac{18.7}{N_{Re}\sqrt{f_g}} \right)} \right]^2$$

Values of f_g are estimated and then f_c is calculated until

f_g and f_c agree to an acceptable tolerance. Using the Drew, Koo and McAdams equation as an initial guess is recommended. After each unsuccessful iteration, the calculated value becomes the assumed value for the next iteration. Also, if more than one pressure loss calculation is to be made as in the case of the iterative procedures discussed in later chapters, then the "converged" value of the previous calculation should be used for the initial guess in the next calculation. Convergence using this method is rapid, normally taking only 2 or 3 iterations. The variation of single-phase friction factor with Reynolds number and relative roughness is shown graphically in Figure 3-5. The Colebrook equation may be applied to flow problems in the smooth, transition and fully rough zones of turbulent flow. For large values of Reynolds number, it degenerates to the Nikuradse equation.

An explicit friction factor equation was proposed by Jain[4], and compared in accuracy to the Colebrook equation. Jain found that for a range of relative roughness between 10^{-6} and 10^{-2} and a range of Reynolds number between 5×10^3 and 10^8 the errors were within $\pm1.0\%$ when compared with the Colebrook equation. The equation gives a maximum error of 3% for Reynolds numbers as low as 2000. The equation is:

$$\frac{1}{\sqrt{f}} = 1.14 - 2 \, \text{Log} \left(\frac{\varepsilon}{d} + \frac{21.25}{N_{Re}^{0.9}} \right) \qquad (3\text{-}15)$$

The determination of the value to use for pipe wall roughness in the friction factor equations is sometimes difficult. It is important to emphasize that ε is not a property that is physically measured. Rather, it is the sand grain roughness that would result in the same friction factor. The only way this can be evaluated is by comparison of the behavior of a normal pipe with one that is sand roughened. Moody has done this and his results, given in Figure 3-6, are still the accepted values. These values should not be considered inviolate and could change significantly because of such things as paraffin deposition, erosion or corrosion. Thus, if measured pressure gradients are available, a friction factor and Reynolds number can be calculated and an effective ε/d obtained from the Moody diagram. This value of ε/d should then be used for future predictions until updated again. If no information on roughness is available, a value of $\varepsilon = 0.0006$ ft is recommended for tubing and line pipe that has been in service for some time.

Example 3-1:

A liquid of specific gravity 0.82 and viscosity of 3 cp (.003 kg/m sec) flows in a 4 in. (101.6 mm) diameter pipe at a velocity of 30 ft/sec (9.14 m/sec). The pipe material is new commercial steel. Calculate the friction

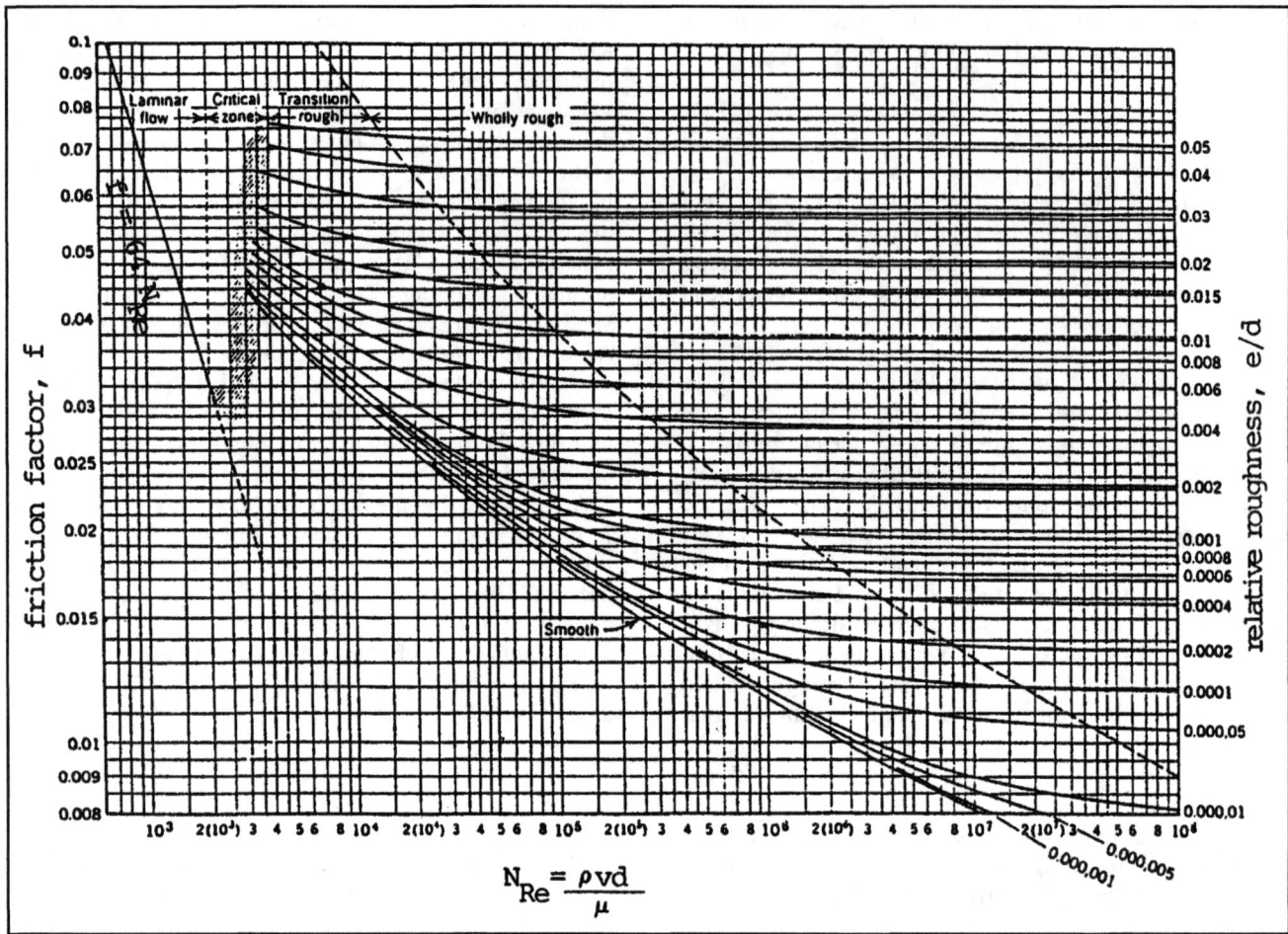

Fig. 3-5. Friction factors for pipe flow.[5]

factor using both the Colebrook equation and the Jain equation.

Solution:

From Figure 3.6, for commercial steel, ε/d = 0.00045. Colebrook Solution: Use the Drew, Koo and McAdams equation for a first guess.

$$N_{Re} = \rho\,vd/\mu = (820)(9.14)(.1016)/.003 = 253,824$$

$$f_g = 0.0056 + 0.5 N_{Re}^{-0.32} = 0.0056 + 0.5 (253,824)^{-0.32}$$

$$f_g = 0.015$$

$$f_c = [1.74 - 2\,Log\,(2\varepsilon/d + 18.7/N_{Re}\sqrt{f_g}]^{-2}$$

$$f_c = [1.74 - 2\,Log\,(2(.00045) + 18.7/253,824\sqrt{.015}]^{-2}$$

$$f_c = 0.0183$$

This value is not close enough to f_g, therefore, another trial is required using f_g = 0.0183:

$$f_c = [1.74 - 2\,Log\,(2(.00045) + 18.7/253,824\sqrt{.0183}\,)]^{-2}$$

$$f_c = 0.0182$$

A third trial using f_g = 0.0182 gives f_c = 0.0182. Jain solution:

$$f = [1.14 - 2\,Log\,(\varepsilon/d + 21.25/N_{Re}^{0.9})]^{-2}$$

$$f = [1.14 - 2\,Log\,(0.00045 + 21.25/(253,824)^{0.9})]^{-2}$$

$$f = 0.01826$$

B. Single-phase Flow

Now that equations and procedures have been presented for evaluating the friction factor in single-phase flow, the pressure gradient equation derived previously can be further developed. Combining Equations 3-8 and 3-11, the pressure gradient equation, which is applicable to any fluid at any pipe inclination angle becomes:

$$\frac{dp}{dL} = \frac{g}{g_c}\rho\,\sin\,\theta + \frac{f\rho v^2}{2g_c d} + \frac{\rho v dv}{g_c dL} \qquad (3\text{-}16)$$

where the friction factor, f, is a function of Reynolds number and pipe roughness. This relationship is shown in

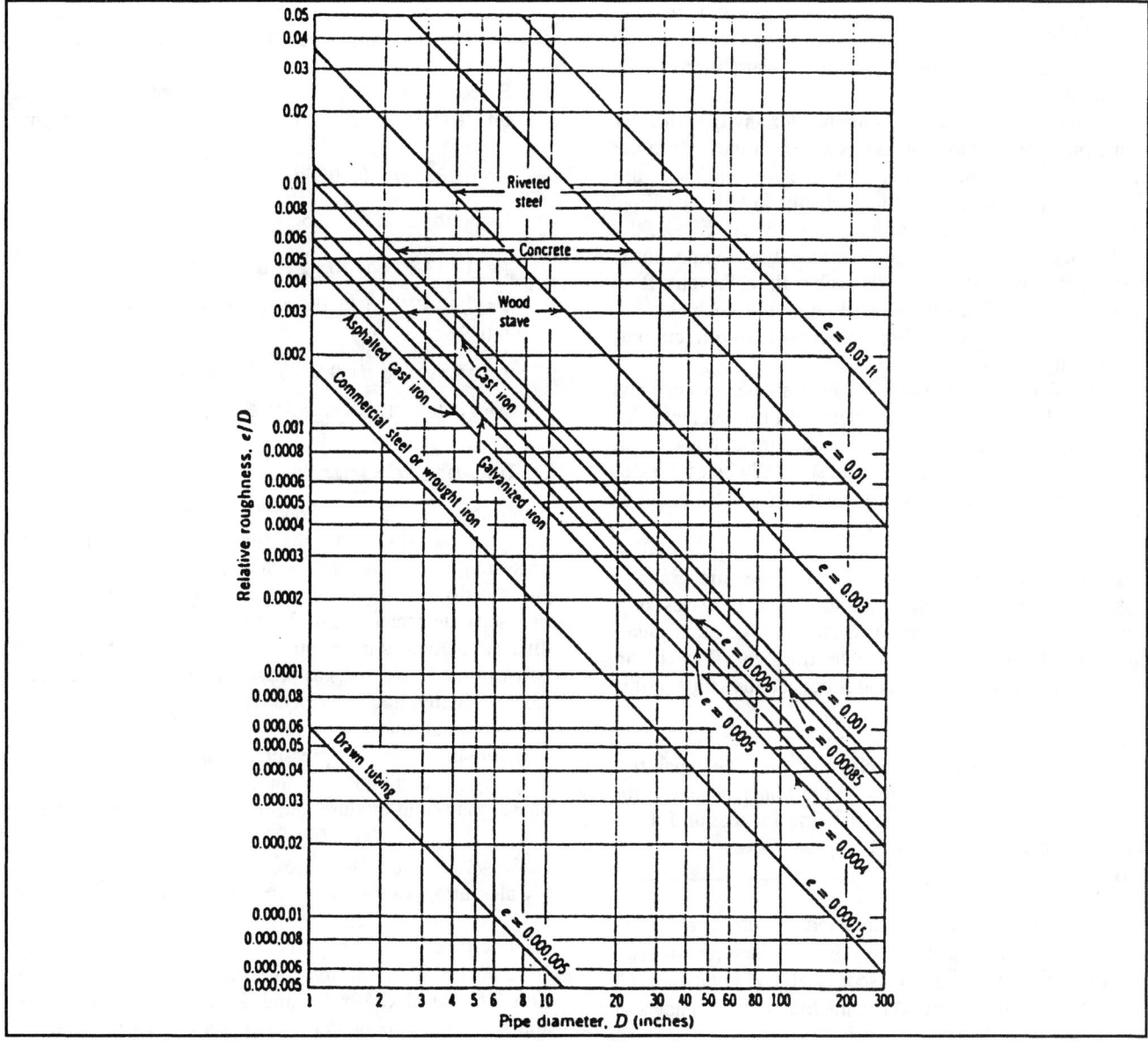

Fig. 3-6. Pipe roughness.[5]

the Moody diagram (Figure 3-5). The total pressure gradient can be considered to be composed of three distinct components, that is:

$$\frac{dp}{dL} = \left(\frac{dp}{dL}\right)_{el} + \left(\frac{dp}{dL}\right)_{f} + \left(\frac{dp}{dL}\right)_{acc} \qquad (3-17)$$

where

$(dp/dL)_{el} = g\rho \sin\theta/g_c$ is the component due to potential energy or elevation change. It is also referred to as the hydrostatic component, as it is the only component that would apply at conditions of no flow.

$(dp/dL)_{f} = f\rho v^2/2g_c d$ is the component due to friction losses.

$(dp/dL)_{acc} = \rho v dv/g_c dL$ is the component due to kinetic energy change or convective acceleration.

Equation 3-16 applies for any fluid in steady state, one-dimensional flow for which f, ρ, and v can be defined. Definition of these variables is what causes most of the difficulty in describing two-phase flow. In two-phase flow, f may be a function of other variables besides the Reynolds number and relative roughness.

Some aspects of the pressure gradient equation as it applies to single-phase flow are discussed to develop a thorough understanding of each component before modifying it for two-phase flow.

The elevation change or hydrostatic component is zero for horizontal flow only. It applies for compressible or incompressible, steady state or transient flow in both vertical and inclined pipes. For downward flow, the sine of the angle is negative, and the hydrostatic pressure increases in the direction of flow.

The friction loss component applies for any type of flow at any pipe angle. It always causes a drop of pressure in the direction of flow. In laminar flow, the friction losses are linearly proportional to the fluid velocity. In turbulent flow, the friction losses are proportional to v^n, where $1.7 \leq n \leq 2$.

The kinetic energy change or acceleration component is zero for constant area, incompressible flow. For any flow condition in which a velocity change occurs, such as compressible flow, a pressure drop will occur in the direction of the velocity increase.

Although single-phase flow has been studied extensively, it still involves an empirically determined friction factor for turbulent flow calculations. The dependence of this friction factor on pipe roughness, which must usually be estimated, makes the calculated pressure gradients subject to considerable error.

Single-phase, incompressible or slightly compressible liquid flow is a trivial solution of the pressure gradient equation. Single-phase compressible flow of gases is a more complex problem to solve and is covered in detail later. Single-phase, compressible transient flow is an extremely complex problem and is beyond the scope of this book.

The preceding descriptions are not meant to be an exhaustive coverage of single-phase flow of Newtonian fluids in pipes. As stated previously, the principal reason for including the material is to form a firm foundation for the more complicated analysis of two-phase flow.

Example 3-2:

Calculate the pressure drop that occurs in a 200 m (656 ft) section of 100 mm (3.94 in.) diameter line pipe when a liquid having a viscosity of 0.05 kg/m–sec (50 cp) and a density of 900 kg/m³ (56.18 lb_m/ft³) flows at a rate of

(a) 3.93 x 10^{-3} m³/sec (0.135 ft³/sec)

(b) 2.355 x 10^{-2} m³/sec (0.83 ft³/sec).

Solution

(a) $v = q/A = 3.93 \times 10^{-3}/\pi \, (0.1)^2/4 = 0.5$ m/sec

$N_{Re} = \rho \, v \, d/\mu = (900)(0.5)(.01)/0.05 = 900$

Since $N_{Re} < 2100$, the flow is laminar, and f is independent of pipe roughness.

$f = 64/N_{Re} = 64/900 = 0.071$

$\Delta p = f \, \rho \, v^2 L/2g_c d = 0.071 \, (900)(0.5)^2(200)/(2)(1)(0.1)$

$\Delta p = 15975$ N/m² = 15.975 kPa (2.31 psi)

(b) $v = q/A = 2.355 \times 10^{-2}/\pi(0.1)^2/4 = 3.0$ m/sec

$N_{Re} = \rho \, v \, d/\mu = 900(3.0)(0.1)/0.05 = 5400$

Since $N_{Re} > 2100$, the flow is turbulent and f depends on Reynolds number and relative roughness. Assume

$\varepsilon = 0.183$ mm (0.0006 ft).

Using the Jain equation for friction factor,

$f = [1.14 - 2 \, Log \, \varepsilon/d + 21.25/N_{Re}^{0.9}]^{-2}$

$f = [1.14 - 2 \, Log \, (0.183/100 + 21.25/(540 \, 0)^{0.9})]^{-2} = 0.0393$

$\Delta p = f \, \rho \, v^2 l/2g_c d = 0.0393 \, (900)(3)^2(200)/(2)(1)(0.1)$

$\Delta p = 3.183 \times 10^5$ N/m² = 318.3 kPa (46.2 psi)

C. Two-phase Flow

Introduction of a second phase into a flow stream complicates the analysis of the pressure gradient equation. The pressure gradient is increased for the same mass flow rate, and the flow may develop a pulsating nature. The fluids may separate because of differences in densities and flow at different velocities in the pipe. A rough interface may exist between the liquid and gas phases. Properties such as densities, velocity, and viscosity, which are relatively simple for individual fluids, become very difficult to determine.

Before modifying the pressure gradient equation for two-phase flow conditions, certain variables unique to a two-phase, gas-liquid mixture must be defined and evaluated.

1. Two-Phase Flow Variables

Calculation of pressure gradients requires values of flow conditions such as velocity, and fluid properties such as density, viscosity, and, in some cases, surface tension. When these variables are calculated for two-phase flow, certain mixing rules and definitions unique to this application are encountered. This section will define and analyze some of the more important properties that must be understood before adapting the previously derived pressure gradient equation for two-phase conditions. In this text, two-phase flow implies gas-liquid flow; however, the liquid phase may include two immiscible fluids such as water and oil. Methods for analysis of a liquid phase that consists of any two components are discussed later.

a. Liquid Holdup. Liquid Holdup H_L, is defined as the fraction of an element of pipe that is occupied by liquid at some instant. That is

$$H_L = \frac{\text{Volume of Liquid in a Pipe Element}}{\text{Volume of the Pipe Element}}$$

Evidently, if the volume element is small enough, the liquid holdup will be either zero or one. It is necessary to be able to determine liquid holdup to calculate such things as mixture density, actual gas and liquid velocities, effective viscosity and heat transfer. In the case of fluctuating flows, such as slug flow, the liquid holdup at a point changes periodically and is taken as the time-averaged value.

The value of liquid holdup varies from zero for single-phase gas flow to one for single-phase liquid flow. Liquid holdup may be measured experimentally by several methods such as resistivity or capacitance probes, nuclear densitometers, or by trapping a segment of the flow stream between quick-closing valves and measuring the volume of liquid trapped.

A value for liquid holdup cannot be calculated analytically. It must be determined from empirical correlations and is a function of variables such as gas and liquid properties, flow pattern, pipe diameter and pipe inclination.

The relative in-situ volume of liquid and gas is sometimes expressed in terms of the volume fraction occupied by gas called gas holdup H_g, or void fraction. Gas holdup is expressed as :

$$H_g = 1 - H_L$$

b. No-Slip Liquid Holdup. No-slip holdup, λ_L, sometimes called input liquid content, is defined as the ratio of the volume of liquid in a pipe element that would exist if the gas and liquid traveled at the same velocity (no slippage) divided by the volume of the pipe element. It can be calculated directly from the known gas and liquid in-situ flow rates from:

$$\lambda_L = \frac{q_L}{q_L + q_g} \tag{3-18}$$

where q_L is the sum of the in-situ oil and water flow rates and q_g is the in-situ gas flow rate. The no-slip gas holdup or gas void fraction is defined as:

$$\lambda_g = 1 - \lambda_L = \frac{q_g}{q_L + q_g}$$

c. Density. All fluid flow equations require that a value of the density of the fluid be available. The density is involved in evaluating the total energy changes due to potential energy and kinetic energy changes. Calculation of density changes as pressure and temperature change requires an equation of state for the fluid under consideration. Equations of state are readily available for single-phase fluids and are presented later. When two immiscible liquids such as oil and water flow simultaneously, the definition of density becomes more complicated. The density of a flowing gas/liquid mixture is very difficult to

evaluate because of the gravitational separation of the phases and the slippage between the phases.

The density of an oil/water mixture may be calculated from the oil and water densities and flow rates if no slippage between the oil and water phases is assumed.

$$\rho_L = \rho_o f_o + \rho_w f_w \tag{3-19}$$

where

$$f_o = \frac{q_o}{q_o + q_w} \tag{3-20}$$

and

$$f_w = 1 - f_o$$

Calculation of the density of a gas/liquid mixture requires knowledge of the liquid holdup. Three equations for two-phase density have been used by various investigators of two-phase flow.

$$\rho_s = \rho_L H_L + \rho_g H_g \tag{3-21}$$

$$\rho_n = \rho_L \lambda_L + \rho_g \lambda_g \tag{3-22}$$

$$\rho_k = \frac{\rho_L \lambda_L^2}{H_L} + \frac{\rho_g \lambda_g^2}{H_g} \tag{3-23}$$

Equation 3-21 is used by most investigators to determine the pressure gradient due to elevation change. Some correlations are based on the assumption of no-slippage and therefore use Equation 3-22 for two-phase density. Equation 3-23 is used by some investigators to define the mixture density used in calculating the friction-loss term and Reynolds number.

d. Velocity. Many two-phase flow correlations are based on a variable called superficial velocity. The superficial velocity of a fluid phase is defined as the velocity that phase would exhibit if it flowed through the total cross sectional area of the pipe alone.

The superficial gas velocity is calculated from:

$$v_{sg} = \frac{q_g}{A} \tag{3-24}$$

The actual area through which the gas flows is reduced by the presence of the liquid to AH_g. Therefore, the actual gas velocity is calculated from:

$$v_g = \frac{q_g}{AH_g} \tag{3-25}$$

where A is the pipe area.

The superficial and actual liquid velocities are similarly calculated from:

$$v_{sL} = \frac{q_L}{A} \tag{3-26}$$

$$v_L = \frac{q_L}{AH_L} \tag{3-27}$$

Since H_g and H_L are less than one, the actual velocities are greater than the superficial velocities.

The two-phase or mixture velocity is calculated based on the total in-situ flow rate from the equation.

$$v_m = \frac{q_L + q_g}{A} = v_{sL} + v_{sg} \tag{3-28}$$

As has been stated previously, the gas and liquid phases may travel at different velocities in the pipe. Some investigators prefer to evaluate the degree of slippage and thus the liquid holdup by determining a slip velocity v_s. The slip velocity is defined as the difference between the actual gas and liquid velocities by:

$$v_s = v_g - v_L = \frac{v_{sg}}{H_g} - \frac{v_{sL}}{H_L} \tag{3-29}$$

Using the previous definitions for the various velocities, alternate forms of the equations for no-slip and actual liquid holdup are:

$$\lambda_L = \frac{v_{sL}}{v_m} \tag{3-30}$$

and

$$H_L = \frac{v_s - v_m + \left[(v_m - v_s)^2 + 4v_s v_{sL} \right]^{1/2}}{2v_s} \tag{3-31}$$

Example 3-3:

Show that for the condition of no-slippage between phases $(v_L = v_g)$, then $H_L = \lambda_L$.

Solution:

By definition,

$$v_L = q_L/A\,H_L, \quad v_g = q_g/A(1-H_L)$$

and since $v_L = v_g$ for no slippage,

$$q_L/A\,H_L = q_g/A(1-H_L)$$

$$q_L\,(1-H_L) = q_g(H_L)$$

$$H_L = q_L/(q_L + q_g) = \lambda_L$$

e. Viscosity. The viscosity of the flowing fluid is used in determining a Reynolds number as well as other dimensionless numbers used as correlating parameters. The concept of a two-phase viscosity is rather nebulous and is defined differently by various investigators. The following equations have been used by various investigators to calculate two-phase, gas/liquid viscosity:

$$\mu_n = \mu_L \lambda_L + \mu_g \lambda_g \tag{3-32}$$

$$\mu_s = \mu_L^{H_L} \times \mu_g^{H_g} \tag{3-33}$$

$$\mu_s = \mu_L H_L + \mu_g H_g \tag{3-34}$$

The viscosity of an oil/water mixture is usually calculated by using the fractions of oil and water flowing in the mixture as weighting factors. The most commonly used equation is

$$\mu_L = \mu_o f_o + \mu_w f_w \tag{3-35}$$

This equation is not valid if an oil/water emulsion is formed.

The viscosities of natural gas, crude oil and water may be estimated from empirical correlations, described in the next section, if measured viscosities are not available.

f. Surface Tension. Correlations for the interfacial tension between water and natural gas and crude oil and natural gas as functions of temperature and pressure are given in the next section. The interfacial tension depends on other fluid properties such as oil gravity, gas gravity and dissolved gas.

When the liquid phase contains both water and oil, the same weighting factors as used for calculating density and viscosity are used. That is:

$$\sigma_L = \sigma_o f_o + \sigma_w f_w \tag{3-36}$$

where

σ_o = oil surface tension, and
σ_w = water surface tension

2. Modification of the Pressure Gradient Equation for Two-Phase Flow

The pressure gradient equation, which is applicable to any fluid flowing in a pipe inclined at a given angle θ from horizontal, was given previously as:

$$\frac{dp}{dL} = \left(\frac{dp}{dL}\right)_{el} + \left(\frac{dp}{dL}\right)_f + \left(\frac{dp}{dL}\right)_{acc} \tag{3-17}$$

a. Elevation Change Component. For two-phase flow the elevation change component becomes

$$\left(\frac{dp}{dL}\right)_{el} = \frac{g}{g_c} \rho_s \sin\theta \tag{3-37}$$

where ρ_s is the density of the gas/liquid mixture in the pipe element. Considering a pipe element that contains liquid and gas, the density of the mixture can be calculated from Equation 3-21. If no slippage between the gas and liquid phases is assumed, the density term is defined by Equation 3-22. Use of Equation 3-21 involves the determination of an accurate value of liquid holdup H_L, whereas the density defined in Equation 3-22 can be calulated from the in-situ gas and liquid flow rates.

b. Friction Component. The friction component becomes:

$$\left(\frac{dp}{dL}\right)_f = \frac{\left(f\rho v^2\right)_f}{2g_c d} \tag{3-38}$$

where f, ρ and v are defined differently by different investigators. The friction component is not analytically predictable except for the case of laminar, single-phase flow. Therefore, it must be determined by experimental means or by analogies to single-phase flow.

The method that has received by far the most attention is the one resulting in two-phase friction factors. Among the most common definitions are the following:

$$\left(\frac{dp}{dL}\right)_f = \frac{f_L \rho_L v_{sL}^2}{2g_c d} \tag{3-39}$$

$$\left(\frac{dp}{dL}\right)_f = \frac{f_g \rho_g v_{sg}^2}{2g_c d} \tag{3-40}$$

$$\left(\frac{dp}{dL}\right)_f = \frac{f_{tp} \rho_f v_m^2}{2g_c d} \tag{3-41}$$

In general, the two-phase friction factor methods differ only in the way the friction factor is determined and to a large extent on the flow pattern. For example, in the mist flow pattern, Equation 3-40, which is based on gas is normally used; whereas in the bubble regime, Equation 3-39, which is based on liquid, is frequently used. The definition of ρ_f in Equation 3-41 can differ widely depending on the investigator.

Most investigators have attempted to correlate friction factors with some form of a Reynolds number. The various Reynolds numbers used to evaluate friction factors are defined when the friction factor correlations are discussed for the individual correlations. One variation, which deserves mention, is that several correlations for predicting vertical flowing pressure losses make use of only the numerator of the Reynolds number.

c. Acceleration Component. The acceleration component for two-phase flow is represented by:

$$\left(\frac{dp}{dL}\right)_{acc} = \frac{\left(\rho v dv\right)_k}{g_c dL} \tag{3-42}$$

The acceleration component is completely ignored by some investigators and ignored in some flow patterns by others. When it is considered, various assumptions are made regarding the relative magnitudes of parameters involved to arrive at some simplified procedure to determine the pressure drop due to kinetic energy change.

From the discussion of the various components contributing to the total pressure gradient, it follows that the principal considerations for developing pressure gradient equations are developing methods for predicting liquid holdup and two-phase friction factor. This is the approach followed by almost all researchers in the study of two-phase flowing pressure gradients.

d. Two-Phase Flow Patterns. Whenever two fluids with different physical properties flow simultaneously in a pipe, there is a wide range of possible flow patterns. By flow pattern, reference is made to the distribution of each phase in the pipe relative to the other phase. Many investigators have attempted to predict the flow pattern that will exist for various sets of conditions, and many different names have been given to the various patterns. Of even more significance, some of the more reliable pressure-loss correlations rely on a knowledge of the existing flow pattern. Also, as a result of the increase in the number of two-phase lines from offshore platforms to onshore facilities, concern has grown regarding the prediction of not only flow pattern, but expected liquid slug sizes and frequencies.

Prediction of flow patterns for horizontal flow is a more difficult problem than for vertical flow. For horizontal flow, the phases tend to separate due to differences in density causing a form of stratified flow to be common. Govier[7] has presented a series of flow pattern descriptions for horizontal air/water flow and vertical air/water flow. These are shown in Figures 3-7 and 3-8 to illustrate the various patterns that can result and also to show that all depend to some extent on the relative magnitudes of v_{sL} and v_{sg}. When flow occurs in a pipe inclined at some angle other than vertical or horizontal the flow patterns take other forms. For inclined upward flow, the pattern is almost always slug or mist. The effect of gravity on the liquid precludes stratification. For inclined downward flow the pattern is usually stratified, mist or annular.

e. Pressure Traverse Calculation. The calculation of a two-phase flowing pressure traverse involves use of an iterative or trial-and-error procedure if temperatures or pipe inclination change with location or distance. In calculating a traverse, the flow conduit is divided into a number of pressure or length increments, and the fluid properties and pressure gradient are evaluated at average conditions of pressure, temperature and pipe inclination in the increment. The accuracy of the pressure traverse calculation increases as the number of increments increases, but so does the number of calculations that must be performed. This presents no problem if a computer is available, but the time involved may be signifi-

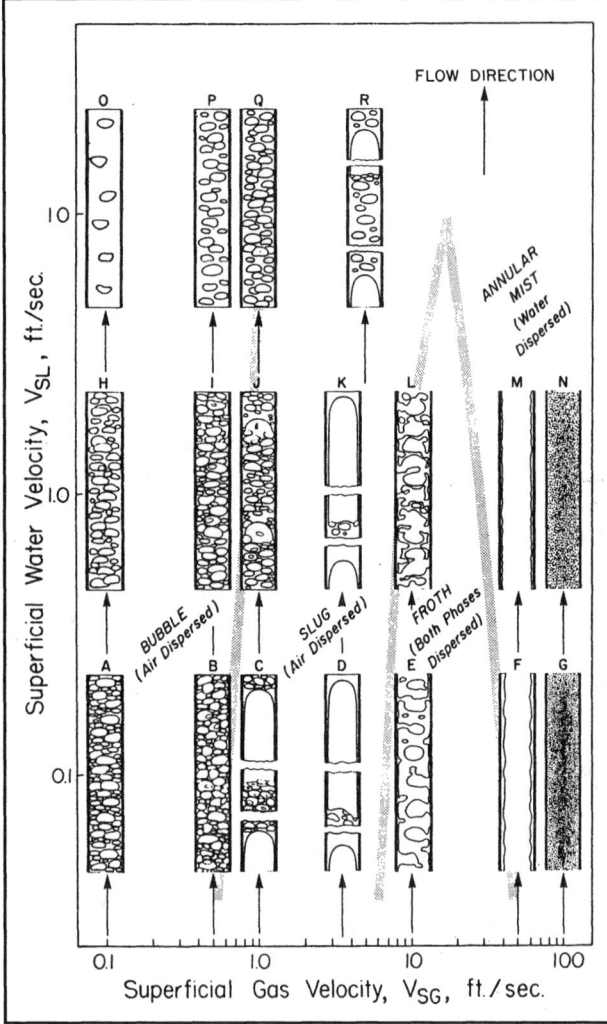

Fig. 3-7. Two-phase vertical flow patterns.

cant for hand calculations. The increments should be smaller at lower pressures where the pressure changes rapidly with distance. For computer calculations, the pressure increments should be no larger than one-tenth of the average pressure in the increment. Algorithms are given both for incrementing on pressure and incrementing on length.

Procedure for Incrementing Pipe Length

1. Starting with the known pressure value, p_1, at location L_1 select a pressure increment, Δp.

2. Starting at the location where the pressure is known, estimate a length increment, ΔL, corresponding to the pressure increment Δp.

3. Calculate the average pressure and, for non-isothermal cases, the average temperature in the increment. Temperature may be a function of location.

4. From laboratory data or empirical correlations, deter-

mine the necessary fluid and PVT properties at conditions of average pressure and temperature determined in step 3.

5. Calculate the pressure gradient, dp/dL in the increment at average conditions of pressure, temperature and pipe inclinations using the appropriate pressure gradient correlation.

6. Calculate the length increment corresponding to the selected pressure increment, $\Delta L = \Delta p/(dp/dL)$.

7. Compare the estimated and calculated values of ΔL obtained in steps 2 and 6. If they are not sufficiently close, estimate a new length increment and go to step 3. Repeat steps 3 through 7 until the estimated and calculated values are sufficiently close.

8. Set $L = L_1 + \Sigma \, \Delta L$ and $p = p_1 + \Sigma \, \Delta p$.

9. If $\Sigma \, \Delta L$ is less than the total conduit length, return to step 2. If $\Sigma \, \Delta L$ is greater than total conduit length, interpolate between the last two values of L to obtain the pressure at the end of the conduit.

This procedure is iterative if either temperature or pipe inclination are functions of length or location. For isothermal, constant pipe inclination cases, steps 2 and 7 may be omitted. This procedure must be used with caution for calculating two-phase pressure traverses for downward flow. In this case, the hydrostatic pressure can increase in the direction of flow, while the friction loss component causes a pressure decrease in the direction of flow. This can create conditions of either a negative or zero pressure gradient, which will cause problems in step 6 of the procedure. The result would be either a negative or infinite length increment. This possibility is eliminated if the following procedure is used. A flow chart for calculating a pressure traverse by dividing the pipe into length increments is shown in Figure 3-9.

Procedure for Incrementing Pressure Drop

1. Starting with the known pressure value, p_1, at location L_1, select a length increment, ΔL.

2. Estimate a pressure increment, Δp, corresponding to the length increment, ΔL.

3. Calculate the average pressure and, for non-isothermal cases, the average temperature in the increment. Temperature may be a function of location.

4. From laboratory data or empirical correlations, determine the necessary fluid and PVT properties at conditions of average pressure and temperature determined in step 3.

5. Calculate the pressure gradient, dp/dL in the increment at average conditions of pressure, temperature,

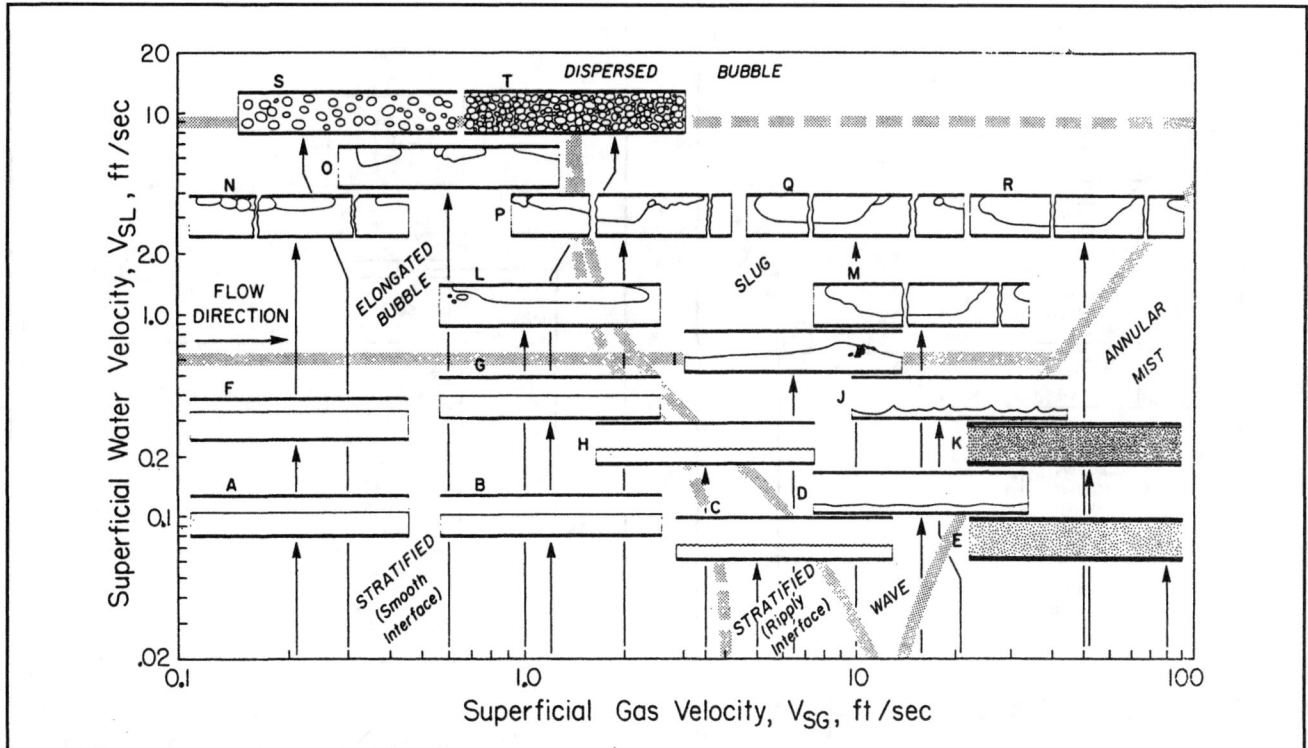

Fig. 3-8. Two-phase horizontal flow patterns.

and pipe inclination, using the appropriate pressure gradient correlation.

6. Calculate the pressure increment corresponding to the selected length increment, $\Delta p = \Delta L \, (dp/dL)$.

7. Compare the estimated and calculated values of obtained in steps 2 and 6. If they are not sufficiently close, estimate a new pressure increment and go to step 3. Repeat steps 3 through 7 until the estimated and calculated values are sufficiently close.

8. Set $L = L_1 + \Sigma \, \Delta L$ and $p = p_1 + \Sigma \, \Delta p$.

9. If $\Sigma \, \Delta L$ is less than the total conduit length, return to step 2.

Using this procedure, the length increments can be selected so that their sum is exactly equal to the total conduit length, and interpolation is not required in the last step. Also, the calculation of a negative or zero pressure gradient in downward flow presents no problem in step 6. This method is always iterative even if temperature and inclination angle are constant, since fluid properties in an increment depend on the unknown pressure. Also, it is not possible to select Δp as some fraction of pressure in the pipe.

If the length increments are set equal, large pressure drops may occur in the low-pressure, high-velocity segments. A flow chart for calculating a pressure traverse by dividing the pipe into pressure drop increments is shown in Figure 3-10.

f. Procedure When Temperature Distribution is Unknown. When a more rigorous method of determining temperature distribution is desired, it is necessary to account for heat transfer to or from the flowing fluids. Heat transfer calculations for two-phase flow can be very important when calculating pressure gradients in geothermal wells, steam-injection wells, wet-gas pipelines in offshore locations or in cold climates, flow of high pourpoint crudes, etc. In general, heat transfer calculations are always preferable to assuming known temperature distribution, but for multicomponent systems, they require that either the inlet or outlet temperature be known.

To perform heat transfer calculations, it is first necessary to change the energy balance equation to a heat balance equation. Combining Equations 3-3 and 3-3a and assuming no work is done on or by the fluid ($dW_s = 0$):

$$dh + \frac{vdv}{g_c} + \frac{g}{g_c}dZ + dq = 0 \qquad (3\text{-}43)$$

If specific enthalpy and the heat-added term are expressed as heat-per-unit mass, then the mechanical energy equivalent of heat constant, J, must be introduced.

$$Jdh + \frac{vdv}{g_c} + \frac{g}{g_c}dZ + Jdq = 0 \qquad (3\text{-}44)$$

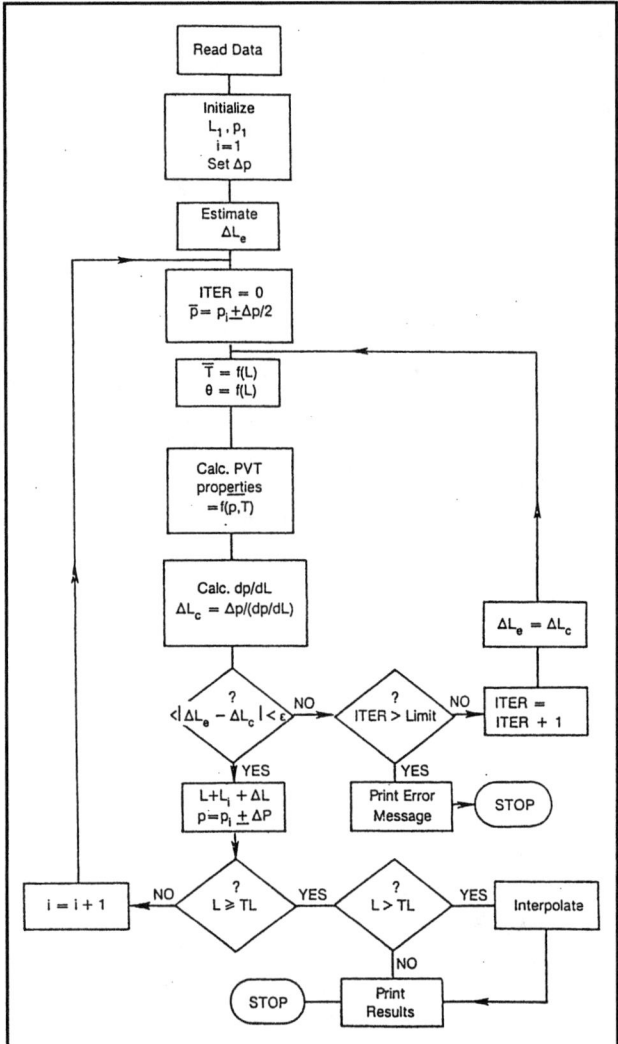

Fig. 3-9. Flow chart for calculating a pressure traverse (incrementing on length).

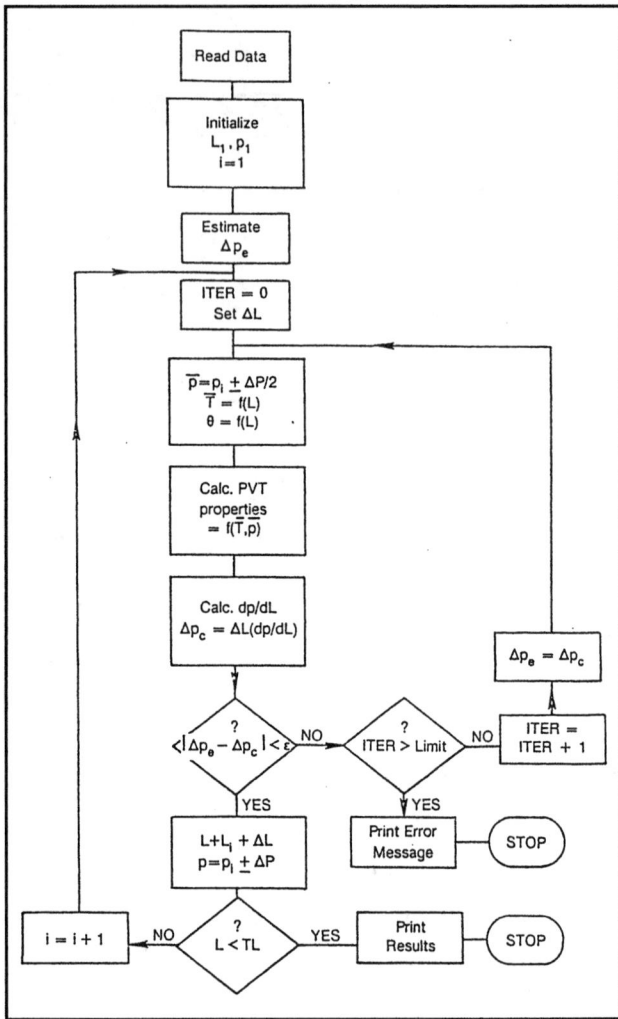

Fig. 3-10. Flow chart for calculating a pressure traverse (incrementing on pressure).

Expressing elevation in terms of pipe length and angle and solving for specific enthalpy gradient:

$$\frac{dh}{dL} = \frac{-dq}{dL} - \frac{g}{g_c}\frac{\sin\theta}{J} - \frac{vdv}{g_c JdL} \qquad (3\text{-}45)$$

The heat added to the system per unit length dq/dL is negative since heat is lost to the surroundings when the fluid temperature is greater than the surrounding temperature. The heat loss gradient can be expressed as:

$$\frac{dq}{dL} = \frac{U(\pi d)}{w_T}\left(\overline{T} - \overline{T}_g\right) \qquad (3\text{-}46)$$

where

$\overline{T}$ = average fluid temperature over dL
$\overline{T}_g$ = average ground or surrounding temperature over dL
U = overall heat transfer coefficient

d = pipe inside diameter
w_T = total mass flow rate.

Heat Transfer Coefficient

The overall heat transfer coefficient can be a combination of several coefficients that depend on the method of heat transfer and the pipe configuration. For unburied pipelines, there will be convective heat losses between the flowing fluids and the pipe wall, conductive losses through the wall and through any insulation or coating material, and conductive losses to the environment. There can also be significant thermal radiation transfer. A complex mixture of heat losses can exist in wellbores due to the variety of materials through which the heat must flow. For example, a well may be cased and the annulus can be cemented, liquid filled or gas filled.

In general, the overall heat transfer coefficient is the reciprocal of the sums of the individual resistances to heat transfer. Consider a buried pipeline.

Then:

$$U = \frac{1}{R_g + R_p + R_f} \qquad (3\text{-}47)$$

where

R_g = resistance to conductive heat transfer from the pipe to the ground

R_p = resistance to conductive heat transfer through the pipe wall and coatings

R_f = resistance to convective heat transfer between the flowing fluids and the pipe wall

For steady-state heat transfer, R_g can be expressed as

$$R_g = \frac{\ln(4D_c / d)}{2k_g / d} \qquad (3\text{-}48)$$

where

D_c = depth from surface to centerline of pipe (Frequently taken as depth of fill on top of pipe)

k_g = thermal conductivity of earth

d = pipe inside diameter

For unsteady state heat transfer, R_g, can be expressed as:

$$R_g = \frac{f(t)}{2k_g / d} \qquad (3\text{-}49)$$

where $f(t)$ is a time-dependent, dimensionless function.

The conductive resistance due to the pipe wall is normally expressed as the steady-state, radial, one-dimensional conduction equation:

$$R_p = \frac{\ln(r_o / r_i)}{2k_p / d} \qquad (3\text{-}50)$$

where

r_o, r_i = outer and inner radii of the pipe

k_p = thermal conductivity of the pipe

R_p can require modifications to account for scale or paraffin buildup on walls or to account for insulation or conrosion protection coatings and wraps.

The convection term, R_f, is extremely complex in two-phase flow, and depends on flow pattern in addition to normally accepted single-phase flow parameters. Heat transfer due to thermal radiation is frequently combined with the convection term. The convective resistance is inversely proportional to the Nusselt factor, N_{Nu}, and the

thermal conductivity of the wetting film on the pipe wall. N_{Nu} in turn depends on the flow characteristic—laminar, transition or turbulent, and is normally correlated with three common dimensionless groups: Reynolds number, Grashof number, and Prandtl number. In turbulent and transition flow, convective heat transfer can be significant so that temperature differences between the fluid and the wall are ignored. In almost all two-phase flow problems, flow is turbulent. A commonly used approximation of the Nusselt factor is:

$$N_{Nu} = 0.027 N_{Re}^{0.8} N_{Pr}^{1/3} \qquad (3\text{-}51)$$

and

$$R_f = \frac{d_i}{k_f N_{Nu}} \qquad (3\text{-}52)$$

where

k_f = thermal conductivity of fluids

N_{Pr} = Prandtl number = $\mu\, c_p / k_f$

C_p = specific heat at constant pressure

For flow in a wellbore, heat transfer is a combination of convection in the casing-tubing annulus and conduction into the earth. The time dependent function in Equation 3-49 for intervals longer than a week can be expressed as:

$$f(t) = -\ln\left(\frac{r_c}{2\sqrt{\infty t}}\right) - 0.290 \qquad (3\text{-}53)$$

where

r_c = outer radius of casing

∞ = thermal diffusivity of the earth

t = time since well began flowing.

Typical values of $f(t)$ range from 0.5 to 3.0.

Equation 3-53 is totally inadequate for short-time, heat-transfer applications or when non-steady flow occurs. This is the case for transient well tests, cyclic steam injection, etc.

In most two-phase design calculations involving buried pipelines or wellbores, R_g is by far the largest of the possible resistances to heat transfer. The result will then be $U = 1/R_g$. If a pipeline is insulated, however, R_p can become the largest term, resulting essentially in $U = 1/R_p$. Seldom is convective heat transfer comparable to the conductive terms. However, for unburied pipelines, the thermal radiation losses can make this term very important.

Magnitudes of overall heat-transfer coefficients can vary widely. Insulated, buried pipelines can have U values as low as 0.1 BTU/(hr°F ft²). Uninsulated, unburied lines can have values higher than 100.

Example 3-4:

Flow in a wellbore is controlled by unsteady heat conduction to the ground. Calculate the overall heat transfer coefficient after 30 days (720 hours) if:

$r_c = 0.5$ ft

$k_g = 1.4$ BTU/(hr ft °F)

$\infty = 0.04$ ft^2/hr.

Solution:

$f(t) = -\ln(r_c/2\sqrt{\infty t}) - 0.290$

$= -\ln(0.5/2\sqrt{(0.04)(720)}) - 0.290 = 2.78$

$R_g = f(t)/2k_g/d$

$= 2.78/(2)(1.4)/(1.0)$

$= 0.992$ hr ft^2 °F/BTU

$U = 1/R_g = 1/0.992 = 1.01$ BTU/hr ft^2 °F

Example 3-5:

Estimate the value of R_f for hot water injection into a well if:

$k_f = 0.4$ BTU/hr ft °F

$d_i = 1.0$ ft

$w_T = 200,000$ lb$_m$/hr

$\mu = 0.3$ cp $= 0.726$ lb$_m$/ft hr

$T = 300$°F

$c_p = 0.5$ BTU/lb$_m$°F

Solution:

$N_{Pr} = \mu c_p/k_f = (0.726)(0.5)/(0.4) = 0.908$

$N_{Re} = (w_T/\mu)(4/\pi d)$

$= (200,000)/(0.726) \cdot 4/\pi(1.0) = 3.51 \times 10^5$

$N_{Nu} = 0.027 N_{Re}^{0.8} N_{Pr}^{1/3} (0.027)(3.51 \times 10^5)^{0.8}(0.908)^{1/3}$

$= 714$

$R_f = d_i/k_f N_{Nu}$

$R_f = 1/(0.4)(714) = 3.50 \times 10^{-3}$ ft^2 hr °F/BTU

Mixture Enthalpy

The total mixture enthalpy of a two-phase, gas-liquid mixture at a given pressure and temperature can be calculated from a knowledge of the in-situ mass or mole fraction of each phase and the enthalpy of each phase. If x is the no-slip, gas, mass fraction (quality), then the no-slip mixture enthalpy is:

$$h = h_L(1-x) + h_g x \qquad (3-54)$$

where

$h, h_L, h_g = $ specific enthalpy of the mixture, liquid and gas, respectively

The quality may be related to the no-slip holdup by:

$$x = \frac{\rho_g \lambda_g}{\rho_L \lambda_L + \rho_g \lambda_g} \qquad (3-55)$$

A simplified flow chart of a computing algorithm for simultaneous pressure and temperature traverse calculations is given in Figure 3-11.

Convergence on both pressure and temperature in a given pipe length increment requires a double iterative procedure. Most efficient results are obtained when converging on the least sensitive variable first. Normally, mixture enthalpy is more sensitive to temperature than to pressure. This is shown in Figure 3-12 in which h (BTU/100-mol) is plotted vs. T for several isobars and for a specific feed composition. Similar curves can be developed for any feed composition using results from vapor-liquid equilibrium flash calculations and empirical enthalpy correlations. The outer loop of the flow chart is thus the temperature loop, and the inner loop is the pressure loop.

III. FLUID PROPERTY CALCULATIONS

The general pressure gradient equation (Equation 3-16) was derived in the previous section, and it was pointed out that all the terms in this equation must be evaluated at in-situ conditions. That is, during the calculation of a pressure traverse, the pressure gradient must be calculated at several points in the piping system at the pressure and temperature existing at this point. This is step 4 in the procedure for calculating a pressure traverse. Also, as will be seen in the following section, it will frequently be necessary to evaluate various in-situ fluid properties or fluid velocities to calculate flow pattern, friction factor and liquid holdup.

The variables in the pressure gradient equation are fluid density, mixture velocity and friction factor. Calculation of these variables for the gas/liquid mixture requires values of the individual components at the conditions of interest. The mixture can be composed of natural gas, crude oil or condensate, and water. Also, evaluation of the friction factor usually requires a value for the viscosity of the individual components at various pressures and temperatures.

The fluid properties required to calculate the necessary parameters could be obtained from a laboratory analysis of samples of the fluids, if these are available. Frequently, a pressure-volume-temperatue (PVT) analysis will have been conducted on the reservoir fluid to obtain data for reservoir engineering calculations. Unfortunately, these

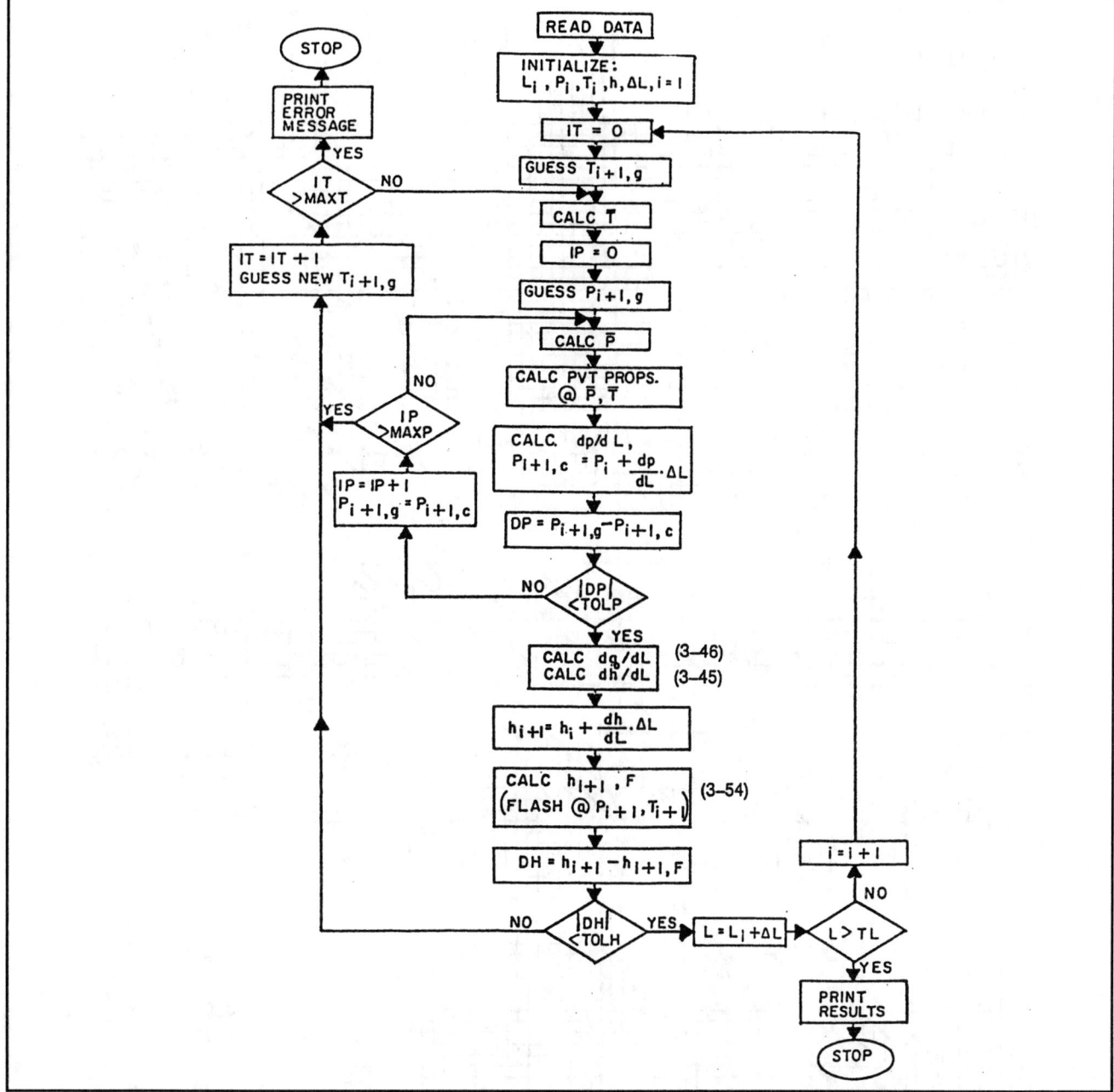

Fig. 3-11. Heat transfer calculation algorithm.

analyses are usually conducted at reservoir temperature only and are not applicable to piping system calculations, since the temperature of the fluids in the piping system is constantly changing.

This fact has made it necessary to resort to empirical fluid property correlations for calculating pressure traverses, to account for temperature change effects. However, if measured values are available at one temperature, these may be used to improve the accuracy of the empirical correlations by forcing the empirical methods to match the measured values at this temperature.

In any design situation, the flow or production rates of the fluids will be known at standard conditions. Also, the fluid densities at standard conditions will usually be known in the form of specific gravities. These must be converted to flow rates or velocities and densities at in-situ conditions before a pressure traverse can be calculated.

The equations necessary for making these conversions will first be given for each of the three fluid components.

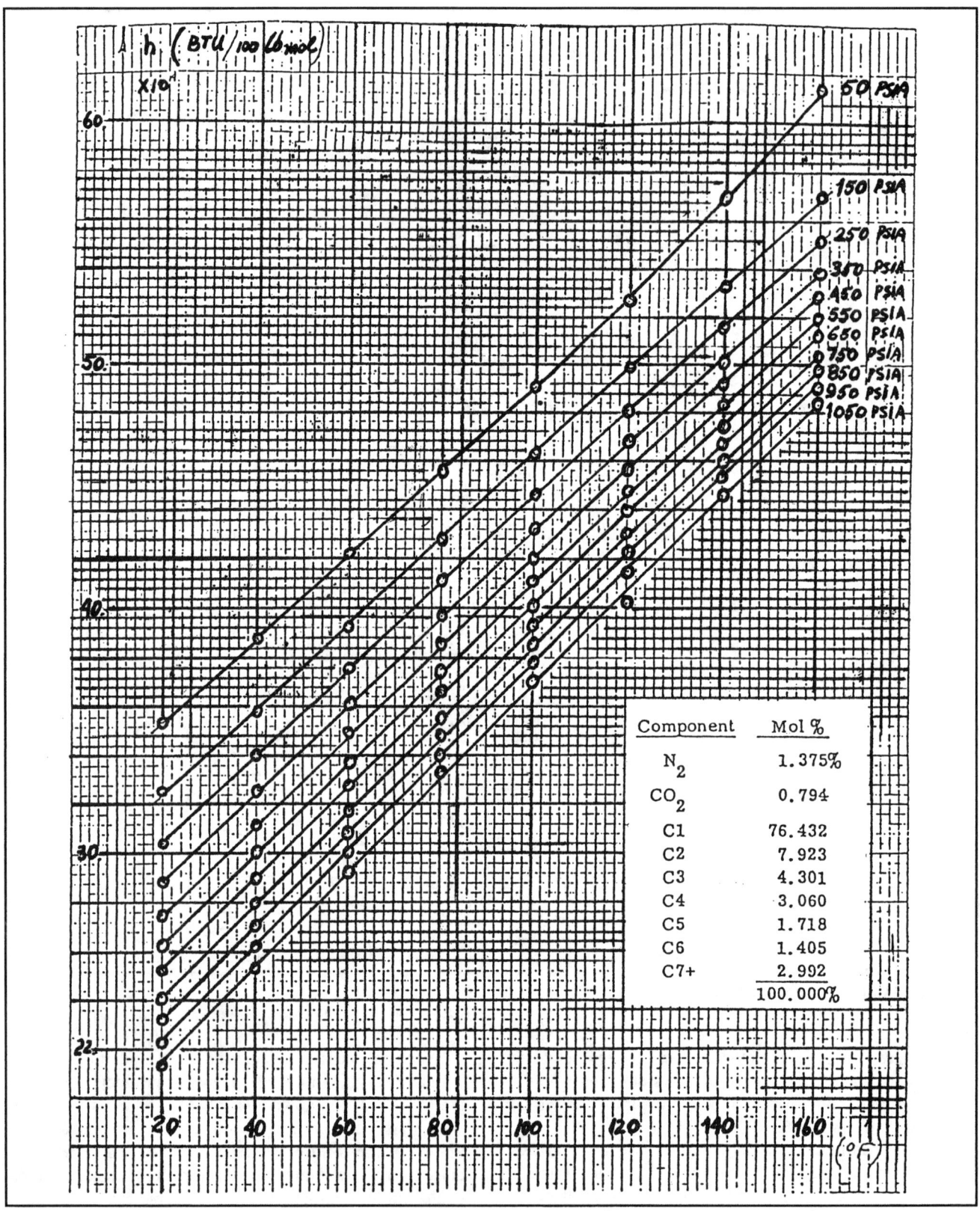

Fig. 3-12. Variation of mixture enthalpy with pressure and temperature.

These equations will require knowledge of various fluid PVT properties, which will then be presented. The equations will be presented in the customary field units, but may be easily converted to metric or SI units if required.

A. Fluid Density

Calculation of a two-phase fluid density requires values for the densities of the gas, oil and water. The equations for converting the densities from specific gravities to in-situ values will be presented.

1. Gas

The specific gravity of a gas is defined as the ratio of the density of the gas to the density of air, both measured at standard conditions of pressure and temperature; that is,

$$\gamma_g = \left(\frac{\rho_g}{\rho_{air}}\right)_{psc,Tsc} = \frac{M_g}{M_{air}} \quad (3\text{-}56)$$

where

ρ_g = gas density,
ρ_{air} = air density,
M_g = gas molecular weight,
p_{sc} = standard pressure (14.7 psia),
T_{sc} = standard temperature (60°F =520°R), and
γ_g = gas gravity

Using the engineering equation of state for a gas, it can be shown that the relationship among gravity, pressure and temprature is:

$$\rho_g = \frac{2.7\gamma_g p}{ZT} \quad (3\text{-}57)$$

where

ρ_g = gas density at p and T, lbm/ft^3,
p = pressure, psia,
T = temperature, °R,
γ_g = gas gravity (air = 1), and
Z = gas compressibility factor.

Methods to estimate Z as a function of p, T and γ_g will be presented subsequently.

2. Oil

The specific gravity of a liquid is defined as the ratio of the density of the liquid to the density of pure water, both measured at standard conditions. That is:

$$\gamma_L = \left(\frac{\rho_L}{\rho_w}\right)_{psc,Tsc} \quad (3\text{-}58)$$

In the petroleum industry, it is common to express the gravity in terms of the API gravity of the oil, or:

$$\gamma_o = \frac{141.5}{131.5 + API} \quad (3\text{-}59)$$

where

γ_o = oil specific gravity, and
API = oil API gravity.

The density of the oil plus any gas dissolved in the oil at the pressure and temperature of interest may be calculated by dividing the weight or mass per stock tank barrel by the in-situ volume that would be occupied by a stock tank barrel of oil and its solution gas.

$$\rho_o = \frac{350\gamma_o + 0.0764\gamma_g R_s}{5.615 B_o} \quad (3\text{-}60)$$

where

ρ_o = oil density, lbm/ft^3,
R_s = solution or dissolved gas, scf/STB,
B_o = oil formation volume factor, bbl/STB,
350 = density of water at s.c., lbm/STB,
0.0764 = density of air at s.c., lbm/scf, and
5.615 = conversion factor, ft^3/bbl

If the pressure and temperature conditions are such that all of the available gas is in solution, that is, the pressure is above the bubblepoint at the temperature of interest, increased pressure will merely compress the liquid and increase its density. For the case of $p \geq p_b$, the oil density is calculated from:

$$\rho_o = \rho_{ob} \text{EXP}\left[C_o\left(p - p_b\right)\right] \quad (3\text{-}61)$$

where

ρ_o = density at p, T,
ρ_{ob} = density at p_b, T,
p = pressure, psia,
p_b = bubblepoint pressure at T, psia, and
C_o = oil isothermal compressibility at T, psi^{-1}
EXP(X) = e^x = $(2.7183)^x$

Correlations for calculating R_s, B_o, C_o and p_b at various conditions will be presented later.

3. Water

For purposes of calculating two-phase flowing pressure gradients, the effect of gas in solution in the water is often ignored since it is very low compared to R_s in the oil. However, the effects of dissolved solids and temperature must be considered. The density of water may be estimated from:

$$\rho_w = \frac{\rho_{wsc}\gamma_w}{B_w} = \frac{62.4\gamma_w}{B_w} \quad (3\text{-}62)$$

where

$$\rho_w = \text{water density at p and T, lbm/ft}^3,$$
$$\rho_{wsc} = \text{density of pure water at s.c.} = 62.4 \text{ lbm/scf},$$
$$\gamma_w = \text{water specific gravity, and}$$
$$B_w = \text{water formation volume factor, ft}^3\text{/scf}.$$

The value of γ_w depends on the dissolved solids in the water. Equations for calculating B_w will be given subsequently.

B. Fluid Velocity

To calculate the in-situ velocities of the gas, oil and water from surface flow rates, the actual volumetric flow rates must first be calculated. The following equations can be used to correct standard flow rates to in-situ flow rates, from which the velocities can be calculated.

1. Gas

The gas-producing rates are usually stated in standard volumes per unit time, such as standard cubic feet per day, scf/day. These may be converted to cubic feet per second at in-situ conditions, from which the superficial velocities in ft/sec can be obtained.

$$v_{sg} = \frac{q_g}{A} = \frac{q_{sc}B_g}{A} \qquad (3\text{-}63)$$

where

$$v_{sg} = \text{superficial gas velocity, ft/sec,}$$
$$q_{sc} = \text{free gas flow rate, scf /sec,}$$
$$B_g = \text{gas formation volume factor, ft}^3\text{/scf, and}$$
$$A = \text{cross-sectional area of the pipe, ft}^2$$

If the gas is in contact with oil in the piping system, the solution gas must be subtracted from the measured separator and stock tank gas before the in-situ velocity is calculated. A convenient equation for making this calculation is:

$$q_g = \frac{q_o (R - R_s) B_g}{86400} \qquad (3\text{-}64)$$

where

$$q_g = \text{gas flow rate, ft}^3\text{/sec,}$$
$$q_o = \text{oil producing rate, STB/day}$$
$$R = \text{producing gas/oil ratio, scf/STB,}$$
$$R_s = \text{solution gas/oil ratio, scf/STB, and}$$
$$B_g = \text{gas formation volume factor, ft}^3\text{/scf.}$$
$$86,400 = \text{seconds/day}$$

$$B_g = \frac{p_{sc}ZT}{T_{sc}p} = \frac{0.0283ZT}{p} \qquad (3\text{-}65)$$

where

$$Z = \text{gas compressibility factor,}$$

$$T = \text{temperature, }^\circ\text{R, and}$$
$$p = \text{pressure, psia.}$$

2. Oil

To calculate the in-situ superficial velocity of the oil, the expanded volume of the oil must be accounted for. That is:

$$v_{so} = \frac{q_o B_o}{A} \qquad (3\text{-}66)$$

where

$$v_{so} = \text{superficial oil velocity, ft/sec}$$
$$q_o = \text{oil flow rate, scf/sec}$$
$$B_o = \text{oil formation volume factor, ft}^3\text{/scf}$$
$$A = \text{pipe area, ft}^2$$

If the oil rate is given in STB/day, the equation becomes:

$$v_{so} = \frac{6.5 \times 10^{-5} q_o B_o}{A} \qquad (3\text{-}66)$$

where

$$q_o = \text{STB/day}$$
$$v_{so} = \text{ft/sec}$$
$$A = \text{ft}^2$$
$$B_o = \text{ft}^3\text{/scf or bbl/STB}$$

3. Water

The in-situ, superficial water velocity is calculated from:

$$v_{sw} = \frac{6.5 \times 10^{-5} q_w B_w}{A} \qquad (3\text{-}68)$$

where

$$q_w = \text{water rate, STB/day}$$
$$B_w = \text{water formation volume factor, ft}^3\text{/scf or bbl/STB}$$
$$A = \text{pipe area, ft}^2$$
$$v_{sw} = \text{superficial water velocity, ft/sec}$$

The superficial liquid velocity is the sum of the oil and water velocities, and the superficial mixture velocity is the sum of the liquid and gas superficial velocities; that is,

$$v_{sL} = v_{so} + v_{sw} \qquad (3\text{-}69)$$

$$v_m = v_{sL} + v_{sg} \qquad (3\text{-}28)$$

C. Empirical Fluid Property Correlations

The most widely used empirical fluid property correlations for estimating the parameters required to calculate densities and velocities will be presented in this section. Many of these correlations were originally published in graphical form only, but where possible the correlations will be given in equation form. This will greatly facilitate calculation of pressure gradients using computers or programmable calculators.

Most of the equations presented here are available for Hewlett-Packard 41-C programmable calculators in the form of a plug-in module called the Petroleum Fluids Pac.[7]

1. Gas Compressibility Factor

The gas compressibility or Z-factor is a function of the pseudoreduced pressure and temperature of the gas. The correlation shown in Figure 3-13, from Standing and Katz[8] gives good values for hydrocarbon gases. Corrections to the pseudocritical pressures and temperatures can be made to account for impurities such as N_2, CO_2, and H_2S. The pseudoreduced values are defined as:

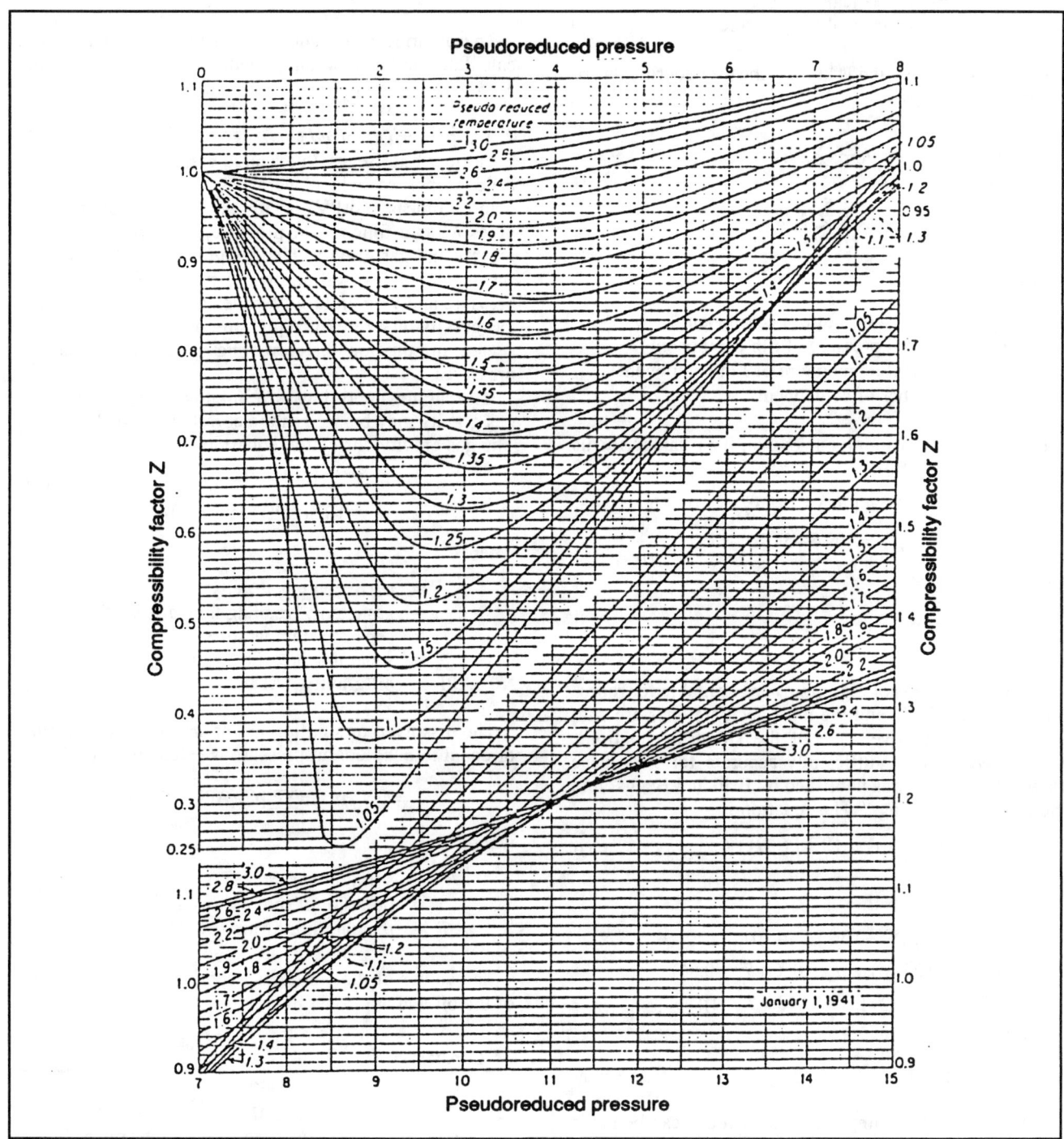

Fig. 3-13. Gas compressibility factor.

$$p_{pr} = \frac{p}{p_{pc}}$$

$$T_{pr} = \frac{T}{T_{pc}} \qquad (3\text{-}70)$$

where

p = pressure of interest
T = temperature of interest
p_{pc} = pseudocritical pressure
T_{pc} = pseudocritical temperature

If the gas composition is known, the pseudocriticals are calculated from:

$$p_{pc} = \sum_{i=1}^{N} y_i p_{ci}$$

$$T_{pc} = \sum_{i=1}^{N} y_i T_{ci} \qquad (3\text{-}71)$$

where

y_i = mol fraction of i th component
p_{ci} = critical pressure of i th component
T_{ci} = critical temperature of i th component
N = number of components

If the gas composition is unknown, the pseudocriticals may be estimated from

$$T_{pc} = 170.5 + 307.3 \gamma_g \qquad (3\text{-}72)$$

$$p_{pc} = 709.6 - 58.7 \gamma_g$$

where

p_{pc} = psia
T_{pc} = °R

Several equations or algorithms are available for reproducing Figure 3-13, and the most accurate ones are trial and error or iterative. One of the simplest equations, which gives values sufficiently accurate for two-phase flow calculations, was published by Brill and Beggs[9] and modified by Standing.[10] The equation is:

$$Z = A + (1 - A)\text{EXP}(-B) + Cp_{pr}^{D} \qquad (3\text{-}73)$$

where

$A = 1.39(T_{pr} - 0.92)^{0.5} - 0.36 T_{pr} - 0.101$

$B = p_{pr}(0.62 - 0.23 T_{pr}) + p_{pr}^2[0.066/(T_{pr} - 0.86)$
$\quad - 0.037] + 0.32 p_{pr}^6 / \text{EXP}[20.723(T_{pr} - 1)]$

$C = 0.132 - 0.32 \log T_{pr}$

$D = \text{EXP}(0.715 - 1.128 T_{pr} + 0.42 T_{pr}^2)$

If the gas contains impurities, corrections can be made to p_{pc} and T_{pc} according to Wichert and Aziz[11] as:

$$T_{pc}' = T_{pc} - \varepsilon, \qquad (3\text{-}74)$$

$$p_{pc}' = \frac{p_{pc} T_{pc}'}{T_{pc} + \varepsilon(B - B^2)}$$

where

ε = $120(A^{0.9} - A^{1.6}) + 15(B^{0.5} - B^4)$
B = mol fraction H_2S
A = mol fraction $CO_2 + B$

These corrected pseudocritical values are then used to calculate the pseudoreduced values for use in Equation 3-73.

2. Solution or Dissolved Gas

Vasquez and Beggs[12] presented correlations for several fluid properties, which were based on data from more than 600 measured PVT analyses. The correlation for R_s gives R_s as a function of pressure, temperature, oil API gravity and gas gravity. The gas gravity resulting from a flash separation of oil and gas depends on the separator pressure and temperature. Vasquez and Beggs based their gas gravity on a reference pressure of 114.7 psia and presented an equation for correcting γ_g for other separator pressures.

If separator conditions are unknown, the uncorrected gas gravity may be used in the correlations for R_s and B_o. The gas gravity correction equation is:

$$\gamma_{gc} = \gamma_g[1.0 + 5.912 \qquad (3\text{-}75)$$
$$\times 10^{-5}(\text{API})T\log(p/114.7)]$$

where

γ_{gc} = corrected gas gravity
γ_g = gas gravity resulting from a separation at p, T
T = separator temperature, °F
p = separator pressure, psia
API = oil gravity, °API

The dissolved or solution gas at any pressure less than or equal to bubblepoint pressure is calculated from:

$$R_s = C_1 \gamma_{gc} p^{C_2} \text{EXP}[C_3(\text{API})/(T + 460)] \qquad (3\text{-}76)$$

where

R_s = solution gas, scf/STB
p = pressure of interests psia
T = temperature of interest, °F

The values of the constants depend on the API gravity of the oil and are given by:

Constant	API $\leq$ 30	API >30
C1	0.0362	0.0178
C2	1.0937	1.1870
C3	25.7240	23.9310

This method for estimating R_s is used in the HP Petroleum Fluids Pac.[7] Other correlations for R_s were published by Standing[13] and Lasater.[14]

If the initial solution gas, $R_{si} = R_{sb}$ is known, Equation 3-76 may be solved for bubblepoint pressure, p_b.

Although frequently ignored in two-phase flow calculations, an equation for calculating gas in solution in water as published by Craft and Hawkins[15] is given.

$$R_{sw} = R_{swp}[1 - XY \times 10^{-4}]$$ (3-77)

where

R_{sw} = gas dissolved in brine, scf/STB
R_{swp} = gas dissolved in pure water, scf/STB
X = $3.471T^{-0.837}$
Y = water salinity, ppm
T = temperature, °F
R_{swp} = $C_1 + C_2p + C_3p^2$

where

C_1 = $2.12 + 3.45 \times 10^{-3}T - 3.59 \times 10^{-5}T^2$
C_2 = $0.0107 - 5.26 \times 10^{-5}T + 1.48 \times 10^{-11}T^2$
C_3 = $-8.75 \times 10^{-7} + 3.9 \times 10^{-9}T - 1.02 \times 10^{-11}T^2$

3. Formation Volume Factor

The formation volume factor of a fluid is a convenient parameter to use for converting from standard volumes to actual or in-situ volumes existing at any pressure and temperature in the system. Equations are given for gas, oil and water.

a. Gas. The gas formation volume factor is defined as the actual volume occupied by a given quantity of gas at some pressure and temperature, divided by the volume which the gas would occupy at standard conditions. It is calculated from:

$$B_g = \frac{p_{sc}ZT}{T_{sc}p}$$ (3-78)

For pressure in psia and temperature in °R, using p_{sc} = 14.7 psia and T_{sc} = 520 °R, Equation 3-78 becomes:

$$B_g = \frac{0.0283ZT}{p}$$

b. Oil. The Vasquez and Beggs method may be used to estimate B_o as a function of γ_g, API, R_s and T. The equation is:

$$B_o = 1 + C_1R_s + C_2(T - 60)(API/\gamma_{gc})$$
$$+ C_3R_s(T - 60)(API/\gamma_{gc})$$ (3-79)

where

B_o = volume/standard volume, e.g. bbl /STB
R_s = solution gas at p, T, scf/STB
T = temperature of interests °F
p = pressure of interest, psia
API = oil API gravity
γ_{gc} = gas gravity

The constants are determined from:

Constant	API ≤ 30	API > 30
C_1	4.677×10^{-4}	4.670×10^{-4}
C_2	1.751×10^{-5}	1.100×10^{-5}
C_3	-1.811×10^{-8}	1.337×10^{-9}

The oil formation volume factor decreases at pressures above the bubblepoint pressure and is calculated from:

$$B_o = B_{ob} \, EXP \, [C_o(p_b - p)]$$ (3-80)

where

B_{ob} = oil FVF at p_b
p_b = bubblepoint pressure, psia
p = pressure of interest, psia
C_o = oil isothermal compressibility, psi⁻¹

c. Water. The equation given in the HP Petroleum Fluids Pac is

$$B_w = B_{wp}(1 + XY \times 10^{-4})$$ (3-81)

where

B_w = formation volume factor for brine in contact with gas, bbl/STB
B_{wp} = FVF for pure water, bbl/STB
Y = water salinity, ppm
X = $5.1 \times 10^8p + (T-60)(5.47 \times 10^{-6} - 1.95 \times 10^{-10}p) + (T-60)^2(-3.23 \times 10^{-8} + 8.5 \times 10^{-13}p)$

$$B_{wp} = C_1 + C_2p + C_3p^2$$ (3-82)

where

C_1 = $0.9911 + 6.35 \times 10^{-5}T + 8.5 \times 10^{-7}T^2$
C_2 = $1.093 \times 10^{-6} - 3.497 \times 10^{-9}T + 4.57 \times 10^{-12}T^2$
C_3 = $-5 \times 10^{-11} + 6.429 \times 10^{-13}T - 1.43 \times 10^{-15}T^2$
T = °F
p = psia

4. Isothermal Compressibility

The isothermal compressibility for oil saturated with gas can be calculated using the following equation presented by Vasquez and Beggs.

$$C_o = \frac{5R_s + 17.2T - 1180\gamma_{gc} + 12.61(\text{API}) - 1433}{p \times 10^5} \quad (3\text{-}83)$$

where

- C_o = oil compressibility, psi^{-1}
- R_s = solution gas/oil ratio, scf/STB
- T = temperature of interest, °F
- p = pressure of interest. psia
- γ_{gc} = gas gravity
- API = oil API gravity

An equation for estimating water isothermal compressibility, ignoring the corrections for dissolved gas and solids is:

$$C_w = (C_1 + C_2 T + C_3 T^2) \times 10^{-6} \quad (3\text{-}84)$$

where

- C_1 = 3.8546 − 0.000134 p
- C_2 = -0.01052 + 4.77 x 10^{-7} p
- C_3 = 3.9267 x 10^{-5} − 8.8 x 10^{-10} p
- T = °F
- p = psia
- C_w = psi^{-1}

The isothermal compressibility for gas is seldom required for two-phase flow calculations. However, it may be calculated from:

$$C_g = \frac{1}{p} - \frac{1}{Z}\frac{\partial Z}{\partial p} \quad (3\text{-}85)$$

where

- C_g = psi^{-1}
- p = psia

5. Viscosity

To calculate the losses due to viscous shear or friction, a value for the viscosity of the fluids is required. Calculation of a Reynolds number always requires viscosity. Equations are presented for the viscosity of oil, both above and below bubblepoint, for water and for natural gas.

a. Oil. Equations for oil viscosity were presented by Beggs and Robinson.[16] For $p \leq p_b$,

$$\mu_o = A\mu_{oD}^B \quad (3\text{-}86)$$

where

- μ_o = oil viscosity at the pressure and temperature of interest, cp
- μ_{oD} = dead or gas-free oil viscosity, cp
- μ_{oD} = $10^x - 1.0$

- x = $YT^{-1.163}$
- Y = 10^Z
- Z = 3.0324 - 0.0203(API)
- A = 10.715 $(R_s + 150)^{-0.515}$
- B = 5 44 $(R_s + 150)^{-0.338}$
- R_s = scf/STB
- T = °F

For Pressures greater than p_b,

$$\mu_o = \mu_{ob}(p/p_b)^m \quad (3\text{-}87)$$

where

- μ_{ob} = viscosity at p_b
- p = pressure of interest
- p_b = bubblepoint pressure
- m = $C_1 p^{C2} \text{EXP}(C_3 + C_4 p)$
- C_1 = 2.6
- C_2 = 1.187
- C_3 = −11.513
- C_4 = −8.98 x 10^{-5}

b. Water. A graphical correlation for water viscosity which was published by Matthews and Russell,[17] has been converted to equation form by Meehan.[18] The correlation accounts for both the effects of pressure and salinity.

$$\mu = \mu_{wD}[1 + 3.5 \times 10^{-2} p^2 (T - 40)] \quad (3\text{-}88)$$

$\mu_{wD} = A + B/T$

A = -4.518 x 10^{-2} + 9.313 x 10^{-7}Y − 3.93 x 10^{-12}Y^2

B = 70.634 + 9.576 x 10^{-10}Y^2

where

- μ_w = brine viscosity at p and T, *cp*
- μ_{wD} = brine viscosity at p = 14.7, T, *cp*
- p = pressure of interest, psia
- T = temperature of interest, °F
- Y = water salinity, ppm

The effects of pressure and salinity are often neglected for two-phase flow calculations. A simpler equation, which considers only temperature effects, was presented by Brill and Beggs.[9]

$$\mu_w = \text{EXP}(1.003 - 1.479 \times 10^{-2} T + 1.982 \times 10^{-5} T^2) \quad (3\text{-}89)$$

where T is in °F and μ_w is in *cp*.

c. Gas. The most widely used method to estimate gas viscosity was presented by Lee *et al.*[77] The equation is applicable to natural gases containing impurities if the corrected Z-factor is used to calculate the value of gas density required in the equation.

$$\mu_g = A \times 10^{-4} \text{EXP}(B\rho_g^C) \quad (3\text{-}90)$$

where

$$A = (9.4 + 0.02M)T^{1.5}/(209 + 19M + T)$$
$$B = 3.5 + 0.01\ M + 986/T$$
$$C = 2.4 - 0.2B$$
$$\mu_g = \text{gas density at } p, T, cp$$
$$\rho_g, = \text{gas density at } p, T, \text{gm}/cc$$
$$M = \text{gas molecular weight}$$
$$T = \text{temperature of interest, °R}$$

The gas density in gm/cc may be calculated from:

$$\rho_g = \frac{0.0433\gamma_g p}{ZT} \qquad (3\text{-}91)$$

where

$$p = \text{psia}$$
$$T = \text{°R}$$
$$\rho_g = \text{gm/cc}$$

6. Interfacial Tension

The interfacial tension existing between the gas and liquid phases has very little effect on two-phase pressure gradient calculations. However, some of the pressure gradient prediction methods require a value for interfacial tension to use in calculating certain dimensionless numbers. Empirical graphs for estimating the gas/oil interfacial tension were presented by Baker and Swerdloff[19] and graphs for gas/water interfacial tension were published by Hough.[20] Regression analysis was used to fit equations to these graphs for specific temperatures. The effect of temperature can be estimated by linear interpolation.

a. Gas/Oil Interfacial Tension. Graphs were presented for dead oil interfacial tension measured at temperatures of 68°F and 100°F. Equations which fit these graphs are:

$$\sigma_{68} = 39 - 0.2571(\text{API}) \qquad (3\text{-}92)$$
$$\sigma_{100} = 37.5 - 0.2571(\text{API}) \qquad (3\text{-}93)$$

where

$$\sigma_{68} = \text{interfacial tension at 68°F, dynes/cm}$$
$$\sigma_{100} = \text{interfacial tension at 100°F, dynes/cm}$$
$$\text{API} = \text{gravity of stock tank oil, °API}$$

It has been suggested that if the temperature is greater than 100°F, the value at 100°F should be used. Also, if $T < 68$, use the value calculated at $T = 68$. For intermediate temperatures, use linear interpolation between the values obtained at 68 and 100°F. That is:

$$\sigma_T = 68 - \frac{(T-68)(\sigma_{68} - \sigma_{100})}{32} \qquad (3\text{-}94)$$

where

$$\sigma_T = \text{interfacial tension at } 68 < T < 100$$

The effect of gas going into solution as pressure is increased on the gas/oil mixture is to reduce the interfacial tension. The dead oil interfacial tension can be corrected by multiplying it by the following correction factor.

$$C = 1.0 - 0.024\ p^{0.45} \qquad (3\text{-}95)$$

where p is in psia.

The interfacial tension at any pressure is then obtained from:

$$\sigma_o = C\sigma_T \qquad (3\text{-}96)$$

The interfacial tension becomes zero at miscibility pressure, and for most systems this will be at any pressure greater than about 5000 psia. Equation 3-95 will give a value of zero at a pressure of 3977 psia. If this occurs, a limiting value of 1 dyne/cm should be used to calculate the dimensionless numbers in the following section.

b. Gas/Water Interfacial Tension. Equations were fitted to graphs of interfacial tension versus pressure at two temperatures. These equations are:

$$\sigma_{w(74)} = 75 - 1.108\ p^{0.349} \qquad (3\text{-}97)$$
$$\sigma_{w(280)} = 53 - 0.1048\ p^{0.637} \qquad (3\text{-}98)$$

The same limitations on temperature as stated for the gas/oil case apply for gas/water interfacial tension for interpolation purposes. That is, for $74 < T < 280$:

$$\sigma_{w(T)} = \sigma_{w(74)} - \frac{(T-74)(\sigma_{w(74)} - \sigma_{w(280)})}{206} \qquad (3\text{-}99)$$

D. Predicting Flowing Temperatures

All the fluid property correlations presented previously require a value of fluid temperature to calculate the required fluid property. The flowing temperature profile in a gas well or an oil well is usually assumed to be linear between the surface temperature and the bottomhole temperature. A linear temperature profile is also usually assumed for surface flowline calculations. The linear assumption for well flow will usually not introduce significant errors if a good value for surface flowing temperature can be obtained. The heat loss from a fluid in a pipe is a function of the mass flow rate in the pipe and will therefore change with a change in producing rate.

An algorithm for coupling pressure and heat loss calculations was presented earlier in this section. The iterative solution was necessary because both the overall heat transfer coefficient and the enthalpy change depend on pressure. If some average heat transfer coefficient can be determined, an approximate temperature profile can be calculated independently of the pressure loss calculation. This will of course be less accurate, but in many cases the amount of data available will not be sufficient to perform the more accurate calculation.

1. Flowing Temperature in Wells

An equation for temperature in a well as a function of location L, as derived by Ramey,[21] can be written as:

$$T_L = T_1 - g_T[L - A(1 - \text{EXP}(-L/A))] \qquad (3\text{-}100)$$

where

$$
\begin{array}{rcl}
T_1 & = & \text{temperature at fluid entry } (L = 0) \\
T_L & = & \text{temperature at location } L, \\
g_T & = & \text{geothermal gradient,} \\
A & = & \text{relaxation distance} = wC_p/\pi dU \\
w & = & \text{mass flow rate,} \\
C_p & = & \text{specific heat of the flowing fluid,} \\
d & = & \text{pipe diameter,} \\
U & = & \text{overall heat transfer coefficient, and} \\
L & = & \text{distance from fluid entry.}
\end{array}
$$

When the equation is written in this form it assumes that the fluid and surroundings temperature are equal at the inlet to the pipe. This will be the case for flowing wells, where T_1 is the reservoir temperature. Also included is the assumption that the heat loss is independent of time. This assumption limits application of Equation 3-100 to wells that have been producing for a considerable length of time.

When multiphase flow is occurring in a well, the variables involved in evaluating the relaxation distance, A, are very difficult to determine, especially the overall heat transfer coefficient U. In view of this fact, Shiu and Beggs[22] developed an empirical method to estimate A based on measured temperature profiles from 270 wells. Using the measured temperatures T_L at various locations L, a value of A for each test was calculated from Equation 3-100. An equation to estimate A was then developed as a function of data which will usually be known. The equation is:

$$A = C_1 w^{C_2} \rho_L^{C_3} d^{C_4} (\text{API})^{C_5} \gamma_g^{C_6} \qquad (3\text{-}101)$$

where

$$
\begin{array}{rcl}
A & = & \text{relaxation distance, ft} \\
w & = & \text{total mass flow rate, lbm/sec} \\
\rho_L & = & \text{liquid (oil and water) density at standard} \\
& & \text{conditions, lbm/ft}^3 \\
d & = & \text{pipe I. D., in.} \\
\text{API} & = & \text{oil gravity, °API} \\
\gamma_g & = & \text{gas gravity (air = 1)}
\end{array}
$$

$$
\begin{array}{rcl}
C_1 & = & 0.0149 \\
C_2 & = & 0.5253 \\
C_3 & = & 2.9303 \\
C_4 & = & 0.2904 \\
C_5 & = & 0.2608 \\
C_6 & = & 4.4146
\end{array}
$$

Equation 3-101 is applicable for flowing oil wells only, although a similar approach could be used for gas wells if insufficient data are available to calculate A. Equation 3-101 has been found to give good results for dry gas wells (no liquid production) by using values for liquid density and oil gravity of 62.4 and 50, respectively.

2. Flowing Temperature in Pipelines

To calculate a temperature profile in a pipeline, it is usually assumed that the temperature of the surroundings is constant. Modification of Equation 3-100 to account for this results in:

$$T_L = T_s + (T_1 - T_s)\,\text{EXP}\,(-L/A) \qquad (3\text{-}102)$$

where T_s is the surroundings temperature and the other variables are defined in Equation 3-100.

For flow of gases, the Joule-Thomson effect may be included, but since this effect depends on pressure, an iterative solution is required. The more rigorous equation is:

$$
\begin{aligned}
T_L = {} & T_s + \mu A(dp/dL) \\
& + [T_1 - T_s - \mu A(dp/dL)]\,\text{EXP}\,(-L/A) \qquad (3\text{-}103)
\end{aligned}
$$

where

$$
\begin{array}{rcl}
\mu & = & \text{Joule-Thomson coefficient, and} \\
dp/dL & = & \text{pressure gradient at } L.
\end{array}
$$

As was discussed earlier the data necessary to calculate the heat transfer coefficient U is seldom available. A simplified approach to estimating flowing temperatures in either wells or pipelines may be used if at least one measured set of inlet and outlet temperatures is available along with one measured flow rate. This approach can be used for oil and gas wells. A procedure is:

1. Using the measured temperatures and flow rates, solve the flowing temperature equation for A. Equation 3-100 applies for wells, while Equation 3-102 applies for pipelines.

2. Considering all the variables in A except flow rate to be constant, solve for the constant.

$$A = wC_p/\pi dU = Cw$$

or

$$C = A/w$$

3. Use this value of C to estimate a value for A for other flow rates.

IV. WELL FLOW CORRELATIONS

One of the most important components in the total well system is the well tubing. As much as 80 percent of the total pressure loss, that is $\overline{p}_R - p_{sep}$, can be consumed in lifting the fluids from the bottom of the hole to the surface. The tubing pressure loss is expressed in Figure 3-1 as $p_{wf} - p_{wh}$. The flow may exist in tubing or in the annulus between the tubing and the casing. The wells may be vertical of can be drilled at large deviation angles, especially in the case of offshore wells or wells drilled in urban areas.

The general pressure gradient equation, which will apply to flow of any fluid in a pipe at any inclination angle, was given as Equation 3-16.

$$\frac{dp}{dL} = \frac{\rho g \sin\theta}{g_c} + \frac{f\rho v^2}{2g_c d} + \frac{\rho v dv}{g_c dL} \qquad (3\text{-}16)$$

If the angle from vertical is used in the equation, it becomes

$$\frac{dp}{dL} = \frac{\rho g \cos\phi}{g_c} + \frac{f\rho v^2}{2g_c d} + \frac{\rho v dv}{g_c dL} \qquad (3\text{-}104)$$

where

ϕ = angle of the well from vertical.

Equation 3-16 was written as the composite of three components in Equation 3-17 as:

$$\left(\frac{dp}{dL}\right)_{total} = \left(\frac{dp}{dL}\right)_{el} + \left(\frac{dp}{dL}\right)_{f} + \left(\frac{dp}{dL}\right)_{acc} \qquad (3\text{-}17)$$

The ranges of contribution of each of these components to the total pressure drop in the well can be seen from the following table, where the contributions are listed as percent of total Δp in the tubing, $p_{wf} - p_{wh}$, for both oil and gas wells.

	Percent of Total Δp	
Component	Oil Wells	Gas Wells
Elevation (Hydrostatic)	70-90	20-50
Friction	10-30	30-70
Acceleration	0-10	0-10

The density of the fluids in oil wells is usually much greater than for gas wells, and since the hydrostatic component depends on liquid holdup, the most important parameter that must be evaluated is the liquid holdup.

In gas wells, the fluid density is smaller, but the gas is usually moving at a relatively high velocity, which generates more friction loss in the pipe. This necessitates having a good value for pipe roughness from which to obtain a friction factor.

The numbers given in the table are of course only approximations, since some oil wells produce at high gas/liquid ratios (GLRs) and some gas wells produce considerable amounts of liquid condensate or water.

Many correlations have been developed in the last 30 or 40 years for predicting two-phase flowing pressure gradients in producing wells. A list of the many methods and a brief review of each can be found in Brown.[23] Some investigators chose to assume that the gas and liquid travel at the same velocity so that the mixture density can be calculated based on the no-slip liquid holdup λ_L (Equations 3-22 and 3-30). In this case a correlation for H_L would not be necessary, and if acceleration is ignored, only a correlation for two-phase friction factor is neccessary. This is, of course, a gross oversimplification of the problem and generally does not give good results.

No methods presently exist for analytically evaluating either liquid holdup or friction factor. Therefore it has been necessary to develop empirical correlations for these two parameters as functions of variables that will be known or can be calculated from known data. This requires an experimental facility from which values of H_L and two-phase friction factor f_{TP} can be measured under a wide range of flow conditions and flow geometries. A general procedure for accomplishing this is described that will aid in the understanding of how the various correlations were developed. An experimental facility is required from which measurements can be made of q_L, q_g, Δp, H_L, and in some cases flow pattern. The experimental data are then obtained by the following procedure:

1. Establish stable flow conditions at particular values of q_L, q_g, pipe diameter, pipe angle, etc.

2. In a test section of length ΔL, measure H_L and Δp. Methods for measuring H_L include nuclear densitometers, capacitance devices, quick closing valves, etc. Flow pattern may be observed if the test section is transparent.

3. Calculate mixture density and elevation component.

$$\rho_s = \rho_L H_L + \rho_g (1 - H_L)$$

$$\left(\frac{dp}{dL}\right)_{el} = \frac{\rho_s g \sin\theta}{g_c}$$

4. Calculate an acceleration component (if it is to be considered) and the friction component.

$$\left(\frac{dp}{dL}\right)_{f} = \frac{\Delta p}{\Delta L} - \left(\frac{dp}{dL}\right)_{el} - \left(\frac{dp}{dL}\right)_{acc}$$

5. Calculate a two-phase friction factor.

$$f_{Tp} = \frac{2g_c d}{\rho v_m^2}\left(\frac{dp}{dL}\right)_{f}$$

6. Change test conditions and return to Step 2. H_L, f_{TP} and flow pattern should be obtained over a wide range of conditions.

7. Develop empirical correlations for H_L, f_{TP} and perhaps flow pattern as a function of variables that will be known for design cases. These variables include v_{sL}, v_{sg}, d, fluid properties, pipe angle, etc.

The well flow pressure gradient methods described in this section will be discussed from the point of view of their development. Some investigators did not measure H_L, some did not measure flow pattern, and others ignored the contribution of the acceleration component. In some cases, seperate correlations for H_L and f_{TP} were developed for each of three separate flow patterns. Use of these methods requires that the flow pattern existing at the location of interest in the well be determined first. This, of course, requires a flow pattern map or some other means of predicting flow patterns.

Several of the most widely used well flow methods will be discussed in this section. The discussions will be limited to:

1. How the experimental data were obtained.

2. How the correlations for H_L and f_{TP} were developed.

Detailed equations and example calculations for each of these methods may be found in References 9 and 23. Computer subroutines for most of the methods are given in Reference 9.

After discussing the development and application of the various methods, several evaluation studies of the methods using measured field data will be described. This will aid the engineer in choosing which method to use for particular conditions existing in a well or field.

The effects of changes in conditions that can exist from field to field or from well to well in a field will be presented. These conditions include variables such as GLR, pipe size, water cut, etc.

The preparation and use of pressure traverse curves for quick estimates of pressure drops in flowing wells will also be described. These curves can often be used for preliminary evaluation of a well or field prior to a more detailed analysis, which usually requires a computer.

A. Poettmann and Carpenter Method

The principal reason for including this method in the discussions is the fact that this was the first serious attempt at solving the multiphase well flow problem. Also, this method was widely used for many years for design of flowing and gas lift wells.

The Poettmann and Carpenter[24] method was developed using measured field data from some 334 flowing wells and 15 continuous flow gas lift wells. The wells were producing through tubing sizes ranging from 2 3/8 in. to 3 1/2 in. Most of the wells were producing at liquid rates less than 500 STB/day at GLRs less than 1500 scf/STB.

Only a correlation for two-phase friction factor was developed since the only measurements made were surface and bottomhole pressures and flow rates. Liquid holdup was not measured, and the wells were not divided into short length increments. The mixture density was calculated using the no-slip holdup, and acceleration was ignored. A plot of the friction factors calculated from the measured field data is shown in Figure 3-14. The correlating parameter for the friction factor was the mass flow rate divided by $\pi/4$ times the pipe diameter or ρvd in lbm/ft-sec.

This simplified approach, in which the energy losses not included in the hydrostatic term or the acceleration term were absorbed in the friction term, was used for many years, mainly because of the difficulty of measuring liquid holdup. In 1961, Baxendell and Thomas[25] extended the friction factor correlation to higher rates and larger pipe sizes using data obtained in Venezuela. Fancher and Brown[26] used the same approach in an

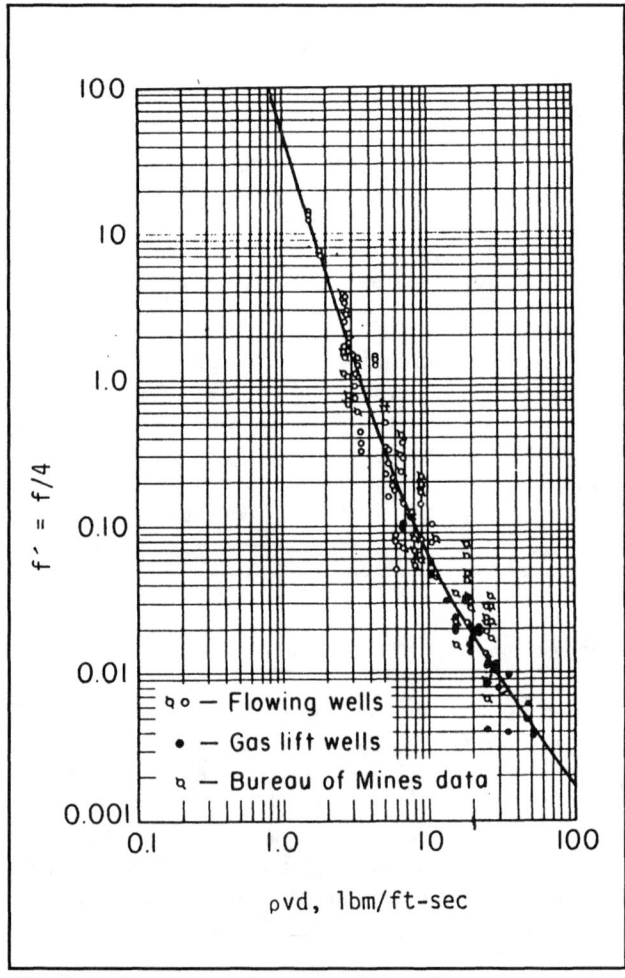

Fig. 3-14. Poettmann-Carpenter friction factor.

attempt to isolate the effects of gas/liquid ratio on the pressure gradient.

These methods, although easy to apply, will give erroneous results when applied to wells that are not producing under conditions very similar to those from which the developing data were obtained.

B. Hagedorn and Brown Method

The Hagedorn and Brown[27] method was developed by obtaining experimental pressure drop and flow rate data from a 1500 ft deep instrumented well. Pressures were measured for flow in tubing sizes ranging from 1-1/4 to 2-7/8 in. O. D. A wide range of liquid rates and gas/liquid ratios was included, and the effects of liquid viscosity were studied by using water and oil as the liquid phase. The oils used had viscosities at stock tank conditions of 10, 35, and 110 cp.

Neither liquid holdup nor flow pattern was measured during the Hagedorn and Brown study, although a correlation for the calculated liquid holdup is presented. The correlations were developed by assuming that the two-phase friction factor could be obtained from the Moody diagram based on a two-phase Reynolds Number. This Reynolds Number requires a value for H_L in the viscosity term. The procedure used for obtaining the calculated H_L is:

1. Measure $\Delta p/\Delta L$.
2. Estimate a value for liquid holdup, H_L^*.
3. Calculate N_{ReTP} and find f_{TP} from the Moody diagram.
4. Calculate $(dp/dL)_f$ and $(dp/dL)_{acc}$.
5. Calculate $(dp/dL)_{el} = \Delta p/\Delta L - (dp/dL)_f - (dp/dL)_{acc}$ and $\rho_s = (dp/dL)_{el}\, g_c/g$.
6. Calculate $H_L = (\rho_s - \rho_g)/(\rho_L - \rho_g)$ and compare with H_L^*. If not close, set $H_L^* = H_L$ and go to Step 3. Continue until convergence is obtained.

The value of H_L obtained is not necessarily the actual liquid holdup, but it is the value required to balance the pressure losses once a friction factor has been selected. Several dimensionless numbers were used to correlate H_L and two secondary correction factors. These dimensionless numbers had been defined earlier by Ros[28] and are given as follows:

$$N_{LV} = v_{SL}\,(\rho_L\,/\,g\sigma)^{0.25}$$
$$N_{gv} = v_{sg}\,(\rho_L\,/\,g\sigma)^{0.25}$$
$$N_d = d\,(\rho_L g\,/\,\sigma)^{0.5} \qquad (3\text{-}105)$$
$$N_L = \mu_L\,(g\,/\,\rho_L\sigma^3)^{0.25}$$

where

N_{LV} = liquid velocity number
N_{gv} = gas velocity number

N_d = diameter number
N_L = liquid viscosity number

The other parameters have been defined previously and the units must be selected so that the numbers will be dimensionless. The three empirical correlations required for obtaining a value of H_L are shown in Figure 3-15.

Two modifications have been made to the original Hagedorn and Brown method that have extended the valid range of application considerably. It was found that for some cases the value calculated for H_L was less than the no-slip holdup λ_L. This is physically impossible in

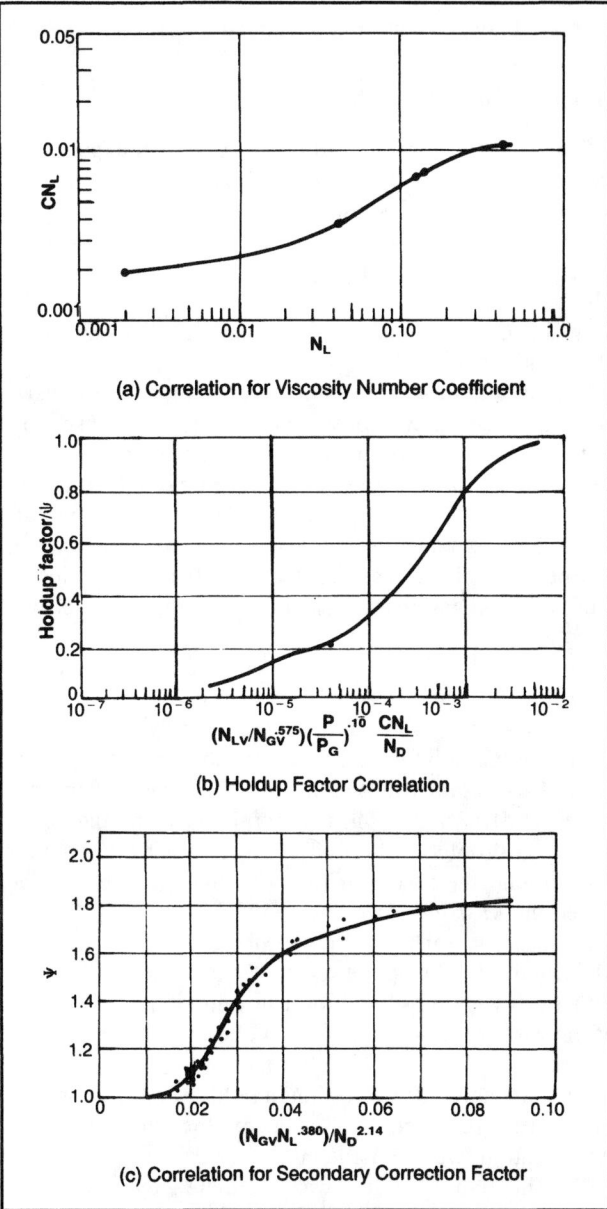

(a) Correlation for Viscosity Number Coefficient

(b) Holdup Factor Correlation

(c) Correlation for Secondary Correction Factor

Fig. 3-15. Hagedorn-Brown holdup Correlations. (a) Correlation for viscosity number coefficient; (b) Holdup factor correlation; (c) Correlation for secondary correction factor.

upward two-phase flow so a lower limit was imposed on the H_L. That is, H_L must be greater than or equal to λ_L.

The second modification involves determining if the flow falls into the bubble-flow pattern as defined by Orkiszewski.[29] If bubble flow does exist, the Griffith[30] correlation is used to determine the pressure gradient in the pipe increment under consideration. The Orkiszewski and Griffith correlations are described in a subsequent section.

The Hagedorn and Brown method has been found to give good results over a wide range of well conditions and is one of the most widely used well flow correlations in the industry. A detailed calculation procedure and example may be found in the appendix.

C. Duns and Ros Method

Duns and Ros[31] published the results of an experimental study of vertical two-phase flow. The experiment, which consisted of some 4000 runs and 20,000 data points, was conducted in a laboratory facility at low pressure using air, oil, and water as the fluid components. The test section was 10 m long and the pipe diameters ranged from 3.2 to 8.02 cm. Some annular flow tests were also conducted.

Liquid holdup was measured with radioactive tracers, and flow pattern was observed through the transparent test section. The experimental work and the preliminary development of the correlations were reported earlier by Ros.[28] Three flow patterns were defined, and a flow pattern map was constructed from which the flow pattern can be determined based on the superficial velocities of the liquid and gas phases. The flow patterns are described as follows:

Region I: The liquid phase is continuous, and the gas moves as discontinuous bubbles or plugs. This region is often referred to as the *Bubble-flow Pattern*.

Region II: Both the liquid and gas phases are discontinuous. This is sometimes called the *Slug-flow Pattern*.

Region III: The gas phase is continuous, and the liquid moves as droplets dispersed in the gas or as an annular ring around the inside of the pipe. This region may be called the *Mist-flow Pattern*.

A transition zone between Regions II and III was also identified. The flow pattern map is shown in Figure 3-16. Equations were presented for determining the bounderies of the various flow patterns as functions of dimensionless numbers.

Separate correlations for liquid holdup and friction factor were presented for each of the flow regions. Acceleration was considered important in Region III only. The liquid holdup was correlated in terms of a dimensionless slip velocity, which was defined as

$$N_s = v_s (\rho_L / g\sigma)^{0.25} \qquad (3\text{-}106)$$

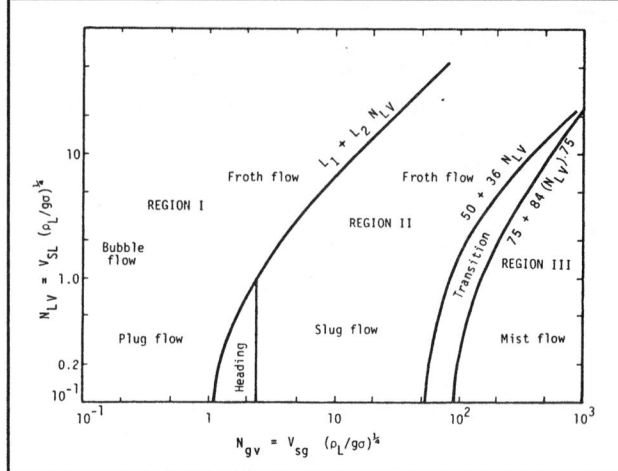

Fig. 3-16. Duns and Ros[31] flow pattern map.

where

$$N_s = \text{dimensionless slip velocity}$$
$$v_s = \text{actual slip velocity}$$

Once v_s is determined, H_L can be calculated from Equation 3-29.

$$v_s = \frac{v_{sg}}{1 - H_L} - \frac{v_{sL}}{H_L} \qquad (3\text{-}29)$$

The slip velocity was considered negligible in Region III, and therefore $H_L = \lambda_L$.

Both the dimensionless slip velocity and the friction factors were correlated as functions of the dimensionless numbers presented earlier in Equation 3-105. The correlations were presented as a series of complex graphs that must be transformed to either equation or tabular form for computer application.

The Duns and Ros method is considered to be applicable over a wide range of well conditions, especially an updated, proprietary version commonly known as the Shell Method. The correlation for Region III, the Mist-flow Pattern, is recommended by both Orkiszewski[29] and Aziz, et al.[32]

D. Orkiszewski Method

Orkiszewski[29] performed a comparison study on some 148 measured well conditions and found that none of the correlations existing at that time (1967) adequately predicted the measured results. He then used the data of Hagedorn and Brown[27] and the field data from the 148 oil well conditions to develop a new correlation to be used in the Bubble- and Slug-flow patterns. He recommended using the Duns and Ros method for Mist-flow.

The flow patterns considered by Orkiszewski are

shown in figure 3-17. Orkiszewski's descriptions of these flow patterns are included.

Bubble Flow

The pipe is almost completely filled with liquid, and the free gas phase is present in small bubbles. The bubbles move at different velocities and, except for their density, have little effect on the pressure gradient. The wall of the pipe is always contacted by the liquid phase.

Slug Flow

The gas phase is more pronounced. Although the liquid phase is still continuous, the gas bubbles coalesce and form plugs or slugs that almost fill the pipe cross section. The gas bubble velocity is greater than that of the liquid. The liquid in the film around the bubble may move downward at low velocities. Both the gas and liquid have significant effects on the pressure gradient.

Transition Flow

The change from a continuous liquid phase to a continuous gas phase occurs. The gas bubbles may join and liquid may be entrained in the bubbles. Although the liquid effects are significant, the gas phase effects are predominant.

Mist Flow

The gas phase is continuous, and the bulk of the liquid is entrained as droplets in the gas phase. The pipe wall is coated with a liquid film, but the gas phase predominantly controls the pressure gradient.

Equations were presented for determining the flow pattern existing under various conditions, and methods for calculating friction factor and two-phase density were presented for the Bubble- and Slug-flow patterns. Acceleration was considered negligible except for the Mist-flow pattern.

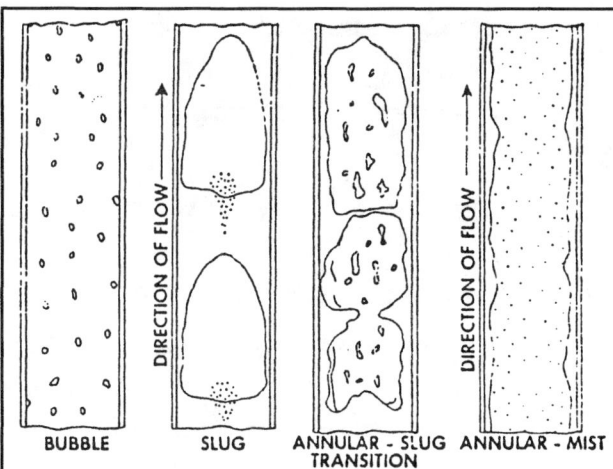

Fig. 3-17. Vertical flow patterns.

In the Slug-flow pattern the liquid density was calculated using a so-called Liquid Distribution Coefficient, rather than the liquid holdup. A distinction was made as to which equations are used to calculate the liquid distribution coefficient depending on whether oil or water was the continuous liquid phase and if the mixture velocity was greater than 10 ft/shsec.

The Orkiszewski method can be computerized and has been widely used in the petroleum industry since its publication. It is applicable over a wide range of well conditions, but in some cases, a mixture density less than the no-slip density will be calculated. This is probably the result of using the Hagedorn and Brown data to develop the equations for Slug-flow. Also, discontinuities in the calculated pressure traverse can occur as the mixture velocity exceeds 10 ft/sec. This results from changing equations for mixture density at this velocity.

E. Aziz, Govier and Fogarasi Method

Aziz, et. al.,[32] proposed what they referred to as a mechanistic model. They proposed a new vertical flow pattern map and presented new equations for calculating the liquid holdup occurring in the Bubble- and Slug-flow patterns. No new equations were proposed for the Annular-mist pattern and the Duns and Ros equations were recommended for this flow pattern.

The flow pattern was correlated with dimensionless numbers which depend primarily on the gas and liquid superficial velocities. The flow pattern map is shown in Figure 3-18. The coordinates, N_x and N_y, are defined from:

$$N_x = v_{sg} \left(\frac{\rho_g}{\rho_{air}} \right)^{0.333} \left(\frac{\rho_L \sigma_w}{\rho_w \sigma_L} \right)^{0.25} \qquad (3\text{-}107)$$

$$N_y = v_{sL} \left(\frac{\rho_L \sigma_w}{\rho_w \sigma_L} \right)^{0.25}$$

where

ρ_g = in-situ gas density,
ρ_L = in-situ liquid density,
σ_L = in-situ gas-liquid interfacial tension,
σ_w = interfacial tension between water and air at s.c.,
ρ_{air} = density of air at s.c., and
ρ_w = density of water at s.c.

The liquid holdup was calculated as a function of a bubble-rise velocity. The bubble-rise velocity was calculated using equations proposed earlier by Zubel, et al.[33]

The new equations were used in a comparison study utilizing measured data from 48 wells. The difference in the accuracy of the new equations and the Orkiszewski method was negligible.

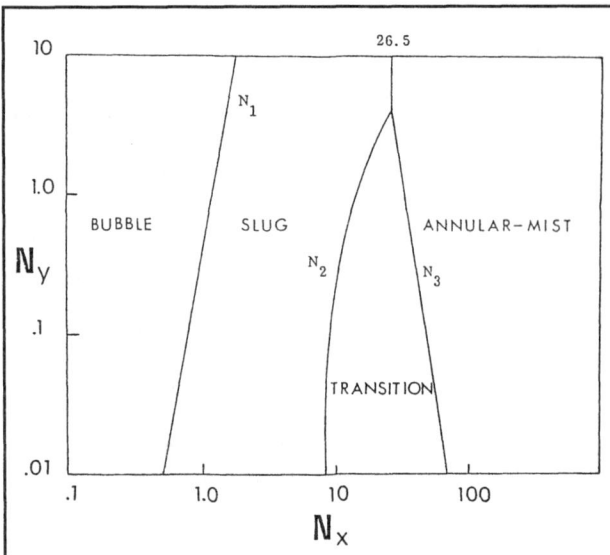

Fig. 3-18. Aziz flow regimes.

F. Chierici, Ciucci and Sclocchi Method

Another modification of the Orkiszewski method was proposed by Chierici, et al.[34] The modification consisted essentially of eliminating the discontinuity inherent in the Orkiszewski equations for mixture density in slug flow. Chierici, et al., found that the fluid physical properties, which must be determined at each step in the pressure traverse calculation, had a considerable effect on the accuracy of the pressure drop calculation. They, therefore, specified which fluid property corelations were to be used with their pressure gradient method.

G. Beggs and Brill Method

The Beggs and Brill[35] correlation was developed from experimental data obtained in a small scale test facility. The facility consisted of 1 in. and 1.5 in. sections of acrylic pipe 90 ft long. The pipe could be inclined at any angle. The parameters studied and their range of variation were: (1) gas flow rate (0 to 300 Mscf/D); (2) liquid flow rate (0 to 30 gal/min); (3) average system pressure (35 to 95 psia); (4) pipe diameter (1 and 1.5 in.); (5) liquid holdup (0 to 0.870); (6) pressure gradient (0 to 0.8 psi/ft); (7) inclination angle (–90° to +90°); and (8) horizontal flow pattern. Fluids used were air and water. For each pipe size, liquid and gas rates were varied so that all flow patterns were observed when the pipe was horizontal. After a particular set of flow rates was set, the angle of the pipe was varied through the range of angles so that the effect of angle on holdup and pressure gradient could be observed. Liquid holdup and pressure gradient were

measured at angles from horizontal of 0, plus and minus 5, 10, 15, 20, 35, 55, 75 and 90 degrees. The correlations were developed from 584 measured tests.

Different correlations for liquid holdup are presented for each of three *horizontal* flow regimes. The liquid holdup that would exist *if the pipe were horizontal* is first calculated and then corrected for the actual pipe inclination angle. The horizontal-flow patterns are illustrated in Figure 3-19. The variation of liquid holdup with pipe inclination is shown in Figure 3-20 for three of the tests.

The holdup was found to be a maximum at approximately +50 degrees from horizontal and a minimum at approximately –50 degrees. The original flow-pattern map has been slightly modified to include a transition zone between the segregated and intermittent flow regimes. The modified flow-pattern map is superimposed on the original in Figure 3-21. A two-phase friction factor

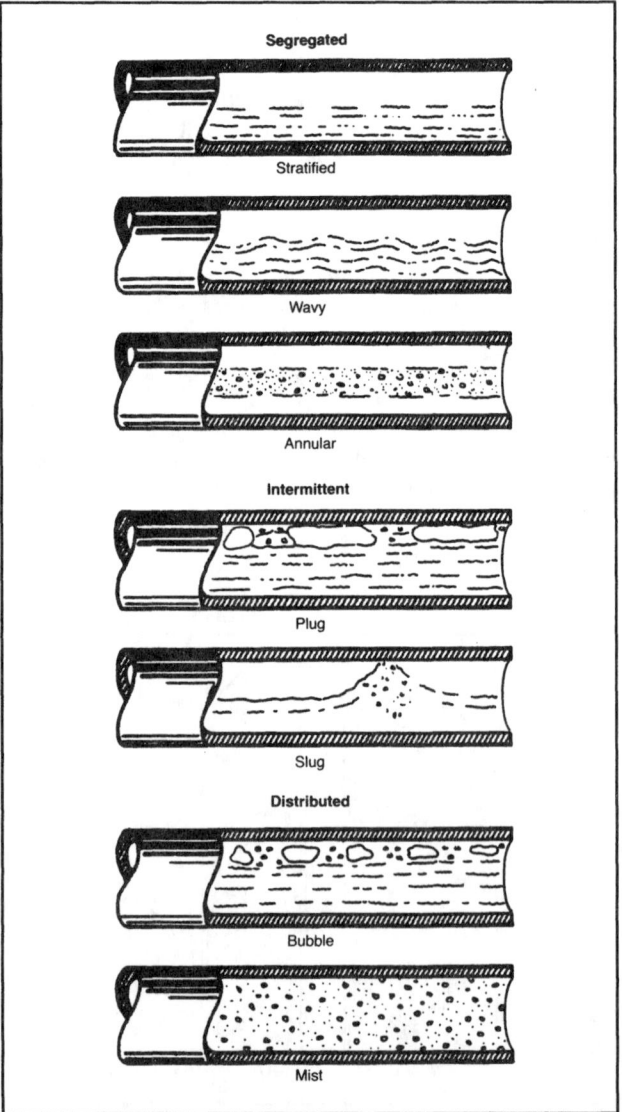

Fig. 3-19. Horizontal flow patterns.

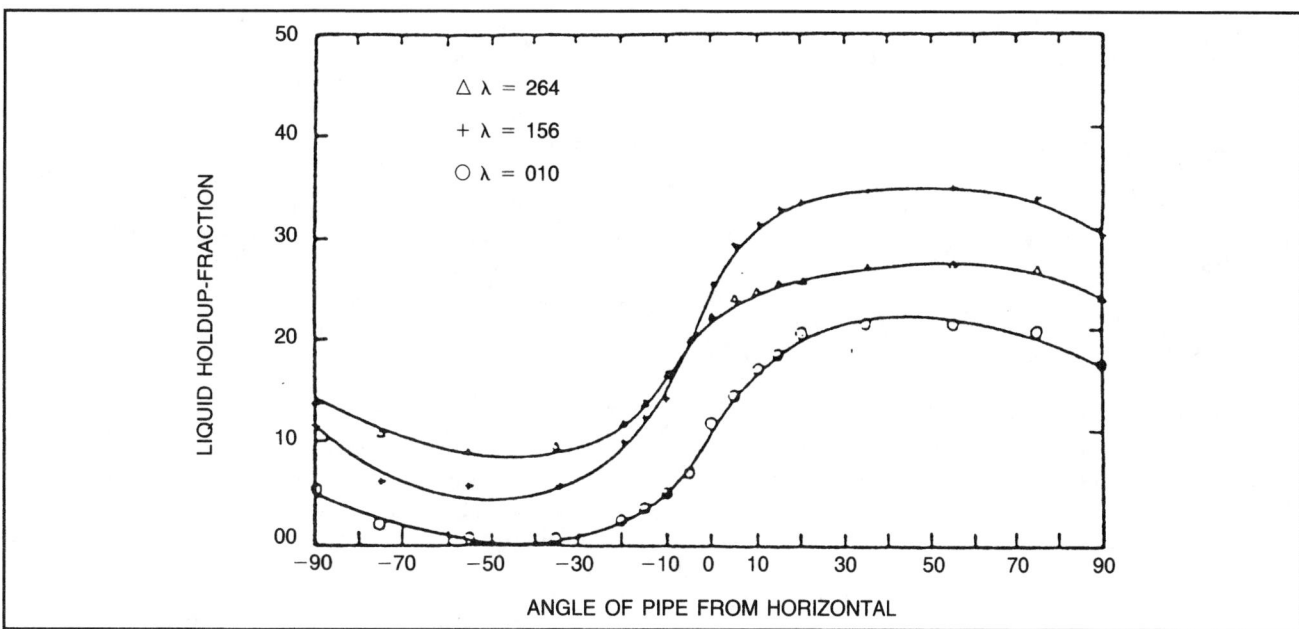

Fig. 3-20. Liquid holdup vs. angle.

is calculated using equations that are independent of flow regime but depend on holdup.

The equations presented by Beggs and Brill apply to flow in a pipe at any angle of inclination, including downward flow. Although the method has been found to slightly over-predict pressure gradients in vertical wells in some cases, it gives good results for pipeline calculations. The fact that this method can be used for pipes at any angle and the fact that it is presented entirely in equation form make it an ideal method for use in handheld, programmable calculators. A program of the Beggs and Brill method for calculating a pressure traverse in wells or pipelines was published by Hein[36, 37] in 1982. The program is for use in the HP-41C calculator and must be used in conjunction with the Petroleum Fluids Pac described earlier.

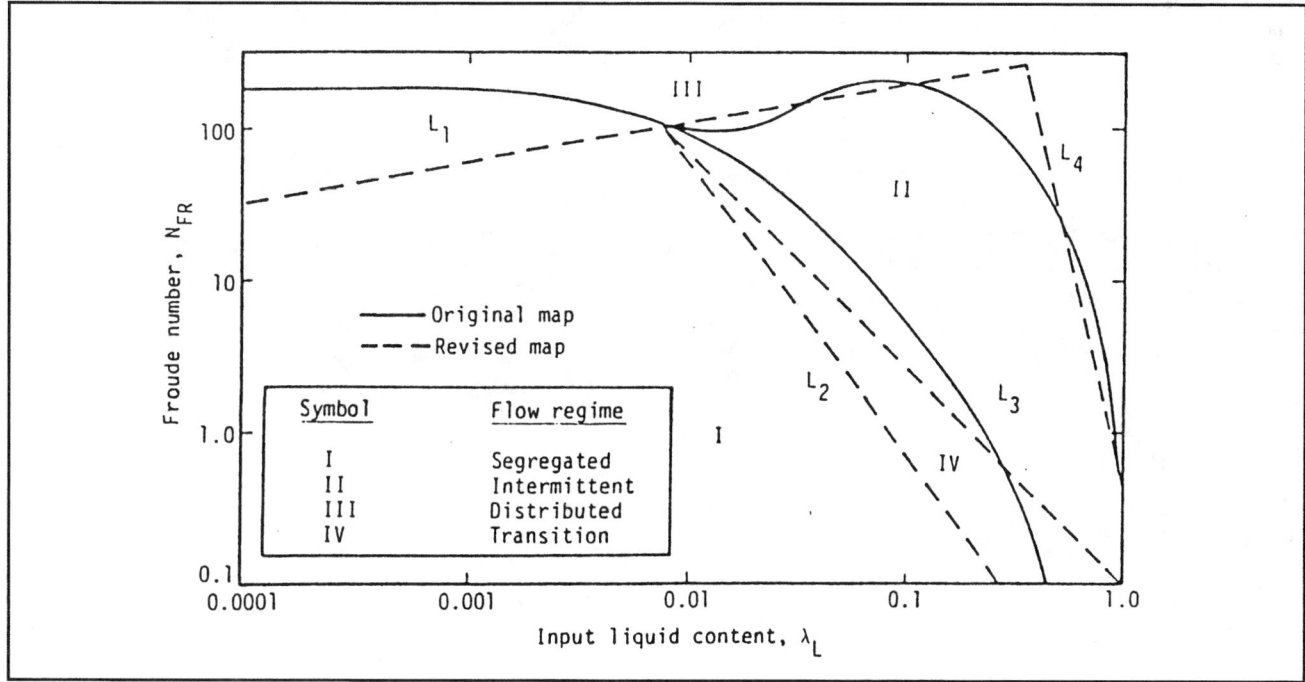

Fig. 3-21. Horizontal flow pattern map.

A detailed calculation procedure and an example utilizing the Beggs and Brill method may be found in the appendix.

H. MONA, Asheim Method

Asheim[78] described a model that was used in a computer program called MONA for two-phase flow calculations. The program was developed at the Norwegian Institute of Technology and was proposed for use in both surface pipe and wells. No experimental data were used in the development. The model does not distinguish among flow patterns.

Liquid holdup was not directly predicted but a linearized functional relationship between gas and liquid velocities was assumed to calculate two-phase density. This relationship, which was similar to earlier work by Nicklen, et al.,[79] requires knowledge of two constants or flow parameters. An extension of the Dukler[55] equation was used to obtain a friction factor. Another adjustable flow parameter was included in the friction factor equation. The program is constructed so that measured field data can be used to adjust and optimize flow parameters that can then be used to make predictions for other flow conditions for cases similar to thoses that were used to calibrate the model. This procedure, of course, could be used in all of the other two-phase flow models described previously. When no calibration data are available, Asheim proposed using values for the flow parameters that were developed by Nicklen.[79] MONA was used to calculate the pressure drop for 50 cases of wells producing from the Ekofisk field in the North Sea. Most of these wells are directionally drilled, and some have fairly high angles of deviation. The results calculated with MONA were compared with those that were obtained using the Baxendell and Thomas[25] method only. None of the more rigorous methods were considered.

Since the proposed model is to be used in surface pipelines, Asheim also used the program to reproduce pressure drops measured in Prudhoe Bay field surface lines. These results were compared with those predicted by both the Beggs and Brill[35] method and a combination of the Dukler[55] and Eaton[54] methods. The new method performed slightly better than either of the other two methods.

I. Hasan and Kabir Method

Hasan and Kabir[80] proposed a model that is to be used especially for directional or deviated wells. The model predicts both flow pattern and pressure gradient. It was based on experimental data obtained from a five in. circular pipe and from annular flow channels for deviation angles up to 32 degrees from vertical. Equations were presented for calculation of flow pattern transitions among several flow patterns for both circular pipes and annular flow. Correlations for gas void fraction were given for the various flow patterns, and friction losses were calculated using "standard charts or correlations in terms of Reynolds number." The proposed model was tested using field data from ten offshore wells. Comparison of the proposed model with that of Beggs and Brill[35] suggested that it performed slightly better. No claim was made as to the new model's superiority over existing methods because of the small data base used in the test.

J. Flow in Annuli

Many situations arise in the petroleum industry in which a well must flow through the annular space between the tubing and casing. Examples of this are dual completions with only one tubing string, wells equipped with a kill string, and gas-lifting high-rate wells in which the gas is injected down the tubing.

None of the previously discussed well flow correlations was developed specifically for annular flow, although Hasan and Kabir[80] included annular flow prediction capability in their model. The correlations developed for circular pipes are sometimes used for annular flow by applying the hydraulic radius concept. This method and a procedure proposed by Cornish,[38] will be discussed in this section.

1. Hydraulic Radius Concept

According to this concept, the diameter of a conduit of circular cross section is equal to four times the hydraulic radius, where the hydraulic radius is defined as the cross sectional area open to flow divided by the wetted perimeter. That is,

$$r_h = \frac{\text{cross sectional area}}{\text{wetted perimeter}} = \frac{\pi d^2/4}{\pi d} = \frac{d}{4}$$

When applied to an annulus, this becomes:

$$r_h = \frac{\pi\left(d_i^2 - d_o^2\right)/4}{\pi(d_i + d_o)} = \frac{d_i - d_o}{4}$$

where

d_o = outside diameter of the tubing
d_i = inside diameter of the casing
r_h = hydraulic radius

Setting the two expressions for r_h equal implies that the correct expression for hydraulic diameter d_h is:

$$d_h = d_i - d_o$$

To calculate superficial velocities, the actual cross sectional area open to flow should be used. Using this concept, any of the previously described well-flow methods can be applied to annular flow.

It is generally assumed that the hydraulic radius concept is valid for annular flow if $d_o/d_i \leq 0.3$ in the case of single-phase flow. This limitation has not been confirmed for two-phase flow.

2. Cornish Method

Cornish[38] presented a method for calculating pressure traverses for annular flow wells in which the liquid flow rate exceeds 5000 STB/day. No experimental data were used to develop the method, and it applied only to cases where there is essentially no-slippage between the gas and liquid phases; that is, for the case where $H_L = \lambda_L$. Cornish assumed this to be the case if his two-phase Reynolds number was greater than 10^5. The Reynolds number is defined as:

$$N_{\mathrm{Re\,TP}} = \frac{w d_h}{\mu_m A}$$

(3-107D)

where

w = mass flow rate
d_h = hydraulic diameter
A = area open to flow
μ_m = mixture viscosity = $\left(\mu_L^{\lambda_L}\right)\left(\mu_g^{\lambda_g}\right)$

The no-slip density was used for the elevation component and the friction factor was obtained from a Moody diagram. The method has not found wide use in the industry because of the limitations stated previously.

K. Evaluation of Correlations Using Field Data

The only meaningful procedure for evaluating the various pressure-gradient prediction methods is by comparison of the pressure drop in a well predicted by the method with actual measured field data. Evaluation studies have been performed by several investigators, but in many cases the study was performed primarily to demonstrate the superior performance of some newly proposed pressure-gradient method.

The results of several evaluation studies are summarized in this section. Also, a table is presented that gives information regarding the development of the well-flow methods evaluated in some of the studies. Table 3-1 summarizes the ranges of data used, the type of data, the pipe sizes and the fluids used to develop the various methods.

TABLE 3-1
Well Flow Correlations

Investigators	Year Presented	Type of Study	Pipe Sizes, in.	Fluids Used	Remarks
Poettmann & Carpenter	1952	Field, Experimental	2, 2.5, 3	oil, water, gas	Correlation developed from well tests with q' > 420 STB/D, GLR < 1500 scf/STB.
Baxendell & Thomas	1961	Field, Experimental	2.5, 3, 3.5	oil, gas	Based on well data from Lake Maracaibo Field. Very high flow rates.
Fancher & Brown	1963	Field, Experimental	2	water, gas	Data from one well, used GLR much higher than Poettman & Carpenter.
Hagedorn & Brown	1963	Intermediate, Experimental	1, 1.25, 1.5	oil, air, water	Data from 1500 ft. experimental well. Used wide range of oil viscosity.
Duns & Ros	1963	Laboratory, Experimental	1.5, 2, 2.5, 3	oil, water, gas	Correlation developed from large number of laboratory data points.
Orkiszewski	1967	Field, Experimental	1, 1.5, 2, 3	oil, water, gas	Utilized some field data and Hagedorn-Brown data. New method for slug flow only.
Aziz, et al.	1972	Theoretical	–	–	Revised Orkiszewski extensions of Griffith-Wallis data. New flow pattern map.
Chierici, et al.	1974	Laboratory, Experimental & Theoretical	0.5, 0.75, 1.0	water, oil, gas	Used Wallis & Nicklen data. New method for slug flow. New flow pattern map.
Beggs & Brill	1973	Laboratory, Experimental	1.0, 1.5	water, air	Method is primarily for inclined flow. Large number of low pressure data points.
Asheim	1986	Theoretical	–	–	Based on work done previously by Dukler.
Hasan & Kabir	1986	Laboratory, Experimental & Theoretical	5, with some annular cases	–	Primarily for directional wells.

TABLE 3-2
Well Flow Comparison Studies

Authors	No. Wells or Tests	GOR Range	Well Depth Range, ft.	Pipe Diameter Range, in.
Brown	35	–	–	2.5-7
Aziz	48	140-10000	4000-12500	–
Espanol	44	140-10000	4300-12500	–
Orkiszewski	148	185-7000	3800-8000	37295
Lawson	726	20-80000	920-12500	37264
Ibe	892	0-78800	1000-12000	1.0-8.8
Rossland	130	480-19000	5000-11000	2.6-5.1
Asheim	50	300-500	6900-10000	3.958-6.184
Hasan	10	–	–	–

The type of field data used in several comparison studies is described in Table 3-2. Results of these studies for the various well-flow correlations compared are given in Table 3-3. The comparison parameters are average percent error and standard deviation of the percent errors. The standard deviation is a measure of the scatter of the calculated errors. Some of the comparison studies failed to calculate the degree of scatter, but presented only average errors. A comparison is, of course, incomplete without some measure of scatter. A positive percent error indicates that the correlation predicted a greater pressure drop than was measured.

TABLE 3-3
Comparison Study Results

Comparison Study Author	Vertical Correlation	Avg. Percent Error	Standard Deviation
Brown	Beggs-Brill	-3.45	7.39
	Orkiszewski	-2.41	16.22
	Aziz, et al.	-9.9	13.95
	No-slip	-6.84	8.68
Aziz, et al.	Aziz, et al.	8.9	14.7
	Orkiszewski	8.9	14.8
	Hagedorn-Brown	-20.5	24.6
	Duns-Ros	-11.1	14.9
Espanol, et al.	Hagedorn-Brown	-24	–
	Orkiszewski	-15.5	–
	Duns-Ros	-16.6	–
Orkiszewski	Orkiszewski	-0.8	10.8
	Duns-Ros	2.4	27
	Hagedorn-Brown	0.7	24.2
Lawson, et al.	Poettmann & Carpenter	107.3	195.7
	Baxendell & Thomas	108.3	195.1
	Fancher & Brown	5.5	36.1
	Duns-Ros	15.4	50.2
	Hagedorn-Brown	1.3	26.1
	Orkiszewski	8.6	35.7
	Beggs-Brill	17.8	27.6
	Aziz, et al.	-8.2	34.7
	Chierici, et al.	42.8	43.9
Ibe	Hagedorn-Brown	1.24	23.3
	Orkiszewski	-0.75	34.4
	Duns-Ros	13.62	32.6
	Beggs-Brill	19.17	31.8
Rossland	Hagedorn-Brown	-3.5	8.5
	Orkiszewski	8.4	28.4
	Duns-Ros	-5.5	12.8
	Beggs-Brill	10.7	15.5
	Poettmann-Carpenter	14	12.3
Asheim	MONA	-2.38	2.92
	Baxendell & Thomas	-13.32	2.96
Hasan, et al.	Hasan, et al.	0.89	10.12
	Beggs-Brill	3.22	11.71

It should be noted that some bias exists in the results of some of the evaluation studies reported in Table 3-3. In the Aziz and Orkiszewski studies, the data used in the comparison were also used to develop the author's correlations. In the Lawson study, of the 726 cases calculated, 346 were from the data used to develop the Hagedorn and Brown method. These tests were later eliminated, and although the accuracy of the Hagedorn and Brown method was diminished, it still gave the best results of all the methods tested.

The Ibe[40] study included data from 300 directional wells, and all of the wells used in the Rossland[41] and Hasan[80] studies were directional.

The results of these evaluation studies emphasize that no one method is best for all well cases. One method may be best in one field while another may be best in another field. Therefore, the best procedure for determining which method to use in a particular field is to gather as much measured field data as possible and make a comparison study using that data.

L. Effects of Variables on Well Performance

During the producing life of a well or field many conditions can change that will affect the well's flowing performance. Also, conditions can change from well to well in a field at a given time, and conditions can certainly vary among fields. Some of these variables that can change are liquid flow rate q_L, gas/liquid ratio GLR, water/oil ratio WOR (or water cut f_w), oil or liquid viscosity μ_L, and tubing size d.

As will be seen in the section on total system analysis, planning for future well performance requires accounting for the change in pressure drop in the tubing $(p_{wf} - p_{wh})$ as these variables change. The effects of changes in these variables can be calculated using any of the well-flow correlations discussed previously.

In this section, the effects of variables will be discussed qualitatively and illustrated graphically. This will lead to a better understanding of changes in well performance as observed in the field. Most of the examples in this section were calculated by Brill, et al.,[42] using the Hagedorn and Brown correlation. Similar results would be obtained using any of the other methods.

Before discussing the changes in individual variables, it will be informative to write the pressure gradient equation in a slightly different form. Ignoring acceleration, Equation 3-16 can be written as:

$$\frac{dp}{dL} = \rho_L H_L + \rho_g (1 - H_L) + \frac{C f \rho_m (q_L + q_g)^2}{d^5} \qquad \text{3-108}$$

or

$$\frac{dp}{dL} = \left(\frac{dp}{dL}\right)_{el} + \left(\frac{dp}{dL}\right)_f$$

where C is a constant depending on the ttllits used. The other terms have been defined previously. This form of the equation will be referred to in the discussion of the effects of changes in variables.

1. Liquid Flow Rate

The effect of increasing liquid rate will be an increase in both H_L and fluid velocity. This will cause an increase in both the hydrostatic and friction terms of Equation 3-108. The effect may be seen graphically in Figure 3-22 that was constructed by choosing some general well conditions and holding everything constant except q_L.

2. Gas/Liquid Ratio

The GLR has more effect on two-phase flowing pressure gradients than any other variable. In a depletion-type field the gas/oil ratio will usually increase with time until late in the life of the reservoir. The GLR may decrease if water cut increases.

The GLR has the most effect on the hydrostatic component of the pressure gradient equation because H_L will decrease as GLR increases. However, the total flow rate will increase, and the friction loss depends on the flow rate squared. This means that as GLR increases, $(dp/dL)_{el}$ decreases but $(dp/dL)_f$ increases. One of the best methods

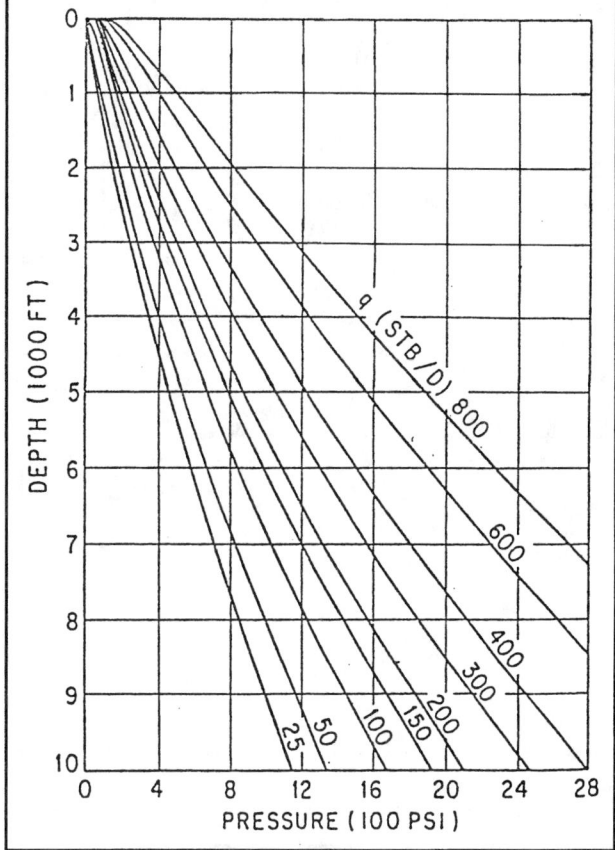

Fig. 3-22. Effect of production rate on pressure gradients.

of artificial lift, i. e., gas/lift, involves the artificial increase of GLR by injecting gas into the tubing string.

The effects of changing GLR can be seen graphically in Figures 3-23 and 3-24. In Figure 3-23, it is observed that as GLR increases, the required flowing bottomhole pressure decreases up to a point. As the GLR increases from 3000 to 5000 scf/STB, the required p_{wf} actually increases. This means that in going from 3000 to 5000 the friction

component has increased more than the hydrostatic component has decreased. The effect of gas rate on the individual components and the total pressure gradient is illustrated in Figure 3-24.

3. Water/Oil Ratio or Water Cut

The total pressure gradient in the well will increase as f_w increases. This results from an increase in liquid density if the water is heavier than the oil and also from a decreasing GLR, since the free gas in the tubing comes primarily from the oil only. These effects can be expressed in Equations 3-19 and 3-109. The effect may be expressed graphically in Figures 3-25 and 3-26. Figure 3-25 shows only the effect of increased liquid density while the total effect is shown in Figure 3-26.

$$\rho_L = \rho_o(1 - f_w) + \rho_w f_w \qquad (3\text{-}19)$$

$$\text{GLR} = \text{GOR}(1 - f_w) \qquad (3\text{-}109)$$

where

ρ_L = liquid density
ρ_o = oil density,

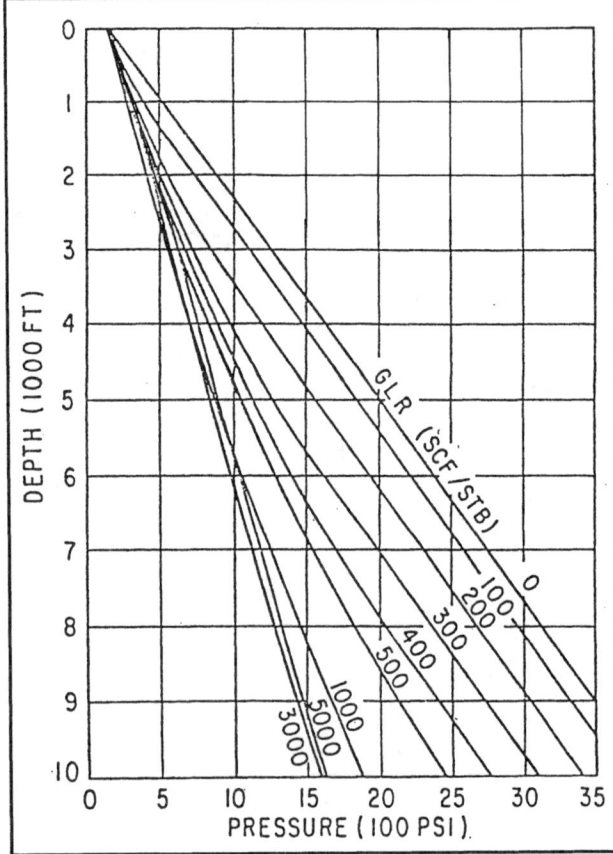

Fig. 3-23. Effect of gas/liquid ratio.

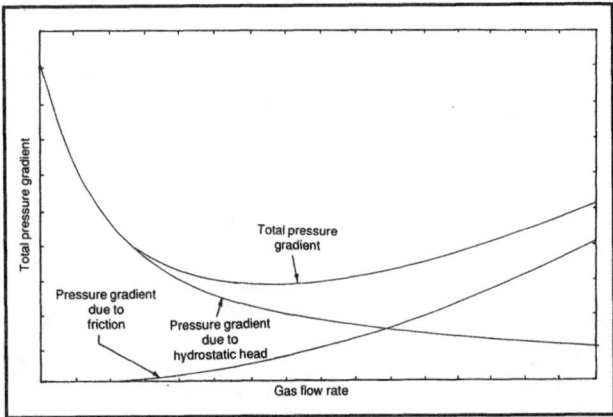

Fig. 3-24. Pressure-drop components in two-phase flow.

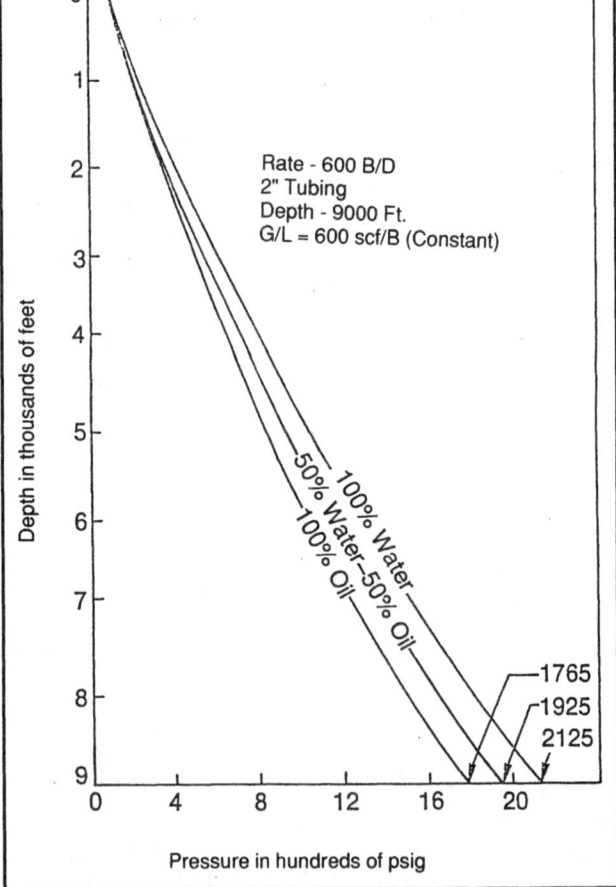

Fig. 3-25. Effect of water cut on required flowing pressure.

ρ_w = water density,

f_w = water cut = $q_w/(q_w+q_o)$,

GLR = $q_g/(q_o + q_w)$, and

GOR = q_g/q_o

4. Liquid Viscosity

The effects of liquid viscosity on pressure drop are very difficult to isolate. This results from the fact that the concept of a gas/liquid mixture viscosity has no physical meaning. The liquid viscosity will affect H_L to some degree and will also increase the shearing stresses in the liquid and, therefore, the friction pressure drop. If an oil/water mixture is present, dispersions or emulsions may form and cause a very large increase in the pressure gradient. At the present time, there is no method to accurately predict the viscosity of an oil/water mixture, much less the viscosity of a gas/oil/water mixture. The viscosity term does not appear explicitly in Equation 3-108 but it is used to calculate a Reynolds number from which the friction factor is determined. It also appears in some of the liquid holdup correlations.

The combined effects of decreasing API gravity and increasing viscosity for a gas/oil mixture are shown qualitatively in Figure 3-27. If water were present, the effects would probably be even more pronounced.

5. Tubing Diameter and Slippage

The selection of the proper tubing size to install in a well is one of the most critical and the most neglected functions of a production engineer. In many cases the tubing size will be selected based on such criteria as what has been used in the past or what is available on the pipe rack. A total system analysis, which combines the reservoir and piping system performance, is required to select the proper tubing size, but the effects of tubing size on velocity and slippage will be discussed here.

As can be seen in Equation 3-108, as d increases, the friction loss and thus the total pressure gradient will decrease up to a point. This can be observed qualitatively in Figure 3-28. However, as the tubing size increases, the velocity of the mixture decreases and eventually the velocity will be too low to lift the liquids to the surface. The well will then begin to load up with liquids and may eventually die. The tubing size at which a well will begin to load or the maximum tubing size which will sustain flow can be determined from a plot such as Figure 3-29.

The effect of declining production rate and, therefore, velocity for a particular tubing size can be shown qualitatively in Figure 3-30. For a particular tubing size, well

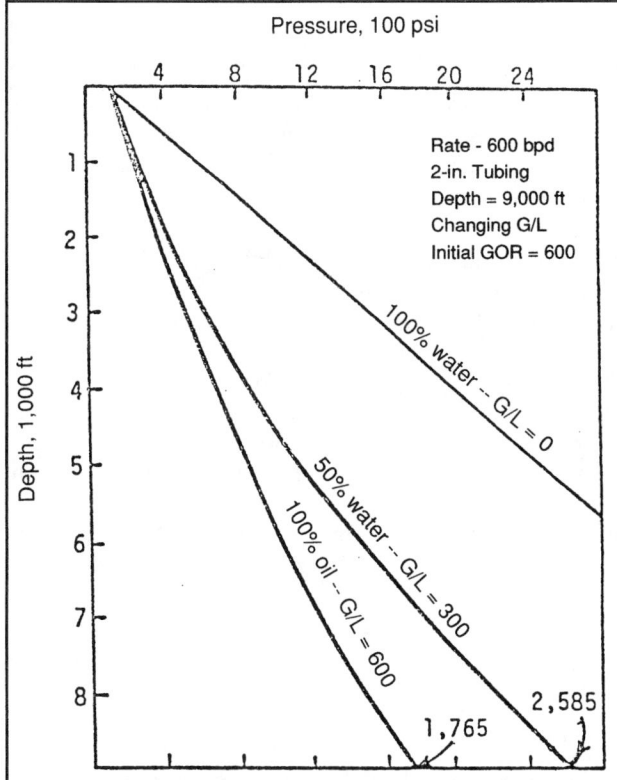

Fig. 3-26. Effect of water cut on required flowing pressure.

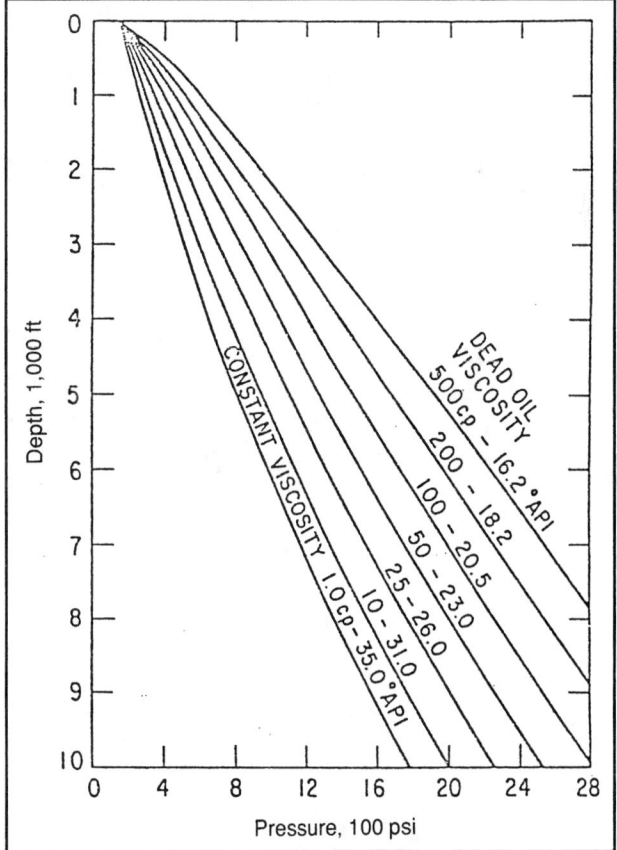

Fig. 3-27. Effect of viscosity.

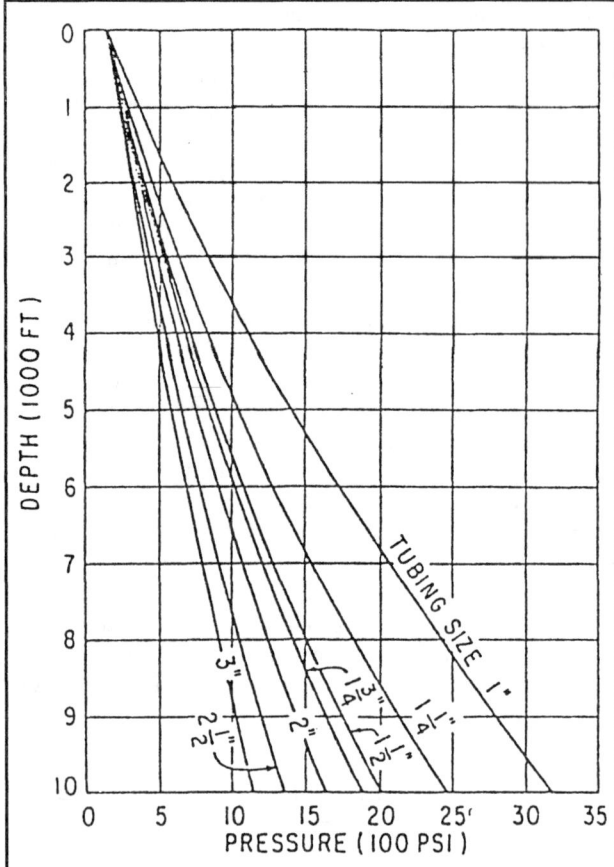

Fig. 3-28. Effect of tubing size.

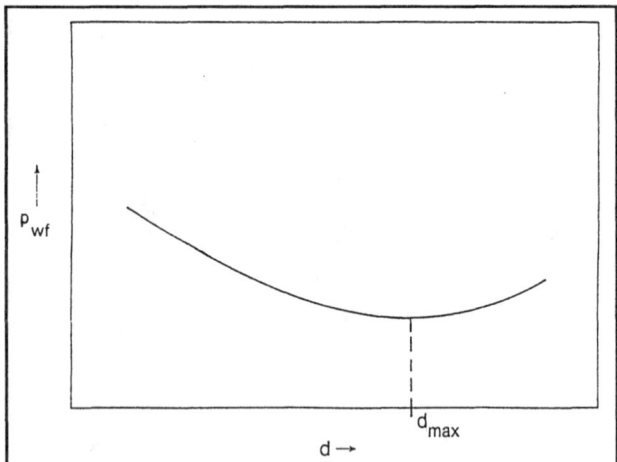

Fig. 3-29. Determining maximum tubing size.

depth, wellhead pressure and gas/liquid ratio, there will exist a minimum production rate that will keep the well unloaded. Figure 3-31 shows the effect of tubing diameter on the minimum rate. This type of information is valuable in determining at what rate a well will begin to load for various tubing sizes.

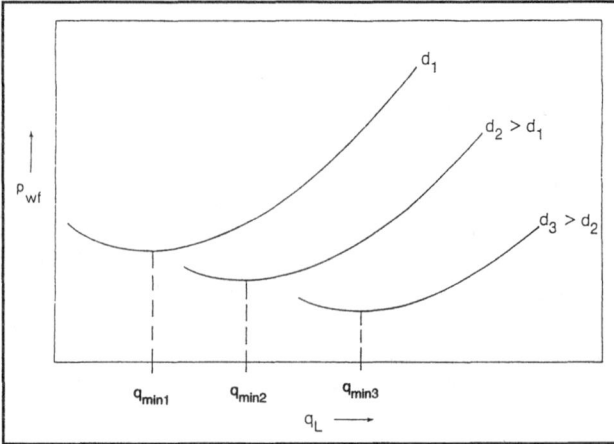

Fig. 3-30. Effect of tubing size on minimum production rate.

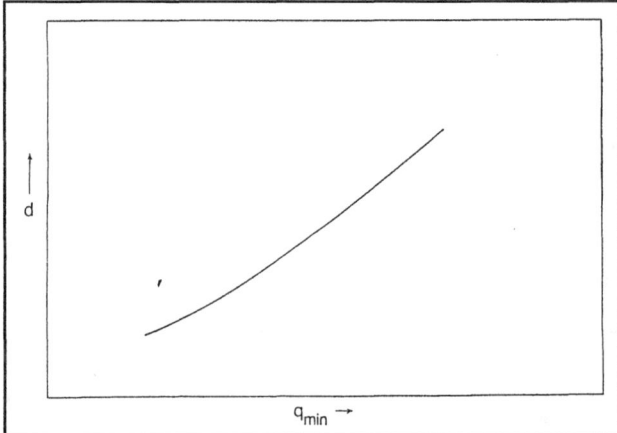

Fig. 3-31. Effect of tubing size on minimum production rate.

The type of information depicted in Figures 3-30 and 3-31 can be calculated using the multiphase flow correlations discussed previously. A simpler procedure for estimating this data for a gas well will be discussed in a subsequent section.

M. Flow in Gas Wells

As was stated previously, the general pressure gradient equation, Equation 3-16, applies for any fluid for which the variables on the right-hand side can be determined. Therefore, a pressure traverse for a gas well can be calculated by dividing the well into short increments, just as was proposed for oil wells. This amounts to the numerical integration of Equation 3-16. This was necessary for two-phase flow case, since no analytical expressions are available for density and velocity.

In the case of single-phase flow, the in-situ density and velocity can be expressed as functions of pressure and temeprature, using the gas equation of state, $pV = ZnRT$.

If acceleration is ignored, and temperature is assumed constant, Equation 3-16 can be integrated over the tubing string to give the following equation.

$$p_{wf}^2 = p_{wh}^2 \text{EXP}(S) \qquad (3\text{-}110)$$

$$+ \frac{C_1 \gamma_g q_{sc}^2 \overline{T} \overline{Z} f (MD)(\text{EXP}(S) - 1)}{S d^5}$$

where

$$S = C_2 \gamma_g (TVD)/(\overline{T}\,\overline{Z})$$

C_1, C_2 = constants depending on units

A set of consistent units is:

p_{wf}, p_{wh}	=	pressure, psia
q_{sc}	=	flow rate, MMscfd
$\overline{T}$	=	average flowing temperature, °R
TVD	=	true vertical depth, ft
MD	=	measured depth, ft
d	=	pipe inside diameter, in.
C_1	=	25
C_2	=	0.0375

In this equation, $\overline{T}$ and $\overline{Z}$ are the average temperature and Z-factor existing in the well, which makes the solution iterative since $Z = f(p)$.

An equation similar to Equation 3-110 was presented by Cullender and Smith.[43] They chose to divide the well into two increments of length *H/2* and evaluate the integral using a series expansion. The Cullender and Smith method was used extensively when most calculations were made by hand, and its use has carried over into computer applications. Actually, if the well is divided into short enough increments, the same results will be obtained from Equations 3-16, 3-110, and the Cullender and Smith method.

Equation 3-110 and the Cullender and Smith method are applicable for essentially dry gas only; that is, a fluid in which the specific gravity of the fluid is constant. These methods have been used for wells producing small amounts of liquid along with the gas by making an adjustment on the gas gravity. The mixture gravity can be estimated from:

$$\gamma_m = \frac{\gamma_g + 4591 \gamma_L / R}{1 + 1123 / R} \qquad (3\text{-}111)$$

where

γ_m	=	mixture gravity (air = 1)
γ_g	=	gas gravity (air = 1)
γ_L	=	liquid gravity (water = 1)
R	=	producing gas/liquid ratio, scf/STB

If the GLR is less than about 10,000, which corresponds to a liquid loading of greater than 100 bbls/MMscf, or if the rate is less than that required to keep the liquids unloaded, the two-phase flow correlations should be used for gas wells.

Another method for taking into account the effects of liquids in gas wells was developed by Gray[44] and is recommended by the API in their manual for sizing subsurface safety valves, Manual 14BM, API 14B. The liquid holdup was correlated with two dimensionless numbers and the no-slip holdup. The accuracy of the Gray method was stated to be questionable if:

1. $v_m > 50$ ft/sec

2. d > 3.5 in.

3. Liquid/gas ratio > 50 bbl/MMscf

4. Water/gas ratio > 5 bbl/MMscf

In practice, the Gray method has been found to give good results for conditions well out of these ranges.

An equation for estimating the minimum gas-producing rate required to keep a well unloaded if water or condensate is being produced was presented by Turner, et al.[45] The minimum rate for a particular tubing size and well-head pressure is calculated from

$$q_{sc(\text{min})} = \frac{3.06 v_{\text{min}} A p_{wh}}{TZ} \qquad (3\text{-}112)$$

where

q_{sc}	=	MMscfd
v_{min}	=	minimum velocity, ft/sec
A	=	area of tubing, ft²
T	=	surface flowing temperature, °R
Z	=	gas deviation factor at T, p_{wh}
p_{wh}	=	wellhead flowing pressure, psia

Two equations were given for v_{min} depending on whether the liquid is water or condensate.

$$v_{\text{min(water)}} = \frac{5.62(67 - 0.0031\,p_{wh})^{0.25}}{(0.0031\,p_{wh})^{0.5}} \qquad (3\text{-}113)$$

$$v_{\text{min(condensate)}} = \frac{4.02(45 - 0.0031\,p_{wh})^{0.25}}{(0.0031\,p_{wh})^{0.5}} \qquad (3\text{-}114)$$

The Turner method was originally stated to be applicable for LGR less than 130 bbl/MMscf, but has been found to give good results for rates as high as 250 bbl/MMscf.

N. Flow in Directional Wells

The general pressure gradient equation applies to directionally drilled wells if the effects of pipe angle on flow pattern and liquid holdup can be accounted for. Equation 3-16 may be written in terms of the pipe angle measured from vertical ϕ rather than the angle from horizontal θ.

$$\frac{dp}{dL} = \frac{\rho g \cos\phi}{g_c} + \frac{f \rho v^2}{2 g_c d} + \frac{\rho v dv}{g_c dL} \qquad (3\text{-}115)$$

The only well-flow methods discussed previously in which the effect of angle on liquid holdup is considered were the Beggs and Brill method and the Hasan and Kabir method.

Inclusion of the pipe angle in the pressure-gradient equation accounts for the fact that the hydrostatic or elevation component acts only over the true vertical depth (TVD) of the well while the friction loss occurs over the entire pipe length or measured depth (MD). Therefore, any of the previously discussed methods can be used in directional wells if the well is divided into increments in which the angle is fairly constant. The feasibility of this approach is demonstrated with the results of the evaluation study by Rossland.[41] All the wells used in this study were directional, but low errors were obtained using the vertical methods.

In the case of directional gas wells, Equation 3-110 may be applied.

O. Use of Prepared Pressure Traverse Curves

Almost all the previously discussed well flow correlations require the use of a computer to calculate a pressure traverse or to calculate the pressure drop occurring in the tubing string for given flow conditions. Before large amounts of money are invested in a project, the most accurate design methods available should be used to design the project. This will almost always involve computer application, but in some cases it is not feasible for the field engineer to conduct an involved computer study. In some cases, it may be advantageous to construct a set of pressure traverse curves for hypothetical values of the variables such as q_L, GLR, d, f_w, etc. These curves can then be used to estimate the pressure drop that would occur in a well producing under similar conditions. Use of traverse or gradient curves will not be as accurate as computer calculations, but the more closely the curves match the actual well conditions, the more accurate the results will be. In this section, the preparation of working curves will be discussed and sources of curves prepared for general conditions will be listed. Application of the curves to well design problems will be outlined, and several example problems will be worked using some general curves. Finally, some of the sources of errors which can occur from using the curves will be discussed.

1. Preparation of Pressure Traverse Curves

The curves are usually prepared in a format similar to Figure 3-32, which was calculated using the Hagedorn and Brown correlation. If a set of curves is to be prepared for a particular field, the correlation used should be the one that most closely matches any field data available. To prepare a curve such as Figure 3-32, the following parameters are selected:

1. Pipe inside diameter, d
2. Liquid flow rate, q_L
3. Water fraction, f_w
4. Average flowing temperature, T
5. Oil, gas, and water gravities

A pressure traverse is then calculated for several values of GLR, starting at zero pressure at zero well depth. The maximum value of GLR used is the one that will give the minimum pressure gradient for the chosen conditions. Figures will be prepared for the full range of pipe sizes, liquid rates and water fractions expected to occur in the field under consideration. The average flowing temperature and fluid properties can be selected from fluid samples taken in the field.

2. Generalized Curves

Sets of gradient or traverse curves, which were prepared using average fluid properties and flowing temperature, are available from several sources. These curves were usually prepared using an oil API gravity of 35°, a gas gravity of 0.6 to 0.7, and a water gravity of 1.0 to 1.10. Use of the general curves may introduce considerable errors unless the actual field conditions happen to correspond to the conditions for which the curves were prepared. For example, one could not expect curves calculated using a 35° API oil to match the flow of a 20° API or 50° API oil in a well. Some sources of the general curves are:

1. *Handbook of Gas-Lift*, U. S, Industries, Inc.[46]
2. *Gas Lift Theory and Practice*, K. E. Brown[47]
3. *Gradient Curves for Well Analysis and Design*, CIM Special Volume 20 (SI Units)[32]
4. *The Technology of Artificial Lift Methods*, Volume 3, K. E. Brown[23]
5. H. D. Beggs & Associates, Petroleum Consultants[26]

Other sets have been prepared by various companies using different correlations, such as the Shell curves, which were prepared using a modified, proprietary version of the Duns and Ros method.

The curves presented in the appendix and used in the example problems in this book were prepared using a computer program developed by Source 5. The Hagedorn and Brown method was used to calculate the pressure gradients. This program can be used to generate curves for any condition using one of six multiphase flow correlations.

3 Application of Traverse Curves

The application of the curves for estimating either a bottomhole flowing pressure from a known flowing wellhead pressure or vice versa will be demonstrated with several example problems in this section.

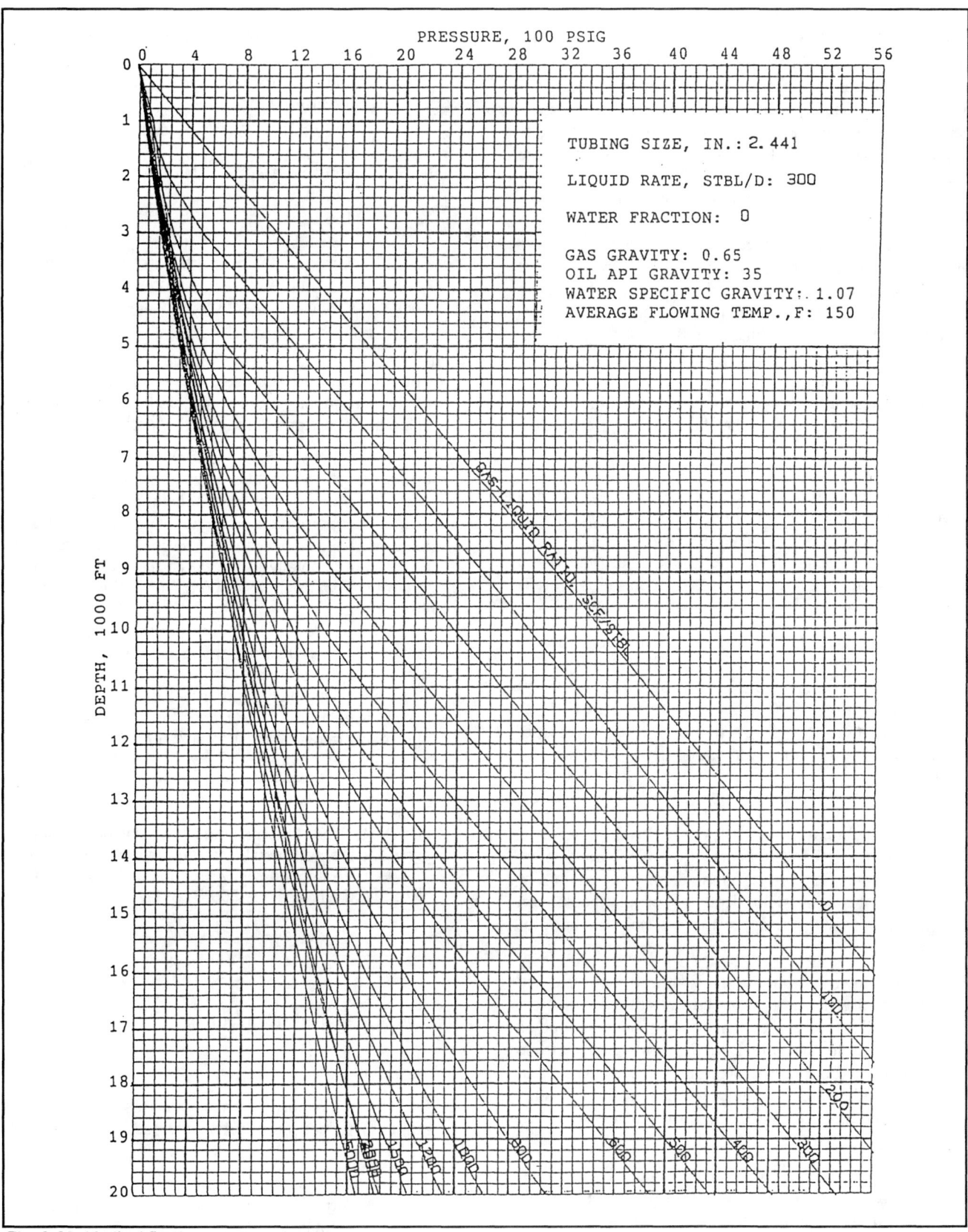

Fig. 3-32. Vertical flowing pressure traverses.

For a particular case the tubing size, liquid flow rate, water fraction and producing gas/liquid ratios will be known or assumed. Therefore, only one curve on a graph will be of interest. Several curves for various GLRs were plotted on the same graph to save space. It has already been demonstrated that the pressure gradient existing at any location in a pipe is highly dependent on the average pressure existing at that location in the pipe. This occurs because the density and therefore, the velocity of the fluid both depend on pressure. This requires that the section of the traverse curve corresponding to the average pressure conditions existing in the well of interest must be used. That is, the location of the top or bottom of the well must be selected at a point on the curve that corresponds to the known wellhead or bottomhole pressure. Therefore, the numbers on the depth axis are reference numbers only. That is, zero depth on the curve does not actually represent the wellhead unless the wellhead pressure is zero. If the actual wellhead pressure is greater than zero, then some number greater than zero on the depth axis will represent the wellhead. This means that some number on the depth axis greater than the actual well depth will represent the location of the bottom of the well. In other words, the curves must be "shifted" vertically to correspond to actual well conditions, unless the actual wellhead pressure is zero. A procedure for estimating an unknown pressure is:

1. Select the chart that most closely corresponds to the known conditions of tubing ID, liquid production rate, and water fraction. In the appendix, charts for water fractions of 0, 0.5 and 1.0 are provided for each tubing size and flow rate. For other water fractions, interpolation would be required.

2. Enter the pressure axis at the known pressure. Proceed vertically from this pressure to the intersection of the appropriate GLR curve. Proceed horizontally to the left to the intersection of the depth axis. This locates the number on the depth axis which represents the equivalent depth of which the known pressure exists, i.e. either the wellhead or bottomhole.

3. If the known pressure is the wellhead pressure, add the actual well depth to the equivalent depth located in Step 2. This represents the axis depth which is equivalent to the actual well depth. If the known pressure is bottomhole pressure, subtract the actual well depth from the number found in Step 2. This gives the axis depth that is equivalent to the actual wellhead pressure.

4. From the point located in Step 3, proceed horizontally to the right to the intersection of the same GLR line. From this point proceed vertically upward to the pressure axis. Read the unknown pressure.

Example 3-6: Finding the Required Flowing Bottomhole Pressure

The following data are known for a particular well:

$$d = 2.441 \text{ in. (2 7/8 tubing)}$$
$$q_L = 1000 \text{ STB/day}$$
$$f_w = 0.50$$
$$GLR = 400 \text{ scf/STB}$$
$$H = 12000 \text{ ft}$$
$$p_{wh} = 160 \text{ psig}$$

Find the flowing bottomhole pressure, p_{wf}, required to lift this fluid to the surface. This could represent one point on an outflow curve if $p_{node} = p_{wf}$.

Solution:

The curve that most closely corresponds to the given conditions must first be selected. Figure 3-33 is selected. The steps in the solution are:

1. Find the equivalent depth corresponding to the known pressure, which is 160 psig for this case. To do this, enter the pressure axis at 160 psig, proceed vertically until the appropriate GLR line, which is 400 scf/STB for this case, is intersected. From this intersection proceed horizontally to the depth axis, which locates the equivalent depth as 1400 ft. This point represents the surface.
2. Add the well depth, 12,000 ft, to the surface equivalent depth of 1400 ft to obtain 13,400 ft. This point represents the bottom of the well.
3. Proceed horizontally and intersect the same GLR line. Proceed vertically upward from this intersection and read the pressure at 12,000 ft as 3320 psig.

Example 3-7:

This example demonstrates the use of the vertical curves to find the maximum permissible wellhead pressure that will result in a required production rate. The well is equipped with 2 7/8 tubing. Other data are:

$$\text{Required oil rate} = 500 \text{ STB/day}$$
$$H = 12,000 \text{ ft}$$
$$GOR = 800 \text{ scf/STB}$$
$$\bar{p}_R = 4000 \text{ psig}$$
$$f_w = 0.5$$
$$J = 5 \text{ STB/day-psi}$$

Find the required value for p_{wh}.

Solution:

1. Find p_{wf} necessary to satisfy the reservoir requirements. This can usually be obtained using Vogel or Fetkovich.

Use the Vogel method and assume the reservoir is saturated, that is $p_b \geq \bar{p}_R$.

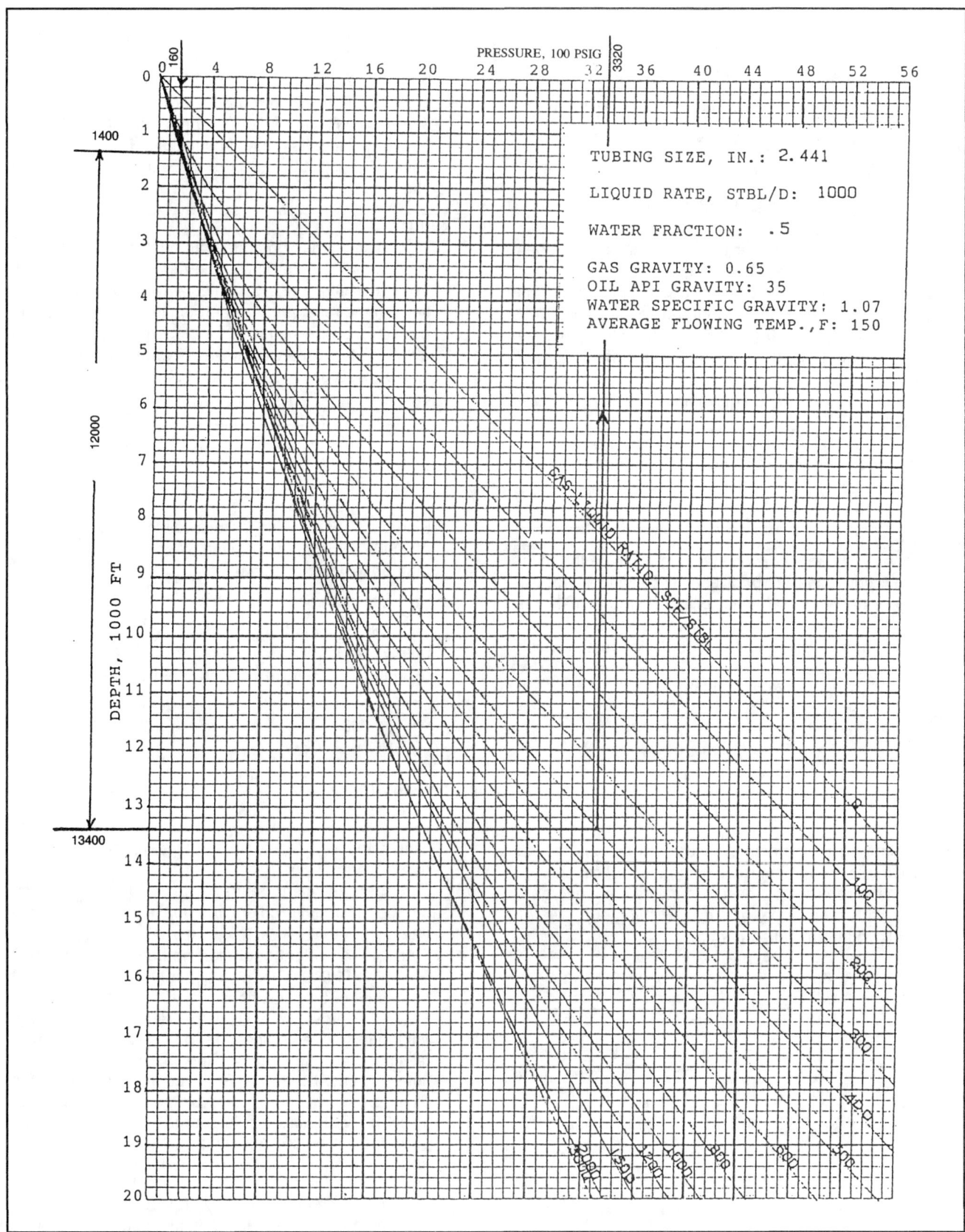

Fig. 3-33. Example 3-6 solution.

$$q_L(max) = \frac{J\bar{p}_R}{1.8} = \frac{5(4000)}{1.8} = 11,111 \text{ STBL/day}$$

Solving Equation 2-33 for p_{wf} (see Example 2-2):

$$p_{wf} = \bar{p}_R[(1.266 - 1.25 q_L / q_{L(max)})^{0.5} - 0.125]$$

$$p_{wf} = 4000[(1.266 - 1.25(1000/11111))^{0.5}$$

$$-0.125]$$

$$p_{wf} = 4000(0.949) = 3800 \text{ psig}$$

2. Select the vertical curve that most closely matches the given conditions. This is Figure 3-34.
3. Locate the known pressure, p_{wf} = 3800, and the equivalent depth. Use the appropriate GLR line. GLR = GOR (1 - f_w) = 800 (1 - 0.5) = 400. This equivalent depth is about 14,700 ft and represents the bottom of the well.
4. Determine the surface equivalent depth as 14,700 − 12,000 = 2700 ft.
5. From the 2700 ft depth, proceed horizontally to the 400 GLR line. Read the pressure at this depth as approximately 360 psig.

Example 3-8:

The vertical curves may be used to estimate the minimum production rate for given well conditions that will prevent the well from loading up. This demonstrates the effect of slippage, as previously discussed. Using the given data, determine the production rate at which the well may load up.

d = 2.992 in. (3-1/2 tubing)
H = 8000 ft
p_{wh} = 240 psig
f_w = 0
GLR = 500 scf/STB

Solution:

The flowing bottomhole pressure will be determined for various production rates using the vertical curves. When the required p_{wf} begins to increase for decreasing q_L, this means that $q_L < q_{min}$. Plotting of p_{wf} vs. q_L will allow determination of q_{min}.

TABLE 3-3C

$q_L = q_o$	Equiv. Surface Depth	Equiv. Bottom Hole Depth	p_{wf}
8000	800	8800	3040
6000	1100	9100	2640
4000	1300	9300	2120
3000	1900	9900	2000
2000	2000	10000	1760
1500	2600	10600	1750
1000	3000	11000	1650
800	3000	11000	1560
600	3100	11100	1510
500	3100	11100	1450
400	3200	11200	1460
300	3600	11600	1600
200	3400	11400	1640

A plot of the values of q_L vs. p_{wf} (Figure 2-35) reveals that p_{wf} reaches a minimum at a production rate of about q_{min} = 500 STB/day. Plotting the data for rates between 200 and 2000 STB/day on an expanded scale would result in more accuracy, but it is obvious that this well will not flow at a rate less than 400-600 STB/day under its present conditions. For other conditions of p_{wh}, GLR, f_w, etc., the value of q_{min} would change.

Example 3-9:

This example demonstrates the application of the vertical curves in determining the minimum GLR that will allow a well to flow at a particular rate for given conditions. If the reservoir is not producing the required volume of gas, the well may be placed on gas lift. From a flowing test on a well, it was determined that $q_{L(max)}$ = 3100 STB/day, FE = 1.0 and $\bar{p}_R$ = 3200 psig. Other data are

d = 1.995 (2-3/8 tubing)
p_{wh} = 400 psig
H = 9400 ft
f_w = 0.5
p_b = 3200 psig

Find the GLR required for this well to produce at a rate of 900 STB/day total fluid. Assume that Vogel's method is valid for f_w = 0.5.

Solution:

1. Find the value of p_{wf} required by the reservoir to inflow 900 STB/day. Solving Vogel's equation (Equation 2-33) for p_{wf}: (See Example 2-2)

$$p_{wf} = \bar{p}_R\left[\left(1.266 - \frac{1.25 q_L}{q_{L(max)}}\right)^{0.5} - 0.125\right]$$

$$p_{wf} = 3200\left[\left(1.266 - \frac{1.25(900)}{3100}\right)^{0.5} - 0.125\right] = 2640$$

2. Select the vertical curve for the appropriate conditions, Figure 3-36.
3. Find the equivalent depth at p_{wh} = 400 psig, assuming the minimum gradient line applies. This is approximately 2500 ft.
4. Find the equivalent depth for the bottomhole as 2500 + 9400 = 11,900 ft. Draw a horizontal line at 11,900 ft across all of the GLR lines.
5. Enter the pressure axis at the required p_{wf} = 2640 psig and draw a vertical line to intersect the horizontal line drawn in Step 4. The intersection of these lines indicates that the GLR required to satisfy the fixed pressure drop is approximately 700 scf/STB. The 700 GLR line would merge with the minimum gradient line before reaching a pressure of 400 psig, so the assumption made in Step 3 is valid. If the GLR line found in Step 5 had not merged at the known p_{wh}, a new surface equivalent depth would be found, and the procedure repeated until the correct GLR was obtained.

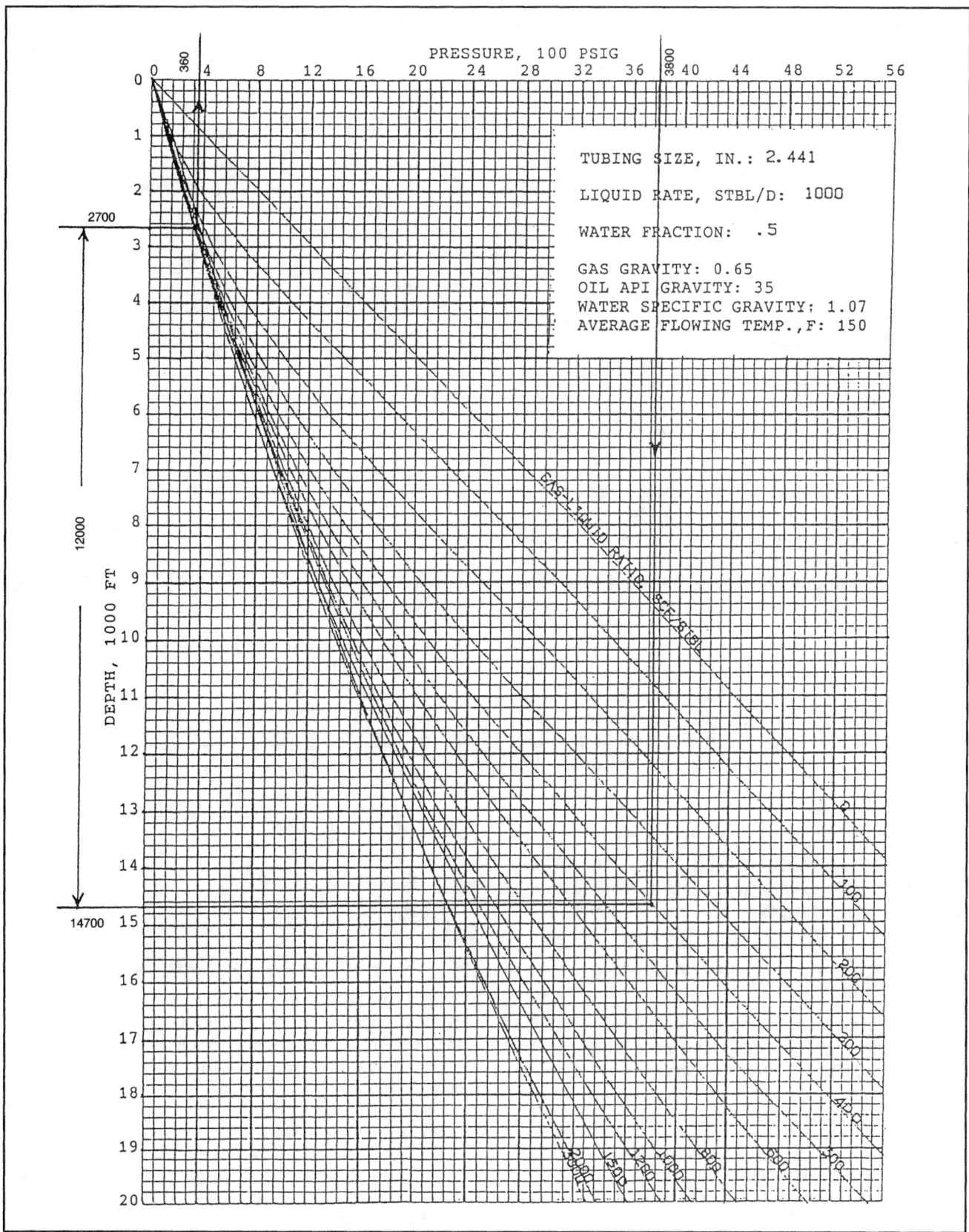

Fig. 3-34. Example 3-7 solution.

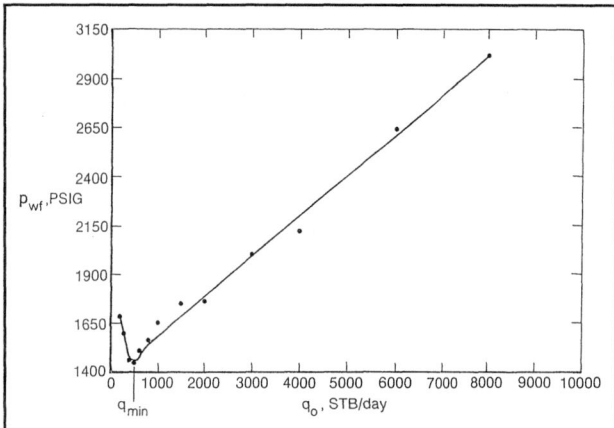

Fig. 3-35. Example 3-8 solution.

The general pressure traverse curves were prepared for vertical wells only in which both the hydrostatic term and friction term act over the total tubing length. A rough approximation of the pressure drop in a directional well may be made by combining vertical and horizontal curves. This is accomplished by using the vertical curves to find the total pressure drop over the TVD and using the horizontal curves to estimate the extra pressure drop due to friction acting over the tubing length MD–TVD. As will be discussed in more detail subsequently, the horizontal curves were prepared assuming an angle of zero from horizontal. Therefore, the hydrostatic term would be zero.

This method should be used only for making very rough preliminary estimates, and a computer calculation should be used in the final design. The procedure is limited to relatively low-pressure wells because the horizontal curves have much lower maximum pressures than the vertical curves.

The following procedure may be used to estimate the flowing bottomhole pressure in a directional well:

1. Using the appropriate vertical curve, find the value of flowing bottomhole pressure that would exist in a vertical well of depth = TVD. Label this pressure as p'_{wf}.

2. Calculate the average pressure in the tubing as $\overline{p} = 0.5\,(p_{wh} + p'_{wf})$.

3. Select a horizontal curve that was prepared for the same pipe size, flow rate, water cut and GLR as was used in Step 1. Enter this curve on the pressure axis at $\overline{p}$ and draw a vertical line to intersect the appropriate GLR line.

4. Determine the pressure gradient, $\Delta p / \Delta L$ existing at this pressure. This can be done by reading the pressure change over a selected length, such as 2000 ft.

5. Calculate the extra pressure drop due to friction over the tubing length not considered in Step 1.

$$\Delta p_f = (MD\text{-}TVD)\Delta p / \Delta L$$

6. Estimate the flowing bottomhole pressure for the directional well as:

$$p_{wf} = p'_{wf} + \Delta p_f$$

The average pressure calculated in Step 2 could then be refined using the p_{wf} calculated in Step 6, and the procedure repeated. However, the accuracy of the method does not warrant this step.

Example 3-10:

Estimate p_{wf} for the directional well described as follows:

q_L	=	1500 STB/day
GLR	=	800 scf/STB
p_{wh}	=	160 psig
MD	=	9000 ft.
f_w	=	0
d	=	1.995 in.
TVD	=	6000 ft.

Solution:

1. p'_{wf} is found to be 1520 psig from Figure 3-37.

2. $\overline{p}$ = 0.5 (160 + 1520) = 840 psig.

3. From Figure 3-38, which is a horizontal curve for the same conditions of q_L, f_w, and d as Figure 3-37, the length corresponding to a pressure of 840 psig is about 5700 ft.

4. Reading on the 800 GLR line, the pressure at 4700 ft is 760, and the pressure at 6700 ft. is 920.

$$\frac{\Delta p}{\Delta L} = \frac{920 - 760}{6700 - 4700} = 0.080 \text{psi/ft}$$

5. Δp_f = (MD – TVD) $\Delta p/\Delta L$= (9000 - 6000)(0.80) = 240 psi

6. p_{wf} = p'_{wf} + Δp_f = 1520 + 240 = 1760 psig

V. PIPELINE FLOW CORRELATIONS

Procedures for calculating the pressure losses occurring in a pipeline are required in the petroleum industry for designing flowlines or gathering lines and for designing long distance pipelines. This section will be concerned primarily with the effect of the flowline on overall well performance, but the correlations discussed also apply to large diameter lines.

The pressure loss in the flowline, expressed as $\Delta p_6 = p_{DSC} - p_{sep}$ in Figure 3-1, can be very small for short flowlines, such as might exist in an offshore situation if the separator is located near the wellhead. Conversely, in many producing areas the distance between the wellhead

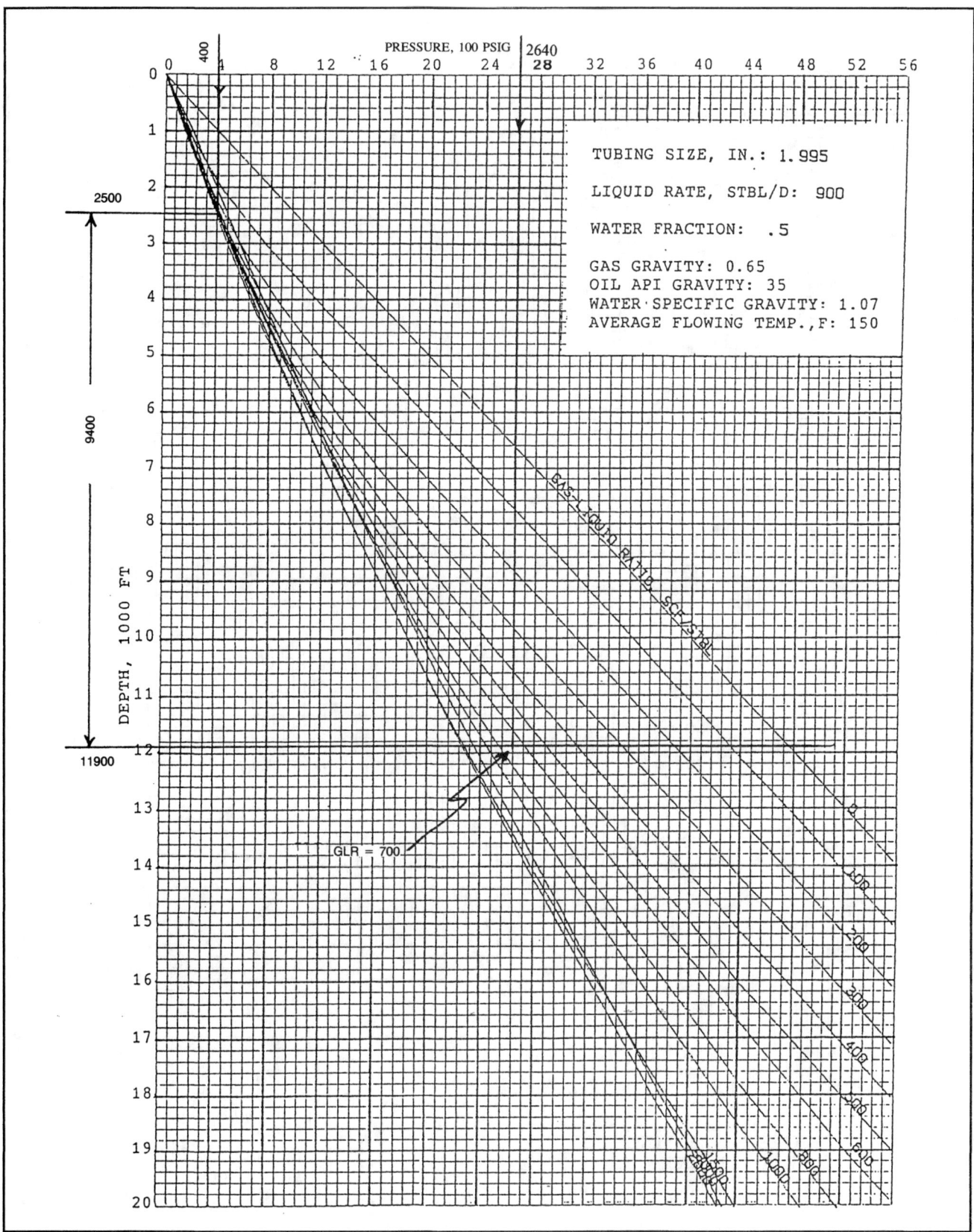

Fig. 3-36. Example 3-9 solution.

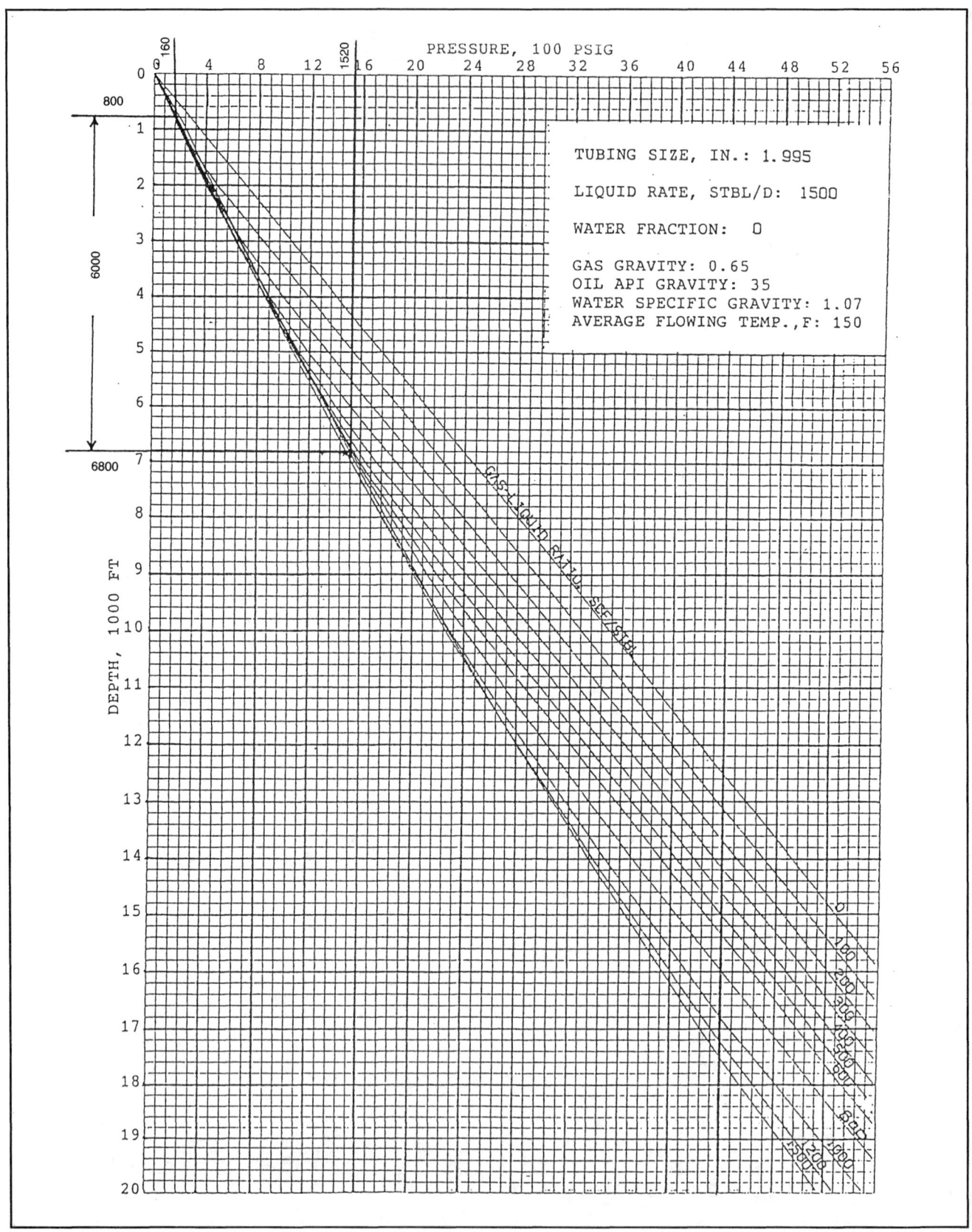

Fig. 3-37. Example 3-10 solution.

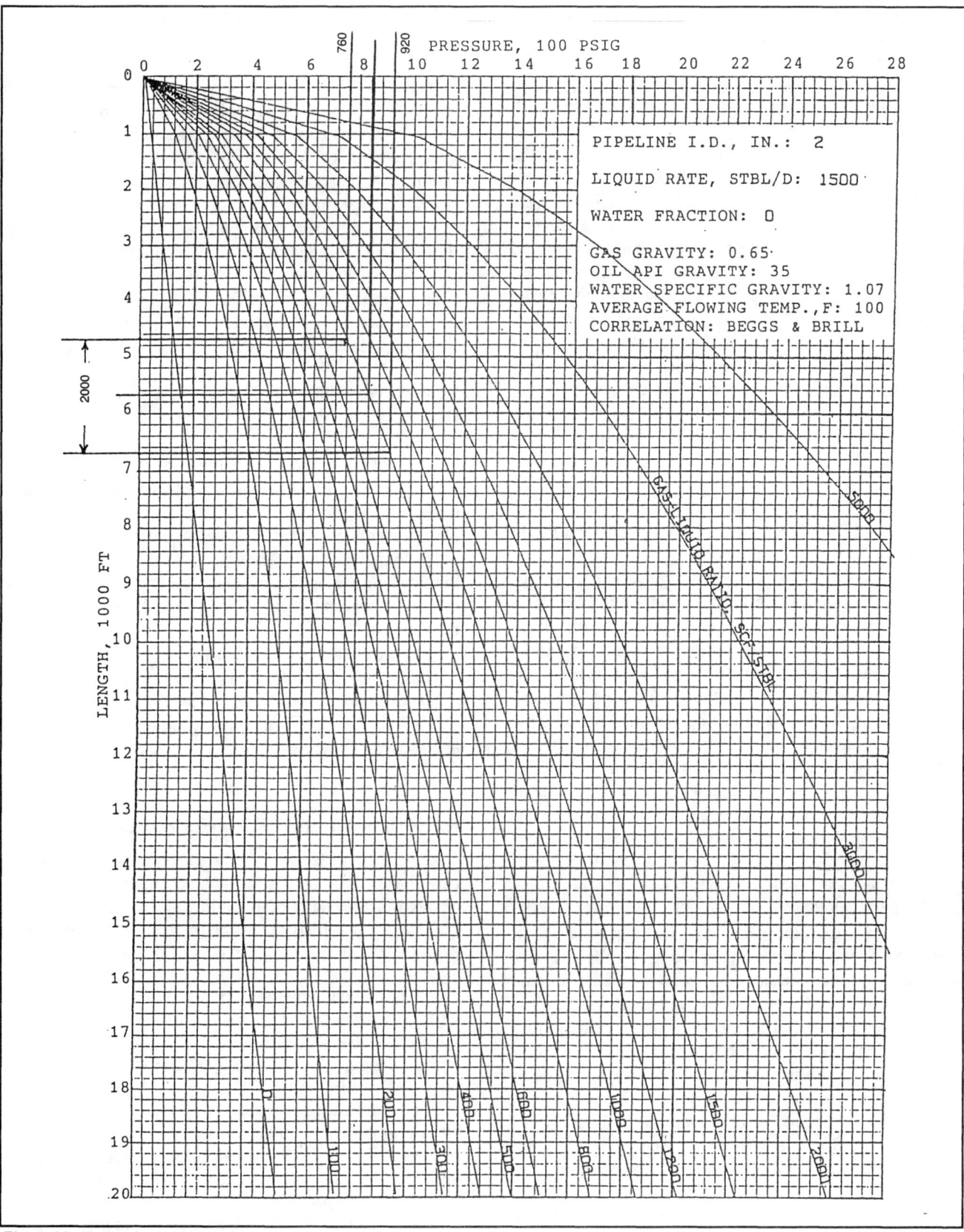

Fig. 3-38. Example 3-10 solution.

and separator may be several miles, and the pressure drop in the flowline might be 20 to 30 percent of the total pressure loss, $\overline{p}_R - p_{sep}$.

The general pressure gradient equation will apply for flow in pipelines, and all three of the components will apply in most cases. Some pipelines can be considered to be essentially horizontal, and in this case the hydrostatic component would be zero. Thus, it is important to have a good correlation for friction factor in pipeline design. The friction factor for multiphase flow can be several times greater than would occur for a single-phase fluid traveling at the same velocity as the two-phase mixture.

Although the hydrostatic or elevation component of the pressure gradient equation is not as important in pipeline flow as it is in wellflow calculations, a substantial part of the total pressure loss in a pipeline can result from lifting the fluids over hills or inclines. Also, it has been found that in most cases, hardly any of the hydrostatic pressure lost in the uphill section of a line is regained in the downhill section. This is true if the flow pattern in the downhill section is segregated, that is, stratified, wavy, or annular. The flow then behaves as open channel flow. Therefore, it is still necessary to be able to predict liquid holdup and flow pattern in pipeline flow.

Some aspects of the pipeline design problem are not related to pressure loss only, but include sizing lines and separation facilities such that the separator will not be overloaded or flooded. Separators are usually designed based on a steady rate of gas and liquid flow. If extra liquid volumes or slugs arrive at the separator periodically, means must be provided to handle the extra liquid. This can be accomplished by installing slug catchers upstream of the separator. Before separation facilities can be sized, the engineer must be able to predict the size or volume of the extra liquid and how fast this volume must be handled. The extra liquid can result from changes in flow conditions, such as adding or deleting wells, which will usually change the liquid holdup in the line. If conditions are changed such that liquid holdup is decreased, then the extra volume of liquid removed from the system will eventually arrive at the separator. Designing for this situation requires accurate holdup prediction methods. This type of problem can arise also during pigging operations.

If flow conditions in a line are such that slug flow exists at the pipeline outlet, liquid and gas will arrive at the separator as alternating slugs of gas and liquid. Even though a separator may be large enough to handle the gas and liquid volumes at steady conditions, a large liquid slug can easily flood the separator. It is, therefore, necessary to be able to predict the horizontal or pipeline flow pattern existing at given conditions. Several flow pattern maps will be discussed in this section, but methods for predicting slugging characteristics, such as slug length,

velocity and frequency, will not be presented. Procedures for accomplishing this can be found in Reference 9.

Many correlations for pipeline flow prediction have been developed over the past 30 or 40 years. A review of many of the correlations and the manner in which they were developed can be found in Reference 23. Since the purpose of this book is to present the best methods to optimize production systems, only the methods most widely used at the present time will be discussed.

A limited number of comparison studies or evaluations of some of the methods has been conducted. The results of some of these studies will be presented. The effects of changes in conditions that can exist in pipeline flow will be discussed from a qualitative viewpoint. These conditions include variables such as GLR, pipe size, liquid rate, etc. The effects of hills or inclines on the pressure loss will be included in this section, and a procedure for separating the friction and elevation losses will be discussed.

As in the section on well flow correlations, the discussions will be limited to how the experimental data were obtained and how the correlations for f_{TP}, H_L, and flow pattern were developed. Detailed equations and an example calculation for the Beggs and Brill method will be included in the appendix. Details of the other methods discussed here and some methods not discussed here can be found in References 9 and 23. Computer subroutines for some of the methods are listed in Reference 9.

The preparation and use of pressure traverse curves for quick estimates of frictional pressure loss in pipelines will also be described. Example problems illustrating the application of these curves will be presented.

A. Horizontal Flow Pattern Prediction

At least eight separate flow patterns that can exist in two-phase horizontal flow have been described. Many different names have been applied to these flow patterns, but the most widely accepted names and descriptions were presented by Alves,[48] and are illustrated in Figure 3-39.

The horizontal flow patterns were classified into only three categories by Degance and Atherton,[49] and these same categories were used by Beggs and Brill, as illustrated in Figure 3-19. The Beggs and Brill flow pattern map is shown in Figure 3-21.

One of the earliest attempts to predict flow pattern by the use of maps was made by Baker in 1958.[50] The coordinates of the map are dimensionless, but the vertical and horizontal axes essentially represent gas velocity and liquid velocity, respectively.

A map that was developed from extensive experimental data was published by Mandhane, et al.[51] in 1974. The

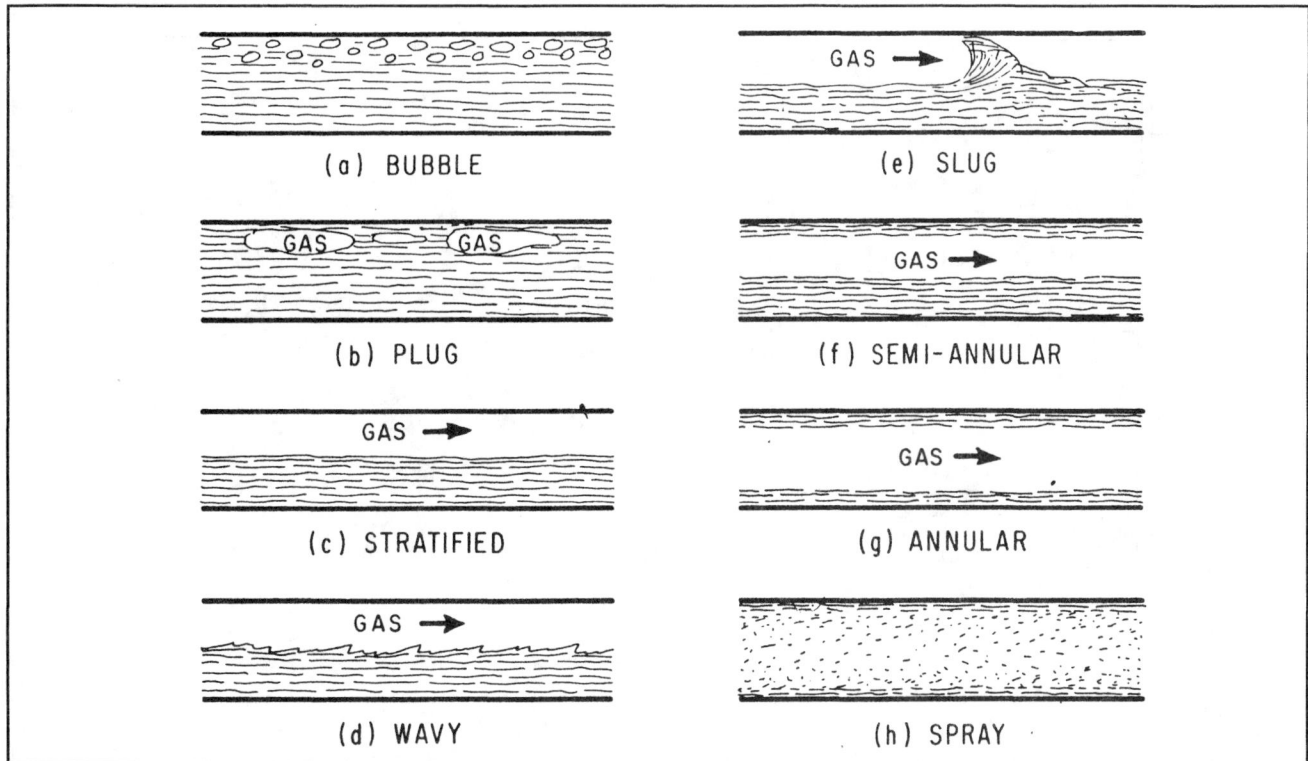

Fig. 3-39. Horizontal flow patterns.

map coordinates are superficial gas and liquid velocities. This is one of the most widely used horizontal flow pattern maps at the present time. The flow patterns and their locations on the Baker and Mandhane maps are illustrated in Figures 3-40 and 3-41.

The flow patterns and maps described above apply for strictly horizontal flow only. It has been observed that stratified or wave flow cannot exist in a pipe that is inclined upward at only a few degrees. When upward inclination occurs, the liquid is held back by gravity

forces and the flow pattern changes to slug. Conversely, if the pipe is inclined downward, stratified flow is predominate, and slug flow will not occur at the conditions predicted by the horizontal maps.

Very few studies have been made to try to include the effect of pipeline angle on the flow pattern. An experimental study performed by Mukherjee[52] addressed this problem, but the maps produced have not been tested using field scale data.

A theoretical study of the inclined two-phase flow pattern problelm was published by Taitel and Dukler.[53] The study involved the prediction of conditions under which the flow pattern would change from stratified to intermittent (slug) or annular. The basis of the equations presented was similarity and dimensional analysis, but several simplifying assumptions were made to make the equations applicable. Also, no limits on the pipe angles for which the method is valid were given.

B. Eaton, et al., Method

The Eaton, et al.,[54] correlations for friction factor and liquid holdup resulted from extensive data that were obtained from a test facility that consisted of two 1700 ft lines. Diameters of 2 in. and 4 in. were utilized, and three liquids were used in each line. The variables studied and their ranges were:

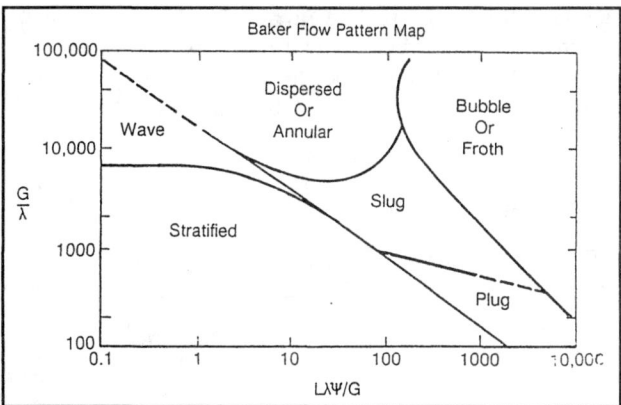

Fig. 3-40. Baker flow pattern map.

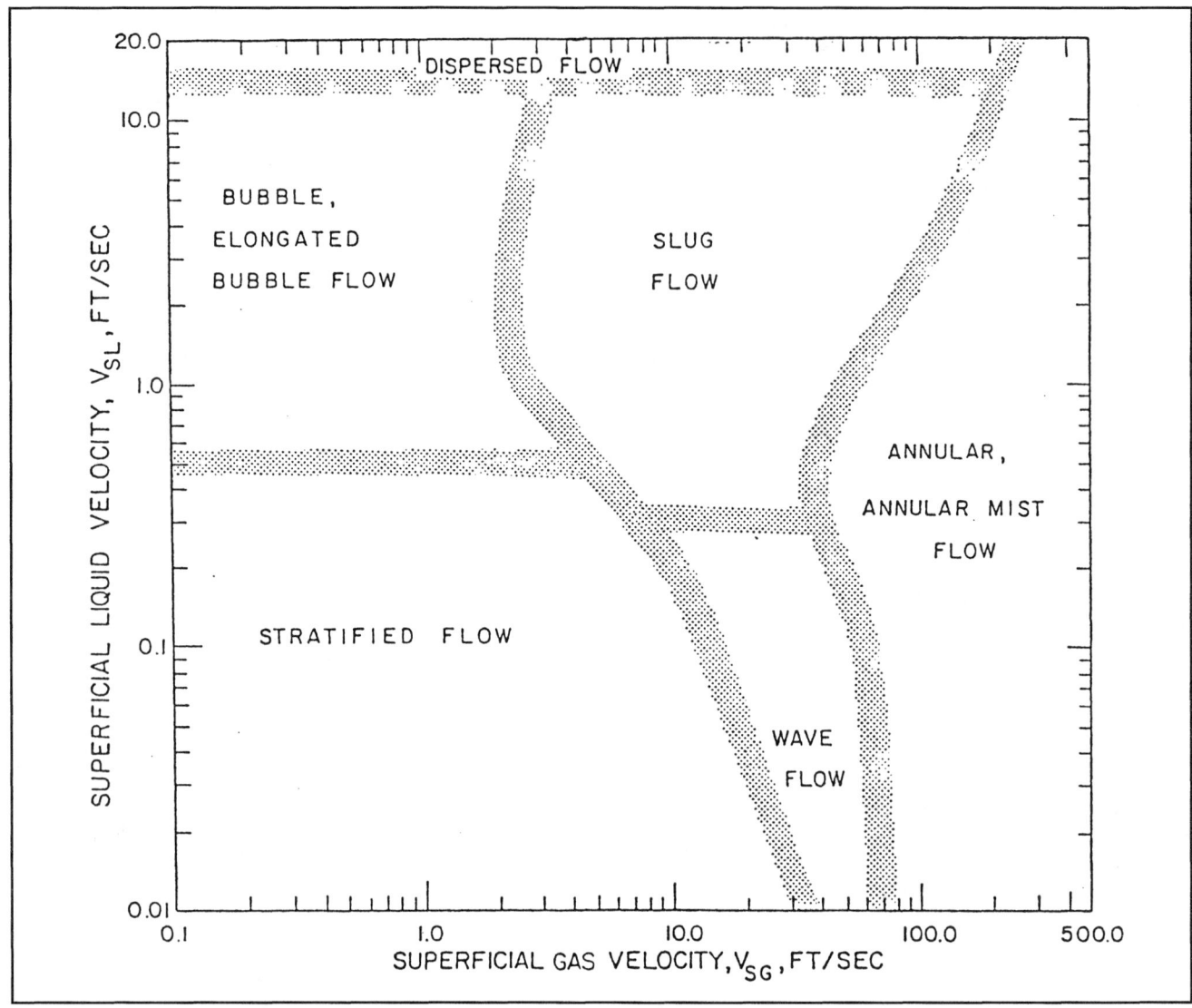

Fig. 3-41. Mandhane flow pattern map.

1. Gas rate, 0 – 10 MMscfd
2. Liquid rate, 50 – 5500 STB/day
3. Liquid viscosity, 1 – 13.5 cp
4. System pressure, 70 – 950 psig

The friction factor and liquid holdup were correlated with dimensionless numbers using regression analysis. Liquid holdup was measured by trapping a segment of the flow stream between quick-closing valves. Flow pattern was not considered in the correlation, and no effect of pipe angle is included.

The friction factor and liquid holdup correlations are shown graphically in Figures 3-42 and 3-43, respectively. The liquid holdup correlation is considered to be one of the best available for horizontal flow, but the friction factor correlation does not degenerate to the single-phase

case as the flow approaches either all liquid or all gas. In the range of low gas-liquid ratios, the friction factor becomes very large. It has been found that the friction factors will be valid if the value of the abscissa correlating group falls between about 10^4 and 10^6.

Detailed calculation procedures and example calculations for the Eaton method can be found in References 9 and 23.

C. Dukler, et al., Method

The American Gas Association sponsored a study to improve methods for predicting pressure drops occurring in two-phase flow pipelines under the direction of Dukler at the University of Houston, and the results were published in a design manual.[56]

The study was conducted by first gathering more than

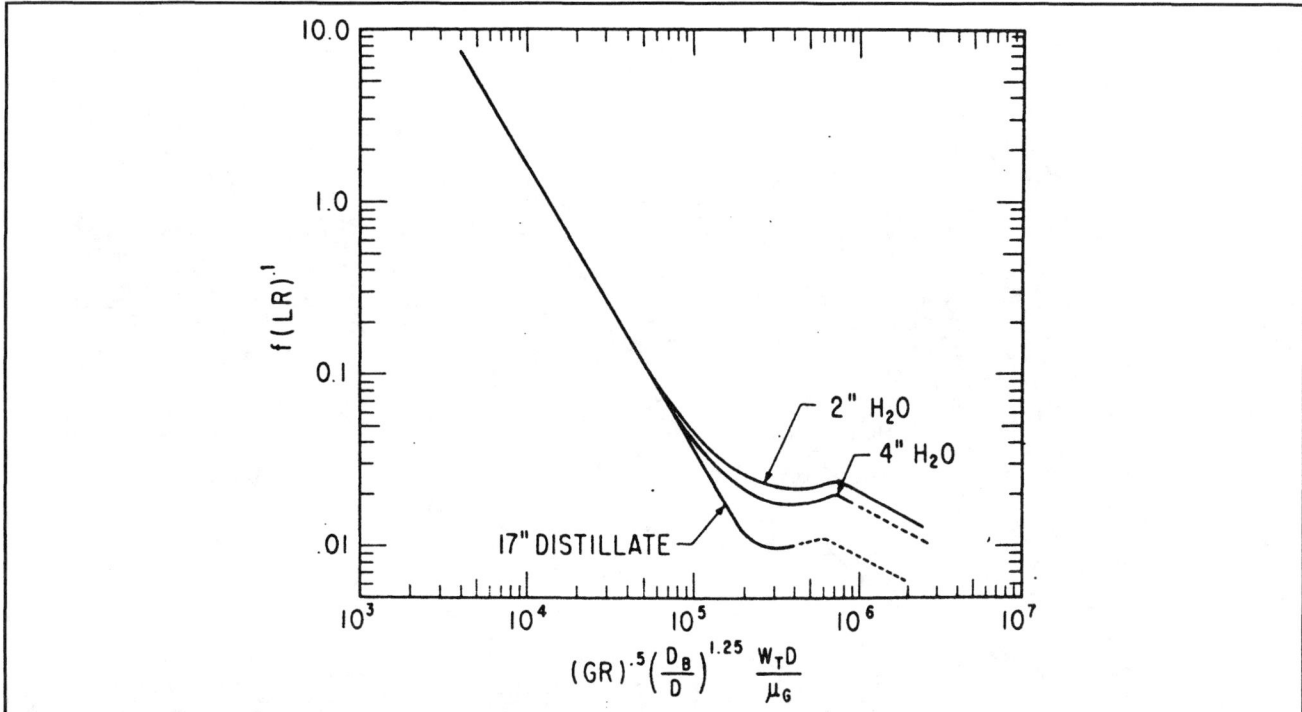

Fig. 3-42. Eaton friction factor correlation.

20,000 experimental data points from both laboratory and field scale facilities. After eliminating the doubtful data, only about 2,600 remained for developing the correlations. Dukler used a combination of dimensional and similarity analysis to arrive at expressions for calculating the frictional pressure loss. A method for predicting in-situ

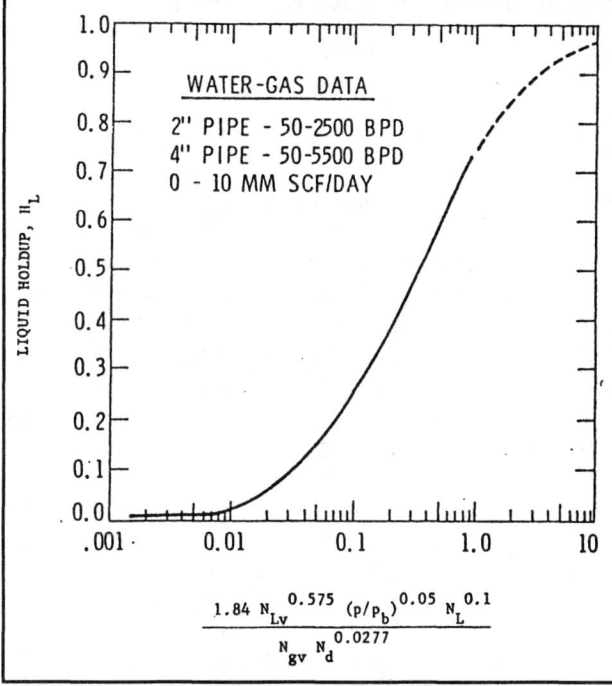

Fig. 3-43. Eaton liquid holdup correlation.

liquid holdup is required because the density term in this component requires a value for H_L.

Liquid holdup was correlated with no-slip holdup λ_L and with a two-phase Reynolds number. Finding a value for H_L is iterative since the Reynolds number includes H_L in the density term. The holdup correlation is shown in Figure 3-44.

A normalized friction factor, from which the two-phase friction factor may be obtained, is illustrated in Figure 3-45.

The Dukler method has been widely used in the petroleum and pipeline industries and gives good results for both small and large diameter pipelines. Although no effect of pipeline inclination is included in the method, it has been successfully combined with a method proposed by Flanigan[56] for hilly terrain pipelines. The Flanigan method will be discussed in a subsequent section.

References 9, 23 and 56 contain detailed calculation procedures and examples using the Dukler method. A computer subroutine that combines Dukler and Flanigan is included in Reference 9.

D. Beggs and Brill Method

The Beggs and Brill method was described earlier in the section on well flow correlations. Although it may be used for pipes at any angle of inclination, its widest application has been in the area of pipeline design as opposed to tubing design. The fact that this method is presented entirely in equation form and therefore requires that no

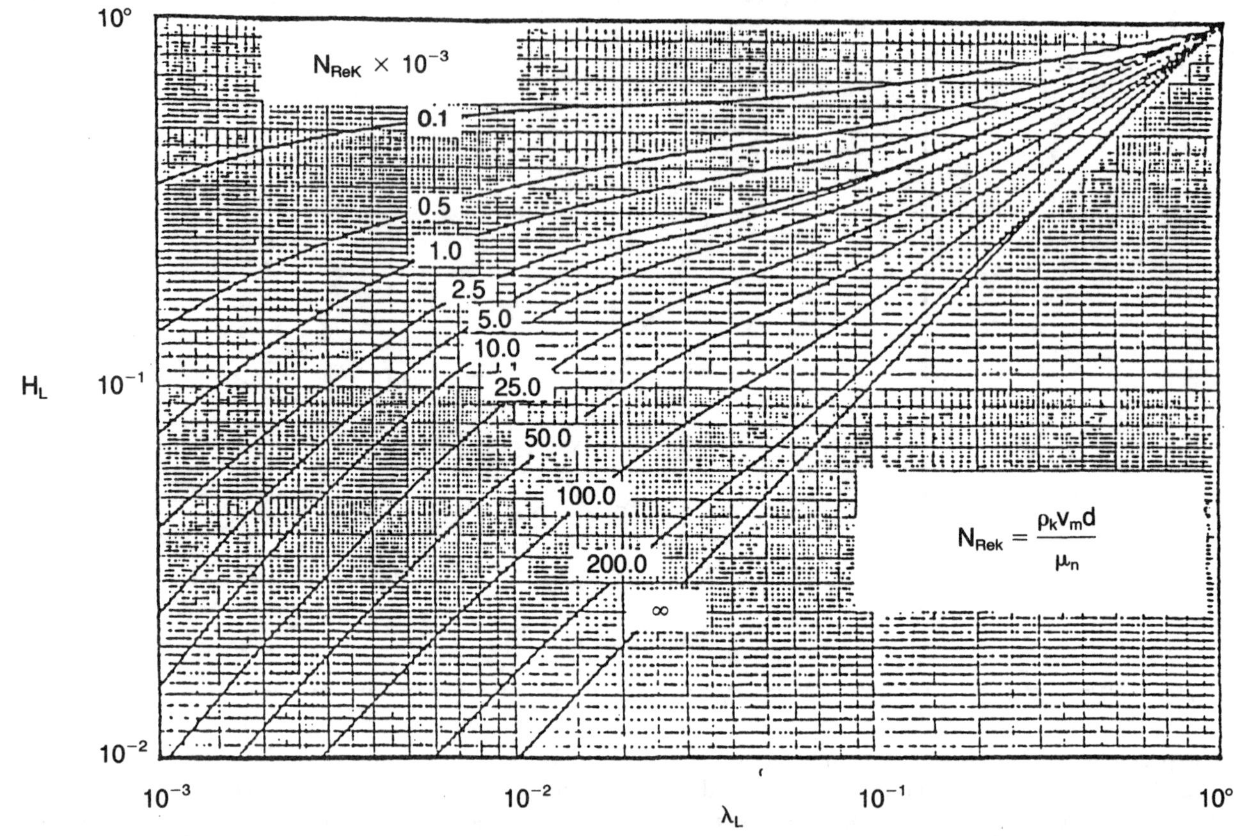

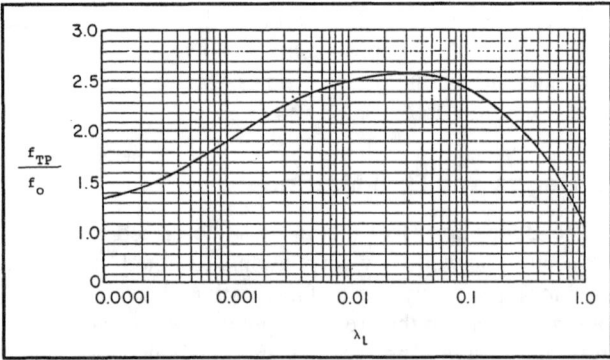

Fig. 3-44. Dukler liquid holdup correlation.

graphs or charts be curve fit for computer or calculator application has increased its acceptance in the industry.

As was mentioned earlier, a program for the HP41CV calculator utilizing the Beggs and Brill method is available.[36] Also, a detailed calculation procedure and example calculations are published in References 9, 23 and 36, as well as in the appendix of this work.

Fig. 3-45. Dukler friction factor correlation.

E. Flanigan Method for Hilly Terrain

A study of the effects of hills on the pressure drop in a two-phase pipeline was conducted by Flanigan.[56] The study was prompted by the observation that a particular gas condensate pipeline that was designed for a total pressure drop of about 30 psi exhibited a gradual increase in pressure drop with time even though the input gas and liquid rates were held fairly constant. Investigation of the source of the extra pressure drop revealed that liquid was accumulating in the low sections of the pipeline and causing an increase in both the elevation or hydrostatic and the friction components. It was found from experimentation that an increase in gas rate at the same liquid rate caused a decrease in the total pressure drop. This was attributed to the fact that the increased gas velocity swept out some of the liquid accumulated in the low sections.

After extensive investigation of this 16 in. pipeline and several other lines in which two-phase flow was occurring, Flanigan developed a method to account for both the increased friction and increased hydrostatic pressure drops. The increase due to friction caused by the presence of the liquid phase was accounted for by a reduced efficiency factor to be used in the Panhandle equation. The

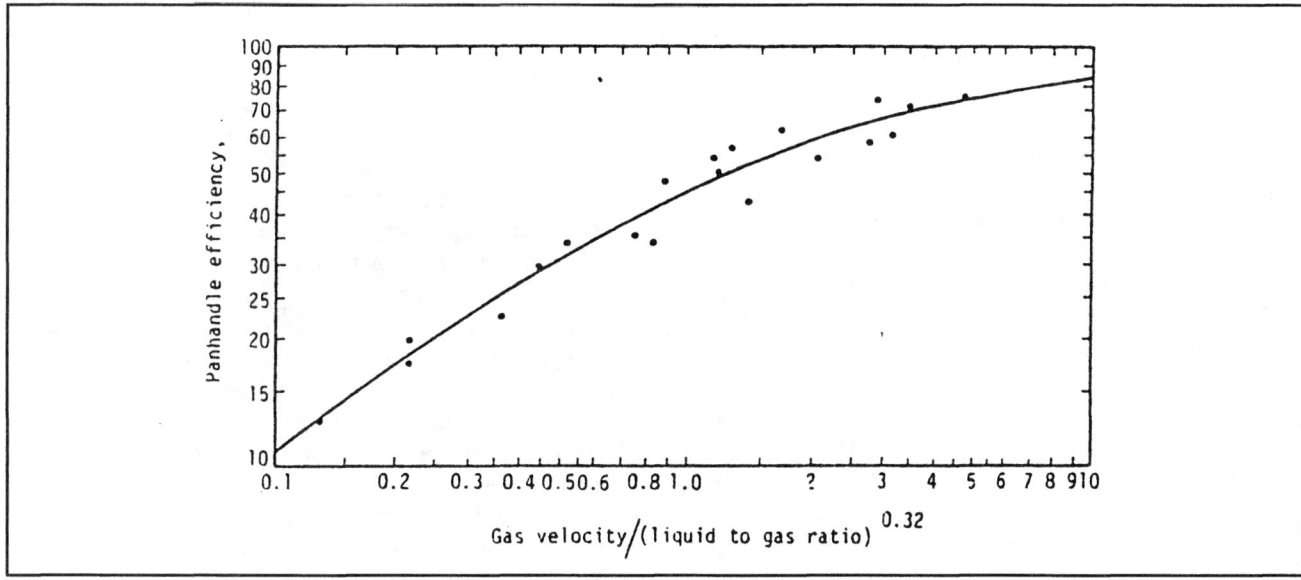

Fig. 3-46. Flanigan efficiency factor.

Panhandle equation will be presented in a subsequent section.

The efficiency factor was correlated with superficial gas velocity and liquid to gas ratio and is illustrated in Figure 3-46. The units to be used in calculating the abscissa value are ft/sec and bbl/MMscf for the gas velocity and liquid-to-gas ratio respectively.

The extra hydrostatic pressure drop due to liquid accumulation in the low sections of a line, which must be added to the friction loss, is calculated from:

$$\Delta p_h = \frac{\rho_L g}{g_c} H_L \sum h_i \qquad (3\text{-}116)$$

where

Δp_h = pressure drop due to hills
ρ_L = liquid density at average pipeline conditions
H_L = holdup factor
h_i = the *vertical* rises of the individual sections of the pipeline.

Flanigan found that neither the angle of inclination of the uphill sections nor the difference in inlet and outlet elevation of the pipeline was important. He also found that recovery of hydrostatic pressure in the downhill sections of the line was negligible.

The holdup factor was found to be a function of superficial gas velocity only. The relationship is shown graphically in Figure 3-47 and may be calculated from:

$$H_L = \frac{1}{1 + 0.3264 v_{sg}^{1.006}} \qquad (3\text{-}117)$$

where v_{sg} is in ft/sec evaluated at average pressure and temperature existing in the pipeline. Since calculation of the average pressure requires knowledge of the inlet and outlet pressures, the calculation of pressure drop is iterative.

The Flanigan equation for calculating the extra pressure drop due to hills has been combined with friction loss methods other than the Panhandle equation. As was mentioned earlier, the American Gas Association pipeline design manual recommends combining Dukler and Flanigan for hilly terrain pipelines. It has also been suggested that the Flanigan equation for hydrostatic pressure drop be combined with friction loss obtained from prepared horizontal pressure traverse curves for quick estimates of pressure drop in hilly terrain pipelines. This procedure will be illustrated with an example in a subsequent section.

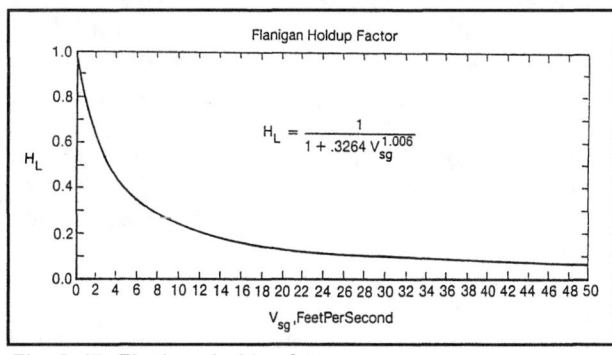

Fig. 3-47. Flanigan holdup factor.

F. Hybrid Model

A pipeline flow model which combines several of the previously published correlations was published by Gregory, et al.[57] A data bank containing 2685 liquid holdup measurements and more than 10,000 pressure drop measurements was used to test existing pipeline correlations. The experimental data points were divided into flow patterns according to Figure 3-41 and the most accurate correlation was selected for each flow pattern. The methods judged to be best for each flow pattern are listed in the following table.

TABLE 3-4
Hybrid Model

Flow Pattern	Holdup Correlation	Friction Loss Method
Bubble	Hughmark (58)	Chenoweth (62)
Stratified	Agrawal (59)	Agrawal
Wave	Chawla (60)	Dukler
Slug	Hughmark	Dukler
Annular	Lockhart (61)	Chenoweth
Dispersed	Beggs and Brill	Lockhart (Modified)

The correlations listed in the table that are not described in this book may be found in the referenced papers. Application of this model requires the use of seven empirical correlations and would therefore require utilization of a computer.

G. MONA, Asheim Method

The model developed by Asheim[78] in Norway and discussed in the section on Wellflow Correlations can also be used for pipeline design. Details of the model can be found in Reference 78.

H. Evaluation of Pipe Flow Correlations

The lack of accurate field data from operating pipelines or flowlines has made the comparison of pipe flow correlations difficult. Although surface and bottomhole pressures are measured in wells for several reasons, once a pipeline is installed, performance measurements are seldom made. Only if the pipeline will not carry its designed rates is it likely to be analyzed. Also, liquid holdup and flow pattern are extremely difficult to measure, and almost no field data exists for these parameters. For these reasons, many of the comparison studies have been based on laboratory scale data, although some comparisons using field data have been published. Some of the published studies will be discussed in this section.

Vohra, et al.[63] made a study using the data that were used to develop the Eaton correlation and the Beggs and Brill data to compare measured and calculated liquid holdup values. Six published correlations were used to calculate the liquid holdup and compare it to the measured value for each test. A description of the experimental data can be found in the referenced papers. The results are reported in Table 3-5 as average percent error and standard deviation. A bias in favor of the Eaton and the Beggs and Brill methods exists since these methods were developed from the test data.

The same experimental data were used by Hernandez, et al.,[64] to compare the friction factor correlations. The values for experimental friction factor were obtained by solving the friction pressure drop component of the general equation for friction factor. The results are reported in Table 3-5, and the same bias exists in favor of the methods from which the data came. In general, the errors in friction factor prediction are larger than those for liquid holdup prediction.

A data bank assembled at the University of Calgary was used in a liquid holdup comparison study by Mandhane, et al.[65] The data bank contained more than 2600 measured holdup data points. These data were used to test eight published liquid holdup correlations. The data were divided into flow patterns based on the map in Figure 3-41, and several measures of accuracy were calculated for each correlation for each flow pattern. Only the overall results are reported in Table 3-5.

A friction-factor comparison study was reported in a later paper[66] by the same authors. The same data bank was used, but more than 10,000 pressure drop points were used. The data breakdown by flow pattern was also made in this study, but only the overall results are reported in Table 3-5.

The data obtained from an operating pipeline, which had an inside diameter of 6.065 in. and a length of almost 20 mi, were used in a comparison study reported by Gregory, et al.[57] The pipeline was not horizontal, and the sum of all the uphill rises (Σh) was 829 ft. The Hybrid Model discussed earlier and three other methods were used to calculate the total pressure loss in the line taking into account the effects of the hills. The results reported in Table 3-5 show good accuracy for all of the correlations tested. The correlation labeled Dukler-Eaton means that the friction loss was calculated using Dukler's method, but the liquid holdup was evaluated using Eaton's correlation.

Fayed and Otten[81] performed tests at 13 different flow conditions, on both 12 in. and 16 in. offshore pipelines. The lines were 1500 ft. long. Liquid flow rates ranged from 44 to 134 STB/day, and gas rates were between 23 and 110 MMscfd. Average pressure in the lines varied between 635 and 1080 psig. Only two correlations were used to calculate the pressure drops and compare them with the measured values. These were the Beggs and Brill method and the Dukler-Eaton method. As reported in Table 3-5, Beggs and Brill underpredicted the

TABLE 3-5
Pipe Flow Comparison Studies

Comparison Study Author	Correlations Compared	Avg. Percent Error	Standard Deviation
Vohra, et. al.	Beggs and Brill	6.0	17.2
(Holdup)	Dukler, et al.	-25.4	25.0
	Eaton, et al.	-3.8	11.4
	Guzhov[67]	29.1	35.9
	Hughmark	16.4	23.9
	Lockhart, et al.	0.7	25.6
	No-slip	-42.1	23.1
Hernandez, et al.	Beggs and Brill	7.7	86.3
(Friction)	Dukler, et al.	7.7	73.3
	Eaton, et al.	42.6	129.1
	Guzhov	80.0	188.1
Mandhane, et al.	Lockhart, et al.	8.0	11.5
(Holdup)	Hoogendorn (68)	9.4	13.9
	Eaton, et al.	12.5	16.8
	Guzhov	8.5	11.7
	Beggs and Brill	10.4	15.1
	Dukler	9.3	13.8
	Hughmark	7.4	11.6
	Chawla	9.5	13.2
Mandhane, et al.	Lockhart, et al.	4.7	117.5
(Friction)	Chisholm (69)	4.7	120.9
	Baker	-7.1	77.8
	Dukler, et al.	-7.9	57.2
	Chawla	4.8	87.3
	Hoogendorn	-19.5	60.4
	Bertuzzi, et al. (70)	-15.8	56.0
	Chenowith, et al.	-2.1	63.5
	Baroczy (71)	-1.7	59.9
	Beggs and Brill	-4.7	57.3
Gregory, et al.	Hybrid Model	-3.1	11.7
(Total Δp)	Dukler-Eaton	-0.4	17.1
	Beggs and Brill	2.0	17.3
	Flanigan	-12.0	18.7
Fyed, et al.	Dukler-Eaton	16.2	12.8
(Total Δp)	Beggs-Brill	-0.4	7.9
Asheim	MONA	-4.5	10.4
(Total Δp)	Beggs and Brill	1.5	15.2
	Dukler-Eaton	14.7	16.6
Osman, et al.	Beggs and Brill	-2.1	27.3
(Total Δp)	Highmark-Dukler	-17.9	32.6
	Dukler	-7.0	33.0
	Hybrid Model	-17.9	33.0

measured pressure drops slightly on the average while Dukler-Eaton overpredicted the measured data.

Osman and El-Feky[82] obtained field data from eight pipelines in a gas-condensate gathering system to compare the predictive accuracy of four design methods. These methods were those of Beggs and Brill, Gregory (Hybrid Model), Hughmark-Dukler[58], and Dukler. The pipeline inside diameters ranged from 4.026 in. to 10.02

in. A wide range of both gas and liquid rates was included. The pipelines were not horizontal, and some were characterized by essentially downhill flow and others by essentially uphill flow. The statistical results of the study are summarized in Table 3-5.

A conclusion that was made in both of the liquid holdup comparison studies was that no available correlation can accurately predict holdup values less than 0.1.

Another conclusion reached in all the studies was that no one method is best for all ranges of parameters. This conclusion is what prompted the development of the Hybrid Model by Gregory, et al. Again, this emphasizes the need for obtaining as much field data as possible before deciding which methods to use for further design in a particular field or for a particular pipeline.

I. Effects of Variables on Pipeline Performance

As was pointed out in the section on well performance, many variables can change from time to time or from location to location in a producing area. The effects of changes in parameters such as line size, gas/liquid ratio, liquid rate and water cut will be discussed briefly in this section. These changes will be illustrated qualitatively by graphs in some cases. When a graph is used, it will have been calculated using one of the pipe-flow correlations discussed previously. Although the results might differ slightly among the methods, the general trend would be the same.

The general pressure gradient equation applies, but for horizontal conditions the elevation component would be zero. To enhance the discussion of the effects of changing conditions, the equation is rewritten here.

$$\frac{dp}{dL} = \frac{Cf\rho_m(q_L + q_g)^2}{d^5}$$

or

$$\frac{dp}{dL} = \left(\frac{dp}{dL}\right)_f$$

1. Liquid Flow Rate

An increase in q_L will cause an increase in the total fluid velocity, thereby increasing the pressure drop due to friction. Thus, the effect is similar to that which will occur in flowing wells. This effect is illustrated in Figure 3-48 for flow at a fixed GLR in a 6 in. line. A common error committed in developing a field is to connect new wells into existing flowlines that are already overloaded. This, of course, increases the wellhead pressure on all the wells tied into the line.

2. Gas/Liquid Ratio

The effect of a change in GLR on pipeline performance depends on whether the line is essentially horizontal or if hills or low sections exist in the line. If the line is horizontal, the increased gas flow will have the opposite effect on pressure drop compared to what will occur in well flow conditions. For a horizontal line, the friction loss will increase approximately as the square of the flow rate. This effect can be seen in Figure 3-49. This means that if the GLR is increased in a gas lift well to decrease

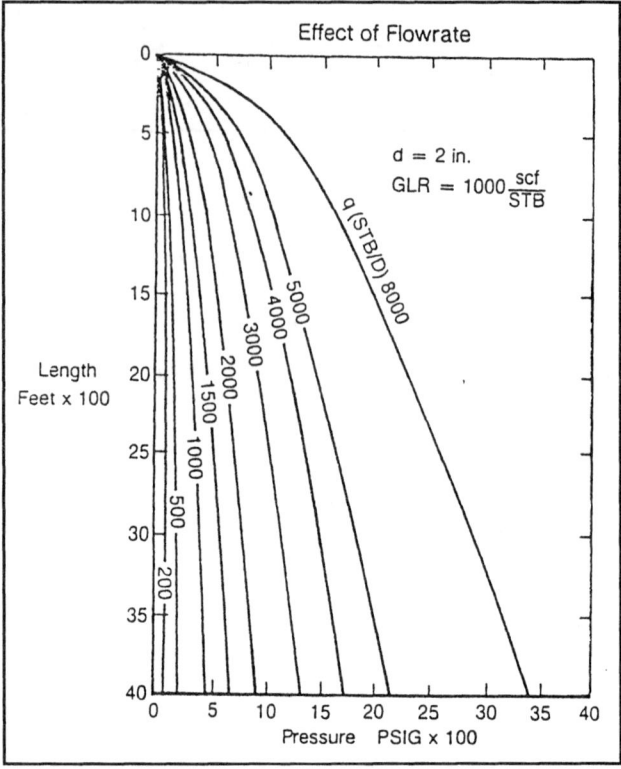

Fig. 3-48. Effect of flow rate.

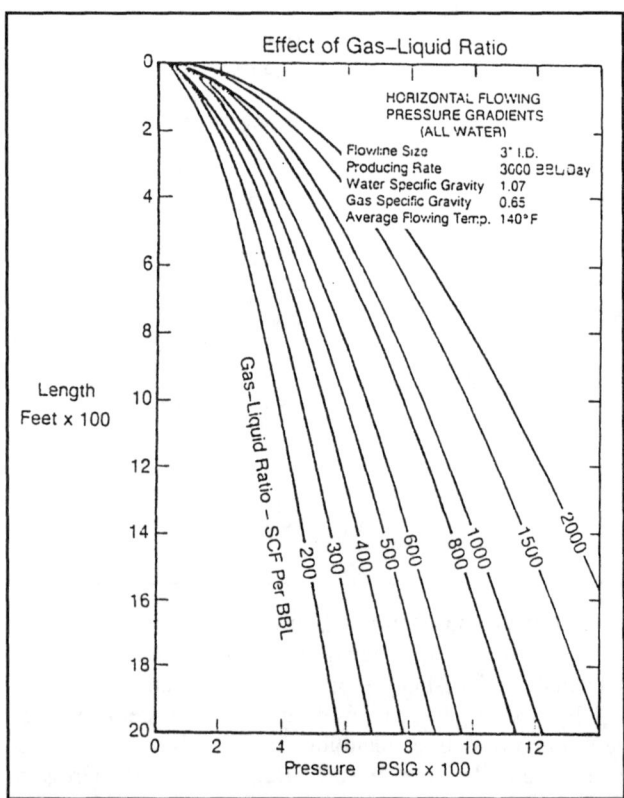

Fig. 3-49. Effect of gas/liquid ratio.

the pressure drop in the tubing, the wellhead pressure will be increased because of the increased Δp in the flowline. However, if the line is not horizontal, an increase in gas velocity will sweep out some of the liquid accumulation in the low sections and may even decrease the overall pressure drop. Therefore, the effect of changing GLR depends on the pipeline profile and would have to be evaluated for each line considered.

3. Water Cut

The effect of changing water cut or WOR is difficult to analyze for pipeline flow. As f_w increases for a fixed gas and liquid rate, the amount of gas in solution R_s will decrease. This happens because the gas is not as soluble in water as it is in oil. This will have the same effect as an increased GLR, as discussed earlier. However, if a very viscous oil is being transported in the line, the effect of water may decrease the pressure drop. If emulsions are formed, the pressure drop may increase several fold.

4. Liquid Viscosity

The effect of oil or liquid viscosity on pressure drop in pipelines during gas/liquid flow cannot be accurately calculated with the present technology. The effective viscosity of the mixture depends on whether dispersions or emulsions are formed and on the degree of "tightness" of the emulsion. Field observations have revealed that the pressure drop does increase with increased viscosity, but all the pipe flow correlations discussed previously merely weight the oil/water viscosity using the water fraction. Also, the viscosity term appears only in the Reynolds number or some other dimensionless group used to find the friction factor.

5. Pipe Diameter

The pipe diameter appears as the fifth power in the denominator of the friction term in Equation 3-108. A decrease in pipe size causes an increase in velocity and, thus, increased frictional pressure drop. This general effect for a horizontal line is illustrated in Figure 3-50. However, if the line is not horizontal, the increased velocity may cause a decrease in liquid holdup or a change in flow pattern, which could decrease the overall pressure drop.

J. Single-phase Gas Flow

The pressure drop occurring in gas lines that are near horizontal can be calculated by means of several equations. All these were derived by integrating the general pressure gradient equation, and the only difference lies in the types of simplifying assumptions made during the integration. Either the Dukler or the Beggs and Brill correlations will degenerate to the single-phase gas case and

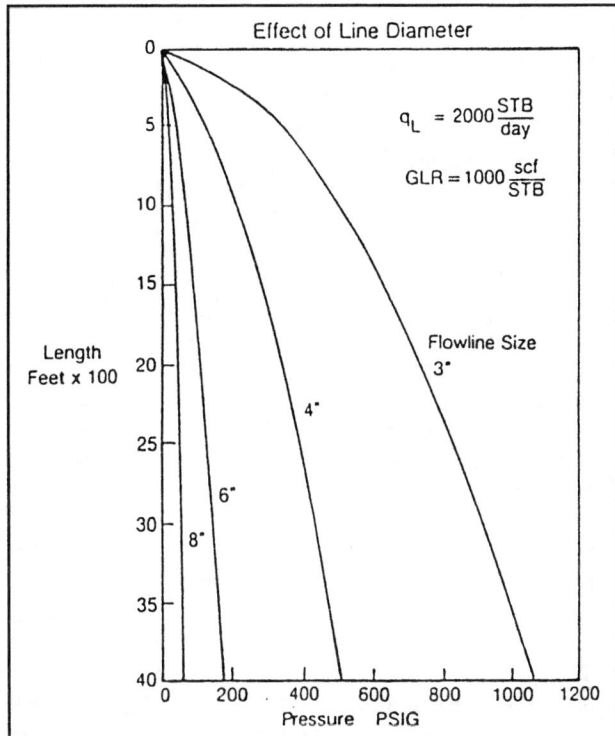

Fig. 3-50. Effect of line diameter.

can, therefore, be used if the line is divided into short increments. Some of the most widely used versions of the gas pipeline equations are presented as follows.

$$q = a_1 E \left(\frac{T_b}{p_b} \right)^{a_2} \left[\frac{p_1^2 - p_2^2}{\overline{T} \overline{Z} L} \right]^{a_3} \left(\frac{1}{\gamma_g} \right)^{a_4} d^{a_5} \qquad (3\text{-}118)$$

where E is the efficiency factor and the values of the a_i constants used in the various equations are tabulated as follows.

Equation	a_1	a_2	a_3	a_4	a_5
Panhandle A	435.87	1.0788	0.5394	0.4604	2.618
Panhandle B	737.0	1.02	0.510	0.490	2.530
IGT	337.9	1.111	0.556	0.4	2.667
Weymouth	433.5	1.0	0.5	0.5	2.667

The following units are to be used in Equation 3-118.

q = ft³/day measured at T_b, p_b
T = °R
p = psia
L = miles
d = inches

The value of the efficiency factor must be estimated from the condition of the pipe inside surface. It usually ranges between about 0.80 and 0.92 for dry gas. An attempt to modify the efficiency factor for the presence of liquids was proposed in Figure 3-46.

K. Use of Prepared Pressure Traverse Curves

The preparation and use of prepared pressure curves for flow in wells was discussed in a previous section. The same methods were used for preparing sets of curves for horizontal two-phase flow, and the same types of errors will result from using the generalized curves.

If several flow conditions are to be analyzed for a particular pipeline, it might be feasible to prepare specific curves for that pipeline, but the pressure profiles would depend on the pipeline profile being considered. Therefore, it would be more feasible to use a computer program to perform the calculations for various conditions. However, for preliminary estimates of the effect of flowlines on the overall well performance, the generalized curves can be used. Since the only curves available were prepared for horizontal flow, if the pipeline under consideration contains hills, some means of accounting for the hydrostatic component must be used. The Flanigan method is suitable for this step.

To the author's knowledge, there are only three sources of curves for horizontal two-phase flow. The Eaton method was used to prepare the curves available from Brown.[47] These curves were calculated for $f_w = 1.0$. The other source, also published by Brown,[23] contains a much larger set of curves that were prepared using the Beggs and Brill method. The curves were calculated for water cuts of zero and 100 percent. The curves included in the Appendix of this book, which were also used in the example problems in this section, were prepared by the author using the Beggs and Brill correlation. The computer program used to prepare the curves can be modified for any possible flow conditions.

Example 3-11:

A well is producing at a rate of 1500 STB/day with a GLR of 600 scf/STB. No water is being produced and the well is located at a distance of 6000 ft from the separator. If the separator pressure is fixed at 120 psig, find the required wellhead pressure for flowline sizes of 2 in. and 3 in.

Solution:

Figure 3-51 represents the conditions for a 2 in. flowline. The same solution procedure as described for the vertical curves will apply for the horizontal curves.

1. Locate the known pressure, $p_{sep} = 120$ psig on the pressure axis and the equivalent length using the 600 GLR line. This length is about 400 ft.
2. Add the line length, 6000 ft, to the outlet equivalent length to obtain 6400 ft. This is the equivalent length corresponding to the upstream conditions.
3. Proceed horizontally and intersect the 600 GLR line. Proceed vertically upward to the pressure axis and

read the upstream pressure as 790 psig.
Following the same procedure on Figure 3-52 for a 3 in. line yields an upstream pressure of about 280 psig.

Example 3-12:

A well that is to produce at a rate of 1500 STB/day with a wellhead pressure of 400 psig is located 5000 ft from the separator. The separator pressure is fixed at 200 psig. The well is producing no water and the GLR is 1000 scf/STB. Find the necessary flowline size required for this well if the terrain between the well and the separator is relatively flat.

Solution:

The solution consists of finding the minimum line size necessary to require an upstream pressure of 400 psig or less. Figures are selected for various line sizes, and the required upstream pressure is found using the same procedure as used in Example 3-11.

Line Size, in.	p_{wh}, psig
2	910
2.5	520
3	360

The results show that a line of size between 2.5 and 3 in. would result in a value of $p_{wh} = 400$ psig. Since only specific pipe sizes are available, the 3-in. line would be selected.

Example 3-13:

The following data apply to a pipeline constructed in a hilly terrain area. Use the horizontal traverse curves and the Flanigan method to find the required upstream pressure p_1.

Length	= 10000 ft	d	= 3 in. I.D.
Σh	= 3000 ft	q_o	= 3000 STB/day
q_w	= 0	GOR	= 800 scf/STB
$\bar{T}$	= 140°F	Oil gravity	= 35° API
γ_g	= 0.65	p_2	= 200 psig

Solution:

The solution will be trial and error or iterative since $\Delta p_h = f(v_{sg})$, $v_{sg} = f(\bar{p})$, and $\bar{p} = 0.5 (p_2 + \Delta p_f + \Delta p_h)$. A suggested solution procedure when using the traverse curves is:

1. Find $p_1{}^*$ using the horizontal curves
2. Estimate $\Delta p_h{}^*$
   For a first guess, use:
   $$\rho_L = 0.433 \, \gamma_L, H_L = 0.2$$
3. Calculate $\bar{p} = 0.5 \, (p_1{}^* + \Delta p_h{}^* + p_2)$
4. Calculate, at $\bar{p}$, $\bar{T}$:
   $$R_s, B_o, \rho_o, Z, B_g, v_{sg}, H_L$$

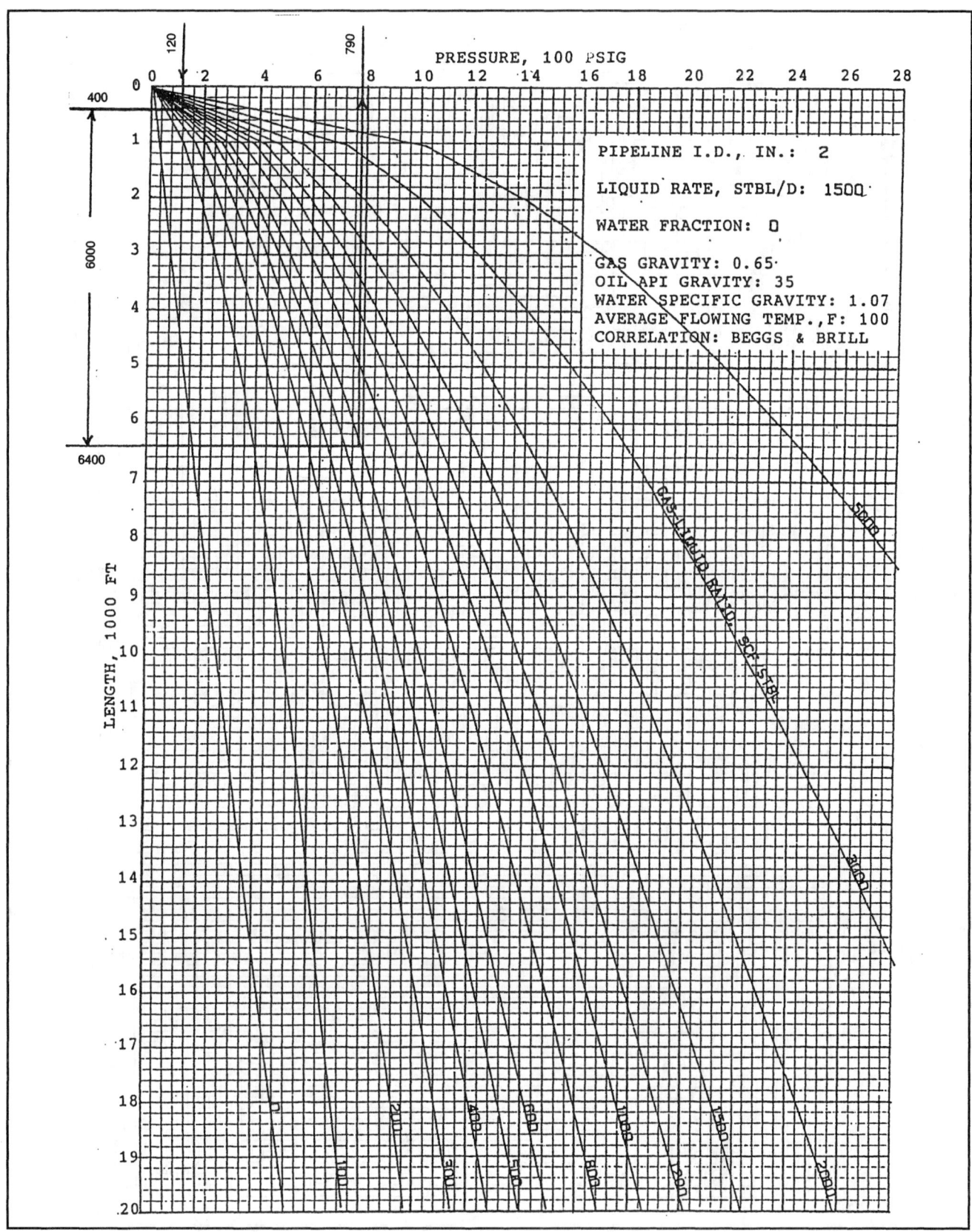

Fig. 3-51. Example 3-11 solution.

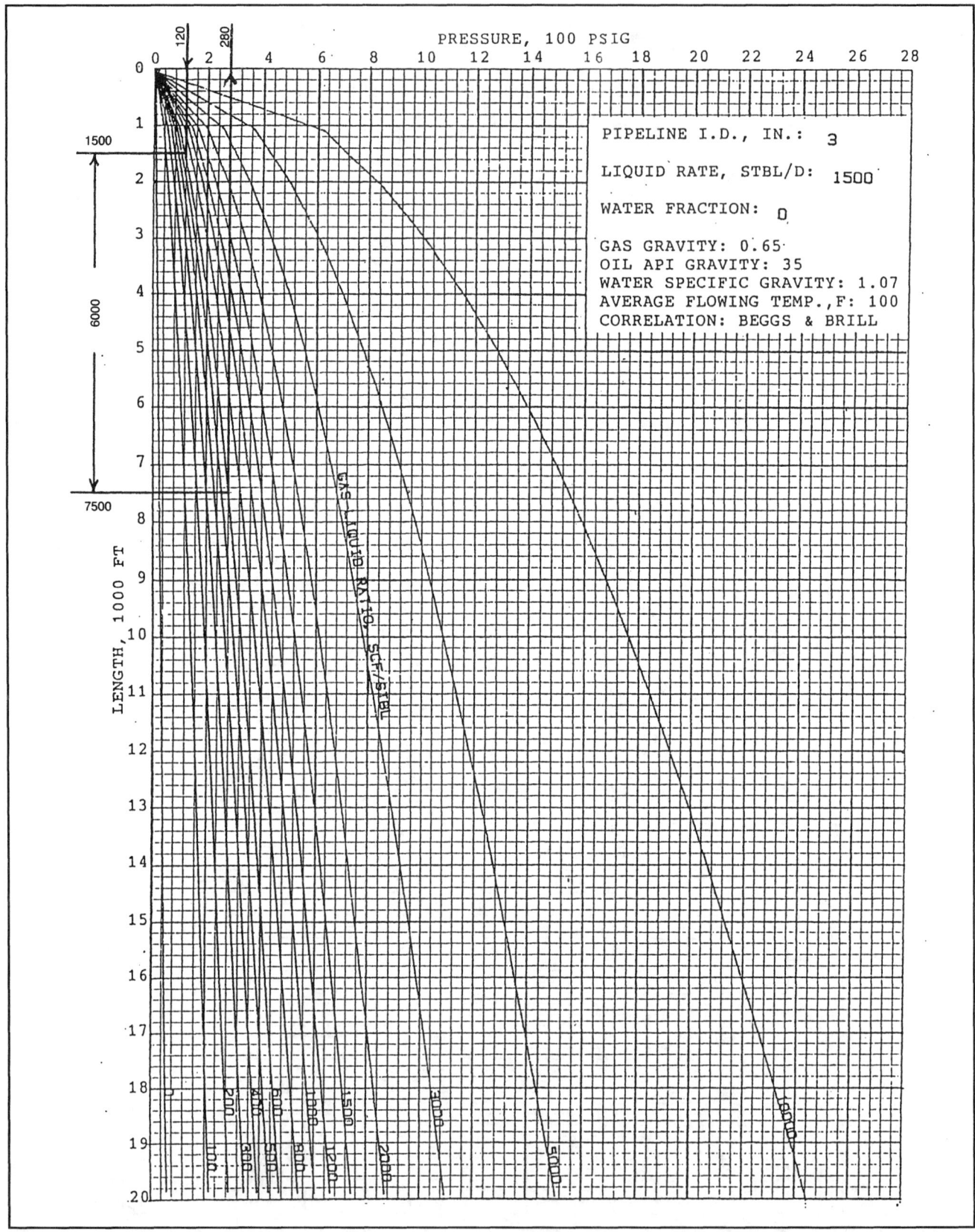

Fig. 3-52. Example 3-11 solution.

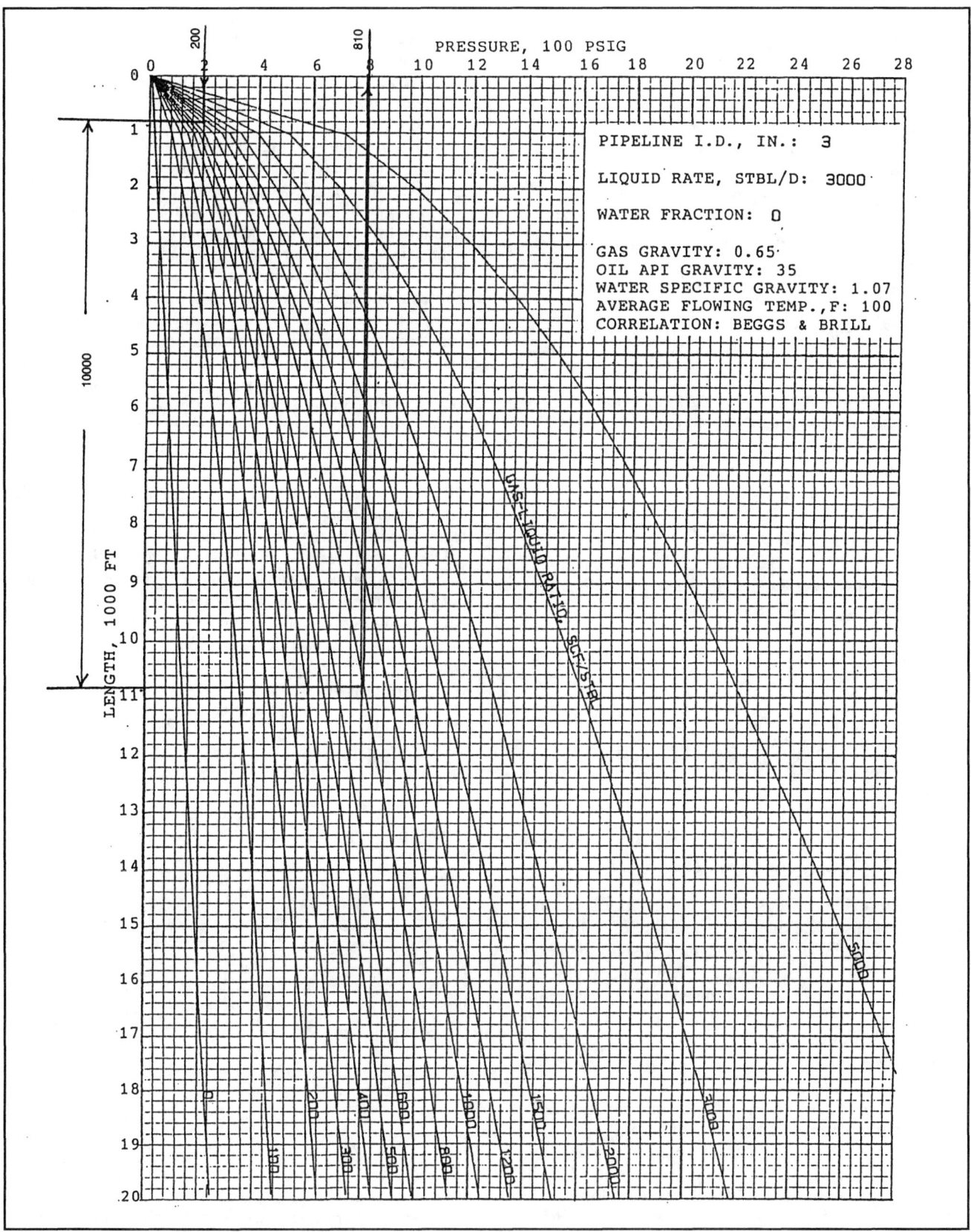

Fig. 3-53. Example 3-13 solution.

5. Calculate $\Delta p_h = \rho_o H_L \Sigma h$
6. Compare Δp_h and $\Delta p_h{}^*$
   If not close, set $\Delta p_h{}^* = \Delta p_h$ and go to Step 3.
7. Calculate $p_1 = p_1{}^* + \Delta p_h$

1. From Figure 3-53, find $p_1{}^* = 810$ psig
2. $\gamma_o = 141.5/(131.5 + 35) = 0.85$
   $\rho_o = 0.433(0.85) = 0.368$ psi/ft
   $\Delta p_h{}^* = \rho_o H_L \Sigma h = 0.368(0.2)(3000) = 220$ psi
3. $\bar{p} = 0.5(810 + 220 + 200) = 615$ psig = 629.7 psia
4. $R_s = 96$ scf/STB (Eq. 3-76)
   $B_o = 1.093$ bbl/STB (Eq. 3-79)
   $\rho_o = 350(.85) + 0.0764(.65)(96)/5.615(1.093)$
   (Eq. 3-60)
   $\rho_o = 49.3$ lb/ft^3 = 0.342 psi/ft
   $Z = 0.930$ (Eq. 3-73)
   $B_g = 0.0257$ ft^3/scf (Eq. 3-65)
   $v_{sg} = q_g/A = q_o(R-R_s)B_g/86400A$ (Eq. 3-64)
   $v_{sg} = 3000(800 - R_s)B_g/86400(0.0491) =$
   $0.7074(800 R_s)B_g$
   $v_{sg} = 0.7074(704)(.0257) = 12.8$ ft/sec
   $H_L = 1/(1 + 0.3264(12.8)^{1.006})$ (Eq. 3-117)
   $H_L = 0.19$
5. $\Delta p_h = 0.342(0.19)(3000) = 195$ psi

   This is not close enough to the initial estimate, $\Delta p_h{}^* = 220$. Set $\Delta p_h{}^* = 195$ psi.

3. $\bar{p} = 0.5(810 + 195 + 200) = 602.5$ psig = 617.2 psia
4. $R_s = 93$, $B_o = 1.092$, $\rho_o = 0.342$, $Z = 0.930$, $B_g = 0.0262$,
   $v_{sg} = 13.1$, $H_L = 0.187$
5. $\Delta p_h = 0.342(.187)(3000) = 192$ psi
6. Since two successive values of Δp_h agree, the solution has converged.
7. $p_1 = 810 + 192 = 1002$ psig

L. Parallel or Looped Pipelines

All the previous discussion regarding pressure drop calculations in surface lines has been with respect to flow through a single pipe of constant inside diameter. However, when the flow capacity of an existing pipeline or gathering system becomes too small, it is common practice to construct another line alongside or parallel to existing lines to increase the flow capacity or reduce the pressure drop. This is sometimes called "looping" the line.

Calculation of the pressure drop occurring in parallel lines is fairly simple in the case of single-phase flow because the flow coming into the branching point will split or divide such that the pressure drop is equal in the two pipes. The two pipes of diameters d_1 and d_2 can be replaced by a single pipe having a diameter of $d_e = (d_1^{2.5} + d_2^{2.5})^{0.4}$. This concept includes the assumption that

there is no change in either friction factor or fluid density between the branches. This concept is inaccurate for two-phase flow because the liquid phase does not always split between the branches in the same ratio as the gas. It has been observed that in many cases all the liquid will go into only one of the branches, while the gas will split in such a manner as to create equal pressure drops in each branch. As has already been discussed, to calculate this pressure drop, the flow rates and gas-liquid ratios existing in a pipe must be known. Therefore, some method to estimate the path of the liquid must be available.

The flow-splitting problem was first discussed in detail by Orange.[83] He observed that in a complex gas distribution system, any condensate that formed in the system eventually reached a single location, even though the gas may have encountered numerous branches upstream of that location. After performing comprehensive studies of the phenomenon, he suggested that if the branch took off less than 20 percent of the total flow, none of the liquid would divert into the lateral. If more than 35 percent of the flow was diverted into the lateral, all the liquid would divert into the lateral. Later studies by Hong[34] and Bergman, et al.[85] revealed that under other flow conditions, particularly high rates at which mist flow might exist, the liquid may also split at the branch. They suggested procedures to estimate the degree of liquid splitting as a function of gas splitting and gas Reynolds number in the main line. Their relationship is shown graphically in Figure 3-54. The Reynolds number for the gas in the main line can be calculated from:
where

$$N_{Re} = \frac{20011 q_{sc}\gamma_g}{\mu d} \qquad (3\text{-}118F)$$

γ_g = gas specific gravity,
q_{sc} = gas flow rate, MMscfd,
d = pipe inside diameter, inches, and
μ = gas viscosity, cp

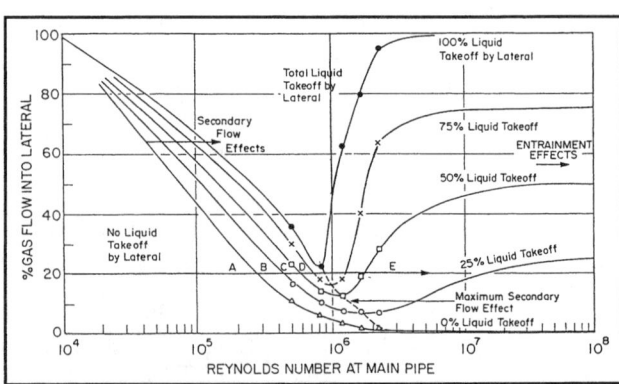

Fig. 3-54. *Generalized chart for route selectively based on Oranje's data.*[85]

The results obtained from this figure have not been extensively tested and may be too simple to actually describe the phenomenon. However, unless more accurate methods are available it can be used to determine the gas and liquid rates in each branch or lateral for nodal analysis purposes.

The following iterative procedure is suggested for calculating the pressure drop occurring across a looped section of a pipeline. It is assumed that the pressure is known at one end of the section where the branches are joined and that this pressure is equal at that point in both branches. The diameters and lengths of the two lines are known. The total gas and liquid rates are also known.

1. Assume a value for the fraction of the gas going into the lateral. This fixes the gas rate in each branch.

2. Use Figure 3-54 to estimate the liquid rate in each branch. This method may not always give the correct liquid split, but it will be close enough for a nodal analysis.

3. Using the gas and liquid rates determined in Steps 2 and 3, calculate the pressure drop in each branch using the appropriate methods.

4. Compare the pressure drops calculated in Step 3. If they are equal, the assumed rates are correct and the pressure drop is correct. If they are not equal, adjust the gas split and go to Step 2. Repeat until convergence on pressure drop is attained.

A more complex analysis of this phenomenon is beyond the scope of this book. More detailed methods for estimating the liquid split can be found in Reference 9.

VI. PRESSURE DROP THROUGH RESTRICTIONS

Although the principal pressure losses in the well system occur in the reservoir, the tubing, and the flowline, the pressure loss in restrictions can be substantial in some wells. The main types of restrictions are:

1. Subsurface safety valves (SSSV's)

2. Surface or bottomhole chokes

3. Valves and fittings

The losses occurring in SSSV's and pipe fittings cannot be avoided, but the pressure drop across a surface choke can be eliminated to obtain the maximum producing capacity from a well.

The flow through a restriction may be either critical (sonic) or subcritical (subsonic). If flow is critical, a pressure disturbance downstream of the restriction will have no effect on either the flow rate through the restriction or the upstream pressure. Since one of the main purposes of a choke is to control flow rate, it will usually be sized so

that critical flow will exist. A rule-of-thumb for distinguishing between critical and subcritical flow states that if the ratio of downstream pressure to upstream pressure is less than or equal to 0.5, then the flow will be critical. This is a closer approximation for single-phase gas than for two-phase flow. Some engineers use either 0.6 or 0.7 as the critical pressure ratio in two-phase flow, although research performed at Tulsa University[87] has shown that in some cases, the ratio must be as low as 0.3 before flow becomes critical.

The purpose of a SSSV is not to control the flow rate, but to shut the well in when wellhead pressure becomes too low. Therefore, they are usually sized for minimum pressure drop and will be operating in subcritical flow.

Procedures for calculating the pressure losses in these three types of restrictions will be presented in this section.

A. Surface Chokes

Equations for estimating the relationship among pressure, flow rate and choke size for both subcritical and critical flow through chokes will be presented for both gas flow and two-phase flow.

1. Gas Flow

A general equation for flow through restrictions can be derived by combining the Bernoulli equation with an equation of state. The irreversible losses are accounted for by use of a discharge coefficient, which depends on the type of restriction. The following equation applies for gas flow in both the critical and subcritical regimes.

For critical flow, the pressure ratio $y = p_2/p_1$ is replaced by the critical pressure ratio y_c.

$$q_{sc} = \frac{C_n(p_1)(d)^2}{\sqrt{\gamma_g(T_1)Z_1}} \sqrt{\left(\frac{k}{k-1}\right)\left(y^{2/k} - y^{(k+1)/k}\right)} \qquad (3\text{-}119)$$

$$C_n = \frac{C_s(C_d)T_{sc}}{p_{sc}}$$

where

q_{sc} = volumetric gas flow rate
C_n = coefficient based on system of units, discharge coefficient and standard conditions
d = I.D. of bore open to gas flow
γ_g = gas specific gravity (air=1.0), dimensionless
k = ratio of specific heats=C_p/C_v, dimensionless
p_1 = upstream pressure, absolute units
p_2 = downstream pressure, absolute units
T_1 = upstream temperature, absolute units
Z_1 = compressibility factor at p_1 and T_1, dimensionless
C_s = coefficient based on system of units
C_d = discharge coefficient (empirical), dimensionless

T_{sc} = standard absolute temperature base, absolute units

p_{sc} = standard absolute pressure base, absolute units

y_c = critical pressure ratio, dimensionless

The pressure ratio at which flow becomes critical depends on the k value for the flowing gas and is given by:

$$y_c = \left(\frac{2}{k+1}\right)^{k/(k-1)} \tag{3-120}$$

TABLE 3-6
Coefficients and Units for Eq. 3-119

Symbol	English System	SI Metric System
q_{sc}	Mscf/d	m³/d
d	in.	mm
p	psia	kPa
T	°R	°K
C_s	27.611	1.6259

The following equation can be used to estimate the relationship among flow rate, upstream pressure and choke size for short restrictions with slightly rounded openings operating in critical flow. The units are obtained from Table 3-6.

$$q_{sc} = \frac{0.487 C_n d^2 p_1}{(T\gamma_g)^{0.5}} \tag{3-121}$$

The value to be used for the discharge coefficient C_d in Equations 3-119 and 3-121 depends on the shape of the opening to the restriction and the length of the restriction as well as the Reynolds number. A value of $C_d = 0.82$ is recommended if no information is available concerning this data.

2. Two-Phase Flow

The following equations may be used to determine the relationship among p_1, q_L and d for gas/liquid flow in the critical regime. These are empirical equations and the coefficient and exponents may vary from field to field or well to well.

$$p_1 = \frac{b q_L R^c}{d^a} \tag{3-122}$$

where

p_1 = upstream pressure, psia
q_L = liquid flow rate, STB/day
R = gas/liquid ratio, scf/STB
d = choke diameter in inches

Values of a, b, and c proposed by different investigators are given in Table 3-7.

TABLE 3-7
Choke Constants

Investigator	a	b	c
Ros (72)	2.00	4.25 x 10⁻³	0.500
Gilbert (73)	1.89	3.86 x 10⁻³	0.546
Baxendell (74)	1.93	3.12 x 10⁻³	0.546
Achong (75)	1.88	1.54 x 10⁻³	0.650

Example 3-14:

Using both the Ros and Gilbert equations, determine the choke size required to obtain a liquid rate of 400 STB/day if wellhead pressure is 900 psia and $R = 600$ scf/STB.

Solution:

Solving Equation 3-122 for d:

$$d = \left[\frac{b q_L R^c}{p_1}\right]^{1/a}$$

Ros:

$$d = \left[\frac{4.25 \times 10^{-3}(400)(600)^{0.5}}{900}\right]^{1/2} = 0.215 \text{ in.}$$

Gilbert:

$$d = \left[\frac{3.86 \times 10^{-3}(400)(600)^{0.546}}{900}\right]^{1/1.89} = 0.218 \text{ in.}$$

Determination of the boundary between critical and subcritical flow for the two-phase case is more complicated than that for single-phase flow. The sonic velocity in a two-phase mixture depends on both the gas and liquid properties. Sachdeva, et al.,[86] presented equations for determining the critical ratio of downstream to upstream pressure at the boundary, and proposed a method for calculating the flow rate through a choke for various pressure conditions. The critical pressure ratio $y_c = p_2/p_1$ is calculated iteratively from:

$$y_c = (N/D)^{k/(k-1)} \tag{3-123}$$

where

$$N = \frac{k}{k-1} + \frac{(1-X_1)\rho_{g1}(1-y)}{X_1\rho_L}$$

$$D = \frac{k}{k-1} + \frac{n}{2} + \frac{n(1-X_1)\rho_{g2}}{X_1\rho_L} + \frac{n}{2}\left[\frac{(1-X_1)\rho_{g2}}{X_1\rho_L}\right]^2$$

X_1 = mass fraction of gas at upstream conditions (quality),
ρ_{g1} = gas density at upstream conditions,
ρ_{g2} = gas density at downstream conditions
ρ_L = liquid density at upstream conditions,
k = ratio of specific heats for the gas, C_p/C_v

$$n = 1 + \frac{X_1(C_p - C_v)}{X_1 C_v + (1 - X_1) C_L} \qquad (3\text{-}124)$$

where

C_L = specific heat of the liquid

Equation 3-123 is dimensionless, and any consistent set of units may be used. It is solved by assuming a value for y and then calculating y_c. Each calculated value of y_c is used for the next estimated y until convergence is reached. A good first guess is $y = 0.5$. The quality is the ratio of the gas mass flow rate to the total mass flow rate. Using field units, it may be calculated from

$$X = \frac{0.0764\gamma_g(R - f_o R_s)}{0.0764\gamma_g(R - f_o R_s) + 5.615(f_o B_o \rho_o + f_w B_w \rho_w)}$$

$$(3\text{-}125)$$

where

γ_g = gas specific gravity,
R = producing GLR, scf/STBL,
f_o = fraction of oil flowing, $q_o/(q_o + q_w)$,
f_w = fraction of water flowing, $(1 - f_o)$,
R_s = solution gas-oil ratio at p_1, T_1, scf/STBO,
ρ_o = oil density at p_1, T_1, lbm/ft^3,
ρ_w = water density at p_1, T_1, lbm/ft^3,
B_o = oil formation volume factor at p_1, T_1,
B_w = water formation volume factor at p_1, T_1,

Sachdeva, et al. also presented an equation to calculate the flow rate through a choke which can be used for both critical and subcritical flow. For field units:

$$q_L = \frac{0.525 C_d d^2}{C_{M2}} \left\{ P_1 \rho_{m2}^2 \left[\frac{(1 - X_1)(1 - y)}{\rho_{L1}} \right. \right.$$

$$\left. \left. + \frac{X_1 k(1 - y^{k-1/k})}{\rho_{g1}(k-1)} \right] \right\}^{0.5}$$

$$(3\text{-}126)$$

where q_L = liquid flow rate, STBL/day,

$$\rho_{m2} = \left[\frac{X_1}{\rho_{g1} y^{1/k}} + \frac{(1-X)_1}{\rho_{L1}} \right]^{-1} \qquad (3\text{-}127)$$

$$c_{m2} = 8.84 \times 10^{-7} \gamma_g(R - f_o R_s)$$

$$+ 6.5 \times 10^{-5}(f_o \rho_o B_o + f_w \rho_w B_w) \qquad (3\text{-}128)$$

where

d = choke inside diameter, inches
p_2 = upstream pressure, psia,
ρ_L = liquid density, lbm/ft^3,
ρ_{g1} = gas density at p_1, T_1, lbm/ft^3, and
X_1 = quality at p_1, T_1 (Equation 3-125)

The fluid properties used to calculate C_{m2} are evaluat-

ed at downstream conditions. If flow is subcritical, that is $y > y_c$, use the actual downstream pressure, p_2. If flow is critical, use $y = y_c$ and $p_2 = y_c p_1$. It was suggested that if an elbow is immediately upstream of the choke the value of C_D is 0.75. If no flow-perturbing effects are upstream, use $C_D = 0.85$.

Example 3-15:

A wellhead choke is installed in a well that is producing oil and gas. Upstream pressure is 1000 psia and downstream pressure is 600 psia. Estimate the oil production rate through the choke under these conditions. Other data are:

GLR	= GOR = 1165 scf/STBO	γ_g	= 0.70
γ_o	= 0.825 = 40° API	d	= 24/64 = 0.38 in.
T_1	= 100°F = 560°R	f_w	= 0
T_2	= 90° F = 550° R	C_p	= 0.537 Btu/lbm-°F
C_v	= 0.414 Btu/lbm-°F	C_L	= 0.55 Btu/lbm-°F
k	= 1.3		

	Upstream Conditions	Downstream Conditions
R_s, scf/STBO	250	140
B_o	1.14	1.08
ρ_o, 1 bm/ft^3	47.3	48.9
ρ_g, 1 bm/ft^3	4.05	2.32
Z	0.834	0.891

Solution:

Calculate critical pressure ratio, y_c:

$$X_1 = $$

$$\frac{0.0764(0.7)(1165 - 250)}{0.0764(0.7)(1165 - 250) + 5.615(1.14)(47.3)} = 0.139$$

$$n = 1 + \frac{0.139(0.537 - 0.414)}{0.139(0.414) + 0.861(0.55)} = 1.032$$

Estimate $y_c = 0.5$:

$$N = \frac{1.3}{1.3 - 1} + \frac{0.861(4.05)(1 - 0.5)}{0.139(47.3)} = 4.599$$

$$D = \frac{1.3}{0.3} + \frac{1.032}{2} + \frac{1.032(0.861)(2.37)}{0.139(47.3)}$$

$$+ \frac{1.032}{2}\left[\frac{0.861(2.37)}{0.139(47.3)} \right]^2$$

$$D = 5.219$$

$$y_c = (4.599/5.219)^{1.3/3} = 0.578$$

Estimated y_c	Calculated y_c
0.5	0.578
0.578	0.556
0.556	0.562
0.562	0.560
0.560	0.560

Since the value of y for the given conditions is $y = p_2/p_1 = 600/1000 = 0.6$, flow is subcritical ($y > y_c$).

Calculate q_L:

$$C_{m2} = 8.84 \times 10^{-7}(0.7)(1165-140)$$
$$+6.5 \times 10^{-5}(48.9)(1.08)$$

$$C_{m2} = 0.0041$$

$$\rho_{m2} = \left[\frac{0.139}{4.05(0.6)^{1/1.3}} + \frac{0.861}{47.3} \right]^{-1}$$

$$= 14.48 \text{ .bm/ft}^2$$

$$q_L = \frac{0.525(0.75)(0.38)^2}{0.0041} \left\{ 1000(14.48)^2 \right.$$
$$\left. \bullet \left[\frac{0.861(0.4)}{47.3} + \frac{0.139(1.3)(0.111)}{4.05(0.3)} \right] \right\}^{0.5}$$

$$q_L = q_o = 979 \text{ STB/day}$$

An equation for calculating flow rates for subcritical flow through multiple orifice valves (MOV) or chokes was presented by Surbey, et al.[87] Most of these valves consist of a stationary disk with two holes and a movable disk with two holes. The size of the opening can be changed by rotating the movable disk. The choke used in the Surbey et al. study was manufactured by the Willis Company. A schematic of the choke is shown in Figures 3-55 and 3-56. The study resulted in a method to modify the discharge coefficient for single-phase liquid flow so that it will apply for two-phase flow through this particular type and size (2 in.) of choke. The equation is:

$$q_L = 34.28 C_{vL} \left(\frac{p_1 - p_2}{\gamma_L} \right)^{0.5} \quad \text{(3-129)}$$

where

q_L	=	flow rate, STBL/day
p_1	=	upstream pressure, psia
p_2	=	downstream pressure, psia
γ_L	=	liquid specific gravity, and
C_{vL}	=	$C_{vtp}/2F_c$
F_c	=	$[\sin(A_1 R^{A_2})]\phi^{A_3} p_1^{A_4/R}$

where

C_{vtp}	=	two-phase discharge coefficient from Figure 3-57,
R	=	gas-liquid ratio, scf/STB,
ϕ	=	angle of choke opening, degrees
A_i	=	values are from Table 3-8

TABLE 3-8
Choke Constants

Constant	Value
A_1	91.9039
A_2	-0.1458
A_3	0.2419
A_4	-20.2600

In some cases, it may be necessary to estimate the pressure drop through a choke in which a single-phase liquid is flowing. This will almost always be subcritical flow since the velocity of sound in a liquid is very large.

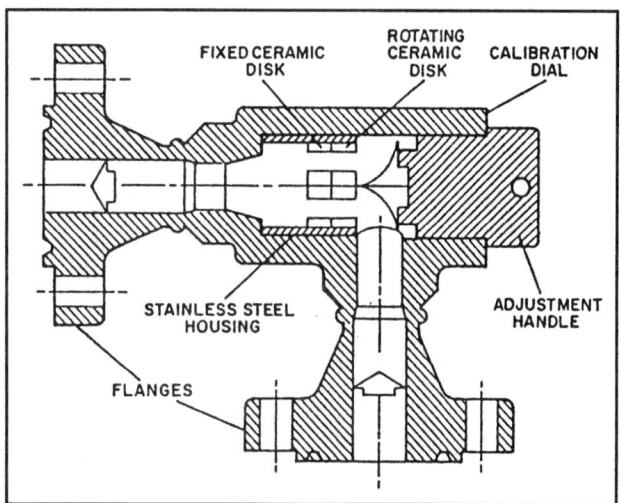

Fig. 3-55. MOV wellhead choke design (after Willis[88]).

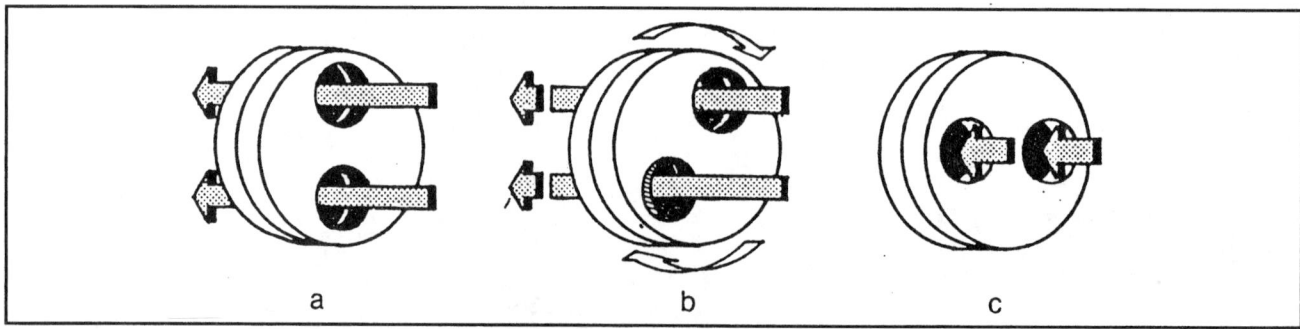

Fig. 3-56. Ceramic choke disk operation (after Willis[88]).

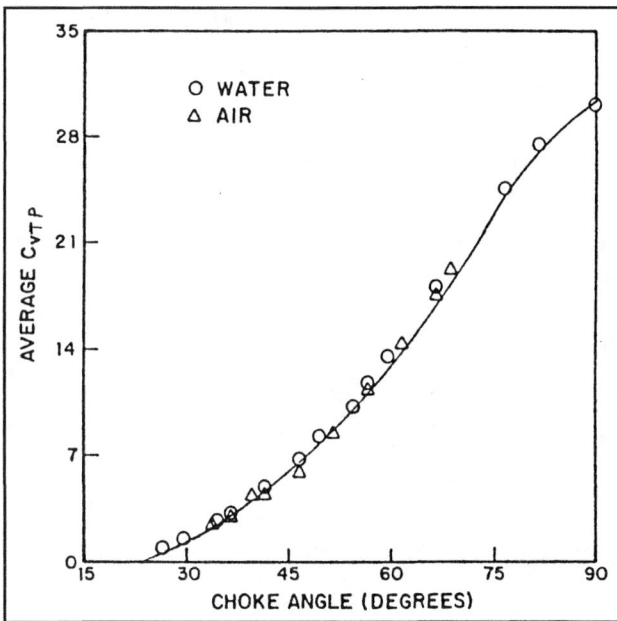

Fig. 3-57. Average C_v vs. choke setting.[87]

The following equation may be used:

$$q_L = 1022.7 C_d d^2 \left(\frac{p_1 - p_2}{\gamma_L} \right)^{0.5} \qquad (3\text{-}131)$$

where

q_L = flow rate in STBL/day,
d = choke diameter, inches,
p = psi, and
γ_L = liquid specific gravity.

Unless C_d is known, a value of 0.85 may be used.

B. Subsurface Safety Valves (SSSV's):

As was stated earlier, the flow through a SSSV will be subcritical and the unknown of interest is usually the pressure drop caused by a SSSV of a particular size. The solution is usually iterative because most of the equations require evaluation of the fluid properties, either at upstream pressure or at average pressure.

1. Gas Flow

An equation published by the API[65] can be used to calculate the pressure drop across a SSSV operating in subcritical flow. The equation is:

$$p_1 - p_2 = \frac{1.048 \times 10^{-6} \gamma_g Z_1 T_1 q_{sc}^2 (1 - \beta^4)}{p_1 d^4 C_d^2 Y^2} \qquad (3\text{-}132)$$

where

p_1 = upstream pressure, psia
p_2 = downstream pressure, psia

γ_g = gas gravity (air = 1)
Z_1 = gas compressibility at p_1, T_1
T_1 = upstream temperature, °R
q_{sc} = gas flow rate, Mscfd
β = beta ratio = d/D
d = bean diameter, in.
D = pipe inside diameter, in.
C_d = discharge coefficient (API suggests 0.9)
Y = expansion factor, dimensionless

The expansion factor determination is iterative and may be calculated from Equation 3-133. Its value ranges between 0.67 and 1.0. For quick estimates, a default value of 0.85 is often used.

$$Y = 1 - [0.41 + 0.35\beta^4] \left(\frac{p_1 - p_2}{kp_1} \right) \qquad (3\text{-}133)$$

where k is the ratio of specific heats of the gas.

2. Two-Phase Flow

A research project was sponsored by the API in 1978 at the University of Tulsa[76] that was designed to improve the equations for sizing SSSV's operating in two-phase subcritical flow. Several of the commercially available SSSV's were used in the experimental phase of the research, and specific equations for discharge coefficient for each valve tested were presented. However, it has been found that in practice a single equation for discharge coefficient will give reasonable results for any type of SSSV. Therefore, only this equation will be presented here. For a more detailed treatment of the problem, reference may be made to a report published by Beggs, et al., in 1980.[76]

The equation for pressure drop is:

$$p_1 - p_2 = \frac{1.078 \times 10^{-4} \rho_n v_m^2}{C_D} \qquad (3\text{-}134)$$

where

ρ_n = no-slip density, lbm/ft³ (Eq. 3-22)
v_m = mixture velocity through the choke, ft/sec, (Eq. 3-28)
p_1 = upstream pressure, psia
p_2 = downstream pressure, psia
C_D = discharge coefficient

$$C_D = C_1 + C_2 N_v + C_3 \beta + C_4 \beta^2 \qquad (3\text{-}135)$$

where

C_1 = 0.233
C_2 = 8.4 X 10⁻⁴
C_3 = 6.672
C_4 = −11.661
N_v = q_g/q_L = $(1 - \lambda_L)/\lambda_L$
λ_L = $q_L/(q_L + q_g)$

$$\beta = d/D$$
d = choke diameter
D = tubing inside diameter

In the previous equations, all the fluid properties necessary for calculating the density and velocities are evaluated at upstream conditions of pressure and temperature. Solving for p_1 from a known p_2 value is, therefore, iterative.

Example 3-16:

A well that is producing through 2 3/8 tubing is equipped with a 0.908 in I.D. SSSV. Pressure and temperature upstream of the SSSV were found to be 615 psia and 140°F respectively. Using the following data calculate the pressure downstream of the SSSV.

q_o = 800 STB/day
q_w = 0
R = 800 scf/STB
Oil gravity = 35°API
γ_g = 0.65

Solution:

The following PVT properties were calculated at p = 615 psia and T = 140°F = 600°R:

R_s = 96 scf/STB

Z = 0.93

B_o = 1.093 bbl/STB

B_g = 0.0257 ft³/scf

$$q'_g = \frac{q_o(R-R_s)B_g}{86400} = \frac{800(800-96)(0.0257)}{86400}$$

q'_g = 0.168 ft³/sec

q'_o = 6.5 x 10⁻⁵ q_oB_o = 6.5 x 10⁻⁵ (800)(1.093)

q'_o = 0.057 ft³/sec

$$\lambda_L = \frac{q'_o}{q'_o + q'_g} = \frac{0.057}{0.057 + 0.168} = 0.253$$

$$\rho_o = \frac{350(.85) + .0764(.65)(96)}{5.615(1.093)} = 49.3 \text{ lbm/ft}^3$$

$$\rho_g = \frac{2.7\gamma_g p}{ZT} = \frac{2.7(.65)(615)}{0.93(600)} = 1.934 \text{ lbm/ft}^3$$

$\rho_n = \rho_o\lambda_L + \rho_g(1-\lambda_L) = 49.3(.253) + 1.934(1 - .253)$

ρ_n = 13.92 lbm/ft³

$N_v = q'_g/q'_o$ = 0.168/0.057 = 2.947

β = 0.908/1.995 = 0.455

C_D = −0.233 + 8.4 x 10⁻⁴ (2.947) + 6.672(0.455)− 11.661(.455)²

C_D = 0.391

$$A = \frac{\pi}{4}d^2 = 0.7854(0.908/12)^2 = 4.497 \times 10^{-3} \text{ ft}^2$$

$$v_m = \frac{q'_o + q'_g}{A} = \frac{0.057 + 0.168}{4\,497 \times 10^{-3}} = 50 \text{ ft/sec}$$

$$p_2 = p_1 - \frac{1.078 \times 10 - 4\rho_n v_m^2}{C_D}$$

$$= 615 - \frac{1.078 \times 10^{-4}(13.92)(50)^2}{0.391}$$

p_2 = 615 − 9.6 = 605.4 psia

C. Valves and Pipe Fittings

The pressure losses occurring through various types of valves and fittings can be approximated by the equivalent length concept. This involves replacing each fitting by an equivalent length of pipe that would produce the same pressure drop as the fitting. This can be expressed in equation form as:

$$\Delta p_f = \frac{fL}{d}\frac{\rho v^2}{2g_c} = K\frac{\rho v^2}{2g_c} \tag{3-136}$$

or

$$\frac{fL}{d} = K$$

where

d = diameter of pipe
f = friction factor for pipe flow
L = length of pipe
K = resistance coefficient depending on the type or size of fitting.

Solving for the length gives:

$$L_e = \frac{Kd}{f} \tag{3-137}$$

An equivalent length, L_e can be calculated for each fitting by using the friction factor calculated for flow in the pipe. All the L_e values can then be added to the actual pipe length for the pressure drop calculation.

Values of the resistance coefficient have been determined for single-phase flow, and it has been found that these values also apply for two-phase flow. The friction factor for two-phase flow is, of course, usually larger than for single-phase flow. The following average values for K can be used to obtain satisfactory results for two-phase flow, although in many cases the relatively small pressure drop in the fittings will be ignored.

Fitting Type	K
Gate Valve	0.15
Elbows	0.2-0.3
Globe Valve	3.0-5.0
Check Valve	6.0-8.0

VII. EROSIONAL VELOCITY

When fluid flows through a pipe at high velocities, it has been found that erosion of the pipe can occur. This is especially true for high capacity gas flow in which the in-situ velocity may exceed 60 to 70 ft/sec. Erosion is not as much of a problem in oil wells, although some high gas-liquid ratio wells may be subject to erosion.

The velocity at which erosion begins to occur cannot be determined exactly, and if some solid particles, such as sand, are in the fluid, erosion may occur at relatively low velocities. The velocity at which erosion may occur has been related to the density of the fluid by the following equation:

$$v_e = \frac{C}{\rho^{0.5}} \tag{3-138}$$

where

v_e = erosion velocity, ft/sec

ρ = fluid density, lbm/ft^3 = $\rho_L \lambda_L + \rho_g (1-\lambda_L)$

C ranges between 75 and 150

A good average value for C has been found to be about 100. If C is set equal to 100 and the gas equation of state is used to express density, Equation 3-138 becomes

$$v_e = \frac{100}{\left[\dfrac{29\,p\gamma_g}{ZRT}\right]^{0.5}}$$

where p, T and Z are the conditions at which the velocity is to be determined.

The equation may be expressed in terms of gas flow rate at standard conditions by:

$$q_e = 1.86 \times 10^5\, A \left(\frac{p}{ZT\gamma_g}\right)^{0.5} \tag{3-139}$$

where

q_e = erosional flow rate, Mscfd

A = area of the pipe, ft^2

p = lowest pressure in the pipe, psia

T = temperature at point where p is determined, °R

Z = gas compressibility factor at p, T

γ_g = gas gravity

Example 3-17:

A gas well is producing through 2 7/8 in. tubing at a wellhead pressure of 800 psia. The wellhead temperature is 140°F and gas gravity is 0.65. Determine the maximum rate at which this well can produce without exceeding the erosional velocity.

Solution:

$A = \dfrac{\pi}{4} d^2 = 0.7854(2.441/12)^2 = 0.032\,\text{ft}^2$

At $p = 800$, $T = 140°F$ the value of Z is 0.91.

$q_e = 1.86 \times 10^5\,(0.032)(800/(.91)(600)(.65))^{0.5}$

$q_e = 8,936$ Mscfd = 8.9 MMscfd

This corresponds to an in-situ velocity of 62.3 ft/sec.

VIII. REFERENCES

1. Drew, T. B., Koo. E. C. and McAdams, W. H.: *Trans. Am. Inst. Chem. Engrs.*, 28, l930.
2. Nikuradse, J.: *Forschungsheft*, p. 301, 1933.
3. Colebrook C. F.: *J. Inst. Civil Engrs.* (London), 1938.
4. Jain, A. K.: "Accurate Explicit Equation for Friction Factor," J. Hydl. Div. ASCE, NoHY5, May, 1976.
5. Moody, L. F.: "Friction Factors for Pipe Flow," *Trans.* ASME, Vol. 66, 1944.
6. Govier, G. W. and Aziz, K.: *The Flow of Complex Mixtures in Pipes*, Van Norstrand Reinhold Co., New York, 1972.
7. HP-41C Petroleum Fluids Pac, Hewlett-Packard, 1000 N. E. Circle Blvd., Corvallis, Ore., 97330.
8. Standing, M. B. and Katz, D. L.: "Density of Natural Gases," *Trans.* AIME, 1942.
9. Brill, J. P. and Beggs, H. D.: *Two-Phase Flow in Pipes*, The Univ. of Tulsa, Tulsa, Okla. 1978.
10. Standing, M. B.: *Volumetric and Phase Behavior of Oil Field Hydrocarbon Systems*, SPE of AIME, 8th Printing, 1977.
11. Wichert, E. and Aziz, K.: "Calculate Z's for Sour Gases," Hydrocarbon Proc., May, 1972.
12. Vasquez, M. and Beggs, H. D.: "Correlations for Fluid Physical Property Prediction," *JPT*, June, 1980.
13. Standing, M. B.: "A Pressure-Volume-Temperature Correlation for Mixtures of California Oils and Gases," API Drlg. & Prod. Practices, 1947.
14. Lasater, J. A.: "Bubble Point Pressure Correlations," *Trans.* AIME, 1958.
15. Craft, B. C, and Hawkins, M. F.: *Applied Petroleum Reservoir Engineering*, Prentice-Hall, NJ, 1959.
16. Beggs, H. D. and Robinson, J. R.: "Estimating the Viscosity of Crude Oil Systems," *JPT*, Sept., 1975.
17. Matthews, C. S. and Russell, D. G.: *Pressure Buildup and Flow Tests in Wells*, SPE Monograph 1, 1967.
18. Meehan. D. N.: "A Correlation for Water Viscosity," *Petr.*

Engr. Int., July, l980.

19. Baker, O. and Swerdloff, W.: "Finding Surface Tension of Hydrocarbon Liquids," *OGJ*, Jan 2, 1956.

20. Hough, E. W.: "Interfacial Tensions at Reservoir Pressures and Temperatures," *Trans.* AIME, 1951.

21. Ramey, H. J.: "Wellbore Heat Transmission," *JPT*, Apr., 1962.

22. Shiu, K. C. and Beggs, H. D.: "Predicting Temperatures in Flowing Oil Wells," *Trans.* AIME, *J. Energy Res. Tech.*, Mar., 1980.

23. Brown, K. E. and Beggs, H. D.: *The Technology of Artificial Lift Methods*, Vol. 1, PennWell Publ. Co., Tulsa, Okla., 1977.

24. Poettmann, F. H. and Carpenter, P. G.: "The Multiphase Flow of Gas Oil and Water Through Vertical Flow Strings," *Drill. & Prod. Practice*, 1952.

25. Baxendell, P. B. and Thomas, R.: "The Calculation of Pressure Gradients for Multiphase Flow in Tubing," *Soc. Pet. Eng. J.*, Mar., 1963.

26. Beggs, H. D.: Lindale, TX, personal communication.

27. Hagedorn, A. R. and Brown, K. E.: "Experimental Study of Pressure Gradients Occurring During Continuous Two-Phase Flow in Small Diameter Vertical Conduits," *JPT*, Apr., 1965.

28. Ros, N. C. J.: "Simultaneous Flow of Gas and Liquid as Encountered in Well Tubing," *JTL*, Oct., 1961.

29. Orkiszewski, J.: "Predicting Two-Phase Pressure Drops in Vertical Pipe," *JPT*, June, 1967.

30. Griffith, P.: "Two-Phase Flow in Pipes," Summer Program, M.I.T., 1962.

31. Duns, H. and Ros, N. C. J.: "Vertical Flow of Gas and Liquid Mixtures in Wells," *Proceedings*, 6th World Petr. Congress, Frankfurt, Germany, 1963.

32. Aziz, K., Govier, G. W., and Forgarasi, M.: "Pressure Drop in Wells Producing Oil and Gas," *J. Cdn. Pet. Tech.*, July-Sept., 1972.

33. Zuber, et al.: "Average Volumetric Concentration in Two-Phase Systems," *J. Heat Transfer*, ASME Trans., Nov., 1965.

34. Chierici, G. L., Ciucci, G. M. and Schlocchi, G.: "Two Phase Flow in Oil Wells, Prediction of Pressure Drop," Ann. Europ. Mtg., April, 1973.

35. Beggs, H. D. and Brill, J. P.: "A Study of Two-Phase Flow in Inclined Pipes," *JPT*, May, 1973.

36. Hein, M.: "3P Flow Analyzer," *OGJ*, Aug. 9, 1982.

37. Hein, M.: "3P Calculator Program for Multiphase Flow Analysis in Pipelines, Casing and Tubing," *OGJ*, Aug. 16, 1982.

38. Cornish, R. E.: "The Vertical Multiphase Flow of Oil and Gas at High Rates," *JPT*, July, 1976.

39. Lawson, J. D., and Brill, J. P.: "A Statistical Evaluation of Methods Used to Predict Pressure Losses for Multiphase Flow in Vertical Oil Well Tubing," Paper SPE 4267, 1973.

40. Ibe, M. C.: "Determination of the Best Combination of Pressure Loss and PVT Property Correlations for Use in Upward Two-Phase Flow," M. S. Thesis, Univ. of Tulsa, 1979.

41. Rossland, L.: "Investigation of the Performance of Pressure Loss Correlations for High Capacity Wells," M. S. Thesis, Univ. of Tulsa, 1979.

42. Brill, J. P. et al.: "Practical Use of Recent Research in Multi-Phase Vertical and Horizontal Flow," *Trans*. AIME, 1966.

43. Cullender, M. H. and Smith, R. V.: "Practical Solution of Gas Flow Equations for Wells and Pipelines with Large Temperature Gradients," *Trans.*, AIME, 1956.

44. Gray, H. E.: "Vertical Flow Correlations in Gas Wells," User Manual, API 14-B SSSV Computer Program.

45. Turner, R. G., Hubbard, M. G., and Dukler, A. E.: "Analysis and Prediction of Minimum Flow Rate for the Continuous Removal of Liquids from Gas Wells," *JPT*, Nov., 1969.

46. "Handbook of Gas-Lift," U. S. Industries Petr. Equip. Div., 1960.

47. Brown, K. E.: *Gas-Lift Theory and Practice*, Petr. Publ. Co., 1965.

48. Alves, G. E.: "Co-Current Liquid-Gas Flow in a Pipe Line Contactor," *Chem. Eng. Prog.*, V. 50, 1954.

49. DeGance, A. E.: and Atherton, R. W.: "Chemical Engineering Aspects of Two-Phase Flow," *Chem Engr.*, Mar. 23, Apr. 20, May 4, July 13, Aug. 10, Oct. 5, Nov. 2, 1970, Feb. 22, 1971.

50. Baker, O.: "Design of Pipelines for Simutaneous Flow of Oil and Gas," *OGJ*, V. 53, 1954.

51. Mandhane, J. M., Gregory, G. A., and Aziz, K.: "A Flow Pattern Map for Gas-Liquid Flow in Horizontal Pipes," *Int. J. Multiphase Flow*, V. 1, 1974.

52. Mukherjee, H.: "An Experimental Study of Inclined Two-Phase Flow," Ph.D. Thesis, Univ. of Tulsa, 1979.

53. Taitel, Y. and Dukler, A. E.: "A Model for Predicting Flow Regime Transitions in Horizontal and Near Horizontal Gas-Liqid Flow," *AICHE J.*, Jan, 1976.

54. Eaton, B. A. et al.: "The Prediction of Flow Patterns, Liquid Holdup and Pressure Losses Occurring During Continuous Two-Phase Flow in Horizontal Pipes," *Trans*. AIME, 1967.

55. Dukler, A. E., et al.: "Gas Liquid Flow in Pipelines, I. Research Results," AGA-API Project NX-28, May, 1969.

56. Flanigan, O.: "Effect of Uphill Flow on Pressure Drop in Design of Two-Phase Gathering Systems," *OGJ*, Mar. 10, 1958.

57. Gregory, G. A., Mandhane, J. M., and Aziz, K.: "Some Design Considerations for Two-Phase Flow in Pipes," C. I. M. Paper No. 374020, May, 1974.

58. Hughmark, G. A.: "Holdup in Gas-Liquid Flow," *Chem. Eng. Prog.*, 68, 1962.

59. Agrawal, S. S., Gregory, G. A., and Govier, G. W.: "An Analysis of Horizontal Stratified Two-Phase Flow in Pipes," *Can. J. CHE.*, 51, 1973.

60. Chawla, J. M.: "Liquid Content in Pipes in Two-Phase Flow of Gas-Liquid Mixtures," *Chimie Ingenieur Technik*, 69, 1969.

61. Lockhart, R. W. and Martinelli, R. C.: "Proposed Correlation of Data for Isothermal Two-Phase Two-Component Flow in Pipes," *Chem. Eng. Prog.*, Jan., 1949.

62. Chenoweth, J. M. and Martin, M. W.: "Turbulent Two-Phase Flow," *Petr. Ref.*, Oct., 1955.

63. Vohra, I. R., et al.: "Comparison of Liquid Holdup Correlations for Gas-Liquid Flow in Horizontal Pipes," SPE 4690, Sept., 1973.

64. Hernandez, F. and Brill. J. P.: "Comparison of Friction Factor Correlations for Gas-Liquid Flow in Horizontal Pipes," SPE 5140, Oct., 1973.

65. API 14B: *Users Manual for API 14B Subsurface Controlled*

Subsurface Safety Valve Sizing Computer Program, API, Washington, D. C., June, 1974.

66. Mandhane, J. M., Gregory, G. A., and Aziz, K.: "Critical Evaluation of Friction Pressure Drop Prediction Methods for Gas-Liquid Flow in Horizontal Pipes," SPE 6036, Oct., 1976.

67. Guzhov, A. I., et al.: "A Study of Transportation in Gas-Liquid Systems," 10th Int. Gas Conf., Hamburg, Germany, 1967.

68. Hoogendorn, C. J.: "Gas-Liquid Flow in Horizontal Pipes," *Chem. Eng. Sci.*, 9, 1959.

69. Chisholm, D.: "A Theoretical Basis for the Lockhart-Martinelli Correlation for Two-Phase Flow," *Int. J. Heat and Mass Transfer*, 10, 1967.

70. Bertuzzi, A. F., et al.: "Simultaneous Flow of Liquid and Gas Through Horizontal Pipe," *Trans. Pet. Soc. AIME*, 17, 1956.

71. Baroczy. C. J.: *Chem. Eng. Prog.*, 62, 1966.

72. Ros, N. C. J.: "An Analysis of Critical Simultaneous Gas-Liquid Flow Through a Restriction and its Application to Flowmetering," Appl. Sci. Res., 9, 1960.

73. Baxendell, P. B.: "Bean Performance—Lake Maracaibo Wells," Internal Company Report, Oct., 1967.

74. Gilbert, W. E.: "Flowing and Gas-Lift Well Performance," *API Drlg. and Prod. Prac.*, 1954.

75. Achong, I.: "Revised Bean Performance Formula for Lake Maracaibo Wells," Internal Company Report, Oct., 1961.

76. Beggs, H. D. and Brill, J. P.: "Development of Methods to Predict Pressure Drop and Closure Conditions for Velocity-Type Subsurface Safety Valves," Report to API OSAPR Committee, Project 10, Feb., 1980.

77. Lee, A. L., et al.: "The Viscosity of Natural Gases," *JPT*, Aug., 1966.

78. Asheim, H: "MONA, An Accurate Two-Phase Well Flow Model Based on Phase Slippage," *SPE Production Engineering*, May, 1986.

79. Nicklin, D. J., Wilkes, J. O., and Davidson, J. F.: "Two-Phase Flow in Vertical Tubes," *Trans*. Inst. Chem Engrs., 1962, 40.

80. Hasan, A. R. and Kabir, C. S.: "Predicting Multiphase Behavior in a Deviated Well," SPE 15449, Presented at 61st Annual SPE Conference, New Orleans, LA, 1986.

81. Fayed, A. S. and Otten, L: "Comparing Measured with Calculated Multiphase Flow Pressure Drop," *OGJ*, Aug. 22, 1983.

82. Osman, M. E. and El-Feky, S. A.: "Design Methods for Two-Phase Pipelines Compared, Evaluated," *OGJ*, Sept, 2, 1985.

83. Oranje, L.: "Condensate Behavior in Gas Pipelines is Predictable," *OGJ*, July 2, 1973.

84. Hong, K. C.: "Flow Splitting a Two-Phase Fluid at a Pipe Tee," SPE 6530.

85. Bergman, D. F., Tec, M. R., and Katz, D.L.: "Retrograde Condensation in Natural Gas Pipelines," Report on Project PR 26-69 of the AGA Pipeline Research Comm., U. of Michigan, 1975.

86. Sachdeva, R., Schmidt, Z., Brill, J. P., and Blais, R. M.: "Two-Phase Flow Through Chokes," SPE 15657, Presented at 61st SPE Fall Conference., New Orleans, LA, 1986.

87. Survey, D. W., Kelkar, B. G., and Brill, J. P.: "Study of Subcricital Flow Through Multiple-Orifice Valves," *SPE Prod. Engr.*, Feb., 1988.

88. "Instructions and Parts List," Willis Oil Tool Co., Cat. 4230A, Long Beach, CA (1971).

4

Total System Analysis

I. INTRODUCTION

The general procedure for applying total system or nodal analysis to a producing well was described in Chapter 1. It was pointed out that methods must be available to calculate the relationship between pressure drop and flow rate for all of the components in the system. The total system, illustrating the various components, was shown in Figure 1-1, which is reproduced as Figure 4-1.

The system analysis procedure requires first selecting a node and calculating the node pressure, starting at the fixed or constant pressures existing in the system. These fixed pressures are usually $\overline{p}_R$ and either p_{wh} or p_{sep}. The node may be selected at any point in the system, and the most commonly selected points are shown in Figure 4-2.

The expressions for the flow into the node and for the flow out of the node can be expressed as:

Inflow

$$p_{inlet} - \Delta p(\text{upstream components}) = p_{node}$$

Outflow

$$p_{outlet} + \Delta p(\text{downstream components}) = p_{node}$$

As was stated earlier, in most cases $p_{inlet} = \overline{p}_R$ and $p_{outlet} = p_{sep}$ or p_{wh}. The two criteria that must be met are:

1. Flow into the node equals flow out of the node.

2. Only one pressure can exist at the node for a given flow rate.

Finding the flow rate and pressure that satisfies the previous requirements can be accomplished graphically by plotting node pressure versus flow rate, as described in Chapter 1. The intersection of the inflow and outflow curves occurs at the rate that satisfies the requirement that the inflow rate equals the outflow rate. This rate will be the producing capacity for the system for a particular set of components. To investigate the effect of changes in any of the components on the producing capacity, new inflow or outflow curves can be generated for each change. If a change is made in an inflow or upstream component only, the outflow curve will not change, and will therefore not require recalculation. Conversely, if the only change made is in a downstream component, the inflow will remain unchanged. This allows isolation of the effect of a change in any component on the total system capacity. This method can be used for determining if existing systems are performing properly and also for designing new systems.

Possible applications of nodal analysis are listed in Chapter 1. Examples of several of these applications will be presented in this chapter, which will illustrate the flexibility of the method. Many of the problems will be worked using the prepared pressure traverse curves for the piping system performance, but the same solution procedures would apply if the calculations were made by a computer. The simplest production system will be considered first as an introduction to the application procedures. More complex and realistic systems will then be considered.

The examples in this chapter will be restricted to flowing wells, either oil or gas. In Chapter 5, the application of systems analysis to artificial lift wells will be presented.

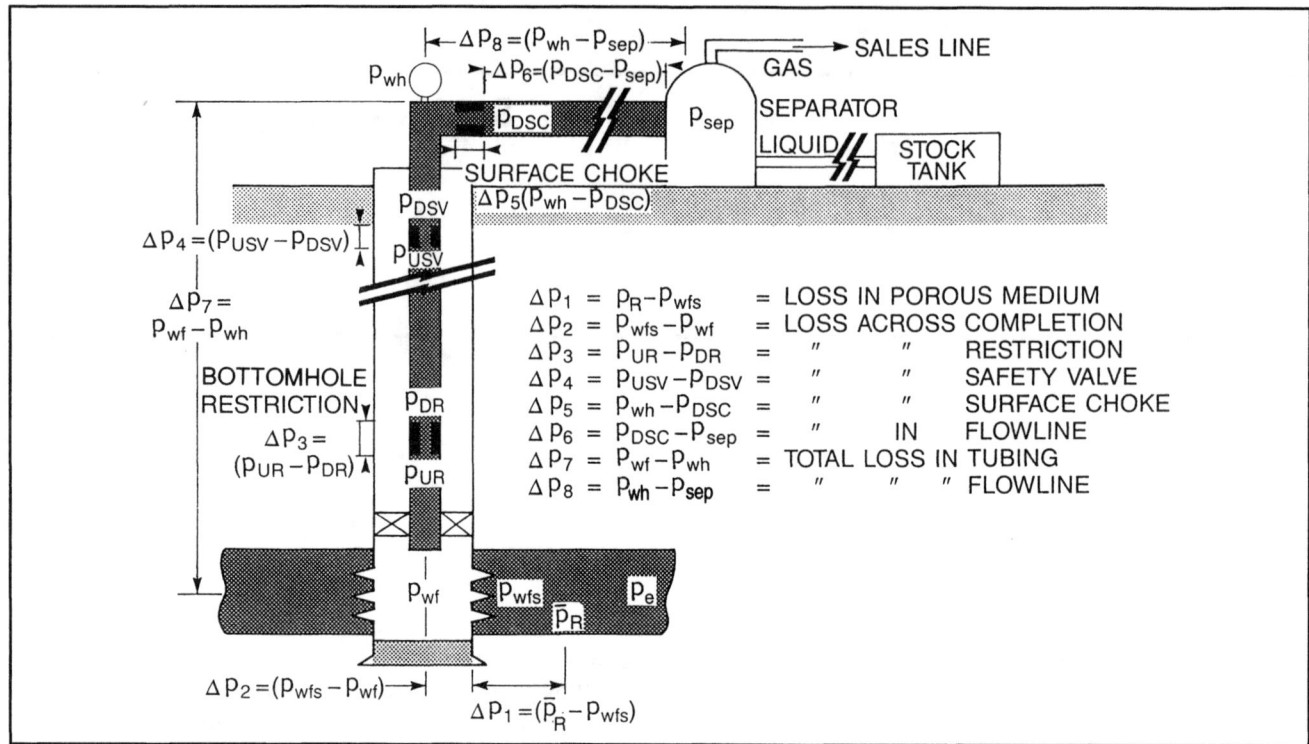

Fig. 4-1. Possible pressure losses in complete system.

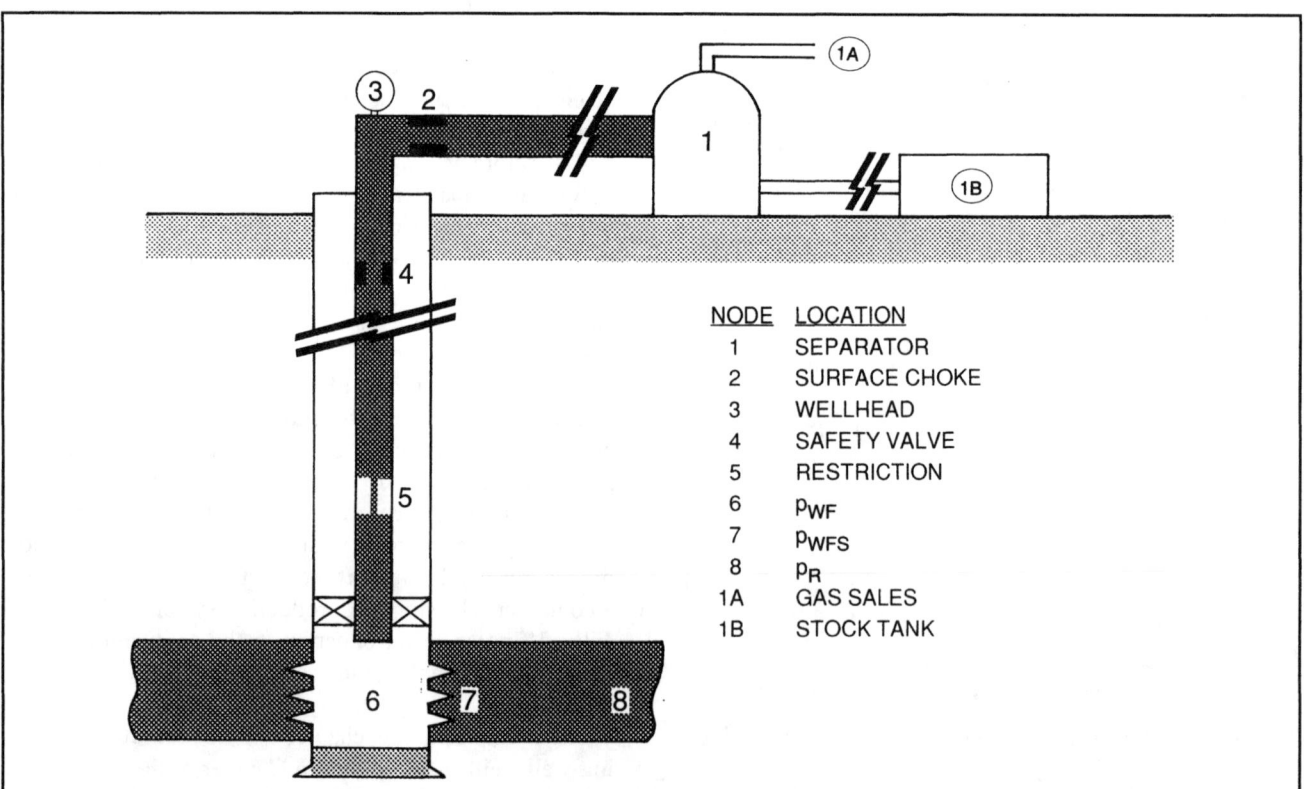

Fig. 4-2. Location of various nodes.

II. TUBING SIZE SELECTION

One of the most important components in the production system is the tubing string. As much as 80 percent of the total pressure loss in an oil well can occur in moving the fluids from the bottom of the hole to the surface.

A common problem in well completion design is to select a tubing size based on totally irrelevant criteria, such as what size tubing is on the pipe rack or what size has been installed in the past. The tubing size selection should be made before a well is drilled, because the tubing size dictates the casing size which dictates the hole size. This is, of course, not possible on an exploratory well because of lack of reservoir data, but once the first well has been drilled, enough data will be available to plan other wells in the field. Selection can also be made using a possible range of expected reservoir characteristics and then refined as more data become available.

There is an optimum tubing size for any well system. Tubing too small will restrict the production rate because of excessive friction loss, while tubing too large will cause a well to load up with liquids and die. A common problem that occurs in completing large capacity wells is to install very large tubing to be "safe." This often results in a decreased flowing life for the wells as reservoir pressure declines and the wells begin to load.

To isolate the effect of tubing size, the wellhead pressure will be considered constant in the following examples. This might be the case for a short flowline discharging into a fixed separator pressure. The node selected will be Node 6 as illustrated in Figure 4-2, so that $p_{node} = p_{wf}$. The expressions for inflow and outflow are:

Inflow

$$\bar{p}_R - \Delta p_{res} = p_{wf}$$

Outflow

$$p_{wh} + \Delta p_{tubing} = p_{wf}$$

___

Example 4-1

Determine the producing capacity of the well described below for nominal tubing sizes of 2-3/8, 2-7/8 and 3-1/2 inches. Other well data are:

$\bar{p}_R = 3482$ psig	$p_b = 3600$ psig
Depth = 10,000 ft	$p_{wh} = 400$ psig
GLR = 400 scf/STB	API = 35°
$\gamma_g = 0.65$	$f_w = 0.5$

Test data:
$q_L = 320$ STB/day, $p_{wf} = 3445$ psig, FE = 1.0
(assumed, since only one test is available)

___

Solution:

Use Vogel's method to calculate the inflow:

$$q_{L(max)} = \frac{q_L}{1 - .2\frac{p_{wf}}{\bar{p}_R} - .8\left(\frac{p_{wf}}{\bar{p}_R}\right)^2}$$

$$= \frac{320}{1 - \frac{.2(3445)}{3482} - \frac{.8(3445)^2}{(3482)^2}}$$

$$q_{L(max)} = 16810 \text{ STB/day}$$

To generate the IPR:

$$q_L = 16810\left[1 - \frac{.2p_{wf}}{3482} - \frac{.8p_{wf}^2}{(3482)^2}\right]$$

Inflow

p_{wf}, psig	q_L, STB/day
3482	0
3000	3930
2500	7464
2000	10442
1500	12866
1000	14735
500	16050
0	16810

The inflow data are plotted on Figure 4-3. It should be pointed out that the inflow data also include implicitly any effects of formation damage or stimulation and perforations. For this case there is insufficient data to consider these effects separately. To calculate the outflow, assume that the prepared pressure traverse curves in the appendix apply to this well. Using the procedure described in Chapter 3, the following data are obtained:

Outflow

q_L, STB/day	1.995 (2 3/8)	2.441 (2 7/8)	2.992 (3 1/2)
400	3200	–	–
600	3280	3160	–
800	3400	3200	–
1000	3500	3250	3130
1500	4400	3400	3200
2000	–	–	3290
2500	–	–	3400

The outflow data are also plotted on Figure 4-3. The flow capacities for the various tubing sizes are read from the intersections of the inflow and outflow curves as:

Tubing I.D., in.	Producing Capacity, STB/day
1.995	800
2.441	1260
2.992	1830

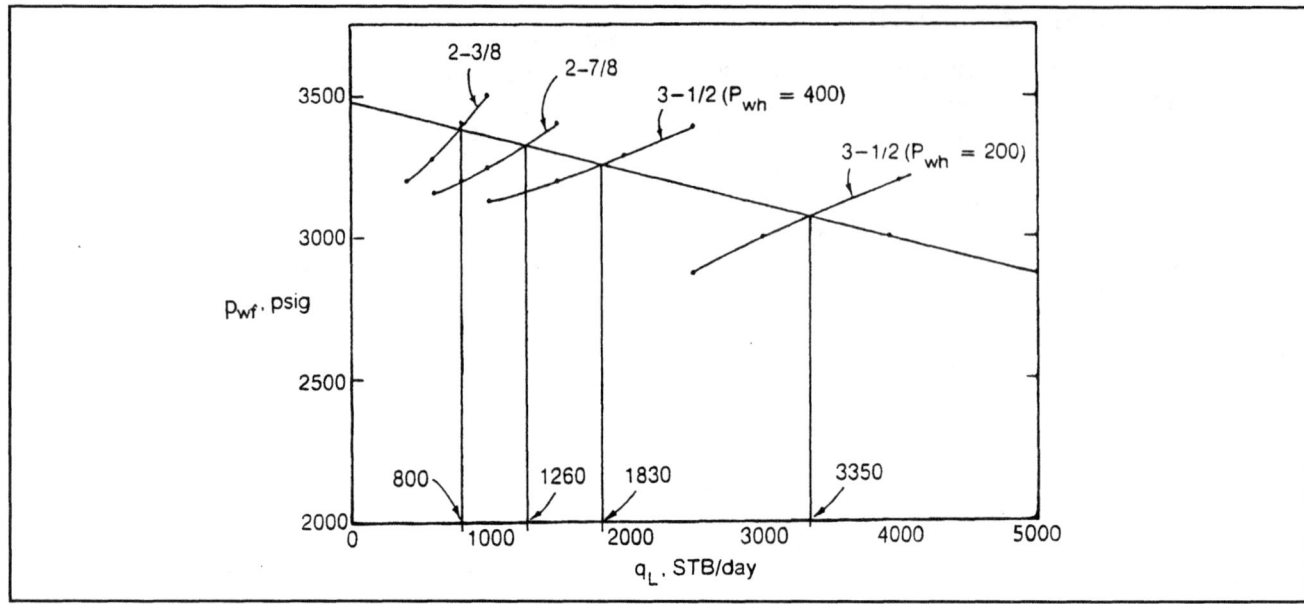

Fig. 4-3. Example 4-1 solution.

The performance of this well is severely restricted by the outflow performance or piping system. Even with the 3-1/2 in. tubing, the well's producing capacity is only about 11 percent of its $q_{L(max)}$. If this well were assigned an allowable rate of 1000 STBO/day, the total liquid rate to obtain this oil rate would be 2000 STBL/day. This could be obtained from this well with the 3-1/2 in. tubing by decreasing the wellhead pressure slightly, but if the water cut increases, other steps would have to be taken to maintain the allowable rate. This could be accomplished by installing larger tubing or placing the well on artificial lift.

The large effect of wellhead pressure on the pressure drop in the tubing is illustrated by decreasing p_{wh} to 200 psig. The effect is illustrated in Figure 4-3 which shows that the producing capacity would be increased from 1830 to 3350 STBL/day for the 3-1/2 tubing.

In many wells, it is necessary to run a small string of tubing in the bottom section of a well if the well is completed with a liner. If the small tubing were run from the surface the producing capacity would be too low, especially if the well is deep. In wells such as these, it is often advantageous to run a tapered tubing string. That is, larger tubing is run from the surface to the top of the liner. The effect of the size of the upper string on producing capacity can be conveniently determined by selecting the point at which the tubing changes size as the node. The inflow will then include the reservoir and the lower section of tubing. The outflow will include the flowline and the upper section of tubing. This situation is illustrated qualitatively in Figures 4-4 and 4-5. The same results

could be obtained by keeping the node at the bottomhole and calculating the tubing pressure drop in two steps.

III. FLOWLINE SIZE EFFECT

The large effect of wellhead pressure on the pressure drop in the tubing was illustrated in the previous example, in which a 200 psi decrease in p_{wh} resulted in an increase in producing capacity of 1520 STB/day. This is caused by the fact that at lower average pressure in the tubing the increased volume of the gas decreases liquid holdup and, thus, the hydrostatic pressure loss. If a well

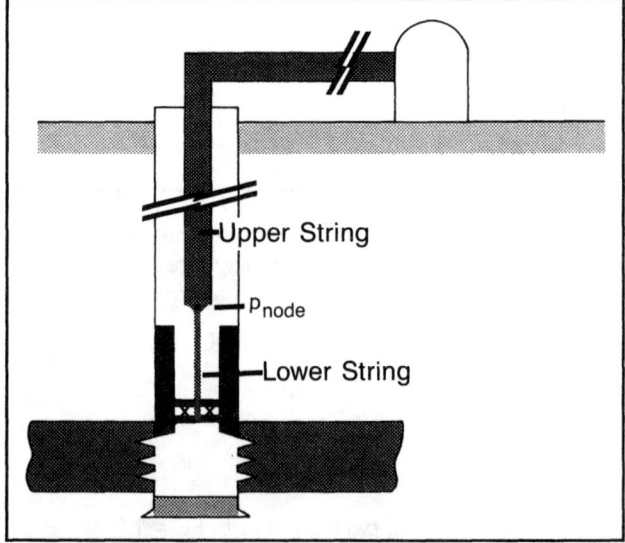

Fig. 4-4. Tapered strings.

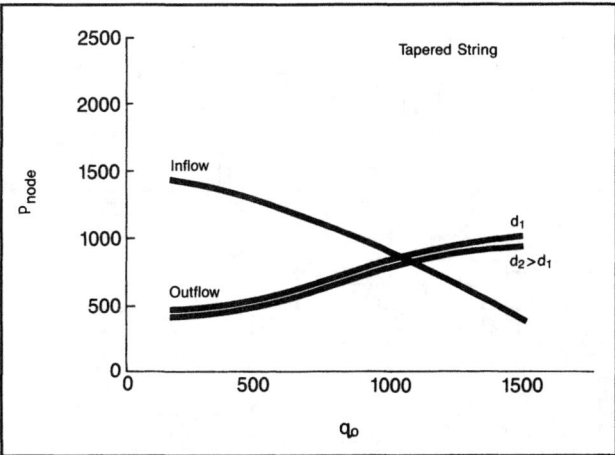

Fig. 4-5. Effect of upper string size.

is producing into a flowline, the wellhead pressure is equal to the sum of the separator pressure and the pressure drop in the flowline, assuming there is no wellhead choke. A common cause of low producing capacity in many wells, especially for wells with long flowlines, is the excessive flowline pressure drop.

Many operators have a tendency to use any size pipe that is convenient or, in some cases, tie two or more wells into a common, small flowline. This can be very detrimental, especially for gas lift wells, because the flowline pressure drop increases as the gas rate increases. This effect will be demonstrated in the section pertaining to artificial lift.

The effect of flowline size will be demonstrated in the form of example problems in this section. The node selected can be either node 6 or node 3, as illustrated in Figure 4-2. As will be seen, node 3 is usually more convenient if the flowline size effect is to be isolated from the tubing effect.

The following example will be solved using both node locations. The effect of separator pressure will be illustrated in the solution utilizing node 3 or $p_{node} = p_{wh}$.

Example 4-2:

The following data pertain to a flowing well that has no surface choke. Calculate the following:

A. The producing capacity as presently equipped using p_{wf} as the node pressure
B. The producing capacity using p_{wh} as the node pressure for the following conditions:
    1. As presently equipped
    2. Flowline size increased to 3 in. I.D.
    3. Separator pressure decreased to 50 psig with present equipment

---

$\overline{p}_R = p_b = 2400$ psig GLR = 800 scf/STB
$p_{sep} = 100$ psig Tubing size = 2,441 in.(2 - 7/8)
Flowline size = 2 in. Tubing depth = 7000 ft
Flowline length = 3000 ft FE = 1.0
$f_w = 0$

Test data: $p_{wf} = 2000$ psig for $q_o = 710$ STB/day

Solution:

Using test data, determine $q_{o(max)}$

$$\frac{p_{wf}}{\overline{p}_R} = \frac{2000}{2400} = 0.833$$

$$q_{o(max)} = \frac{710}{1 - .2(.833) - .8(.833)^2}$$

$$= 2556 \text{ STBO/day}$$

A. ($p_{node} = p_{wf}$)
Calculate the inflow data using Vogel's method:

$$q_o = 2556\left[1 - \frac{.2p_{wf}}{2400} - \frac{.8p_{wf}^2}{(2400)^2}\right]$$

Inflow

p_{wf}	q_o
2400	0
2000	710
1500	1438
1000	1988
500	2361
0	2556

These results are plotted in Figure 4-6. The outflow is determined from

$$p_{sep} + \Delta p_{flowline} + \Delta p_{tubing} = p_{wf}$$

When using the traverse curves, the following procedure is used:
1. For various flow rates, find p_{wh} using the pipeline curves (horizontal) and p_{sep}.
2. For each flow rate and the p_{wh} found in Step 1, use the wellflow (vertical) curves and find p_{wf}.

Outflow

q_o	p_{wh}	p_{wf}
900	3808	1450
1200	510	1720
1500	640	2000

A plot of the inflow and outflow data on figure 4-6 results in a producing capacity of 1175 STBO/day and a value of $p_{wf} = 1700$ psig.

To analyze the effect of changing either flowline size or separator pressure, the entire outflow calculation would have to be repeated, even though the tubing remains the same. This can be avoided by selecting p_{wh} as the node pressure.

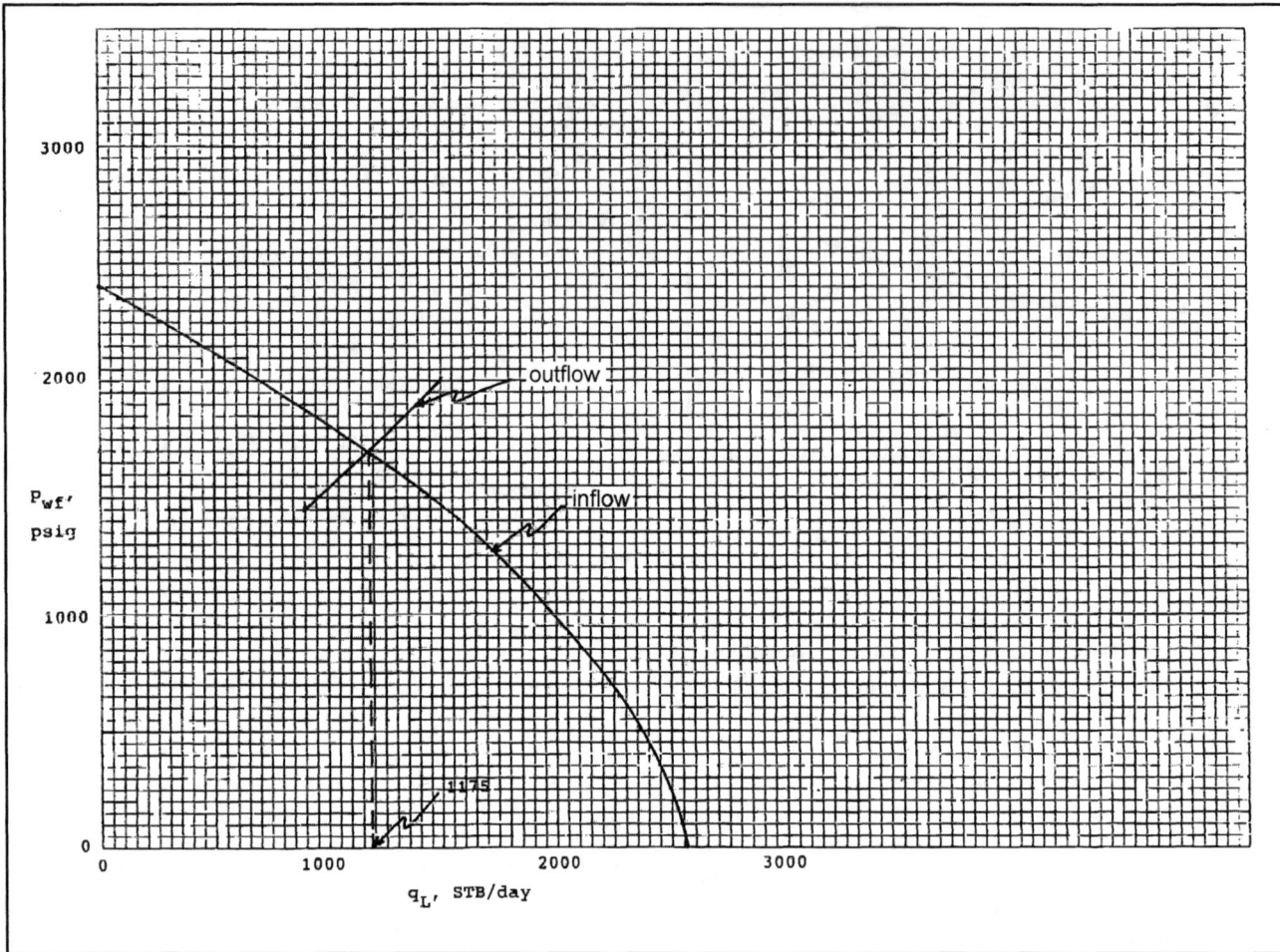

Fig. 4-6. Example 4-2A solution.

B. ($p_{node} = p_{wh}$)

Inflow
$$\bar{p}_R - \Delta p_{res} - \Delta p_{tubing} = p_{wh}$$
Outflow
$$p_{sep} + \Delta p_{flowline} = p_{wh}$$

To expedite the inflow calculation using the traverse curves, it is convenient to solve Vogel's equation for p_{wf} and select flow rates. This results from the fact that the traverse curves are available for specific flow rates. The procedure for generating the inflow data is:

1. Select various flow rates and calculate p_{wf} using Vogel.

2. For each flow rate and p_{wf} determined in Step 1, use the vertical traverse curves to find p_{wh}.

From Vogel:

$$p_{wf} = \bar{p}_R \left[\left(1.266 - \frac{1.25\, q_o}{q_{o(max)}} \right)^{0.5} - 0.125 \right]$$

$$p_{wf} = 2400 \left[\left(1.266 - \frac{1.25\, q_o}{2556} \right)^{0.5} - 0.125 \right]$$

Inflow

q_o	p_{wf}	p_{wh}
900	1880	640
1200	1677	480
1500	1451	240

The wellhead pressures obtained from the horizontal curves in Solution A will be used for plotting the outflow for the 2 inch line.

Outflow

	Outflow, p_{wh}		
q_o	$d = 2,$ $p_{sep} = 100$	$d = 3,$ $p_{sep} = 100$	$d = 2,$ $p_{sep} = 50$
900	380	140	365
1200	510	190	500
1500	640	230	610

The following producing capacities were obtained from Figure 4-7 for the three configurations considered in Solution B:

Flowline Diameter	Separator Pressure	Capacity
2	100	1175
2	50	1180
3	100	1500

These results indicate that the effect of reducing the separator pressure is small compared to the effect of increasing flowline size. This results from the fact that as average pressure in the flowline is decreased in a constant area pipe, the fluid must move faster because of its expansion. This creates more frictional pressure drop. This may not apply if the flowline is in a hilly terrain area, since the increased velocity may decrease the pressure drop caused by the hills.

IV. EFFECT OF STIMULATION

The systems analysis approach can be used to estimate the improvement in well capacity due to fracturing or acidizing. Even though the reservoir capacity may be increased considerably by stimulation; in some cases the well's actual producing capacity increase may be small due to restrictions in the outflow. Before a decision is made on what steps to take to increase the producing

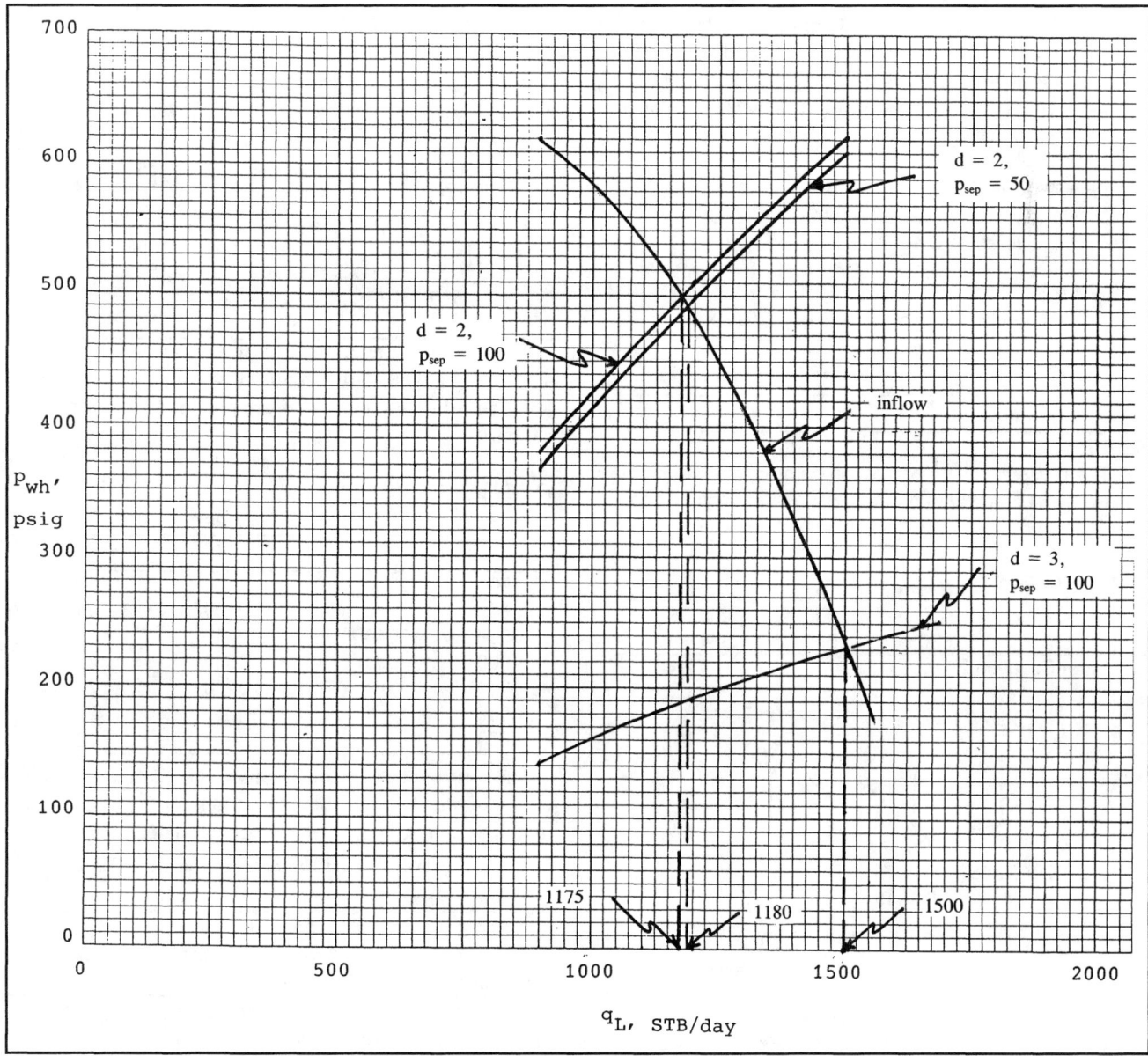

Fig. 4-7. Example 4-2B solution.

capacity, the exact cause of the low productivity should be determined. This can be accomplished only through a total system analysis. Large sums of money are often wasted on workovers because the wrong component of the well system is changed.

The following example illustrates the effect of stimulating a well and how the benefits of an effective stimulation can be nullified by small tubing.

Example 4-3:

Two stabilized tests were conducted on a well that is presently equipped with 2-7/8 in. tubing. Using the following data, determine:

A. The flow efficiency of the well at present.
B. The producing capacity at present conditions for 2-7/8, 3-1/2, and 4 in. tubing if wellhead pressure is maintained at 400 psig.
C. The producing capacity for the conditions stated in Part B if the flow efficiency is increased to 1.3 by stimulation.

$\bar{p}_R = 3482$ psig	GLR = 800 scf/STB
$f_w = 0$	Depth = 10,000 ft
Tubing diameter = 2.441 in.	

Test Data:

Test	p_{wh}, psig	q_o, STB/day
1	920	1000
2	630	2000

Solution:

The present flow efficiency can be calculated from the two tests if the bottomhole pressures for each test are determined using vertical curves or correlations.

Test	p_{wh}	q_o	p_{wf}
1	920	1000	3240
2	630	2000	2990

From Equation 2-50,

$$FE = \frac{2.25\left(\left(1-\frac{p_{wf1}}{\bar{p}_R}\right)q_2 - \left(1-\frac{p_{wf2}}{\bar{p}_R}\right)q_1\right)}{\left(1-\frac{p_{wf1}}{\bar{p}_R}\right)^2 q_2 - \left(1-\frac{p_{wf2}}{\bar{p}_R}\right)^2 q_1} \qquad (2\text{-}50)$$

where

$1 - 3240/3482 = 0.0695$
$1 - 2990/3482 = 0.1413$

$$FE = \frac{2.25[(0.0695)(2000) - (0.1413)(1000)]}{(0.0695)^2(2000) - (0.1413)^2(1000)}$$

$FE = 0.5$

Calculate $q_{o(max)}^{FE=1}$ and generate an IPR for the two FE conditions.

Using Equation 2-46 and Test 1 data:

$$q_o = q_{o(max)}^{FE=1}\left[1.8(FE)\left(1-\frac{p_{wf1}}{\bar{p}_R}\right) - 0.8(FE)^2\left(1-\frac{p_{wf1}}{\bar{p}_R}\right)^2\right]$$

$$q_{o(max)}^{FE=1} = \frac{1000}{1.8(.5)(.0695) - 0.8(.5)^2(.0695)^2}$$

$$= 16238\,\frac{STB}{day}$$

$$q_o = 16238\left[1.8(FE)\left(1-\frac{p_{wf}}{\bar{p}_R}\right) - 0.8(FE)^2\left(1-\frac{p_{wf}}{\bar{p}_R}\right)^2\right]$$

Inflow

p_{wf}	q_o (FE = 0.5)	q_o (FE = 1.3)
3482	0	0
3000	1961	4839
2500	3863	8970
2000	5631	12195
1500	7266	14515
1000	8767	15930
500	10134	–
0	11367	–

These values are plotted as inflow curves on Figure 4-8. Using a wellhead pressure of 400 psig and a GLR = 800, the following p_{wf} values are obtained from the vertical curves.

Outflow

	p_{wf}, psig		
q_o	d = 2.441	d = 2.992	d = 3.476
1000	2220	–	–
2000	2570	2100	–
3000	3040	2440	2160
4000	3600	2680	2320
5000	–	3160	2480
6000	–	–	2710
8000	–	–	3680

The outflow curves for the three tubing sizes are plotted on Figure 4-8. The flow capacities for the various tubing sizes and flow efficiencies are tabulated as follows.

	Producing Capacity, STB/day		
Flow Efficiency	2.441	2.992	3.476
0.5	2600	3610	4330
1.3	3160	4670	6550
Improvement	560	1060	2220

The improvements in producing capacity for the various tubing sizes are also tabulated in the previous table. Notice that the improvement is minimal for the 2-7/8 tub-

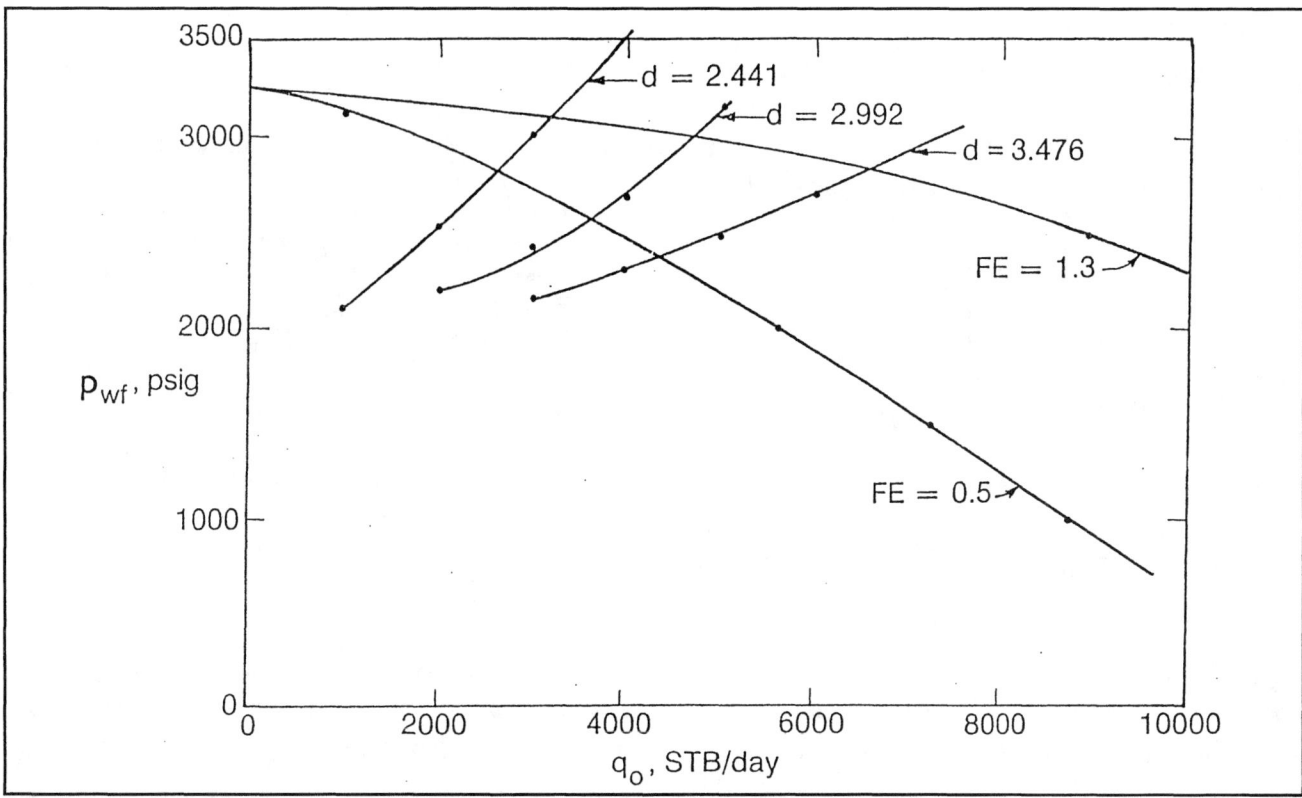

Fig. 4-8. Example 4-3 solution.

ing, even though the flow efficiency was increased by a factor of 2.6. Had this well been equipped with 2-3/8 tubing, which is not uncommon, the improvement would have been negligible.

V. SYSTEMS ANALYSIS FOR WELLS WITH RESTRICTIONS

The analyses performed previously were based on wells that had no restrictions in the outflow segment. Many wells will be equipped with surface chokes, and most offshore wells and wells located in urban areas will be equipped with subsurface safety valves. A surface choke will usually be sized such that flow through the choke is critical, while flow through a SSSV will be subcritical.

In this section, examples will be presented that show the effect of the restriction and the location of the restriction on the producing capacity of the well. Also, the most convenient node location for each analysis will be specified.

A. Surface Chokes

Most flowing wells and some artificial lift wells will be equipped with surface chokes to control the producing

rate, the downstream pressure or pressure surges. These chokes are usually located at the wellhead, but in some cases they may be located near the separator. The location can have a considerable effect on the well's producing capacity, especially if the well has a long flowline.

The following example will be worked by selecting the node at the wellhead, Node 3 in Figure 4-2. This will permit analysis of different surface conditions without recalculating the inflow for each condition change. The analysis will assume that the choke is sized such that critical flow will exist, that is, upstream pressure twice the downstream pressure.

Example 4-4:

The following well is to be equipped with a surface choke operating in critical flow. Determine the well's producing capacity and the choke size required for the following conditions:
1. No choke
2. Choke at wellhead
3. Choke at separator

Well Depth = 7000 ft	Flowline Length = 30 00 ft
Tubing Size = 1.995 in.	Flowline Size = 2 in.
$\bar{p}_R$ = 2500 psia	GLR = 500 scf/STB
f_w = 0	p_{sep} = 120 psig
$c = 0.0023 \dfrac{STB}{day\text{-}psia^{2^n}}$	$n=0.85$

Solution:

Using p_{wh} as the node pressure, the inflow expression will be identical for all three cases. That is,

$$\bar{p}_R - \Delta p_{res} - \Delta p_{tubing} = p_{wh}$$

To calculate Δp_{res}, assume several flow rates and calculate p_{wf} using Fetkovich's method:

$$p_{wf} = \left[\bar{p}_R^2 - \left(\frac{q}{C}\right)^{1/n} \right]^{0.5}$$

$$p_{wf} = \left[(2500)^2 - \left(\frac{q}{0.0023}\right)^{1/.85} \right]^{0.5}$$

Inflow

q_o, STB/day	p_{wf}, psia	p_{wh}, psig (vertical curves)
400	2188	560
600	1973	430
800	1716	280
1000	1397	100

The inflow curve is plotted on Figure 4-9.

1. For no choke in the well, the outflow expression is

$$p_{sep} + \Delta p_{flowline} = p_{wh}$$

The horizontal curves are used to find p_{wh} for various flow rates and $p_{sep} = 120$ psig.

2. For a choke in critical flow at the wellhead, the pressure downstream of the choke, p_d, is found using the horizontal curves, and will be identical to the p_{wh} values found in Step 1. However, to assure that the choke is in critical flow, the wellhead pressure will be equal to twice the p_d value. That is,

$$(p_{sep} + \Delta p_{flowline})2 = p_{wh}$$

3. For a choke located at the separator, the pressure just upstream of the choke, which is also the outlet pressure for the flowline, will be equal to twice the separator pressure. The wellhead pressures for various flow rates can then be found using the horizontal curves. That is,

$$2p_{sep} + \Delta p_{flowline} = p_{wh}$$

The values obtained for the various conditions are tabulated as follows and plotted on Figure 4-9. The producing capacities are also tabulated.

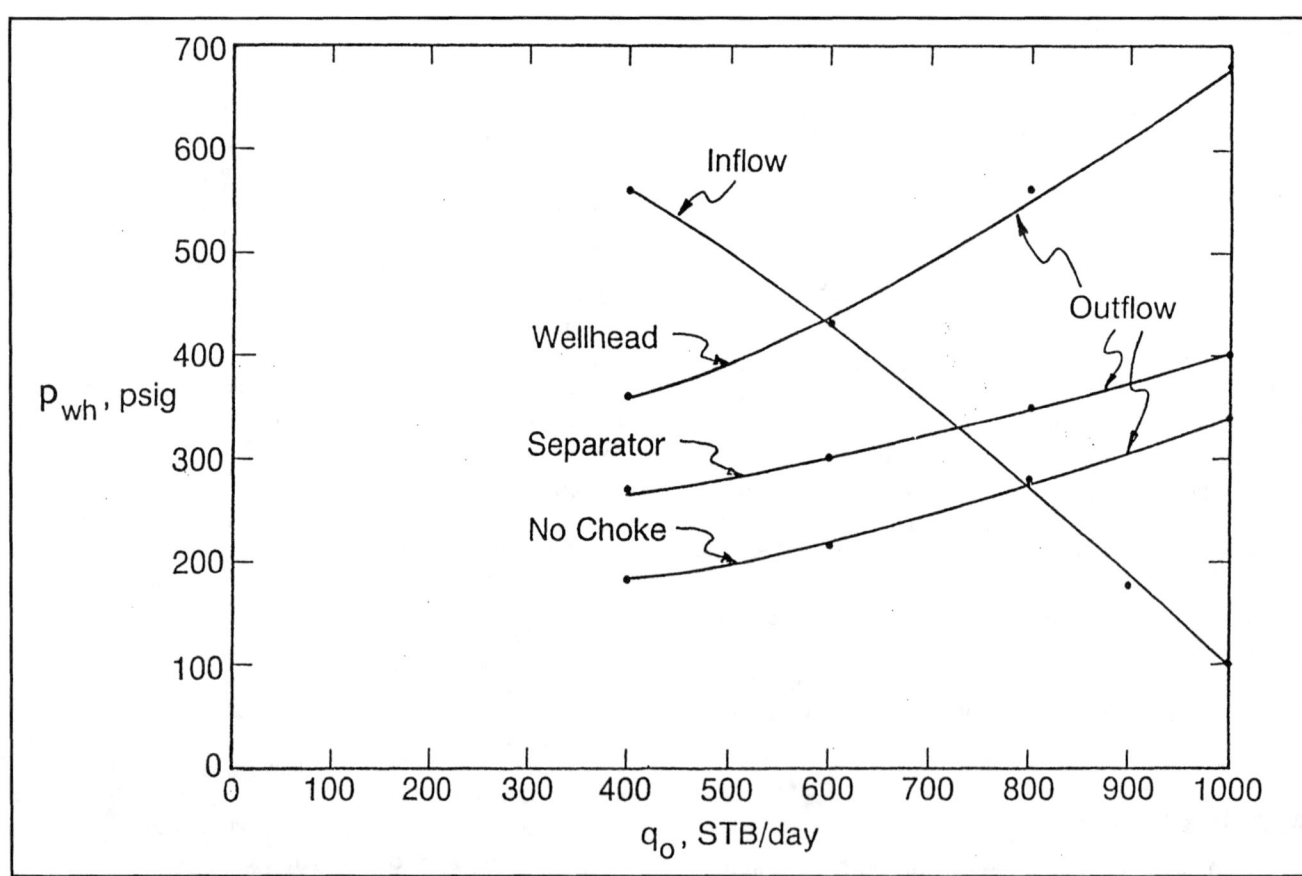

Fig. 4-9. Example 4-4 solution.

Outflow

	p_{wh}, psig		
q_o, STB/day	No Choke	Wellhead	Separator
400	180	360	270
600	215	430	300
800	280	560	350
1000	340	680	400

Choke Location	p_1	Producing Capacity	Choke Size, 64ths in.
No choke	275	800	–
Wellhead	435	595	22
Separator	240	730	34

The choke sizes were calculated using the Ros coefficients in Equation 3-122. The pressure upstream of the choke, p_1, is read from Figure 4-9, at the inflow-outflow intersection.

$$d = \left[\frac{0.0045(q_o)(\text{GLR})^{0.5}}{p_1} \right]^{0.5}$$

For the case of the choke at the wellhead,

$$d = \left[\frac{0.0045(595)(500)^{0.5}}{449.7} \right]^{0.5}$$
$$= 0.35 \text{ in.} = 22/64 \text{ in.}$$

Comparison of the results reveals that locating the choke at the separator rather than at the wellhead increases the producing capacity by 135 STB/day or about 23 percent. There are two reasons for this effect.

1. A lower downstream pressure is doubled to obtain the choke upstream pressure, and

2. The average pressure in the flowline is higher, resulting in less frictional pressure drop.

B. Subsurface Safety Valves

Analysis of the effect of a SSSV in the tubing can be conducted in essentially the same manner as illustrated previously. However, the SSSV will be operating in subcritical flow, and, therefore, the pressure drop across the valve must be calculated. Also, if Node 4 is chosen, the outflow will include the section of tubing above the SSSV. It is convenient to choose the node pressure as the pressure just downstream of the SSSV, since the equations for pressure drop across the SSSV depend on upstream conditions of pressure and temperature. The inflow and outflow expressions are:

Inflow

$$\bar{p}_R - \Delta p_{res} - \Delta p_{(tubing\ below)} - \Delta p_{SSSV} = p_{node}$$

Outflow

$$p_{sep} + \Delta p_{flowline} + \Delta p_{(tubing\ above)} = p_{node}$$

Using this analysis, the outflow curve will not change for different SSSV sizes, and only the pressure drop across the SSSV will change in the inflow calculation. A systems analysis plot that would result for various size SSSVs is shown in figure Figure 4-10. The outflow curve for no SSSV is shown also.

VI. EVALUATING COMPLETION EFFECTS

Nodal system analysis is a convenient method to use in comparing various well completion schemes, such as perforating density and total perforated interval. Methods for calculating the pressure drop across the completion were presented earlier for open hole, perforated, and gravel pack completions.

As was discussed earlier, the completion pressure drop, $p_{wfs} - p_{wf}$, may be included in the reservoir pressure drop component, or it may be isolated to compare effects of various completion methods. If the completion effect is combined with the reservoir effect, the system analysis would be identical to the examples presented earlier, where p_{wf} was selected as the node pressure. A different inflow curve would result for each completion scheme, such as number of perforations used. This is illustrated qualitatively in Figure 4-11.

If gravel-packed completions are being considered, it is advantageous to isolate the pressure drop across the gravel pack. This is necessary so that the critical pressure drop, usually about 300 psi, is not exceeded. This is accomplished by treating the completion or gravel pack as an independent component and plotting pressure drop across the gravel pack versus flow rate.

To analyze a gravel-pack completion, the system is

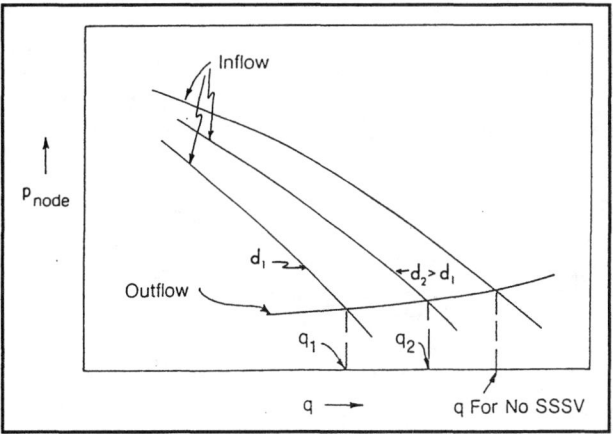

Fig. 4-10. Subsurface safety valve effect.

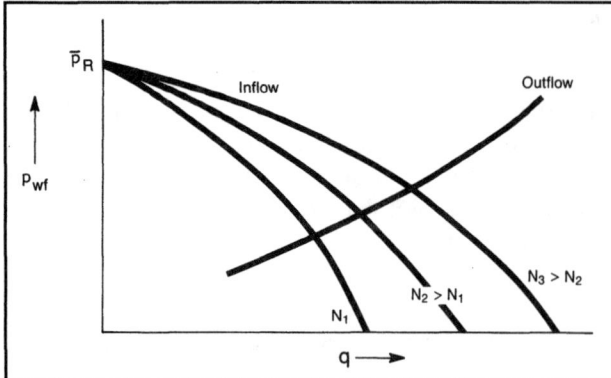

Fig. 4-11. Effect of perforating density.

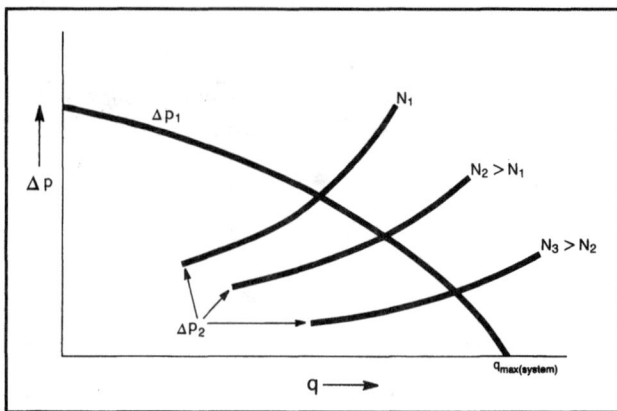

Fig. 4-13. Gravel-pack analysis, producing capacity.

divided at the wellbore. The node pressure for the inflow is p_{wfs}, while the node pressure for the outflow is p_{wf}. That is,

Inflow

$$\bar{p}_R - \Delta p_{res} = p_{wfs}$$

Outflow

$$p_{sep} + \Delta p_{flowline} + \Delta p_{tubing} = p_{wf}$$

Both p_{wfs} and p_{wf} are determined for various flow rates and plotted versus flow rate, as illustrated in Figure 4-12. The intersection of these curves gives the producing capacity that would result if no pressure drop across the gravel pack occurred. The required p_{wfs} may be calculated using equations for oil or gas reservoirs.

The pressure drop available for overcoming the gravel pack's resistance to flow for rates lower than the maximum system rate can be read from Figure 4-12. These are designated as Δp_1. Values of Δp_1 versus q are plotted as illustrated in Figure 4-13.

The pressure drop occurring across the gravel pack for various flow rates can be calculated as a function of the number of perforations, perforation size, perforation length, and gravel permeability, using Equations 2-111 or 2-113. These pressure drops, designated as Δp_2 are also plotted on Figure 4-13. The intersection of the Δp_1 and Δp_2 curves gives the producing capacity and pressure drop across the gravel pack for various completion schemes. This permits determination of the maximum producing rate allowed for any number of perforations to keep Δp below the critical value.

Example 4-5:

A drill stem test was conducted on an oil well to be completed in an unconsolidated formation. It was determined that the well must be completed by gravel packing. Using the following data, determine the producing capacity for perforating densities of 4, 8 and 12 shots per foot. Also determine the maximum producing rate for each perforating density if the maximum pressure drop across the gravel pack is 300 psi. Assume that there is no compacted zone around the perforations.

From DST and PVT Analysis:

$k_o = 100$ md	$\bar{p}_R = 3200$ psig
$S' = 0$	$f_w = 0$
$\mu_o = 0.803$ cp	$p_b = 2200$ psig
$B_o = 1.248$ bbl/STB } assume constant	
$\rho_o = 45$ lbm/ft3	
GOR = 400 scf/STB	$T_R = 180°$ F
API = 35°	$\gamma_g = 0.65$

Completion Data:

$r_w = 6$ in.	$r_e = 750$ ft
$h = 30$ ft	Screen diameter = 4.5 in.
Depth = 7000 ft	$p_{wh} = 200$ psig
Tubing I.D. = 3.958 in.	Gravel permeability = 45 darcies
Perforation diameter = 0.5 in.	Casing I.D. = 8.921 in.

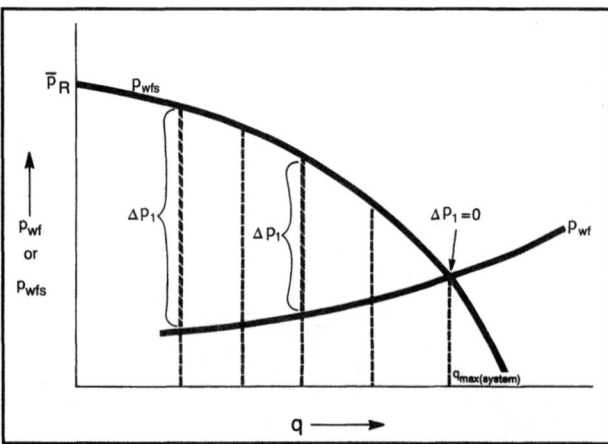

Fig. 4-12. Gravel-pack analysis, system pressure drop.

Solution:

$$J = \frac{0.00708 k_o h}{\mu_o B_o \left[\ln\left(\frac{.472 r_e}{r_w}\right) + S' \right]}$$

$$J = \frac{0.00708(100)(30)}{(.803)(1.248)\left[\ln\left(\frac{.472(750)}{.5}\right) + 0 \right]}$$

$$= 3.23 \frac{STB}{D\text{-}psi}$$

$$q_b = J(\bar{p}_R - p_b) = 3.23(3200 - 2200)$$

$$= 3230 \frac{STB}{D}$$

$$q_o = q_b + \frac{J p_b}{1.8}\left[1 - \frac{.2 p_{wfs}}{p_b} - \frac{.8 p_{wfs}^2}{p_b^2} \right] \qquad (2\text{-}38)$$

$$q_{o(max)} = q_b + \frac{J p_b}{1.8} = 3230 + \frac{3.23(2200)}{1.8}$$

$$q_{o(max)} = 3230 + 3948 = 7178 \frac{STB}{D}$$

Solving Equation 2-38 for p_{wfs}:

$$p_{wfs} = p_b\left[\left(1.2656 - \frac{2.25(q_o - q_b)}{J p_b} \right)^{0.5} - .125 \right]$$

$$p_{wfs} = 2200\left[\left(1.2656 - \frac{2.25(q_o - 3230)}{3.23(2200)} \right)^{0.5} - .125 \right]$$

This equation is valid for $p_{wfs} < p_b$.
For $p_{wfs} \geq p_b$, use

$$p_{wfs} = \bar{p}_R - \frac{q_o}{J} = 3200 - \frac{q_o}{3.23}$$

Inflow

q_o	p_{wfs}
1000	2890
2000	2581
3000	2271
4000	1950
5000	1572
6000	1096

The values of p_{wf} for the outflow are read from the vertical traverse curves for 4-1/2 tubing, using $p_{wh} = 200$ psig.

Outflow

q_o	p_{wf}
1000	1340
2000	1400
3000	1500
4000	1560
5000	1650
6000	1750

Plotting both p_{wfs} and p_{wf} versus q_o on Figure 4-14 indicates a producing capacity of 4800 STB/day for no perforation pressure drop.

Read $\Delta p_1 = p_{wfs} - p_{wf}$ for various values of q_o. This is the pressure drop available to overcome the perforation pressure drop at these flow rates. These values may be obtained by subtracting the required outflow pressure from the required inflow pressure at the same flow rates, or read from Figure 4-14.

q_o	Δp_1
1000	1550
2000	1180
3000	770
4000	390
4800	0

Plot this data as Δp versus q_o on Figure 4-15.

Use Equation 2-111 to determine values of $\Delta p_2 = p_{wfs} - p_{wf}$ for the gravel pack.

$$\Delta p_2 = p_{wfs} - p_{wf} = A_G q_o + B_G q_o^2 \qquad (2\text{-}112)$$

$$A_G = \frac{282.4 \mu_o B_o L}{K_G N r_p^2}$$

$$B_G = \frac{9.20 \times 10^{-14} \beta_G B_o^2 p_o L}{N^2 r_p^4}$$

$$\beta_G = \frac{1.47 \times 10^7}{k_G^{0.55}}$$

$$\beta_G = \frac{1.47 \times 10^7}{(45000)^{0.55}} = 4.06 \times 10^4 \text{ ft}^{-1}$$

$$r_p = 0.25 \text{ in.} = 0.0208 \text{ ft}$$

$$L = r_w - r_{csg} = 6 - \frac{8.921}{2}$$

$$= 1.54 \text{ in.} = 0.128 \text{ ft}$$

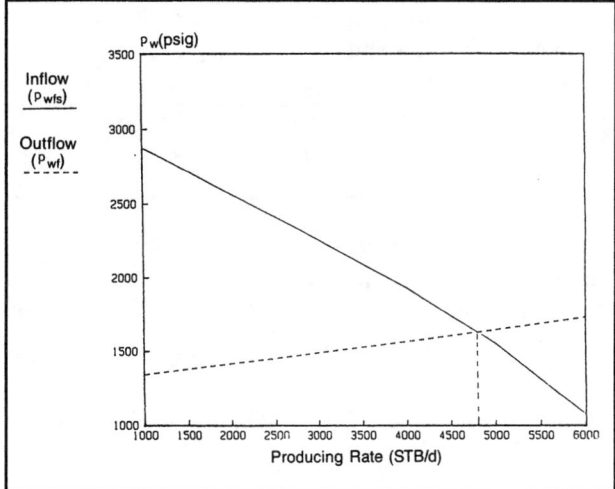

Fig. 4-14. Example 4-5 solution.

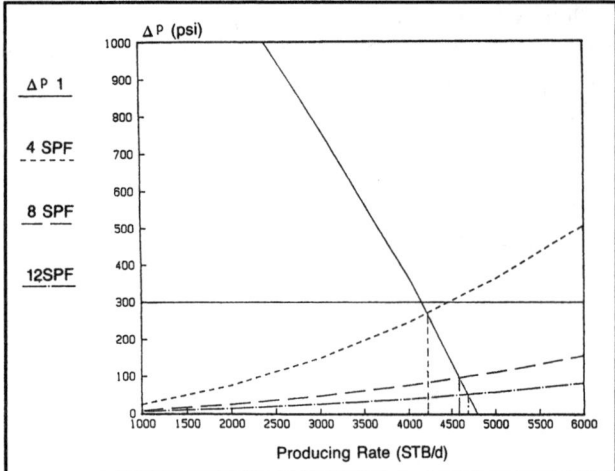

Fig. 4-15. Example 4-5 solution.

$$A_G = \frac{282.4(.803)(1.248)(.128)}{(45000)\,N(.0208)^2} = \frac{1.861}{N}$$

$$B_G = \frac{9.20\times10^{-14}(4.06\times10^4)(1.248)^2(45)(.128)}{N^2(.0208)^4}$$

$$= \frac{0.179}{N^2}$$

$$\Delta p_2 = 1.861\frac{q_o}{N} + 0.179\left(\frac{q_o}{N}\right)^2$$

	Δp_2, psi		
	N = 120	N = 240	N = 360
q_o	(4SPF)	(8SPF)	(12SPF)
2000	80	30	15
3000	158	50	30
4000	260	80	43
5000	390	120	60

Plotting both Δp_1 and Δp_2 versus q_o on Figure 4-15 for the three perforating densities indicates the following producing capacities and corresponding pressure drops:

SPF	q_o	Δp	q_o for $\Delta p = 300$
4	4200	260	4500
8	4600	100	–
12	4700	40	–

These results indicate that the 300 psi limit on Δp would not be exceeded if all the data are correct. However, if fewer than four shots per foot are open, the 300 psi could be exceeded.

VII. NODAL ANALYSIS OF INJECTION WELLS

All the previous examples in this chapter have dealt with flowing production wells. Artificial lift wells will be analyzed in Chapter 5. The production engineer must sometimes design completion configurations or analyze performance of various types of injection wells. These wells may be used for injecting water or some other fluid for enhanced recovery projects or they could be gas injection wells operating in gas storage reservoirs. Nodal analysis may be performed on injection wells by selecting the node at bottom-hole such that the inflow to the node will include the injection pump or compressor and the piping system, while the outflow will consist of the perforations and reservoir. For example, if gas from a compressor is being injected into a well, the inflow and outflow expressions would be:

Inflow

$$p_{comp} - \Delta p_{\,flowline} - \Delta p_{\,tubing} = p_{\,wf}$$

Outflow

$$\bar{p}_R + \Delta p_{res} = p_{wf}$$

This type of analysis could be used to determine the effects on injection rate of various compressor pressures, flowline sizes or tubing sizes. The effect of tubing size on injection rate will be illustrated by an example. For this example, it is assumed that wellhead pressure is constant so that the inflow will include only the pressure drop in the tubing. That is,

Inflow

$$p_{wh} + \Delta p_{el} - \Delta p_f = p_{wf}$$

Equation 3-109 may be used to calculate p_{wf} for various rates. The outflow performance may be calculated using the backpressure equation for gas wells. That is,

Outflow

$$\bar{p}_R + \Delta p_{res} = p_{wf}$$

where

$$q_{inj} = c(p_{wf}^2 - \bar{p}_R^2)^n$$

$$p_{wf}^2 = \bar{p}_R^2 + \left(\frac{q}{C}\right)^{1/n}$$

Example 4-6:
Using the following data, determine the rate at which gas can be injected into this well for 2-3/8 in., 2-7/8 in. and 3 1/2 in. tubing.

$\bar{p}_R$ = 2000 psia	Injection p_{wh} = 4000 psia
γ_g = 0.7	T_{wh} = 150°F
T_{res} = 150°F	Well depth, H = 10000 ft
ϵ = 0.0018 in.	

From a previous injection test:

$$C = 2 \times 10^{-5} \text{ MMscfd/psia}^{2n}, \; n = 0.86$$

Solution:

The inflow performance can be calculated from:

$$p_{wf}^2 = p_{wh}^2 \, EXP(S) - \frac{25\gamma_g \, \overline{T} \, H(EXP(S)-1) \, f \, \overline{Z} \, q_{inj}^2}{Sd^5}$$

where

$$S = 0.0375\gamma_g H/(\overline{TZ})$$

$$f = \left[1.14 - 2\log\left(\frac{\epsilon}{d} + \frac{21.25}{N_{Re}^{0.9}}\right) \right]^{-2}$$

$$N_{Re} = 20011\gamma_g \, q_{inj}/\mu_g d$$

The solution for p_{wf} for any injection rate will be iterative since $\overline{Z}$ depends on the average of p_{wf} and p_{wh}. The iterative procedure will be illustrated for one tubing size and one rate. If a computer is available, the tubing can be divided into short increments of length, and the solution will be more accurate. The procedure for hand calculations is:

1. Assume a value for p_{wf}.
2. Calculate $\overline{p} = (p_{wh} + p_{wf})/2$.
3. Calculate $\overline{Z}$ and $\overline{\mu}_g$ at $\overline{p}, \overline{T}$.
4. Calculate p_{wf} and compare with the assumed value. If not close, use the calculated p_{wf} as next estimate and go to Step 2.

The procedure will be illustrated by calculating the bottomhole pressure for the 2-3/8 in. (1.995 I.D.) tubing for an injection rate of 4 MMscfd. For this rate:

$$N_{Re} = \frac{20011(0.7)(4)}{1.995\mu_g} = \frac{28086}{\mu_g}$$

$$f = \left[1.14 - 2\log\left(\frac{.0018}{1.995} + \frac{21.25}{N_{Re}^{0.9}}\right) \right]^{-2}$$

$$S = \frac{0.0375(0.7)(10000)}{(150+460)\overline{Z}} = \frac{0.4303}{\overline{Z}}$$

$$p_{wf}^2 = (4000)^2 EXP(S)$$

$$-\frac{25(0.7)(150+460)(10000)(EXP(S)-1)f\,\overline{Z}(4)^2}{S(1.995)^5}$$

$$p_{wf}^2 = (4000)^2 EXP(S)$$

$$-\frac{5.405 \times 10^7 (EXP(S)-1)f\,\overline{Z}}{S}$$

Assumed							Calculated
p_{wf}	$\overline{p}$	$\overline{Z}$	μ	N_{Re}	f	S	p_{wf}
4000	4000	0.879	0.025	1.1×10^6	0.019	0.489	4991
4991	4496	0.921	0.027	1.0×10^6	0.019	0.467	4930
4930	4465	0.918	0.027	1.0×10^6	0.019	0.469	4934
4934	4467	0.918	0.027	1.0×10^6	0.019	0.469	4934

This procedure was followed for the various injection rates and tubing sizes to produce the following table of inflow pressures and rates:

Inflow

Injection Rate, MMscfd	Inflow p_{wf}, psia		
	2-3/8 tubing	2-7/8 tubing	3-1/2 tubing
0	5050	5050	5050
4	4934	5010	5035
8	4580	4890	4995
12	3945	4685	4925
16	2910	4390	4830
20	1200	3985	4700
24	0	3455	4540
28	0	2735	4350
32	0	–	4120

The reservoir or outflow performance is calculated from:

$$p_{wf} = \left[\overline{p}_R^2 + (q/C)^{1/n} \right]^{0.5}$$

$$p_{wf} = \left[(2000)^2 + (q/2 \times 10^{-5})^{1.1628} \right]^{0.5}$$

Outflow

Injection Rate, MMscfd	Outflow p_{wf}, psia
0	2000
4	2336
8	2695
12	3039
16	3363
20	3671
24	3964
28	4245
32	4514

The intersections of the various inflow curves with the outflow curve in Figure 4-16 represent the injection rates possible for the 3 tubing sizes. The results are tabulated in the following table:

Tubing size, inches	Injection Rate, MMscfd
2-3/8	14.6
2-7/8	21.5
3-1/2	29.0

This well could also be analyzed for other wellhead pressures. In gas storage operations, the static reservoir pressure will increase as gas is injected, and this would cause an upward shift in the outflow curve in Figure 4-16. This would result in a decreasing injection rate with time, as the intersection of the inflow and outflow curves would shift to the left. The change in injection rate with time could be determined by using a procedure similar to that discussed in Sections VIII and IX in this chapter.

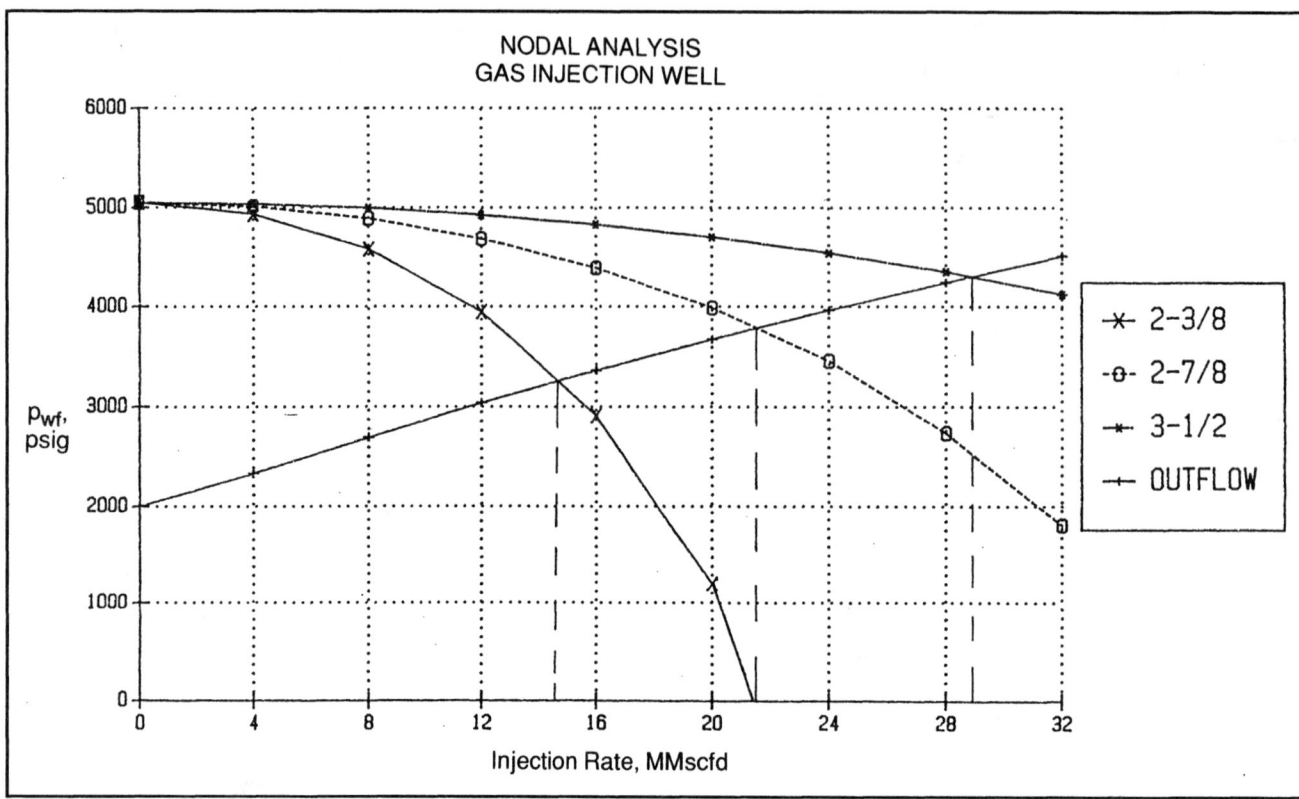

Fig. 4-16. Gas injection rate.

VIII. EFFECT OF DEPLETION

As the pressure in a reservoir declines from depletion, both the inflow and outflow conditions can change. The changes occurring in the reservoir inflow capacity were discussed in Chapter 2, and methods for preparing IPRs for the future were presented.

It is very likely that the outflow conditions will also change with depletion or time, especially in the case of naturally flowing wells. Once a well is placed on artificial lift, the outflow conditions can be held fairly constant. The principal parameters that will change with depletion are gas/oil or gas/liquid ratio and water cut. The effects that these parameters have on the pressure drop in the tubing were discussed in Chapter 3. The producing gas/oil ratio will increase in any reservoir in which the pressure declines below bubblepoint pressure, and the water cut will increase with time or depletion if a water drive is active or if water is being injected in the reservoir for pressure maintenance purposes.

Before any development planning or economic calculations can be performed, it is necessary to be able to predict the producing rate of a well or field as a function of time. The effect of changing reservoir conditions on the producing capacity of a well is considered in this section. Relating this changing performance to time will be discussed subsequently.

Example 4-7:

The following data pertaining to a well producing from a reservoir with a weak water drive were obtained from a material balance calculation and a PVT analysis. The reservoir pressure at the present time is 2250 psig and from a current test $q_{o(max)}$ = 1257 STB/day. Using Standing's method for generating future IPRs, determine the oil producing capacity of the well at the various reservoir pressures given. Plot the producing capacity versus reservoir pressure.

API = 35°	p_{wh} = 80 psig
Depth = 5000 ft	Tubing size = 2-3/8 in.
$q_{o(max)p}$ = 1257 STB/day	$\bar{p}_{RP}$ = 2250 psig
γ_g = 0.7	

$\bar{p}_R$	μ_o, cp	B_o, bbl/STB	k_{ro}	GOR, scf/STB	f_w
2250	3.11	1.173	0.815	400	0
1800	3.59	1.150	0.685	600	0
1500	3.70	1.130	0.580	800	0
1000	3.80	1.110	0.430	1000	0.5

Solution:

To generate the future IPR data, the following relationships will be used to predict $q_{L(max)F}$:

$$q_{L(max)F} = q_{L(max)p} \left[\frac{\bar{p}_{RF} f(\bar{p}_{RF})}{\bar{p}_{RP} f(\bar{p}_{RP})} \right] \quad \text{(2-72)}$$

where

$$f(\bar{p}_R) = k_{ro} / \mu_o B_o$$

$\bar{p}_R$	$f(p_R)$	$p_{RF} f(p_R)$	$q_{L(max)}$
2250	0.223	501.8	1257
1800	0.166	298.8	749
1500	0.139	208.5	522
1000	0.103	103.0	258

Vogel's equation is now used to generate the IPR data for both the present and future times.

$$q_{LF} = q_{L(max)F} \left[1 - \frac{0.2\, p_{wf}}{\bar{p}_{RF}} - \frac{0.8\, p_{wf}^2}{\bar{p}_{RF}^2} \right] \quad \text{(2-73)}$$

Inflow

p_{wf}	$\bar{p}_R = 2250$	$\bar{p}_R = 1800$	$\bar{p}_R = 1500$	$\bar{p}_R = 1000$
		q_L, STB/day		
2250	0	–	–	–
2000	239	–	–	–
1800	412	0	–	–
1600	570	142	–	–
1500	642	208	0	–
1200	837	383	171	–
1000	947	481	267	0
800	1040	564	348	85
600	1118	632	413	153
400	1180	686	464	204
200	1227	725	501	239
0	1257	749	522	258

The node pressure for the analysis is selected as p_{wf}. Therefore, the data in the previous table represent the inflow for the well. These are plotted on Figure 4-17 as p_{wf} versus q_L.

The vertical traverse curves will be used to generate the outflow data for this example. A different outflow curve will have to be prepared for each inflow curve because the gas/liquid ratios and water cuts are changing. It is assumed that tubing size and wellhead pressure remain constant. The outflow data are tabulated as follows and plotted on Figure 4-17.

Outflow

q_L, STB/day	$\bar{p}_R = 2250$	$\bar{p}_R = 1800$	$\bar{p}_R = 1500$	$\bar{p}_R = 1000$
100	–	–	–	600
200	–	–	–	660
300	–	–	510	730
400	–	–	640	–
500	–	730	710	–
700	–	810	–	–
800	950	840	–	–
1000	1040	–	–	–
1200	1120	–	–	–

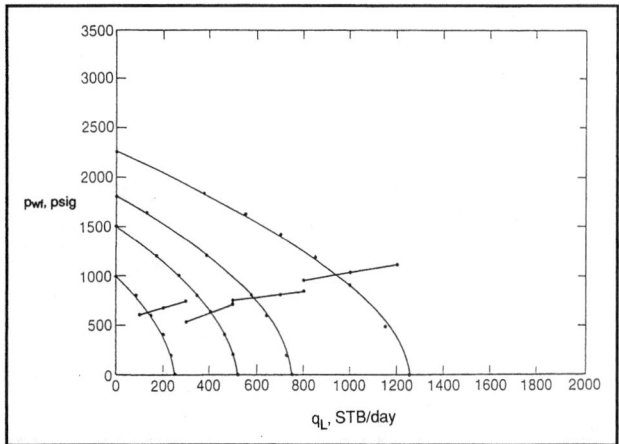

Fig. 4-17. Example 4-7 solution.

The intersections of the inflow and outflow curves for the same conditions of $\bar{p}_R$ give the producing capacity for those conditions. These are

$\bar{p}_R$	q_L	q_o
2250	940	940
1800	590	590
1500	410	410
1000	140	70

A plot of these results is presented in Figure 4-18. A similar analysis can be performed for gas wells using the procedures outlined in Chapter 2 to generate future IPRs. The outflow conditions are not as likely to change for a gas well unless it is producing from a water-drive reservoir, in which case the liquid loading may increase as $\bar{p}_R$ declines. If this occurs, the minimum gas rate necessary to keep the tubing unloaded will eventually be reached, and the tubing size may have to be reduced to keep the well flowing. This situation is illustrated qualitatively in Figure 4-19. The minimum gas rate that will keep the well unloaded for a particular tub-

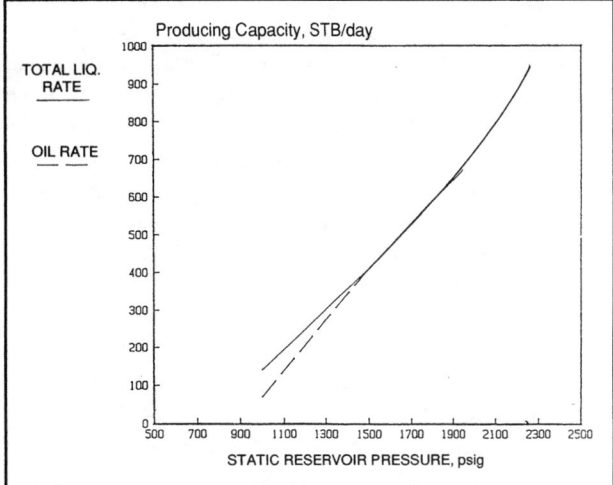

Fig. 4-18. Example 4-7 solution.

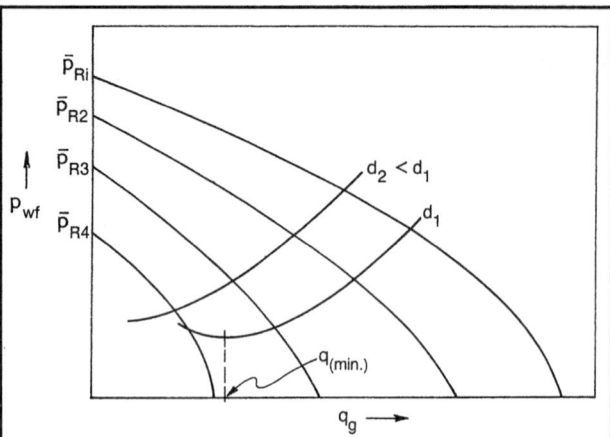

Fig. 4-19. Future gas well performance.

ing size and wellhead pressure can also be estimated using Equation 3-112.

IX. RELATING PERFORMANCE TO TIME

Methods for calculating the producing capacity of a well as a function of static reservoir pressure were presented in the previous section. This analysis procedure could be extended to include an entire reservoir by preparing a graph of producing capacity q_L versus $\bar{p}_R$ for each well producing from the reservoir. The reservoir producing capacity at a particular value of $\bar{p}_R$ would then be the sum of the individual well capacities, $Q_L = \Sigma q_L$. These calculations will be more accurate if a mathematical reservoir model other than the material balance model is used. This will also increase the accuracy of the outflow calculations, since producing gas/oil ratios and water cuts will be more accurate.

This type of analysis could be used to estimate at what value of $\bar{p}_R$ the producing capacity or deliverability will have declined to some particular value. However, this type of information also needs to be known as a function of time to facilitate development planning or to make economic evaluations. For example, oil wells may have to be worked over or placed on artificial lift, or gas wells may require installation of a compressor when certain producing capacities can no longer be met. Purchasing this equipment requires knowledge of when it will be needed. The timing of the expenditure of money is required for any economic evaluation of a project or for comparison of projects that require investment.

A general procedure for developing a relationship among cumulative production, reservoir pressure, producing capacity and time will be described in this section. Although a simplified example will be presented, more complex problems will usually require computer calculations.

The following procedure may be used to relate reservoir and well performance to time:

1. Using data from a material balance or other reservoir model, determine $\bar{p}_R$, GOR, f_w, etc., versus cumulative production, N_p. Typical behavior of these parameters is illustrated in Figure 4-20.

2. Construct inflow-outflow curves, similar to Figure 4-21 for each well. This step was illustrated in Example 4-7.

3. Construct a graph of producing capacity $Q_L = \Sigma q_L$ versus $\bar{p}_R$, as illustrated in Figure 4-22 and Example 4-7.

4. Select a small increment of production, ΔN_p, and determine the average value of static reservoir pressure, $\bar{p}_{R(avg)}$ that existed during this producing interval. The smaller the ΔN_p increments, the more accurate the analysis will be.

5. Using the value of $\bar{p}_{R(avg)}$ determined in Step 4, enter the graph of Q_L versus $\bar{p}_R$ and determine the average producing capacity, $Q_{L(avg)}$ at this value of $\bar{p}_{R(avg)}$.

6. Calculate the time increment required to produce the cumulative production increment. That is, $\Delta t = \Delta N_p / Q_{L(avg)}$.

7. Repeat Steps 4 through 6 and plot $\bar{p}_{R(avg)}$, and $N_p = \Sigma \Delta N_p$ versus $t = \Sigma \Delta t$, to obtain a graph such as Figure 4-23. The time at which the producing capacity reaches some minimum value can then be determined.

Example 4-8:

The well described in Example 4-7 is producing from an 80-acre drainage area. Using the following cumulative production versus pressure data, determine when the oil-producing capacity of the well will decline to 150 STB/day.

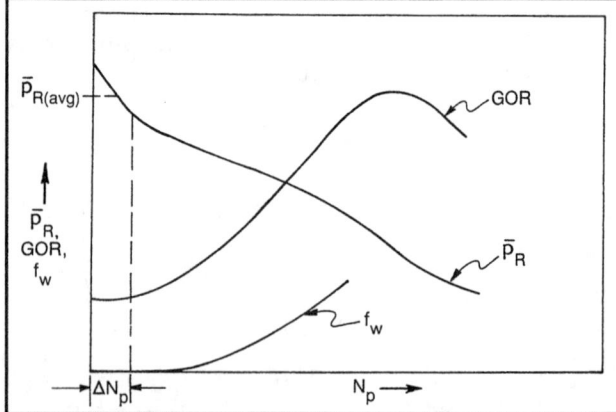

Fig. 4-20. Reservoir performance.

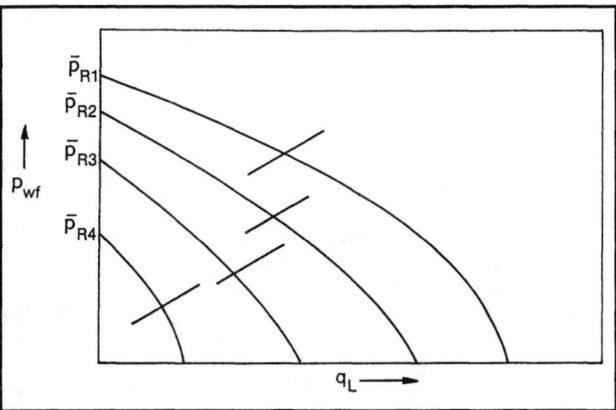

Fig. 4-21. Determining producing capacity.

$\bar{p}_R$, psig	N_p, STB
2250	0
1800	80,000
1500	150,000
1000	300,000

Solution:

A plot of $\bar{p}_R$ versus N_p is shown in Figure 4-24. The following table was constructed using the procedure outlined previously. Oil-producing capacity, $Q_{o(avg)}$, was obtained from Figure 4-18 for the values of $p_{R(avg)}$ corresponding to production increments of 50,000 STB.

$\Delta N_p \times 10^{-3}$	$N_p \times 10^{-3}$	$\bar{p}_{R(avg)}$	$Q_{o(avg)}$	Δt, days	t, days
50	50	2100	760	66	66
50	100	1950	680	74	140
50	150	1630	500	100	240
50	200	1430	380	132	372
50	250	1260	260	192	564
50	300	1100	150	333	897
50	350	930	50	1000	1897

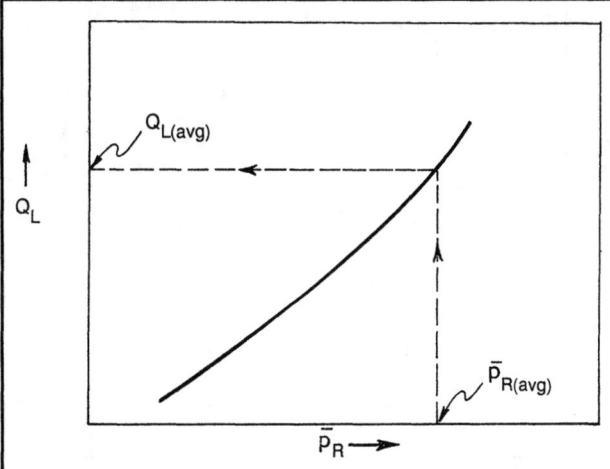

Fig. 4-22. Field producing capacity.

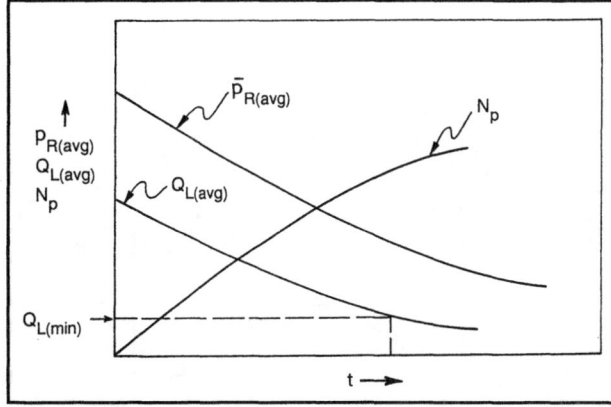

Fig. 4-23. Performance versus time.

The data are plotted on Figure 4-25. Entering the graph at a value of Q_o = 150 STB/day, a time of 930 days is obtained. At that time the cumulative oil produced is approximately 300,000 STB and the reservoir pressure is 1030 psig.

X. ANALYZING MULTIWELL SYSTEMS

The concepts discussed for applying total system or nodal analysis to single wells can also be applied to the analysis of multiwell systems, including entire fields. The procedure will be illustrated qualitatively by referring to the simple system shown in Figure 4-26. In this case, a change made in any component in the system would affect the producing capacity of the total system. Some of the changes that could be considered are:

1. Working over individual wells.

2. Placing some wells on artificial lift.

3. Adding new wells to the system.

4. Shutting in some of the existing wells.

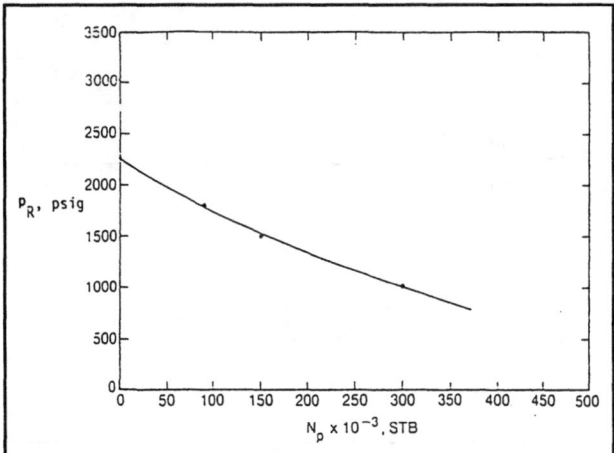

Fig. 4-24. Example 4-8 solution.

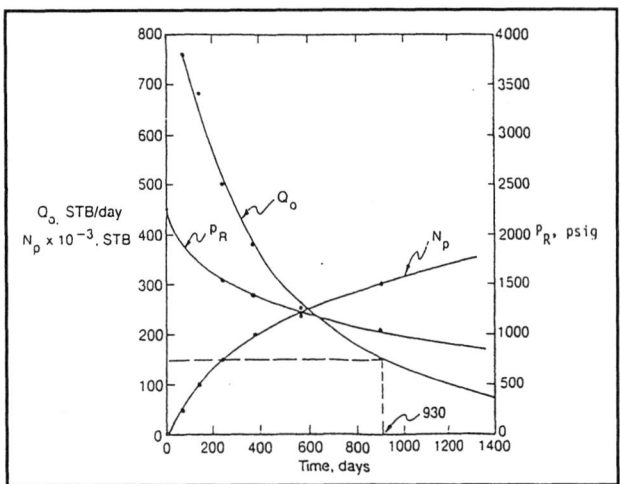

Fig. 4-25. Example of 4-8 solution.

5.   Changes in producing characteristics with time.

6.   Effect of surface line sizes.

7.   Installation of pumps or compressors.

8.   Effect of the final outlet pressure, p_D.

The location of the node for the final analysis must be selected at a point such that there is no further commingling of flow streams downstream of the node.

For the system shown in Figure 4-26, this would be at either point C or point D. Intermediate nodes must be selected at any point where flow streams commingle upstream of the final node, points A and B in the system shown. The analysis must begin at the reservoir pressure that is independent of rate and end at some final outlet pressure that is also independent of rate.

To illustrate the procedure, consideration will be given to either changing the pressure at point D or looping the surface line between points C and D. The node will be

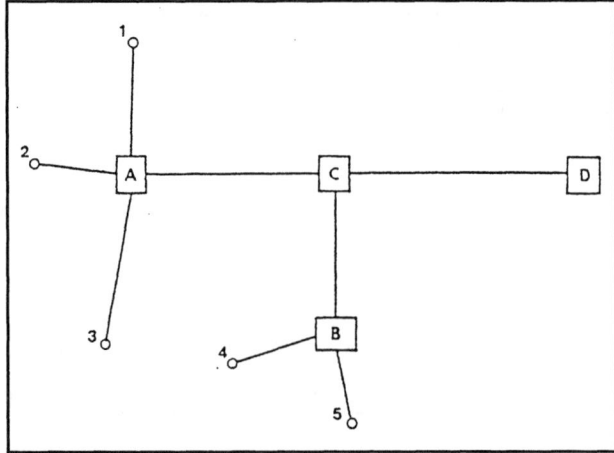

Fig. 4-26. Multiwell system.

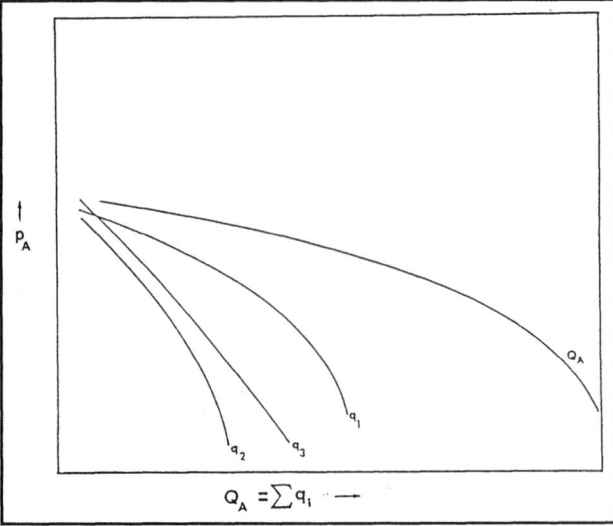

Fig. 4-27. Inflow to Point A.

selected at point C. Intermediate nodes must be considered at points A and B.

The inflow to point A will be calculated from:

$$p_A = \bar{p}_R - \Delta p_{res} - \Delta p_{tubing} - \Delta p_{flowline}$$

This expression would be evaluated for each well feeding into point A for a range of producing rates. This would result in a plot such as illustrated in Figure 4-27. A similar plot for the pressure behavior at point B can be constructed by considering wells 4 and 5. This is illustrated in Figure 4-28. The gas/liquid ratios and water fractions used in calculating the pressure drops in the piping system to this point would be those corresponding to the individual wells, since no commingling of well streams has occurred.

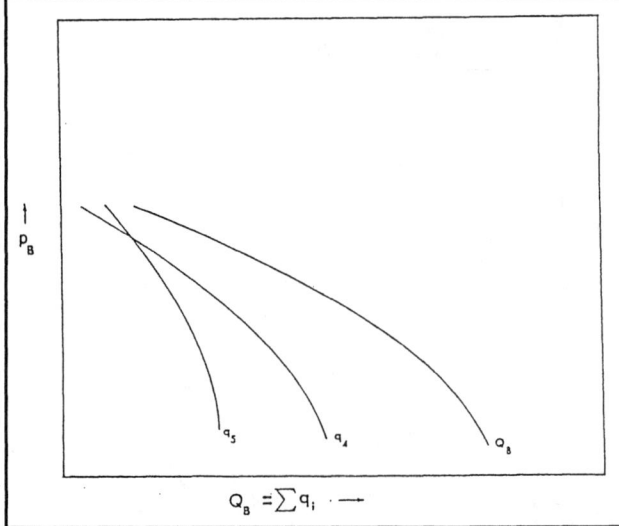

Fig. 4-28. Inflow to Point B.

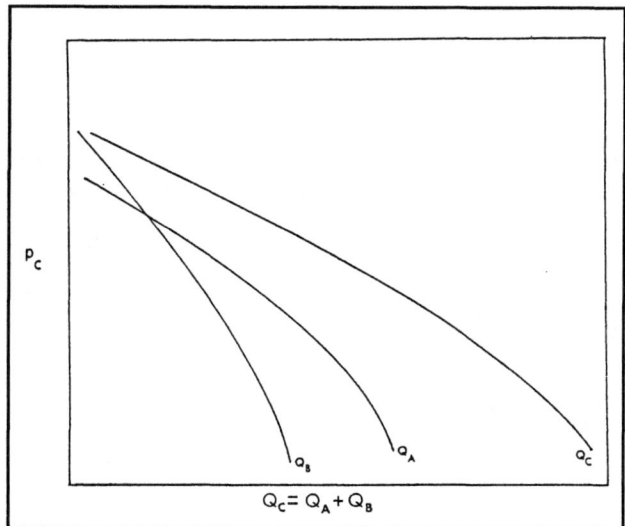

Fig. 4-29. Inflow to Point C.

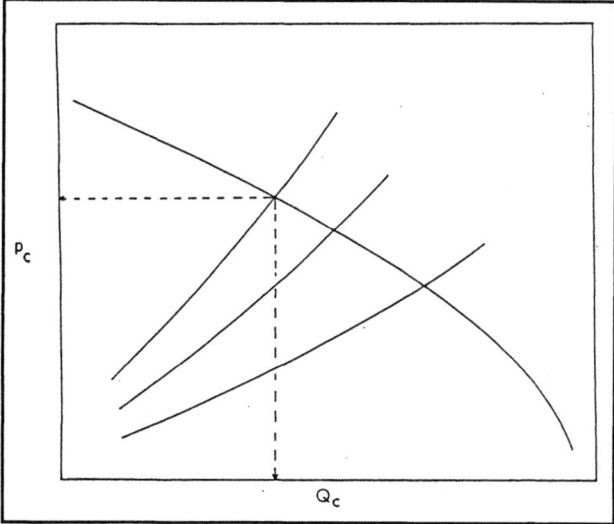

Fig. 4-30. System capacity.

Moving downstream to the next point at which commingling occurs, point C, the inflow expressions for the flows coming from points A and B are:

$$p_C = p_A - \Delta p_{AC} \quad \text{and}$$
$$p_C = p_B - \Delta p_{BC}$$

This will result in a relationship between pressure at point C and the inflow rate into point C as illustrated in Figure 4-29. The calculation of the pressure drops between points A and C, Δp_{AC}, and between points B and C, Δp_{BC}, is complicated by the fact that the GLR and water fractions are functions of rate if the individual wells have different values of these parameters. In this case, the correct GLR and f_w for each commingled rate are calculated using:

$$\text{GLR}_{AC} = \frac{\Sigma q_i \text{GLR}_i}{Q_A}$$

and

$$f_{wAC} = \frac{\Sigma q_i f_{wi}}{Q_A}$$

Similar expressions are used to determine these values for calculating Δp_{BC}.

The expression for the outflow from point C is

$$p_C = p_D + \Delta p_{CD}$$

Calculation of Δp_{CD} for various rates would again require determining the correct GLR and f_w corresponding to each $Q_C = Q_A + Q_B$. A change in either p_D or the line size between points C and D would result in different outflow curves and thus different system capacities, as shown in Figure 4-30. To determine the effect of these changes on individual well performance, the pressure at point C corresponding to an intersection on Figure 4-30 can be used to move upstream to points A and B and thus determine the individual well rates.

The procedure outlined previously will also apply if some of the wells are under choke control or on artificial lift. If a well is flowing through a choke in critical flow, the well's rate will be constant unless the pressure downstream of the choke is increased to the point at which critical flow no longer occurs. The producing rate for a well on a sucker rod pump may be independent of the pressure at the commingling point, but its rate will affect the pressure at the commingling point and thus the producing capacity of other wells feeding into the same point. The producing capacity of wells on gas lift or electrical submersible pumps would be affected by the well-head pressure or the pressure at the commingling point. Analysis of wells on artificial lift will be discussed in Chapter 5.

5

Artificial Lift Design

I. INTRODUCTION

In Chapter 4 it was shown that as the pressure in a reservoir declines from depletion the producing capacity of the wells will decline. The decline is caused by both a decrease in the reservoir's ability to supply fluid to the wellbore, and, in some cases, an increase in the pressure required to lift the fluids to the surface. That is, both inflow and outflow conditions may change.

The only way in which the inflow can be kept high, once the well has been stimulated to reduce reservoir pressure drop to a minimum, is by pressure maintenance or secondary recovery. This will eventually be initiated in most oil reservoirs, but methods are available to reduce the flowing wellbore pressure by artificial means, that is, to modify the outflow performance of the well.

All the methods presented earlier for calculating reservoir performance, or generating IPRs, apply equally well to either flowing or artificial lift wells. The reservoir inflow performance depends on p_{wf} and is completely independent of what methods are employed to obtain a particular value of p_{wf}. Therefore, no new procedures are required for reservoir performance in analyzing artificial lift wells.

Analysis of the outflow, however, is changed considerably from what was presented earlier for some types of artificial lift methods. For other types, the outflow analysis will change very little, only requiring addition of a pressure increase term in the outflow expression. The analysis is frequently expedited by choosing the node location as the point at which the artificial lift energy is introduced into the system. This point is usually very close to the reservoir, but for certain types of artificial lift analysis, it may be some distance above the reservoir.

The four most commonly used artificial lift methods are sucker rod or beam pumping, gas lift, submersible pumping, and hydraulic pumping. For a thorough discussion of each type of system, its frequency of application and relative advantages and disadvantages, reference should be made to Brown.[1] Descriptions of the methods discussed in this chapter will be given as the analysis procedures are presented.

II. CONTINUOUS FLOW GAS LIFT

The operation of a continuous gas lift well is very similar to that of a naturally flowing well. Gas is continuously injected into the tubing through a gas lift valve at a fixed depth and the increased gas/liquid ratio from the valve to the surface decreases the hydrostatic pressure gradient in the tubing, thus decreasing p_{wf}. The only difference between this type of operation and a flowing well is that the gas-liquid ratio changes at some point in the tubing for the gas lift well. The depth at which the operating gas lift valve can be located depends on the gas injection pressure available. The more pressure available the deeper the injection point can be. Also, as the depth of injection is increased, less injection gas is required to achieve the same bottomhole pressure. A simplified schematic and pressure traverse for a gas lift well is shown in Figure 5-1. Other valves are required above the working valve in order to unload the well, and the design and location of these valves will be discussed later.

The design of a continuous flow gas lift system consists of essentially two parts: (1) determination of the performance of the well once it is unloaded and in stabilized operation, and (2) spacing and pressure setting of the upper gas lift valves used in unloading the well.

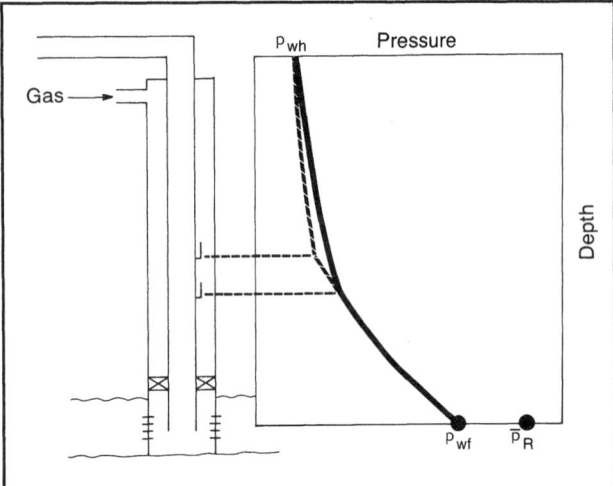

Fig. 5-1. Gas lift well schematic.

Determination of the well's stabilized performance will be discussed first.

In Figure 5-1, there are essentially two variables that can be controlled by the designer for a given tubing size and wellhead pressure. These are the depth at which the gas is injected and the volume of gas that is injected. There can be constraints on these values that must be taken into account.

For example, the amount of surface gas injection pressure controls the depth to injection. As was illustrated in Figure 5-1, the deeper the injection depth, the higher the pressure in the tubing at the point of injection. The pressure in the annulus at the injection point must be between 50 and 150 psi greater than the pressure in the tubing to be able to inject the gas. Therefore, to inject at a deeper point in the well, more injection gas pressure is required.

As will be seen subsequently, there is an optimum injection gas volume for a well that will result in a maximum liquid production rate. If this volume of gas is not available, the well will produce at a lower rate. If several gas lift wells in a field are utilizing a limited volume of injection gas, nodal analysis can be used to determine the optimum volume of gas to allocate to the various wells.

A. Well Performance

The performance of a gas lift well can be analyzed using system nodal analysis in the same manner as was discussed in Chapter 4 for a flowing well. Any convenient node may be selected depending on which parameter is being analyzed. If the effect of the injected gas volume is being analyzed, it may be convenient to select the working valve as the node, particularly if wellhead pressure remains constant. In this case, the node pressure would be the pressure in the tubing at the valve p_v, and the inflow and outflow expressions would be:

Inflow

$$\bar{p}_R - \Delta p_{res} - \Delta p(\text{tubing below valve}) = p_v$$

Outflow

$$p_{wh} + \Delta p(\text{tubing above valve}) = p_v$$

A plot of p_v versus q_L would yield the producing capacity at the intersection of the inflow and outflow curves. A change in injected GLR would not affect the inflow curve.

If the effect of depth of injection were being analyzed, it would be more convenient to select p_{wf} as the node pressure, in which case only the outflow curve would change with injection depth change. This is sometimes used to determine which valve is the working valve in an operating gas lift well.

If a gas lift well has a flowline of considerable length, the wellhead pressure will not be constant, but will increase as the gas injection rate is increased. This results from the increased friction loss in the flowline. Therefore, an excessive gas injection rate can actually cause an increase in p_{wf} and thus decrease the well's producing capacity. Systems analysis can be used to determine the optimum injected gas volume from the standpoint of maximum producing capacity. The optimum injection rate from an economic standpoint can then be determined.

A general procedure for determining the optimum gas rate will be outlined and illustrated graphically. The node will be selected at the reservoir. The inflow and outflow expressions are:

Inflow

$$\bar{p}_R - \Delta p_{res} = p_{wf}$$

Outflow

$$p_{sep} + \Delta p_{flowline} + \Delta p(\text{tubing above valve}) + \Delta p(\text{tubing below valve}) = p_{wf}$$

With p_{wf} as the node pressure, the inflow will be independent of injected GLR, but the flowline pressure drop and the pressure drop in the tubing above the valve will change as injected GLR changes. Therefore, a different outflow curve would be obtained for each injected GLR.

The formation GLR must be used to calculate Δp (tubing below valve), and the total GLR must be used above the valve. As the injected GLR becomes too large, the increase in piping system pressure drop due to friction will exceed the decrease in the hydrostatic pressure in the tubing above the valve. This is illustrated in Figure 5-2.

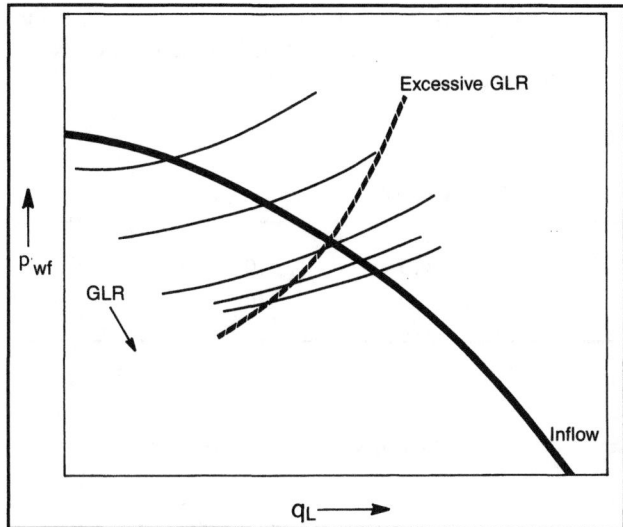

Fig. 5-2. Gas lift well analysis.

The intersections of the inflow and outflow curves give the liquid production rate corresponding to each injected GLR. The required volume of gas to be injected can then be calculated, and a plot of liquid production rate versus gas injection rate can be constructed. This is illustrated in Figure 5-3. The maximum liquid rate obtainable for this well and the corresponding gas injection rate can be read from this plot. Injection at a rate higher than this would actually result in a decreased liquid rate for an increased expenditure of money. This is not an uncommon occurrence in gas lift operations.

As can be observed in Figure 5-3, the slope of the plot ($\Delta q_L / \Delta q_{inj}$) becomes smaller as the maximum liquid rate is approached. That is, a large increase in gas injection rate is required to obtain a small increase in liquid production rate. Depending on the cost to compress the gas and the income from the sale of the oil, the economic optimum gas injection rate may be considerably lower

than that required to obtain the maximum liquid rate. This is illustrated qualitatively on Figure 5-3, but the actual value will depend on the individual well being considered. It can, of course, change if the price of oil or the cost of compression changes.

The liquid production rate that can be expected from any limited available gas volume can also be estimated from a plot such as Figure 5-3.

The calculations required to produce the plots in Figures 5-2 and 5-3 were based on a particular injection depth. As was discussed earlier, the possible injection depth is controlled by the available injection gas pressure. In most cases, the maximum liquid production rate is obtained by injecting the gas as deep as possible, possibly just above the packer. This could, however, require a high surface pressure if the well is deep. The effect of injection depth and, therefore, injection pressure required on liquid production rate can be determined by repeating the procedure illustrated in Figure 5-3 for various injection depths. This would result in a different plot of q_L versus q_{inj} for each injection depth, as illustrated in Figure 5-4. Once the liquid production rate and gas injection rate are determined, the pressure existing in the tubing at the injection depth can be determined. The required surface operating pressure for the injected gas can then be estimated from:

$$p_{so} = (p_v + \Delta p_v)e^{-x} \qquad (5\text{-}1)$$

where

$$
\begin{aligned}
p_{so} &= \text{surface operating pressure, psia} \\
p_v &= \text{tubing pressure at the gas lift valve, psia} \\
\Delta p_v &= \text{pressure drop across the valve, psi} \\
e &= \text{natural log base,} \\
X &= 0.01875\, D_v\, \gamma_g / \overline{T}\, \overline{Z}, \\
D_v &= \text{true vertical depth to injection point, ft,} \\
\gamma_g &= \text{specific gravity of injected gas,}
\end{aligned}
$$

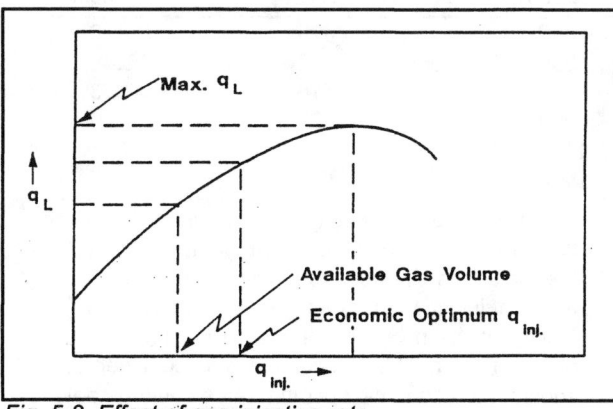

Fig. 5-3. Effect of gas injection rate.

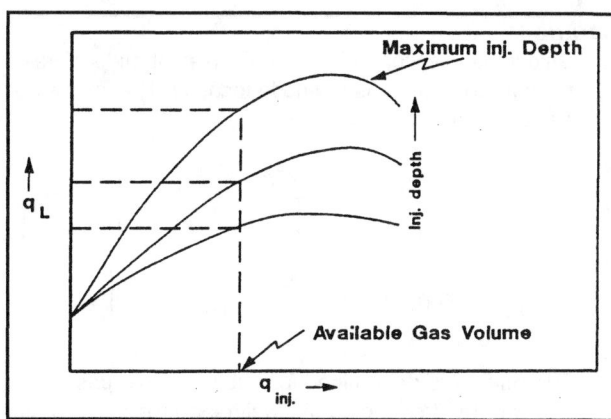

Fig. 5-4. Effect of injection depth.

$\overline{T}$ = average temperature in the annulus, °R
$\overline{Z}$ = average gas compressibility factor.

Solution of Equation 5-1 is iterative since $\overline{Z}$ depends on the unknown pressure p_{so}. It also assumes no friction loss in the annulus. If required injection rates are high or the annular area is small, friction may have to be considered. The procedure for calculating pressure drop for this case was discussed in Chapter 3.

The previous discussion assumed that parameters such as separator pressure, flowline size, tubing size, water cut, formation GLR and static reservoir pressure were fixed. The effects of changes in any of these can be studied for a well on gas lift by using the procedures described previously for flowing wells. For example, if a well is to be placed on gas-lift (and side pocket mandrels have not previously been installed), the tubing will have to be pulled. This presents an excellent opportunity to optimize tubing size. Also, since the flowline will have to transport additional fluid when gas lift is installed, it is often economically feasible to either replace or loop the existing flowline.

Example 5-1:

An oil well that has died due to declining gas/oil ratio is to be placed on gas lift. From a test conducted on the well before it died, it was determined that $q_{o(max)}$ is 1200 STBO/day. Using this test data and the following information, determine the maximum rate at which this well will produce, and prepare a plot of oil production rate versus gas injection rate.

Depth to mid-perforations = 6000 ft
Gas injection depth = 6000 ft
Tubing I.D. = 1.995 in.
Flowline length = 4000 ft
p_{sep} = 100 psig
$\overline{p}_R$ = 1500 psig
Flowline I.D. = 2 in.
Formation GLR = 100 scf/STBO
f_w = 0

Solution:

Select the injection point, which is also at midperforations for this case, as the node location. The inflow is calculated from:

$$p_{wf} = \overline{p}_R \left[\left(1.266 - \frac{1.25q_o}{q_{o(max)}} \right)^{0.5} - 0.125 \right]$$

$$p_{wf} = 1500 \left[\left(1.266 - \frac{1.25q_o}{1200} \right)^{0.5} - 0.125 \right]$$

The outflow is determined for each possible gas injection rate by first finding the wellhead pressure necessary to satisfy the flowline requirements for various liq-

uid rates, and then using these wellhead pressures to find the node pressure necessary to satisfy the tubing and flowline requirements.

Inflow

q_o, STB/day	p_{wf}, psig
300	1277
400	1195
500	1107
600	1013

Outflow

Flowline Requirements

Injected GLR	Total GLR	q_o=300	q_o=400	q_o=500	q_o=600
300	400	150	170	200	230
500	600	160	190	230	260
700	800	170	210	250	300
900	1000	190	230	280	320
1100	1200	200	250	300	360
1400	1500	220	280	340	400

Tubing Requirements

Injected GLR	Total GLR	q_o=300	q_o=400	q_o=500	q_o=600
300	400	1040	1150	1260	1360
500	600	890	1000	1110	1210
700	800	840	960	1070	1180
900	1000	830	950	1060	1180
1100	1200	820	940	1070	1190
1400	1500	810	950	1080	1220

The inflow and six outflow curves are plotted on Figure 5-5. The producing capacities for the various injected GLRs are read from the intersections of the inflow and outflow curves. The oil production rate, injected GLR, and gas injection rates are tabulated as follows:

Injected GLR, scf/STBO	Oil Rate, STBO/day	Injection Rate, Mscfd
300	420	126
500	497	249
700	518	363
900	522	470
1100	515	567
1400	511	715

A plot of the oil production rate for various gas injection rates is shown in Figure 5-6. It can be seen that the maximum oil rate is approximately 520 STBO/day and that injection at a rate greater than about 360 Mscfd results in very little increase in oil rate. If the profit on oil and the gas compression cost were known, the economic optimum rates could be determined.

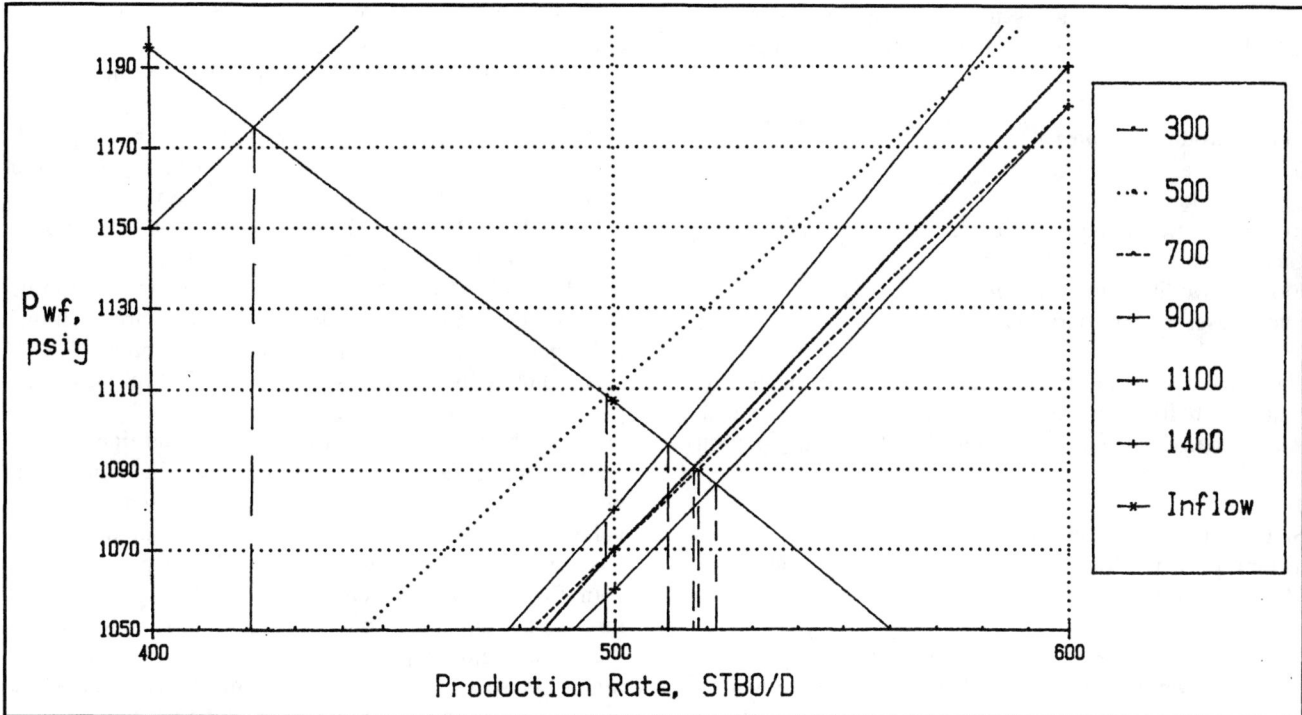

Fig. 5-5. Example 5-1 results.

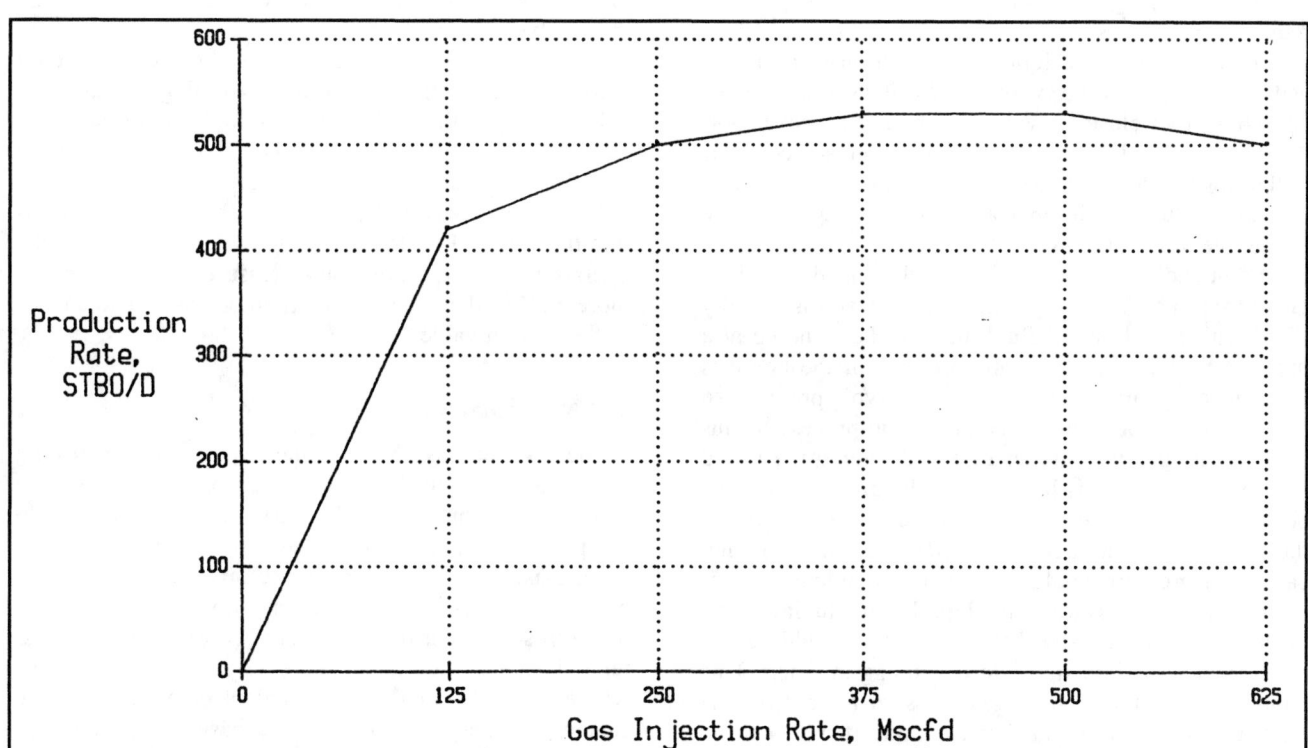

Fig. 5-6. Example 5-1 results.

Once the procedures discussed above have been followed, the following information is known for a particular well:

Liquid production rate: q_L,
Injection gas liquid ratio, GLR_{inj},
Injection depth, D_{inj},
Required bottomhole flowing pressure, p_{wf},
Gas injection surface pressure, p_{so},
Flowing wellhead pressure, p_{wh}, and
Static reservoir pressure, $\overline{p}_R$.

Knowledge or an estimate of all these parameters is required before the unloading valves can be spaced. Additional information required for valve spacing includes:

Static gradient of load fluid, g_s,
Tubing pressure during unloading period, p_{wh}, and
Maximum temporary gas pressure for "kickoff," p_{ko}.

As can be observed from the discussion on performance of a continuous flow gas lift well, once the well is producing in a stable condition, gas is being injected at a continuous rate at only one point in the tubing. Therefore, only one gas lift valve is open, and the required surface operating pressure is known. However, unless a much higher pressure source of gas is available to unload kill fluids from the well to get it into this stable operating condition, other gas lift valves at shallower depths are required. The necessity for having the unloading valves will first be illustrated, followed by a discussion of procedures for spacing the valves. As it is necessary to be able to open or close the valves from the surface, a short discussion of valve mechanics will be presented. This will allow the designer to set the pressure in the dome of the valve so that it will open and close at the proper time during the unloading process.

The unloading process is described using illustrations taken from the API Design Manual.[3] The well is initially filled with liquid or kill fluid that will be standing at a depth such that the hydrostatic pressure of the liquid is equal to or greater than the static reservoir pressure so that no flow is occurring. The unloading process begins by injecting gas into the annulus and forcing the liquid into the tubing. This is illustrated in Figure 5-7A. As can be seen from this figure, if there were only one valve in the well, that is, the bottom or operating valve, enough casing gas pressure would have to be available to support the hydrostatic pressure of the liquid in the tubing down to the depth of the valve before the valve could be uncovered and gas injection into the tubing initiated. This would require a very high gas pressure for deep wells loaded with heavy liquid. For example, if gas is to be injected at a depth of 10,000 ft and the kill fluid gradient

is 0.47 psi/ft, a hydrostatic pressure of 4700 psig would exist in the tubing at this depth. However, once the valve is uncovered and gas is injected into the tubing, the weight of the fluid in the tubing will be reduced considerably, and much lower gas pressure will be required. Thus, the purpose of the unloading valves is to reduce the casing gas pressure required to unload the well.

In Figure 5-7A, fluid from the annulus is being transferred into the tubing through the open valves and U-tubed by injection gas pressure being exerted on top of the liquid column in the annulus. All valves are open initially because of the high pressure in the annulus.

As shown in Figure 5-7B, the fluid level has been pushed down below the top valve and injection gas is entering the tubing through this valve. This lightens the weight of the fluid in the tubing above the top valve, which also reduces the pressure in the tubing at the lower valves. This allows the liquid level in the annulus to be pushed farther down with the available gas pressure.

In Figure 5-7C, injection gas is entering the tubing through the top and second valves. With the fluid level in the casing below the depth of the second valve, the tubing pressure is less than the casing pressure at valve depth, and gas enters the tubing through the second valve.

In Figure 5-7D, the top valve has closed and all other valves are still open. The third and bottom valves are not uncovered. Before the top valve will close, the casing pressure must decrease slightly. This will occur if more gas is passing into the tubing than is being injected into the casing. The second valve must remain open until the third valve is uncovered.

In Figure 5- 7E, the second and third valves are uncovered and injection gas is entering the tubing through both valves, causing the casing pressure to decrease further.

In Figure 5-7F, the top and second valves are closed and the third and bottom valves are open. The bottom valve is not uncovered and can, therefore, pass no gas into the tubing. For this illustration, the third valve is the operating valve. If more casing pressure were available or more gas could be injected, it might be possible to uncover the bottom valve.

B. Valve Spacing

There are at least three distinct types of valve spacing problems encountered in the design of continuous flow installations. One case is that in which the valves are to be spaced, and pressure-charged and run with the tubing in an existing well. A second case involves running side pocket mandrels in a well that may not be placed on gas lift until some later time. This case is frequently encountered in areas such as offshore or in remote locations where pulling tubing is inconvenient or expensive. The spacing of the mandrels must be based on anticipated future conditions. A third case is one in which the open-

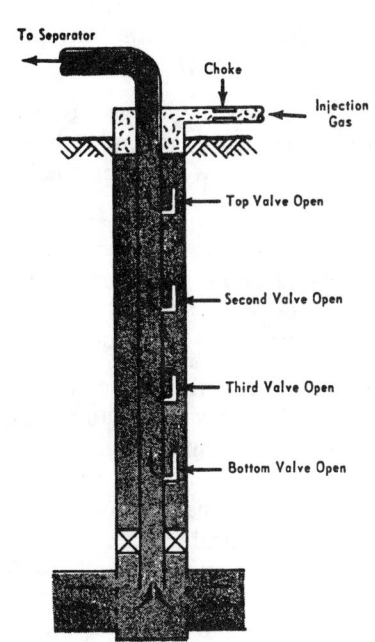

(A) Fluid from casing being transferred into tubing through all valves and u-tubed by injection gas pressure to surface.

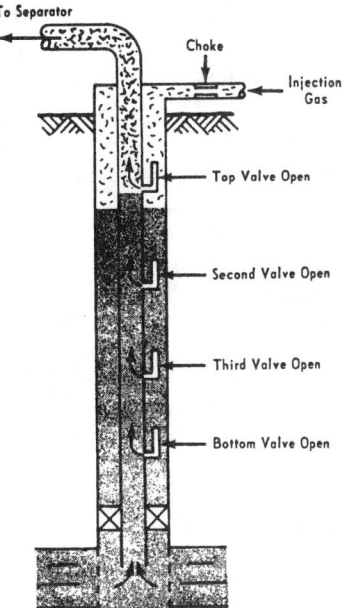

(B) Fluid in tubing being aerated to surface by injection gas through top valve as fluid in annulus is transferred into tubing through lower valves.

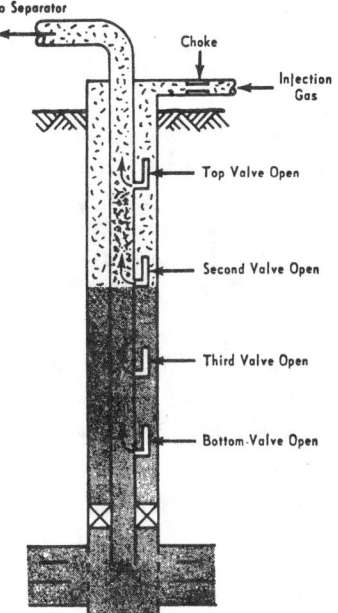

(C) Injection gas entering tubing through top and second valve immediately after second valve uncovered.

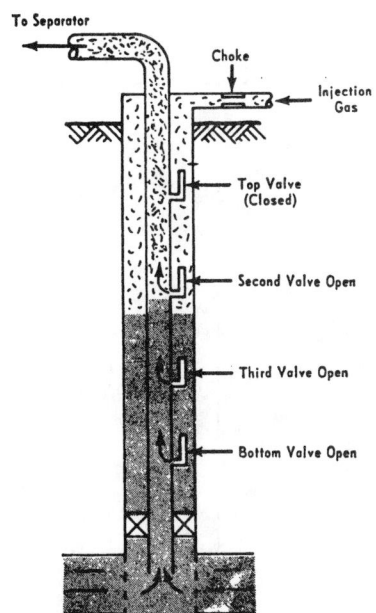

(D) Fluid in tubing being aerated to surface by injection gas through second valve as fluid in annulus is transferred into tubing through third and bottom valves.

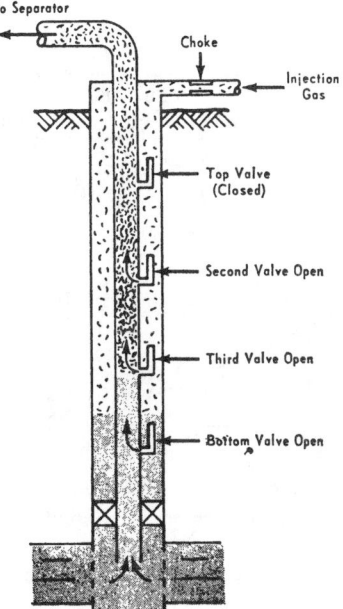

(E) Injection gas entering tubing through second and third valves immediately after third valve is uncovered.

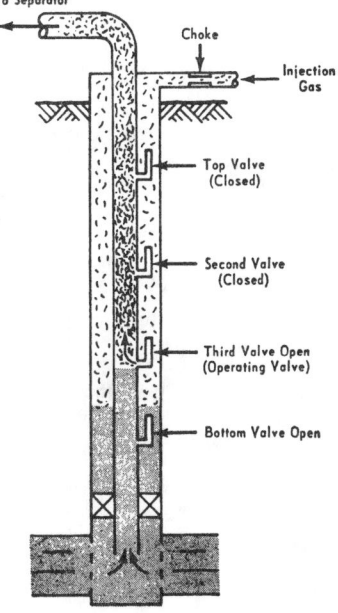

(F) Producing rate equals capacity of tubing from third valve for available injection pressure. Therefore, bottom valve cannot be uncovered.

Fig. 5-7. Unloading sequence.

ing and closing conditions must be calculated for valves to be run in existing mandrels. The mandrels may have been spaced some time earlier, and the existing conditions may be considerably different from the anticipated conditions. Only the first case will be discussed in detail in this book.

The unloading valve design process can be considered to consist of two parts: determining the required depth to each valve or spacing of the valves, and calculating the required pressure setting for each valve to be made at the surface before the valve is run into the well. The spacing and pressure settings must be such as to accomplish the following:

1. It must be possible to displace liquid from the casing into the tubing down to the desired operating depth with the available gas pressure, and

2. It must be possible to open any valve under producing conditions without opening the valve above it.

The spacing will be illustrated qualitatively and graphically first. Then safety factors will be introduced and detailed design procedures and examples will be presented. The procedure is illustrated graphically by making a plot of pressure versus depth, such as in Figure 5-8.

1. Construct the casing pressure traverse by locating the available surface casing pressure at zero depth and extending this line downward, taking into account the weight of the gas.

2. Starting at the flowing bottomhole pressure required to inflow the design production rate, plot the tubing flowing pressure traverse upward using the formation GLR.

3. Locate the intersection of the tubing pressure line and the casing pressure line as the point of balance. This is the depth at which gas could be injected if no pressure drop across the valve were required.

4. Locate the tubing flowing wellhead pressure at zero depth, and connect this point with a line from the point of balance. This represents the tubing pressure traverse above the point of injection based on the design liquid rate and the formation plus the injected gas rate. This is the theoretical condition that would exist in the well after it is unloaded and stabilized, assuming no pressure drop across the operating gas lift valve.

5. Starting at the surface tubing pressure, extend a line downward based on the kill or load fluid gradient. This represents the pressure in the tubing before any

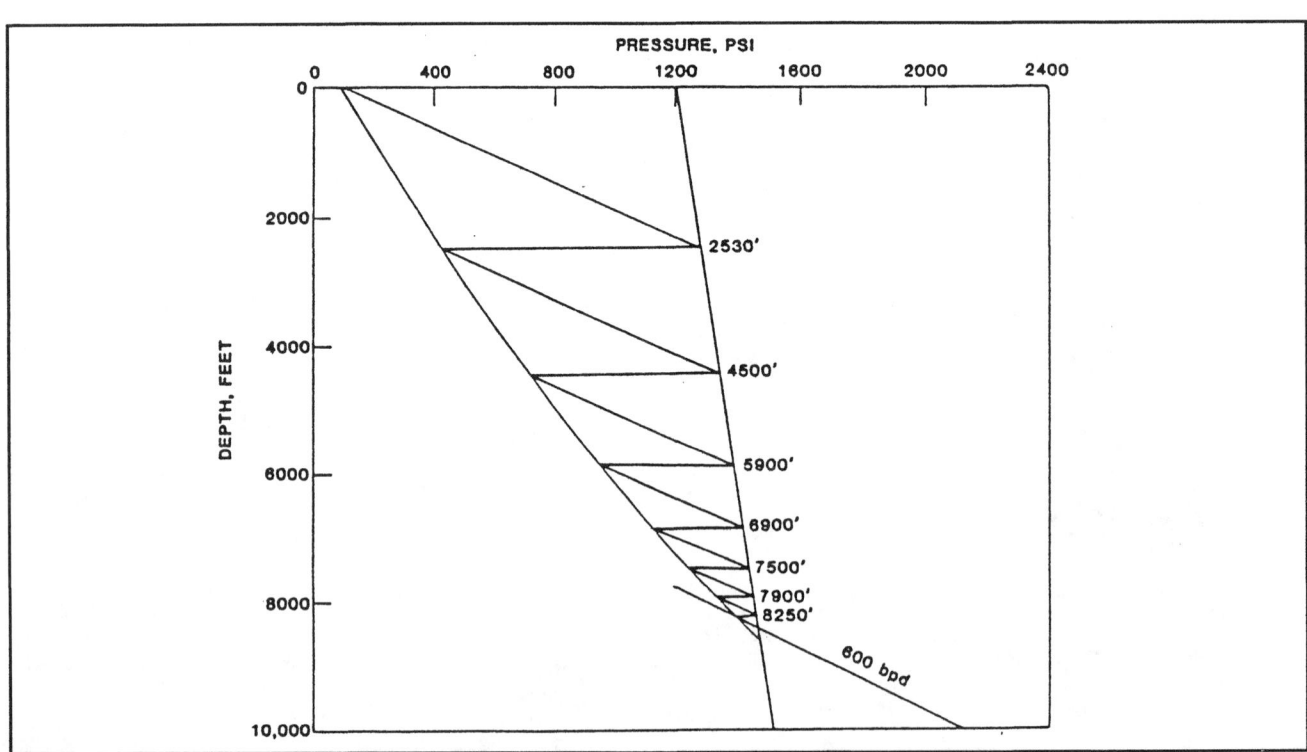

Fig. 5-8. Graphical solution for valve spacing.

gas has entered the tubing. The intersection of this line represents the depth at which the pressures in the tubing and casing are equal as the load fluid is being U-tubed from the casing into the tubing. This is the location of the top valve. When gas is injected into the tubing at this point, the tubing gradient above the valve will then be represented by the flowing gradient line.

6. From the depth of the top valve on the flowing gradient line, extend a line based on the load fluid gradient downward to the intersection with the casing pressure line. This locates the second valve. Repeat this procedure until the difference between the casing and tubing pressures is about 50 psi. Figure 5-8 represents this procedure for a hypothetical case.

Most design methods include several safety factors, and these factors are recommended for the following reasons: (1) an error in well data that would affect injection gas volume or injection gas pressure at depth; (2) an error in valve operating temperature for temperature sensitive valves; (3) a slight error in setting the valve opening pressure in a tester, particularly if the valve opening pressure must be set in the field; and (4) to overcome the load rate of the gas lift valve, which means that a given psi increase in pressure is required to obtain sufficient stem travel in a gas lift valve.

Several safety factors that are incorporated in the design of a continuous flow installation as discussed in Reference 3 are:

1. A pressure differential across the valve of 50 psi is recommended and used for determining valve depths in most installations. Generally, only a few psi differential would be required theoretically if all assumed data were absolutely correct.

2. The maximum tubing pressure at valve depth, used to calculate the reopening pressure of a valve, is based on the maximum flowing tubing pressure at this depth while lifting from the next lower valve with the assumed pressure differential across the lower valve. The minimum flowing tubing pressure at valve depth is opposite the upper gas lift valves at the instant a lower valve is uncovered. However, gas lifting may occur from the lower valve depth with the flowing tubing pressure equal to, or near, this maximum flowing tubing pressure.

3. A pressure gradient based on a flowing load fluid gradient curve is used to locate the valve depths. This condition should not exist for the lower valves when the flowing BHP is less than the static BHP and reservoir fluid feed-in occurs.

4. The total producing GLR is assumed to be the injection GLR during unloading operations for the tra-verse above the point of gas injection. In most installations, some formation gas will be produced before the unloading operations are completed.

5. Generally, the unloading traverse for locating the valve depths is based on the desired producing rate and injection gas volume available to unload the well. The actual unloading producing rate can be controlled by varying the injection gas volume. Decreasing the choke size in the injection gas line will decrease the injection gas volume and producing rate during liquid transfer from casing to tubing. Therefore, for high capacity wells, a minimum fluid gradient traverse for a lower producing rate may exist until the deliverability of the well approaches the desired producing rate.

Not all the above safety factors will be recommended for all installation design methods. Although many of these factors may be used, the gas lift valve manufacturers' literature seldom emphasizes this fact. Experience dictates that certain safety factors should be incorporated in the design calculations to assure unloading.

One means of including a safety factor in the design, frequently called the Camco or Winkler method, is illustrated in Figure 5-9. The safety factor consists of reducing the casing pressure required to open the valve successively for each lower valve in the well. The casing pressure available for U-tubing at each valve is reduced, and therefore, the valves must be spaced closer together. As can be seen from Figure 5-9 the point at which a 50 psi differential between the casing and tubing pressure is reached is at a shallower depth in the well. This means that with the same available surface gas injection pressure, the well would produce less liquid. When this method is used, the location of the point of injection becomes trial and error. That is, the total reduction in casing pressure depends on the number of valves required to reach the point of injection. The spacing of the valves, and therefore the number required, depends on the flowing tubing pressure traverse, which depends on production rate and point of injection. The production rate depends on flowing bottomhole pressure, which depends on the point of injection.

Another means of including safety factors is called the design gradient method and is illustrated in Figure 5-10. Some pseudo flowing wellhead pressure higher than the expected wellhead pressure is selected. This pseudo wellhead pressure is generally selected as the expected flowing wellhead pressure plus 20 percent of the difference between surface casing and tubing pressure. The method is frequently referred to as the 20 percent design gradient method. A line is drawn between this pseudo wellhead pressure and the point of injection. The valve spacing is then carried out as if the pressure in the tubing during unloading were represented by the pseudo pressure line.

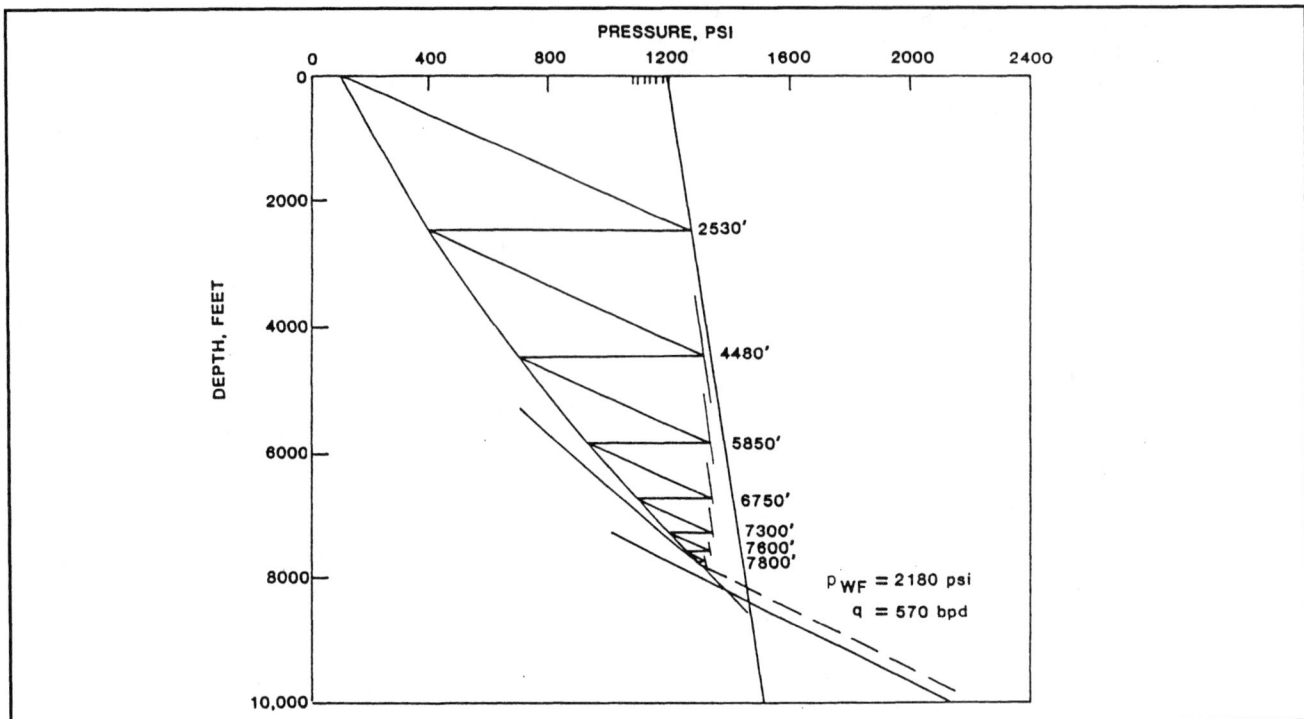

Fig. 5-9. Example design using casing drop of 20 psi.

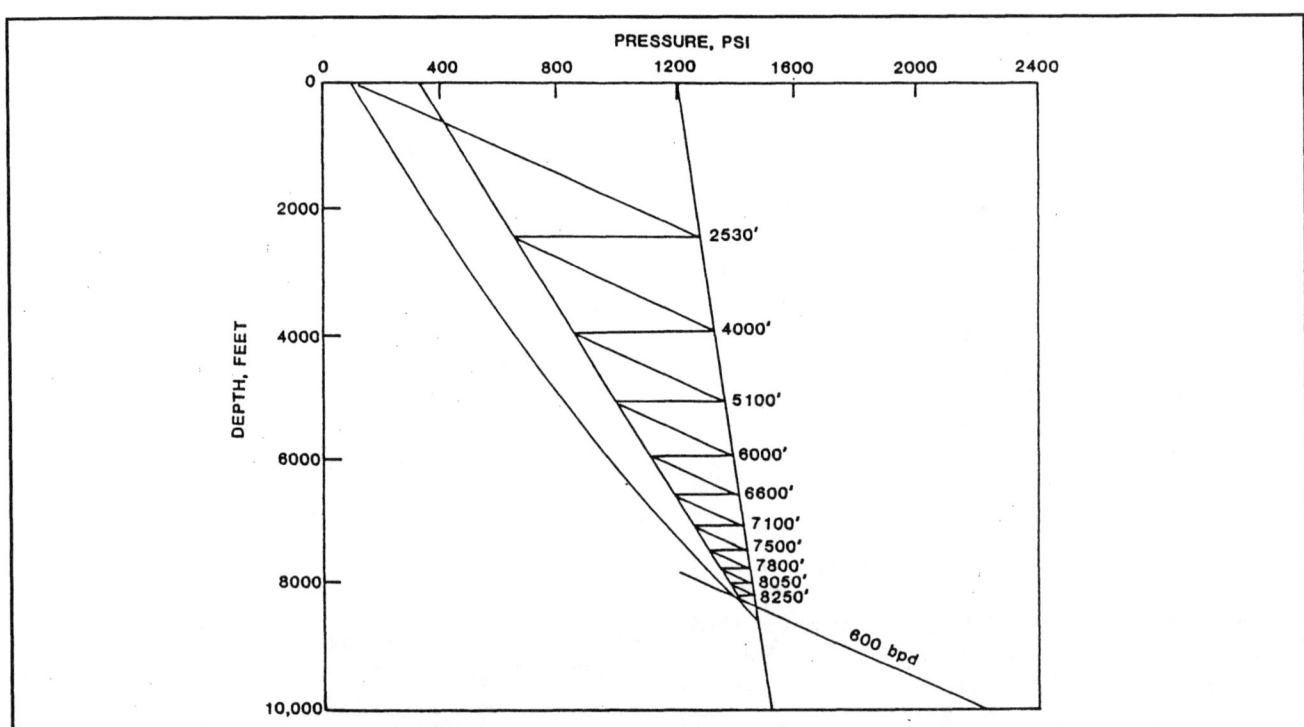

Fig. 5-10. Variable gradient design.

Using this method, full casing pressure is available at the depth of injection, and location of the point of injection is not iterative.

Only two methods of incorporating safety factors into the design have been described. Most gas lift valve manufacturers have their own methods, and some companies use a combination of the two described previously. It is also common practice to install additional valves below the anticipated initial point of injection to be able to handle changing well conditions. Some designers also recommend bracketing the design injection point with several valves above and below this point to handle changing conditions.

1. Gas Lift Valve Performance

As was discussed earlier, one of the requirements for gas lift valve design is that one must be able to open any valve without opening the valve above it in the well. Selection of the pressure at which to charge the bellows of a valve before running it into the hole requires some knowledge as to how the valve responds to various pressures and temperatures, particularly the casing pressure and tubing pressure. A very brief discussion of the most commonly used valve is given here. For a more comprehensive discussion, see the API Gas Lift Design Manual.[3] A well may be equipped with either nonretrievable or retrievable valves. That is, they may be mounted on the outside of the tubing or may be run in side pocket mandrels inside the tubing. Figures 5-11 and 5-12 illustrate both of these cases. The operation of the valves is independent of the mounting but the temperature existing at the valve during unloading and producing operations may be different. For retrievable valves, the temperature is usually considered to be the flowing fluid temperature. For nonretrievable valves, the temperature is considered to be earth temperature. The fact that neither of these may be correct is one of the reasons that safety factors are required in the spacing of the valves. Estimates of the flowing temperature at any depth may be made using either the correlation given in Chapter 3 or Figure 5-13.

Most gas lift valves can be placed into one of two broad categories for analysis.

These categories are called Injection Pressure or casing pressure operated valves, and Production Pressure or fluid operated valves. The two types are shown in Figures 5-14 and 5-15. Only one type of valve will be considered in this book, the unbalanced pressure charged valve that is primarily responsive to injection or casing pressure, Figure 5-14.

The design of a valve spacing program for a continuous-flow gas lift well requires the calculation of several pressures. The pressures at which a valve will close and reopen downhole are essential in spacing the valves. It is necessary to calculate the required bellows or dome pressure existing downhole that controls the opening and closing pressures. Since the bellows must be pressured up or charged at the surface, this bellows pressure must be converted to standard conditions. To be sure that the

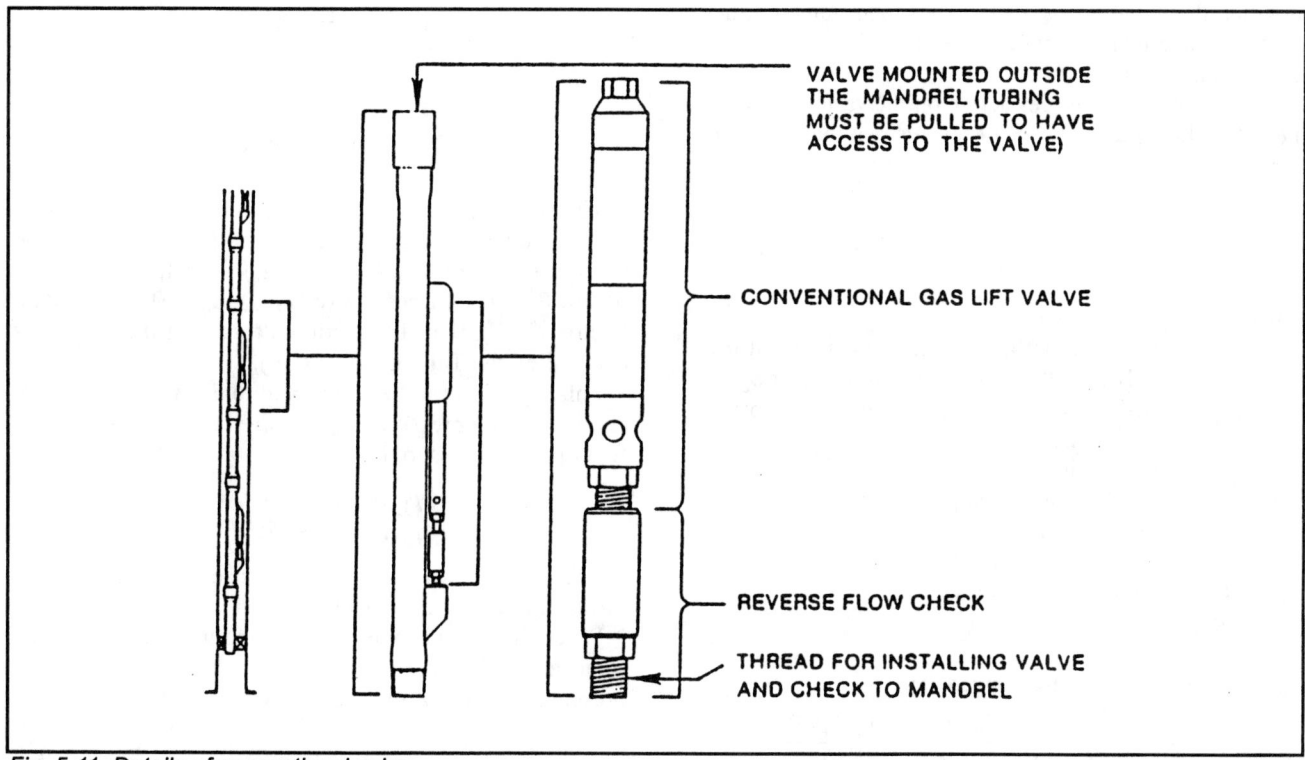

Fig. 5-11. Details of conventional valve.

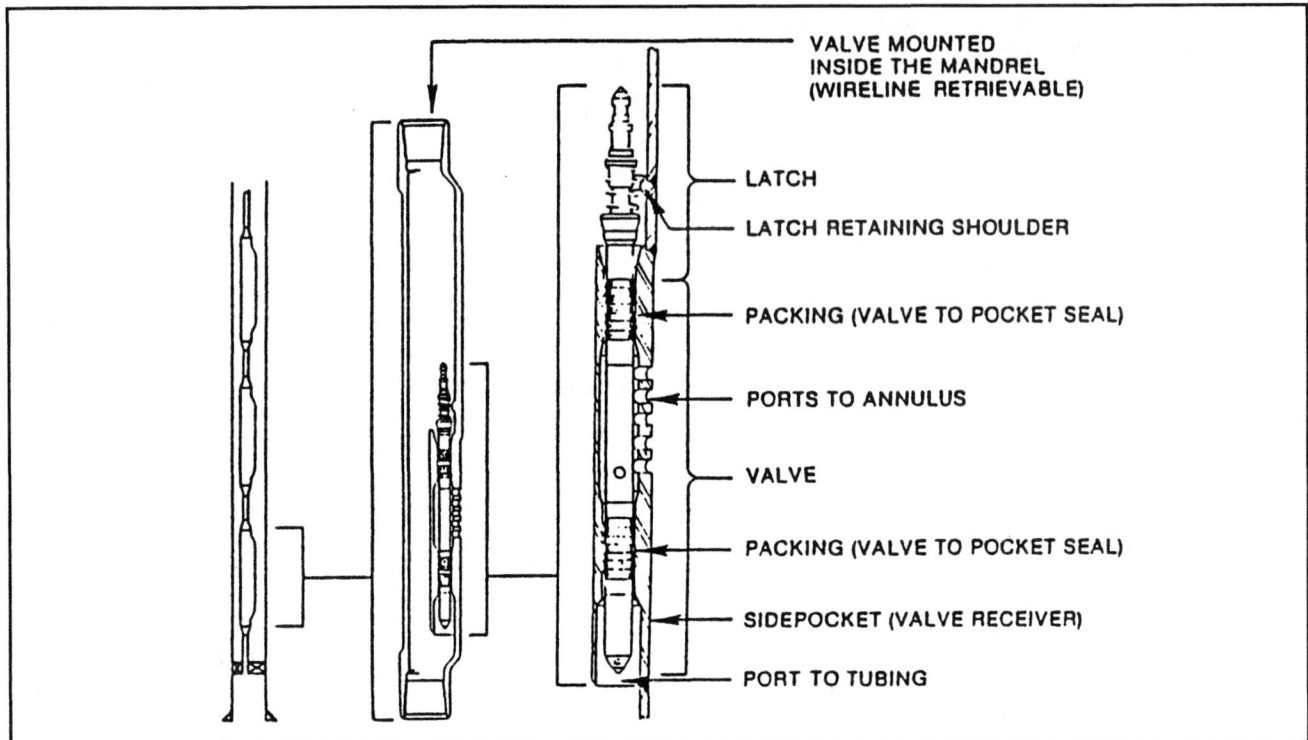

Fig. 5-12. Details of wireline retrievable valve.

valve will open at the correct pressure downhole, the valve is first tested at the surface at a temperature of 60°F. This testing is done in a test rack and the pressure at which the valve opens under these conditions is called the test rack opening pressure. A schematic of a test rack is shown in Figure 5-16.

An analysis of the forces acting on the various areas of the valve shown in Figure 5-14 reveals that the valve will open when the casing pressure p_1 is equal to:

$$p_1 = \frac{p_{bt} - p_2(A_p / A_b)}{1 - (A_p / A_b)} = \frac{p_{bt} - Rp_2}{1 - R} \qquad (5\text{-}2)$$

where

p_1 = injection or casing pressure existing at the valve depth. This is also called p_o or p_c.

p_{bt} = bellows or dome pressure at valve depth. Also called p_{dt}.

p_2 = production or tubing pressure at valve depth. Also called production pressure p_{pt} or tubing pressure p_t.

A_p = area of valve port or seat.

A_b = area of the bellows.

R = A_p/A_b

Equation 5-2 can be rearranged as:

$$p_1 = \frac{p_{bt}}{1 - (A_p / A_b)} - \frac{p_2(A_p / A_b)}{1 - (A_p / A_b)} = \frac{p_{bt} - p_2 R}{1 - R}$$

or

$$p_1 = \frac{p_{bt}}{1 - R} - p_{pe}$$

where

$$p_{pe} = \frac{Rp_2}{1 - R} = \text{Production Pressure Effect, } p_{pe}$$

This represents the amount that the opening pressure (p_1) is reduced as a result of the assistance from p_2. The ratio $R/(1-R)$ or $(A_p/A_b)/(1-A_p/A_b)$ is called the Production Pressure Effect Factor (p_{PEF}) or the Tubing Effect Factor (TEF). The Production Pressure Effect (p_{PE}) is also called the Tubing Effect (TE).

This equation can be rearranged to determine the valve charge (dome) pressure (p_{bt}) required to obtain the specified opening pressure (p_1).

$$\begin{aligned} p_{bt} &= p_1(1 - A_p / A_b) + p_2(A_p / A_b) \\ &= p_1(1 - R) + p_2 R \end{aligned} \qquad (5\text{-}3)$$

The dome pressure (p_{bt}) in this case is at the temperature of the valve in the well.

When the valve is open, both p_1 and p_2 are acting on the same area (A_b), and, therefore, the theoretical injection pressure or casing pressure at which the valve closes is

$$p_1 = p_{bt} \qquad (5\text{-}4)$$

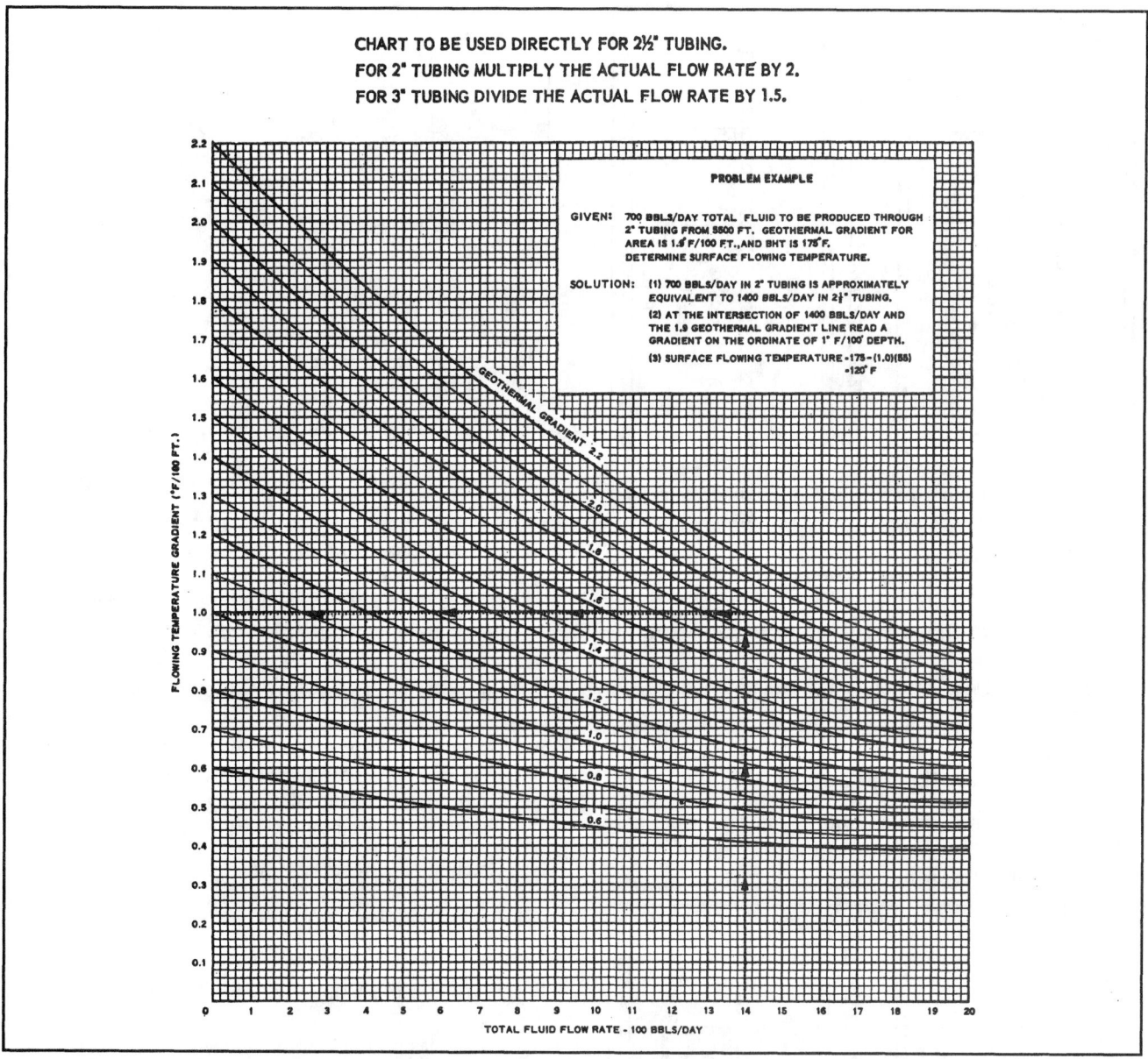

CHART TO BE USED DIRECTLY FOR 2½" TUBING.
FOR 2" TUBING MULTIPLY THE ACTUAL FLOW RATE BY 2.
FOR 3" TUBING DIVIDE THE ACTUAL FLOW RATE BY 1.5.

PROBLEM EXAMPLE

GIVEN: 700 BBLS/DAY TOTAL FLUID TO BE PRODUCED THROUGH 2" TUBING FROM 5500 FT. GEOTHERMAL GRADIENT FOR AREA IS 1.8 F/100 FT., AND BHT IS 175°F. DETERMINE SURFACE FLOWING TEMPERATURE.

SOLUTION: (1) 700 BBLS/DAY IN 2" TUBING IS APPROXIMATELY EQUIVALENT TO 1400 BBLS/DAY IN 2½" TUBING.

(2) AT THE INTERSECTION OF 1400 BBLS/DAY AND THE 1.8 GEOTHERMAL GRADIENT LINE READ A GRADIENT ON THE ORDINATE OF 1° F/100' DEPTH.

(3) SURFACE FLOWING TEMPERATURE = 175 − (1.0)(55) = 120° F

Fig. 5-13. Flowing temperature gradient for different flow rates, geothermal gradients, and tubing sizes.[3]

When the valve is in the test rack, as in Figure 5-16, p_2 is equal to zero and the test rack opening pressure is given by

$$p_{vo} = \frac{p_b @ 60°F}{1 - (A_p / A_b)} = \frac{p_b}{1 - R} \qquad (5\text{-}5)$$

Tables 5-1 and 5-2 list the values of A_p and A_b for two manufacturer's gas lift valves. The spread of a gas lift valve is defined as the difference between the opening and closing pressure of the valve.

2. Otis Design Procedure

The procedure that was described qualitatively earlier, sometimes called the 20 percent design gradient method, was first used by Otis Engineering. It has the advantage that all the available gas injection pressure at depth can be used in spacing the valves, and, therefore, gas can be injected deeper for a fixed available pressure. Also, the point of injection can be determined before the valves are spaced. The following procedure may be used for locating the valves and determining the pressure settings for a fixed production rate.

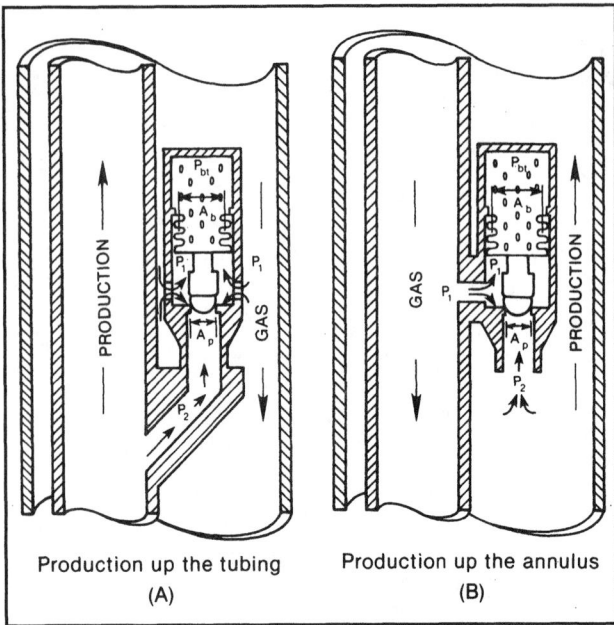

Fig. 5-14. Injection pressure operated valves.[3]

I. Select Depth to Operating Valve (Injection Depth)

A. Plot kick-off (p_{ko}) and operating casing pressure (p_{so}) versus depth.

B. Determine p_{wf} required for desired production rate using the Inflow Performance methods.

C. Plot the flowing gradient from p_{wf} upward using gradient curves for appropriate d_T, q_L, and formation gas/liquid ratio, GLR$_f$.

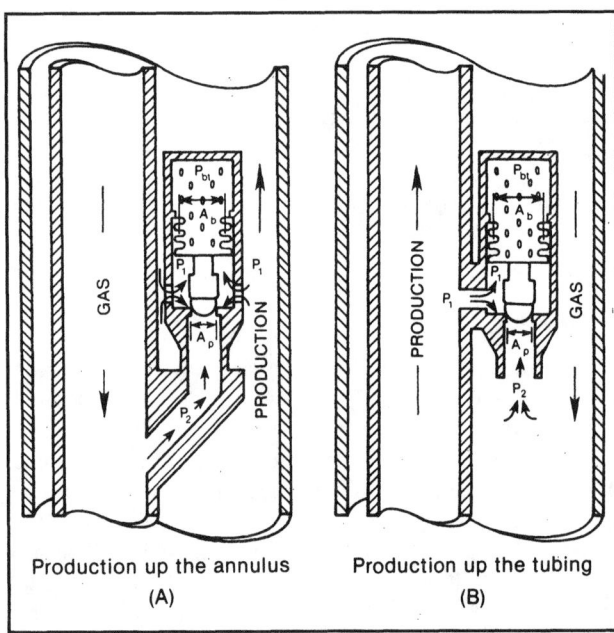

Fig. 5-15. Production pressure operated valves.[3]

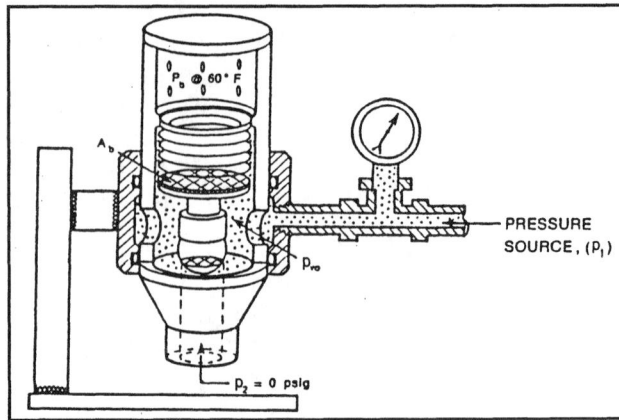

Fig. 5-16. Test rack.[3]

D. Locate the intersection of the operating casing pressure line with the flowing gradient line. This is the depth at which the casing and tubing pressures are balanced or equal.

E. Subtract the operating differential pressure, usually 100 psi, from the pressure at the intersection found in D. The depth at which this pressure occurs on the flowing gradient line is the operating valve depth.

II. Determine the Necessary Injection Gas Rate

A. Using the two-phase gradient curves, determine the GLR line which connects the pressure at the injection point to the wellhead flowing pressure, p_{wh}. This gives the required total GLR, GLR$_T$.

B. Calculate the necessary injection gas rate.

$$q_{gi} = q_L(\text{GLR}_T - \text{GLR}_f).$$

III. Determine the Depths to the Unloading Valves

A. Draw a design gradient line.
   1. Locate the design surface pressure,

$$p_{whd} = p_{wh} + 0.20(p_{so} - p_{wh}).$$

   2. Connect this point to the point of injection.

B. Draw a kill fluid gradient line (usually 0.5 psi/ft) from p_{wh} to intersect the p_{ko} line. This locates the top valve.

C. From the top valve depth on the design gradient line, draw a kill fluid gradient line to intersect the p_o line. This is the depth to the second valve. Repeat this procedure until the operating valve depth is reached.

IV. Select the Port Size for the Operating Valve.

(This is necessary to get $R = A_p/A_b$). The port size should be large enough to pass the required gas injection

TABLE 5-1

SPECIFICATIONS FOR CAMCO PRESSURE OPERATED GAS LIFT VALVES

Type Valve	A_b Effective Bellows Area sq. in.	Port Size I.D. in.	Area of Port with Bevel sq. in.	A_p/A_b	$(1-A_p/A_b)$	$TEF = \left[\dfrac{A_p/A_b}{(1-A_p/A_b)}\right]$ Tubing Effect Factor
J50, JR50	0.12	3/32	0.0077	0.06	0.94	0.07
JP50		1/8	0.0133	0.11	0.89	0.12
		5/32	0.0204	0.17	0.83	0.21
		13/64	0.0340	0.28	0.72	0.39
		11/32	0.0955	0.80	0.20	4.00
AK, AKR	0.31	1/8	0.0133	0.043	0.957	0.045
BK, BKF		3/16	0.0291	0.094	0.906	0.104
BKR, BP		1/4	0.0511	0.164	0.836	0.196
J40, J41		9/32	0.0643	0.207	0.793	0.261
JR40, PK		5/16	0.0792	0.255	0.745	0.342
		3/8	0.1134	0.365	0.635	0.575
CP, J20	0.77	3/16	0.0291	0.038	0.962	0.040
JR20, R20		1/4	0.0511	0.067	0.933	0.072
R21, R25		5/16	0.0792	0.104	0.896	0.116
R28, R29		3/8	0.1134	0.148	0.852	0.174
		7/16	0.1538	0.201	0.799	0.252
		1/2	0.2002	0.262	0.738	0.355

NOTE: The valves are grouped according to their bellows size. A specific valve type may not be available in all port sizes shown for a given bellows. This is particularly true for valves with a cross-over seat.

1. The maximum port size for the AK series valves is 9/32-inch I.D.
2. Valves with cross-over seats and their corresponding maximum I.D. port sizes are as follows:

AKR-2: 5/32-inch	BKF: 1/4-inch
BKF-3: 3/16-inch	BKF-10: 1/4-inch
BKR-1: 1/4-inch	JR20: 3/16-inch
JR40: 3/16-inch	JR50: 3/32-inch
R25: 5/16-inch	R28: 5/16-inch

3. The specifications apply to the pilot section only of a pilot valve.
4. Port areas were calculated based on the nominal I.D. plus 0.005-inch for the bevel.

rate corrected for temperature at the valve. The corrected gas rate is

$$q_{gi} = \frac{q_{gsc}(T_v + 460)}{520}$$

V. Determine the Required Dome Pressures for all the Valves.

A. Read the opening pressure (p_o) at valve depth from the p_o line.

B. Read the design tubing load (p_t) at a valve depth from the design gradient line.

C. Calculate the closing or dome pressure (p_{bt}) at valve depth.

$$p_{bt} = p_o(1 - R) + p_t R,$$

where $R = A_p/A_b$.

D. Calculate the dome pressure required at surface conditions to give p_{bt} at the valve temperature.

$$p_b = C_t(p_{bt})$$

C_t is obtained from Table 5-3.

E. Calculate the test rack opening pressure

$$p_{vo} = p_b/(1 - R)$$

Example 5-2:

Using the following data, determine:
1. Depth to point of injection.
2. Required gas injection rate.
3. Valve spacing and test rack opening pressures.

Depth to mid-perforations = 7500 ft.
Oil gravity = 35°API
Gas gravity = 0.65
Water fraction = 0
Formation GOR = 200 scf/STB
p_{wh}=100 psig
T_{wh} = 100°F

TABLE 5-2

SPECIFICATIONS FOR OTIS SPREADMASTER PRESSURE-OPERATED GAS LIFT VALVES

R VALUES

Bellows Area and Seat Area Relationships for Otis Spreadmaster Valves

$$R = A_p/A_b$$

Where:

A_p = Area of spread control seat—in^2
A_b = Effective area of bellows—in^2
A_b for 1" O.D. Valves—.32 in^2
A_b for 1-1/2" O.D. Valves—.77 in^2

Diameter of Spread Control Seat—in.	For 1" O.D. Valves R	1 – R	For 1-1/2" O.D. Valves R	1 – R
3/16	.0863	.9137	.0359	.9641
1/4	.1534	.8466	.0638	.9362
9/32	.1942	.8058	—	—
5/16	.2397	.7603	.0996	.9004
11/32	.2900	.7100	—	—
3/8	.3450	.6550	.1434	.8566
7/16	.4697	.5303	.1952	.8048
1/2	—	—	.2562	.7438
9/16	—	—	.3227	.6773

TYPE S OTIS SPREADMASTER GAS LIFT VALVES WITH PILO-PORT TYPE NS OTIS CONVENTIONAL VALVE

PILO-PORT DIAMETER INCHES

Valve O.D. Inches	Lift Port Diam. Inches	1/8 8/64 .125	1/16 12/64 .187	1/4 16/64 .250	9/32 18/64 .281	5/16 20/64 .312	11/32 22/64 .343	3/8 24/64 .375	7/16 28/64 .437	1/2 32/64 .500	9/16 34/64 .562	Thread Connection Inches	Valve Length Inches
1.000	7/16	221NS1017	221NS1018	221NS1001	221NS1003	221NS1005	221NS1004					1/2 NPT	14 7/8
1.500	9/16			221NS1524		221NS1513		221NS1512	221NS1514			1/2 NPT	20 1/4
1.500	1/4			221 NS1525		221 NS1509		221NS1508	221NS1507	221NS1510	221NS1511	1/4 NPT	20 1/4

TYPE RS OTIS WIRELINE RETRIEVABLE VALVE

PILO-PORT DIAMETER INCHES

Valve O.D. Inches	Lift Port Diam. Inches	1/8 8/64 .125	1/16 12/64 .187	1/4 16/64 .250	9/32 18/64 .281	5/16 20/64 .312	11/32 22/64 .343	3/8 24/64 .375	7/16 28/64 .437	1/2 32/64 .500	9/16 34/64 .562	Packing O.D. In. Upper	Lower	Valve Length Inches
1.000	7/16	221 RS1014	221 RS1015	221 RS1001	221 RS1003	221 RS1005	221 RS1004					1.032	1.032	13 1/8
1.500	8/16			221RS1518		221RS1513		221 RS1512	221RS1514			1.562	1.500	20 13/16
1.500	1/4			221RS1519		221RS1509		221 RS1508	221RS1507	221RS1510	221RS1511	1.562	1.500	20 13/16

(Courtesy Otis Engineering, Dallas, Texas)

TABLE 5-3
Temperature Correction Factors for Nitrogen Based on 60°F

°F	C_t	°F	C_t	°F	C_t	°F	C_t	°F	C_t	°F	C_t
61	.998	101	.919	141	.852	181	.794	221	.743	261	.698
62	.996	102	.917	142	.850	182	.792	222	.742	262	.697
63	.994	103	.915	143	.849	183	.791	223	.740	263	.696
64	.991	104	.914	144	.847	184	.790	224	.739	264	.695
65	.989	105	.912	145	.845	185	.788	225	.738	265	.694
66	.987	106	.910	146	.844	186	.787	226	.737	266	.693
67	.985	107	.908	147	.842	187	.786	227	.736	267	.692
68	.983	108	.906	148	.841	188	.784	228	.735	268	.691
69	.981	109	.905	149	.839	189	.783	229	.733	269	.690
70	.979	110	.903	150	.838	190	.782	230	.732	270	.689
71	.977	111	.901	151	.836	191	.780	231	.731	271	.688
72	.975	112	.899	152	.835	192	.779	232	.730	272	.687
73	.973	113	.898	153	.833	193	.778	233	.729	273	.686
74	.971	114	.896	154	.832	194	.776	234	.728	274	.685
75	.969	115	.894	155	.830	195	.775	235	.727	275	.684
76	.967	116	.893	156	.829	196	.774	236	.725	276	.683
77	.965	117	.891	157	.827	197	.772	237	.724	277	.682
78	.963	118	.889	158	.826	198	.771	238	.723	278	.681
79	.961	119	.887	159	.825	199	.770	239	.722	279	.680
80	.959	120	.886	160	.823	200	.769	240	.721	280	.679
81	.957	121	.884	161	.822	201	.767	241	.720	281	.678
82	.955	122	.882	162	.820	202	.766	242	.719	282	.677
83	.953	123	.881	163	.819	203	.765	243	.718	283	.676
84	.951	124	.879	164	.817	204	.764	244	.717	284	.675
85	.949	125	.877	165	.816	205	.762	245	.715	285	.674
86	.947	126	.876	166	.814	206	.761	246	.714	286	.673
87	.945	127	.874	167	.813	207	.760	247	.713	287	.672
88	.943	128	.872	168	.812	208	.759	248	.712	288	.671
89	.941	129	.871	169	.810	209	.757	249	.711	289	.670
90	.939	130	.869	170	.809	210	.756	250	.710	290	.669
91	.938	131	.868	171	.807	211	.755	251	.709	291	.668
92	.936	132	.866	172	.806	212	.754	252	.708	292	.667
93	.934	133	.864	173	.805	213	.752	253	.707	293	.666
94	.932	134	.863	174	.803	214	.751	254	.706	294	.665
95	.930	135	.861	175	.802	215	.750	255	.705	295	.664
96	.928	136	.860	176	.800	216	.749	256	.704	296	.663
97	.926	137	.858	177	.799	217	.748	257	.702	297	.662
98	.924	138	.856	178	.798	218	.746	258	.701	298	.662
99	.923	139	.855	179	.796	219	.745	259	.700	299	.661
100	.921	140	.853	180	.795	220	.744	260	.699	300	.660

$$C_t = \frac{\text{Gas Lift Valve Dome Pressure at 60 °F}}{\text{Gas Lift Valve Dome Pressure at Well Temperature}}$$

T_{res} = 182°F

Tubing I.D. = 1.995 in.

p_{so} = 870 psig

p_{ko} = 920 psig

Δp across operating valve = 100 psi

Injected gas surface temperature = 100°F

Load fluid gradient = 0.5 psi/ft

Static liquid level is at surface

Well to be unloaded to p_{wh} = 100 psig

Desired production rate = 600 STBO/day

Valves will be 1 in. retrievable.

From a previous test:

$\bar{p}_R$ = 2000 psig, q_o = 383 STB/day for p_{wf} = 1850 psig

I.

A. Using Figure 5-18, determine the gas pressure at 7000 ft:

p_s	p at 7000 ft
920	1100
870	1050

B. Using Vogel's Method to determine p_{wf} required for a liquid rate of 600 STBO/day:

From test:

$p_{wf} / \bar{p}_R = 1850/2000 = .925$

$q_{o(max)}$ = 383/[1–.2(.925) –.8(.925)²] = 2935 STB/day

p_{wf} = 2000[1.266–1.25(600/2935)).⁵–.125] = 1760 psig

C. Intersection is at 5400 ft (from graph, Figure 5-17)

D. Operating valve depth = 5100 ft (from graph)

II.

A. Using the pressure traverse curves for 1.995 tubing and 600 STBO/day establishes the required GLR above the operating valve as 400 scf/STB.

B. q_{gi} = 600(400–200) = 120,000 scf/day

III.

A. p_{whd} = 100 + 0.20(770) = 254 psig

B, C. The valve depths are found from the graph as:

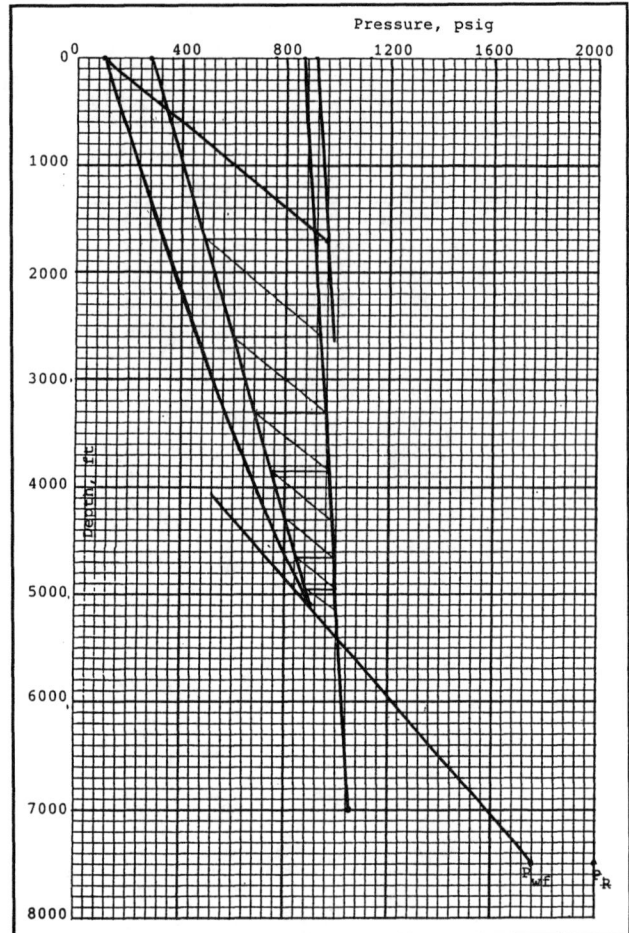

Fig. 5-17. Example 5-2 solution.

Valve No.	Depth
1	1700
2	2600
3	3300
4	3850
5	4300
6	4660
7	4950
8	5150

IV.

A port size must be selected which will pass gas at a rate of 120 Mscfd with an upstream pressure of 1000 psig and a downstream pressure of 900 psig. The temperature at the operating valve is

$$T_v = 100 + \frac{(182-100)}{7500}(5150) = 156°F$$

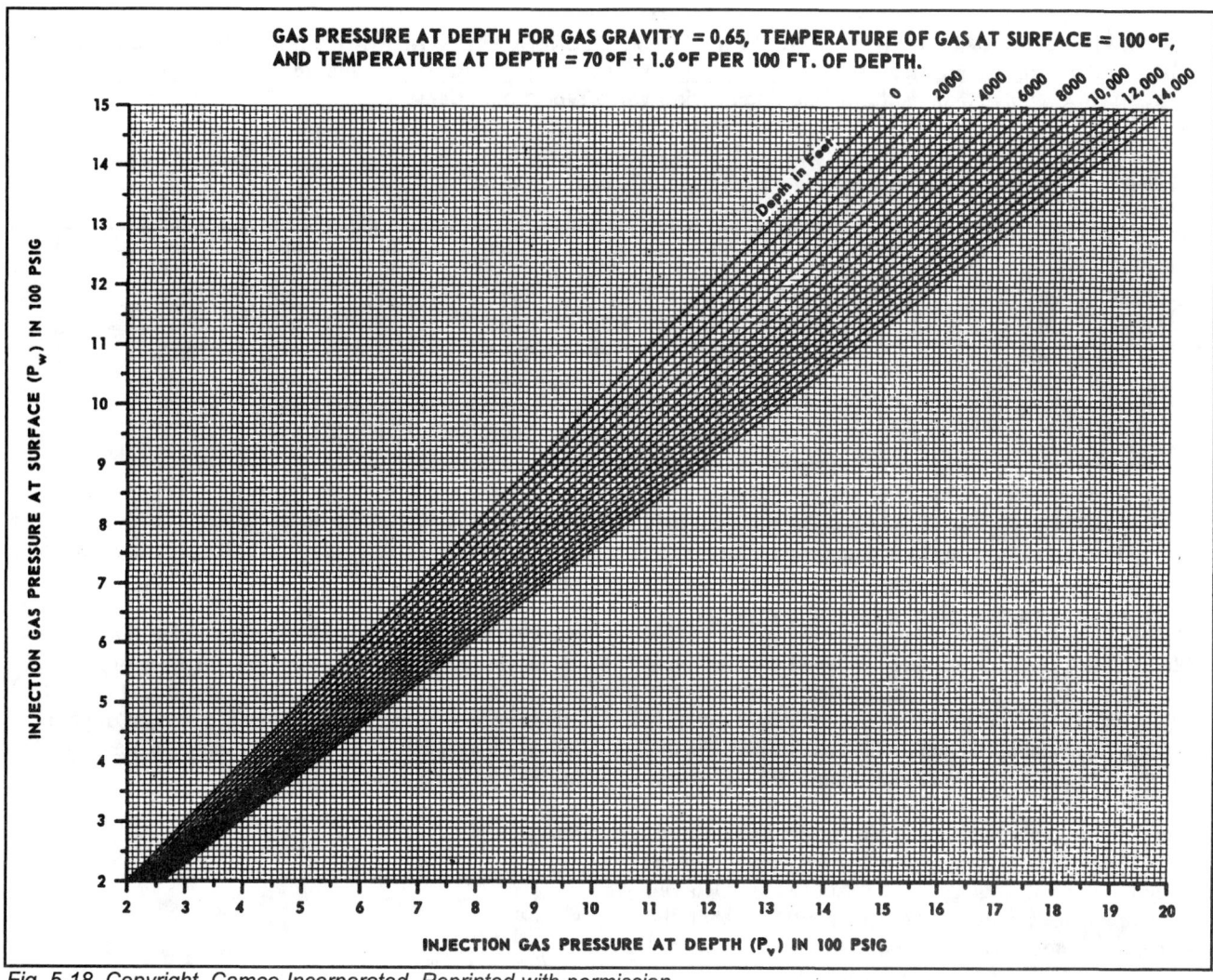

GAS PRESSURE AT DEPTH FOR GAS GRAVITY = 0.65, TEMPERATURE OF GAS AT SURFACE = 100 °F, AND TEMPERATURE AT DEPTH = 70 °F + 1.6 °F PER 100 FT. OF DEPTH.

Fig. 5-18. Copyright, Camco Incorporated. Reprinted with permission.

The corrected gas volume is:

$$q_{gi} = \frac{120(156 + 460)}{(60 + 460)} = 142 \ Mscfd \qquad (5\text{-}5C)$$

From the choke capacity chart (Fig. 5-22), a 16/64 orifice will pass gas at a rate of 250 Mscfd. Although a smaller port size will suffice, select the 0.25 in. port for possible changing conditions. From Table 5-2: $R = .1534$, $(1 - R) = .8466$ $A_b = 0.32$ in.^{2} Use the same port size for all the valves.

V.

For the top valve:

A. $p_o = 970$

B. $p_t = 490$

C. $p_{bt} = p_o(1 - R) + p_t R = 970(.8466) + 490(.1534)$
    $p_{bt} = 896$ Psig

D. From Table 5-3: $C_t = .886$ for $T_v = 120°F$. The table for valve setting temperatures of 60°F was used. Some companies use 80°F.

$$p_b = C_t(p_{bt}) = .886(896) = 794 \ psig$$

E. $p_{vo} = p_b/(1 - R) = 794/0.8466 = 938 \ psig$

Table 5-4 summarizes the design:

The example problem illustrated the design for a vertical well. If a directionally drilled well is to be placed on gas lift, the same design procedure can be used if the flowing pressure traverses are calculated based on the measured depth and the measured depths are converted to true vertical depths for plotting. This is convenient because the load fluid and

TABLE 5-4

OPENING PRESSURE CALCULATIONS FOR SINGLE-ELEMENT, UNBALANCED NITROGEN CHARGED GAS LIFT VALVES

Company <u>Example 5-2</u> Field _____ Lease _____ Well No. _____

Valve Manufacturer <u>Otis Engineering</u> Valve Type <u>1 inch</u> Date _____

Valve No.	L ft.	Port Size in.	A_p/A_b	p_o@L psig	Design p_t@L psig	p_{bt} psig	T_v@L °F	C_t	p_b psig	p_{vo} psig
1	1700	.25	.1534	970	490	896	120	.886	794	938
2	2600			935	600	884	128	.872	771	911
3	3300			955	680	913	136	.860	785	927
4	3850			970	750	936	142	.850	796	940
5	4300			980	800	952	147	.842	802	947
6	4660			990	847	968	151	.836	809	956
7	4950			1000	882	982	154	.832	817	965
8	5150			1005	906	990	156	.829	821	970

the injected gas are under static conditions and are, therefore, unaffected by hole angle.

III. SUBMERSIBLE PUMP SELECTION

Many high-volume wells are equipped with electric submersible pumps (ESP) to lift the liquid and decrease the flowing bottomhole pressure. A submersible pump is a multistage centrifugal pump that is driven by an electric motor located in the well below the pump. Electric power is supplied by means of a cable from the surface.

Submersible pumps are available for production rates ranging from about 300 to 60,000 bbls/day and are ideal for high water-cut, low gas/liquid ratio wells. Detailed design procedures will not be presented here, but the application of nodal systems analysis in determining the size and power requirements of a submersible pump will be described. A comprehensive discussion of submersible pump selection can be found in Brown.[1]

The pump and motor are suspended on the tubing at a certain depth in the well. The annulus is either vented or tied into the well's flowline, so that as much gas as possible is separated from the liquid before it enters the pump, In some cases, a centrifugal separator will be placed between the pump and motor for obtaining maximum gas-liquid separation. Some manufacturers claim up to 90 percent gas separation with a downhole separator. A schematic of a well equipped with a submersible pump is given in Figure 5-25, along with the pressure traverse in the well. An illustration of a typical pump assembly is presented in Figures 5-26 and 5-27.

To perform a nodal analysis on a submersible pumping well, the node is selected at the pump. The pump can be handled as an independent component in the system in a manner similar to that used in analyzing gravel-packed completions. The node pressure is either the pump intake pressure p_{up} or the pump discharge pressure p_{dn}. The pres-

sure gain that the pump must generate for a particular producing rate is $\Delta p = p_{dn} - p_{up}$. These pressures and their locations are illustrated in Figure 5-25. The pressure traverse below the pump will be calculated based on the formation gas/liquid ratio and the casing size. The traverse in the tubing above the pump will be based on the gas/liquid ratio entering the pump and the tubing size. If no information is available regarding the amount of gas separated, it may be assumed to be about 50 percent. The inflow and outflow expressions are:

Inflow

$$\bar{p}_R - \Delta p_{res} - \Delta p_{csg} \text{(below pump)} = p_{up}$$

Outflow

$$p_{sep} + \Delta p_{flowline} + \Delta p_{tub} \text{(above pump)} = p_{dn}$$

The following procedure may be used to estimate the pressure gain and power required to achieve a particular producing capacity.

Inflow

1. Select a value for liquid producing rate q_L.

2. Determine the required p_{wf} for this q_L using the reservoir performance procedures described in Chapter 2.

3. Determine the pump suction pressure p_{up} using the casing diameter and the total producing GLR to calculate the pressure drop below the pump.

4. Repeat for a range of liquid producing rates and plot p_{up} vs. q_L as illustrated in Figure 5-28.

Outflow

1. Select a value for q_L.

2. Determine the appropriate GLR for tubing and flowline pressure drop calculations.

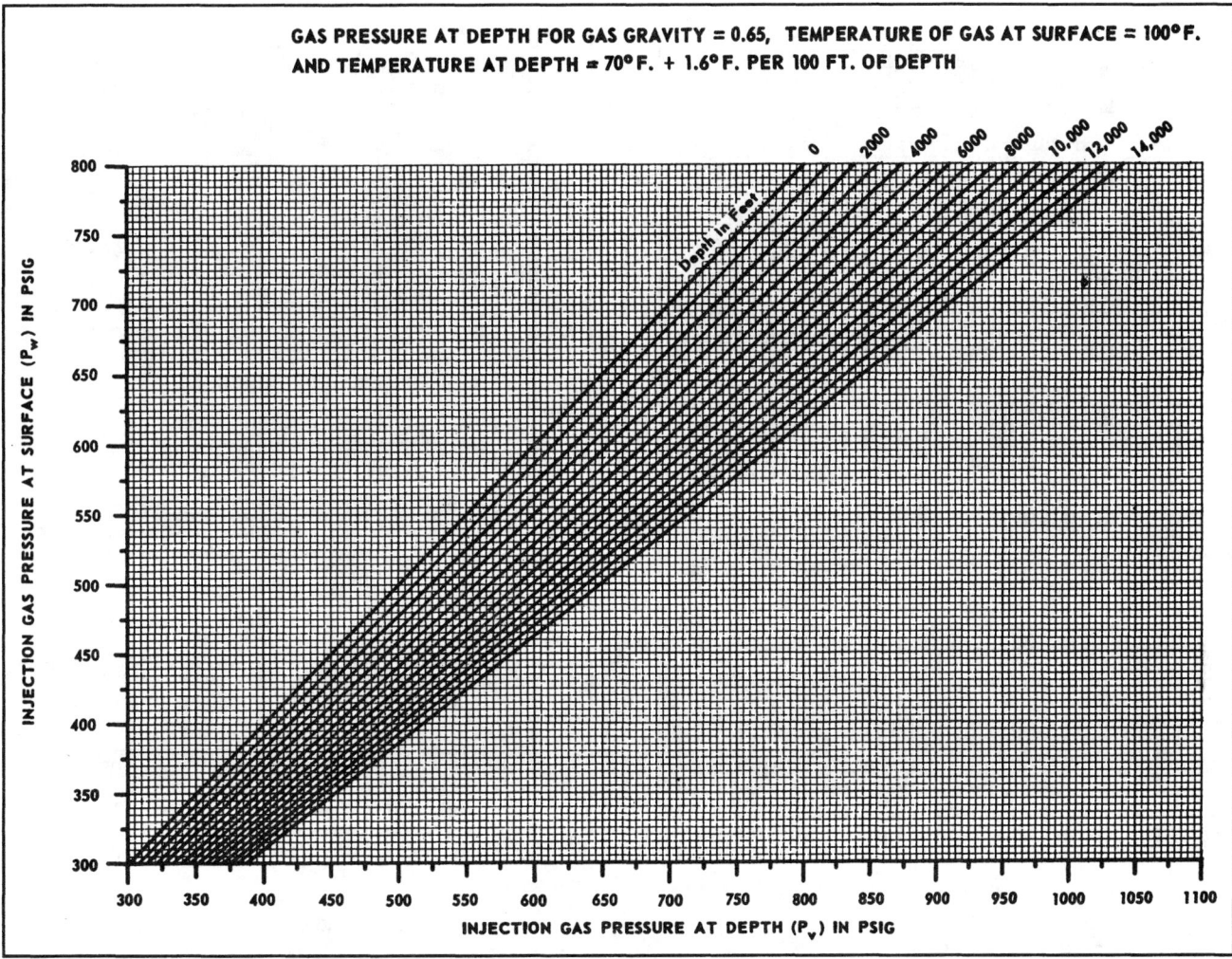

GAS PRESSURE AT DEPTH FOR GAS GRAVITY = 0.65, TEMPERATURE OF GAS AT SURFACE = 100°F.
AND TEMPERATURE AT DEPTH = 70°F. + 1.6°F. PER 100 FT. OF DEPTH

Fig. 5-19. Copyright, Camco Incorporated. Reprinted with permission.

a. Determine p_{up} and fluid temperature at the pump at this q_L value from inflow calculations.

b. Determine dissolved gas R_s, at this pressure and temperature.

c. Estimate fraction of free gas E_s, separated at the pump. This will be dependent on whether or not a downhole separator is to be used. If not, use $E_s = 0.5$.

d. Calculate the GLR downstream of the pump from $GLR_{dn} = (1 - E_s)(R_{total} - f_o R_s)$

where

R_{total} = total producing gas/liquid ratio,

R_s = solution gas/oil ratio at suction conditions, and

f_o = fraction of oil flowing.

3. Determine p_{dn} using GLR_{dn} to calculate the pressure drop in the tubing and the flowline if the casinghead gas is vented. If the casing is tied into the flowline,

the total GLR will be used to determine the pressure drop in the flowline.

4. Repeat for a range of q_L and plot p_{dn} vs. q_L on the same graph.

5. Select various producing rates and determine the pressure gain Δp required to achieve an intersection of the inflow and outflow curves at these rates. The suction and discharge pressures can also be determined for each rate.

6. Calculate the power requirement, pump size, number of stages, etc., at each producing rate.

The required horsepower may be calculated from

$$HP = 1.72 \times 10^{-5} \Delta p(q_o B_o + q_w B_w)$$

where

HP = horsepower required,

Δp = pressure gain, psi,

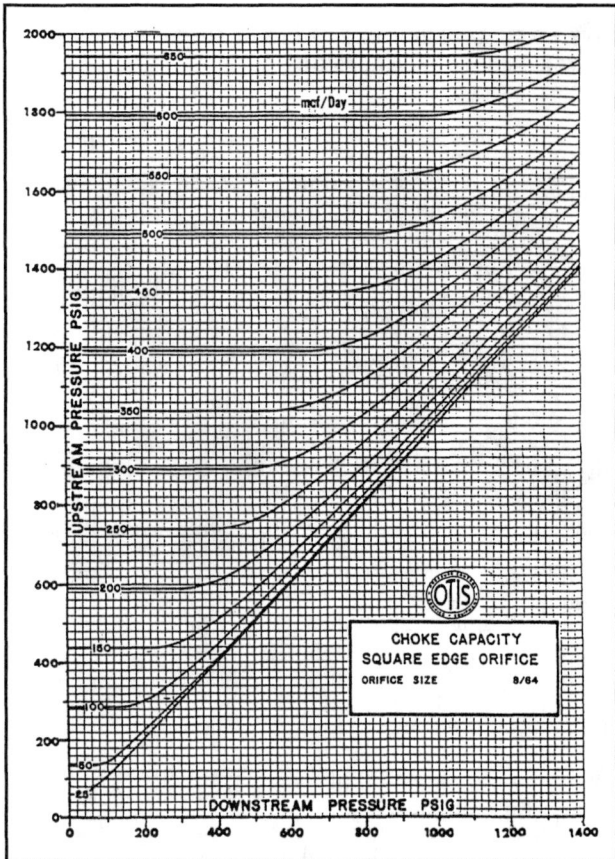

Fig. 5-20. (Courtesy Otis Engineering, Dallas, Texas.)

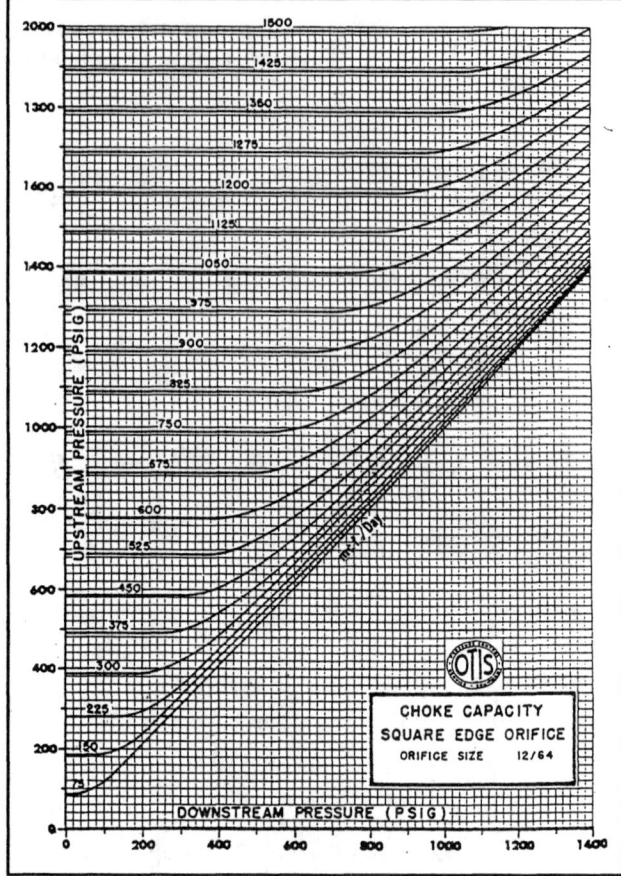

Fig. 5-21. (Courtesy Otis Engineering, Dallas, Texas.)

q_o = oil rate, STB/day,

q_w = water rate, STB/day,

B_o = oil formation volume factor at suction conditions, bbl/STB, and

B_w = water formation volume factor at suction conditions, bbl/STB.

The pressure gain can be converted to head gain if necessary for pump selection. This is accomplished by dividing the pressure gain by the density of the fluid being pumped. The actual plotting of the data is not required if the pump is to be selected for specific rates, as all the necessary information is calculated before plotting.

Other variables that could be analyzed using systems analysis include pump setting depth, effect of a downhole separator and pump speed.

Example 5-3:

The well described in Example 4-1 is to be equipped with an ESP set at a depth of 7000 ft from the surface. Assume that the 2-7/8 (2.441) in. tubing is run to 7000 ft and that the casing I.D. below the pump is 6.366 in. If one-half of the free gas is separated at the pump, determine the horsepower and head gain required if the well

is to produce at a total liquid rate of 3000 STBL/day.

$\bar{p}_R$ = 3482 psig	p_b = 3600 psig
Depth to mid-perfs = 10,000 ft	p_{wh} = 400 psig
R_{total} = 400 scf/STBL	f_w = 0.5
$q_{L(max)}$ = 16810 STBL/day	T_{pump} = 200°F
γ_g = 0.65	γ_o = 35° API
γ_w = 1.07	

Solution:

The required p_{wf} to inflow 3000 STBL/day is:

$$p_{wf} = 3482 \left[\left(1.266 - \frac{1.25(3000)}{16810} \right)^{0.5} - 0.125 \right]$$

$$= 3120 \, \text{psig}$$

Using the vertical pressure traverse curve for flow in the casing, entering at a pressure of 3120 psig on the 400 GLR line and moving upward 3000 ft gives a pump suction pressure of about 2040 psig. Use Equation 3-76 to determine R_s at suction conditions:

$$R_s = 0.0178(0.65)(2054.7)^{1.187} \, \text{EXP}[23.931 \, (35)/(200 + 460)] = 352 \, \text{scf/STBO}$$

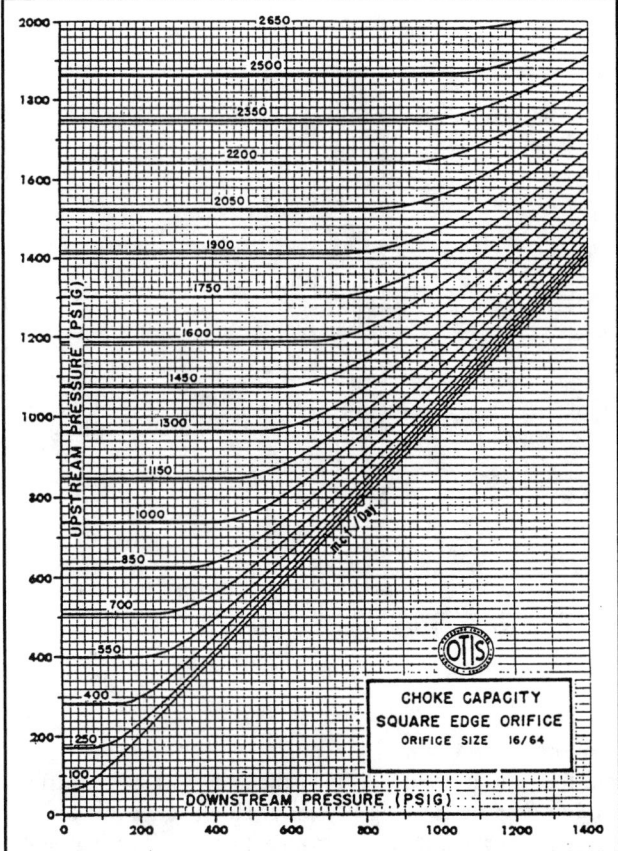

Fig. 5-22. (Courtesy Otis Engineering, Dallas, Texas.)

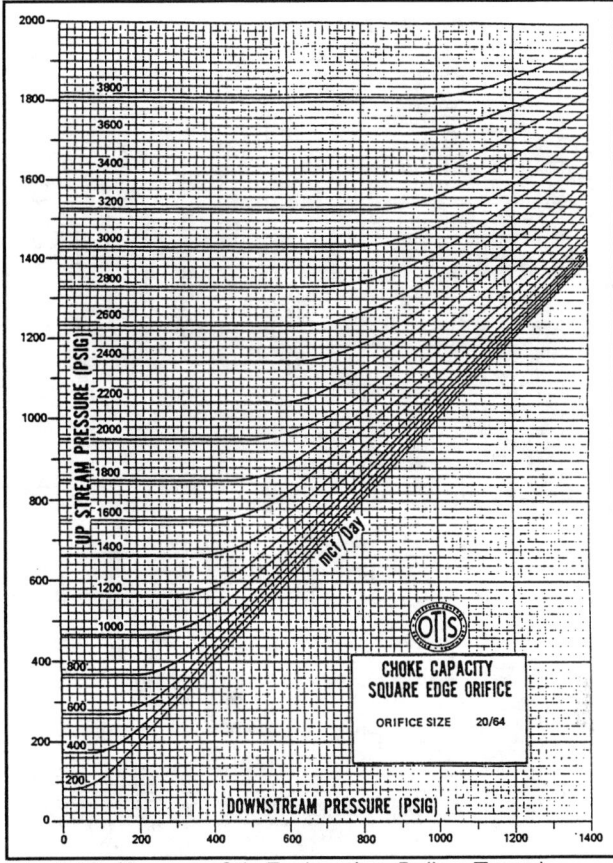

Fig. 5-23. (Courtesy Otis Engineering, Dallas, Texas.)

Calculate the gas which will go through the pump and the tubing, GLR_{dn}:

$GLR_{dn} = (1-E_s)(R_{total} - f_o R_s) = (1 - 0.5)(400 - 0.5(352))$
= 112 scf/STBL

GLR_{dn} = 112 scf/STBL
Using the vertical pressure traverse curve for flow in 2.441 I.D. tubing at a rate of 3000 STBL/day and $f_w = 0.5$, enter at p_{wh} = 400 psig and using the GLR = 100 line, find the pump discharge pressure to be approximately 3200 psig at 7000 ft. The required pressure gain across the pmp is therefore $\Delta p = 3200 - 2040$ = 1160 psi.
Calculate B_o for R_s = 352 scf/STBO at 200°F:
From Equation 3-79:

$B_o = 1.0 + 4.67 \times 10^{-4} (352) + 1.1 \times 10^{-5} (200 - 60)(35/0.65) + 1.337 \times 10^{-9} (200 - 60)(352)(35/0.65)$
B_o = 1.251 bbl/STB
Assume B_w = 1.0
Use Equation 3-60 to calculate the oil density:

$$\rho_o = \frac{350(0.845) + 0.0764(0.65)(352)}{5.615(1.251)}$$

ρ_o = 45 lbm/ft^3

$\rho_L = \rho_o f_o + \rho_w f_w = 45(0.5) + 62.4(1.07)(0.5) =$
55.9 lbm/ft^3

$$\rho_L = \frac{55.9}{144} = 0.388 \text{ psi/ft}$$

$HP = 1.72 \times 10^{-5} \Delta p (q_o B_o + q_w B_w)$

$HP = 1.72 \times 10^{-5} (1160)[1500(1.251) + 1500(1.0)] =$
67 hp

$$\text{Head gain} = \frac{1160 \text{ psi}}{0.388 \text{ psi/ft}} = 2990 \text{ ft}$$

IV. SUCKER ROD OR BEAM PUMPING

Although nodal systems analysis is not as widely applied to the analysis of wells equipped with sucker rod pumps, the effects of pump setting depth or liquid level can be determined using this method. Also if the casing annulus is tied into the well's flowline, the producing rate will have a direct effect on casing pressure and thus bottomhole flowing pressure.

Sucker rod pumping is the most widely used artificial lift method. That is, more artificial lift wells are equipped

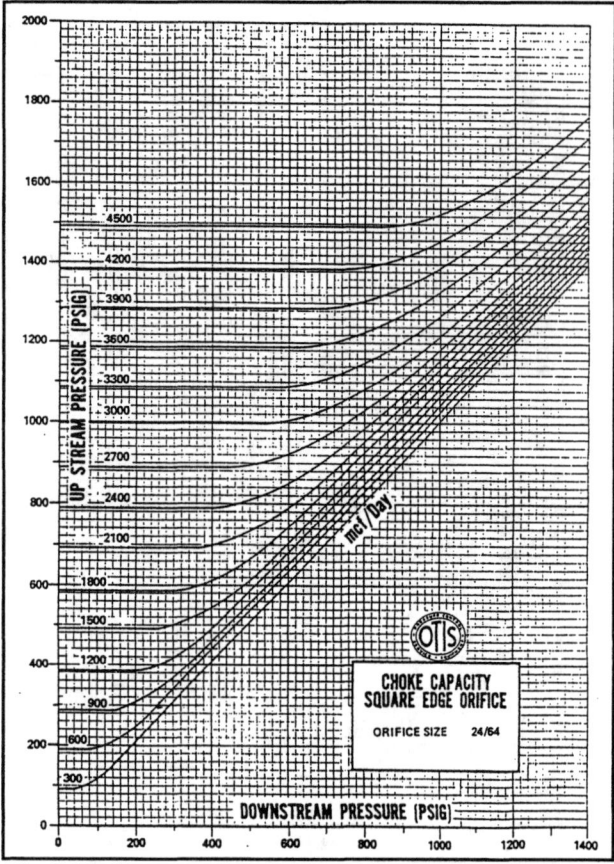

Fig. 5-24. (Courtesy Otis Engineering, Dallas, Texas.)

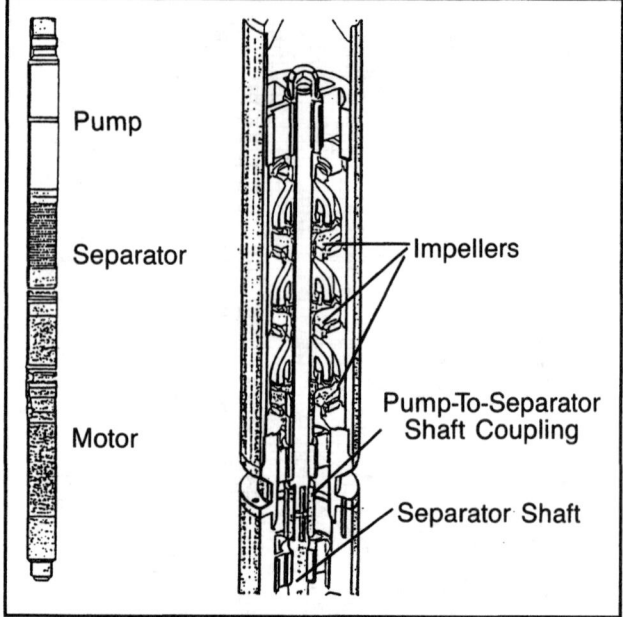

Fig. 5-26. Submersible pump assembly.

with rod pumps than any other type of artificial lift method. This does not mean that more oil is produced by rod pumping, since many rod-pumped wells produce at very low rates. A discussion of the relative popularity of the various methods can be found in Brown.[1]

A sucker rod pump is a positive displacement pump, and thus the pressure drop in the tubing does not affect the bottomhole flowing pressure. However, p_{wf} is affected by the surface casing pressure, the pump setting depth and the working liquid level, usually referred to as the fluid level. A schematic of a rod-pumped well and the corresponding traverse are shown in Figures 5-29 and 5-30. A more detailed representation of the pump and the operating sequence is shown in Figure 5-31. A method for

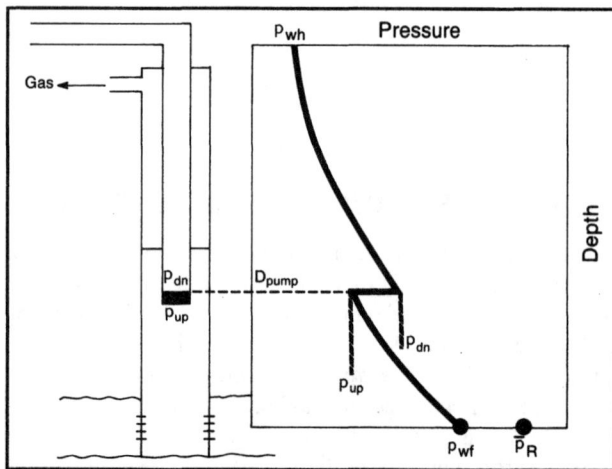

Fig. 5-25. Submersible pump schematic.

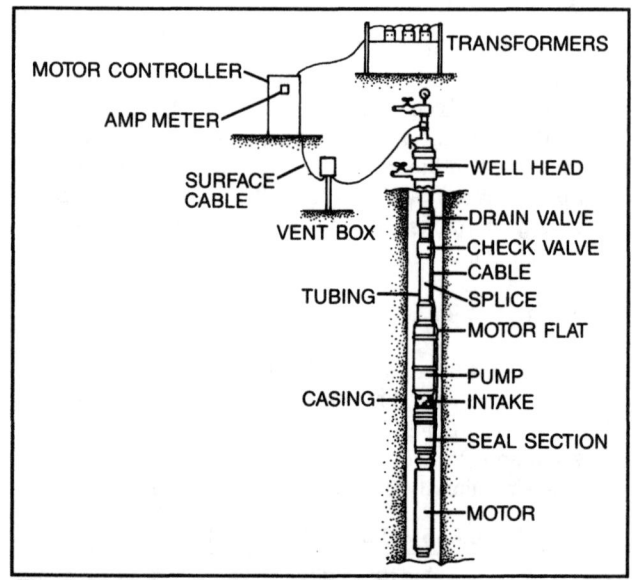

Fig. 5-27. Submersible pump installation.

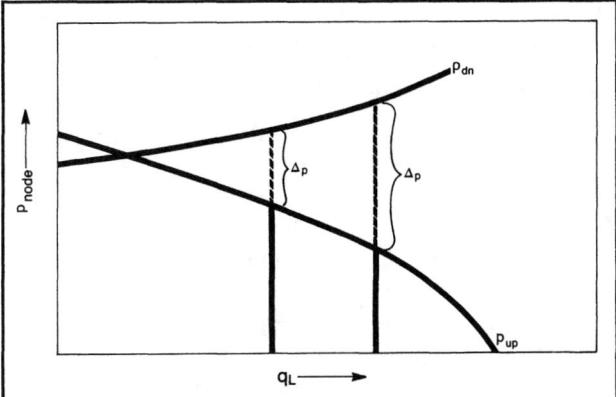

Fig. 5-28. Submersible pump selection.

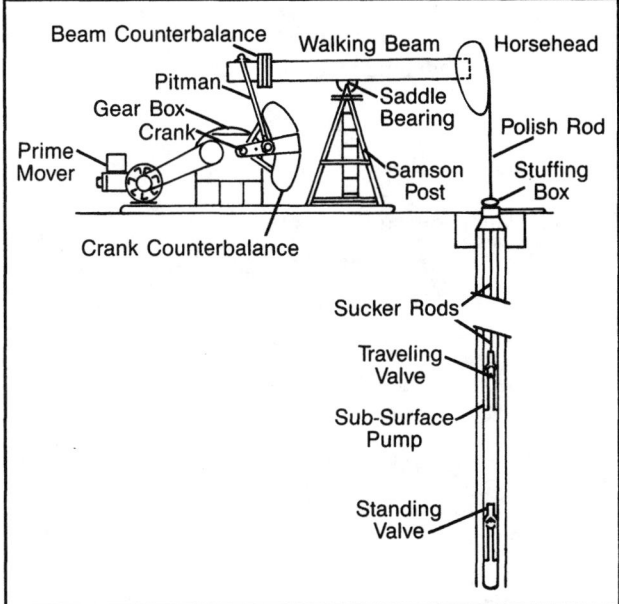

Fig. 5-29. Beam pumping unit.

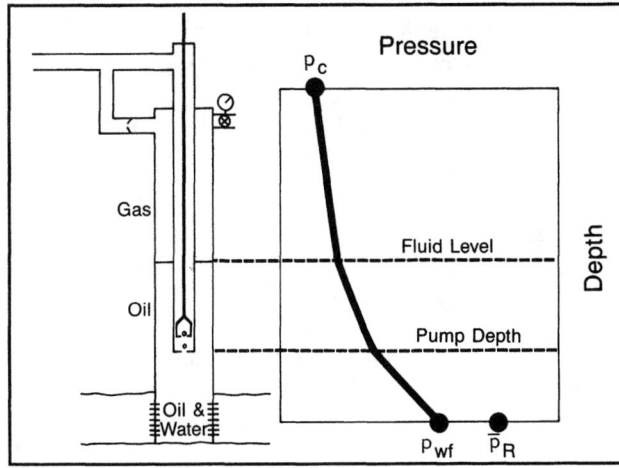

Fig. 5-30. Rod pumping well.

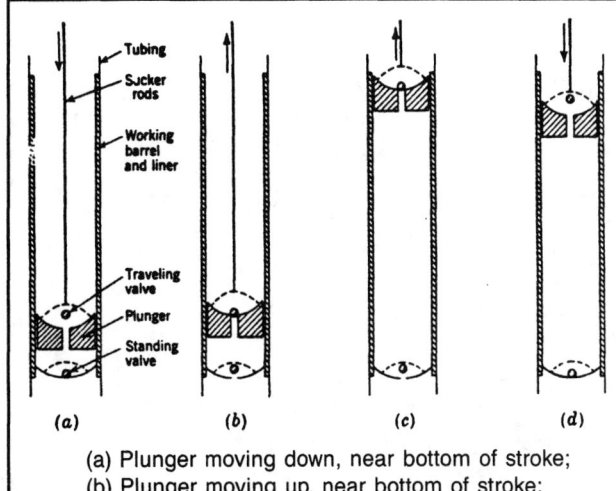

(a) Plunger moving down, near bottom of stroke;
(b) Plunger moving up, near bottom of stroke;
(c) Plunger moving up, near top of stroke;
(d) Plunger moving down, near top of stroke.

Fig. 5-31. The pumping cycle.

measuring the fluid level in a well is illustrated and described in Figures 5-32 and 5-33.

If the node pressure selected is p_{wf} the inflow and outflow expressions for a well which is being rod pumped are:

Inflow

$$\overline{p}_R - \Delta p_{res} = p_{wf}$$

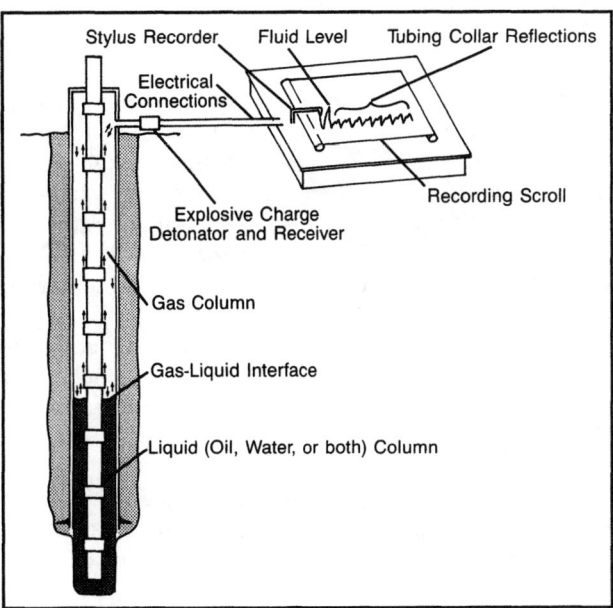

Fig. 5-32. Measuring well fluid level.

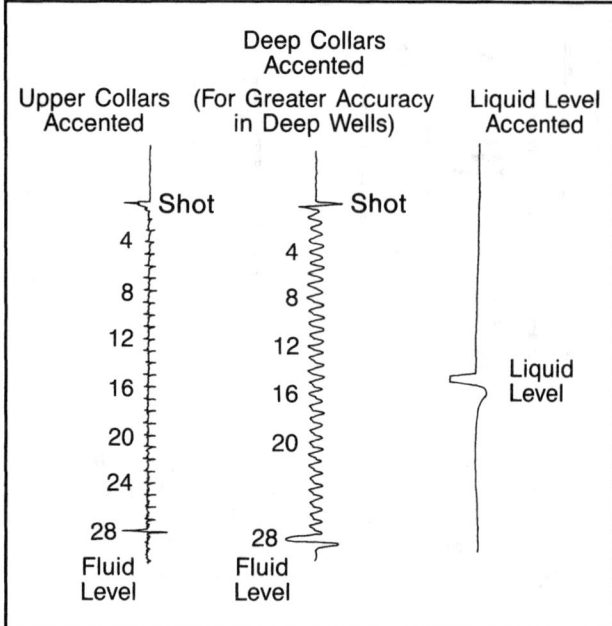

Fig. 5-33. Reflections of sound waves.

Outflow

Constant casing pressure, p_c

$p_c + \Delta p_{gas} + \Delta p$(fluid above pump) $+ \Delta p$(fluid below pump) $= p_{wf}$

Casing tied into flowline

$p_{sep} + \Delta p_{flowline} + \Delta p_{gas} + \Delta p$(above pump) $+ \Delta p$(below pump) $= p_{wf}$

The pressure drop in the gas is usually calculated using the hydrostatic pressure only and may be calculated using equations given in Chapter 3. The fluid above the pump will usually be oil only with gas bubbling through it. The fluid below the pump will be a mixture of oil, water and gas if the well is producing any water. In most cases, any pressure drop due to friction is ignored since the fluid will usually be moving at a low velocity in the casing below the pump.

The effect on the density of the fluid of the gas bubbling through the fluid and escaping into the annulus can be estimated using an empirical curve presented by Gilbert.[2] This is often referred to as Gilbert's "S" curve and is shown in Figure 5-34.

The gas rate q in the group of terms on the vertical axis is the amount of gas bubbling through a column of liquid occupying an area A. The pressure in this group of terms is the average pressure in the liquid column. The liquid hydrostatic gradient is multiplied by F_x and used to cal-

culate the pressure change in the liquid column. Therefore, the pressure calculation is iterative.

p_{wf} Calculation, Pumping Well

The following nomenclature will be used in presenting a procedure for calculating p_{wf}:

$$
\begin{aligned}
p_c &= \text{surface casing pressure,} \\
p_F &= \text{pressure at the fluid level,} \\
p_p &= \text{pressure at the pump,} \\
p_w &= \text{pressure at oil-water interface,} \\
D_F &= \text{depth to fluid level,} \\
D_p &= \text{depth to pump,} \\
D_{perf} &= \text{depth to perforation mid-point,} \\
D_{ow} &= \text{depth to oil-water interface,} \\
\rho_o &= \text{oil density,} \\
\rho_L &= \text{oil-water mixture density,} \\
\rho_{og} &= \text{oil density with gas bubbling through,} \\
\rho_{Lg} &= \text{oil-water-gas mixture density,} \\
f_w &= \text{producing water fraction.}
\end{aligned}
$$

Procedure:

1. Measure p_c and fluid level D_F.

2. Calculate p_F using the gas weight from
$$ p_F = p_c \text{EXP}(0.01875 \gamma_g D_F / \overline{TZ} $$

   This will be trial and error, since $\overline{Z} = f((p_F + p_c)/2)$.

3. Calculate p_p. If the well is pumped down, that is if $D_F = D_p$, then $p_p = p_F$.
$$ p_p = p_F + \rho_{og}(D_p - D_F) $$

   To determine ρ_{og}:
   a. Estimate $\rho_o = 0.433 \gamma_o$
   b. Estimate $F^*_x = 0.95$
   c. Calculate $\rho^*_{og} = \rho_o F_x$
   d. Calculate $p^*_p = p_F + \rho_{og}^* (D_p - D_F)$
   e. Calculate $\overline{p} = (p_F + p^*_p)0.5$
   f. Determine R_s, B_o and ρ_o at $\overline{p}$, $\overline{T}$
   g. Calculate $F_y = q_g / a(p)^{0.4}$ and find F_x from Figure 5-34.
   h. Compare F_x and F^*_x. If not close, set $F^*_x = F_x$ and go to step C. Continue until $F_x = F^*_x$.

4. Calculate $p_{wf} = p_p + \rho_{Lg}(D_{perf} - D_p)$, where
$$
\begin{aligned}
\rho_{Lg} &= \rho_L F_x, \\
\rho_L &= \rho_o(1 - f_w) + \rho_w f_w
\end{aligned}
$$

A value for F_x is found iteratively using the procedure described in Step 3.

The two-phase flow methods may also be used to find p_{wf} by assuming a very low liquid flow rate in the fluid column above the pump.

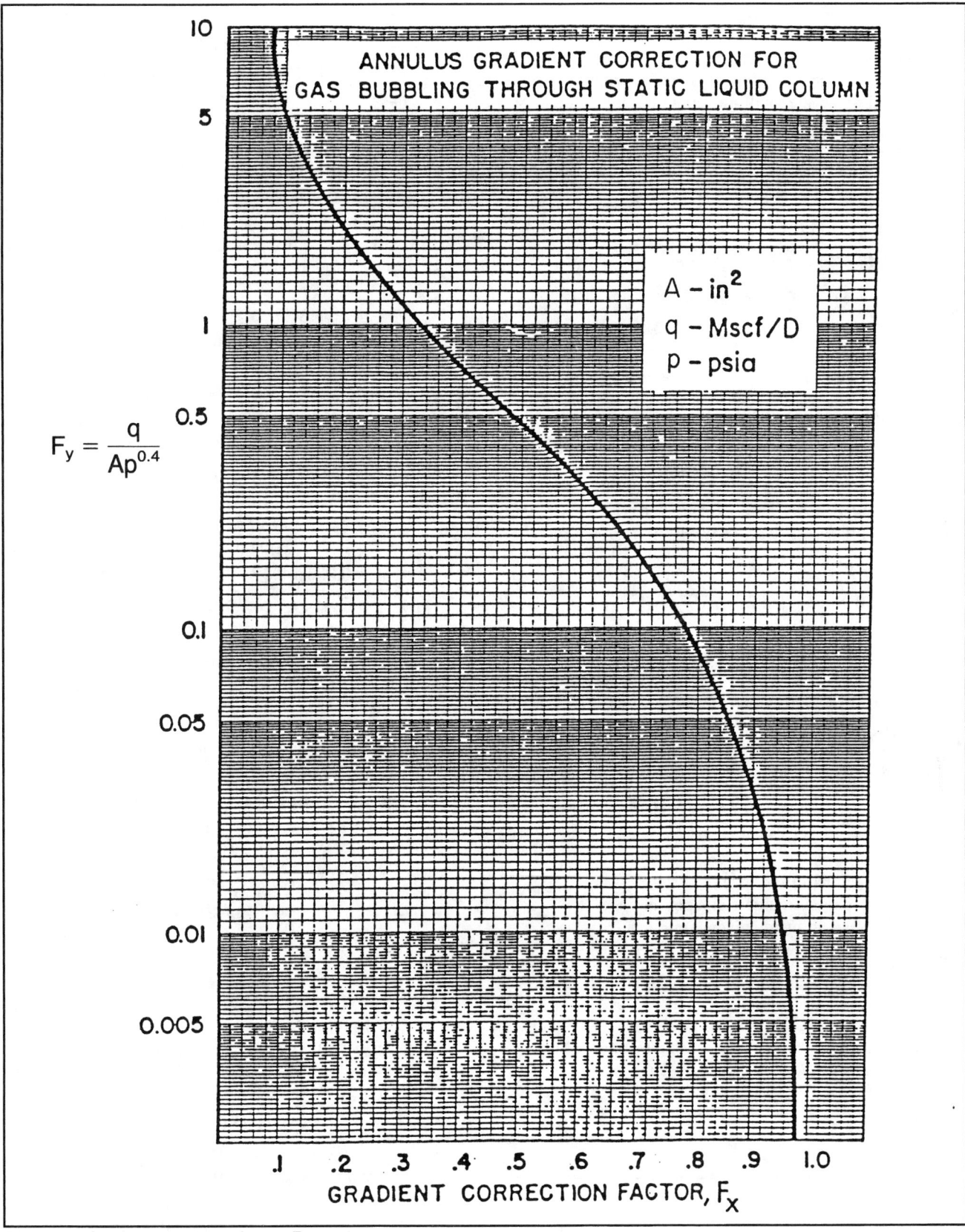

Fig. 5-34. Gradient correction factor, F_x.

p_{ws} Calculation, Shut-in Well

When the well is shut in, no gas will be bubbling through the liquid, and, therefore, F_x will not be required. However, R_s and B_o must be calculated to determine p_o. Also, the oil and water will be separated due to density difference, and the depth to the oil-water interface D_{ow} must be estimated.

Procedure

1. Measure p_c and F_L.

2. Calculate p_F.

3. Calculate the pressure at the oil-water interface.

$$p_w = p_F + \rho_o(D_{ow} - D_F)$$

To determine ρ_o:
   a. Estimate $\rho^*_o = 0.433\ \gamma_o$
   b. Calculate $p^*_w = p_F + \rho^*_o(D_{ow} - D_F)$
   c. Calculate $\bar{p} = 0.5(p_F + p^*_w)$
   d. Calculate R_s, B_o, and ρ_o at $\bar{p}$, $\bar{T}$
   e. Compare ρ_o and ρ^*_o. If not close, set $\rho^*_o = \rho_o$ and go to Step b. Repeat until $\rho_o \cong \rho^*_o$.

4. Calculate p_{ws}

$$p_{ws} = p_w + \rho_w(D_{perf} - D_{ow})$$

where
$$\rho_w = 0.433\ \gamma_w.$$

The depth to the oil-water interface cannot be measured, but it can be calculated by using the pumping and static fluid levels and the producing water fraction. The equation is

$$D_{ow} = D_{perf} - f_w(D_{perf} - D_p) - (D_{Fp} - D_{Fs})(1 - (d_T/d_c)^2)$$

where
D_{Fp} = pumping fluid level,
D_{Fs} = shut-in fluid level,
d_T = tubing outside diameter, and
d_c = casing inside diameter

Example 5-4:

Using the following data, calculate the flowing bottom-hole pressure for this well.

q_o = 70 STB/day	D_{perf} = 7350 ft
D_p = 6200 ft	q_w = 83 STB/day
d_T = 2.375 in.	d_c = 4.95 in.
API = 37°	γ_g = 0.7
T_R = 187°F	T_p = 170°F
γ_w = 1.0	T_s = 100°F
q_g = 25.2 Mscfd	

Solution:

1. $p_c = 120 + 14.7 = 134.7\ \text{psia}, D_F = 6000\ \text{ft}$

2. Estimate $\bar{Z}^* = 0.9$
   $p_F = 134.7\ \text{EXP}[0.01875(.7)(6000)/(595)\bar{Z}^*]$
   $p_F = 134.7\ \text{EXP}(0.132/\bar{Z}^*) = 134.7\ \text{EXP}(0.132/0.9)$
   $p_F = 156\ \text{psia}$
   $\bar{p} = (134.7 + 156)0.5 = 145.4$
   At $\bar{p} = 145.4$ and $\bar{T} = 595$, $\bar{Z} = 0.98$
   $p_F = 134.7\ \text{EXP}(0.132/.98) = 154\ \text{psia}$
   $\bar{p} = (134.7 + 154)0.5 = 144.4$, $\bar{Z} = 0.98$
   Since $\bar{Z}^* = \bar{Z}$, $p_F = 154\ \text{psia} = 139\ \text{psig}$

3. Calculate p_p :
   $\gamma_o = 141.5/(131.5 + 37) = 0.84$
   a. $\rho_o^* = 0.433(.84) = 0.364\ \text{psi/ft}$
   b. Estimate $F_x = 0.95$
   c. $\rho_{og}^* = 0.364(.95) = 0.345$
   d. $p_p^* = 139 + 0.345(6200 - 6000) = 208\ \text{psig}$
   e. $\bar{p} = (139 + 208)0.5 = 173.5\ \text{psig} = 188.2\ \text{psia}$
      $\bar{T} = (187 + 170)0.5 = 179°\text{F} = 639°\text{R}$
   f. $R_s = 25\ \text{scf/STB}, B_o = 1.081$

   $$\rho_o = \frac{350(.84) + .0764(.7)(25)}{5.615(1.081)} = 48.66\ \text{lb/ft}^3$$

   $\rho_o = 0.338\ \text{psi/ft}$

   g. $A = \frac{\pi}{4}(4.95^2 - 2.375^2) = 14.8\ \text{in}^2$

   $$F_y = \frac{q}{A\bar{p}^{0.4}} = \frac{25.2}{14.8(188.2)^{.4}} = 0.21$$

   $F_x = 0.68\ \text{(Figure 5.34)}$
   h. $F_x \neq F_x^*$, therefore, $F_x^* = 0.68$
      $\rho_{og}^* = 0.338(.68) = 0.230$
      $p_p^* = 139 + .230(200) = 185$
      $\bar{p} = (139 + 185)(0.5) = 162\ \text{psig} = 176.7\ \text{psia}$
      $R_s = 24, B_o = 1.081, \rho_o = 0.338$

      $$F_y = \frac{q}{A\bar{p}^{0.4}} = 0.21 \qquad F_x = 0.68$$

   Since $F_x = F_x^*$, convergence is attained, therefore,
   $p_p = 185\ \text{psig}$

4. Calculate p_{wf} :
   $f_w = q_w/(q_w + q_o) = 83/(83 + 70) = 0.54$
   a. $\rho_L = 0.433(.54) + 0.338(.46) = 0.389\ \text{psi/ft}$
   b. Estimate $F_x^* = 0.68$
   c. $\rho_{Lg} = 0.389(.68) = 0.265$
   d. $p_{wf}^* = 185 + 0.265(7350 - 6200) = 489\ \text{psig}$
   e. $\bar{p} = (185 + 489)0.5 = 337\ \text{psig} = 352\ \text{psia}$
   f. Assume change in ρ_o is negligible

   g. $A = \frac{\pi}{4}(4.95)^2 = 19.2\ \text{in}^2$

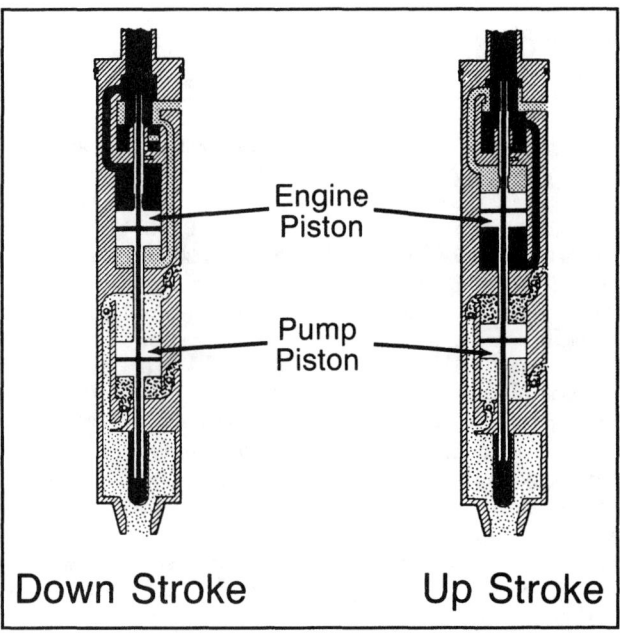

Fig. 5-35. Pump and engine assembly.

$$F_y = \frac{25.2}{19.2(352)^{0.4}} = 0.13$$

$$F_x = 0.74$$

h. $F_x \neq F_x{}^*$, therefore, set $F_x{}^* = 0.74$

$$\rho_{Lg} = 0.389(.74) = 0.288$$

$$p_{wf}{}^* = 516$$

$$\bar{p} = 351\,\text{psig} = 365\,\text{psia}$$

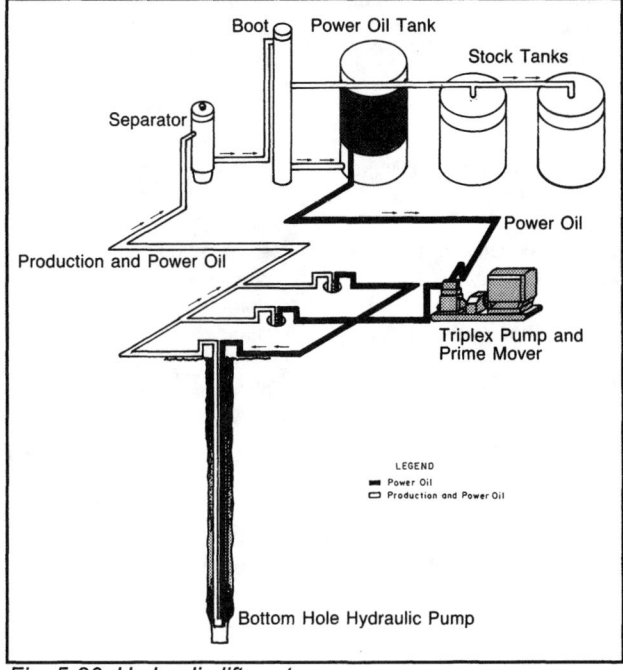

Fig. 5-36. Hydraulic lift system.

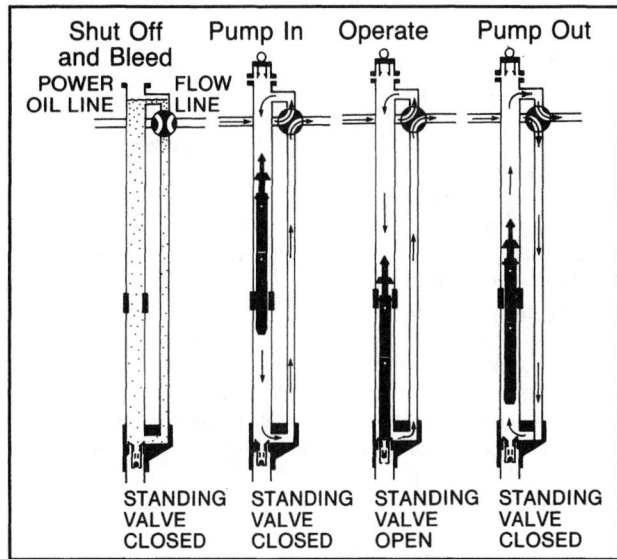

Fig. 5-37. Installing and retrieving the pump.

$$F_y = 0.21 \qquad F_x = 0.74$$

Since $F_x = F_x{}^*$, convergence is attained, therefore,

$$p_{wf} = 516\,\text{psig}$$

V. HYDRAULIC PUMPING

Another artificial lift method that employs a positive displacement pump is the hydraulic pumping system. In this system, the power is transmitted to a subsurface hydraulic motor that is coupled to a pump similar to a sucker rod pump. The power is transmitted by hydraulic or power fluid that is pumped from the surface down an extra string of tubing. If the gas from a hydraulically pumped well is vented, the systems analysis would be identical to that for a rod-pumped well. A schematic of the pump and motor assembly is illustrated in Figure 5-35. The power fluid system is shown in Figure 5-36 and the method used for installing and retrieving the pump without the use of extra equipment is illustrated in Figure 5-37.

VI. SUMMARY

A few of the most widely used artificial methods were discussed, and the application of systems analysis to the design of some of these methods was illustrated. Many other methods exist for decreasing the flowing bottom-hole pressure in a well to increase its producing rate. Some of these are plunger lift, intermittent gas lift, jet pumping, and hydraulically powered centrifugal pump. Many of these methods are discussed in Reference 1.

Regardless of the methods used to move the fluids from the wellbore to the separator, an analysis that considers

TABLE 5-5
Relative Advantages of Artificial Lift Systems
S. G. Gibbs -Nabla Corporation

Rod Pumping	Hydraulic Pumping	Electric Submersible Pumping	Gas Lift
*Relatively simple system design.	*Not so depth limited–can lift large volumes from great depths (500 BPD from 15000 ft.). Have been installed to 18000 ft.	*Can lift extremely high volumes (20000 BPD+ in shallow wells with large casing. Currently lifting ±120,000 B/D from water supply wells in Middle East with 600 HP units.	*Can handle large volume of solids with minor problems.
*Units easily changed to other wells with minimum cost.	*Crooked holes present minimal problem.	*Unobtrusive in urban locations.	*Handles large volume in high P.I. wells (continuous lift) (50000 BLPD+).
*Efficient, simple, and easy for field people to operate.	*Unobtrusive in urban locations.	*Simple to operate.	*Fairly flexible–convertible from continuous to intermittent to chamber or plunger lift as well declines.
*Applicable to slim holes and multiple completions.	*Power source can be remotely located.	*Easy to install downhole pressure sensor for telemetering pressure to surface via cable.	*Unobtrusive in urban locations.
*Can pump a well down to very low pressure (depth and rate dependent).	*Analyzable.	*Crooked holes present no problem.	*Power source can be remotely located.
*System usually is naturally vented for gas separation and fluid level soundings.	*Flexible–can usually match displacement to well's capability as well declines.	*Applicable offshore.	*Easy to obtain downhole pressures and gradients.
*Flexible–can match displacement rate to well capability as well declines.	*Can use gas or electricity as power source.	*Corrosion and scale treatment easy to perform.	*Lifting gassy wells is no problem.
*Analyzable.	*Downhole pumps can be circulated out in free systems.	*Availability in different sizes.	*Sometimes servicable with wireline unit.
*Can lift high temperature and viscous oils.	*Can pump a well down to fairly low pressure.	*Lifting cost for high volumes generally very low.	*Crooked holes present no problem except for retrieving and running valves by wire line.
*Can use gas or electricity as power source.	*Applicable to multiple completions.		*Corrosion is not usually as adverse.
*Corrosion and scale treatments easy to perform.	*Applicable offshore.		*Applicable offshore.
*Applicable to pump-off control if electrified.	*Closed system will combat corrosion.		
*Availability of different sizes.			

Relative Disadvantages of Artificial Lift Systems
S. G. Gibbs -Nabla Corporation

Rod Pumping	Hydraulic Pumping	Electric Submersible Pumping	Gas Lift
*Obtrusive in urban locations.	*Not easy for field personnel to troubleshoot.	*System is depth limited (10,000 ft.±) due to cable cost and inability to install enough power downhole.	*Lift gas is not always available.
*Heavy and bulky in offshore operations.	*Difficult to obtain valid well tests in low volume wells.	*Gas and solids production are troublesome.	*Not efficient in lifting small fields or one-well leases.
*Susceptible to paraffin problems.	*Requires two strings of tubing for some installations.	*Not easily analyzable unless good engineering "know-how."	*Difficult to lift emulsions and viscous crudes.
*Tubing cannot be internally coated for corrosion.	*Problems in treating power water where used.	*Lack of production rate flexibility.	*Not efficient for small fields or one-well leases if compression equipment is required.
*Crooked holes present a friction problem.	*Power oil systems are fire hazard.	*Casing size limitation.	*Gas freezing and hydrate problems.
*High solids production is troublesome.	*Large oil inventory required in power oil system which detracts from profitability.	*Should not be set below fluid entry.	*Problems with dirty surface lines.
*Gassy wells usually lower volumetric efficiency.	*High solids production is troublesome.	*Not applicable to multiple completions.	*Difficulty in retrieving valves in highly deviated wells.
*Is depth limited, primarily due to rod capability.	*Operating costs are sometimes higher.	*Only applicable with electric power.	
	*Unusually susceptible to gas interference; usually not vented.	*High voltages (1000V±) are necessary.	
	*Vented installations are more expensive because of extra tubing required.	*Impractical in low volume wells and in wells with deep lift.	
	*Treating for scale below packer is difficult.	*Expensive to change equipment to match declining well capability.	
		*Cable causes problems in handling tubulars.	
		*Cables deteriorate in high temperatures.	

the components of the system as completely separate and unrelated entities cannot adequately describe the total system. Total system or nodal analysis is necessary to optimize the performance of any oil- or gas-producing well.

The choice of the type of artificial lift system for a well or field depends on many factors, such as well depth, availability of gas, production rates required, hole deviation, etc. Some of the relative advantages and disadvantages of the various methods are listed in Table 5-5, as published by Gibbs.[2]

VII. REFERENCES

1. Brown, K. E.: *The Technology of Artificial Lift Methods, Vol. 2*, PennWell Publ. Co., Tulsa, Okla., 1980.
2. Gibbs S. G.: "Predicting the Behavior of Sucker Rod Pumping Systems," *JPT*, July, 1963.
3. *Gas Lift*, Book 6 of the Vocational Training Series, American Petroleum Institute, Dallas, Texas, 1984.

Nomenclature

		Dimensions			Dimensions
A	area	L^2	G_p	cumulative gas produced	L^3
B_g	gas formation volume factor		ΔG_p	gas produced during an interval	L^3
B_o	oil formation volume factor		h	thickness (general and individual	
B_{ob}	oil formation volume factor at			bed)	L
	bubble-point conditions		H	enthalpy (always with phase or	
B_t	total (two-phase) formation			system subscripts)	mL^2/t^2
	volume factor		H_L	liquid holdup	
B_w	water formation volume factor		H_g	gas void fraction	
C_f	formation (rock) compressibility	Lt^2/m	$J(PI)$	productivity index	L^4t/m
c_g	gas compressibility	Lt^2/m	J_s	specific productivity index	L^3t/m
c_o	oil compressibility		k	absolute permeability	L^2
c_{pr}	pseudoreduced compressibility		k_g	effective permeability to gas	L^2
c_w	water compressibility	Lt^2/m	k_o	effective permeability to oil	L^2
C	coefficient of gas-well back-		k_{rg}	relative permeability to gas	
	pressure curve	$L^{3-2n}t^{4n}/m^{2n}$	k_{ro}	relative permeability to oil	
C	concentration	various	k_{rw}	relative permeability to water	
C_L	condensate or natural gas liquids		k_w	effective permeability to water	L^2
	content	various	K	equilibrium ratio (y/x)	
d	diameter	L	ln	natural logarithm, base e	
D	depth	L	log	common logarithm, base 10	
E	efficiency		L	length	L
E_v	volumetric efficiency		m	mass	m
f	fraction		m	slope	various
f	friction factor		M	molecular weight	m
f_w	producing water fraction		n	exponent of back-pressure curve,	
F	force	mL/t^2		gas well	
F_{wo}	instantaneous producing water-		n	total moles	
	oil ratio		n_j	moles of component j	
g	acceleration of gravity	L/t^2	N	initial oil in place in reservoir	L^3
g_c	conversion factor in Newton's		N_p	cumulative oil produced	L^3
	Second Law of Motion	mL/Ft^2	N_{Re}	Reynolds number	
G	total initial gas in place in		ΔN_p	oil produced during an interval	L^3
	reservoir	L^3	p	pressure	m/Lt^2

		Dimensions
p_a	atmospheric pressure	m/Lt^2
p_b	bubblepoint (saturation) pressure	m/Lt^2
p_c	critical pressure	m/Lt^2
p_{cf}	casing pressure, flowing	m/Lt^2
p_{cs}	casing pressure, static	m/Lt^2
p_d	dewpoint pressure	m/Lt^2
p_D	dimensionless pressure	
p_e	external boundary pressure	m/Lt^2
p_i	initial pressure	m/Lt^2
p_{pc}	pseudocritical pressure	m/Lt^2
p_{pr}	pseudoreduced pressure	
p_r	reduced pressure	
p_{sc}	pressure, standard conditions	m/Lt^2
p_{sp}	separator pressure	m/Lt^2
p_{wh}	tubing pressure, flowing	m/Lt^2
p_{ts}	tubing pressure, static	m/Lt^2
p_w	bottomhole pressure, general	m/Lt^2
p_{wf}	bottomhole pressure, flowing	m/Lt^2
p_{wfs}	bottomhole pressure, sandface	m/Lt^2
p_{ws}	bottomhole pressure, static	m/Lt^2
$\bar{p}$	average pressure	m/Lt^2
$\bar{p}_R$	pressure, reservoir average	m/Lt^2
p_c	capillary pressure	m/Lt^2
q	production rate or flow rate	L^3/t
q_D	dimensionless production rate	
q_g	gas production rate	L^3/t
q_o	oil production rate	L^3/t
q_w	water production rate	L^3/t
r	radial distance	L
r_d	radius of drainage	L
r_e	external boundary radius	L
r_s	radius of well damage or stimulation (skin)	L
r_w	well radius	L
r_{wa}	apparent or effective wellbore radius (includes effects of well damage or stimulation)	L
R	producing gas/oil ratio	
R	universal gas constant (per mole)	mL^2/t^2T
R_p	cumulative gas/oil ratio	
R_s	solution gas/oil ratio (gas solubility in oil)	
R_{sb}	solution gas/oil at bubble-point conditions	
R_{si}	initial solution gas/oil ratio	
R_{sw}	gas solubility in water	
s	skin effect	
S	saturation	
S_g	gas saturation	
S_{gc}	critical gas saturation	
S_{gr}	residual gas saturation	
S_L	total (combined) liquid saturation	
S_o	oil saturation	

		Dimensions
S_{or}	residual oil saturation	
S_w	water saturation	
S_{ws}	critical water saturation	
t	time	t
t_D	dimensionless time	
t_s	time for stabilization of a well	t
T	temperature	T
T_c	critical temperature	T
T_f	formation temperature	T
T_{pc}	pseudocritical temperature	T
T_{pr}	pseudoreduced temperature	
T_r	reduced temperature	
T_R	reservoir temperature	T
T_{sc}	temperature, standard conditions	T
v	specific volume	L^3/m
v	velocity	L/t
V	volume	
V_b	bulk volume	
V_p	pore volume	
w	mass flow rate	m/t
W	initial water in place in reservoir	L^3
W	water (always with identifying subscripts)	various
W	work	mL^2/t^2
W_e	cumulative water influx (encroachment)	L^3
W_p	cumulative water produced	L^3
ΔW_e	water influx (encroachment) during an interval	L^3
ΔW_p	water produced during an interval	L^3
x	mole fraction of a component in liquid phase	
y	mole fraction of a component in vapor phase	
z	gas deviation factor (compressibility factor, $z = pV/nRT$)	
Z	elevation referred to datum	L
α	alpha angle	
γ	gamma specific gravity	
γ_g	gamma gas specific gravity	
γ_o	gamma oil specific gravity	
γ_w	gamma water specific gravity	
Δ	delta difference	
θ	theta angle	
λ_g	lambda gas void fraction (no-slip)	
λ_L	lambda liquid holdup (no-slip)	
μ	mu viscosity	m/Lt
μ_g	mu gas viscosity	m/Lt
μ_o	mu oil viscosity	m/Lt
μ_w	mu water viscosity	m/Lt

			Dimensions				Dimensions
ρ	rho	density	m/L^3	ρ_w	rho	water density	m/L^3
ρ_L	rho	liquid density	m/L^3	σ	sigma	surface tension	
ρ_g	rho	gas density	m/L^3			(interfacial tension)	m/t^2
ρ_o	rho	oil density	m/L^3	ϕ	phi	porosity	

Appendix
A

Two-phase Flow Correlation Examples Hagedorn and Brown Method

The Hagedorn and Brown method, ignoring acceleration, requires solution of Equation 3-16 for each increment into which the well is divided.

$$\frac{dp}{dh} = \frac{g}{g_c}(\rho_m \cos\phi) + \frac{f\rho_f v_m^2}{2g_c d} \qquad \text{(A-1)}$$

Empirical correlations are presented for determining the mixture density, ρ_m and the friction factor f. The parameters in Equation A-1 are defined by

$$
\begin{aligned}
\rho_m &= \rho_L H_L + \rho_g(1 - H_L), \\
\rho_L &= \text{liquid density}, \\
\rho_g &= \text{gas density} \\
H_L &= \text{liquid holdup (fraction of pipe occupied} \\
&\quad \text{by liquid)}, \\
\phi &= \text{angle of well or segment from vertical}, \\
v_m &= v_{sL} + v_{sg}, \\
v_{sL} &= \text{superficial liquid velocity} = q_L/A_p, \\
v_{sg} &= \text{superficial gas velocity} = q_g/A_p, \\
A_p &= \text{area of flow string} \\
d &= \text{flow string I.D.}, \\
\rho_f &= \rho_n^2 / \rho_m \\
\rho_n &= \rho_L\lambda_L + \rho_g(1 - \lambda_L), \\
&= v_{sL}/v_m.
\end{aligned}
$$

The friction factor is calculated using the Jain equation or found from the Moody diagram using the pipe relative roughness and the following Reynolds number:

$$N_{\text{Rem}} = \frac{\rho_n v_m d}{\mu_m}$$

where
$$\mu_m = (\mu_L)^{H_L} (\mu_g)^{(1-H_L)},$$

$$
\begin{aligned}
\mu_L &= \text{liquid viscosity}, \\
\mu_g &= \text{gas viscosity}.
\end{aligned}
$$

Determination of H_L requires use of three empirical correlations. These are presented in Fig. 3-15a, b, and c. To determine H_L from these figures, the following dimensionless numbers must be evaluated from known data:

$$
\begin{aligned}
N_{LV} &= v_{sL}[\rho_L/g\sigma]^{.25}, \\
N_{gv} &= v_{sg}[\rho_L/g\sigma]^{.25}, \\
N_d &= d[\rho_L g/\sigma]^{.5}, \\
N_L &= \mu_L[g/\rho_L\sigma^3]^{.25}
\end{aligned}
$$

where σ = gas-liquid surface tension. These equations are valid for any consistent set of units. For field units, the equations are

$$
\begin{aligned}
N_{Lv} &= 1.938 \, v_{sL}(\rho_L/\sigma)^{.25} \\
N_{gv} &= 1.938 \, v_{sg}(\rho_L/\sigma)^{.25}, \\
N_d &= 120.872 \, d \, (\rho_L/\sigma)^{.5}, \\
N_L &= 0.15726 \, \mu_L \, (1.0/\rho_L\sigma^3)^{.25}
\end{aligned}
$$

where
$$
\begin{aligned}
v_{SL}, v_{sg} &= \text{ft/sec}, \\
\rho_L &= \text{lbm/cu ft}, \\
\sigma &= \text{dynes/cm}, \\
d &= \text{ft}, \\
\mu_L &= \text{centipoise}.
\end{aligned}
$$

The procedure for finding H_L is:

1. Calculate N_L

2. Find CN_L from Fig. 3-15a

3. Calculate

$$\bar{X}_{H_L} = \frac{N_{Lv}(CN_L)p^{0.1}}{N_d N_{gv}^{0.575} p_a^{0.1}}$$

where p_a = base pressure (14.7 psia)

4. Find

$$\frac{H_L}{\psi} \text{ from Fig. 3-15b.}$$

5. Calculate

$$\bar{X}_\psi = \frac{N_{gv} N_L^{0.38}}{N_d^{2.14}}$$

6. Find ψ from Fig. 3-15c

7. Calculate $H_L = \psi(H_L/\psi)$

A constraint on liquid holdup is $H_L \geq \lambda_L$.

Once H_L is determined, N_{Re} and thus f can be calculated. The pressure gradient can then be calculated.

This is Step 5 in the Procedure for Calculating a Pressure Traverse that was presented earlier. All the fluid properties and velocities used in the above equations are evaluated at the average pressure and temperature in the tubing increment.

Example A-1:

During the calculation of a pressure traverse in a gas well producing liquid, the following conditions were determined at the average pressure and temperature in the pipe increment.

$\bar{p}$ = 1500 psia	v_{sg} = 30 ft/sec
$\bar{T}$ = 180 °F	v_{sL} = 5 ft/sec
d = 2.992 in.	ε = 0.0006 ft
μ_g = 0.012 cp	ρ_L = 50 lbm/cu ft
μ_L = 0.45 cp	ρ_g = 8 lbm/cu ft
σ = 25 dynes/cm	

Using the Hagedorn and Brown method, determine the pressure gradient.

Solution:

Before finding H_L and f, some preliminary calculations are made:

v_m = $v_{sL} + v_{sg}$ = 5 + 30 = 35 ft/sec

λ_L = 5/35 = 0.143

ρ_n = 50 (0.143) + 8 (1 - 0.143) = 14 lb/ft³

A_p = $\pi d^2/4$ = 0.7854 (0.249)² = 0.0487 ft²

ρ_L/σ = 50/25 = 2

N_{Lv} = 1.938 (5) (2)·25 = 11.52

N_{gv} = 1.928 (30) (2)·25 = 69.14

N_d = 120.872 (0.249) (2)·5 = 42.56

N_L = 0.15726 (0.45) [1/(50) (25)³]·25 = 0.0024

Determine H_L

1. N_L = 0.0024
2. From Fig. 3-15a, CN_L = 0.002
3. $\bar{X}_H = \dfrac{11.52\,(0.002)\,(1500)^{0.1}}{42.56\,(69.14)^{0.575}\,(14.7)^{0.1}} = 7.53 \times 10^{-5}$
4. From Fig. 3-15b, $H_L / \psi = 0.29$
5. $\bar{X}\psi = \dfrac{69.14\,(0.0024)^{0.38}}{(42.56)^{2.14}} = 2.28 \times 10^{-3}$
6. From Fig. 3-15c, $\psi = 1.0$
7. $H_L = 1.0\,(0.29) = 0.29$

$\rho_m = 50\,(0.29) + 8\,(1 - .29) = 20.18$ lb/ft³

$\rho_f = (14)^2/20.18 = 9.71$ lb/ft³

$\mu_m = (0.45)^{0.29}\,(0.012)^{(1-0.29)} = 0.034$ cp

$N_{Rem} = \dfrac{1488\,(14)\,(35)\,(0.249)}{0.034} = 5.29 \times 10^6$

$\dfrac{\varepsilon}{d} = \dfrac{0.0072}{2.992} = 0.0024$

From Fig. 3-5 or Eq. 3-15

$f = 0.025$

$\dfrac{dp}{dh} = 20.18 + \dfrac{0.025\,(9.71)\,(35)^2}{2\,(32.2)\,(0.249)} = 20.18 + 18.54$

$\dfrac{dp}{dh} = 38.72$ lb/ft³ = 0.269 psi/ft

Beggs and Brill Method

The Beggs and Brill method requires the determination of the flow pattern that would exist in the pipeline if the pipe were horizontal. Different equations are used to calculate liquid holdup for each flow pattern. The flow patterns defined are shown in Fig. 3-19.

Determination of the correct flow regime requires calculating several dimensionless numbers, including a two-phase Froude number.

The following variables are used to determine which flow regime would exist if the pipe were in a horizontal position. This flow regime is a correlating parameter and gives no information about the actual flow regime unless the pipe is horizontal.

$$N_{FR} = \frac{v_m^2}{g\,d}$$

$$\lambda_L = \frac{v_{sL}}{v_m}$$

$$L_1 = 316\,\lambda_L^{0.302}$$

$$L_2 = 0.0009252\,\lambda_L^{-2.4684}$$

$$L_3 = 0.10\,\lambda_L^{-1.4516}$$

$$L_4 = 0.5\,\lambda_L^{-6.738}$$

The horizontal flow regime limits are:

Segregated:

Limits: $\lambda_L < 0.01$ and $N_{FR} < L_1$
or $\lambda_L \geq 0.02$ and $N_{FR} < L_2$

Transition:

Limits: $\lambda L \geq 0.01$ and $L_2 < N_{FR} \leq L_3$

Intermittent:

Limits: $0.01 \leq \lambda_L < 0.4$ and $L_3 < N_{FR} \leq L_1$
or $\lambda_L \geq 0.4$ and $L_3 < N_{FR} \leq L_4$

Distributed:

Limits: $\lambda_L < 0.4$ and $N_{FR} \geq L_1$
or $\lambda_L \geq 0.4$ and $N_{FR} > L_4$

When the flow falls in the transition regime, the liquid holdup must be calculated using both the segregated and intermittent equations and interpolated using the following weighting factors.

H_L (transition) $= A \times H_L$ (segregated) $+ B \times H_L$ (intermittent) where:

$$A = \frac{L_3 - N_{FR}}{L_3 - L_2}$$

$$B = 1 - A$$

The same equations are used to calculate liquid holdup for all flow regimes. The coefficients and exponents used in the equations are different for each flow regime.

The liquid holdup depends on flow regime and is calculated from

$$H_{L(\phi)} = \psi H_{L(0)}$$

where $H_{L(0)}$ is the holdup which would exist at the same conditions in a horizontal pipe and ψ is the inclination correction factor.

$$H_{L(0)} = \frac{a\lambda_L^b}{N_{FR}^c}$$

where a, b, and c are determined for each flow pattern from Table A-1.

TABLE A-1
Flow Pattern

Flow Pattern	a	b	c
Segregated	0.98	0.4846	0.0868
Intermittent	0.845	0.5351	0.0173
Distributed	1.065	0.5824	0.0609

The value calculated for $H_{L(0)}$ is constrained by

$$H_{L(0)} \geq \lambda_L$$

The factor for correcting the holdup for the effect of pipe inclination is given by:

$$\psi = 1 + C\,[\sin(1.8\phi) - 0.333 \sin^3(1.8\phi)]$$

where ϕ is the actual angel of the pipe from horizontal, and

$$C = (1 - \lambda_L)\ln[(d)(\lambda_L)^e(N_{LV})^f(N_{FR})^g]$$

where d, e, f, and g are determined for each flow condition from Table A-2.

TABLE A-2
Intermittent Uphill

Flow Pattern	d	e	f	g
Segregated uphill	0.011	−3.768	3.539	−1.614
Intermittent uphill	2.96	0.305	−0.4473	0.0978
Distributed uphill	No correction		$C = 0$, $\psi = 1$	$H_L \neq f(\phi)$
All flow patterns downhill	4.70	-0.3692	0.1244	-0.5056

with with the restriction that $C \geq 0$.

Once $H_{L(\phi)}$ is determined, the two-phase density is calculated from

$$\rho_s = \rho_L H_L + \rho_g H_g$$

where $H_g = 1 - H_L$.

The pressure gradient due to elevation change is then

$$\left(\frac{dp}{dZ}\right)_{el} = \frac{g}{g_c}(\rho_s \sin\phi)$$

The pressure gradient due to friction is

$$\left(\frac{dp}{dZ}\right)_f = \frac{f_{tp}\rho_n v_m^2}{2g_c d}$$

where

$$\rho_n = \rho_L\lambda_L + \rho_g\lambda_g$$

$$f_{tp} = f_n\left(\frac{f_{tp}}{f_n}\right)$$

The no-slip friction factor f_n is determined from the Moody diagram (Fig. 3-5) or from Equation 3-15 using the following Reynolds number:

$$N_{Re} = \frac{\rho_n v_m d}{\mu_n}$$

where

$$\mu_n = \mu_L \lambda_L + \mu_g \lambda_g$$

The ratio of the two-phase to no-slip friction factor is calculated from:

$$\frac{f_{tp}}{f_n} = e^s$$

where:

$$S = [\ln(y)]/\{-0.0523 + 3.182\ln(y) \\ -0.8725[\ln(y)^2] + 0.01853[\ln(y)]^4\}$$

and

$$y = \frac{\lambda_L}{\left[H_{L(\phi)}\right]^2}$$

The value of S becomes unbounded at a point in the interval $1 < y < 1.2$; and for y in this interval, the function S is calculated from:

$$S = \ln(2.2y - 1.2)$$

Although the acceleration pressure gradient is very small except for high velocity flow, it should be included for high flow rates.

$$\left(\frac{dp}{dZ}\right)_{acc} = \left[\frac{\rho_s v_m v_{sg}}{g_c p}\right]\frac{dp}{dz}$$

If we define an acceleration term as

$$E_k = \frac{\rho_s v_m v_{sg}}{g_c p}$$

the total pressure gradient can be calculated from:

$$\frac{dp}{dZ} = \frac{\left(\frac{dp}{dz}\right)_{el} + \left(\frac{dp}{dz}\right)_f}{1 - Ek}$$

Example A-2:

Using the following data for a hilly terrain pipeline, calculate the outlet pressure using the Beggs and Brill method.

$q'_o = 7140$ STB/day p_1(inlet) = 425 psia
$q'_g = 25.7$ MMcf/day $\bar{T} = 90\ °F$ $\mu_g = 0.70$
$d = 12$ in. $\gamma_o = 0.83 = 40\ °API$

Divide the pipeline into two sections. Section 1 rises 300 ft in one mile. Section 2 drops 300 ft in 3000 ft.

Solution:

Section 1

1. Estimate Δp^* and calculate $\bar{p}$
   $\Delta p^* = 30$ psi, $\bar{p} = 425 - 30/2 = 410$ psia
2. From fluid property correlations, at 410 psia and 90 °F:

   $R_s = 96$ scf/STB $\sigma_o = 19.6$ dyne/cm

   $B_o = 1.047$ $Z = 0.925$

   $\mu_o = 2.4$ cp $\mu_g = 0.0105$ cp

3. Calculate flow rates and densities

   $$\rho_o = \frac{350\gamma_o + 0.0764\,R_s\gamma_g}{5.615\,B_o}$$
   $$= \frac{350(0.83) + 0.0764(96)(0.7)}{5.615(1.047)}$$

$\rho_o = 50.29$ lbm/cu ft

$$\rho_g = \frac{2.7\,p\,\gamma_g}{ZT} = \frac{2.7(410)(0.7)}{0.925(550)} = 1.52\ \text{lbm/cu ft}$$

$q_o = 6.49\times10^{-5}\ q'_o\,B_o = 6.49\times10^{-5}(7140)(1.047)$

$q_o = 0.485\ \text{ft}^3/\text{sec}$

$$q_g = \frac{3.27\times10^{-7}\,Z(q'_g - q'_o R_s)T}{p}$$

$$q_g = \frac{3.27\times10^{-7}(0.925)(25.7\times10^6 - 7140(96))(550)}{410}$$

$q_g = 10.1\ \text{ft}^3/\text{sec}$

4. Calculate the in-situ superficial velocities
   $v_{sL} = q_L/A = 0.485/0.7854\,(1)^2 = 0.617$ ft/sec
   $v_{sg} = q_g/A = 10.1/0.7854 = 12.87$ ft/sec
   $v_m = v_{sL} + v_{sg} = 0.617 + 12.87 = 13.49$ ft/sec

5. Determine the flow pattern
   $$\lambda_L = \frac{v_{sL}}{v_m} = \frac{0.617}{13.49} = 0.0457$$

   $$N_{FR} = v_m^2/gd = (13.49)^2/(32.2)(1) = 5.65$$

   $$N_{Lv} = 1.938\,v_{sL}\left[\frac{\rho_L}{\sigma_L}\right]^{.25} = 1.51$$

   $L_1 = 316\,\lambda_L^{0.302} = 124$

   $L_2 = 0.009252\,\lambda_L^{-2.4684} = 1.86$

   $L_3 = 0.10\,\lambda_L^{-1.4516} = 8.82$

Since $\lambda_L > .01$ and $L_2 < N_{FR} \le L_3$, flow pattern is TRANSITION, therefore interpolation is required.

6. Calculate liquid holdup
   a. *Segregated*

   $$H_{L(0)} = \frac{0.98(\lambda_L)^{0.4846}}{(N_{FR})^{0.0868}} = 0.189$$

   $C = (1 - \lambda_L)\ln[0.011(\lambda_L)^{-3.768}(N_{Lv})^{3.539}(N_{FR})^{-1.614}]$
   $\quad = 5.516$

   $\phi = \arcsin(300/5280) = 3.257\ °$

   $\psi = 1 + C[\sin(1.8\phi) - \sin^3(1.8\phi)/3] = 1.56$

   $H_{L(\phi)}$ (segregated) $= H_{L(0)}\psi = 0.189(1.56) = 0.295$

   b. *Intermittent*

   $$H_{L(0)} = \frac{0.845(\lambda_L)^{0.5351}}{N_{FR}^{0.0173}} = 0.157$$

   $C = (1 - \lambda_L)\ln[2.96(\lambda_L)^{0.305}(N_{Lv})^{-0.4437}(N_{FR})^{0.0978}]$
   $\quad = 0.1246$

   $\psi = 1 + C(0.1015) = 1.013$

   $H_{L(\phi)}$ (intermittent) $= 0.157(1.013) = 0.159$

   $$A = \frac{L_3 - N_{FR}}{L_3 - L_2} = \frac{8.82 - 5.65}{8.82 - 1.86} = 0.455$$

   $B = 1 - A = 0.545$

$H_L \text{(transition)} = A \times H_L \text{(seg.)} + B \times H_L \text{(int.)}$
$= 0.455(0.295) + 0.545(0.159) = 0.221$

7. Calculate the actual and no-slip densities

$\rho_s = \rho_L H_L + \rho_g H_g + 50.29(0.221) + 1.52(0.779)$
$\rho_s = 12.3 \text{ lbm/cu ft}$
$\rho_n = \rho_L \lambda_L + \rho_g \lambda_g = 50.29(0.0457) + 1.52(0.9543)$
$\rho_n = 3.75 \text{ lbm/cu ft}$

8. Calculate the friction factor

$N_{Ren} = \dfrac{1488 \, \rho_n v_m d}{\mu_L \lambda_L + \mu_g \lambda_g}$

$= \dfrac{1488(3.75)(13.49)(1)}{2.4(0.0457) + 0.0105(0.9543)} = 6.29 \times 10^5$

$fn = 0.0126$

$y = \dfrac{\lambda_L}{[H_L(\phi)]^2} = \dfrac{0.0457}{(0.221)^2} = 0.936$

$X = \ln y = -0.066$

$S = X/[-0.0523 + 3.182\,X - 0.8725\,X^2$
$\quad + 0.01853\,X^4] = 0.248$

$f_{tp} = f_n \, \text{EXP}(S) = 0.0126(1.28) = 0.016$

9. Calculate the pressure gradient

$\dfrac{dp}{dL} = \dfrac{\dfrac{g}{g_c}\rho_s \sin\phi + \dfrac{f_{tp}\rho_n v_m^2}{2\,g_c\,d}}{1 - \dfrac{\rho_s v_m v_{sg}}{g_c p}}$

$\dfrac{dp}{dL} = \dfrac{12.3(0.057) + \dfrac{0.016(3.75)(13.49)^2}{64.4(1)}}{1 - \dfrac{12.3(13.49)(12.87)}{32.2(410)(144)}}$

$\dfrac{dp}{dL} = \dfrac{0.70 + 0.17}{0.999} = 0.871 \text{ psf/ft} = 0.0061 \text{ psi/ft}$

10. Calculate the pressure drop

$\Delta p = \left(\dfrac{dp}{dL}\right)\Delta L = 0.0061(5280) = 32.2 \text{ psi}$

This is close enough to the estimated Δp^* of 30 psi for hand calculations. If a computer program was being used, the calculated Δp would be taken as the new estimated Δp^* and the procedure would be repeated until the error was smaller. The pressure at the end of Section 1 is $425 - 32.2 = 393.8$ psia. The same procedure must be followed to calculate the pressure at the end of Section 2.

Section 2

1. Estimate $\Delta p^* = 0$, $\bar{p} = 392.8$ psia
2. $R_s = 90$ \quad $\mu_g = 0.0105$ cp
   $B_o = 1.046$ \quad $\sigma_o = 19.6$ dyne/cm
   $\mu_o = 2.4$ cp \quad $Z = 0.925$

3. $\rho_o = \dfrac{350(0.83) + 0.0764(90)(0.7)}{5.615(1.046)}$
   $= 50.28 \text{ lbm/cu ft}$

$\rho_g = \dfrac{2.7(392.8)(0.7)}{0.925(550)} = 1.46 \text{ lbm/cu ft}$

$q_o = \dfrac{0.485(1.046)}{1.047} = .484 \text{ ft}^3/\text{sec}$

$q_g = \dfrac{3.27\times10^{-7}(0.925)(25.7)\times10^6 - 7140(90)(550)}{392.8}$
$= 10.61 \text{ ft}^3/\text{sec}$

4. $v_{sL} = 0.484/0.785 = 0.616 \text{ ft/sec}$
   $v_{sg} = 10.61/0.785 = 13.52 \text{ ft/sec}$
   $v_m = 0.616 + 13.52 = 14.14 \text{ ft/sec}$

5. $\lambda_L = 0.616/14.14 = .04356$

$N_{FR} = (14.14)^2/(32.2) = 6.21$

$N_{Lv} = \dfrac{1.51(0.616)}{0.617} = 1.51$

$L_1 = 316(0.04356)^{0.302} = 122.7$

$L_2 = 2.11 \qquad L_3 = 9.45$

Since $\lambda_L > .01$ and $L_2 < N_{FR} < L_3$, flow pattern is TRANSITION.

6.

a. *Segregated*

$H_{L(0)} = \dfrac{0.98(.04356)^{0.4846}}{(6.21)^{0.0868}} = 0.183$

$C = (1-\lambda_L)\ln(4.7\lambda_L^{-.3692}N_{Lv}^{.1244}N_{FR}^{-.5056}) = 1.75$

$\phi = \arcsin(300/3000) = -5.74°$

$\psi = 1 + C(\sin(1.8\phi) - \sin^3(1.8\phi)/3)$
$\psi = 1 - 1.75(0.177) = 0.69$
$H_{L(\phi)} = 0.183(0.69) = 0.126$

b. *Intermittent*

$H_{L(0)} = \dfrac{0.945(.04356)^{0.5351}}{(6.21)^{0.0173}} = 0.153$

$C = 1.75, \qquad \psi = 0.69$
$H_{L(\phi)} = 0.153(0.69) = 0.106$

$A = \dfrac{9.45 - 6.21}{9.45 - 2.11} = 0.44 \quad B = 0.56$

$H_L \text{(transition)} = 0.44(0.126) + 0.56(0.106) = 0.115$

7. $\rho_s = 50.28(0.115) + 1.46(1 - 0.115) = 7.07 \text{ lbm/cu ft}$
   $\rho_n = 50.28(0.04356) + 1.46(1 - 0.04356) = 3.59 \text{ lbm/cu ft}$

8. $N_{Re_n} = \dfrac{1488\,(3.59)\,(14.14)}{2.4\,(0.04356) + 0.0105\,(0.9564)}$

$= 6.59 \times 10^5$

$f_n = 0.0125 \quad y = \dfrac{0.04356}{(0.115)^2} = 3.29$

$X = \ln\ y = 1.192$

$S = 0.319$

$f_{tp} = 0.0125\ \text{EXP}\,(0.319) = 0.017$

9. $\dfrac{dp}{dL} = \dfrac{7.07\,(-0.10) + \dfrac{(0.017)\,(3.59)\,(14.14)^2}{64.4\,(1)}}{0.999}$

$= \dfrac{-7.07 + 0.189}{0.999}$

$\dfrac{dp}{dL} = -0.518\ \text{psf/ft} = -0.0036\ \text{psi/ft}$

10. $\Delta p = (-0.0036)\,(3000) = -10.8$ psi

The estimated Δp^* was zero. Iteration through two more trials gives a pressure drop of –8.6 psi. The pressure at the outlet end of the pipe is then 425 – 32.2 + 8.6 = 401.4 psia.

Appendix
B

Pressure Traverse Curves

These pressure traverse (gradient) curves are included for making estimates of pressure drops occurring in both vertical and horizontal pipes in which two-phases are flowing. The accuracy of the values obtained from these curves will decrease as the actual conditions diverge from the conditions used to prepare the curves.

Vertical Curves

The vertical curves were prepared for water cuts of 100%, 50%, and 0%.

Tubing O.D., in.	Tubing I.D., in.	Liquid Rates, STB/day
2-3/8	1.995	50–1500
2-7/8	2.441	100–3000
3-1/2	2.992	300–6000
4-1/2	3.958	500–10000

Horizontal Curves

The horizontal curves were prepared for 100% oil flowing.

Pipe I.D., in.	Liquid Rates, STB/day
2.0	100–2000
2.5	100–3000
3.0	400–10000
3.5	600–10000
4.0	800–10000
5.0	5000–10000
6.0	5000–10000

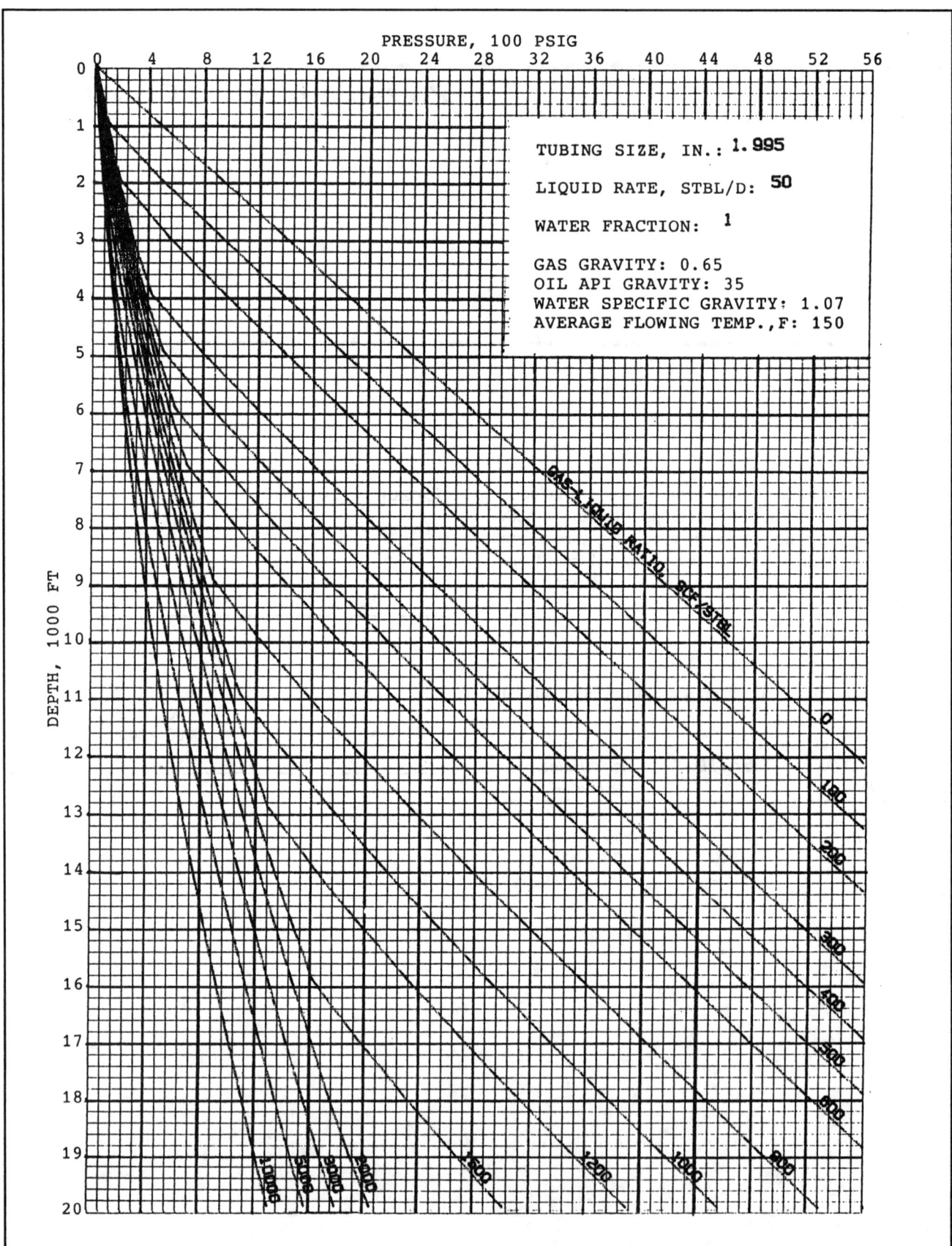

PRESSURE, 100 PSIG

DEPTH, 1000 FT

TUBING SIZE, IN.: 1.995

LIQUID RATE, STBL/D: 50

WATER FRACTION: 1

GAS GRAVITY: 0.65
OIL API GRAVITY: 35
WATER SPECIFIC GRAVITY: 1.07
AVERAGE FLOWING TEMP.,F: 150

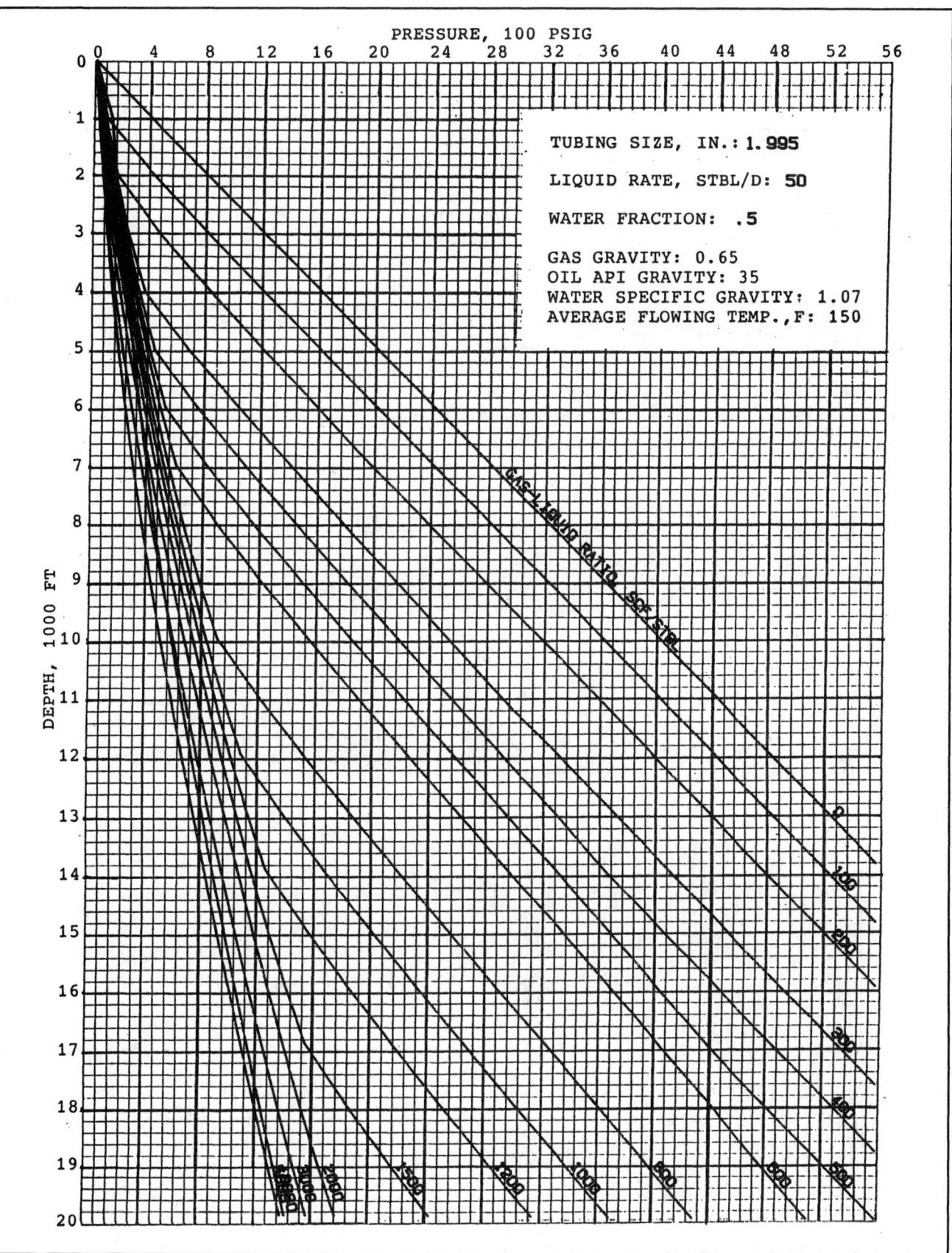

PRESSURE, 100 PSIG

DEPTH, 1000 FT

TUBING SIZE, IN.: **1.995**

LIQUID RATE, STBL/D: **50**

WATER FRACTION: **.5**

GAS GRAVITY: 0.65
OIL API GRAVITY: 35
WATER SPECIFIC GRAVITY: 1.07
AVERAGE FLOWING TEMP.,F: 150

GAS-LIQUID RATIO, SCF/STBL

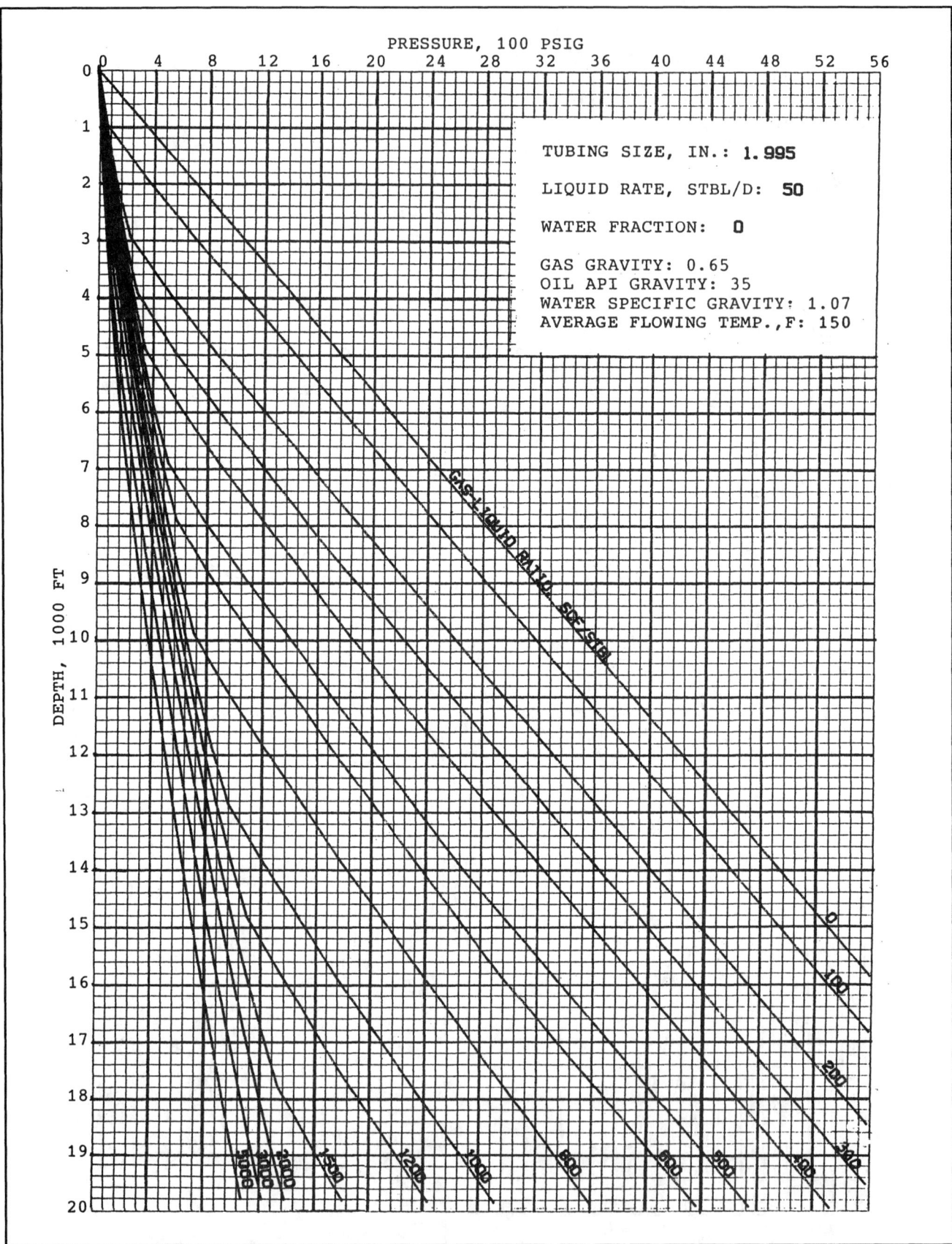

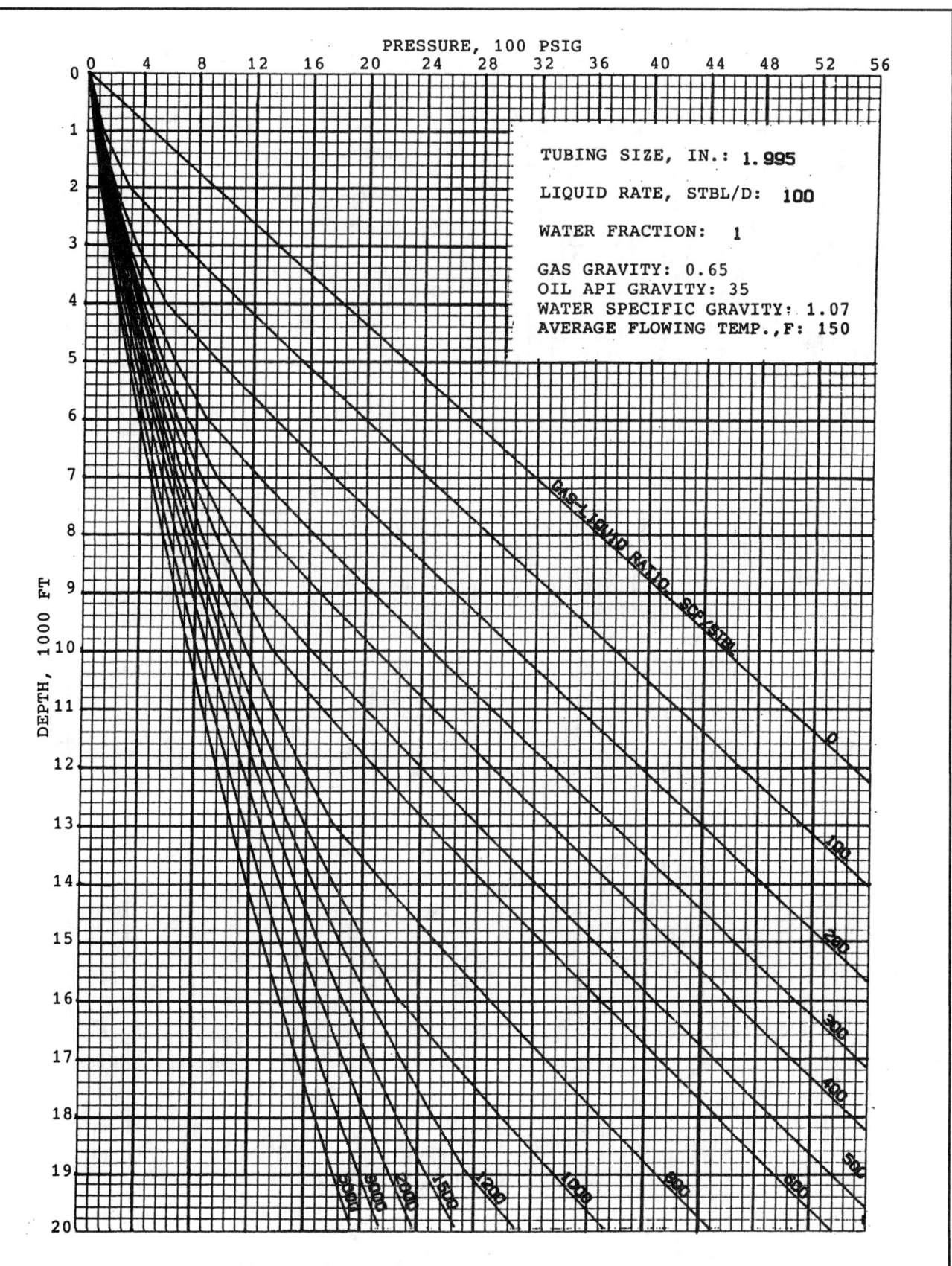

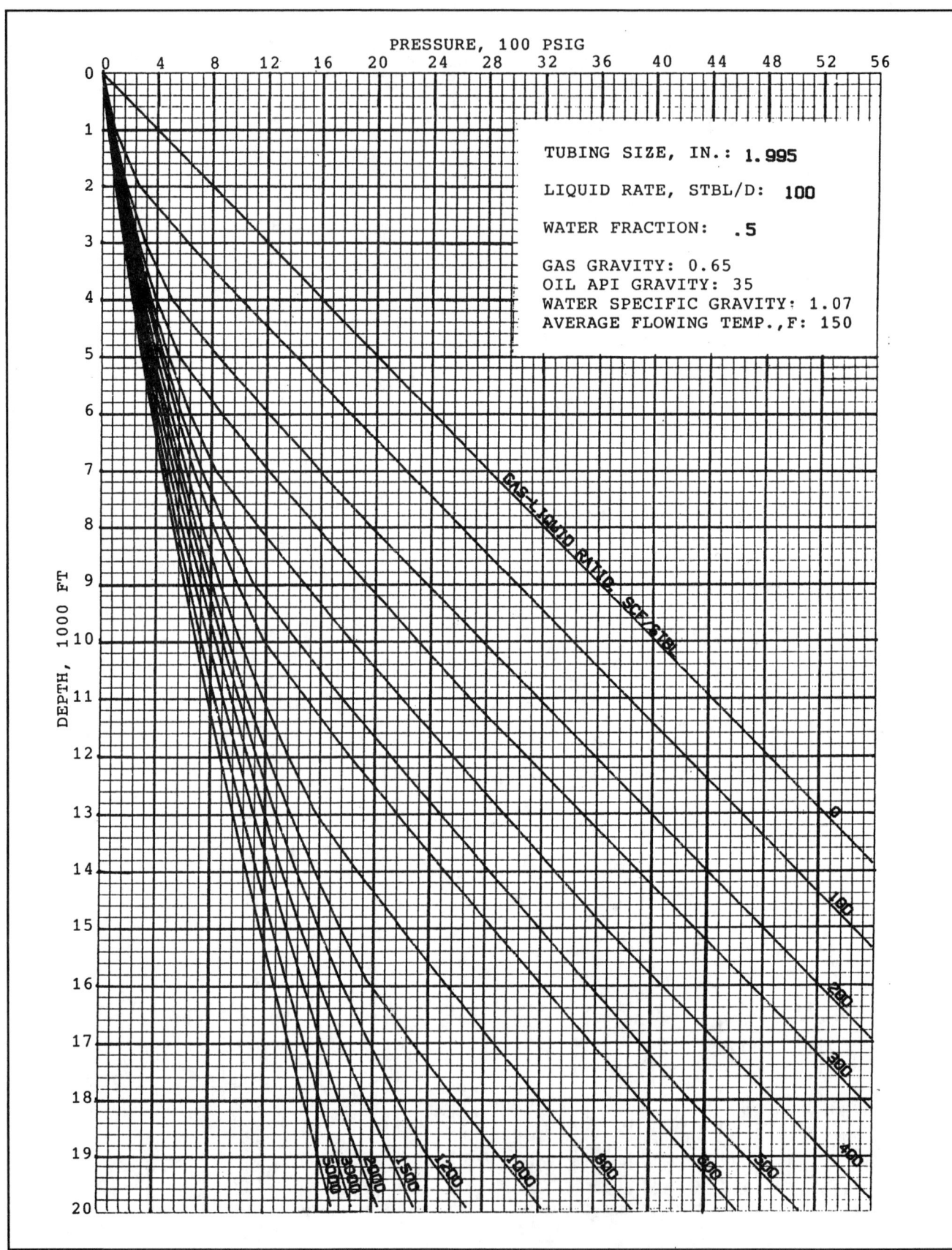

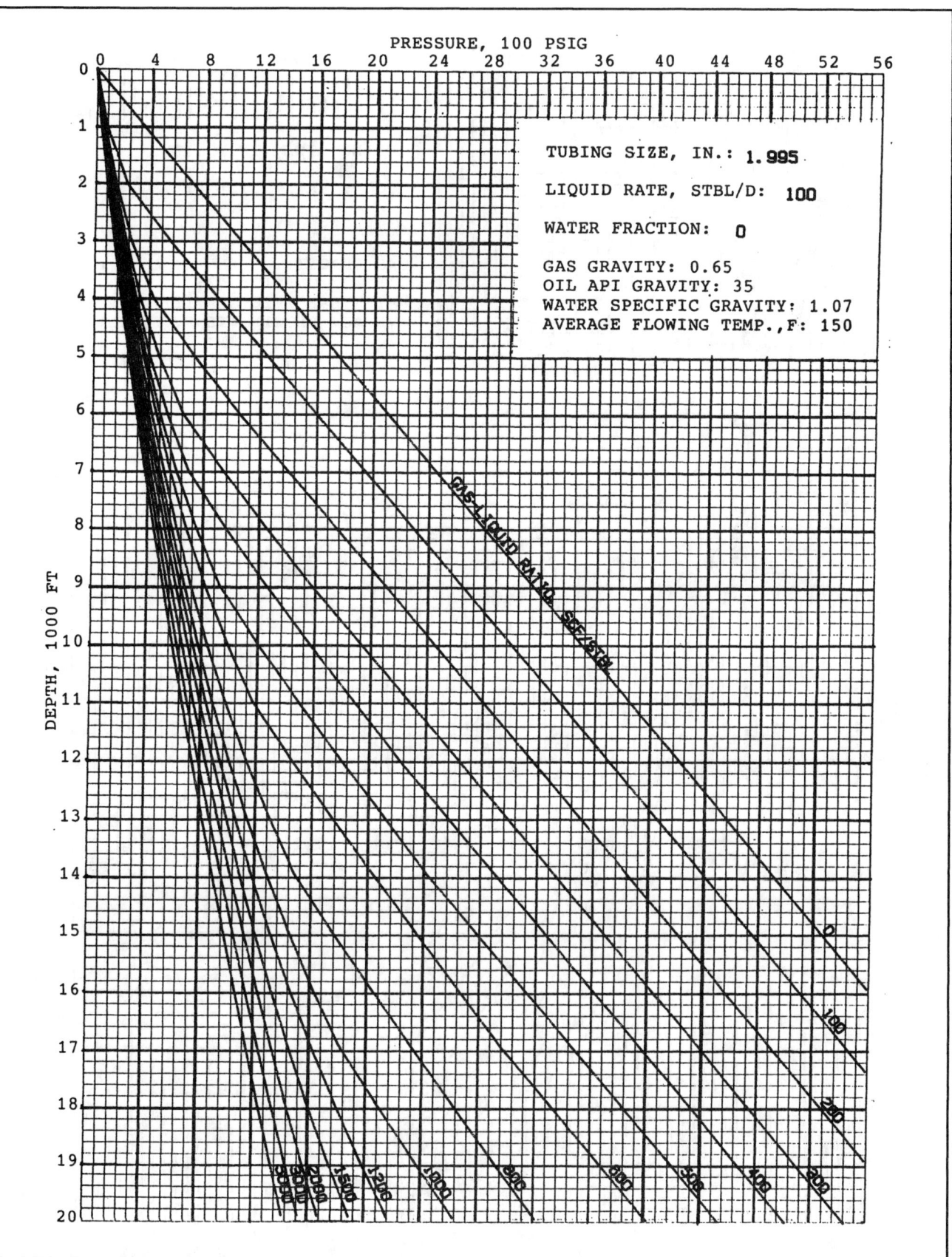

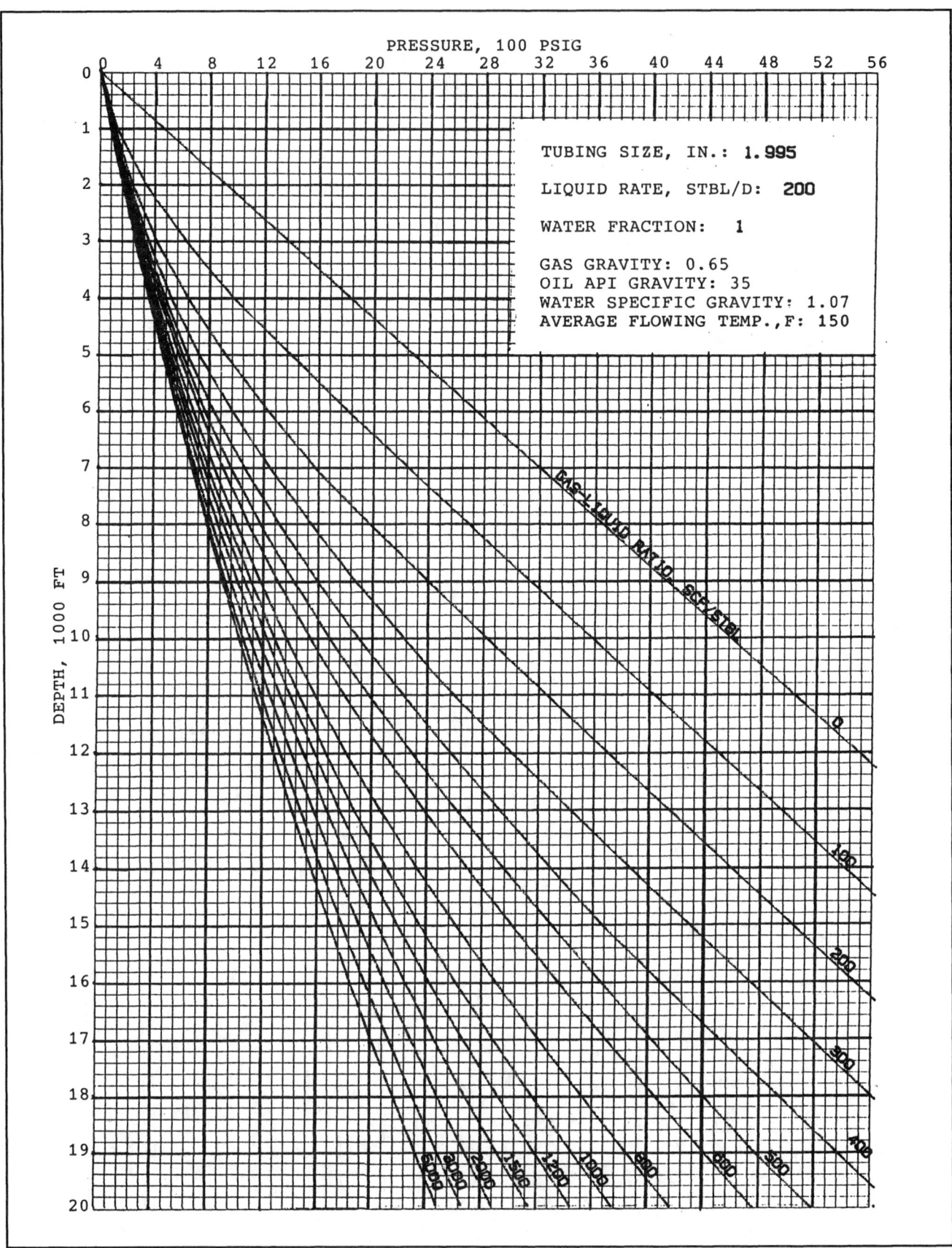

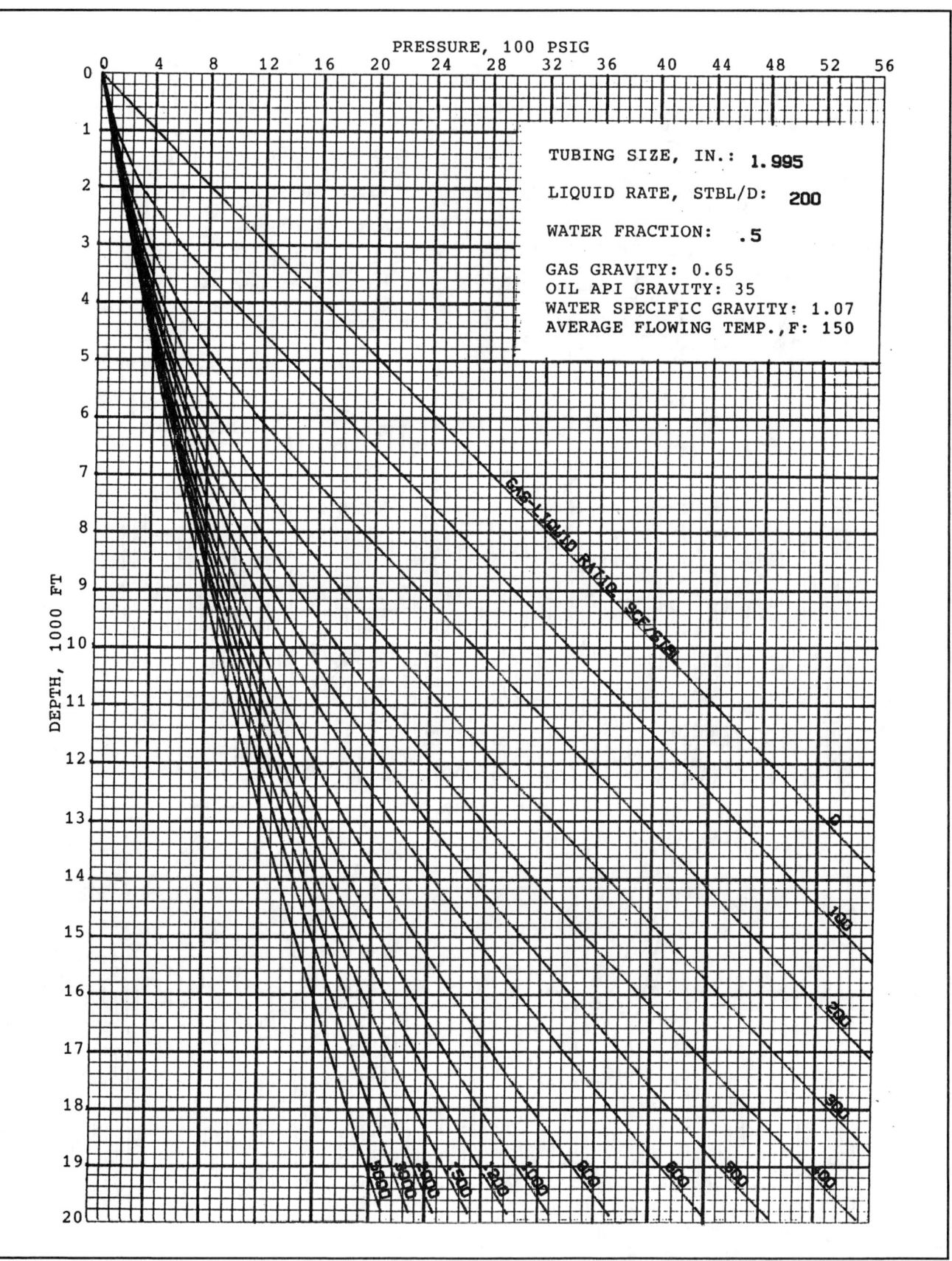

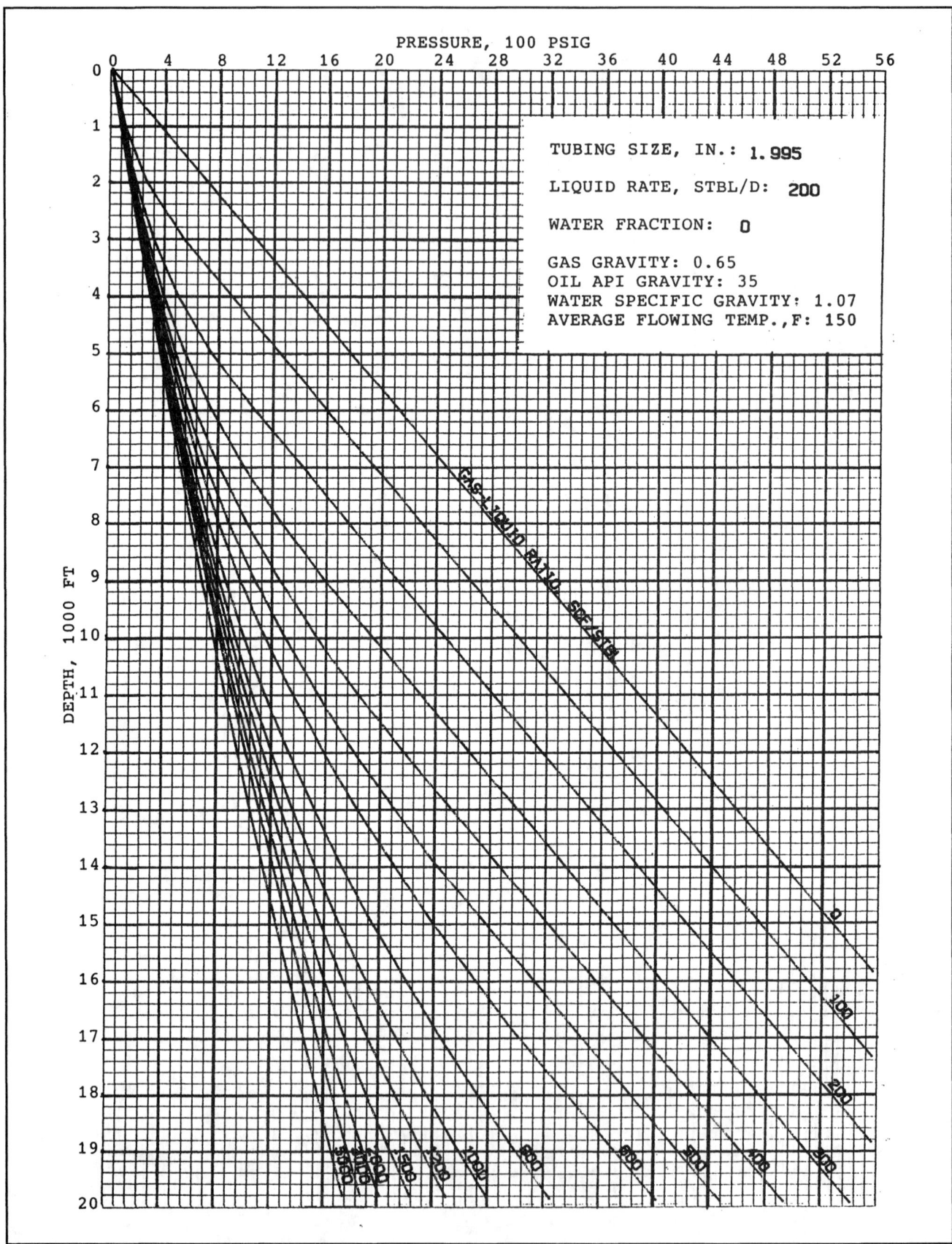

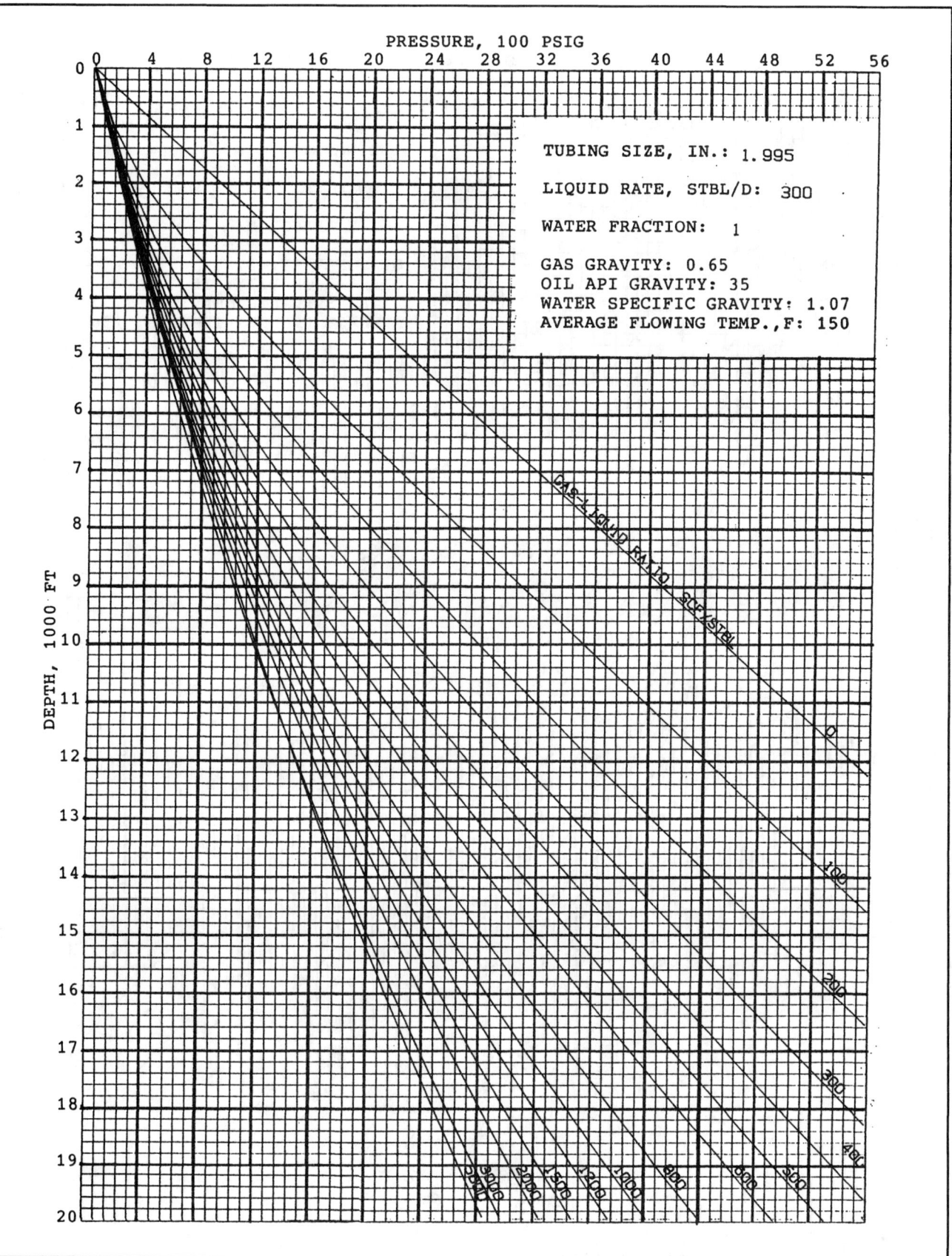

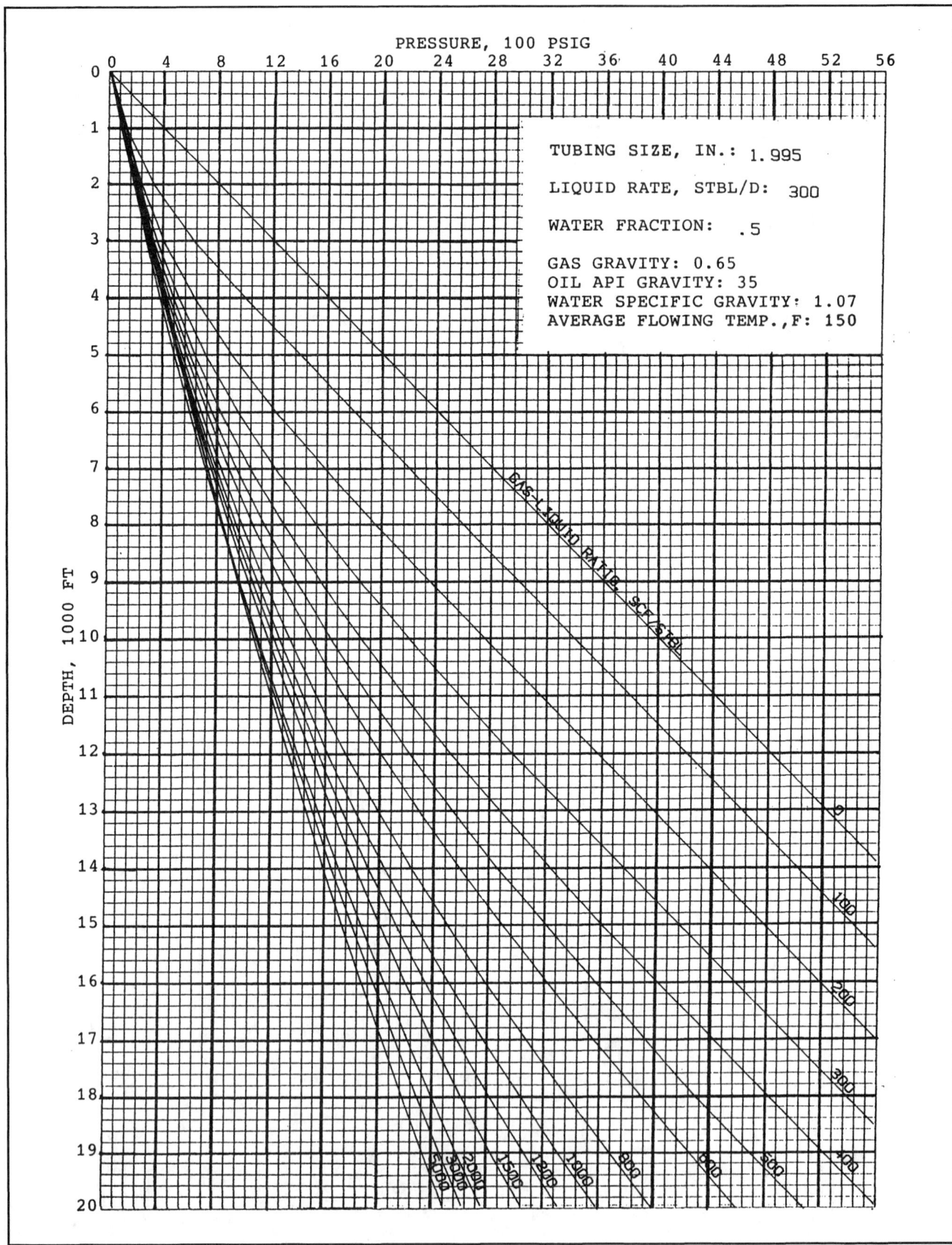

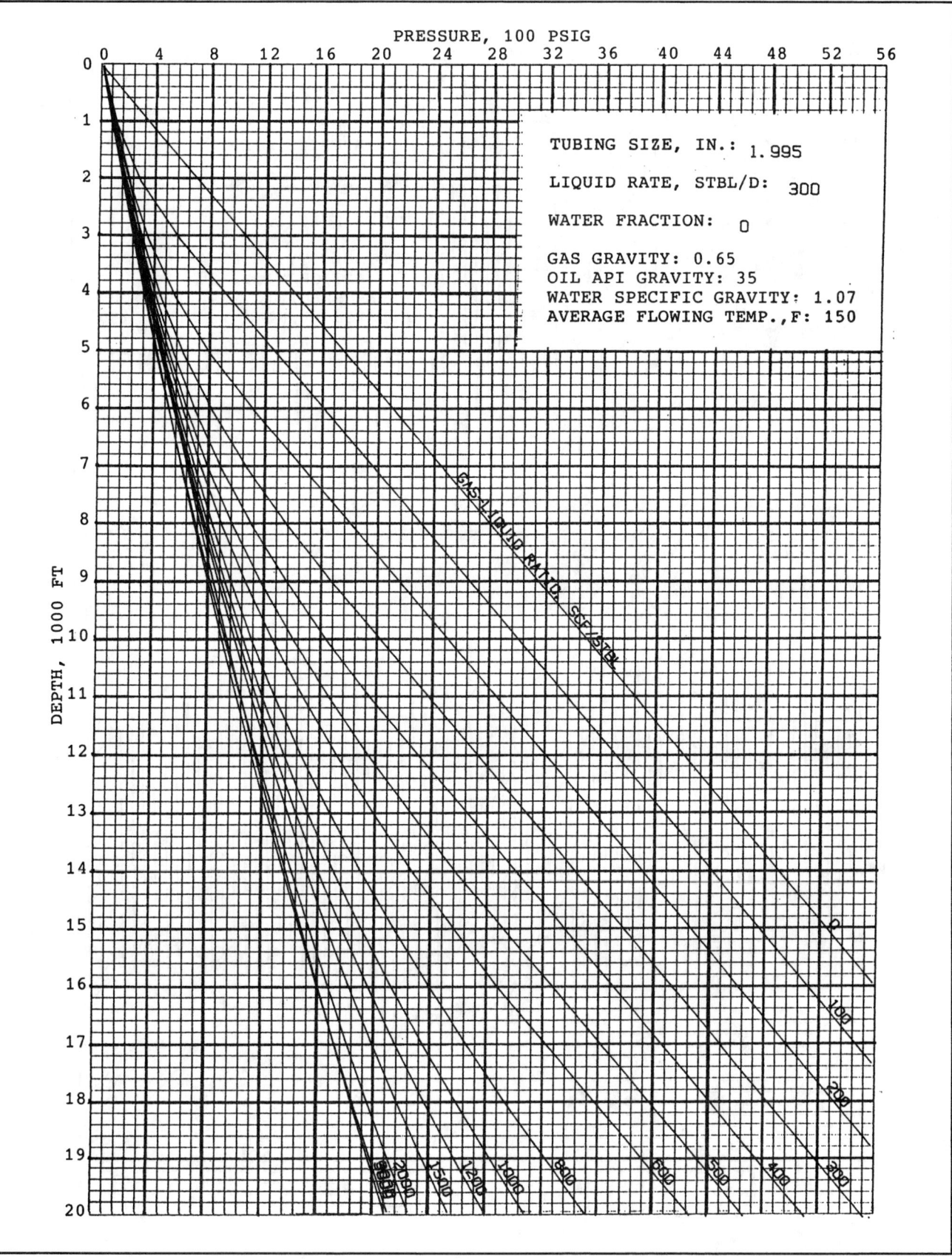

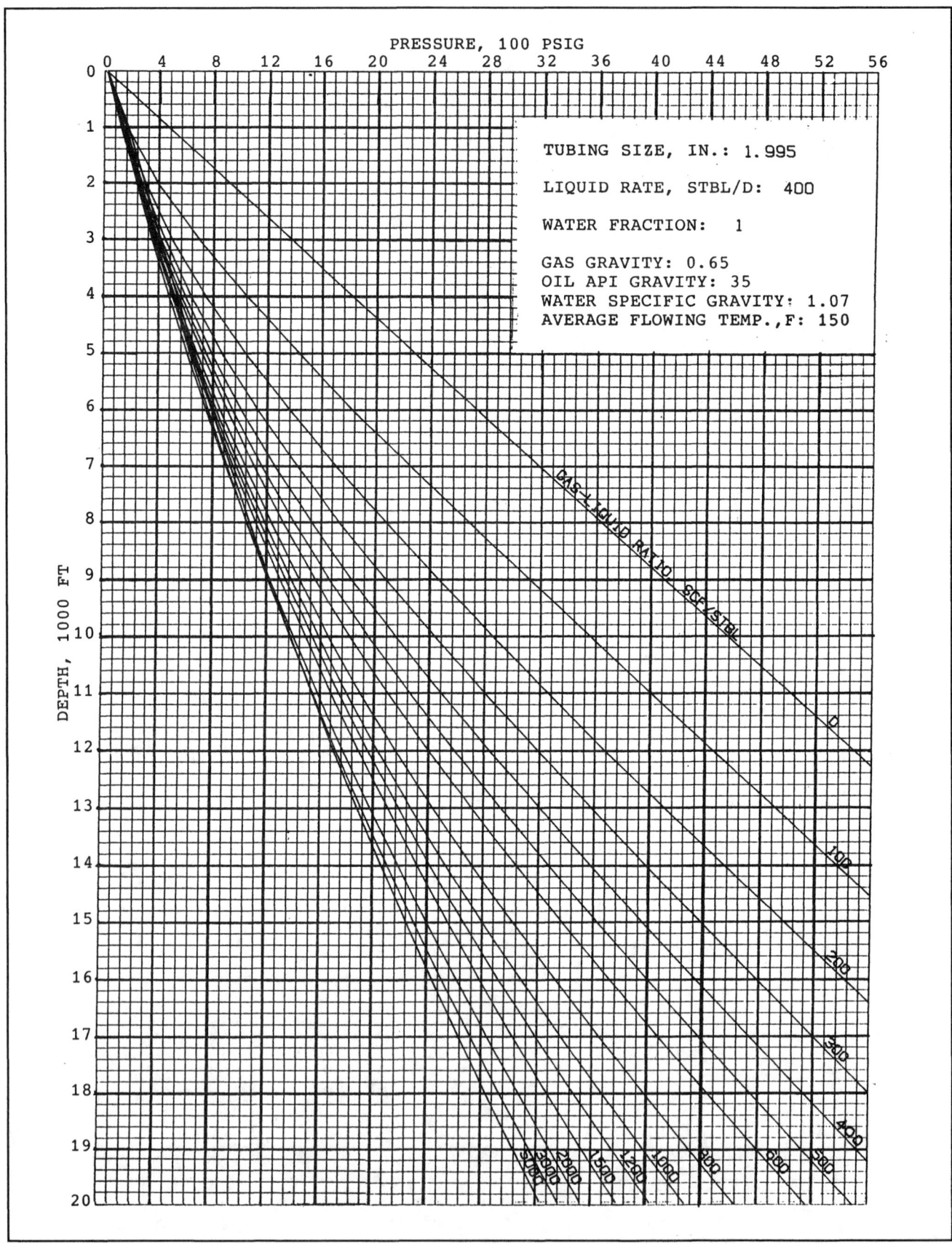

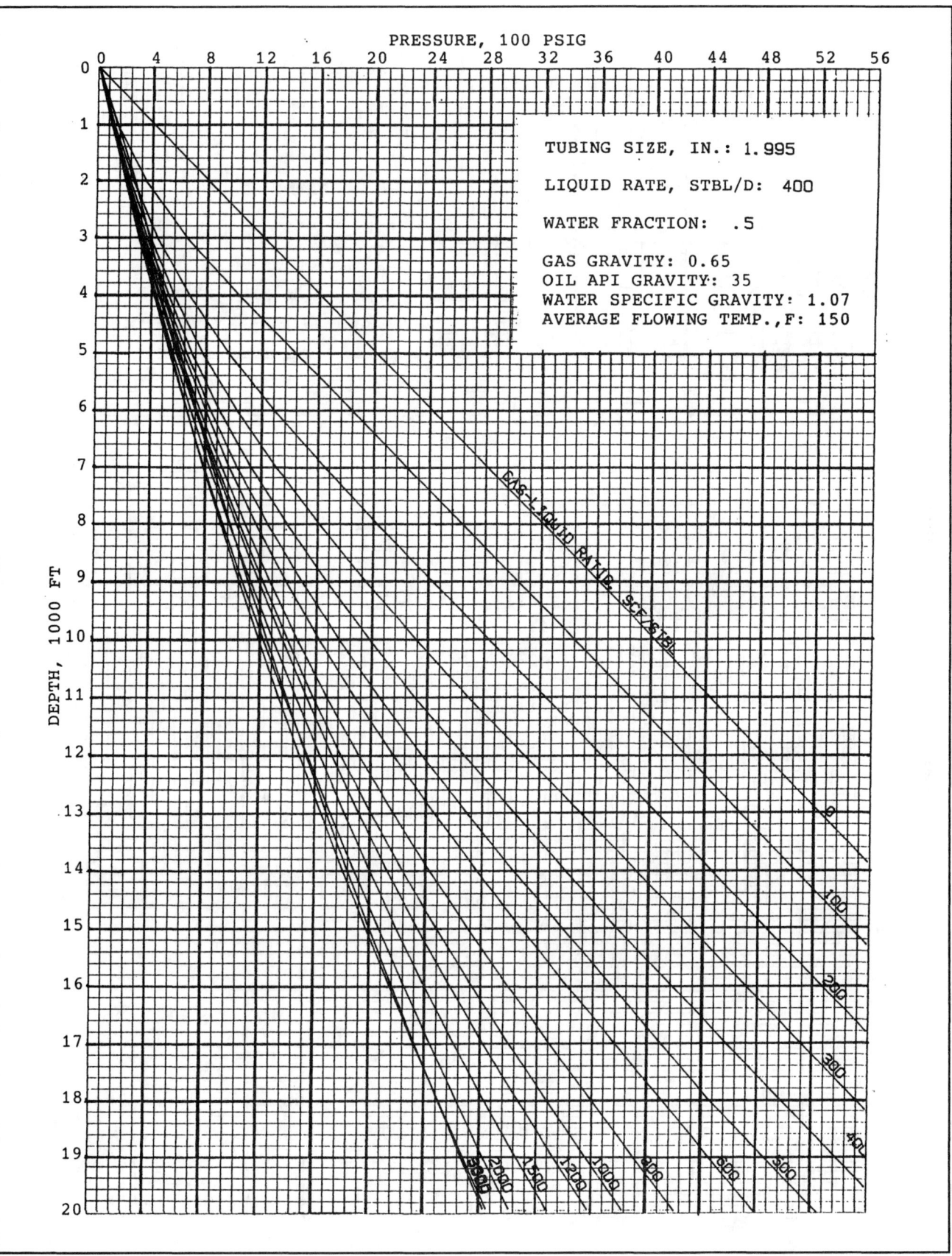

PRESSURE, 100 PSIG

TUBING SIZE, IN.: 1.995

LIQUID RATE, STBL/D: 400

WATER FRACTION: .5

GAS GRAVITY: 0.65
OIL API GRAVITY: 35
WATER SPECIFIC GRAVITY: 1.07
AVERAGE FLOWING TEMP.,F: 150

DEPTH, 1000 FT

GAS-LIQUID RATIO, SCF/STBL

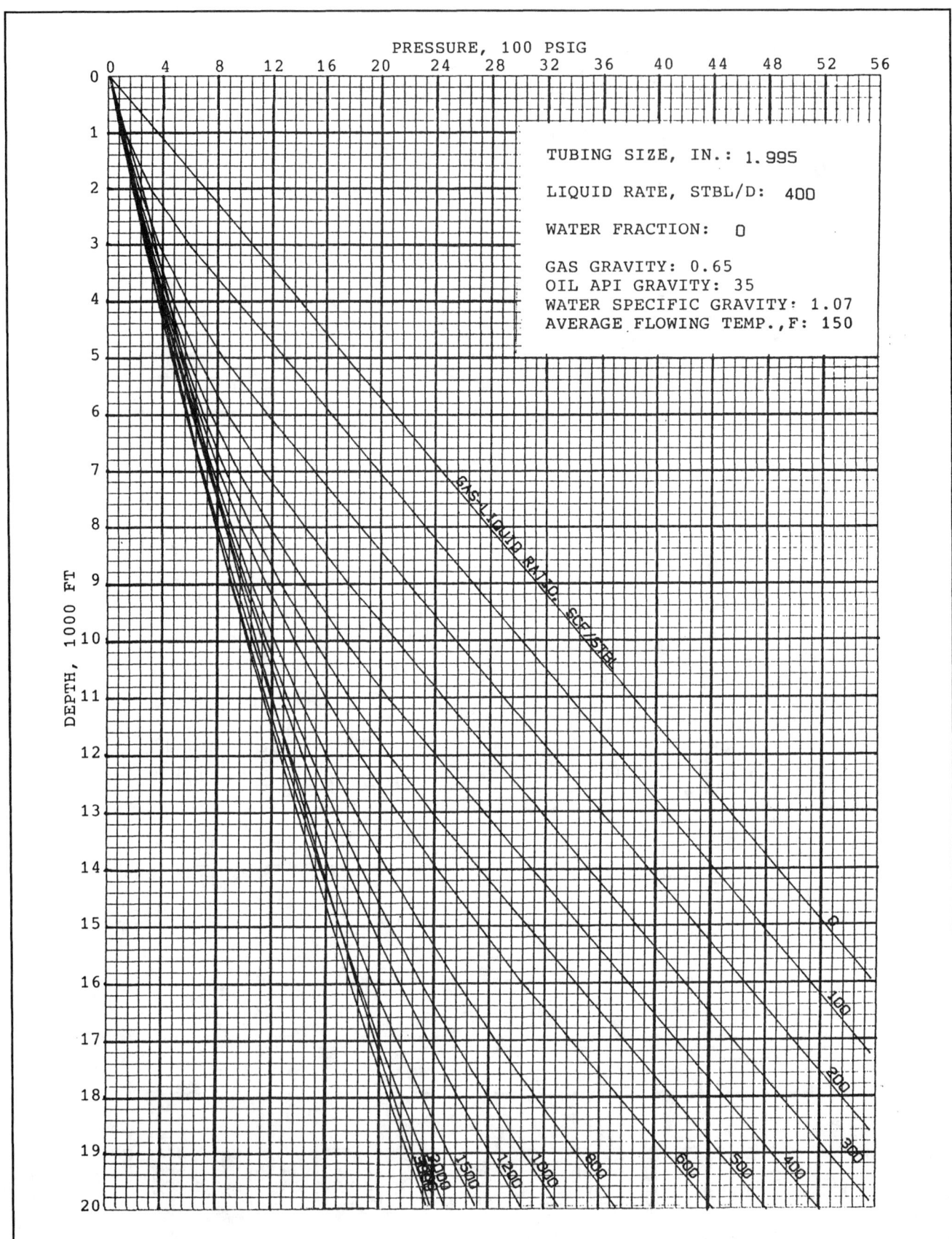

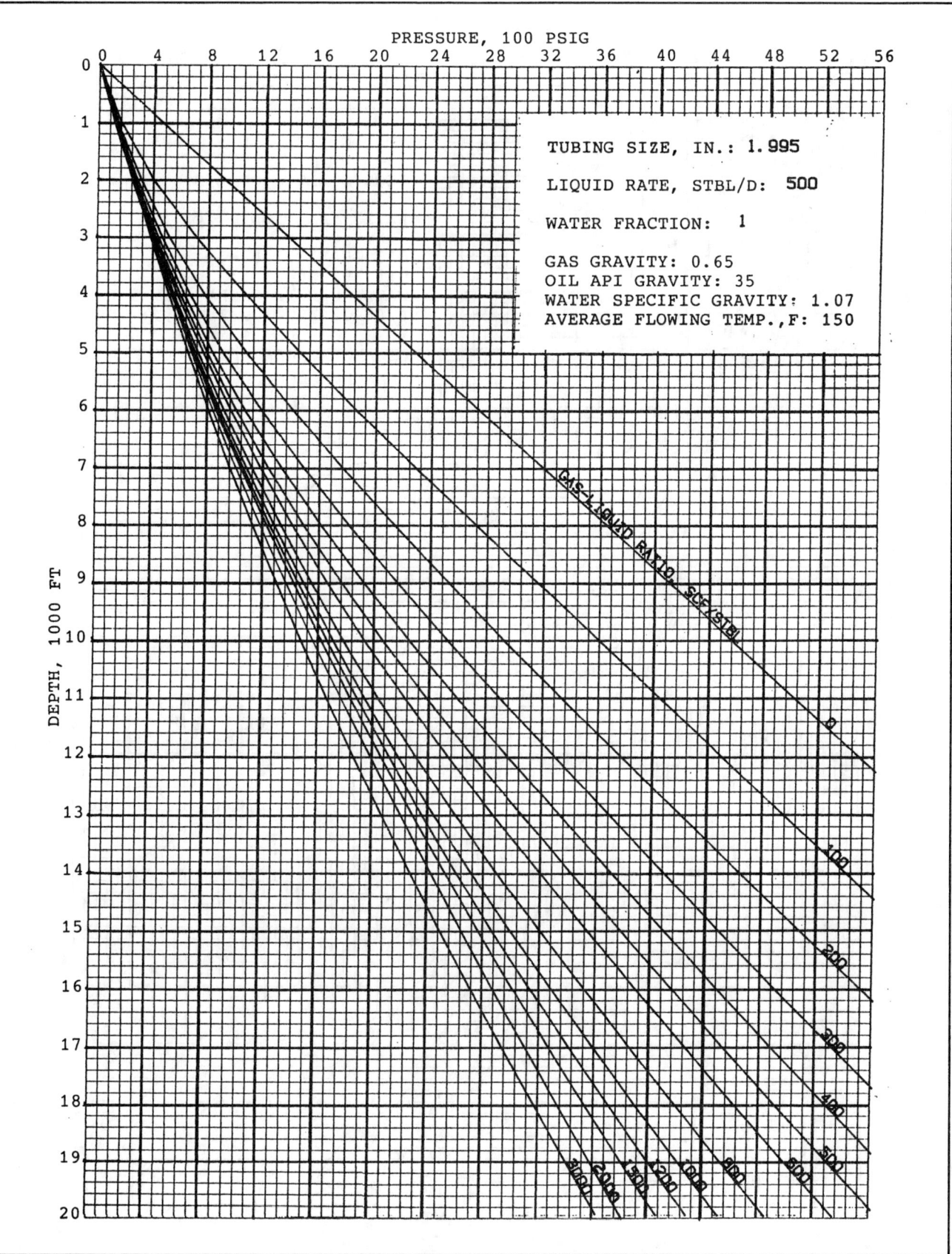

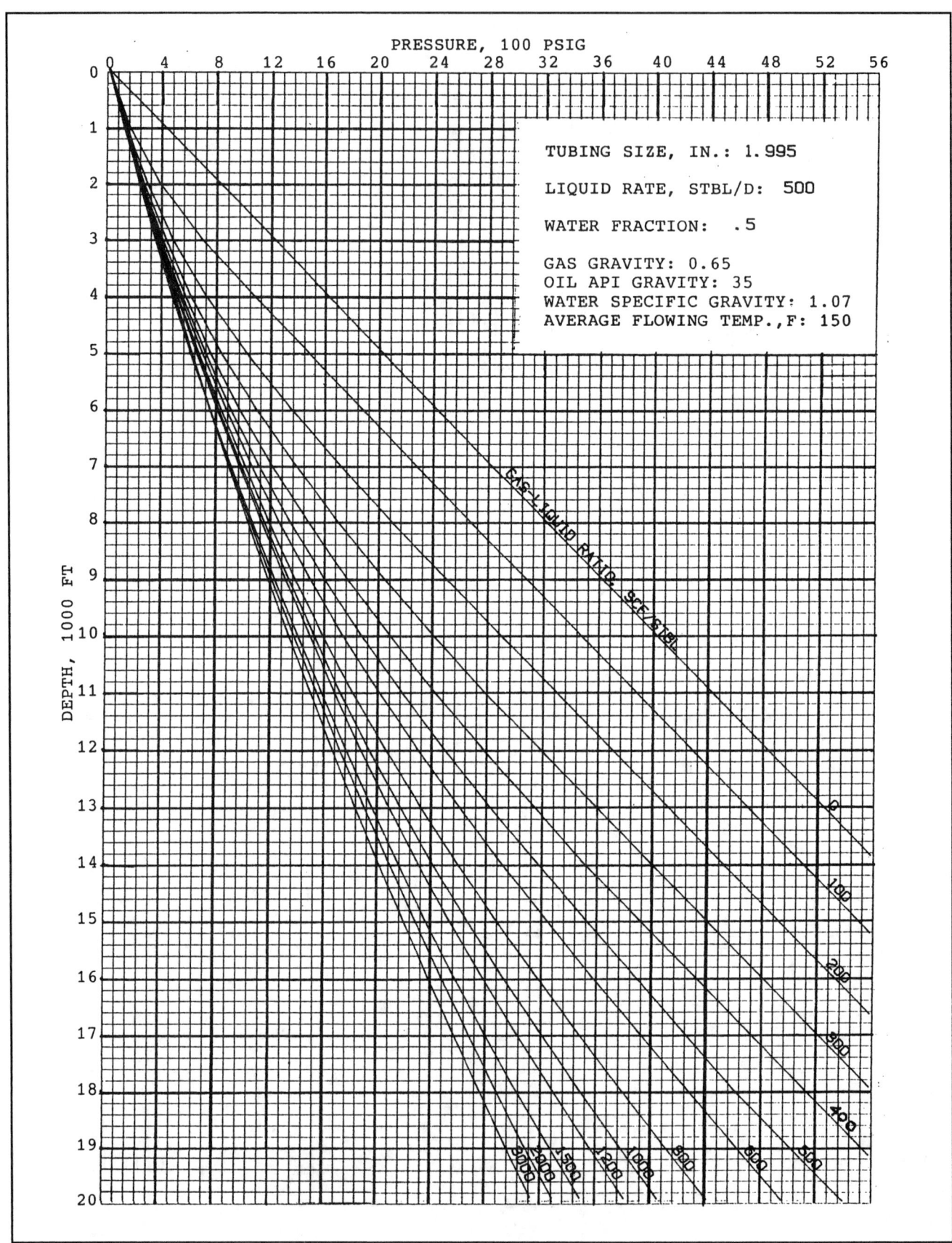

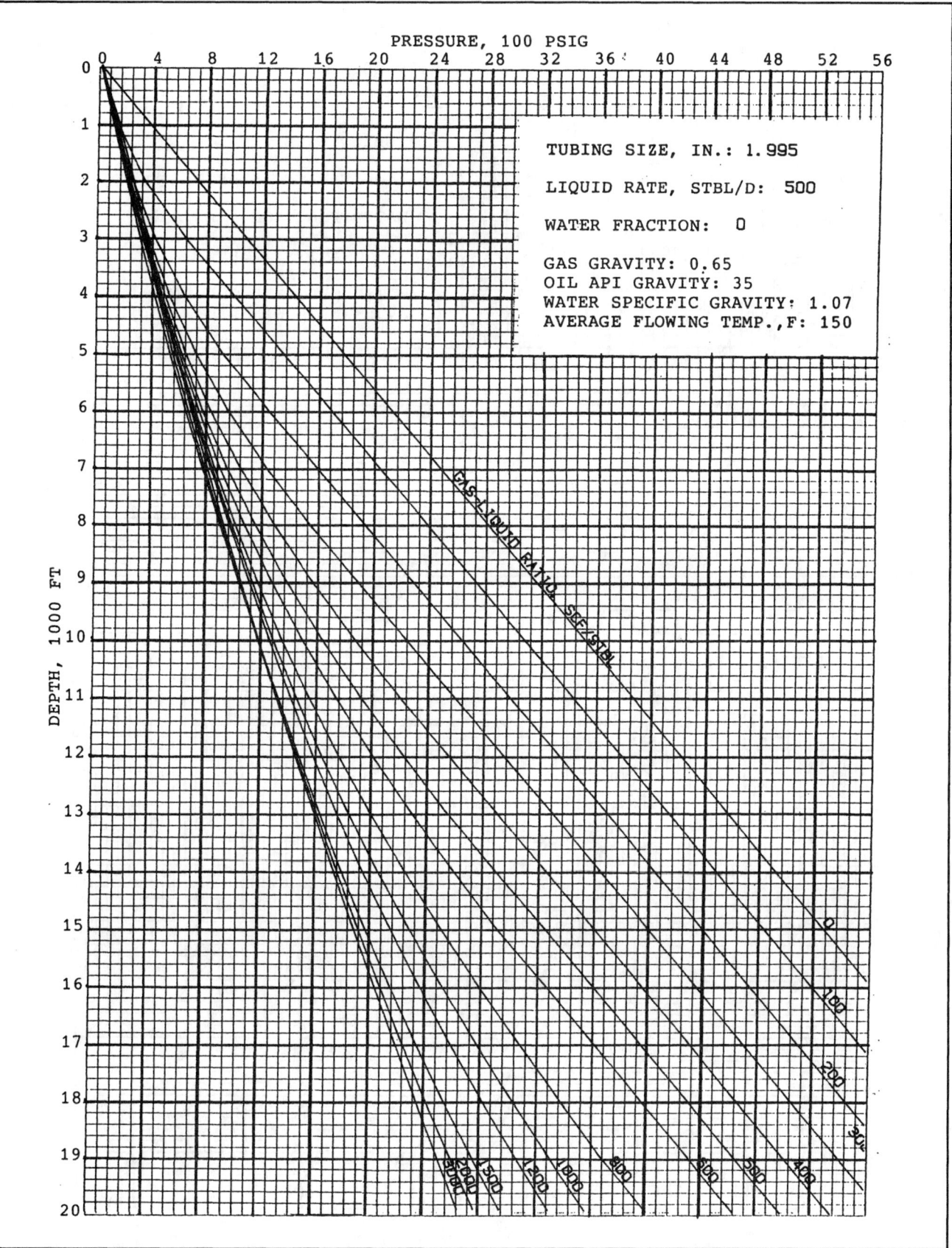

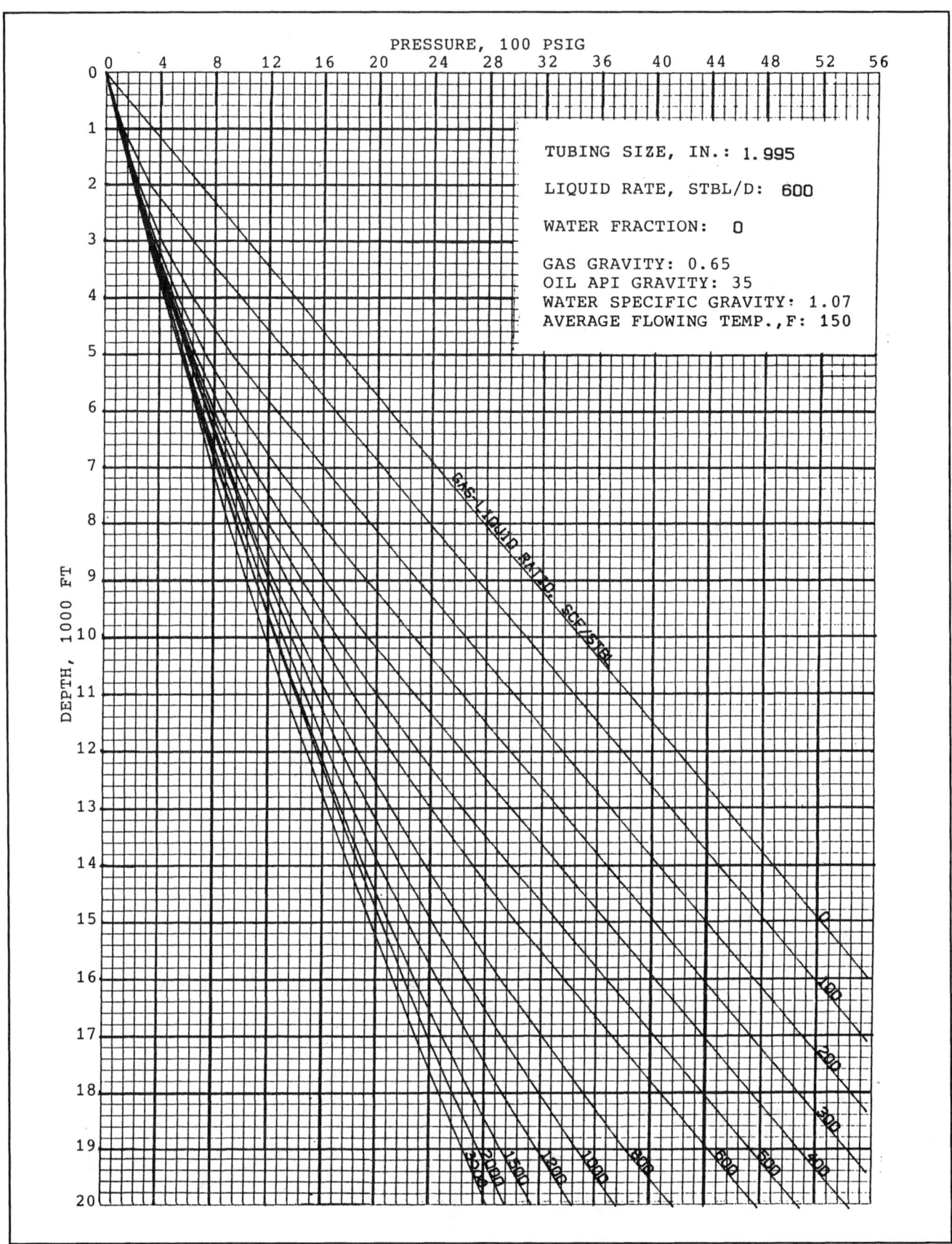

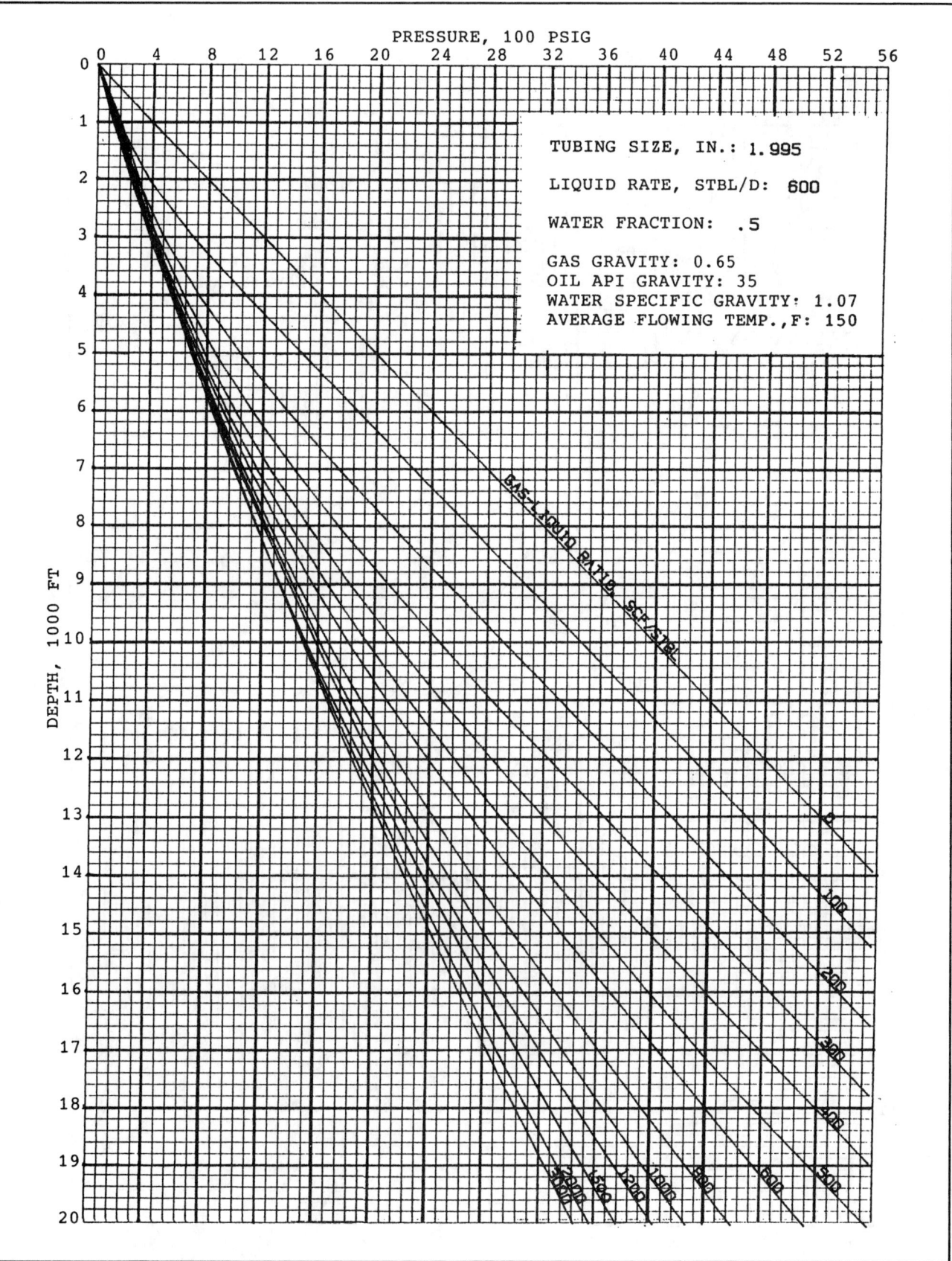

TUBING SIZE, IN.: 1.995

LIQUID RATE, STBL/D: 600

WATER FRACTION: .5

GAS GRAVITY: 0.65
OIL API GRAVITY: 35
WATER SPECIFIC GRAVITY: 1.07
AVERAGE FLOWING TEMP.,F: 150

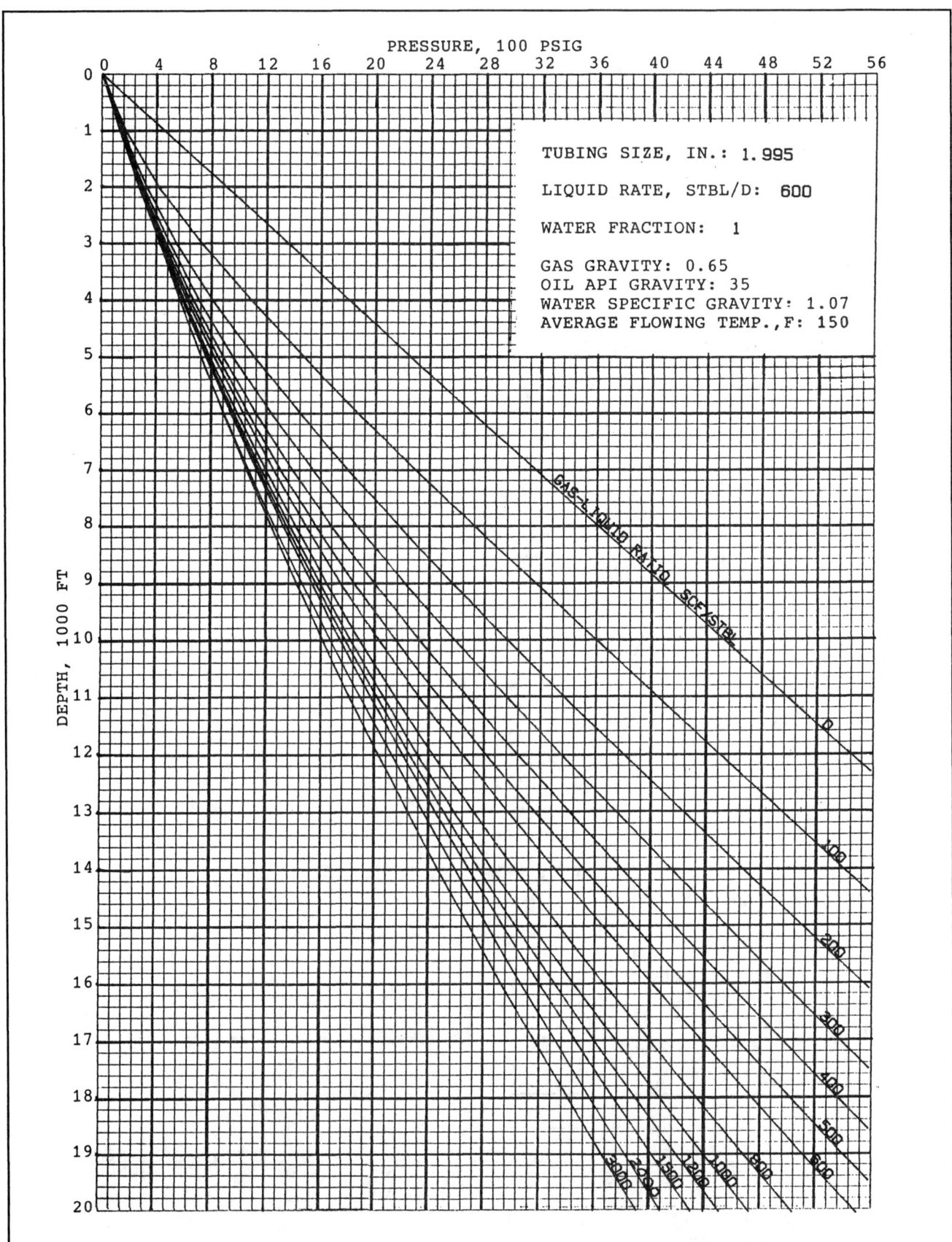

PRESSURE, 100 PSIG

DEPTH, 1000 FT

TUBING SIZE, IN.: 1.995

LIQUID RATE, STBL/D: 600

WATER FRACTION: 1

GAS GRAVITY: 0.65
OIL API GRAVITY: 35
WATER SPECIFIC GRAVITY: 1.07
AVERAGE FLOWING TEMP.,F: 150

GAS-LIQUID RATIO, SCF/STBL

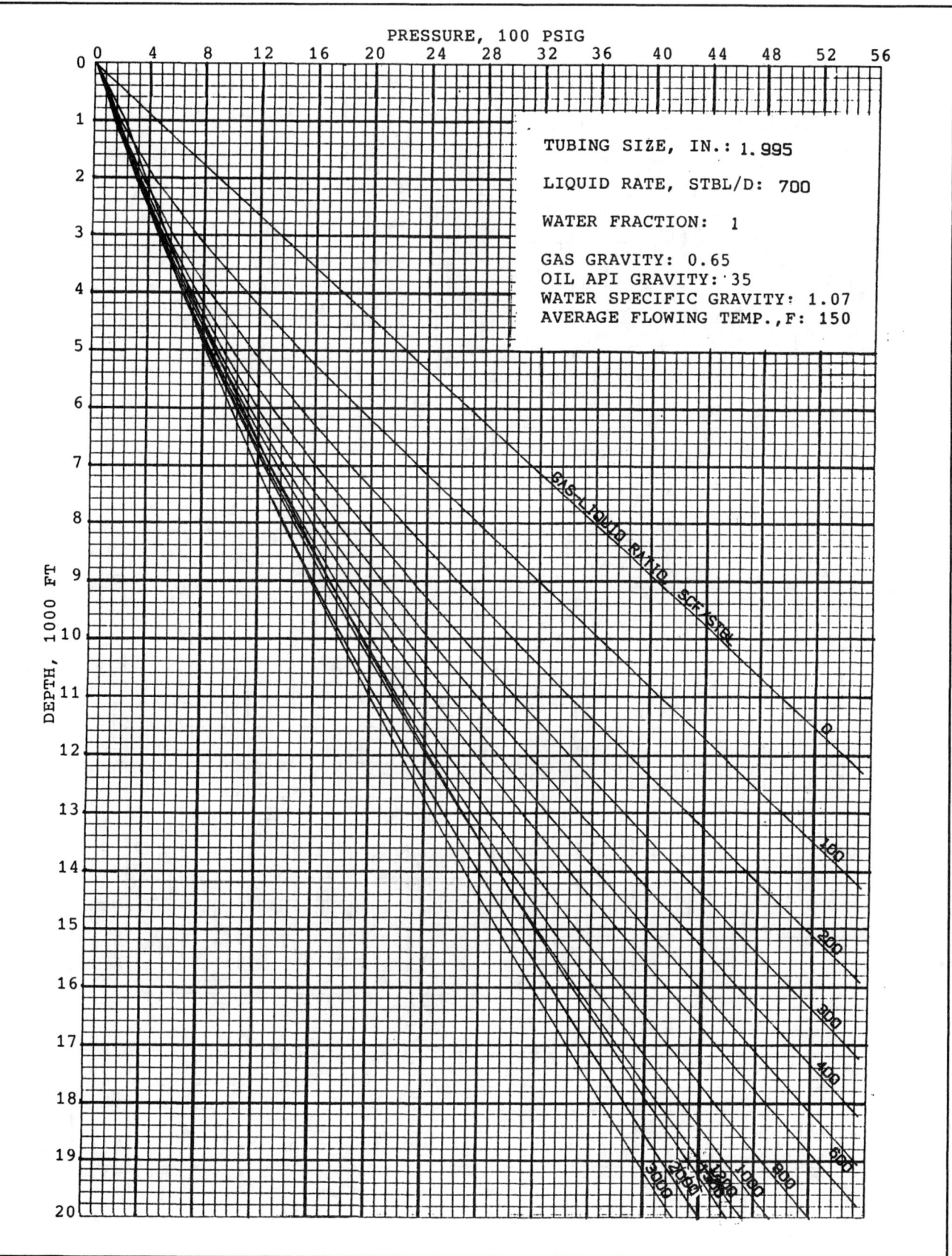

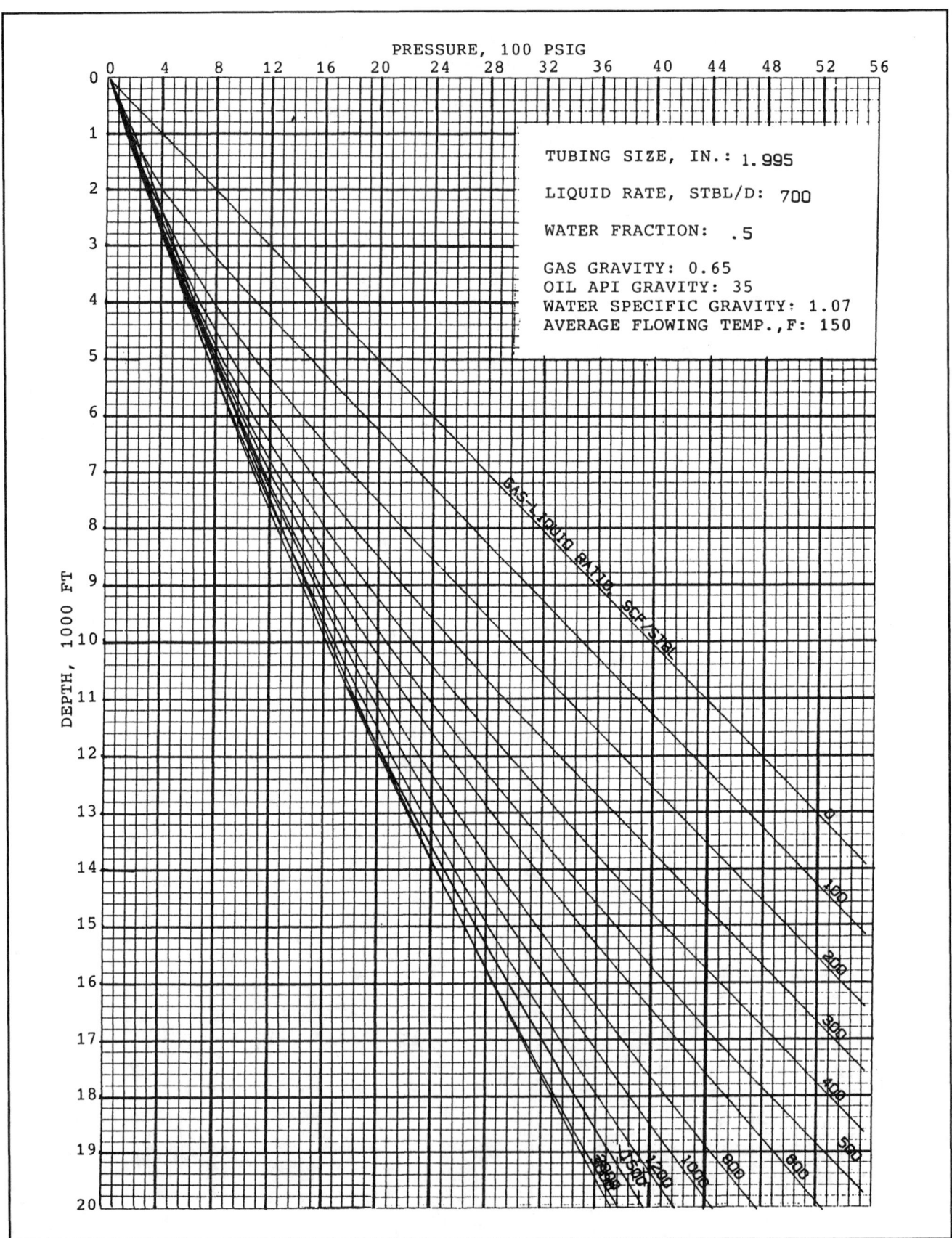

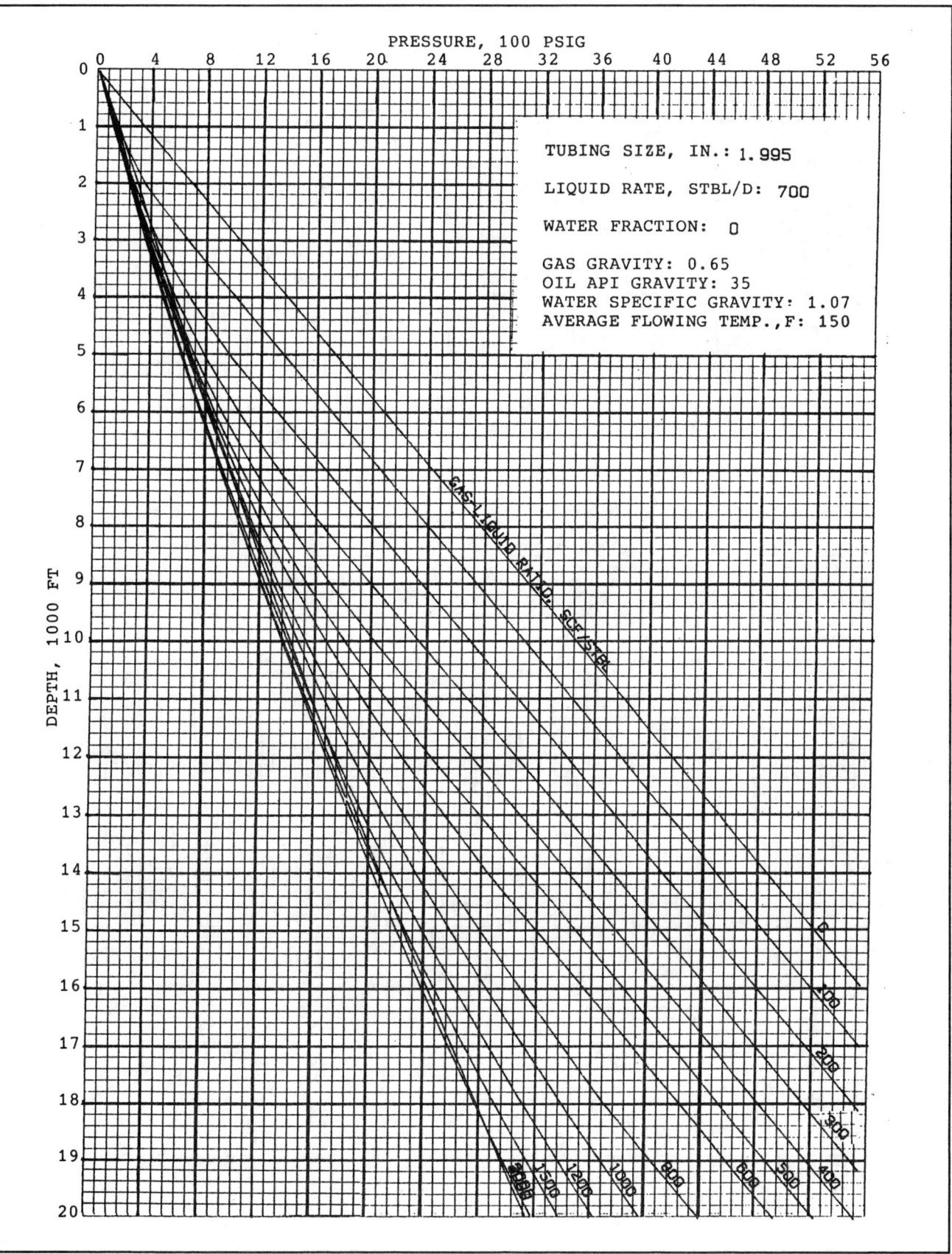

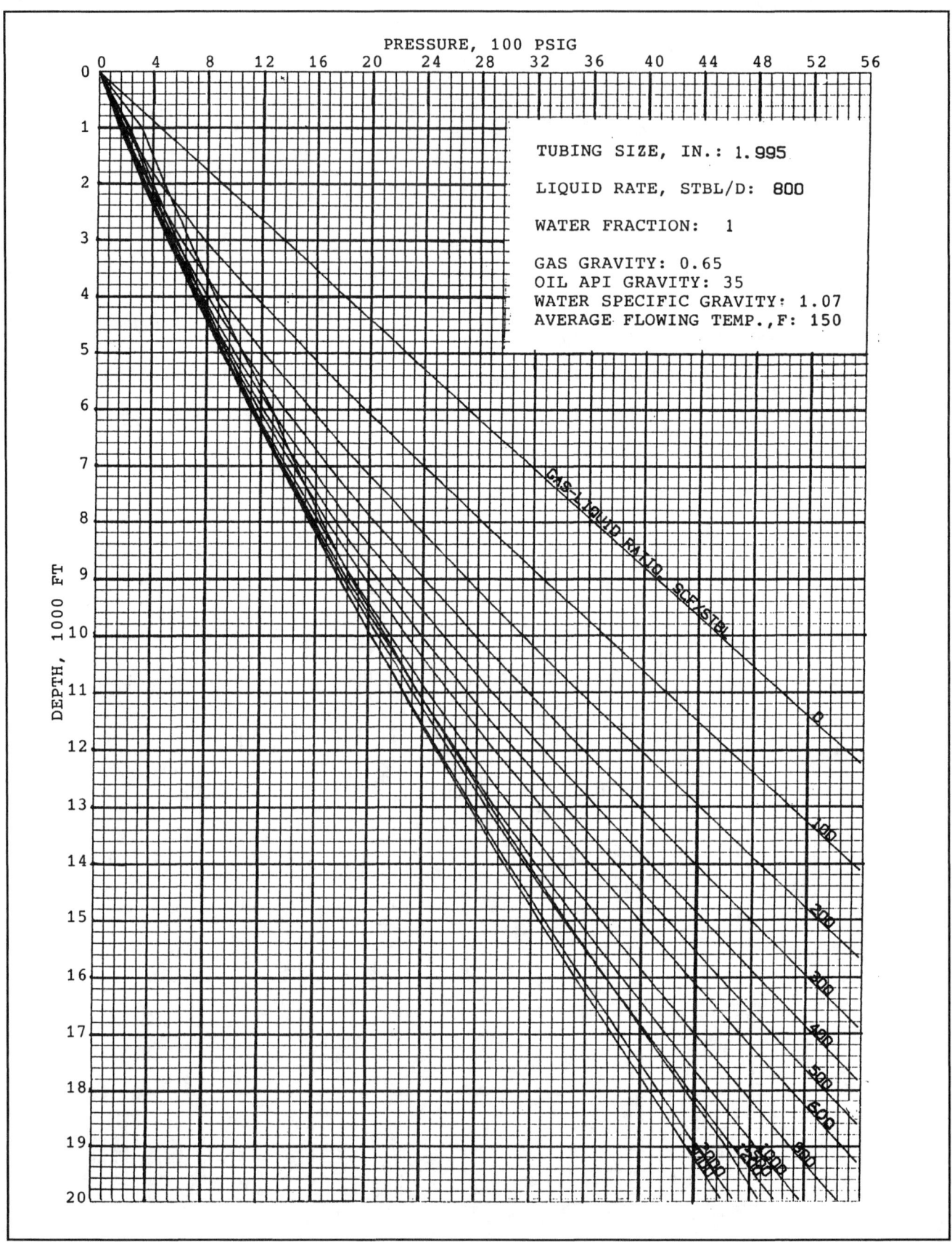

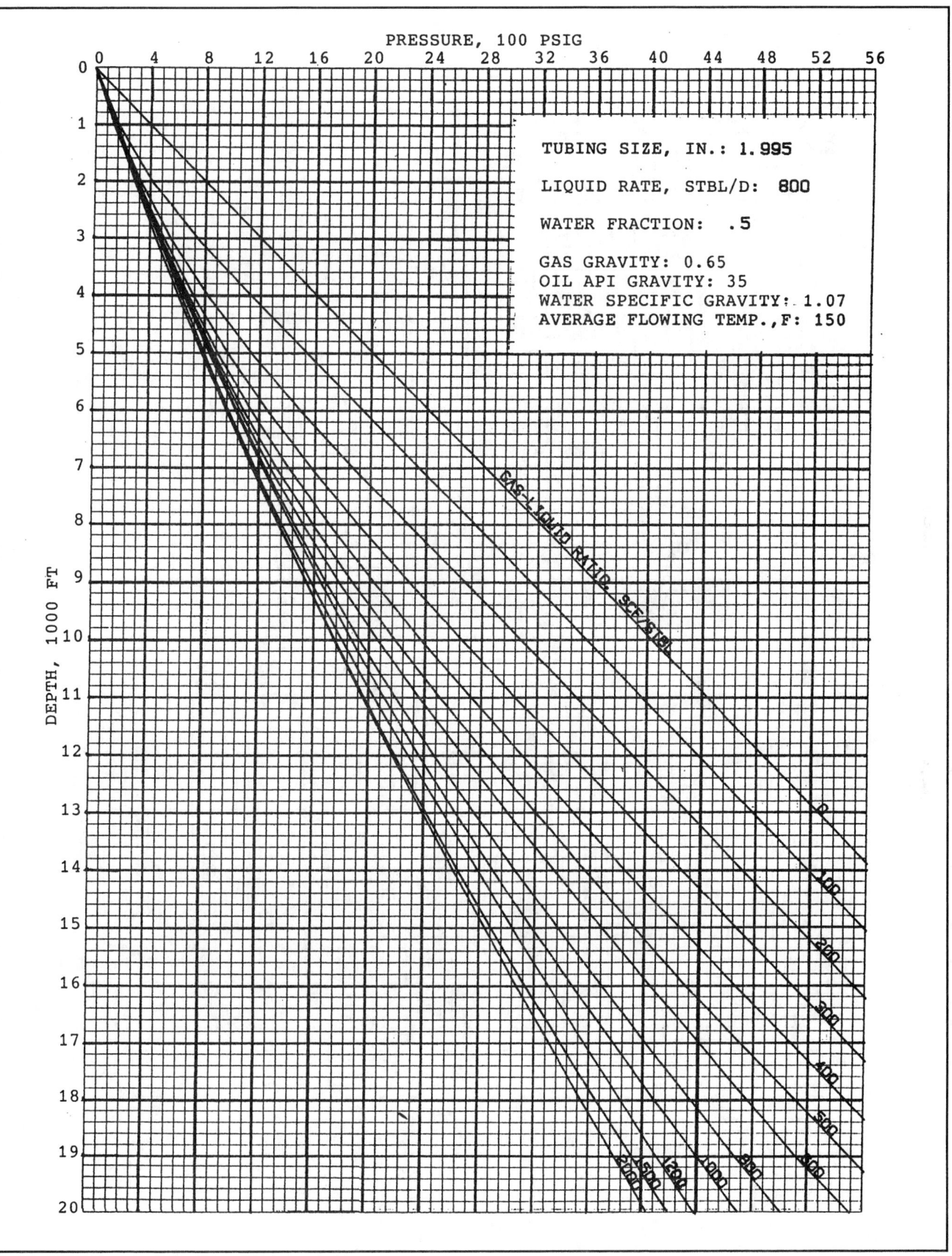

PRESSURE, 100 PSIG

DEPTH, 1000 FT

TUBING SIZE, IN.: 1.995

LIQUID RATE, STBL/D: 800

WATER FRACTION: .5

GAS GRAVITY: 0.65
OIL API GRAVITY: 35
WATER SPECIFIC GRAVITY: 1.07
AVERAGE FLOWING TEMP.,F: 150

GAS-LIQUID RATIO, SCF/STBL

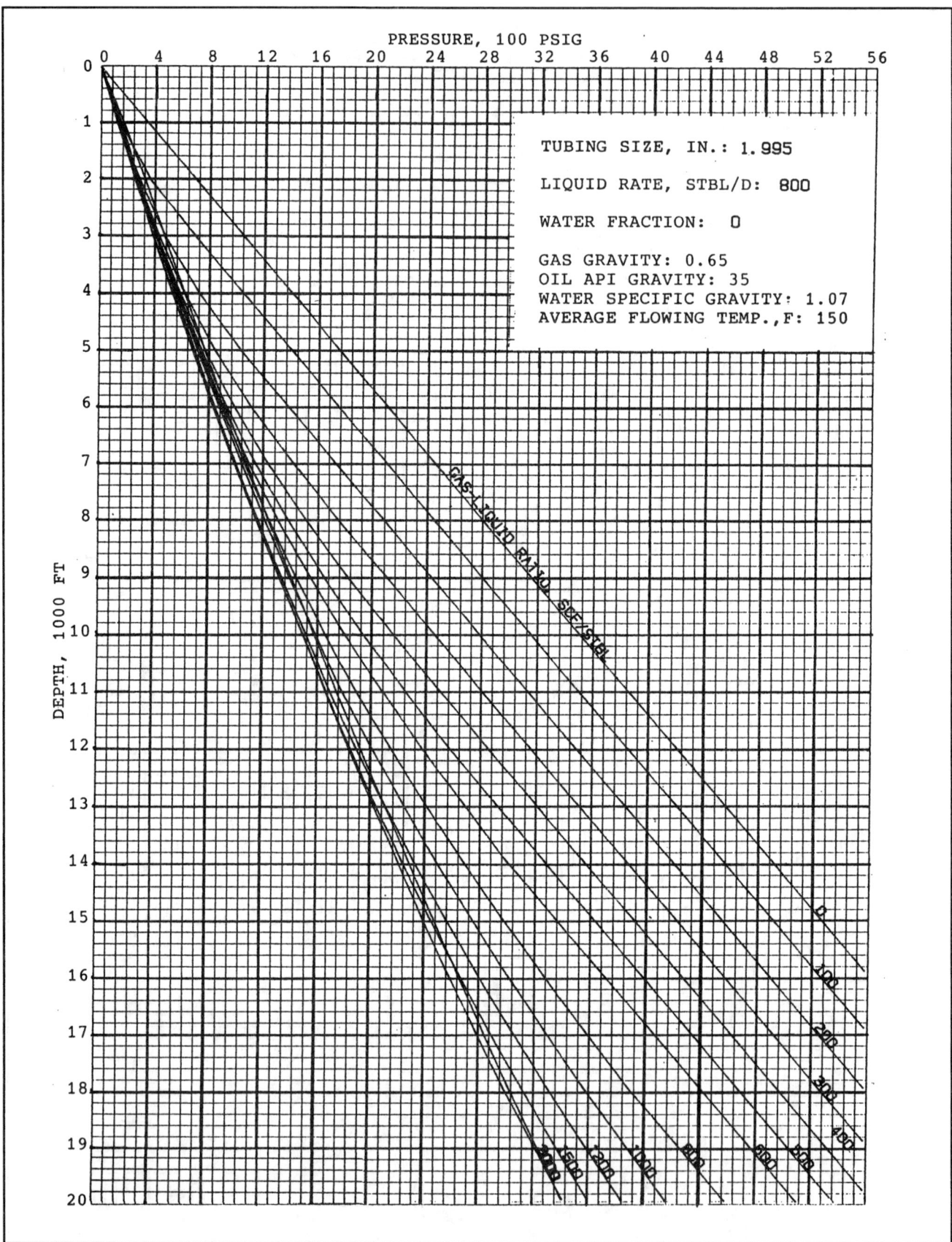

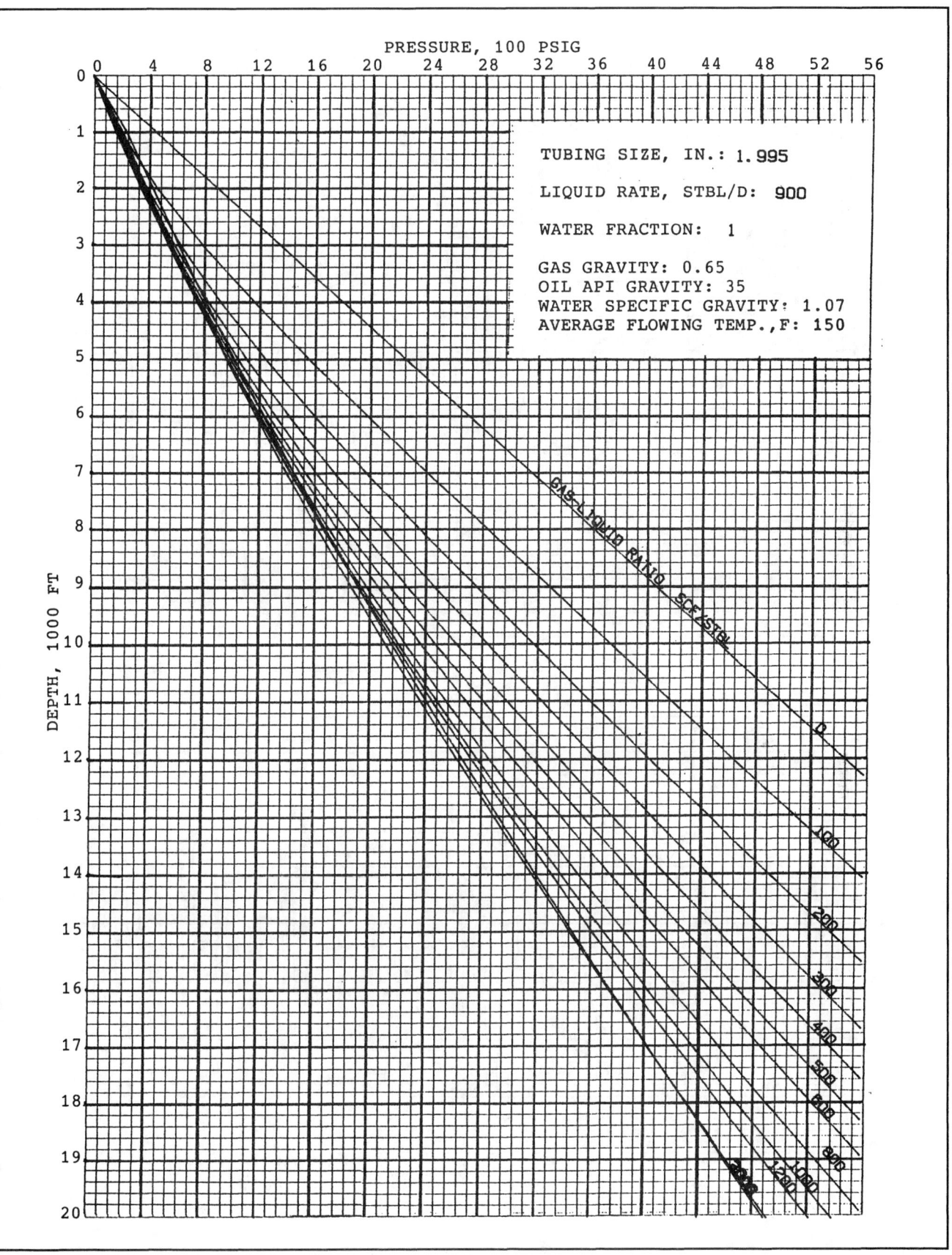

PRESSURE, 100 PSIG

TUBING SIZE, IN.: 1.995

LIQUID RATE, STBL/D: 900

WATER FRACTION: 1

GAS GRAVITY: 0.65
OIL API GRAVITY: 35
WATER SPECIFIC GRAVITY: 1.07
AVERAGE FLOWING TEMP.,F: 150

DEPTH, 1000 FT

GAS-LIQUID RATIO SCF/STBL

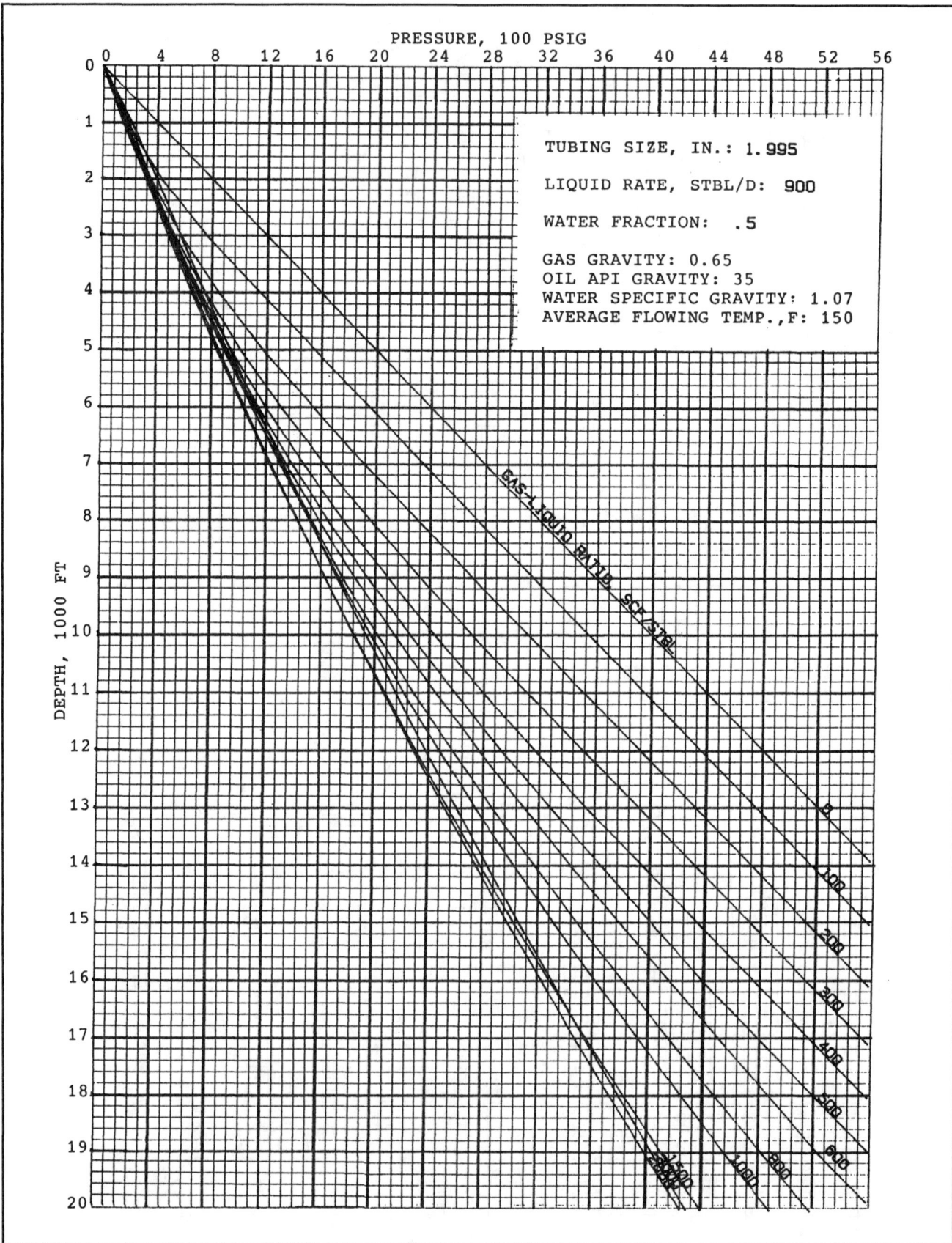

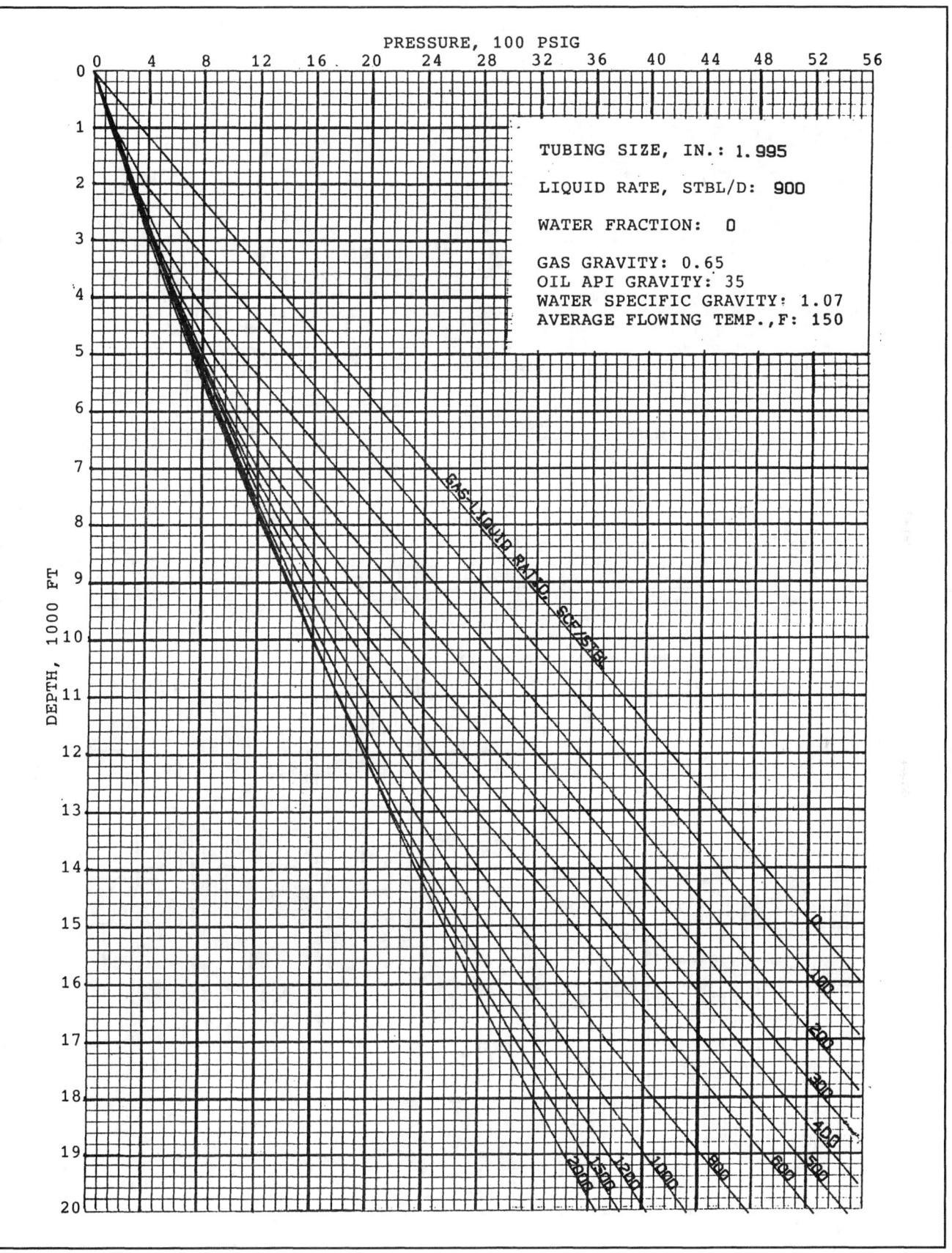

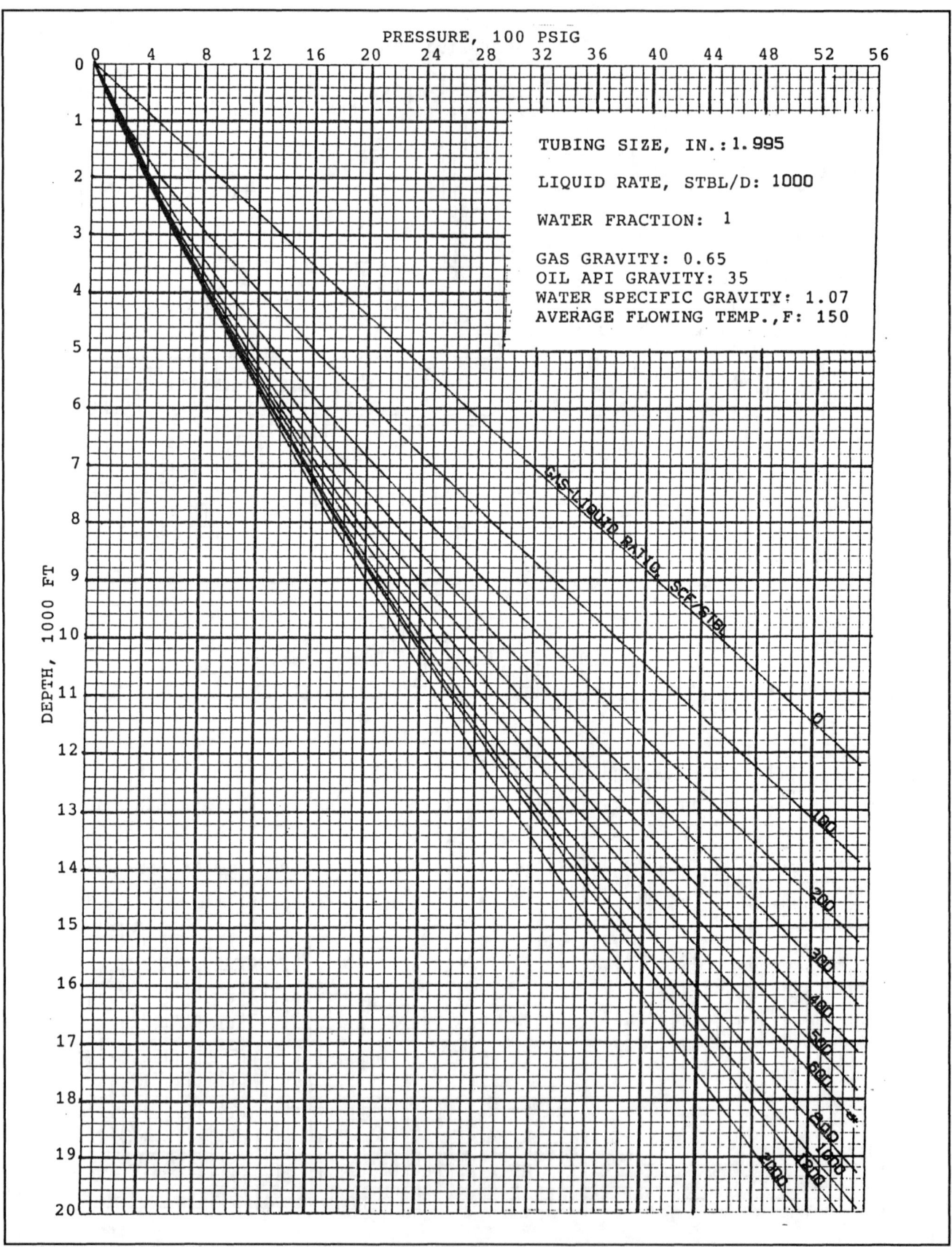

PRESSURE, 100 PSIG

DEPTH, 1000 FT

TUBING SIZE, IN.: 1.995

LIQUID RATE, STBL/D: 1000

WATER FRACTION: 1

GAS GRAVITY: 0.65
OIL API GRAVITY: 35
WATER SPECIFIC GRAVITY: 1.07
AVERAGE FLOWING TEMP., F: 150

GAS-LIQUID RATIO, SCF/STBL

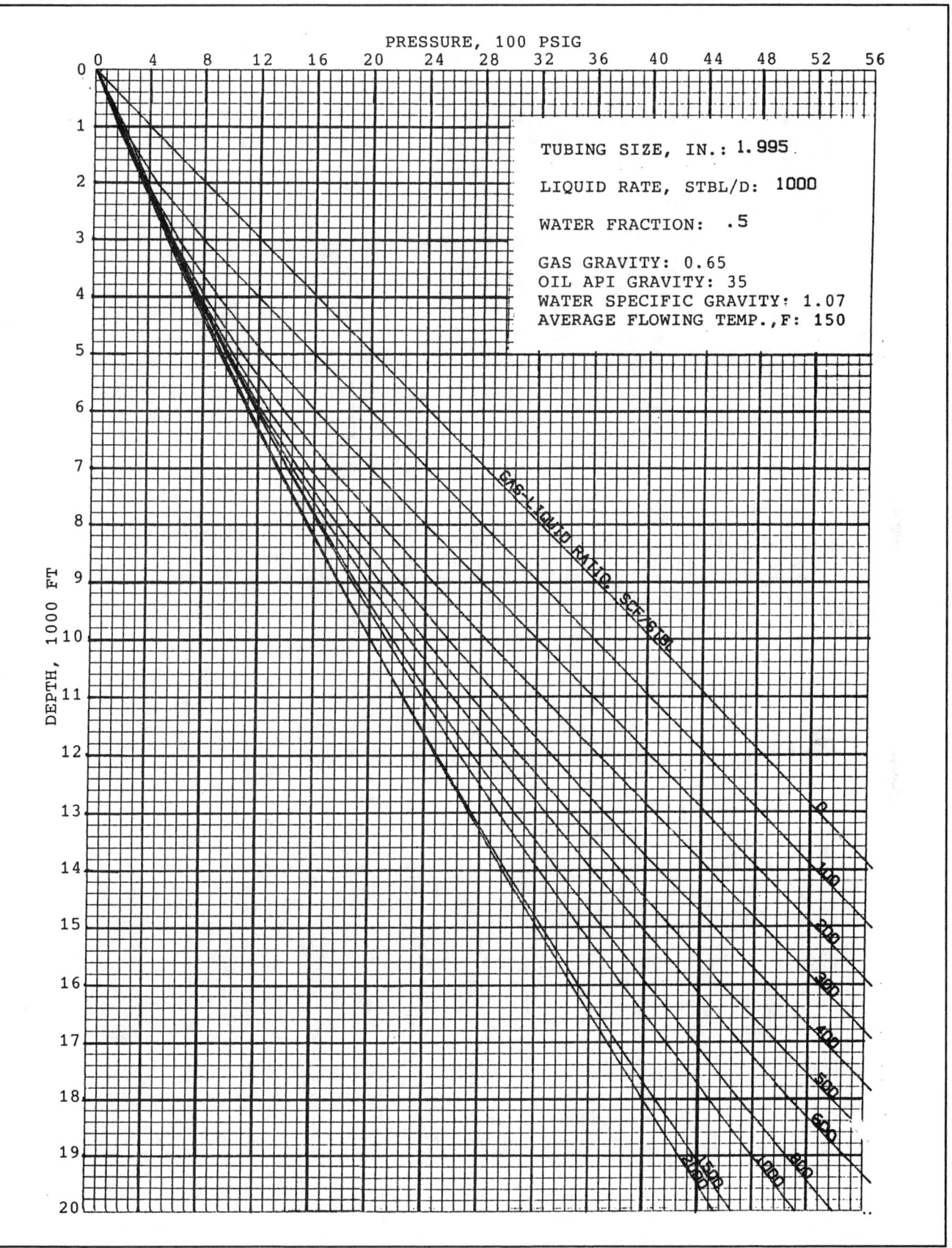

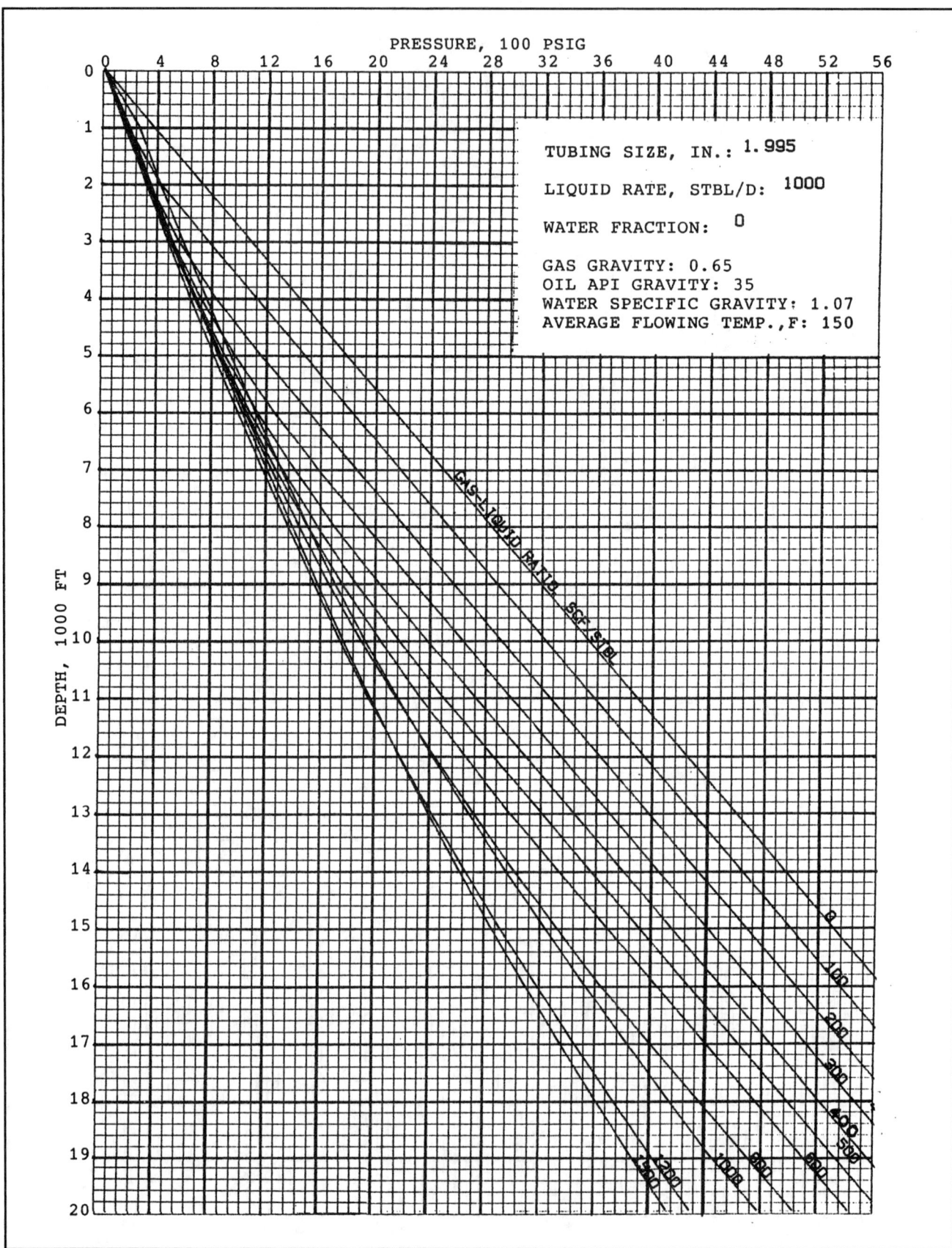

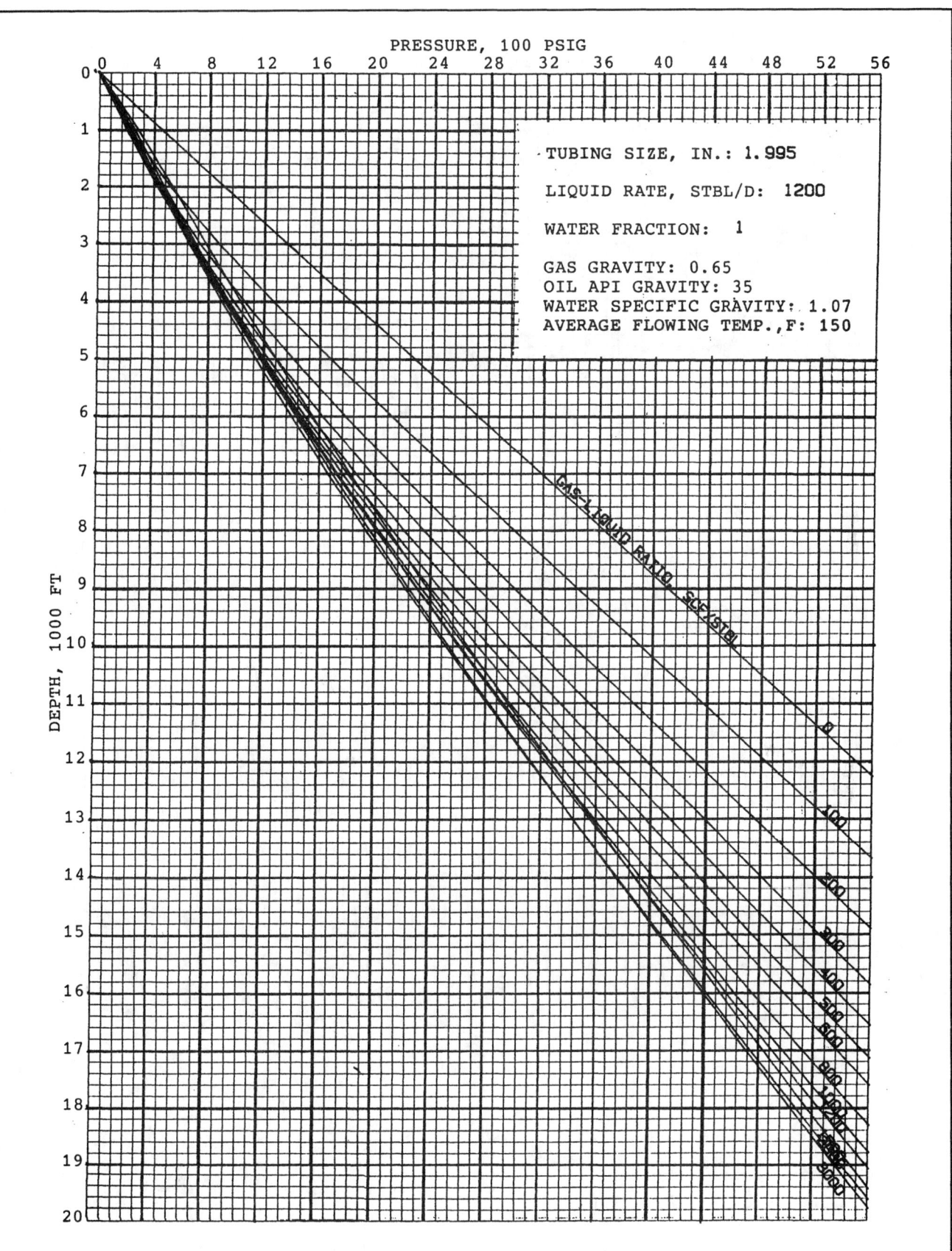

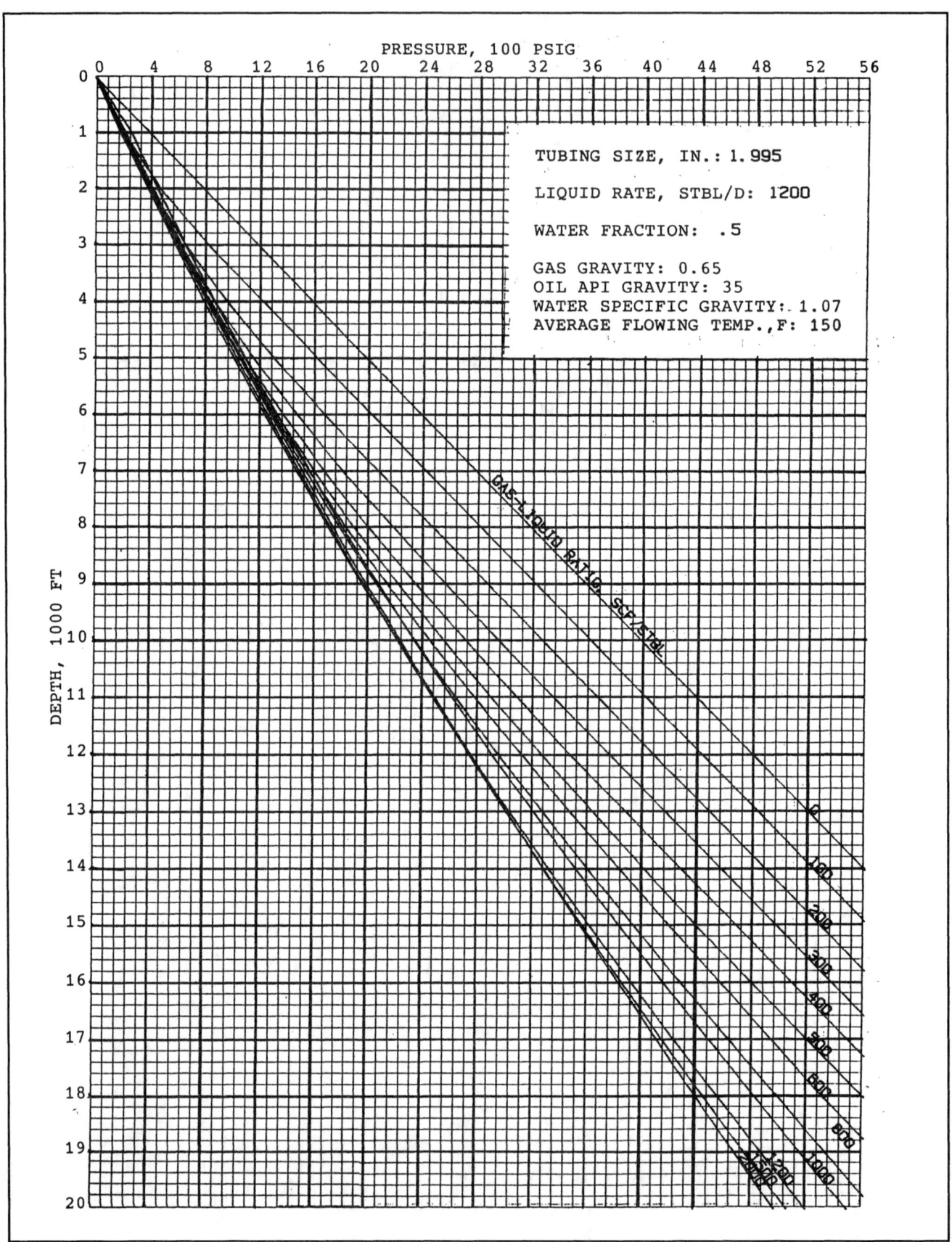

PRESSURE, 100 PSIG

TUBING SIZE, IN.: 1.995

LIQUID RATE, STBL/D: 1200

WATER FRACTION: .5

GAS GRAVITY: 0.65
OIL API GRAVITY: 35
WATER SPECIFIC GRAVITY: 1.07
AVERAGE FLOWING TEMP., F: 150

GAS-LIQUID RATIO, SCF/STBL

DEPTH, 1000 FT

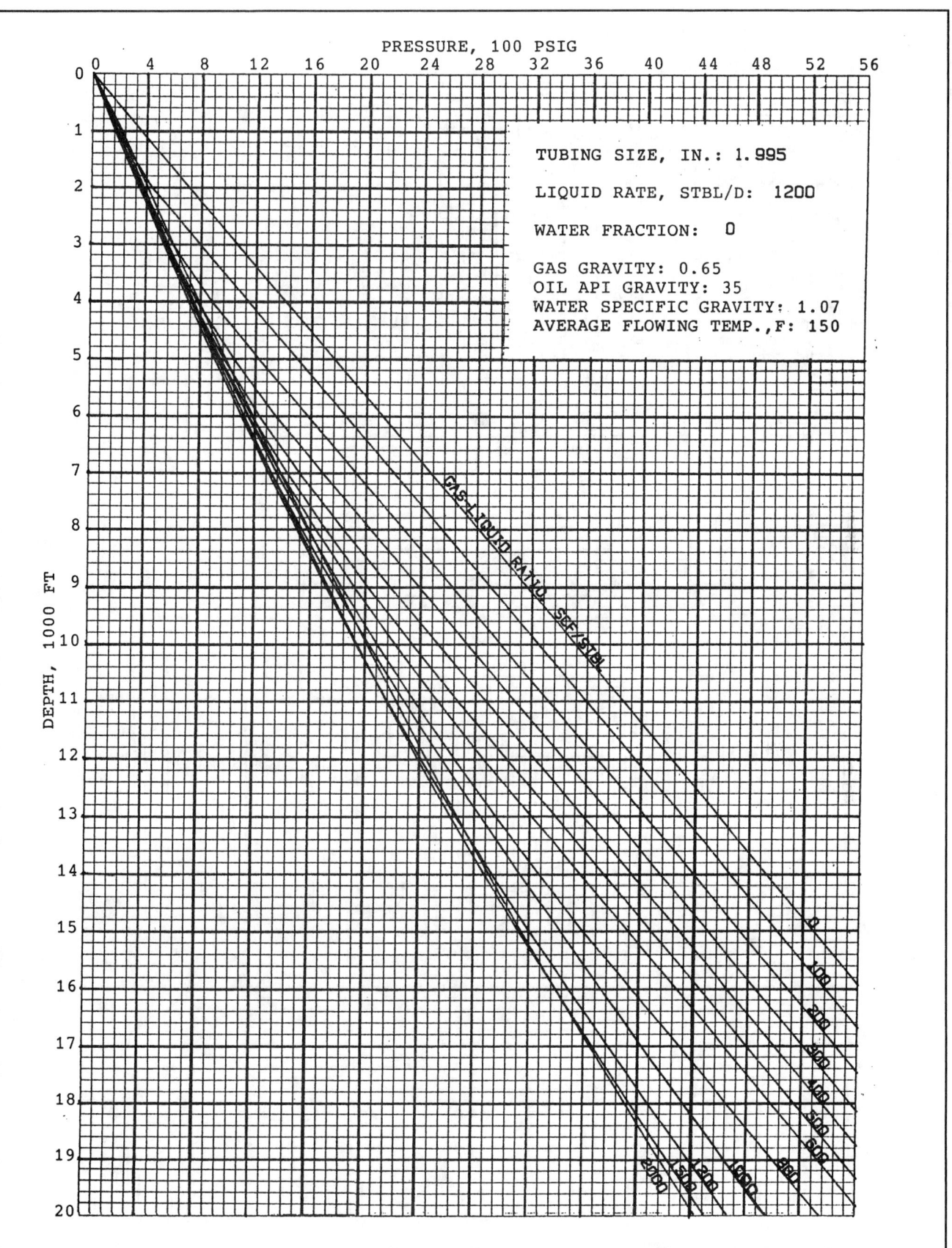

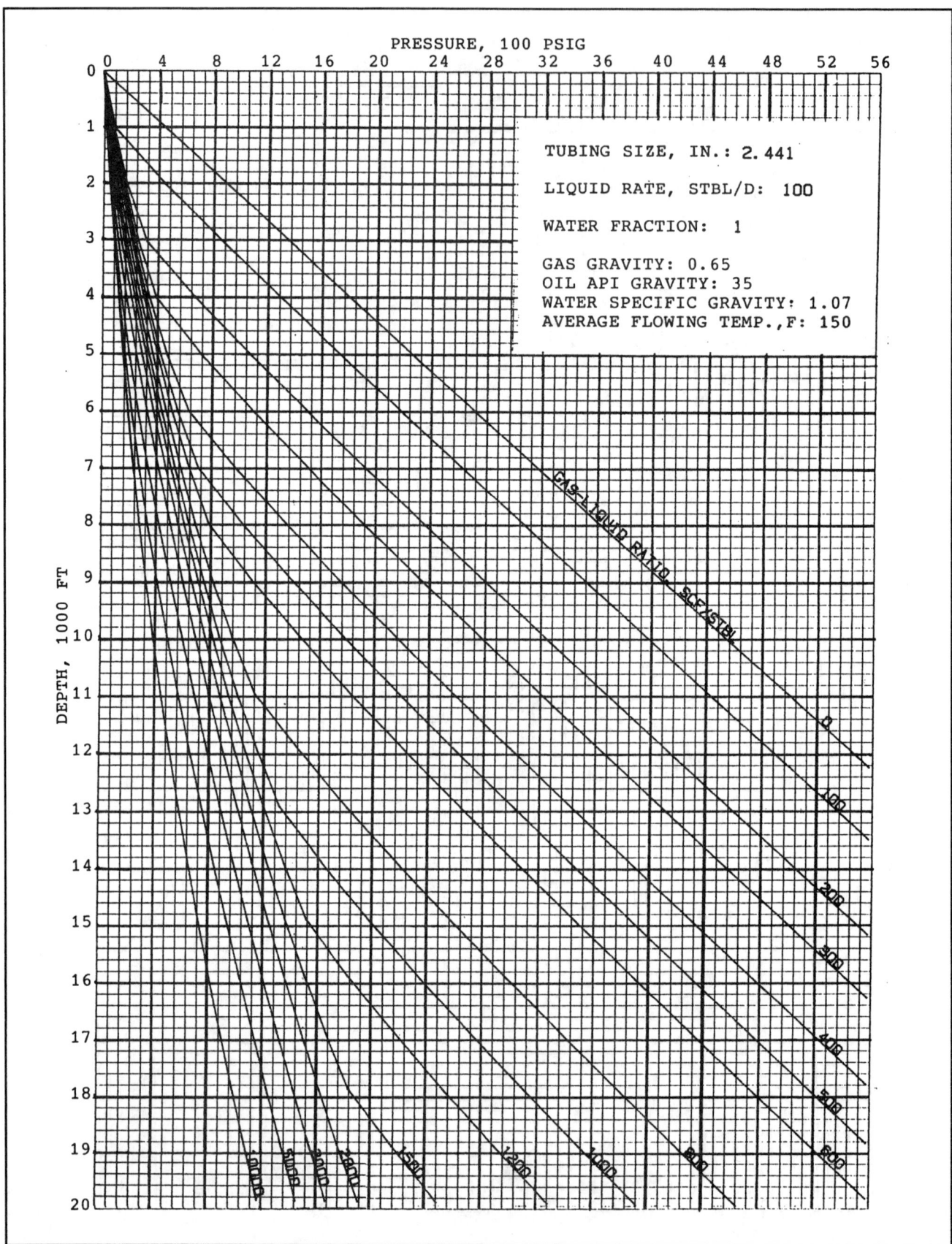

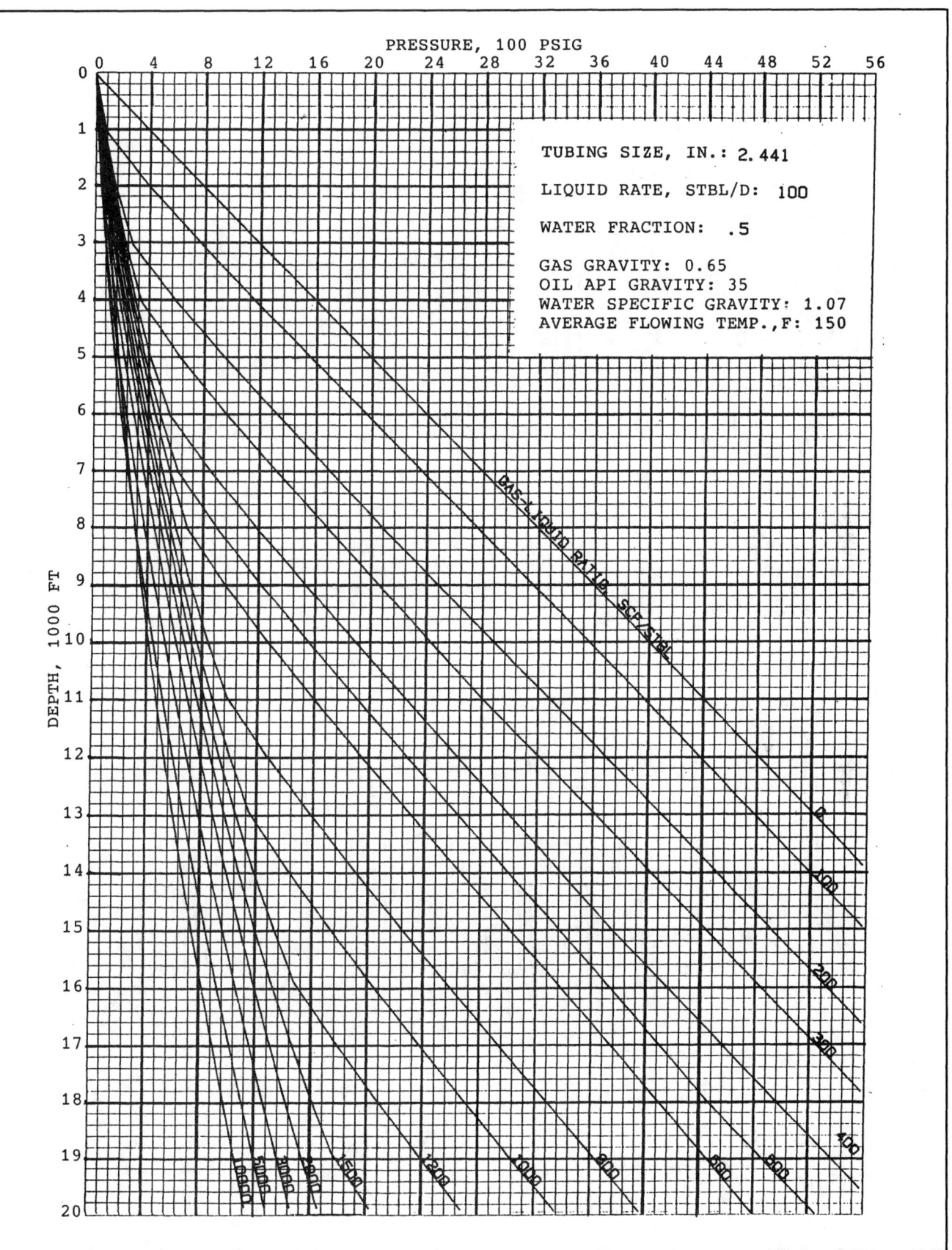

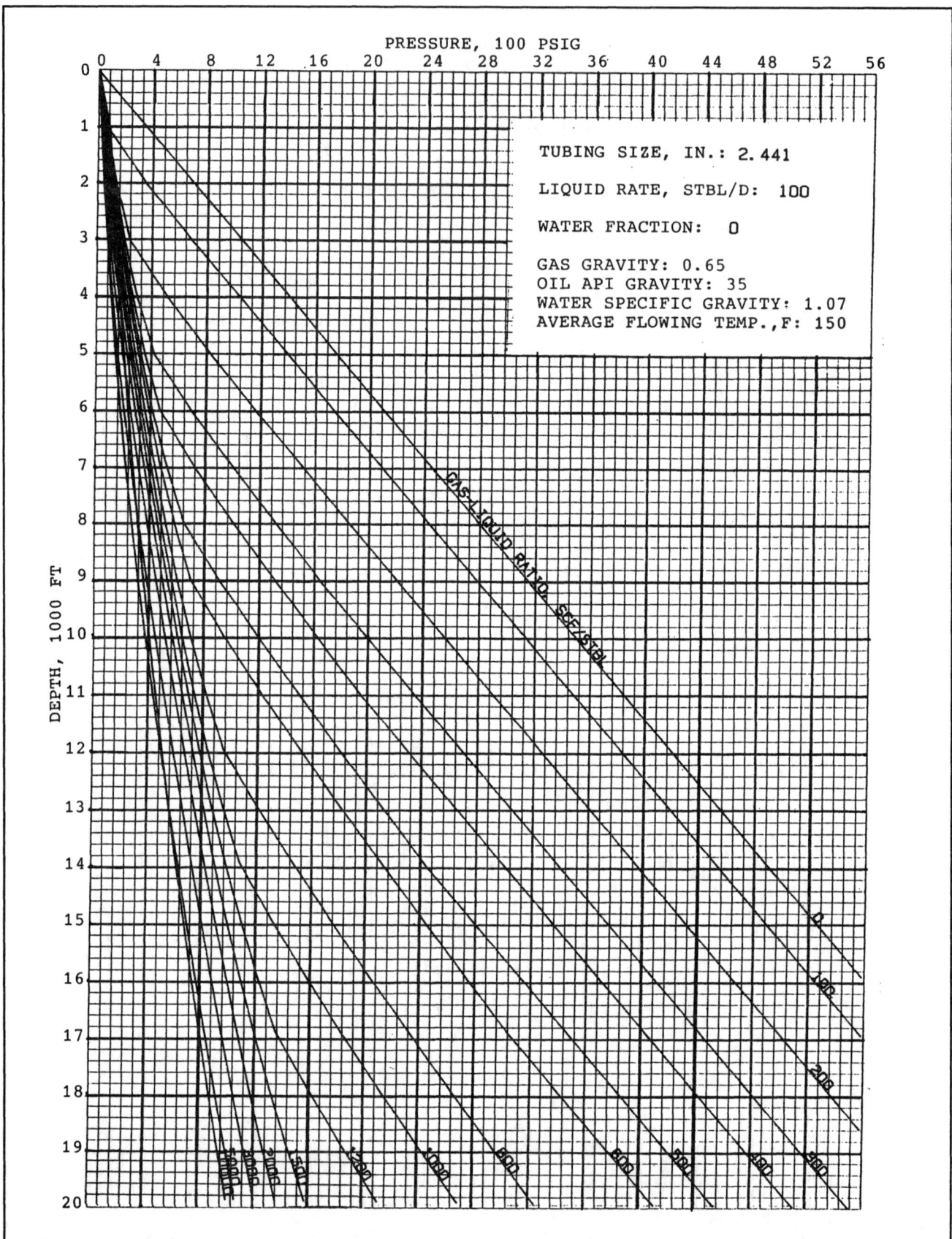

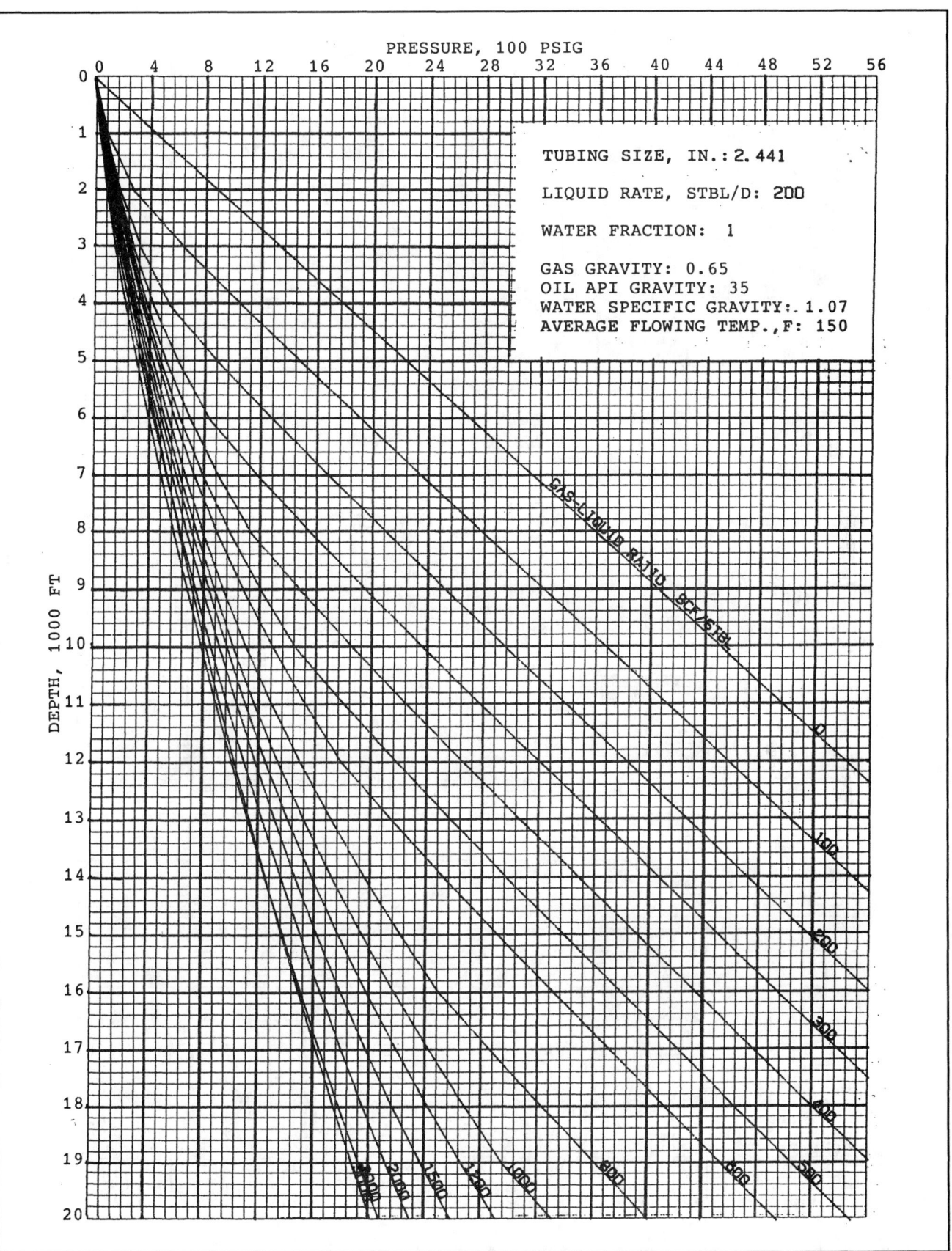

TUBING SIZE, IN.: 2.441

LIQUID RATE, STBL/D: 200

WATER FRACTION: 1

GAS GRAVITY: 0.65
OIL API GRAVITY: 35
WATER SPECIFIC GRAVITY: 1.07
AVERAGE FLOWING TEMP.,F: 150

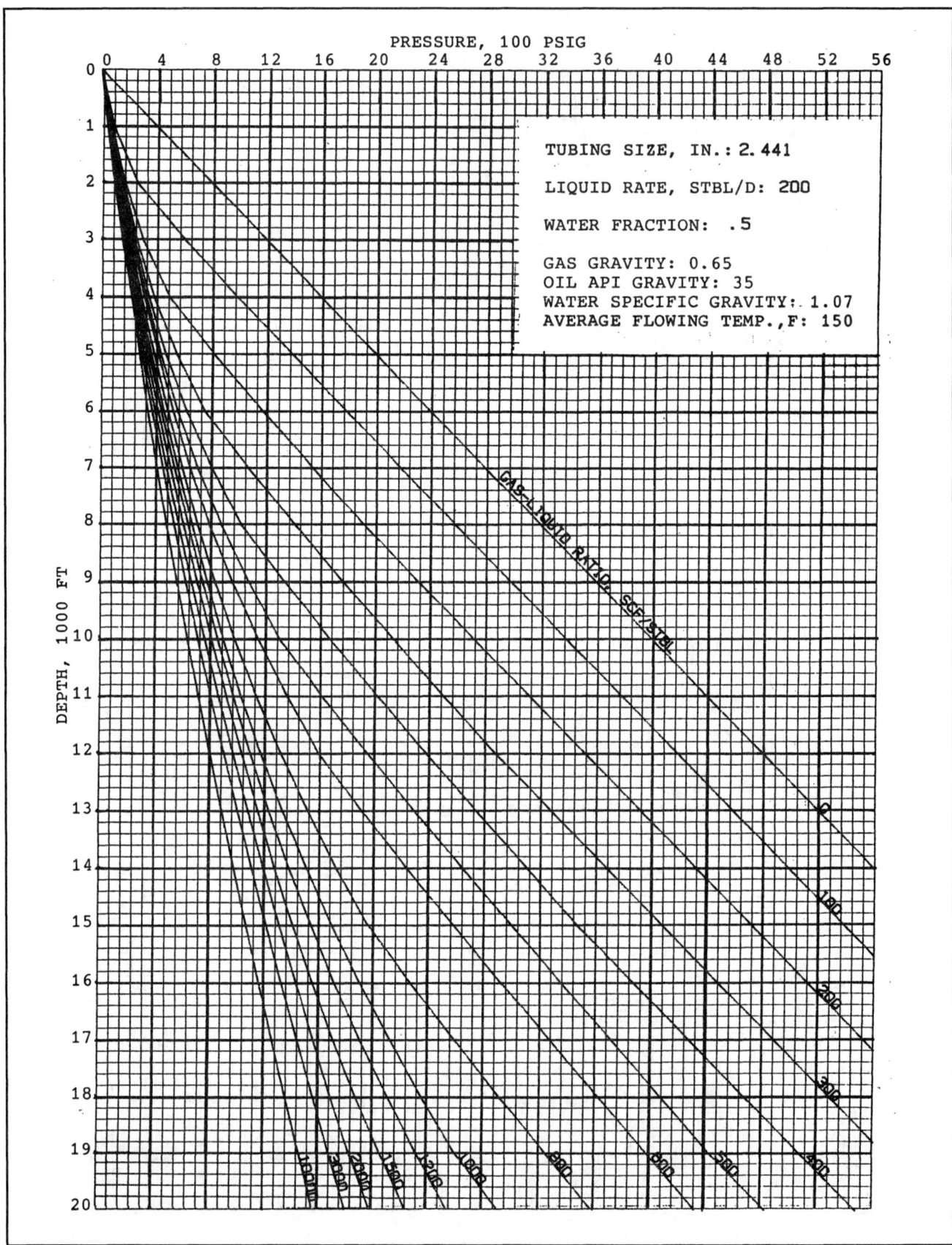

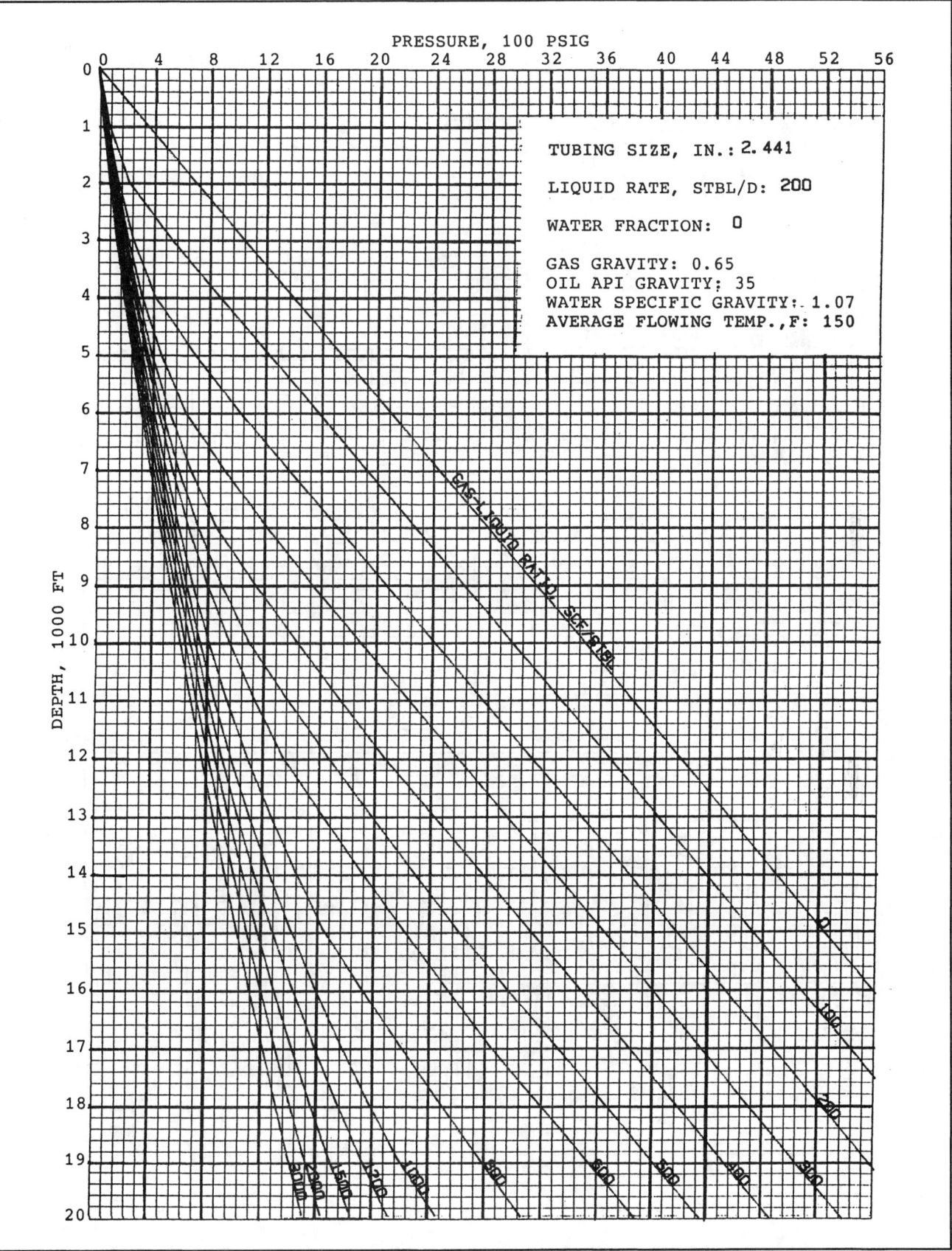

PRESSURE, 100 PSIG

TUBING SIZE, IN.: 2.441

LIQUID RATE, STBL/D: 200

WATER FRACTION: 0

GAS GRAVITY: 0.65
OIL API GRAVITY: 35
WATER SPECIFIC GRAVITY: 1.07
AVERAGE FLOWING TEMP., F: 150

DEPTH, 1000 FT

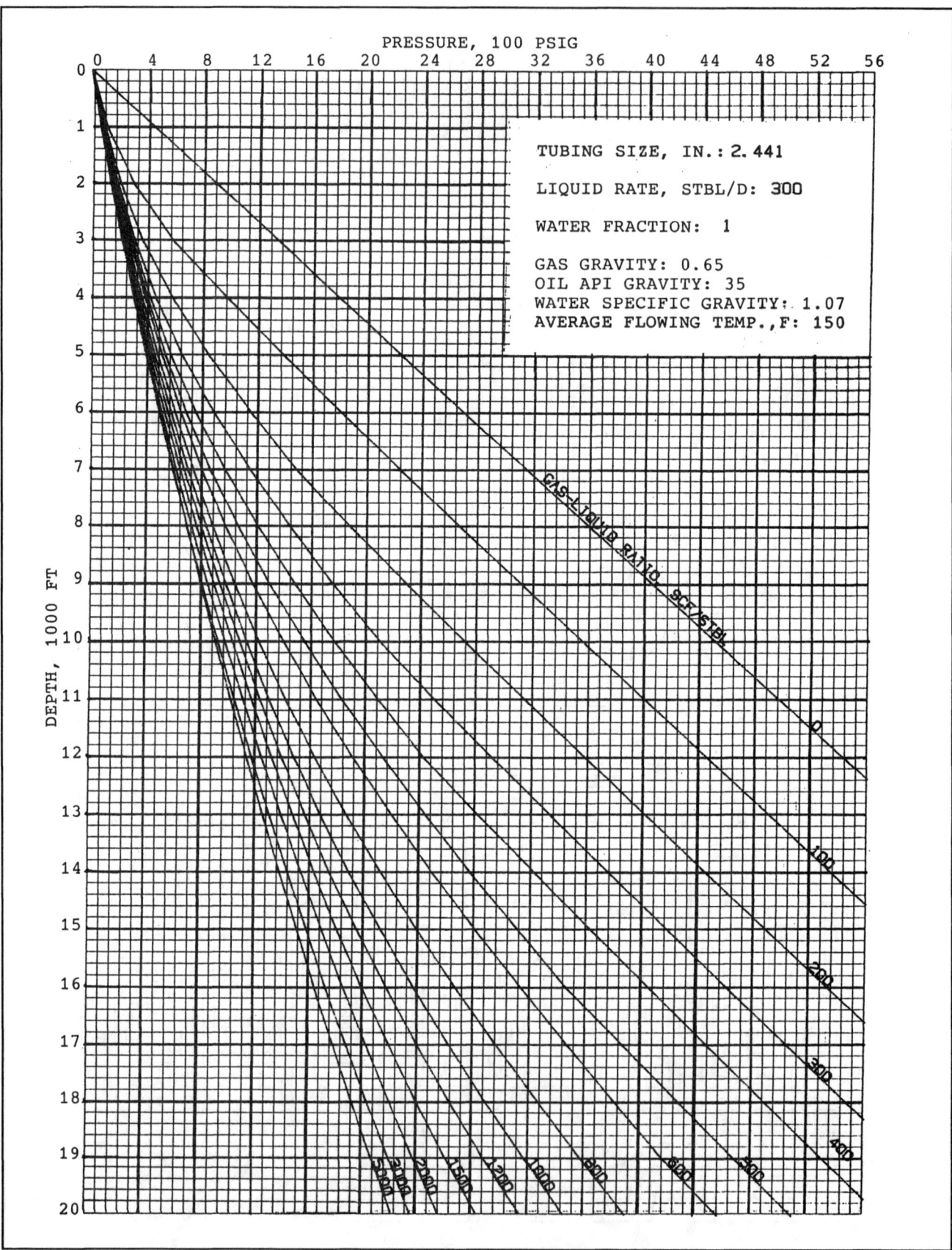

PRESSURE, 100 PSIG

DEPTH, 1000 FT

TUBING SIZE, IN.: 2.441

LIQUID RATE, STBL/D: 300

WATER FRACTION: 1

GAS GRAVITY: 0.65
OIL API GRAVITY: 35
WATER SPECIFIC GRAVITY: 1.07
AVERAGE FLOWING TEMP.,F: 150

GAS-LIQUID RATIO, SCF/STBL

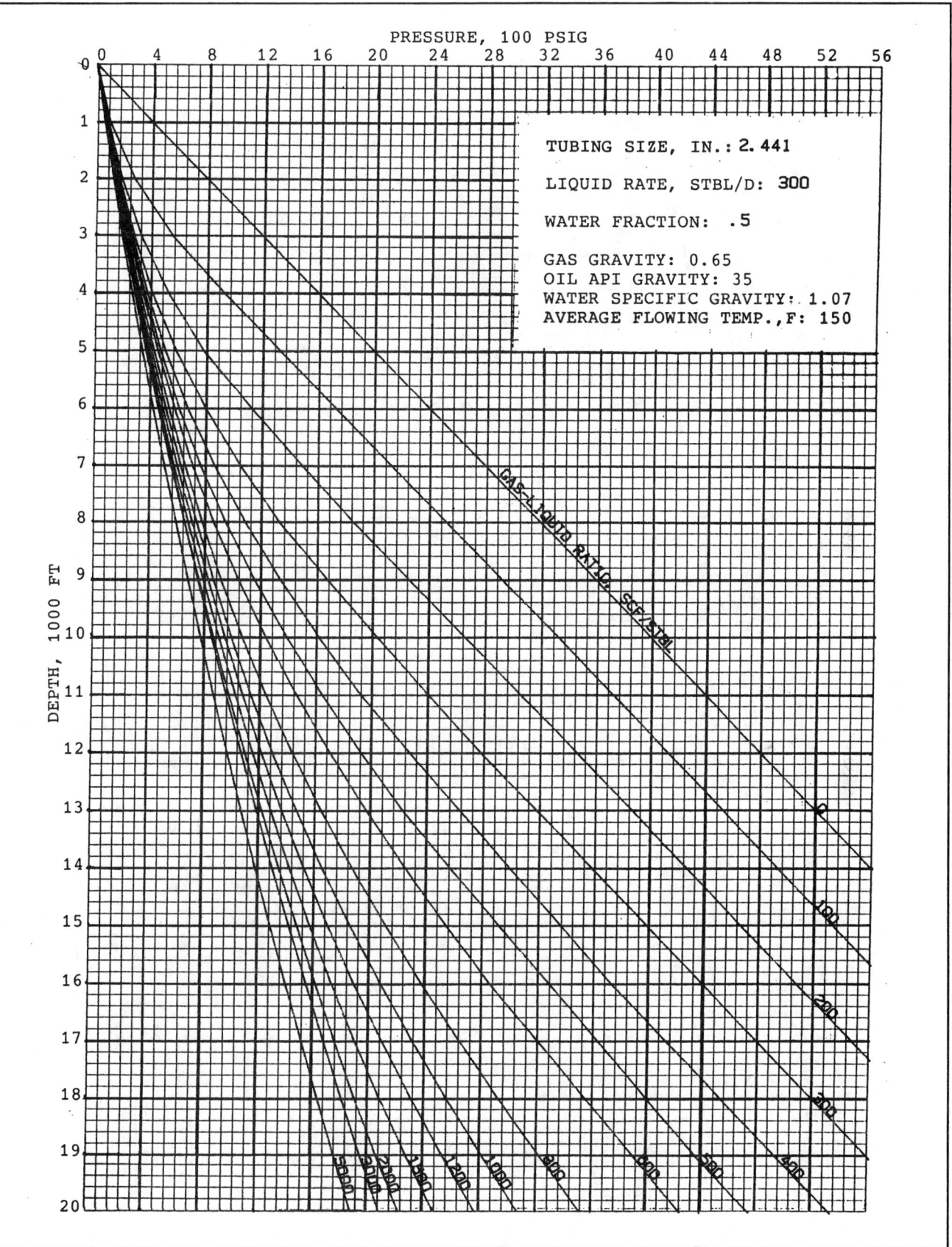

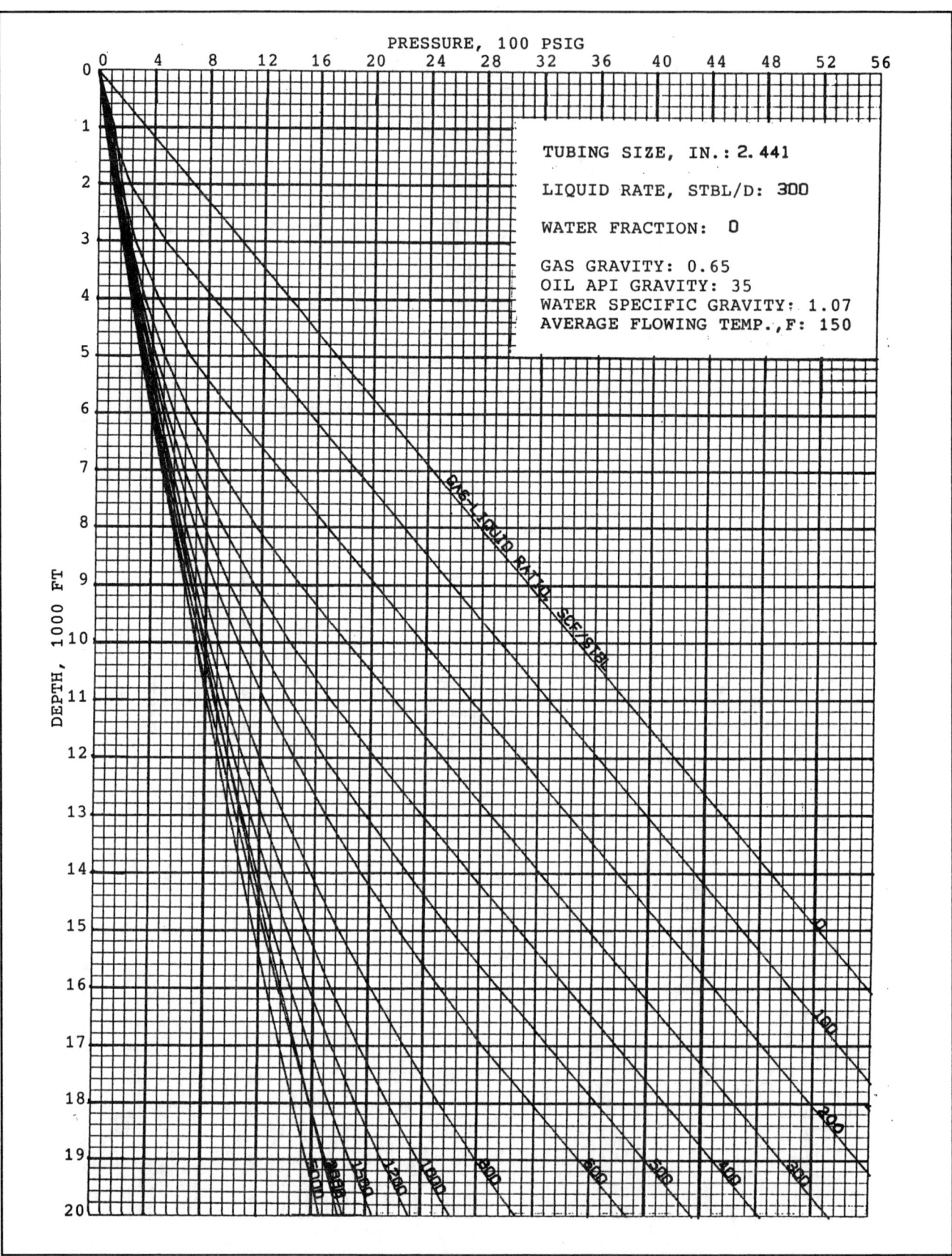

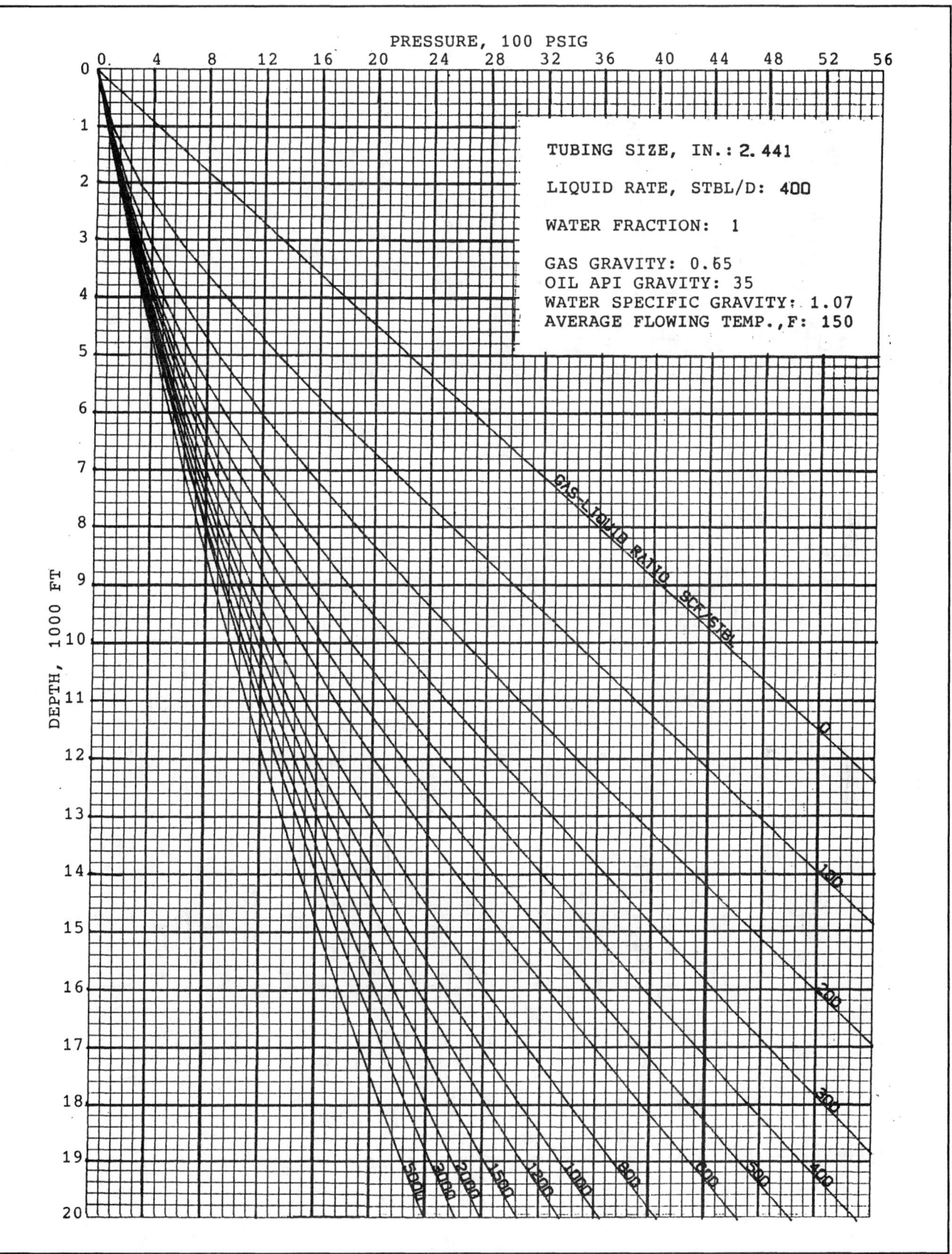

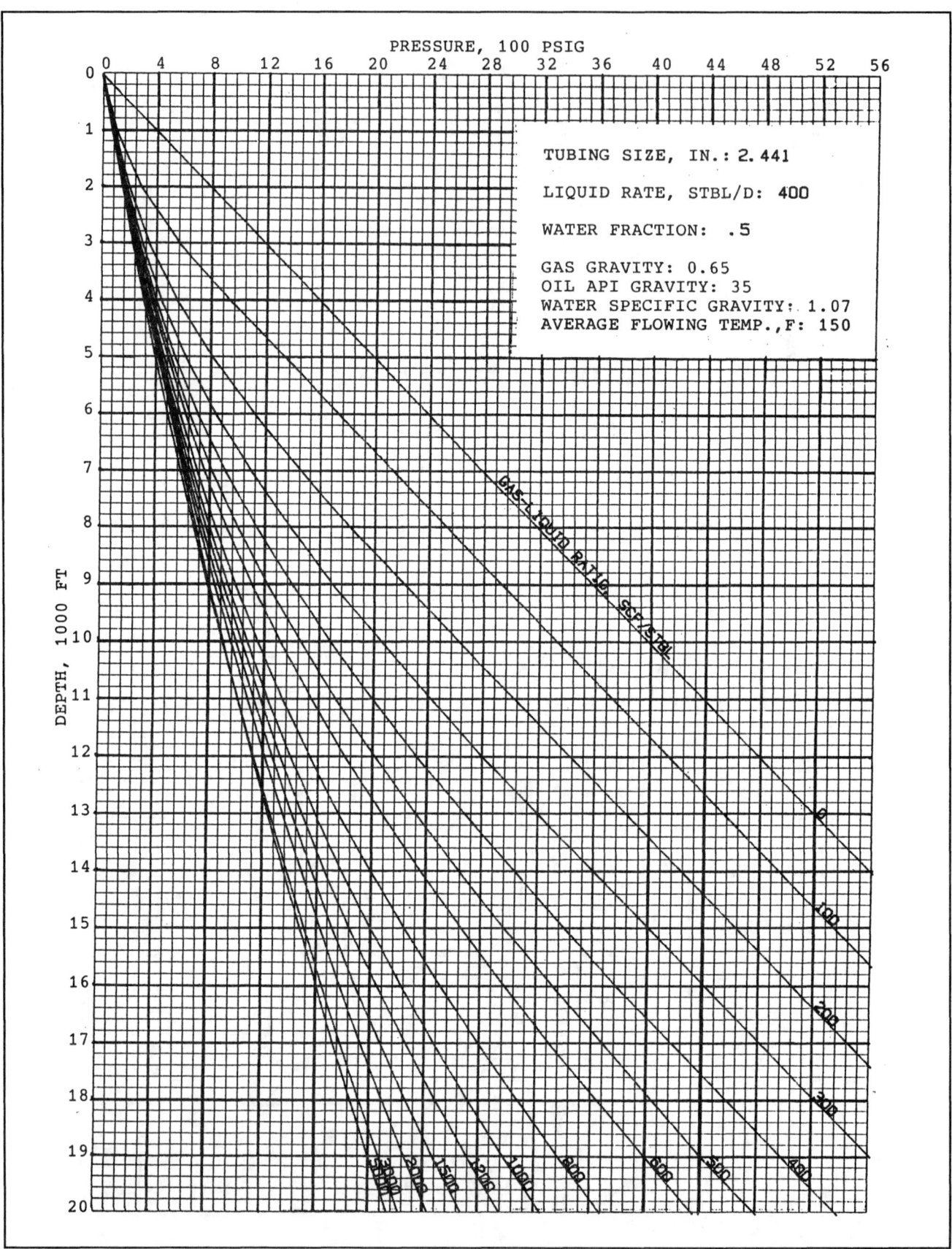

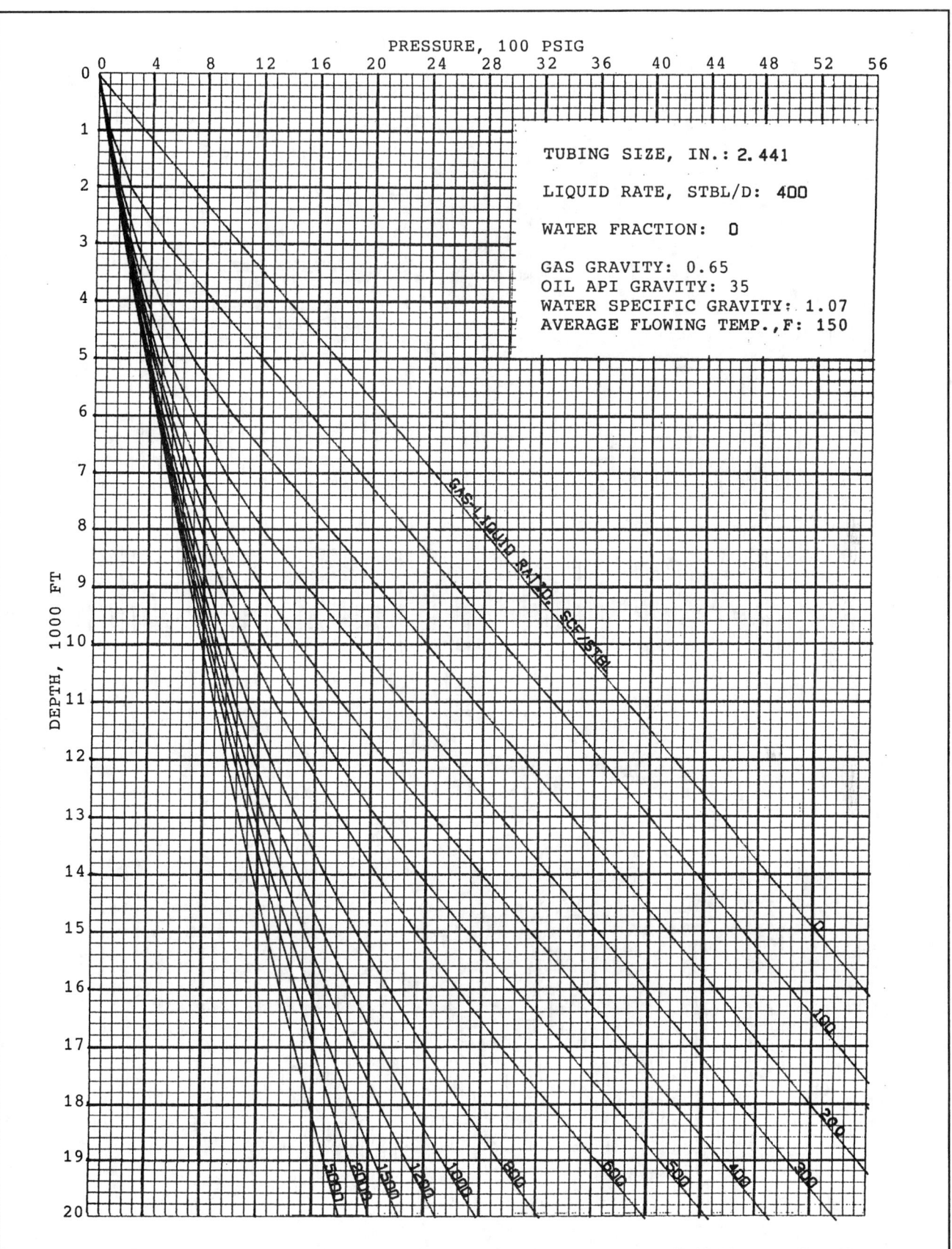

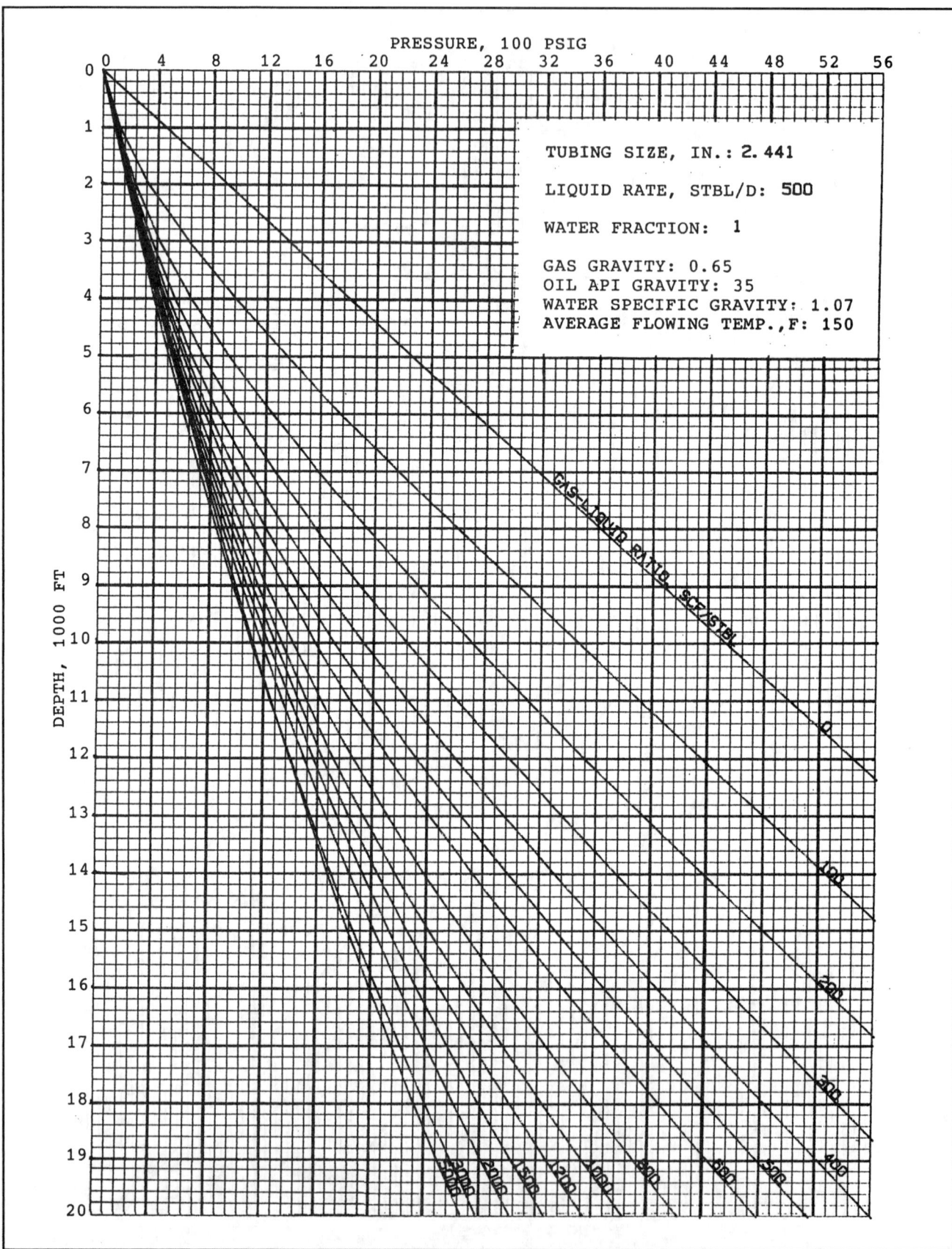

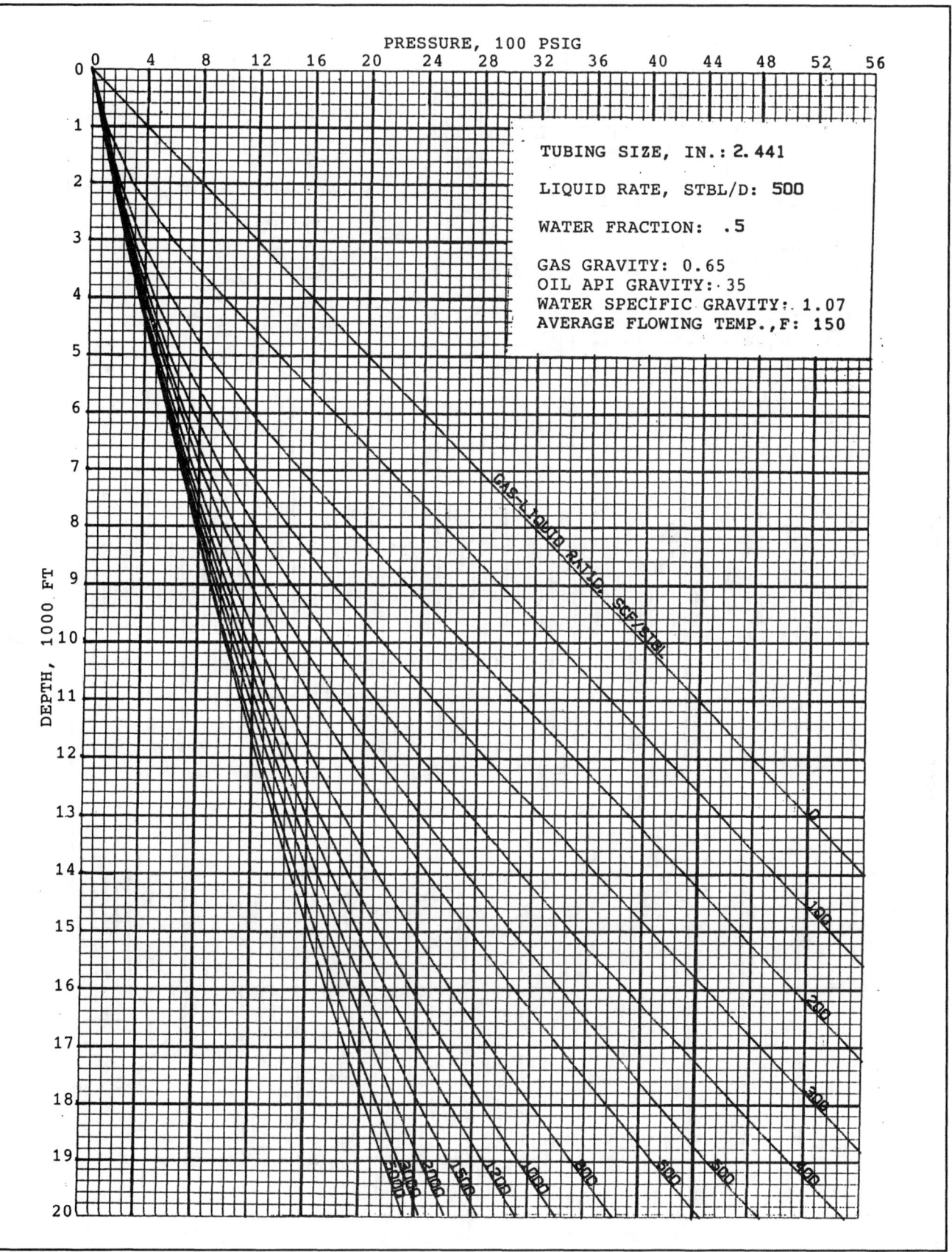

PRESSURE, 100 PSIG

DEPTH, 1000 FT

TUBING SIZE, IN.: 2.441

LIQUID RATE, STBL/D: 500

WATER FRACTION: .5

GAS GRAVITY: 0.65
OIL API GRAVITY: 35
WATER SPECIFIC GRAVITY: 1.07
AVERAGE FLOWING TEMP., F: 150

GAS-LIQUID RATIO, SCF/STBL

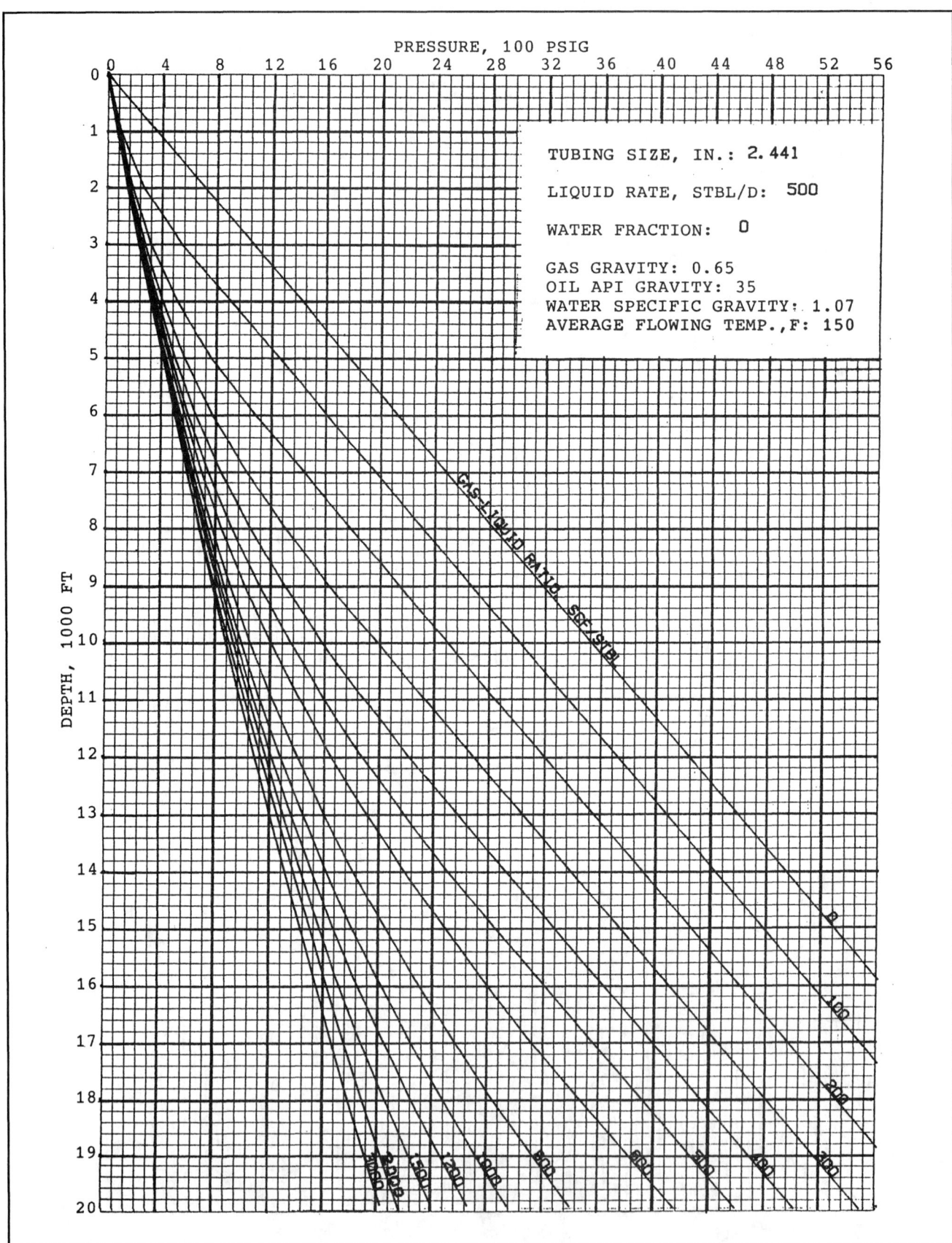

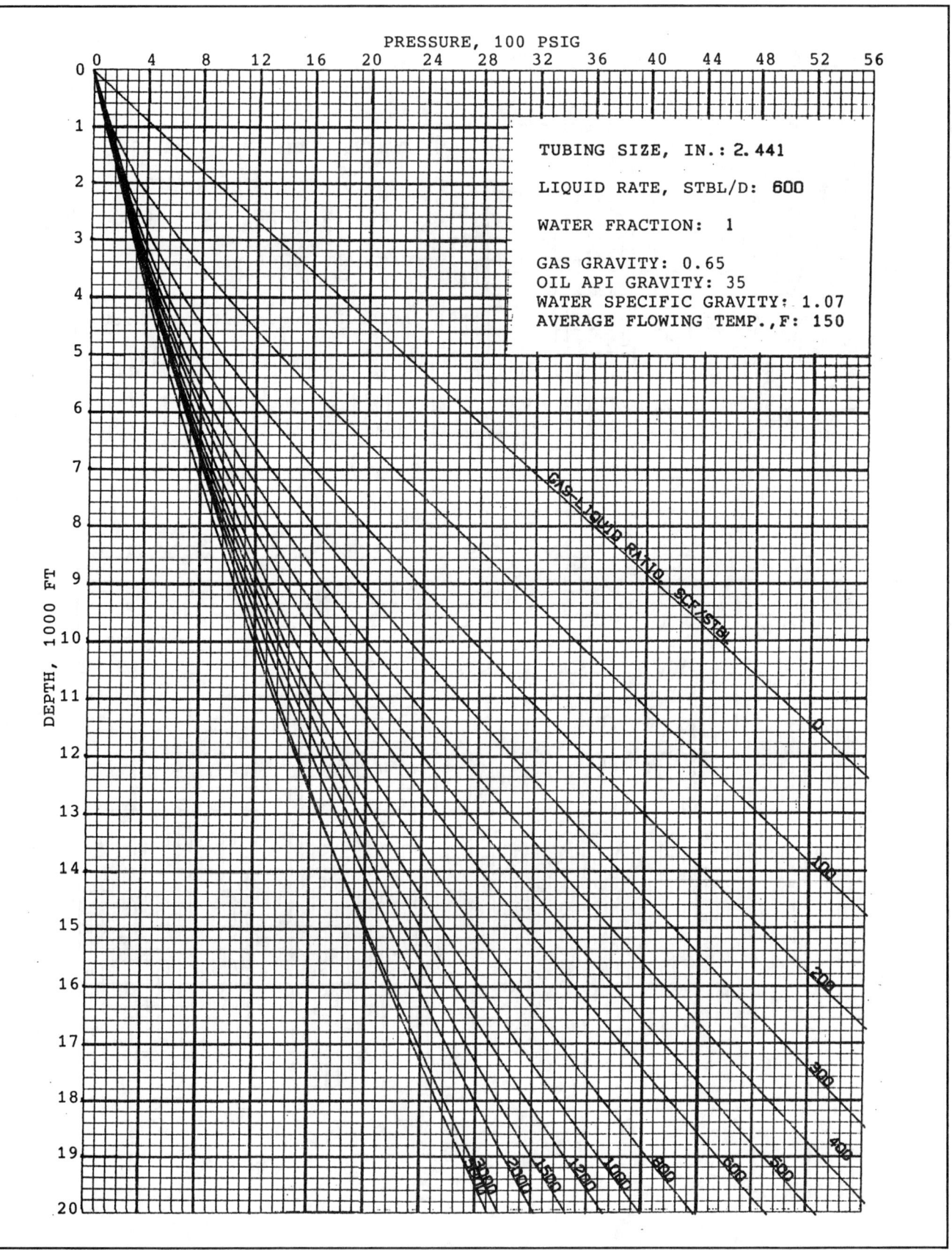

PRESSURE, 100 PSIG

TUBING SIZE, IN.: 2.441

LIQUID RATE, STBL/D: 600

WATER FRACTION: 1

GAS GRAVITY: 0.65
OIL API GRAVITY: 35
WATER SPECIFIC GRAVITY: 1.07
AVERAGE FLOWING TEMP., F: 150

DEPTH, 1000 FT

GAS-LIQUID RATIO SCF/STBL

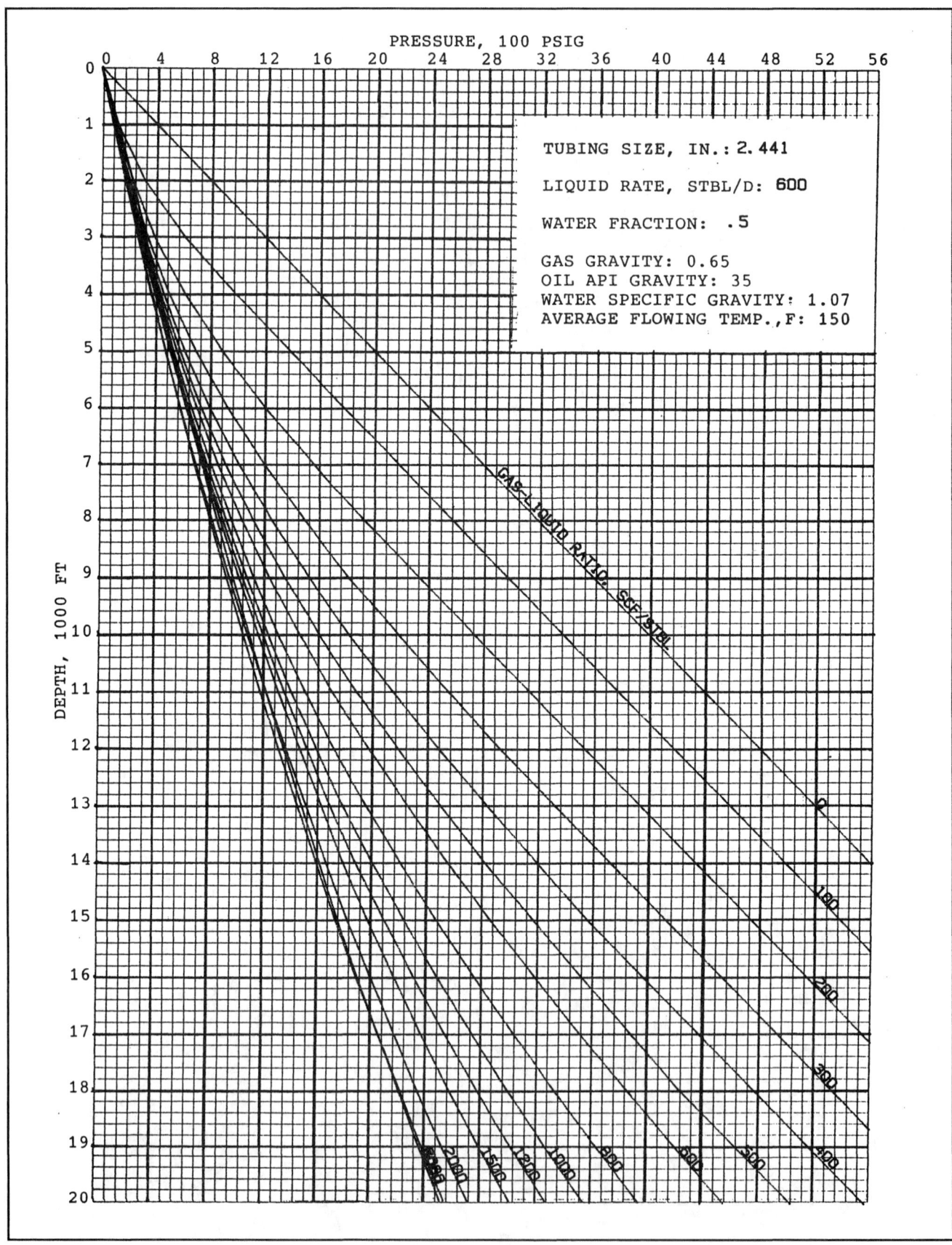

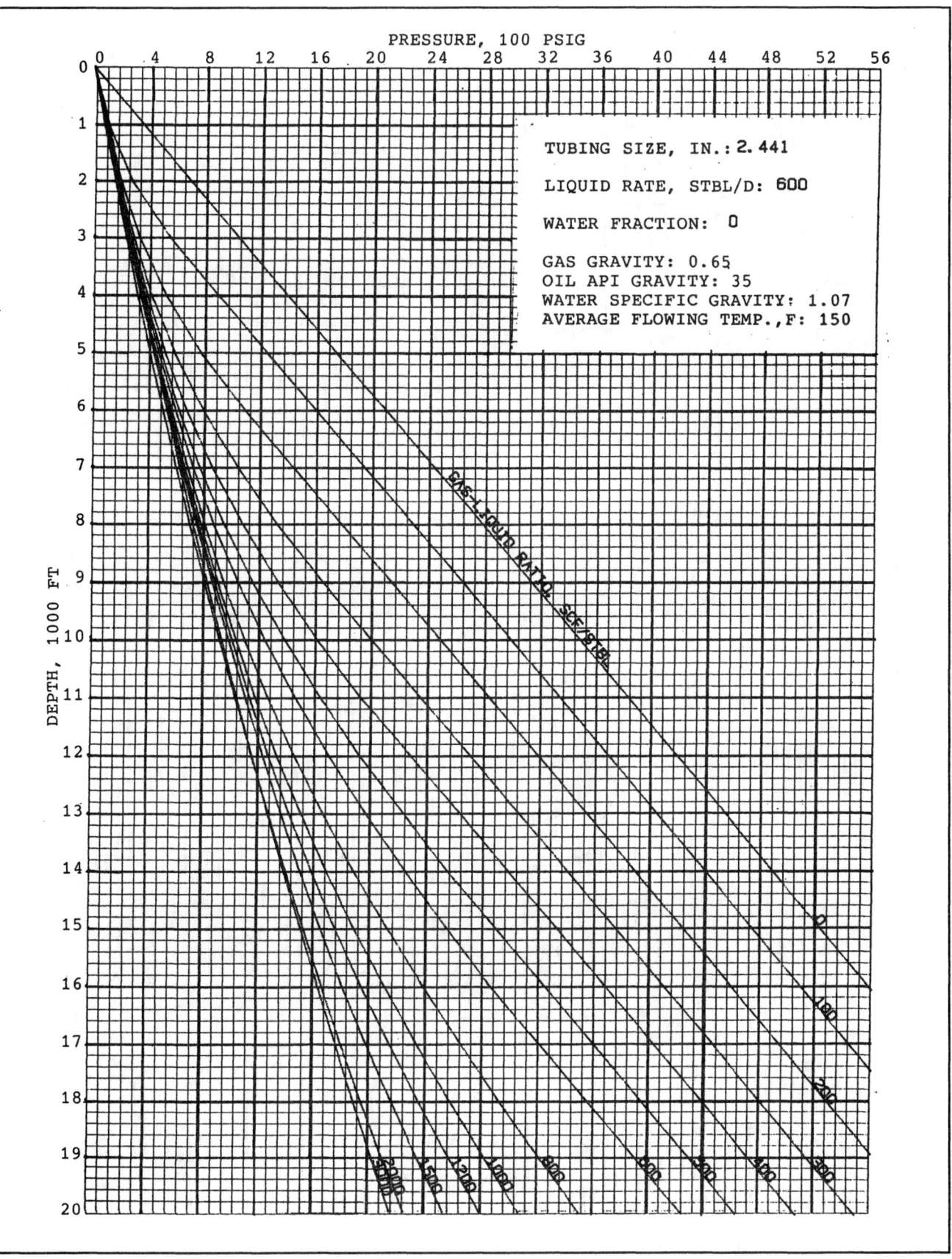

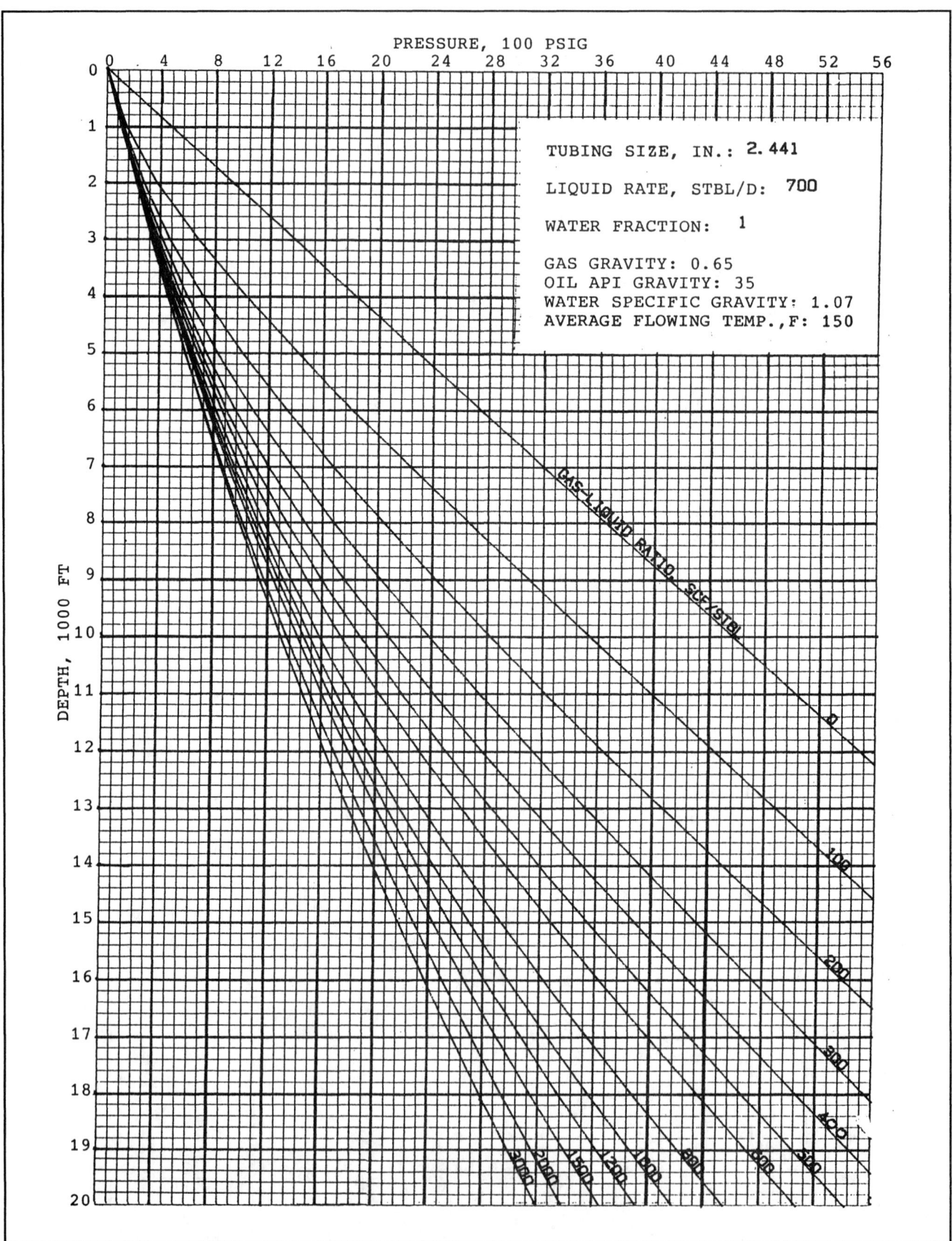

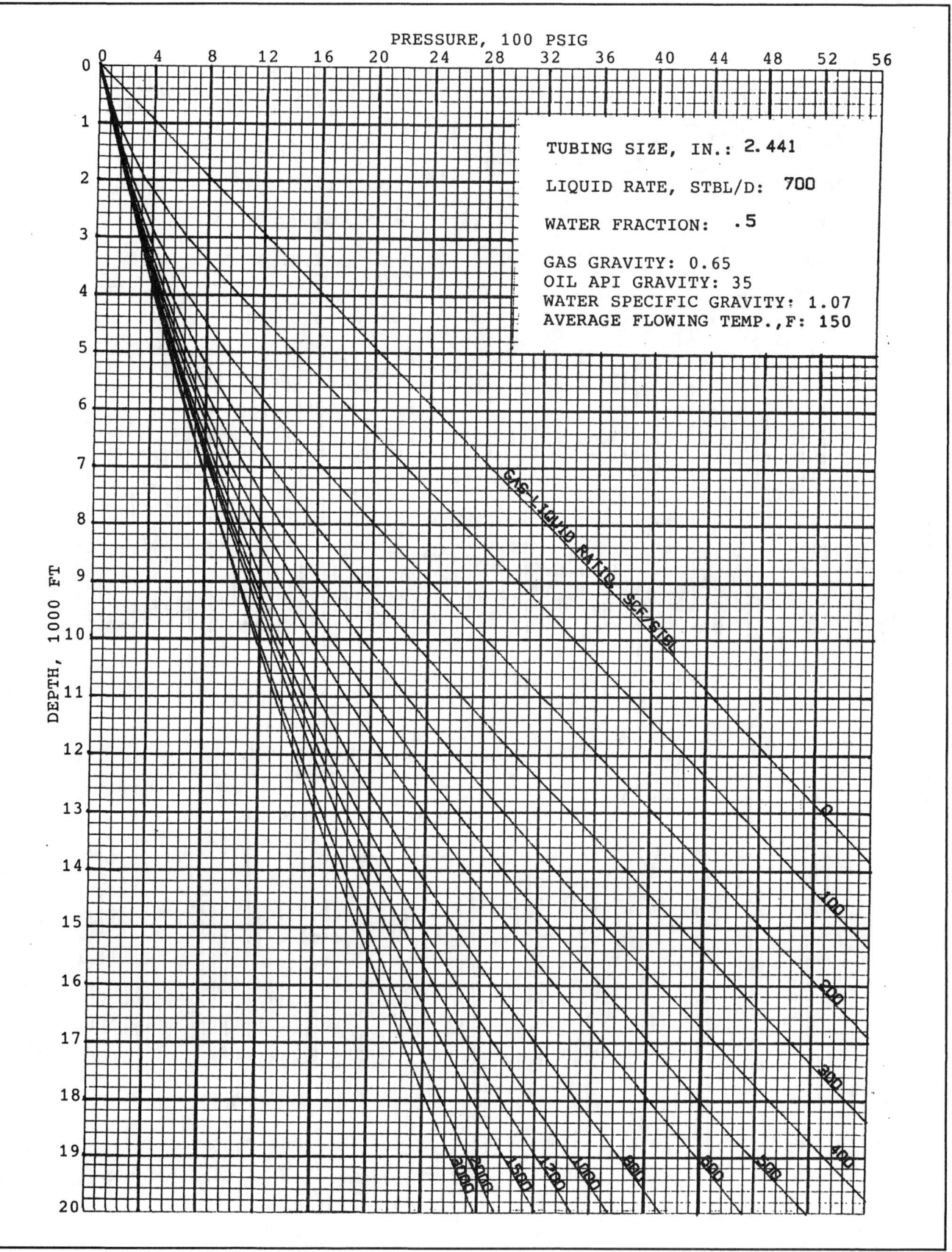

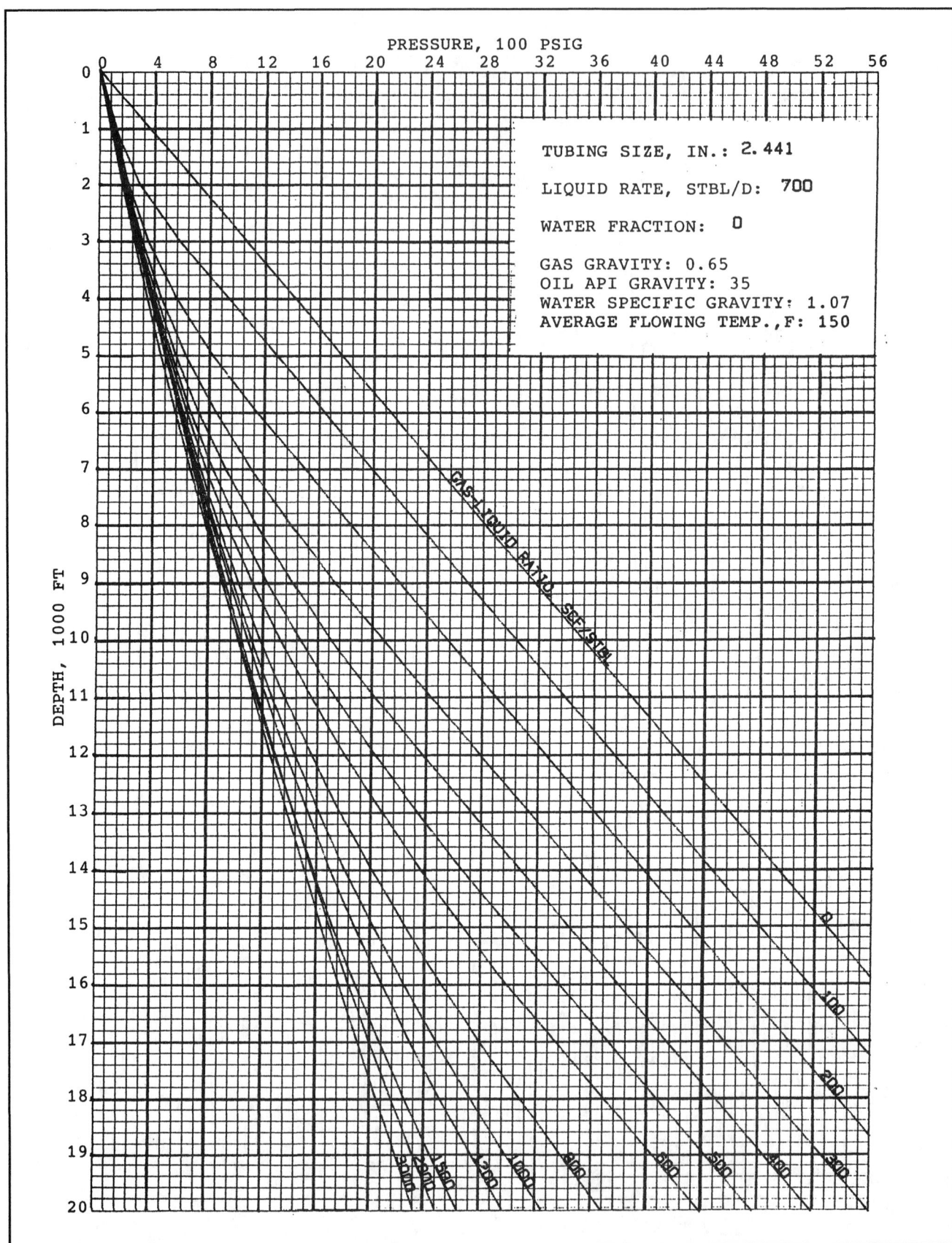

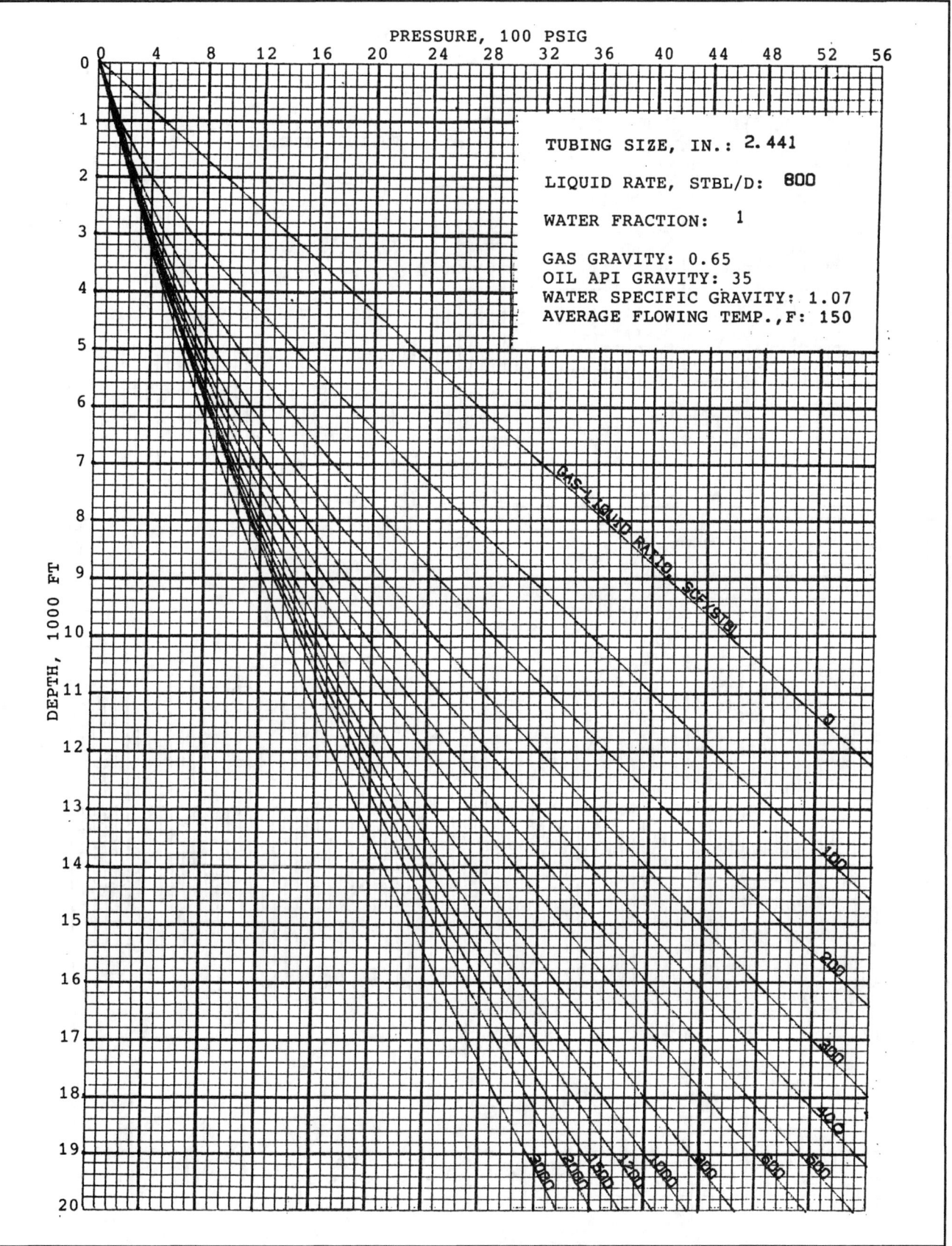

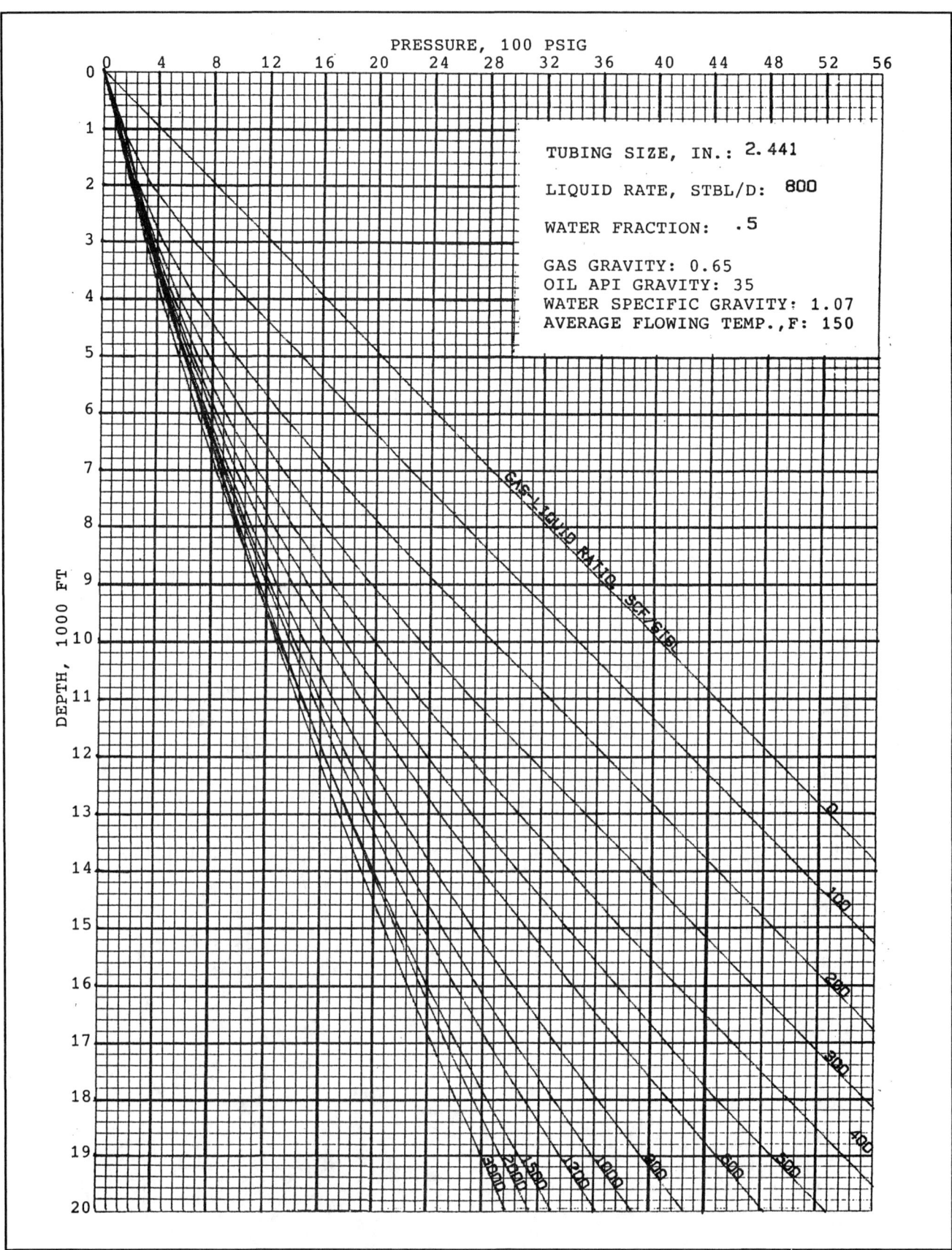

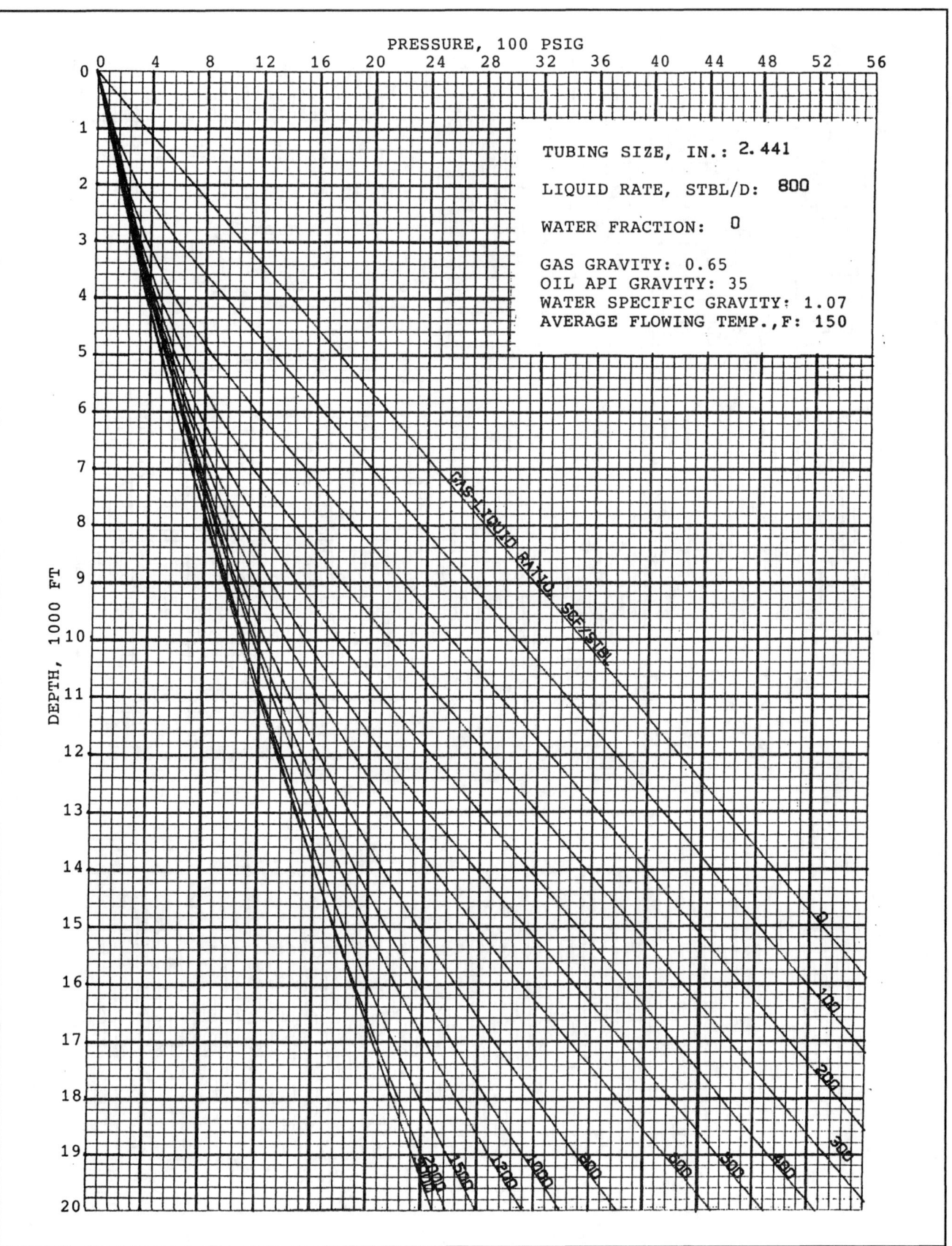

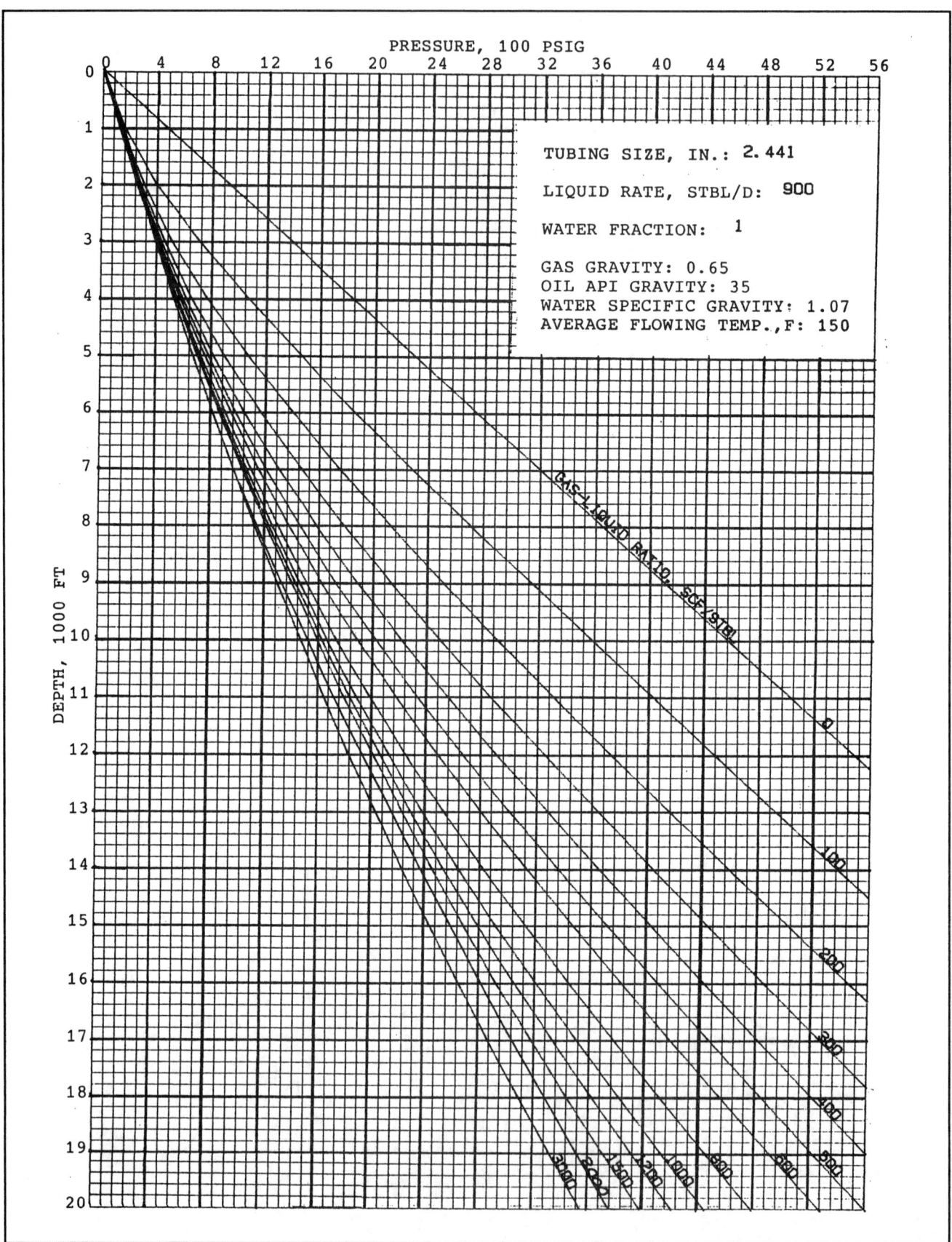

PRESSURE, 100 PSIG

TUBING SIZE, IN.: 2.441

LIQUID RATE, STBL/D: 900

WATER FRACTION: 1

GAS GRAVITY: 0.65
OIL API GRAVITY: 35
WATER SPECIFIC GRAVITY: 1.07
AVERAGE FLOWING TEMP.,F: 150

DEPTH, 1000 FT

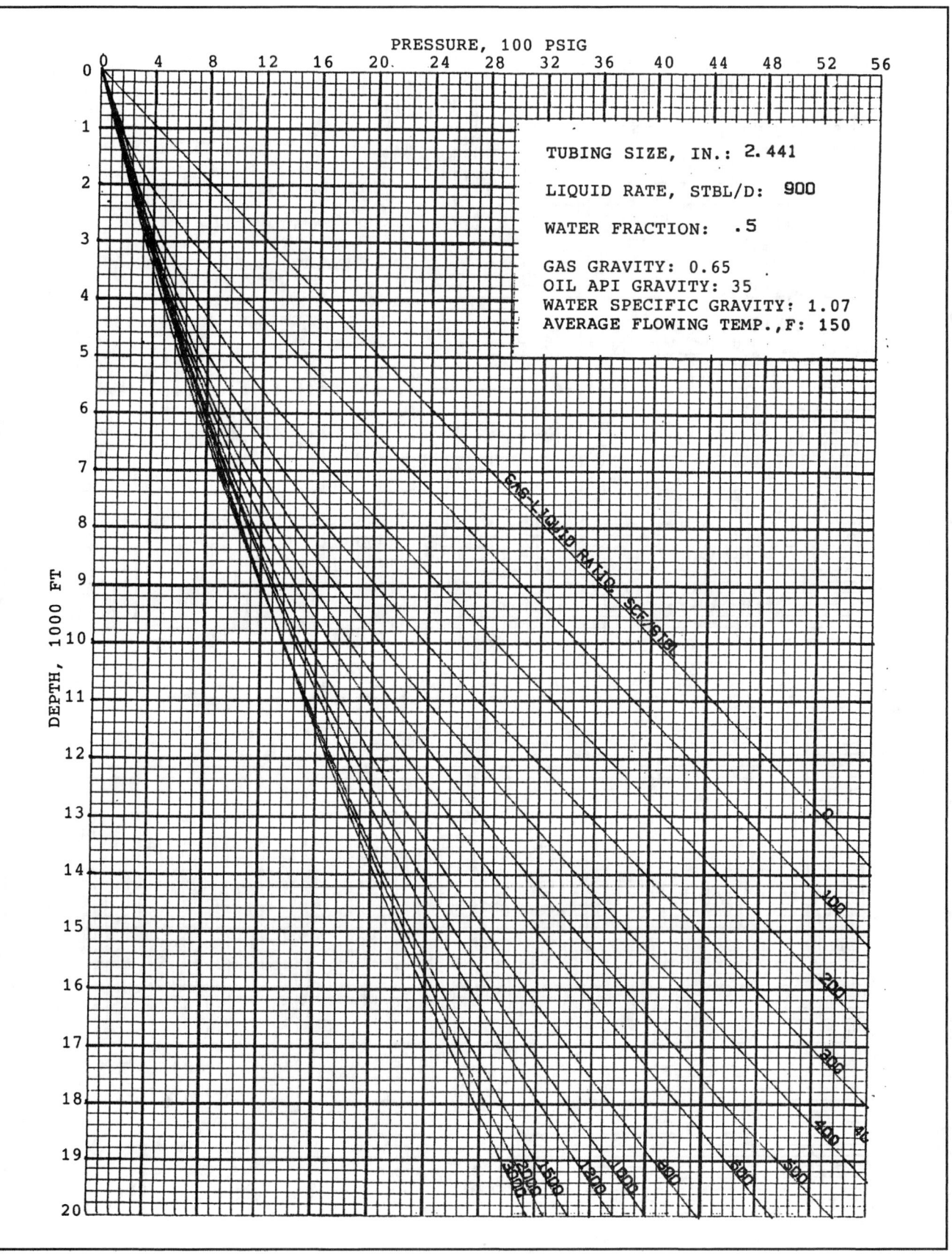

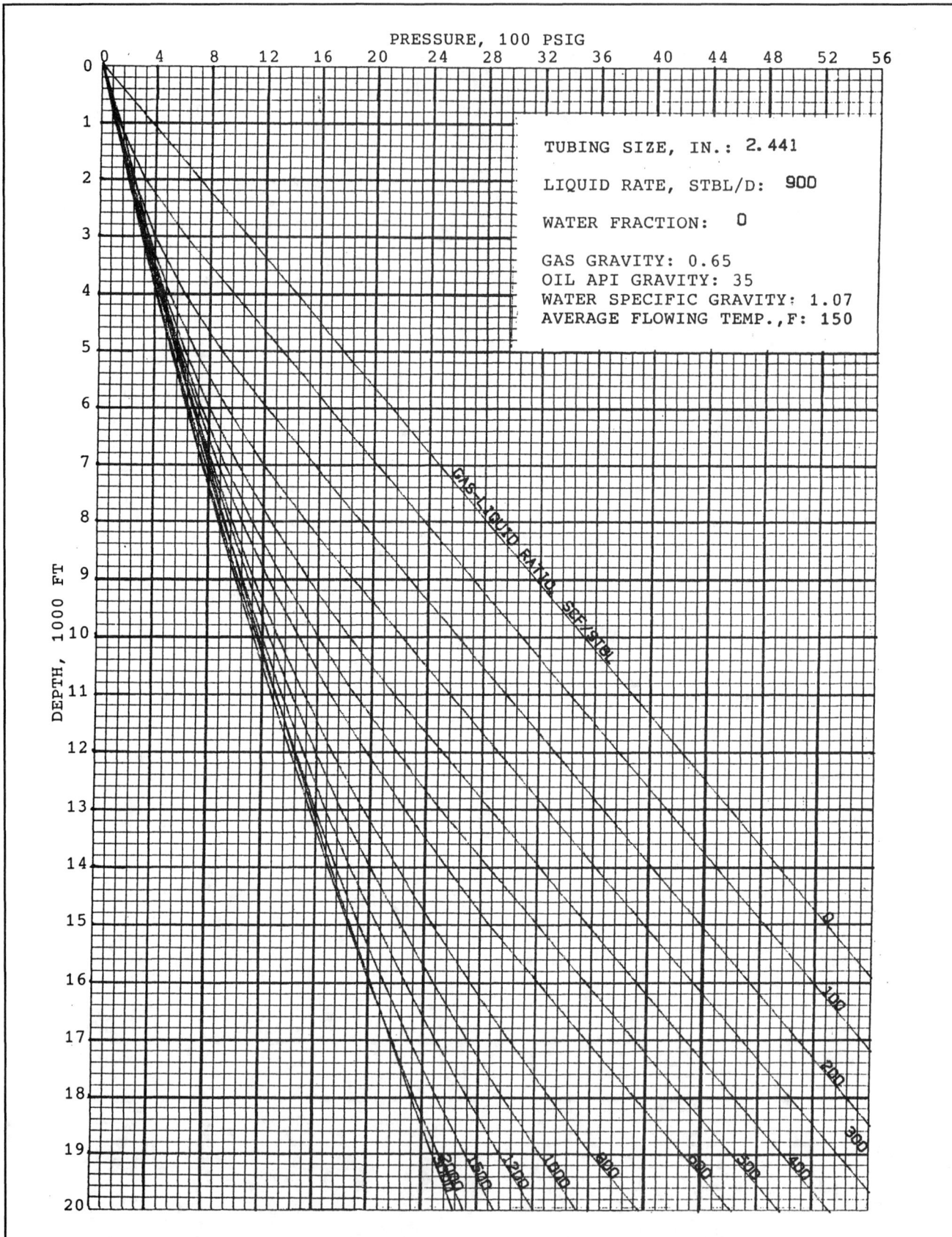

PRESSURE, 100 PSIG

TUBING SIZE, IN.: 2.441

LIQUID RATE, STBL/D: 900

WATER FRACTION: 0

GAS GRAVITY: 0.65
OIL API GRAVITY: 35
WATER SPECIFIC GRAVITY: 1.07
AVERAGE FLOWING TEMP.,F: 150

GAS-LIQUID RATIO, SCF/STBL

DEPTH, 1000 FT

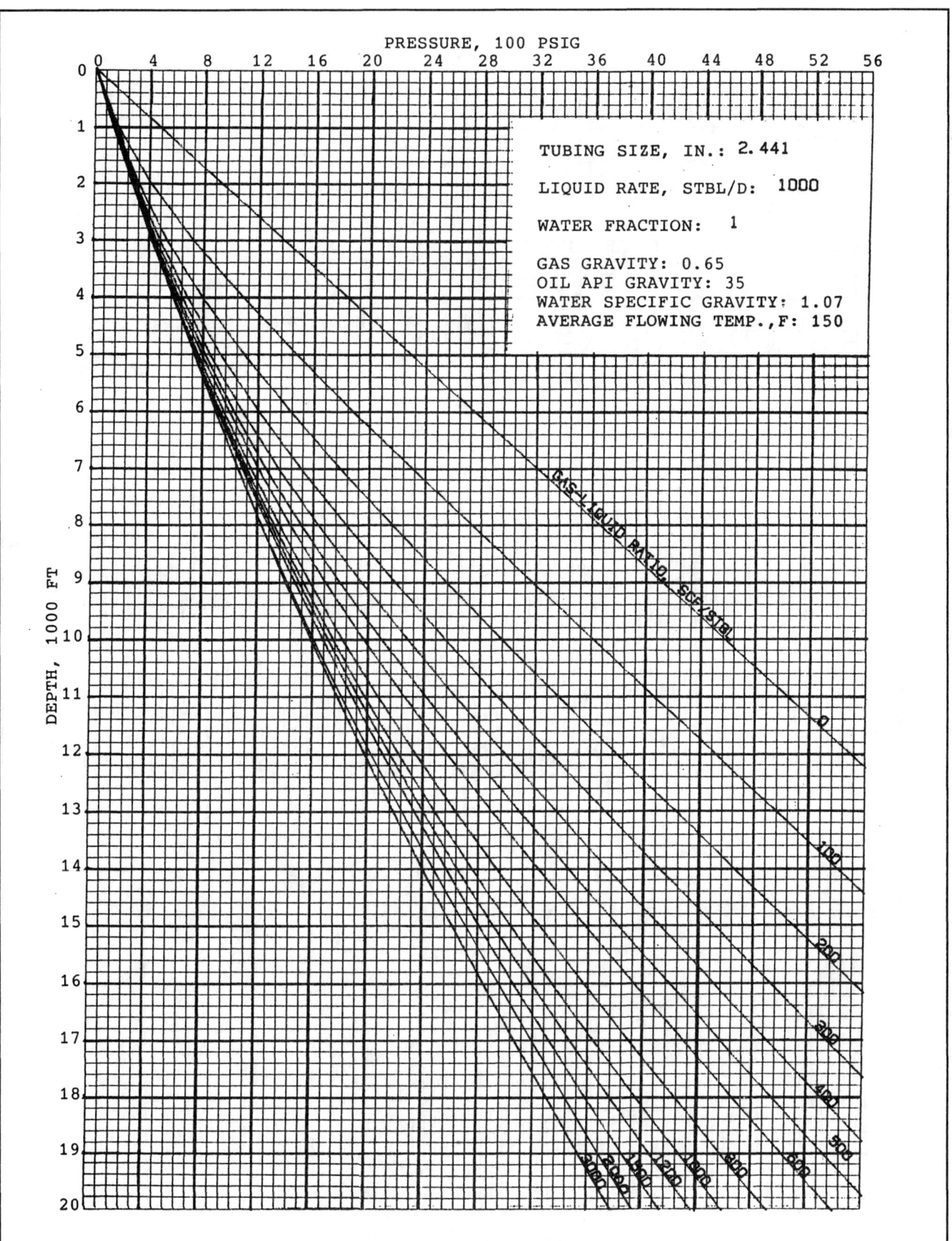

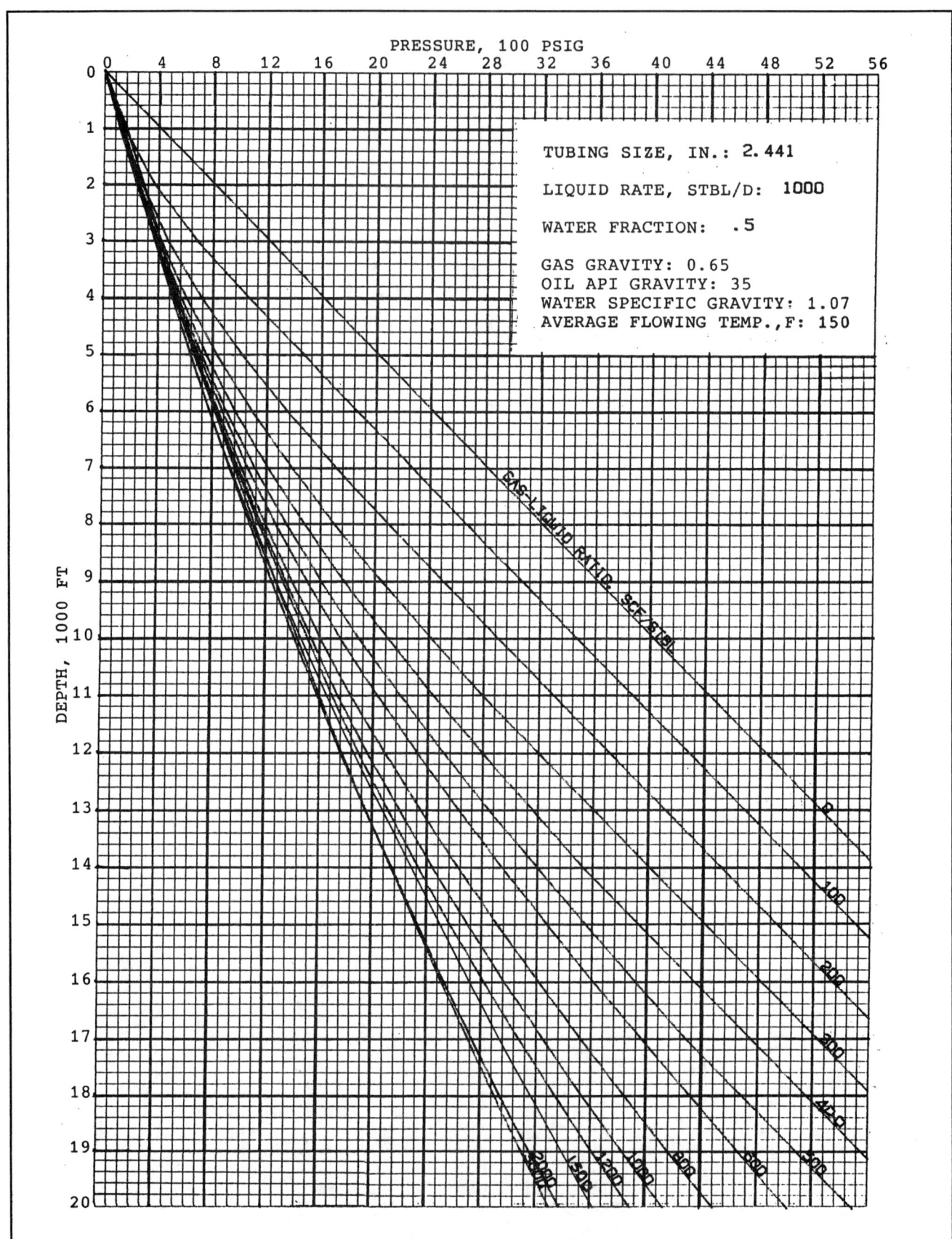

PRESSURE, 100 PSIG

DEPTH, 1000 FT

TUBING SIZE, IN.: **2.441**

LIQUID RATE, STBL/D: **1000**

WATER FRACTION: **.5**

GAS GRAVITY: 0.65
OIL API GRAVITY: 35
WATER SPECIFIC GRAVITY: 1.07
AVERAGE FLOWING TEMP.,F: 150

GAS-LIQUID RATIO, SCF/STBL

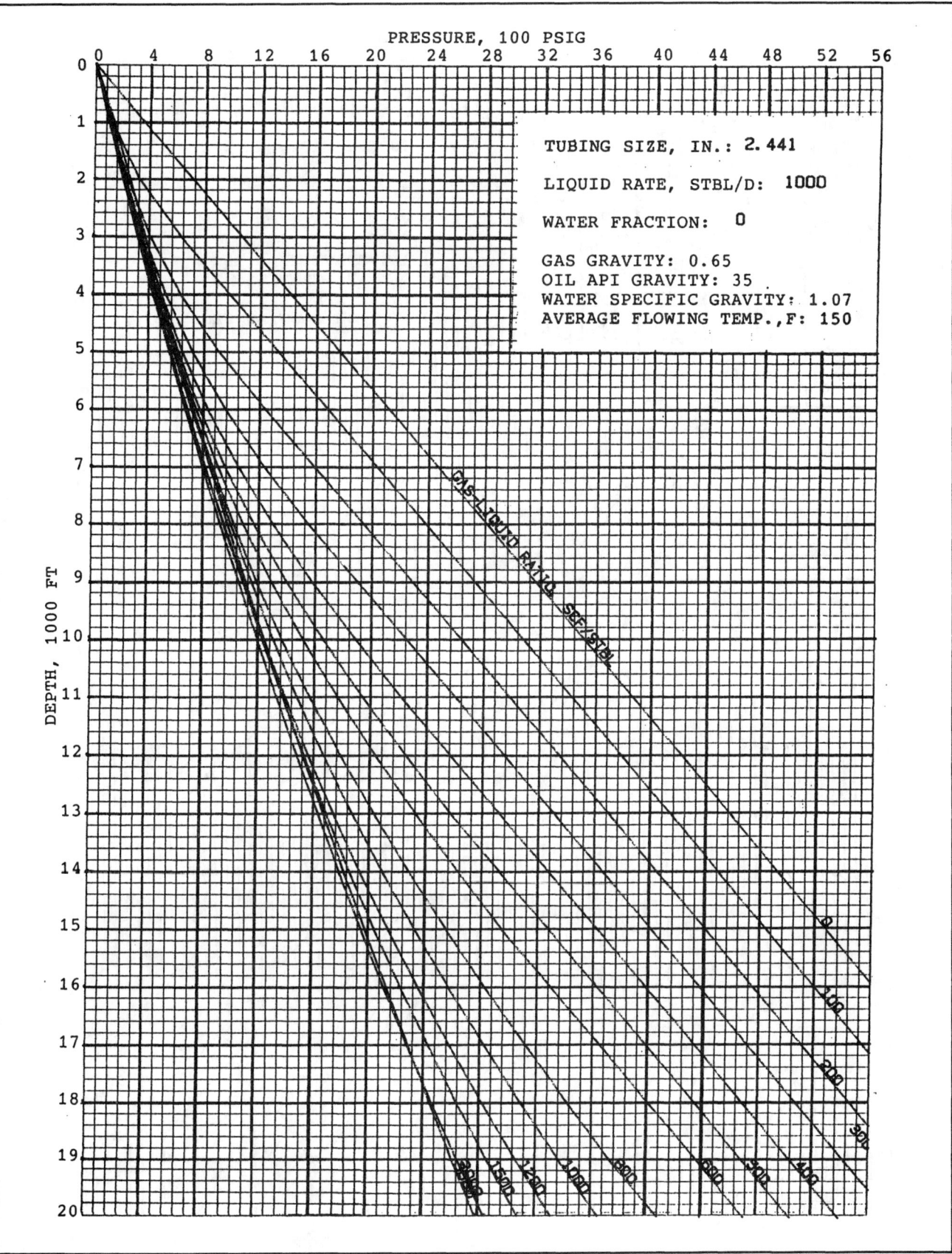

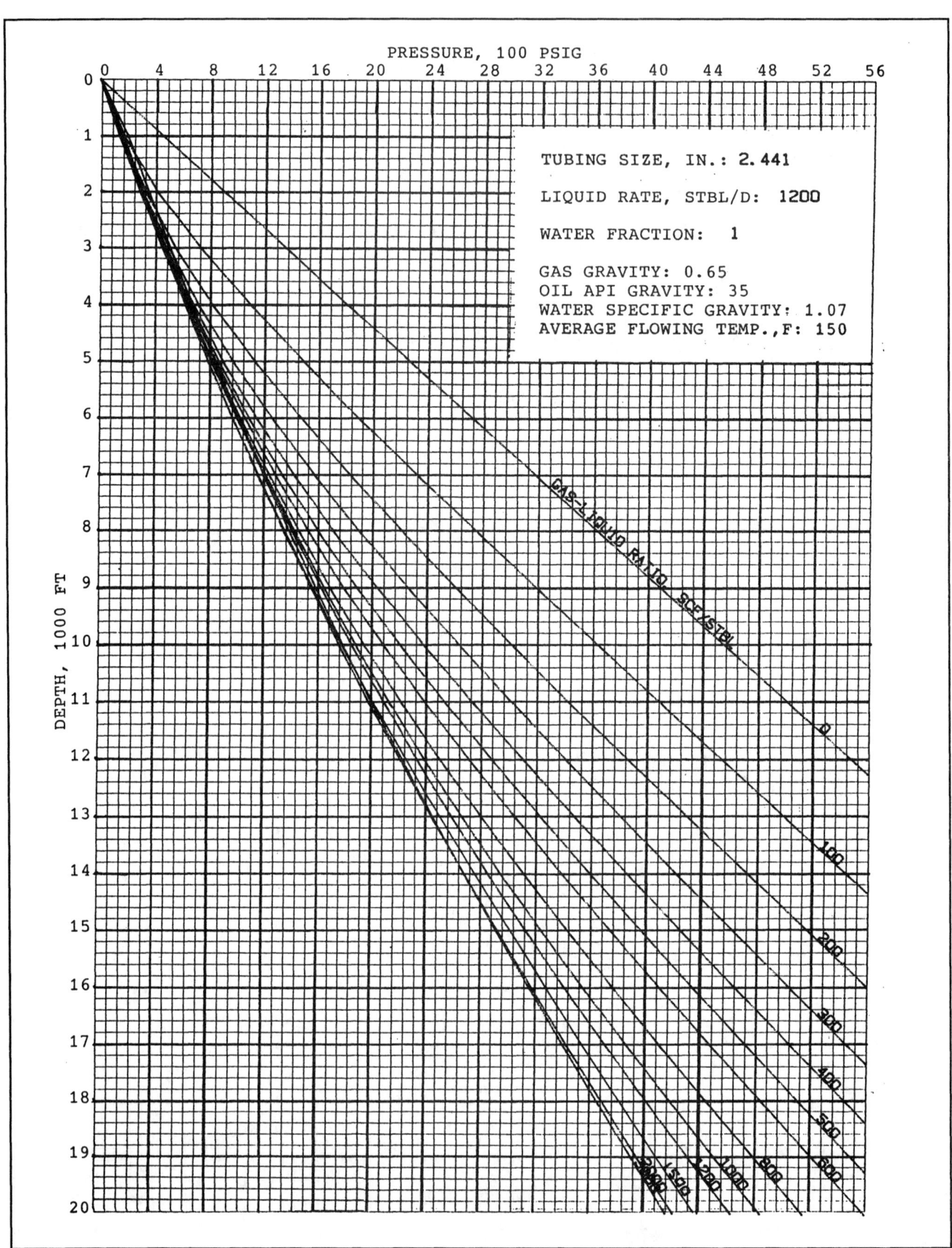

PRESSURE, 100 PSIG

TUBING SIZE, IN.: **2.441**

LIQUID RATE, STBL/D: **1200**

WATER FRACTION: **1**

GAS GRAVITY: 0.65
OIL API GRAVITY: 35
WATER SPECIFIC GRAVITY: 1.07
AVERAGE FLOWING TEMP.,F: 150

DEPTH, 1000 FT

GAS-LIQUID RATIO, SCF/STBL

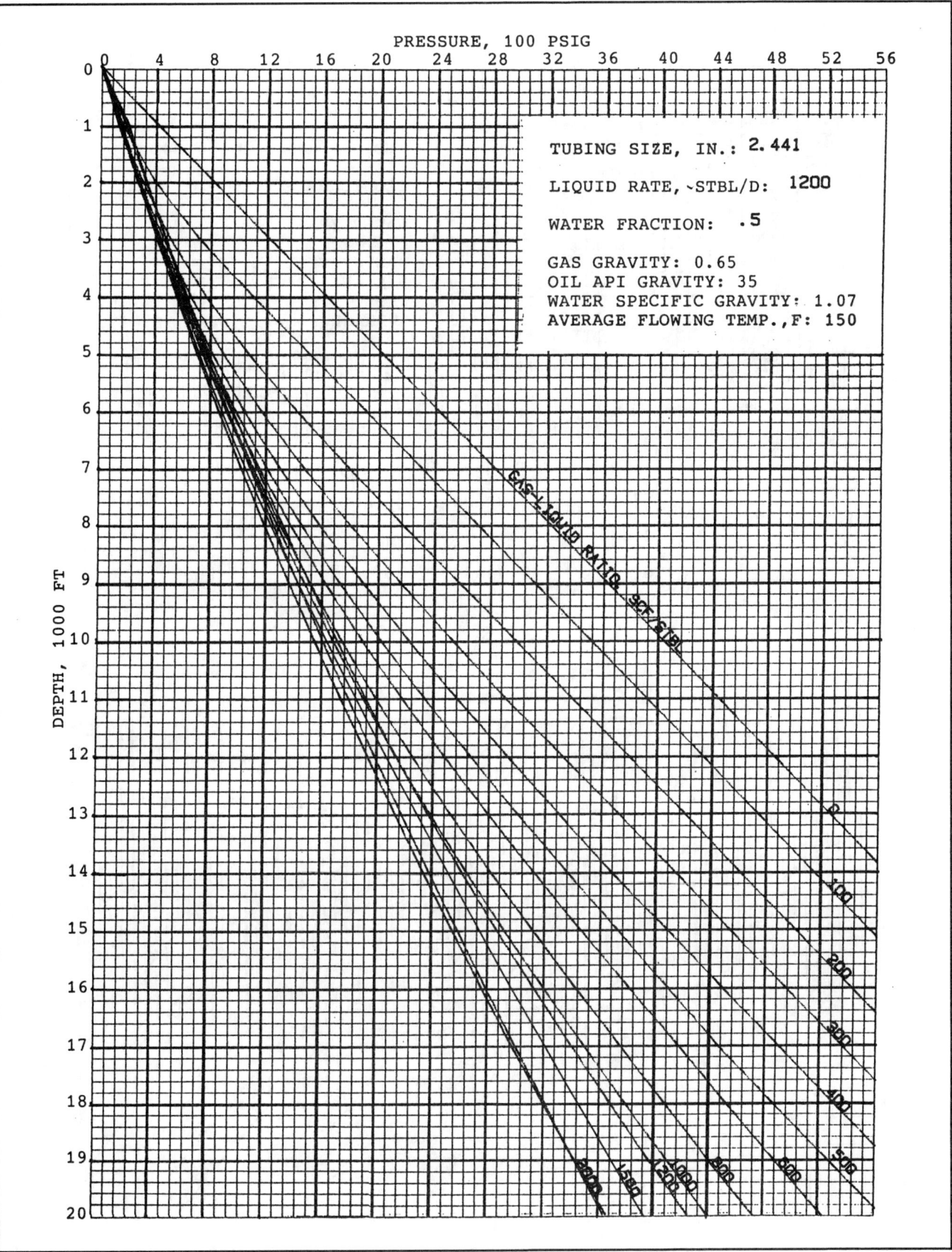

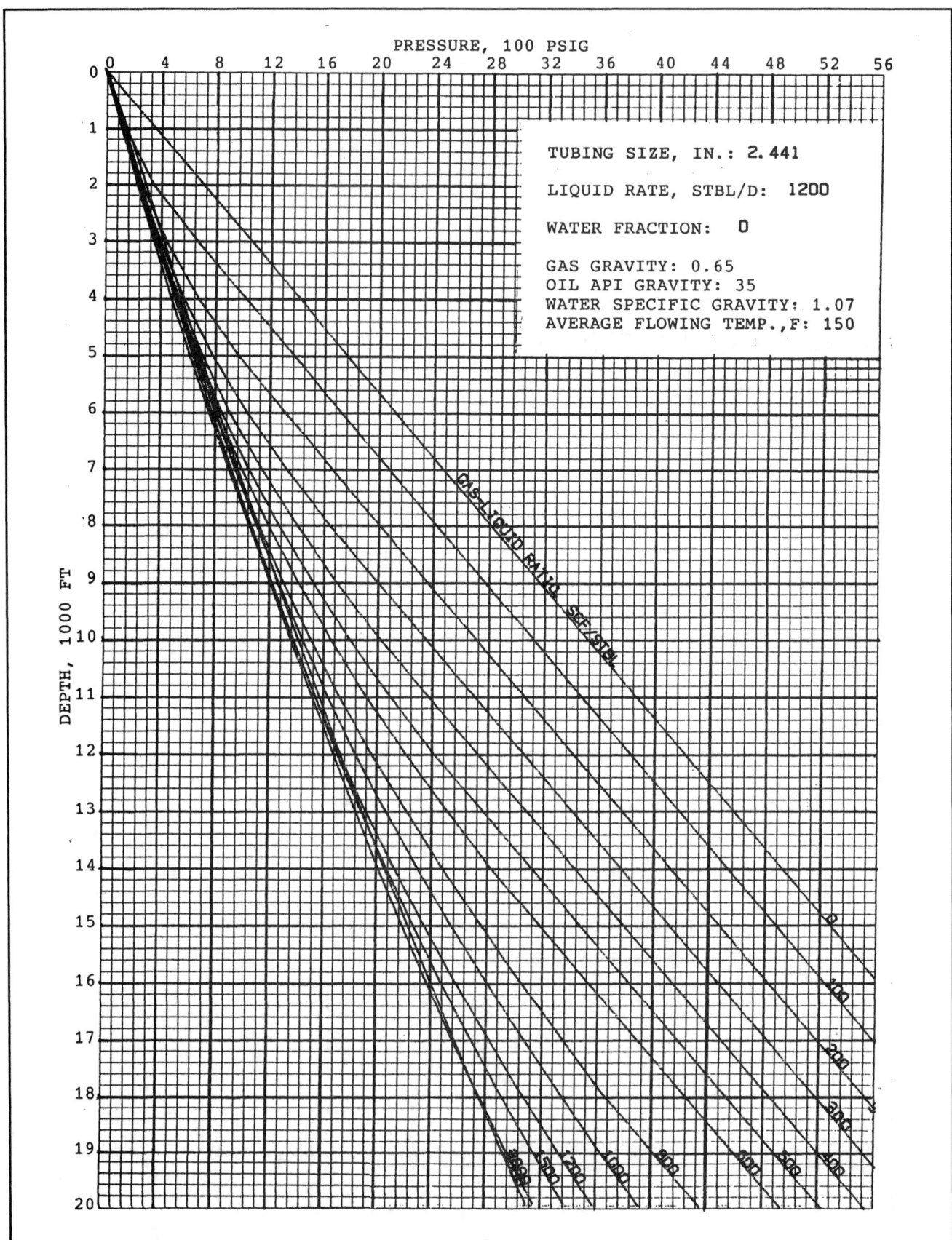

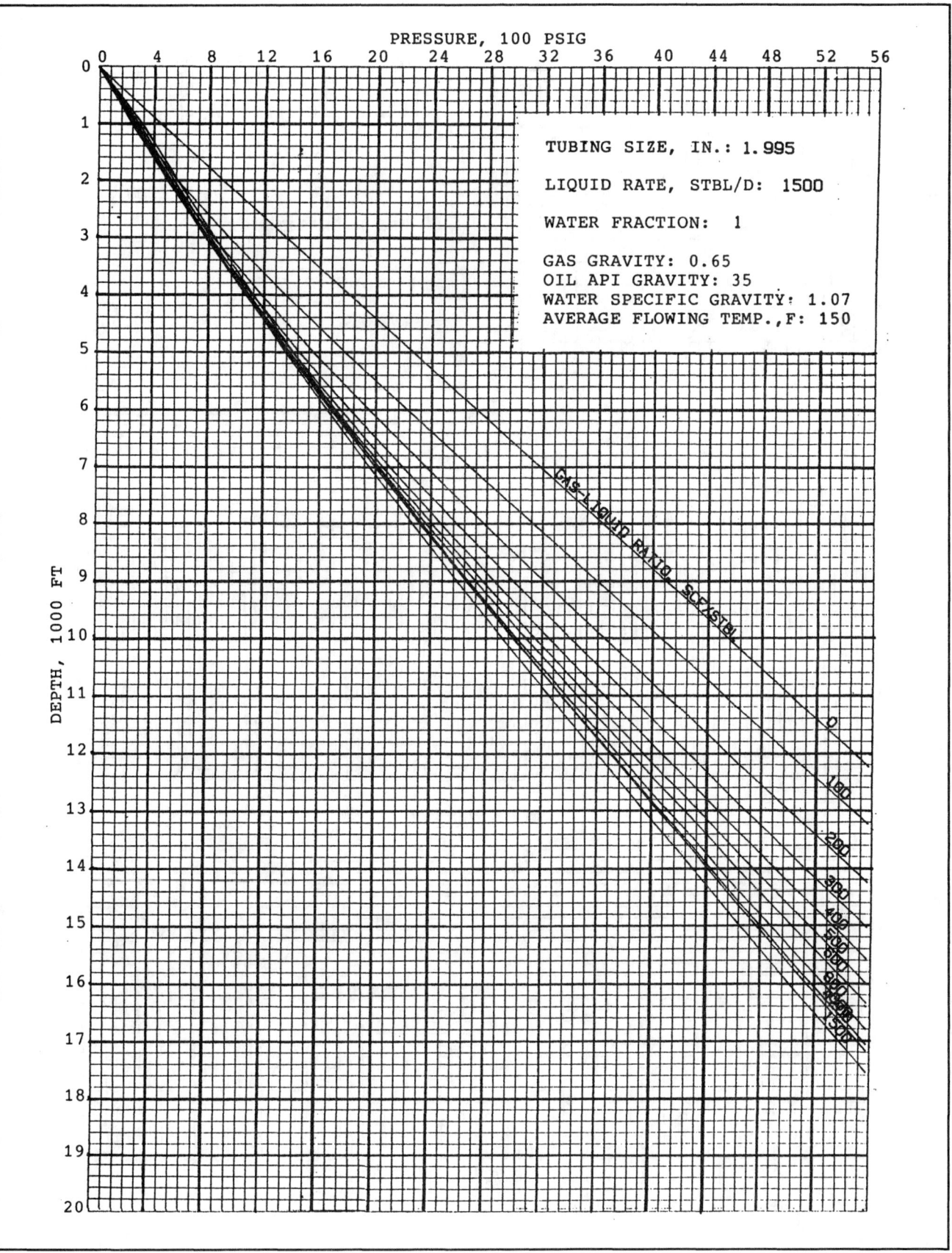

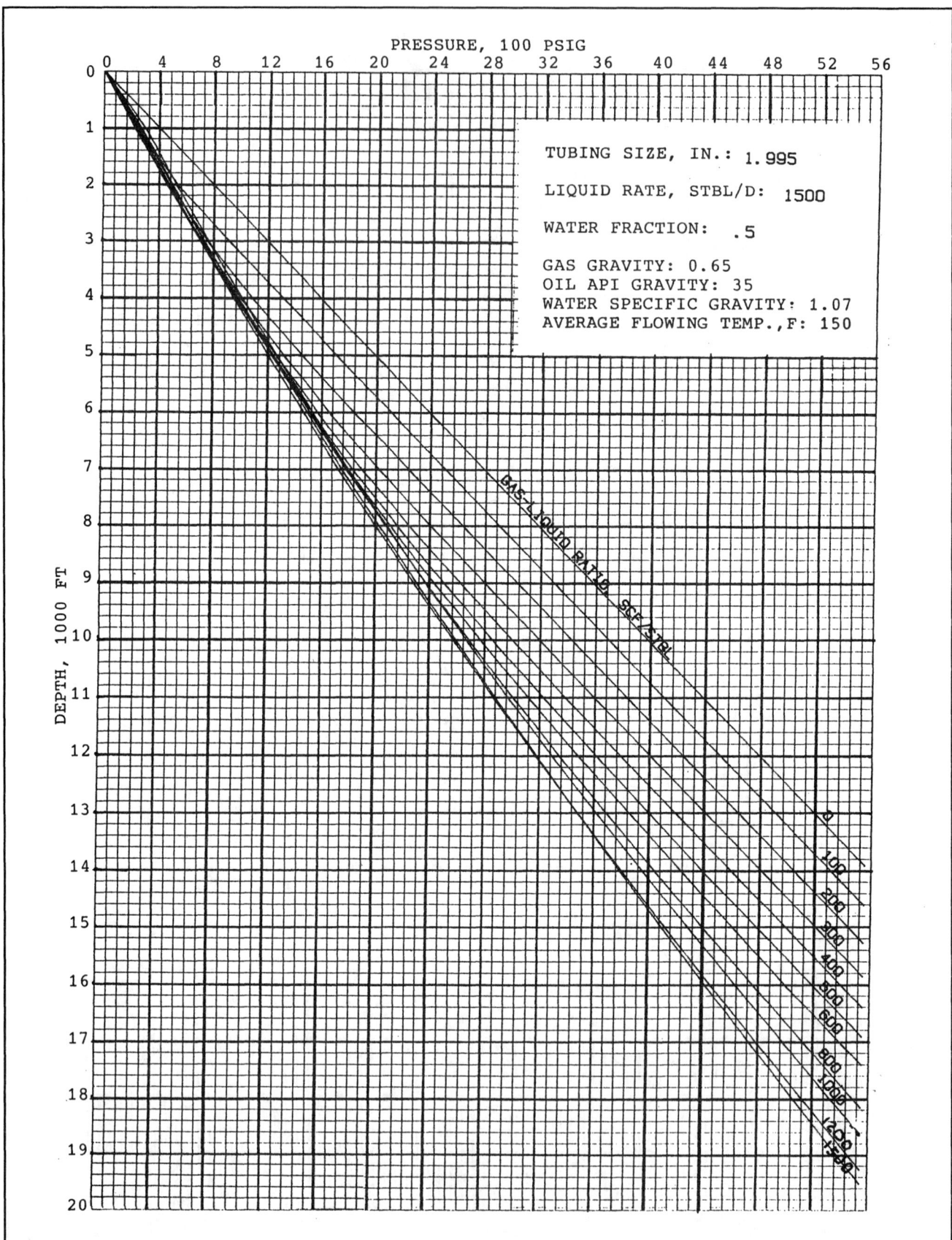

PRESSURE, 100 PSIG

DEPTH, 1000 FT

TUBING SIZE, IN.: 1.995

LIQUID RATE, STBL/D: 1500

WATER FRACTION: .5

GAS GRAVITY: 0.65
OIL API GRAVITY: 35
WATER SPECIFIC GRAVITY: 1.07
AVERAGE FLOWING TEMP.,F: 150

GAS-LIQUID RATIO, SCF/STBL

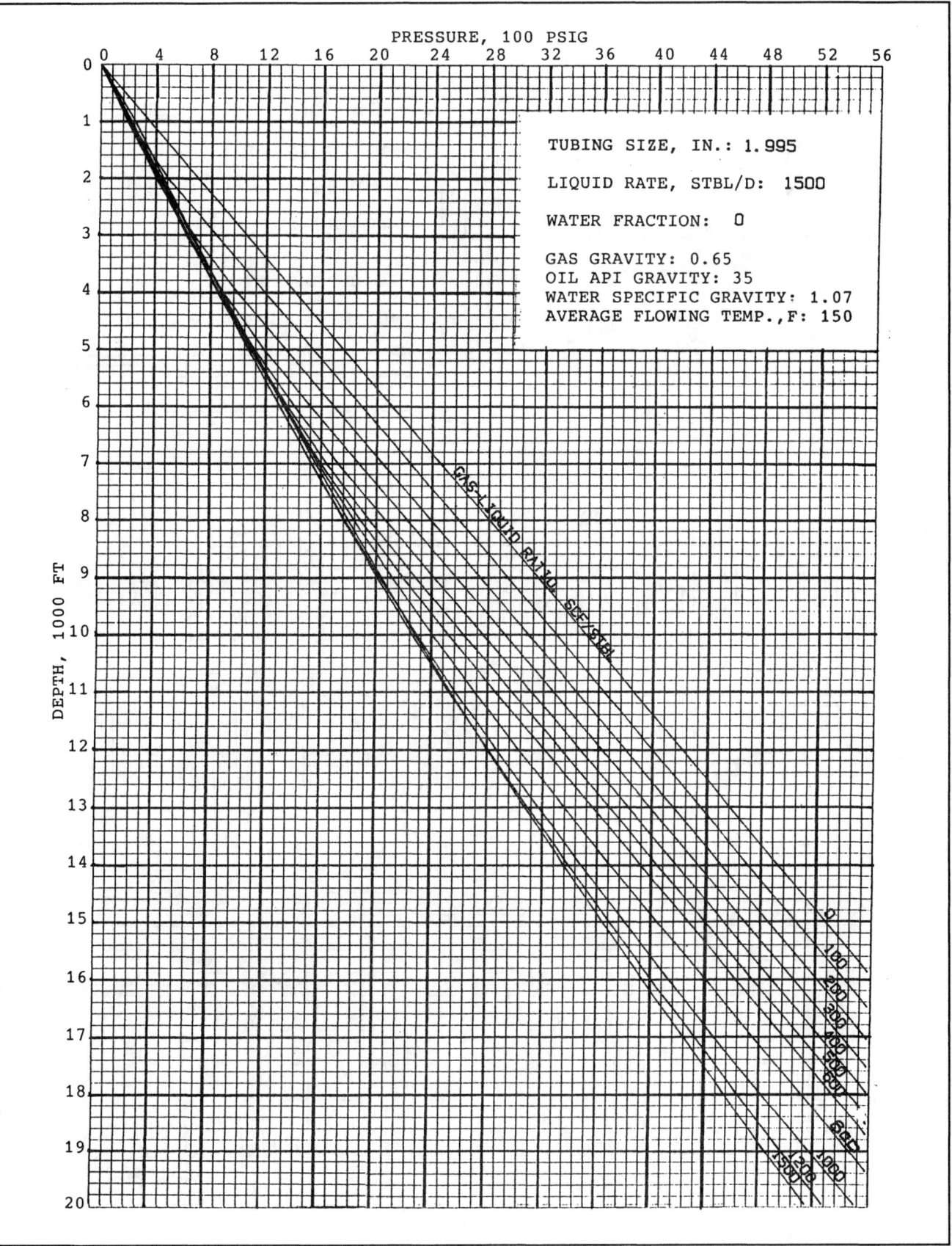

TUBING SIZE, IN.: 1.995

LIQUID RATE, STBL/D: 1500

WATER FRACTION: 0

GAS GRAVITY: 0.65
OIL API GRAVITY: 35
WATER SPECIFIC GRAVITY: 1.07
AVERAGE FLOWING TEMP.,F: 150

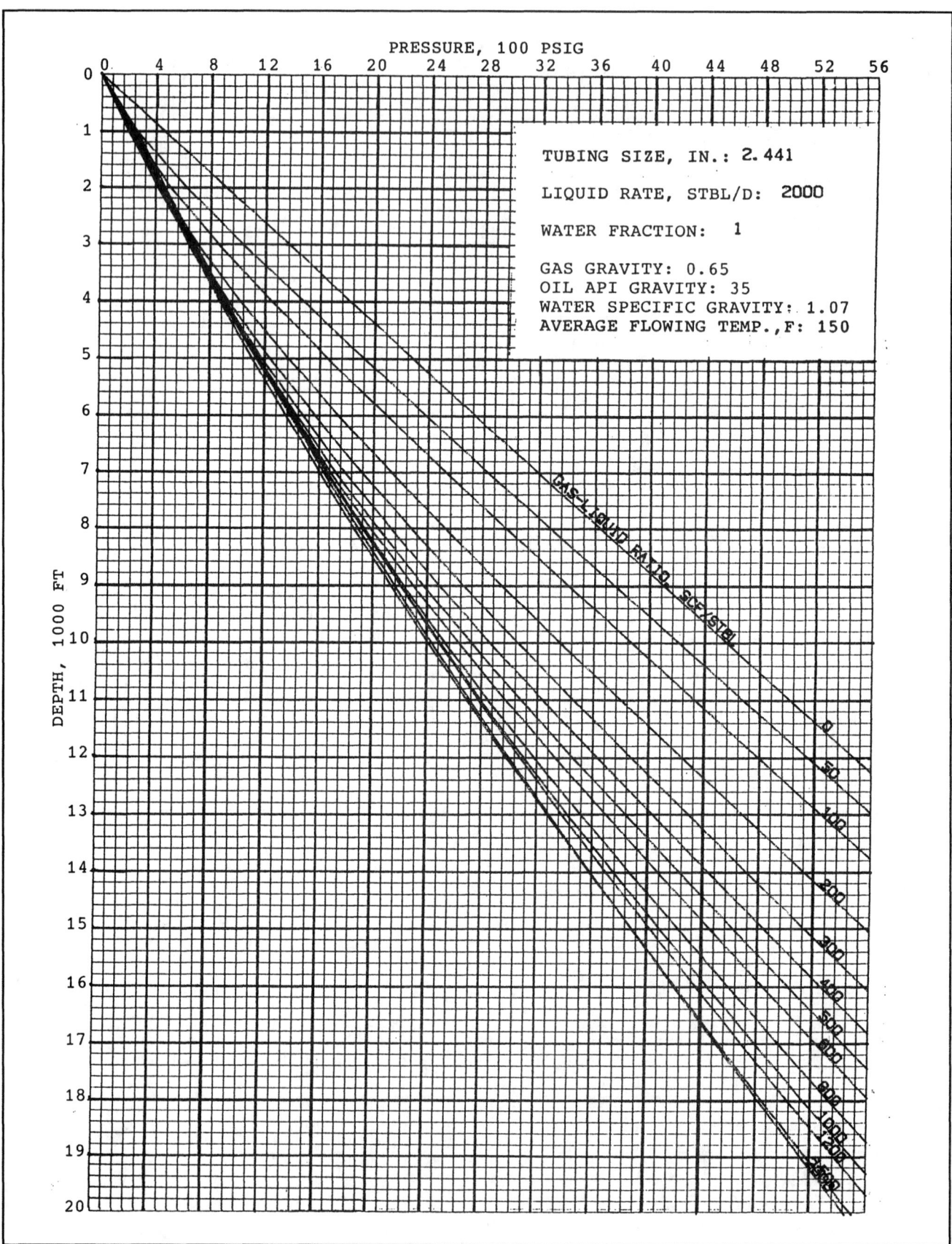

PRESSURE, 100 PSIG

DEPTH, 1000 FT

TUBING SIZE, IN.: 2.441

LIQUID RATE, STBL/D: 2000

WATER FRACTION: 1

GAS GRAVITY: 0.65
OIL API GRAVITY: 35
WATER SPECIFIC GRAVITY: 1.07
AVERAGE FLOWING TEMP.,F: 150

GAS-LIQUID RATIO, SCF/STBL

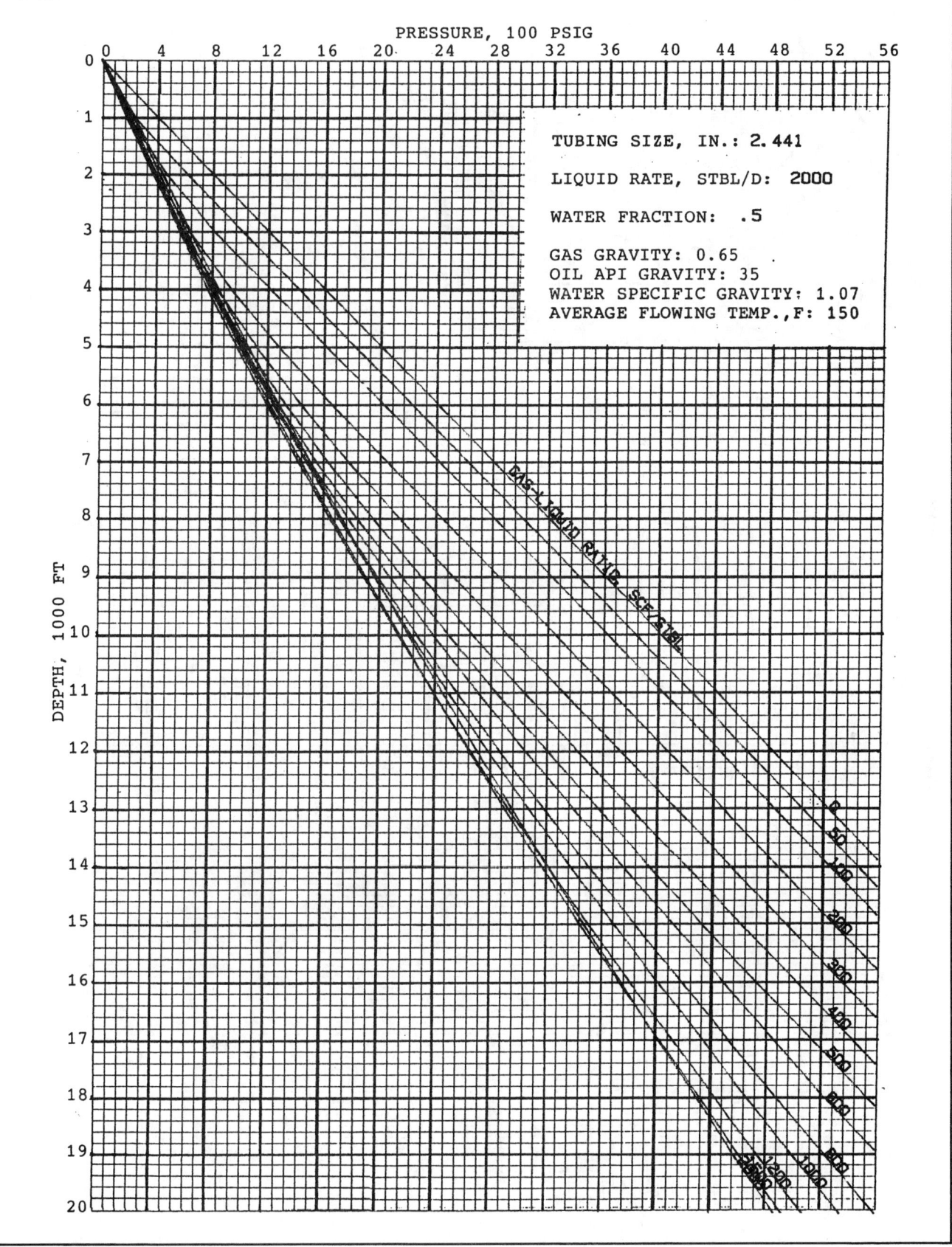

TUBING SIZE, IN.: 2.441

LIQUID RATE, STBL/D: 2000

WATER FRACTION: .5

GAS GRAVITY: 0.65
OIL API GRAVITY: 35
WATER SPECIFIC GRAVITY: 1.07
AVERAGE FLOWING TEMP.,F: 150

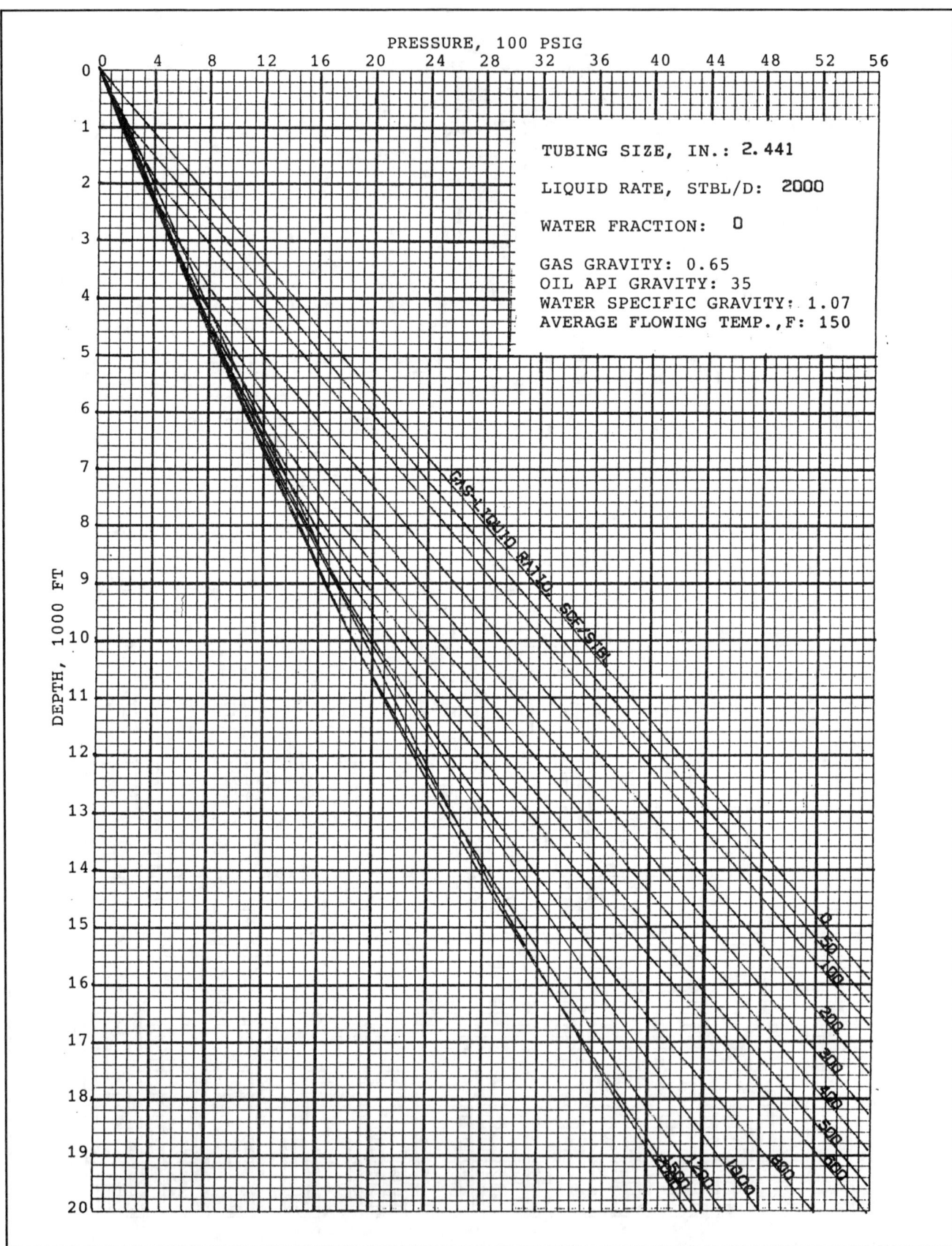

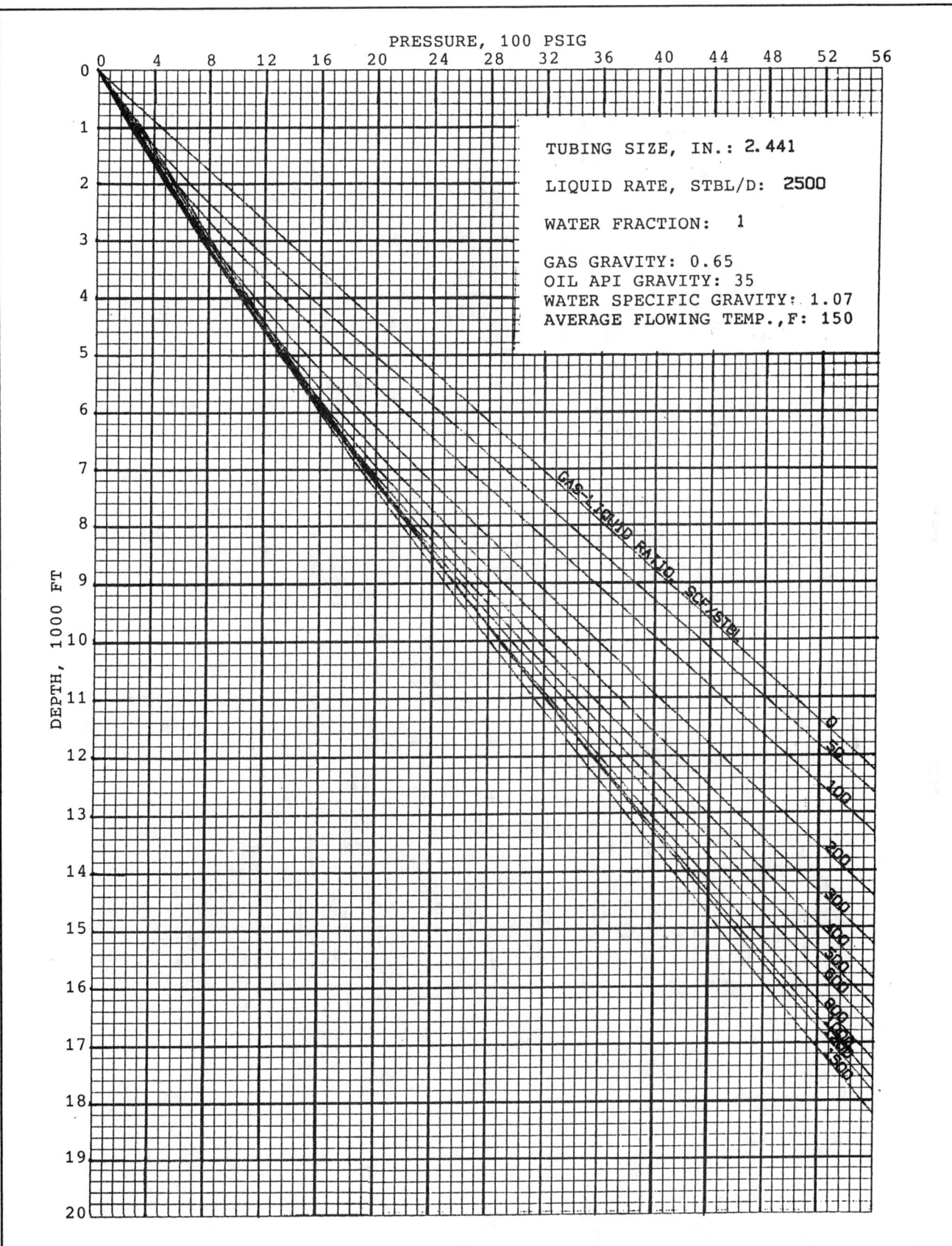

PRESSURE, 100 PSIG

DEPTH, 1000 FT

TUBING SIZE, IN.: 2.441

LIQUID RATE, STBL/D: 2500

WATER FRACTION: 1

GAS GRAVITY: 0.65
OIL API GRAVITY: 35
WATER SPECIFIC GRAVITY: 1.07
AVERAGE FLOWING TEMP.,F: 150

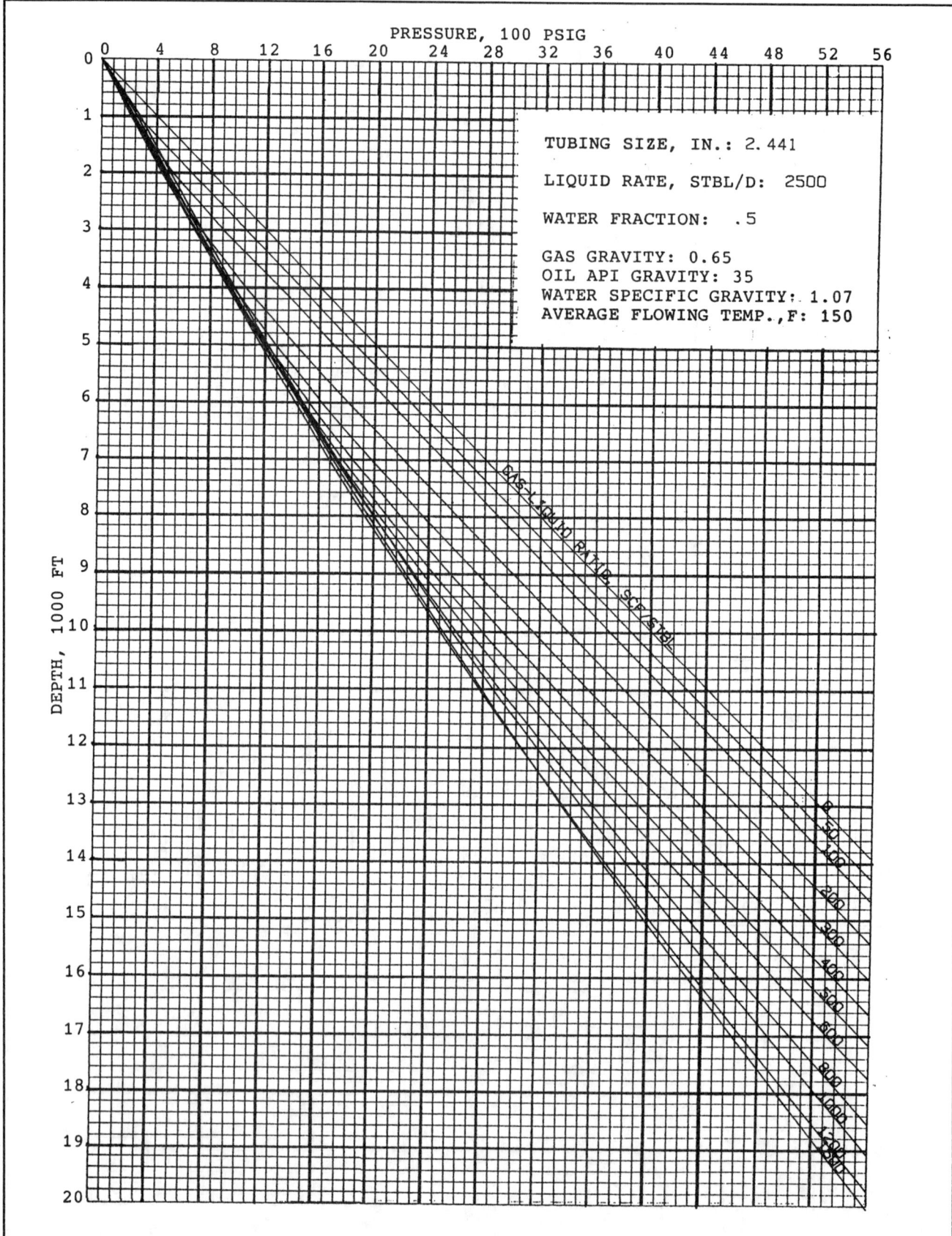

PRESSURE, 100 PSIG

DEPTH, 1000 FT

TUBING SIZE, IN.: 2.441

LIQUID RATE, STBL/D: 2500

WATER FRACTION: .5

GAS GRAVITY: 0.65
OIL API GRAVITY: 35
WATER SPECIFIC GRAVITY: 1.07
AVERAGE FLOWING TEMP.,F: 150

GAS-LIQUID RATIO, SCF/STBL

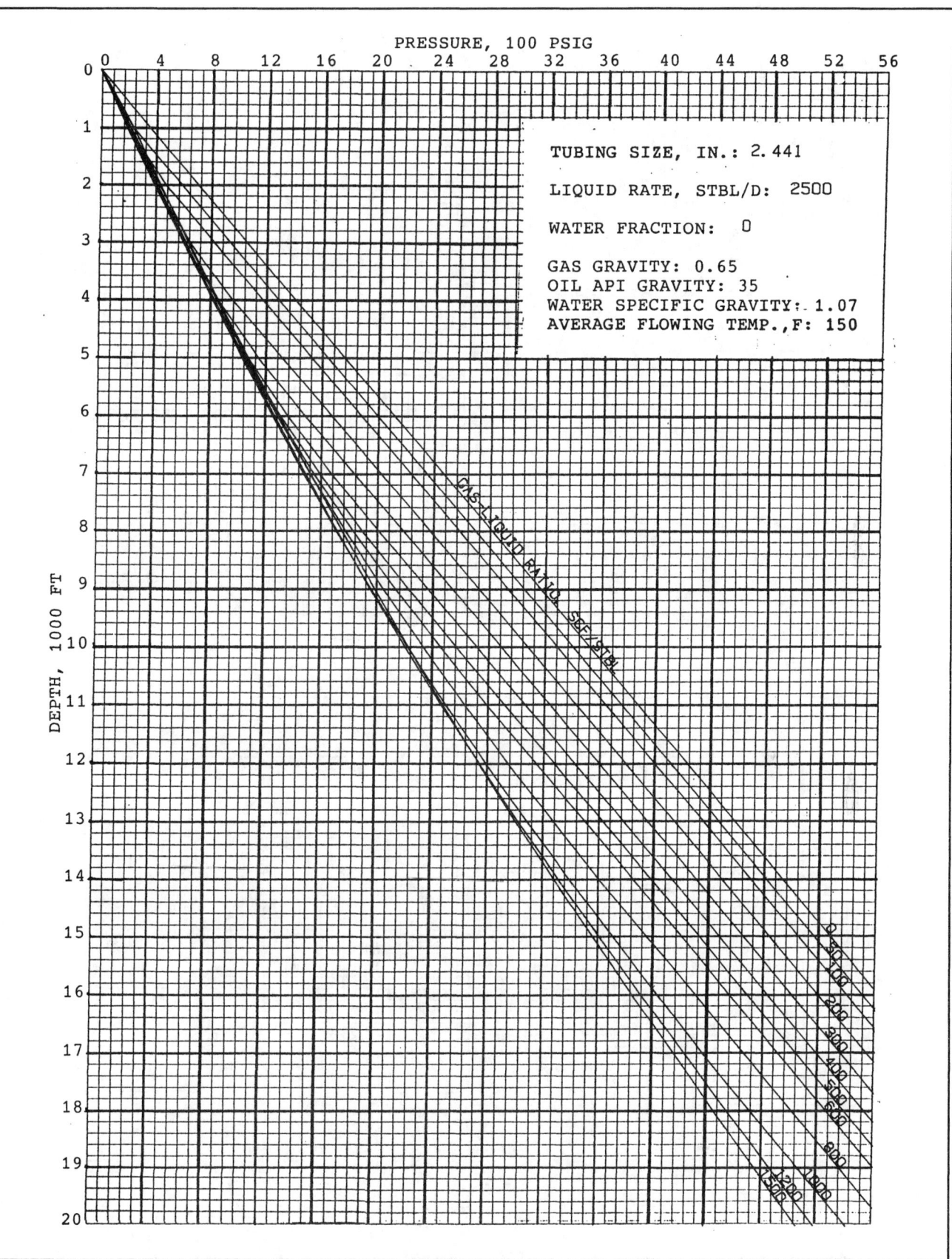

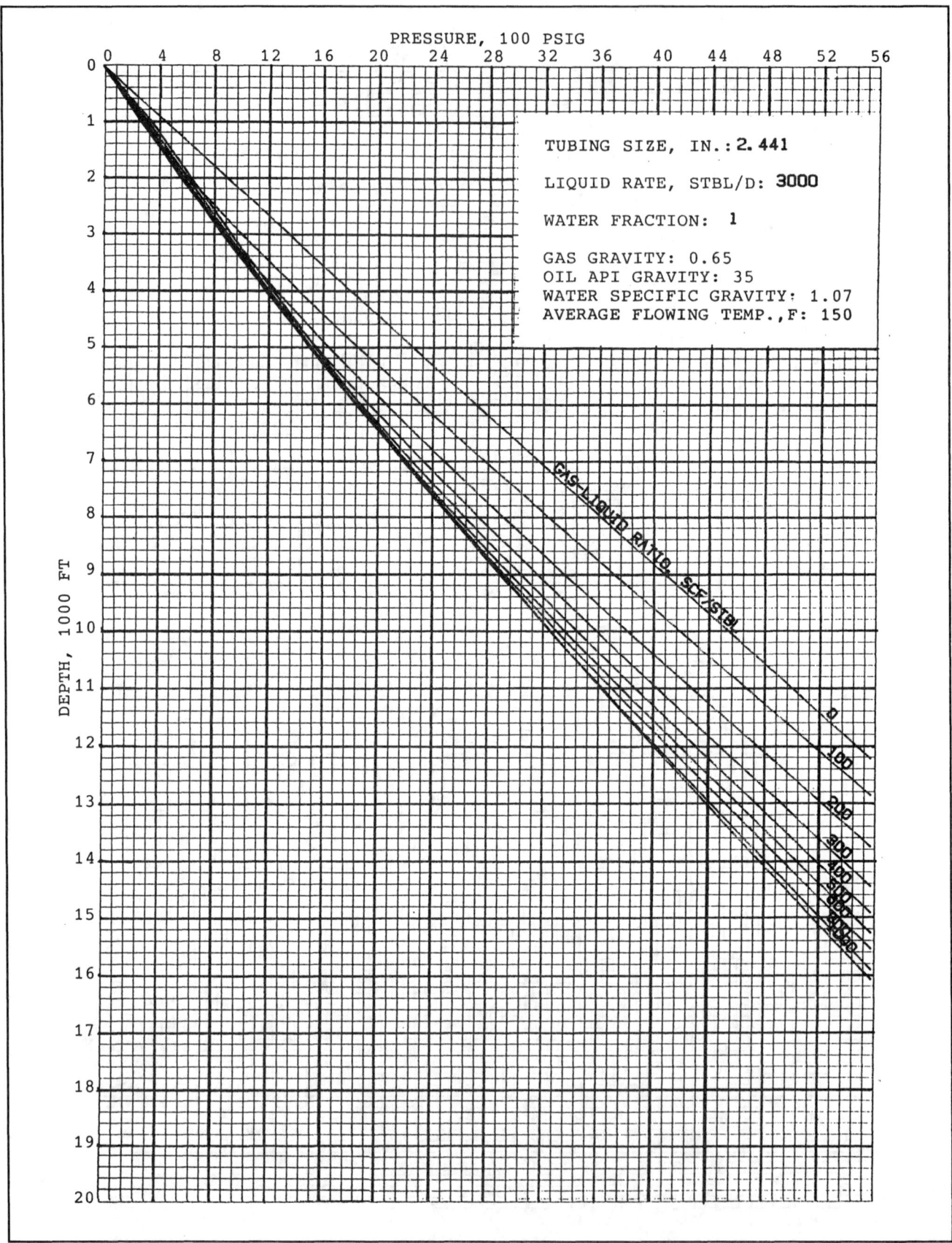

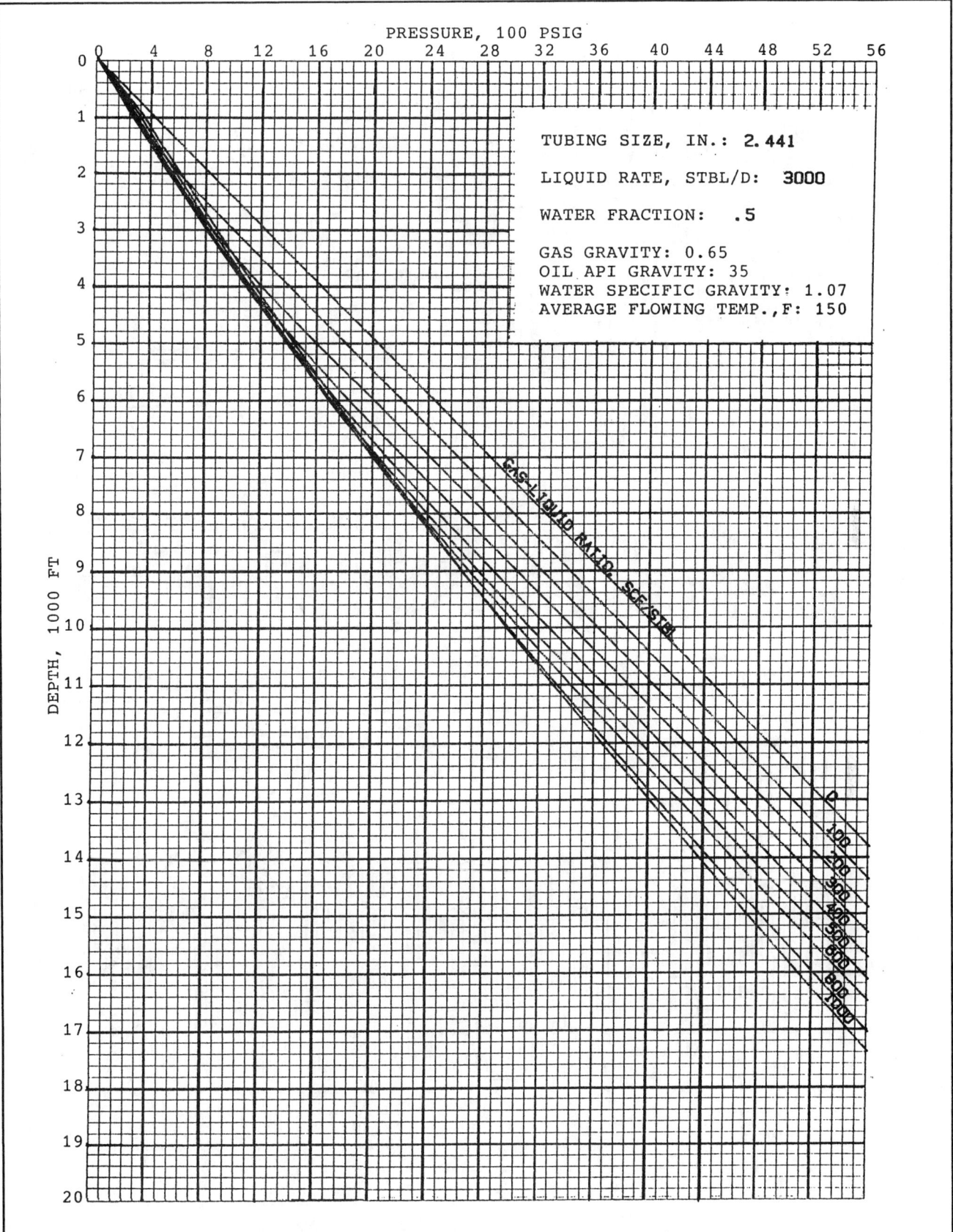

PRESSURE, 100 PSIG

DEPTH, 1000 FT

TUBING SIZE, IN.: **2.441**

LIQUID RATE, STBL/D: **3000**

WATER FRACTION: **.5**

GAS GRAVITY: 0.65
OIL API GRAVITY: 35
WATER SPECIFIC GRAVITY: 1.07
AVERAGE FLOWING TEMP.,F: 150

GAS-LIQUID RATIO, SCF/STB

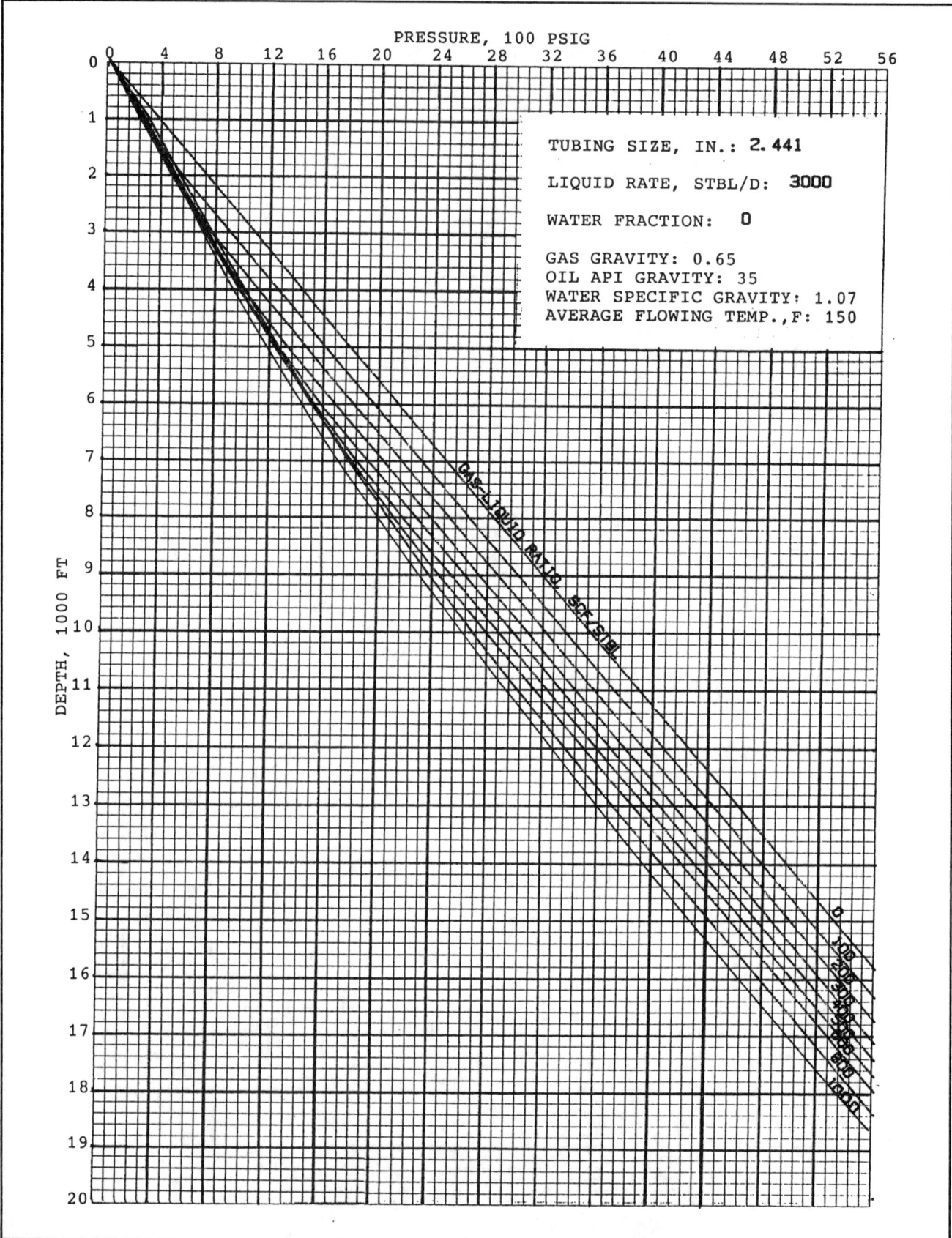

PRESSURE, 100 PSIG

DEPTH, 1000 FT

TUBING SIZE, IN.: **2.441**

LIQUID RATE, STBL/D: **3000**

WATER FRACTION: **0**

GAS GRAVITY: 0.65
OIL API GRAVITY: 35
WATER SPECIFIC GRAVITY: 1.07
AVERAGE FLOWING TEMP.,F: 150

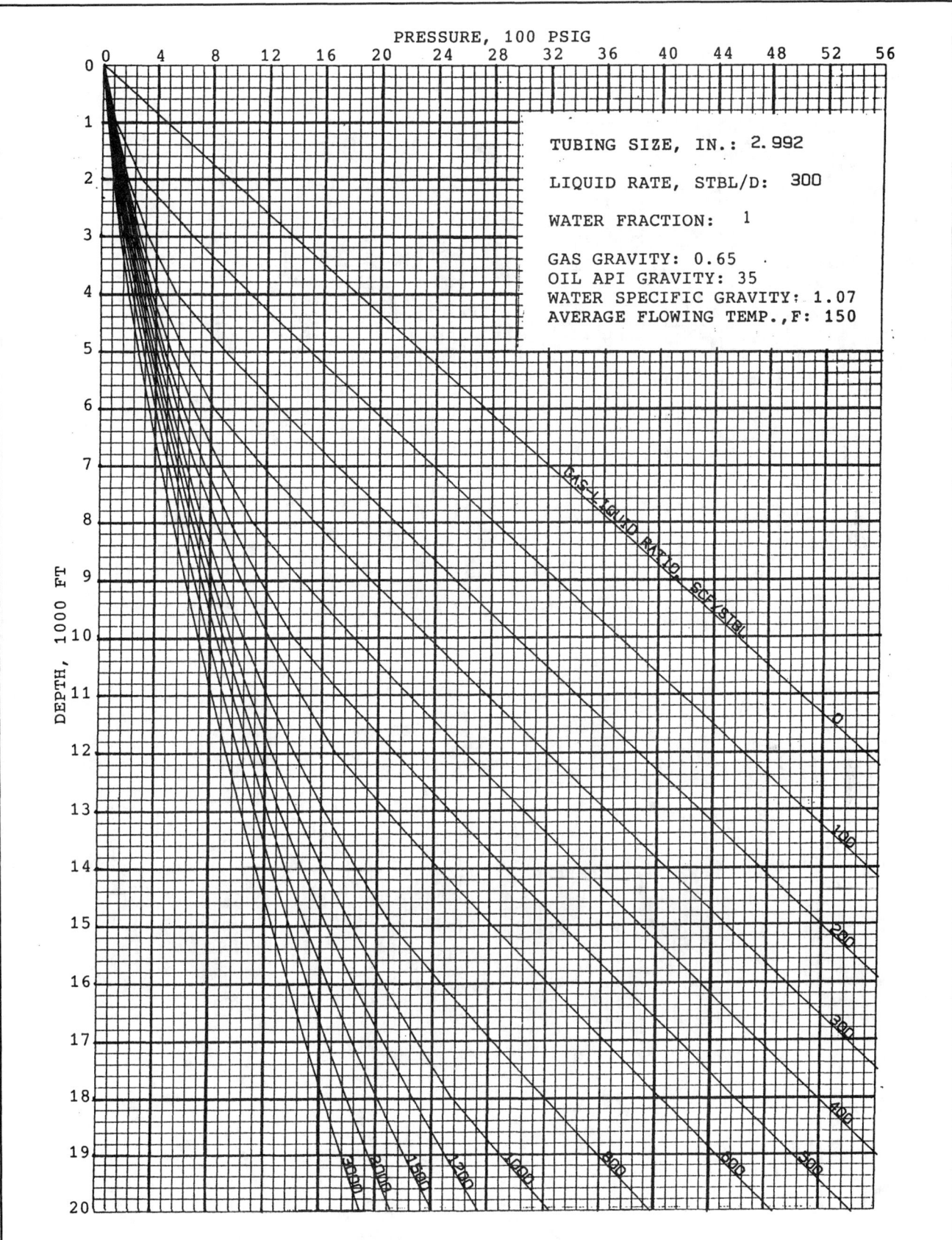

TUBING SIZE, IN.: 2.992

LIQUID RATE, STBL/D: 300

WATER FRACTION: 1

GAS GRAVITY: 0.65
OIL API GRAVITY: 35
WATER SPECIFIC GRAVITY: 1.07
AVERAGE FLOWING TEMP., F: 150

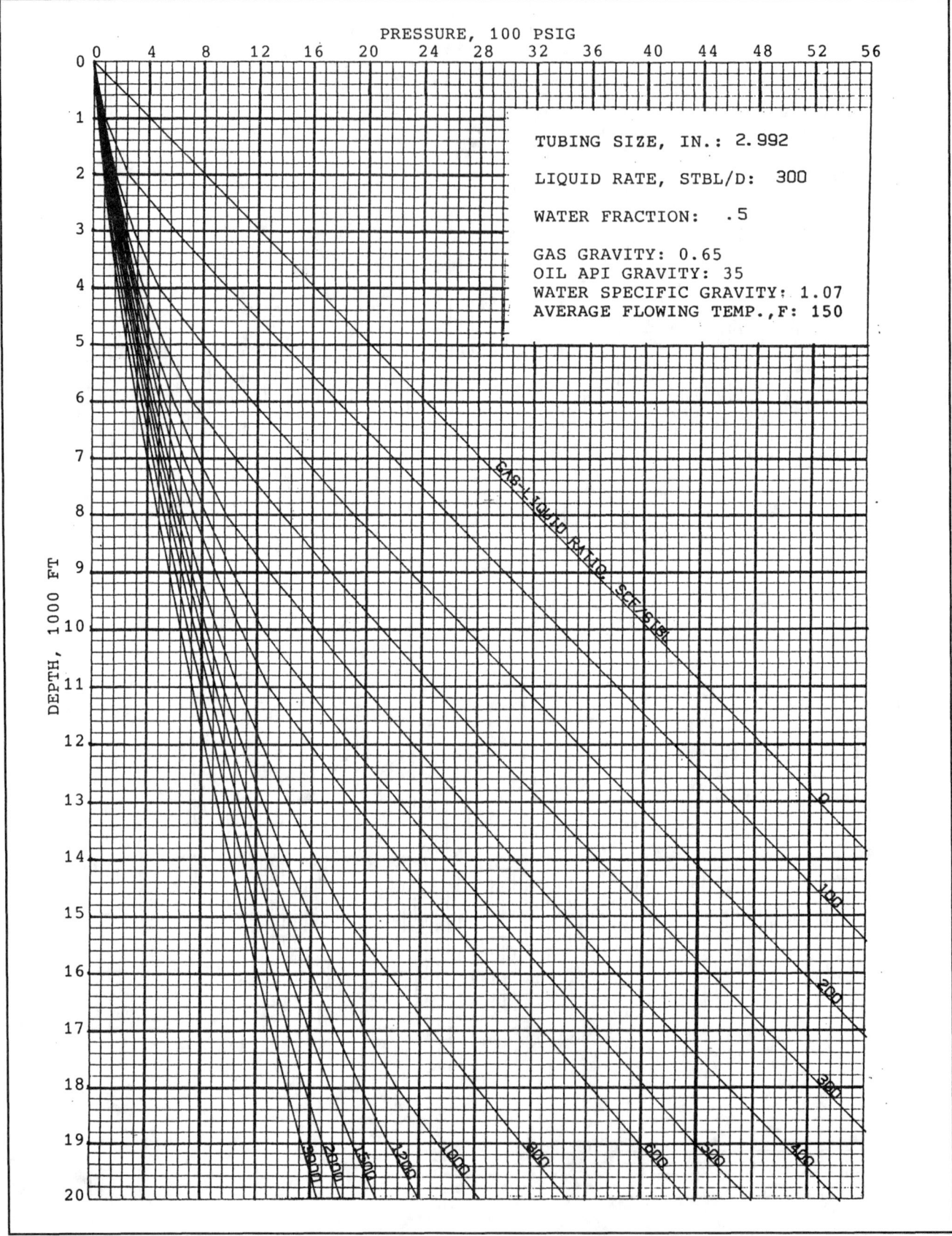

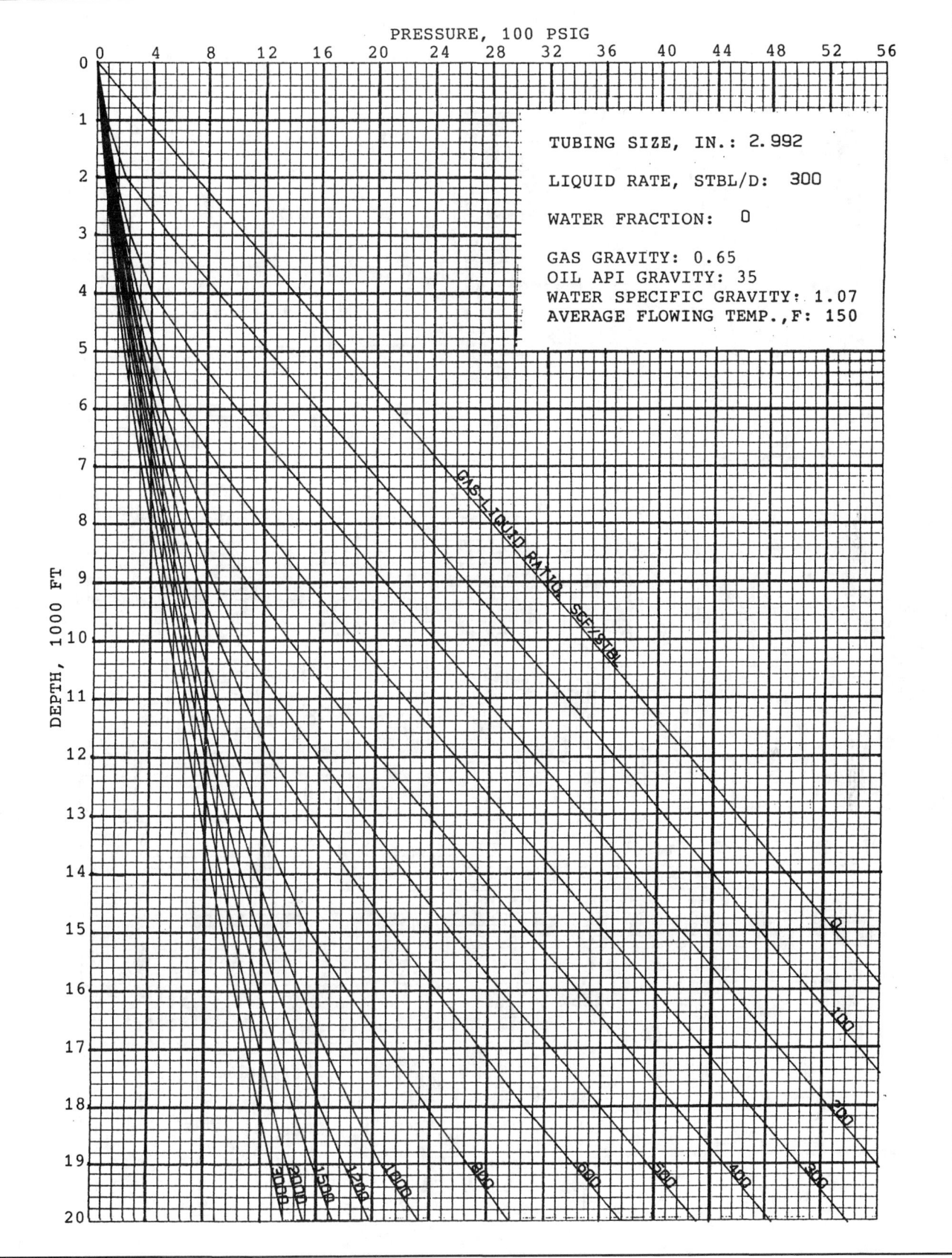

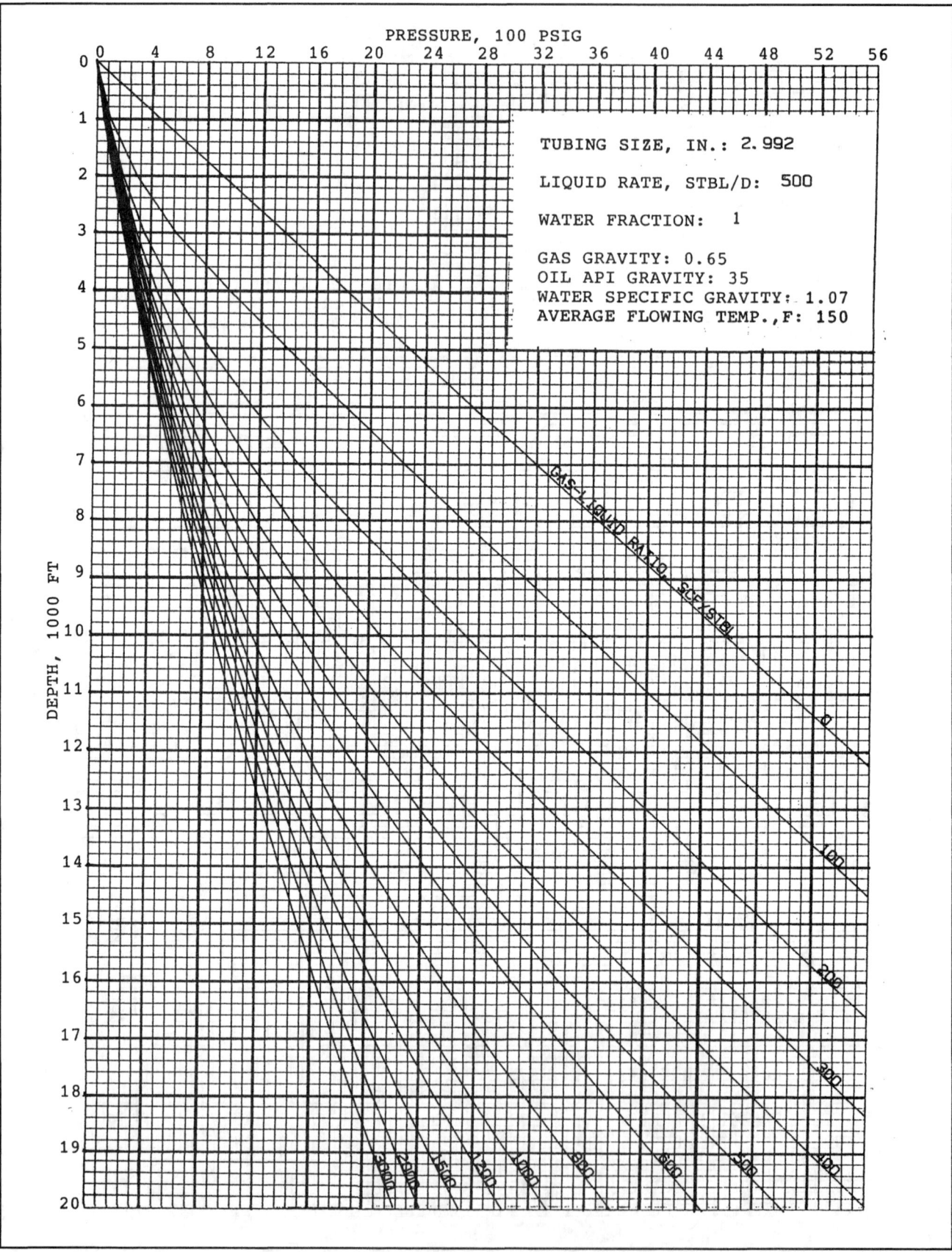

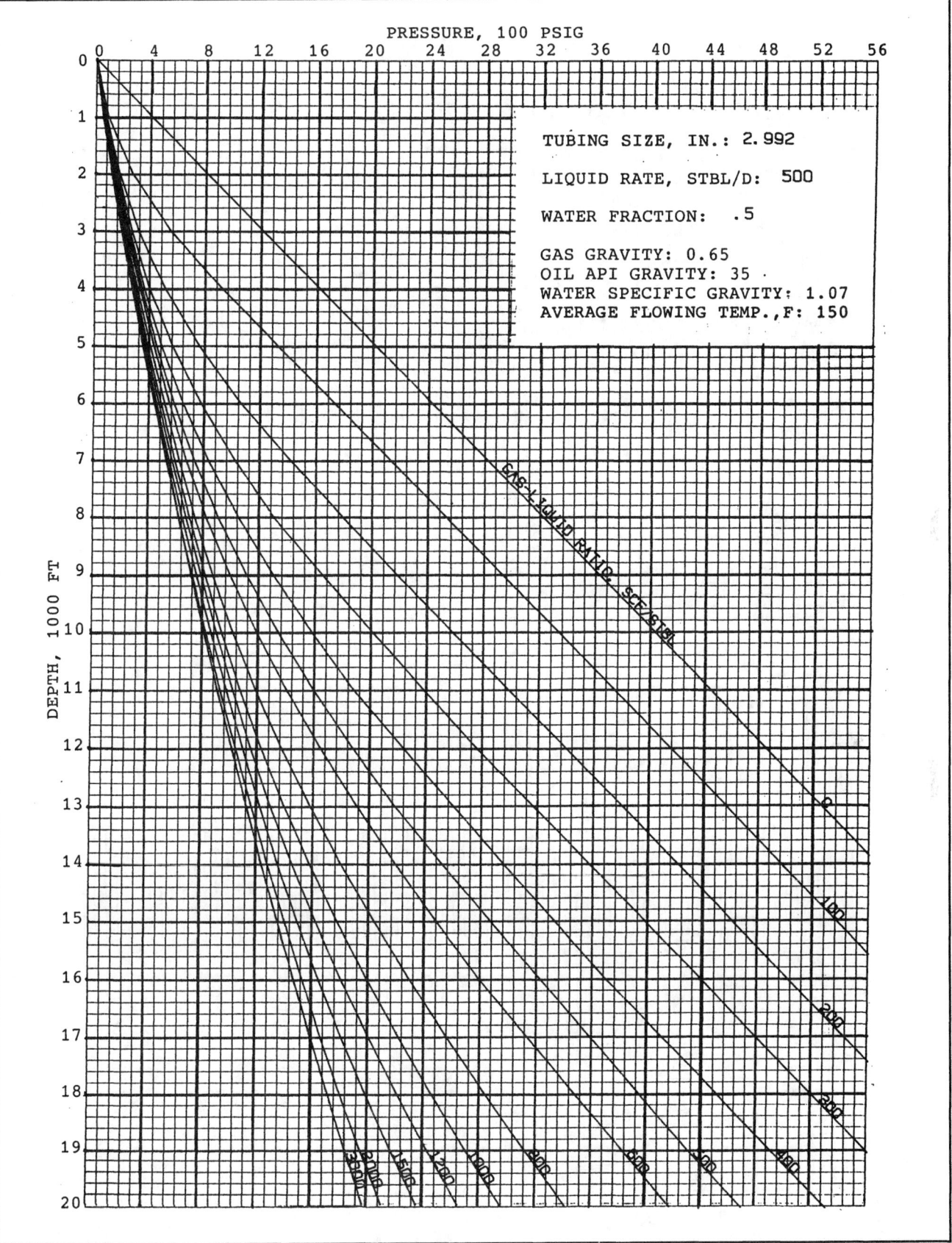

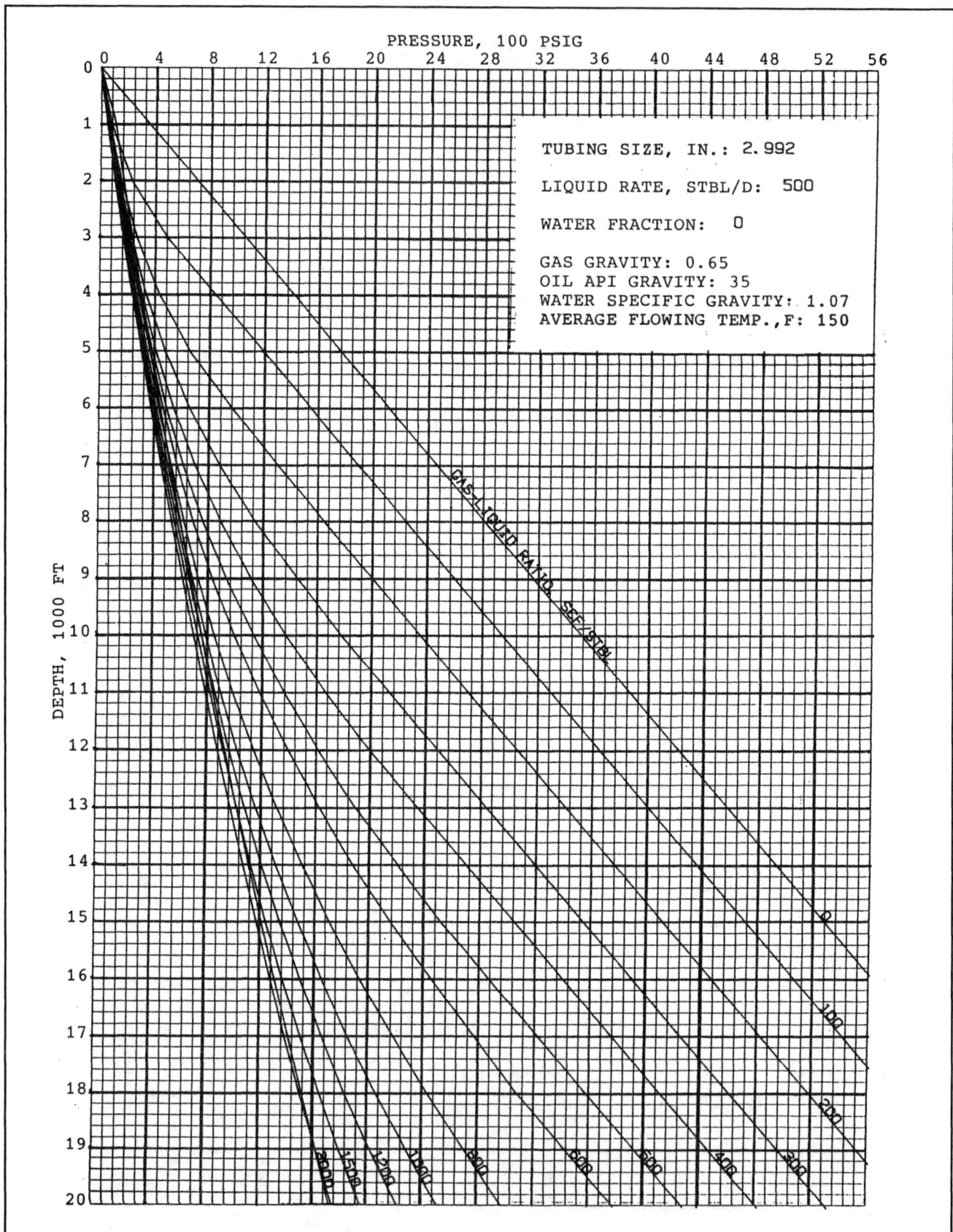

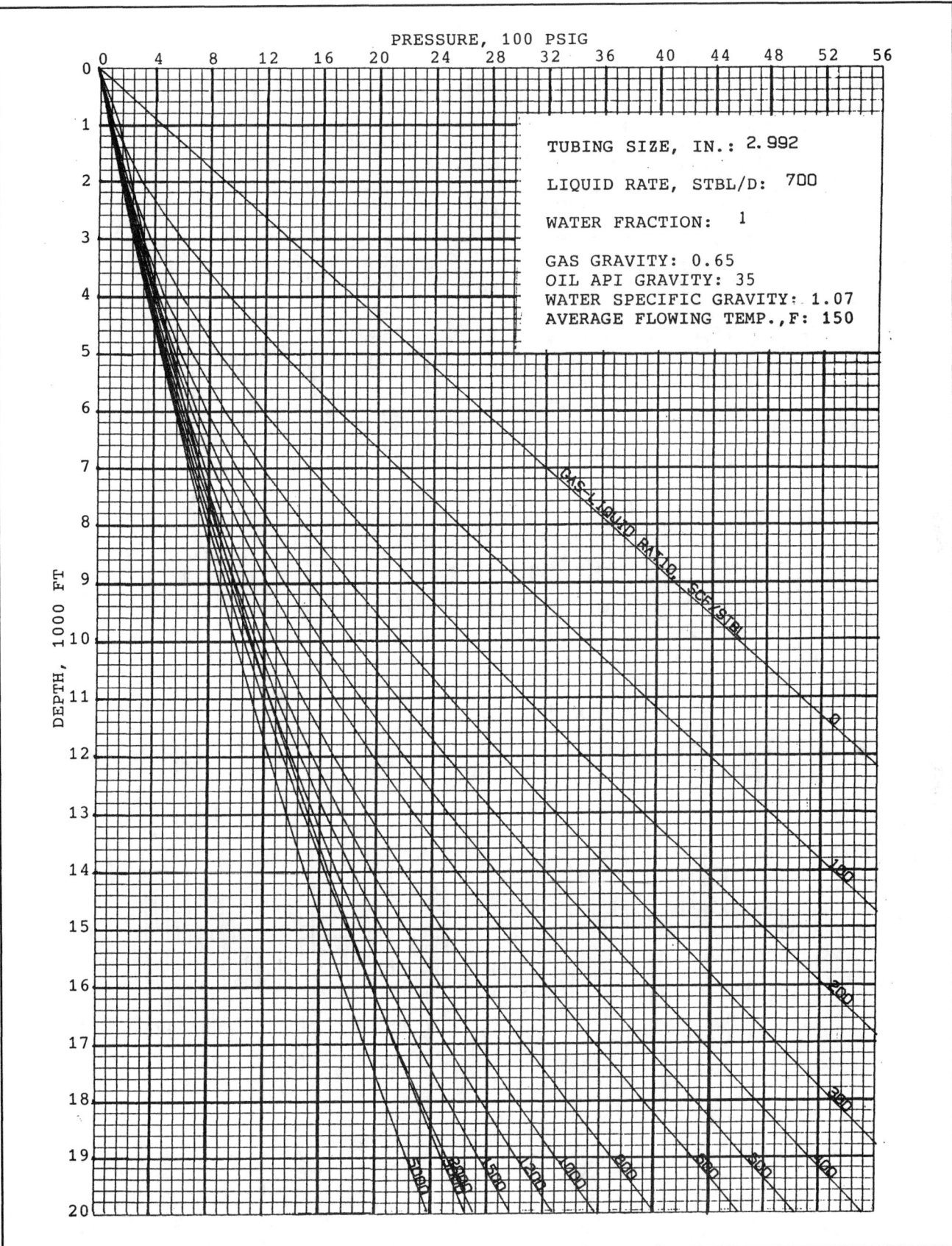

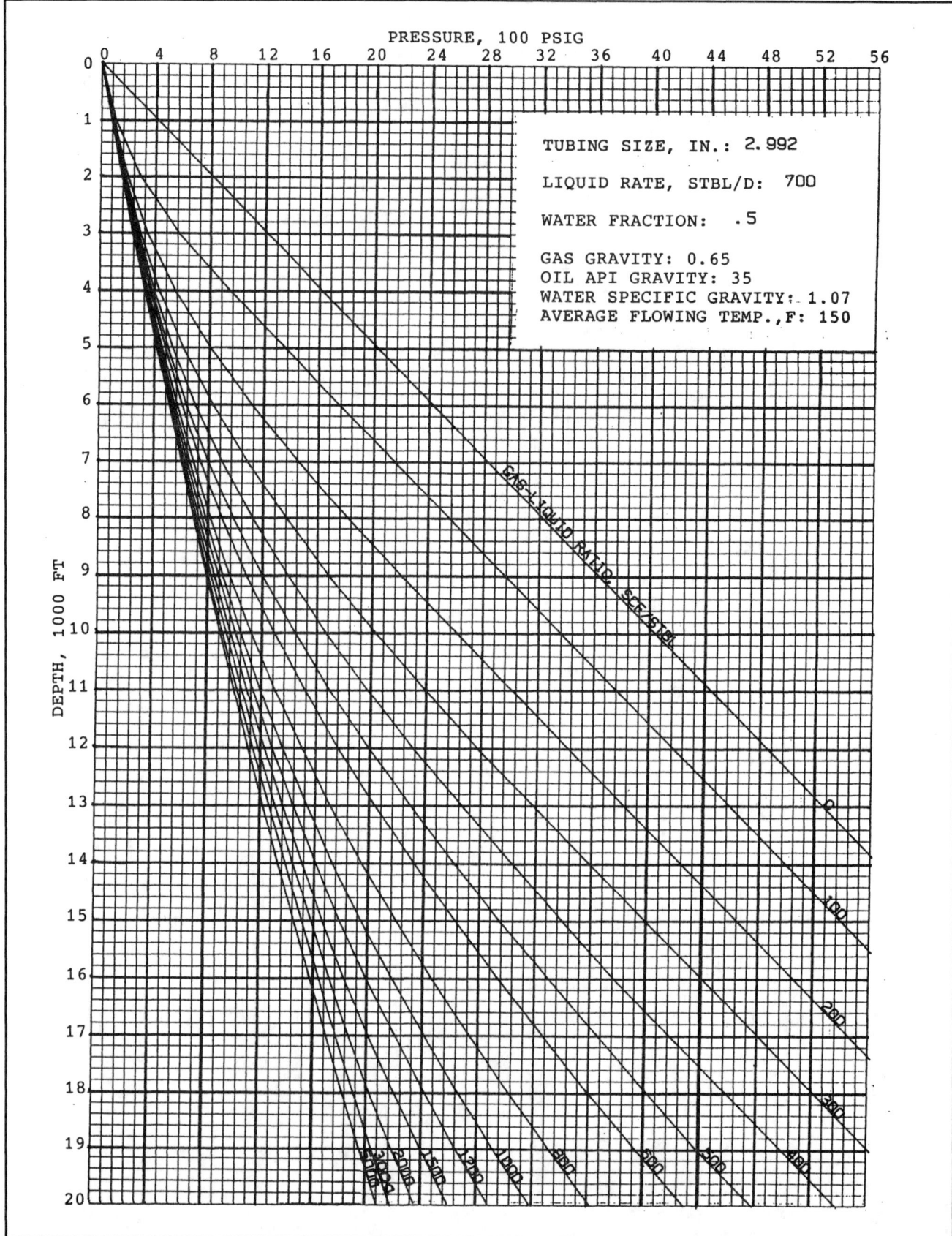

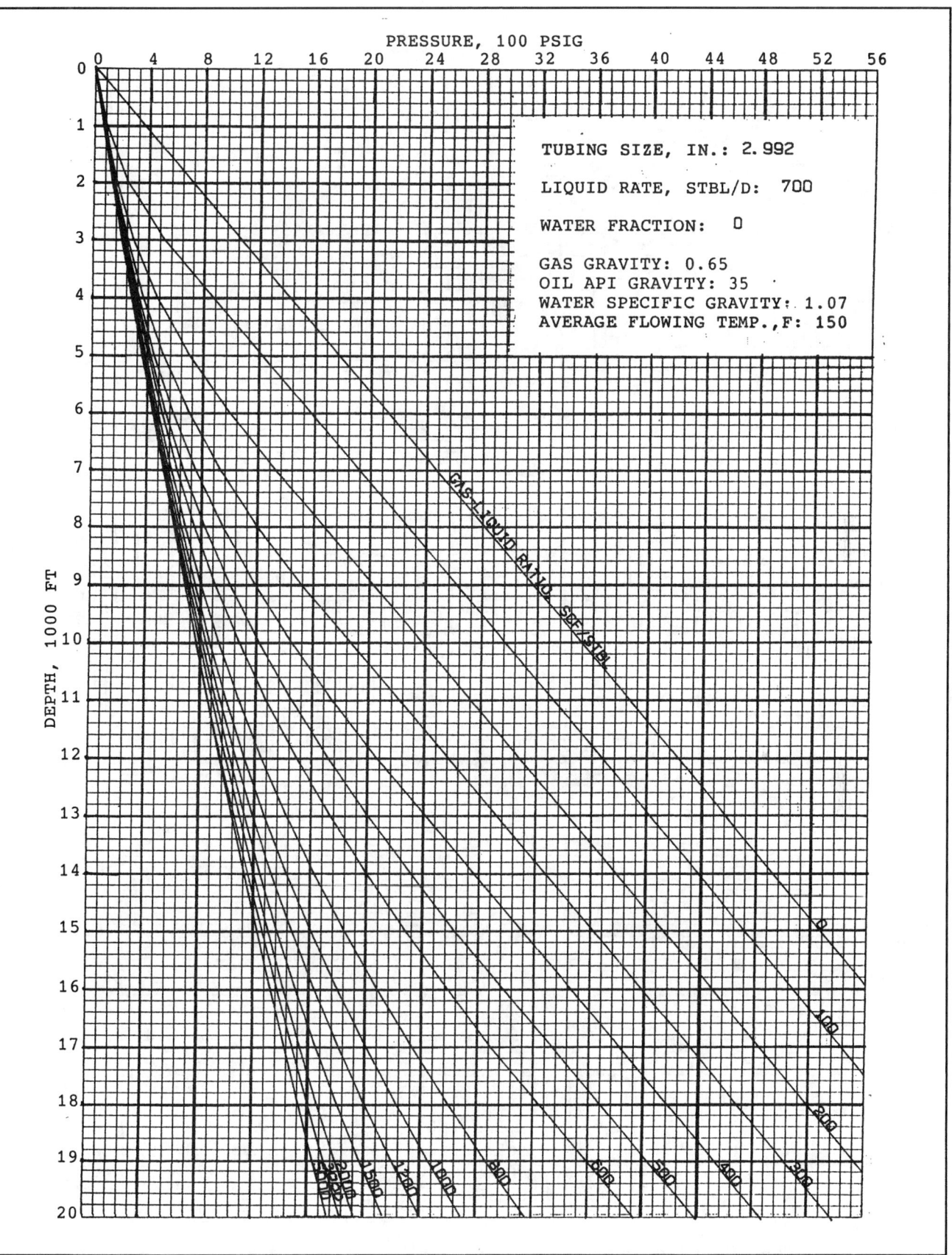

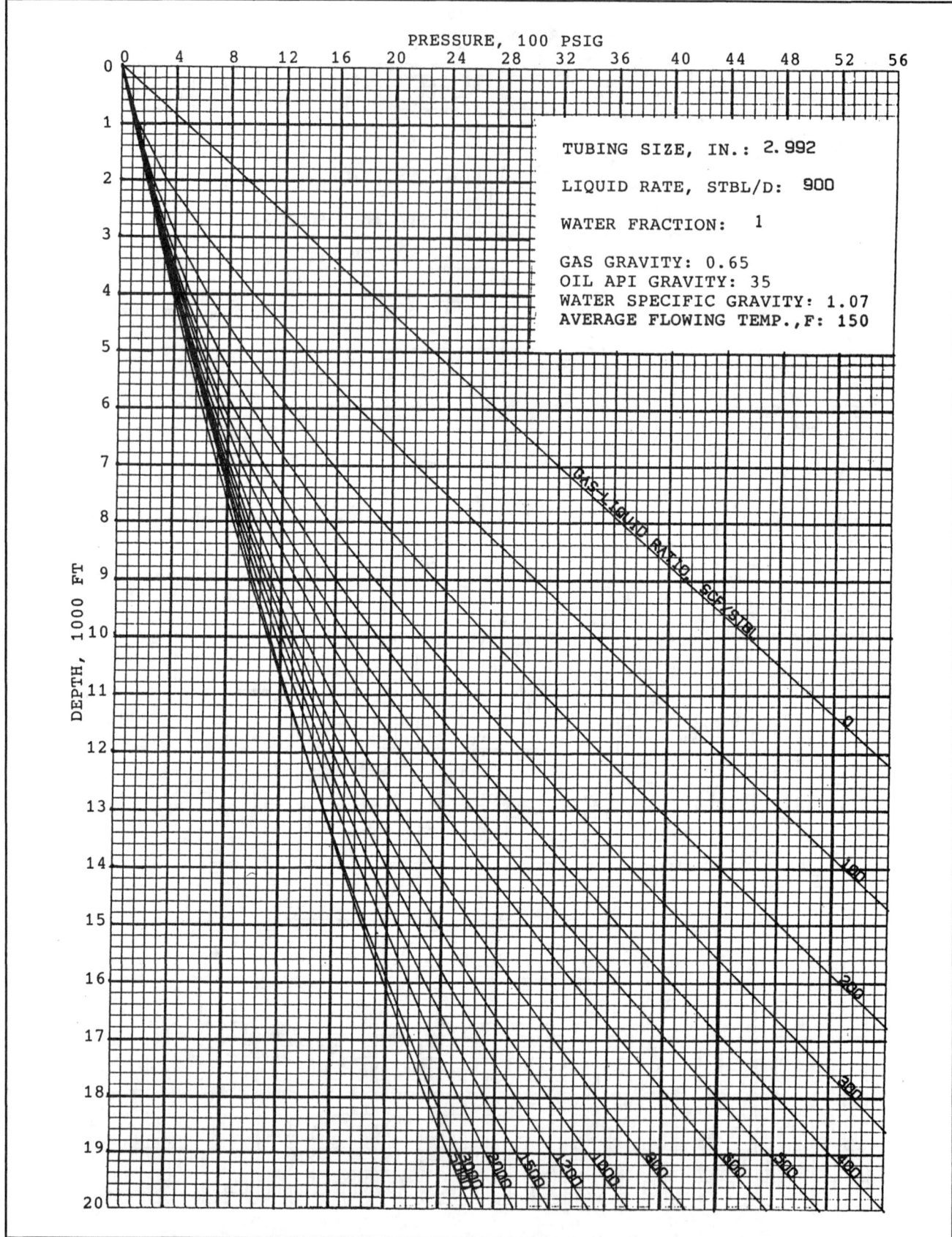

PRESSURE, 100 PSIG

TUBING SIZE, IN.: 2.992

LIQUID RATE, STBL/D: 900

WATER FRACTION: 1

GAS GRAVITY: 0.65
OIL API GRAVITY: 35
WATER SPECIFIC GRAVITY: 1.07
AVERAGE FLOWING TEMP.,F: 150

DEPTH, 1000 FT

GAS-LIQUID RATIO, SCF/STBL

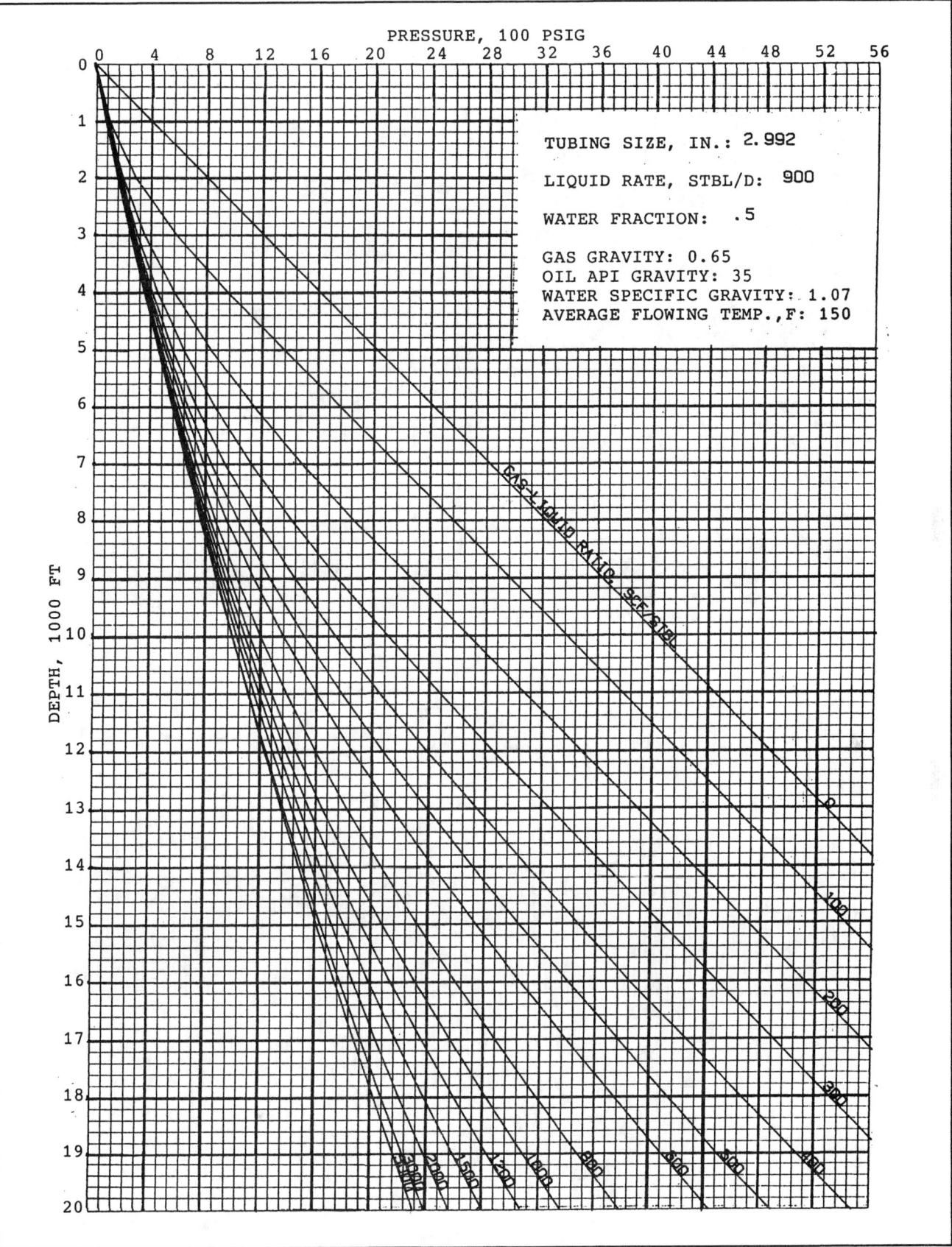

PRESSURE, 100 PSIG

TUBING SIZE, IN.: 2.992

LIQUID RATE, STBL/D: 900

WATER FRACTION: .5

GAS GRAVITY: 0.65
OIL API GRAVITY: 35
WATER SPECIFIC GRAVITY: 1.07
AVERAGE FLOWING TEMP.,F: 150

DEPTH, 1000 FT

GAS-LIQUID RATIO, SCF/STBL

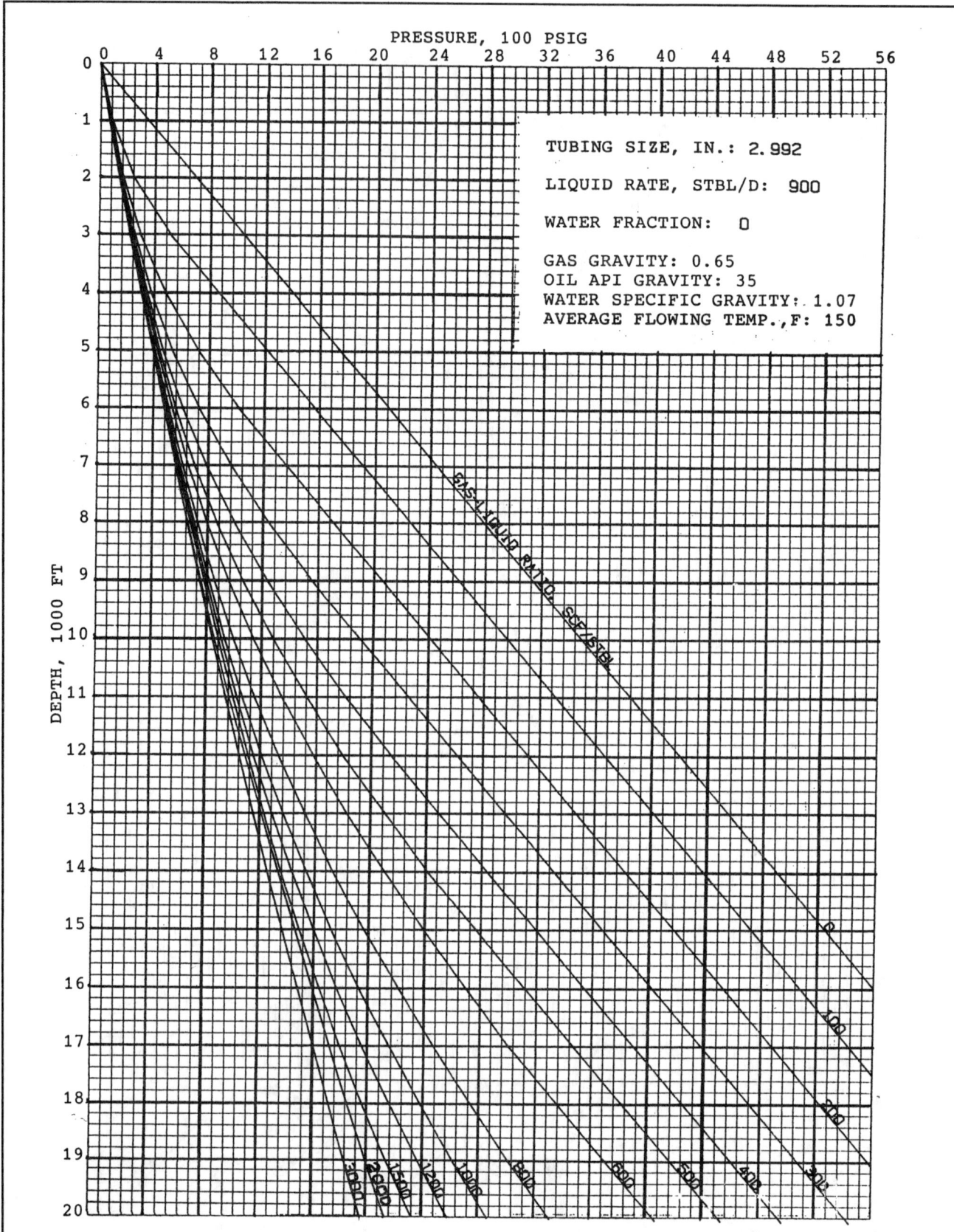

PRESSURE, 100 PSIG

DEPTH, 1000 FT

TUBING SIZE, IN.: 2.992

LIQUID RATE, STBL/D: 900

WATER FRACTION: 0

GAS GRAVITY: 0.65
OIL API GRAVITY: 35
WATER SPECIFIC GRAVITY: 1.07
AVERAGE FLOWING TEMP., F: 150

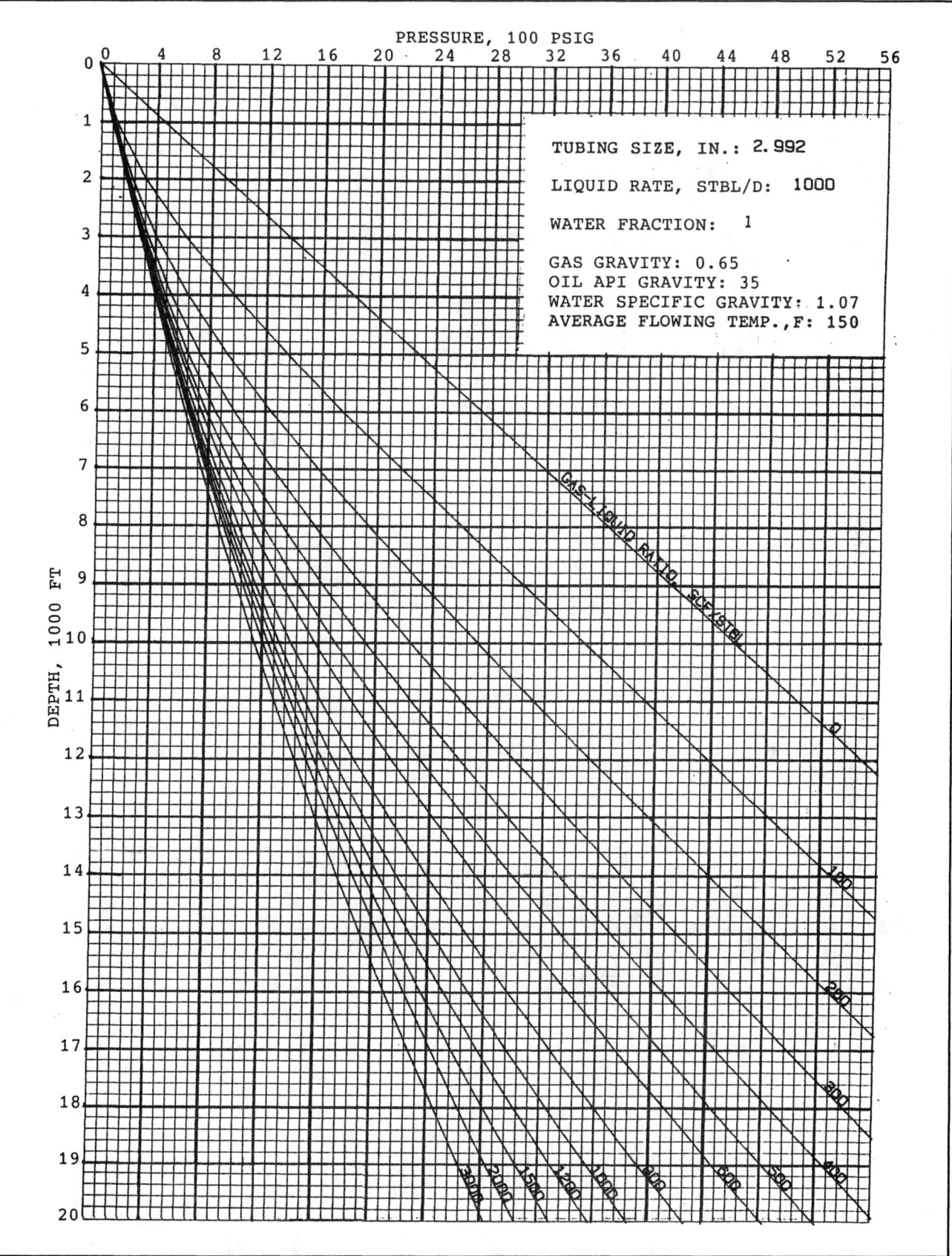

PRESSURE, 100 PSIG

DEPTH, 1000 FT

TUBING SIZE, IN.: 2.992

LIQUID RATE, STBL/D: 1000

WATER FRACTION: 1

GAS GRAVITY: 0.65
OIL API GRAVITY: 35
WATER SPECIFIC GRAVITY: 1.07
AVERAGE FLOWING TEMP.,F: 150

GAS-LIQUID RATIO SCF/STBL

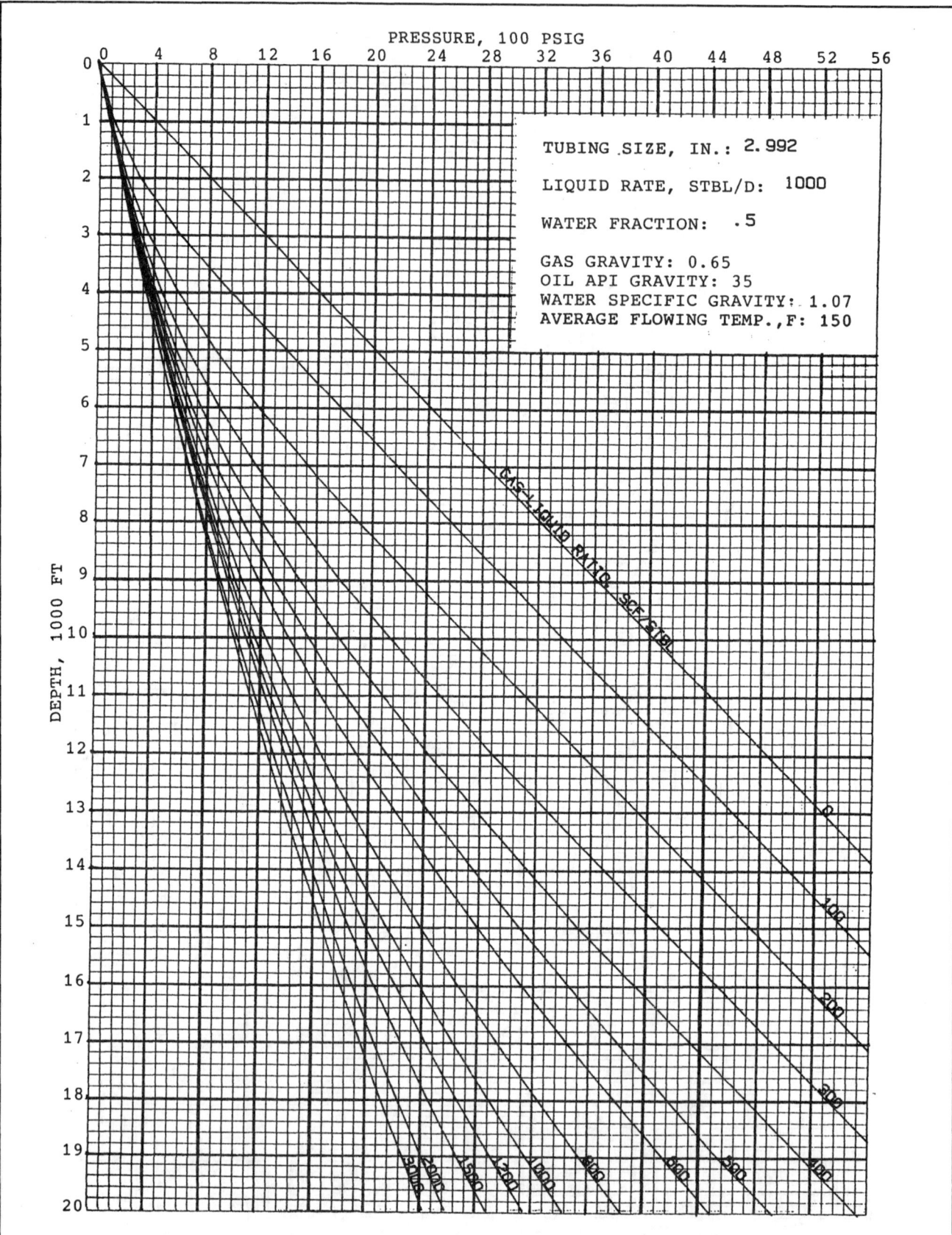

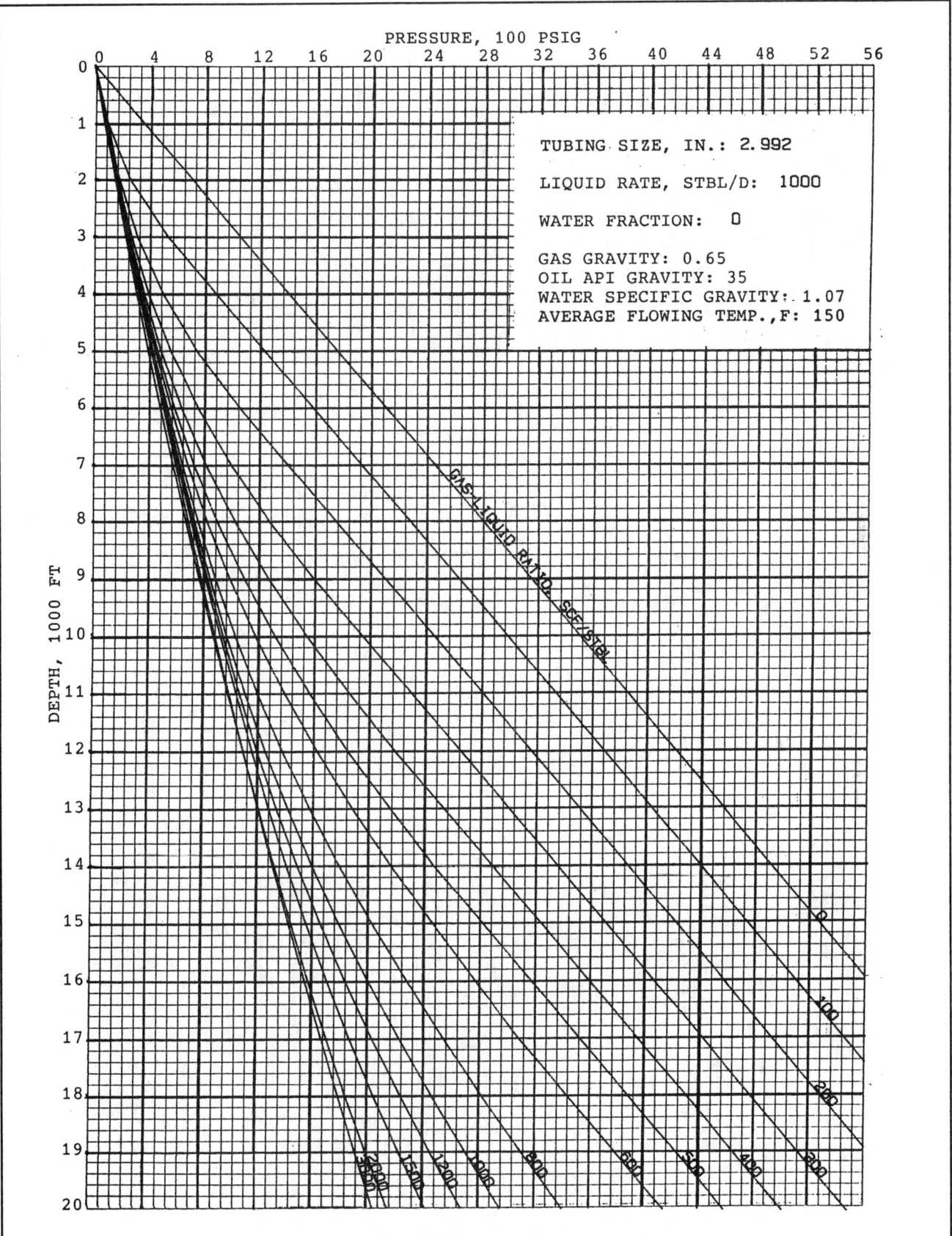

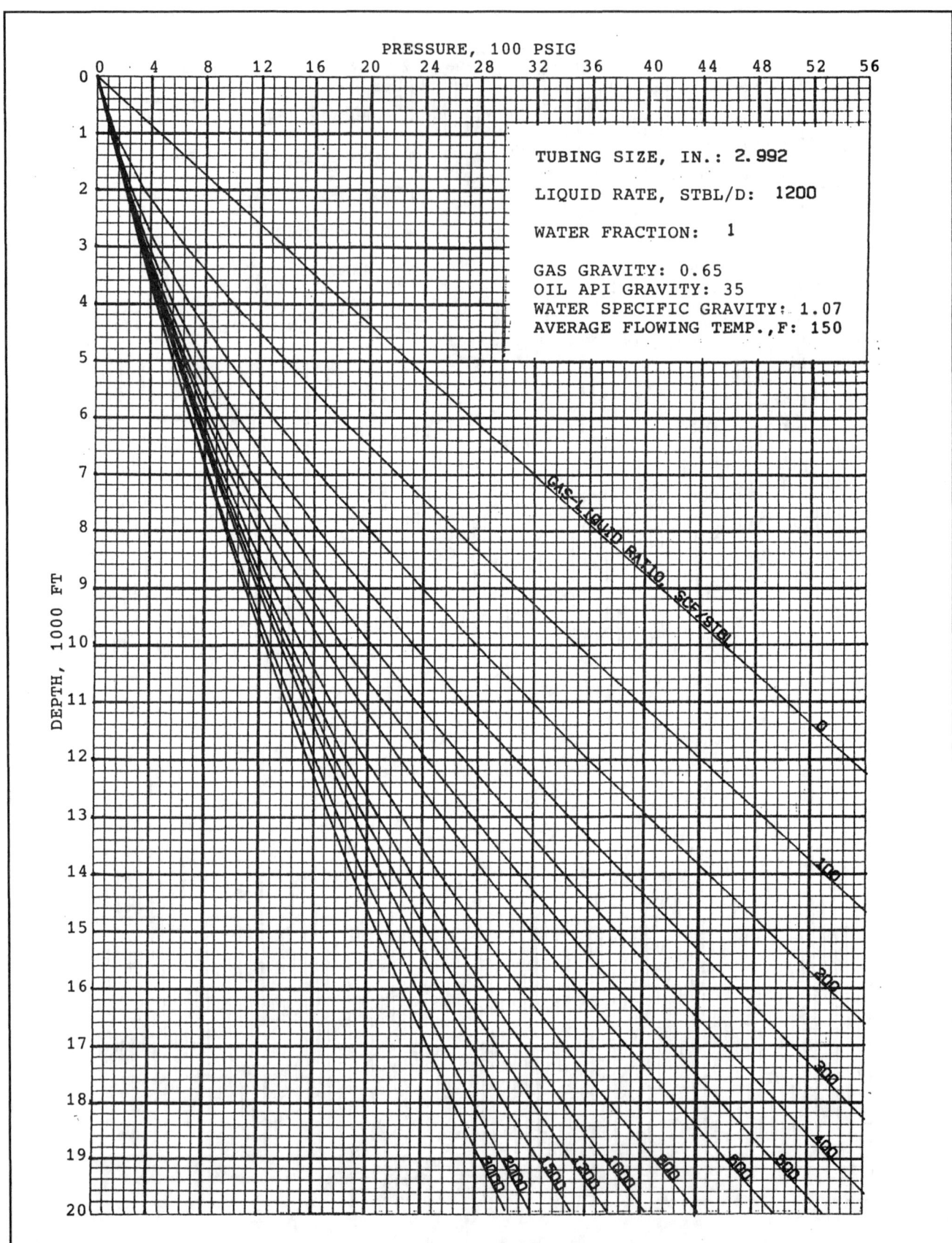

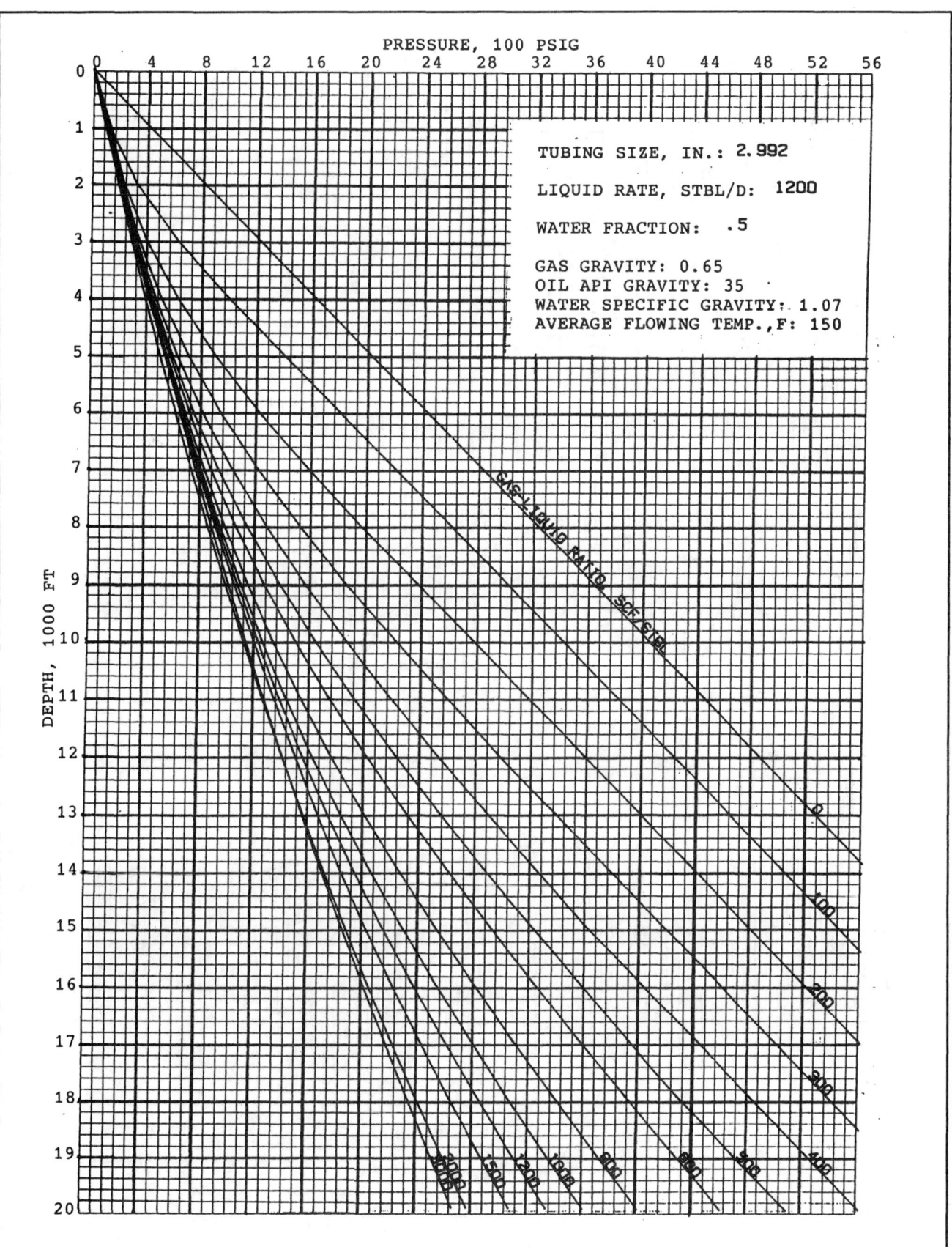

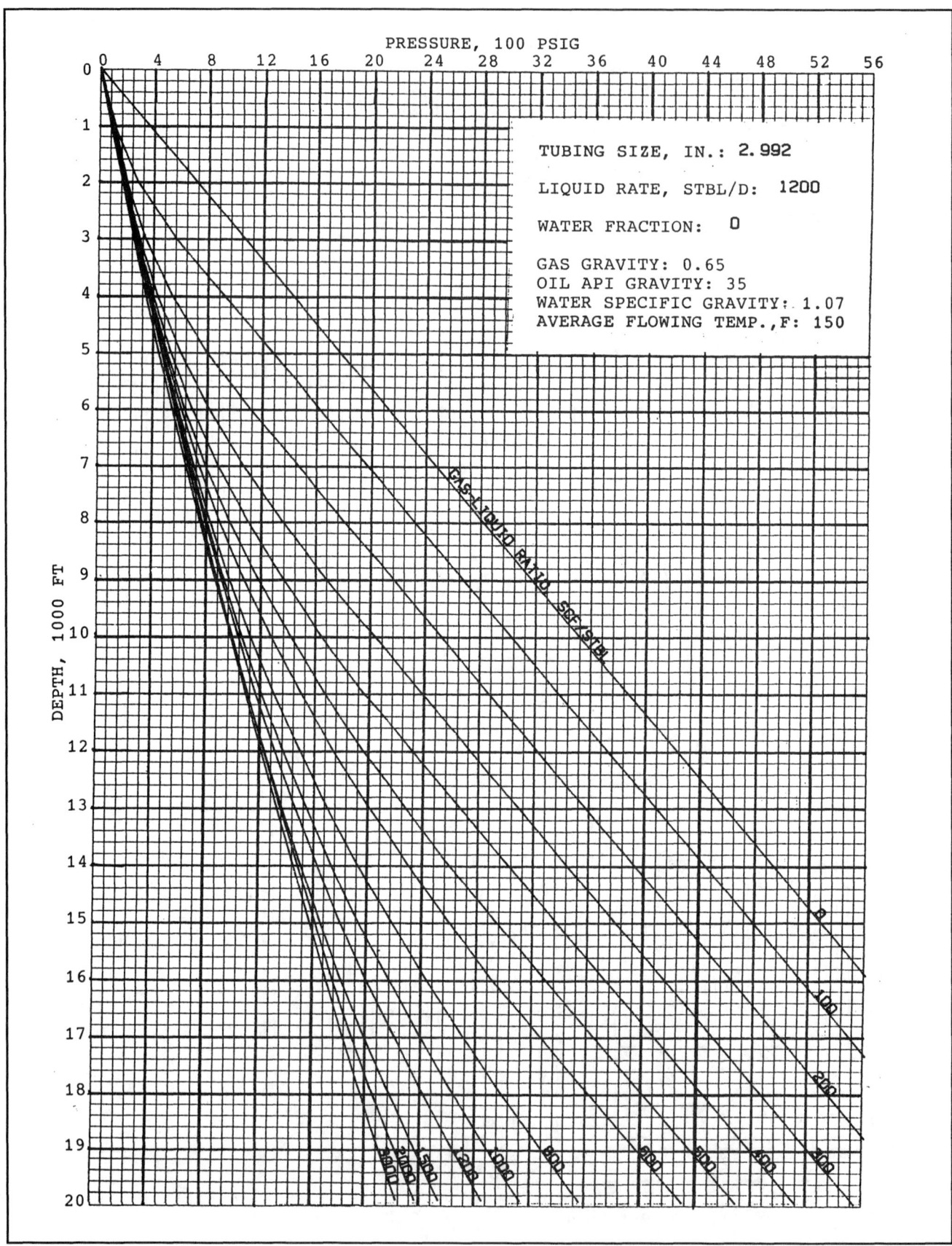

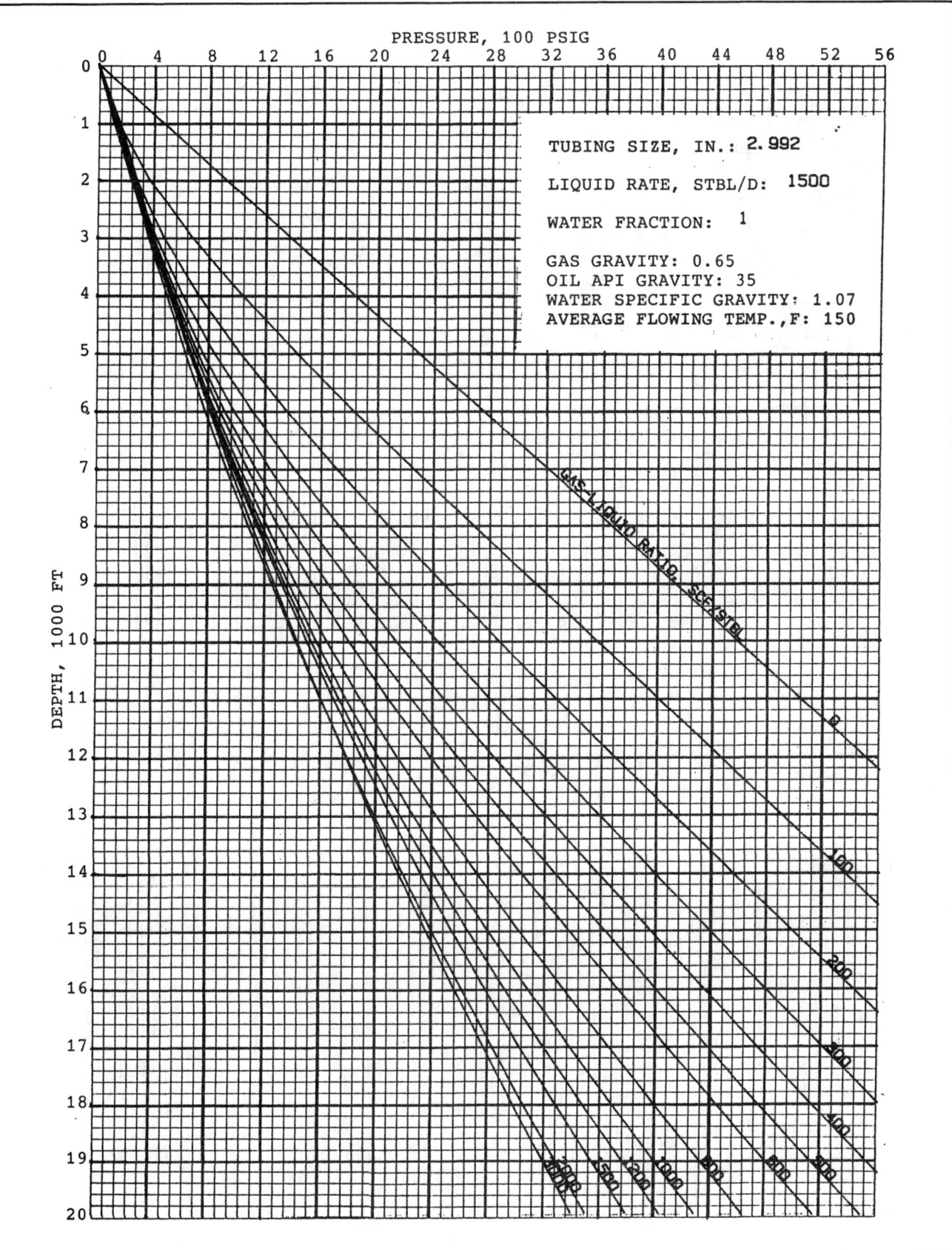

PRESSURE, 100 PSIG

TUBING SIZE, IN.: 2.992

LIQUID RATE, STBL/D: 1500

WATER FRACTION: 1

GAS GRAVITY: 0.65
OIL API GRAVITY: 35
WATER SPECIFIC GRAVITY: 1.07
AVERAGE FLOWING TEMP.,F: 150

GAS-LIQUID RATIO, SCF/STBL

DEPTH, 1000 FT

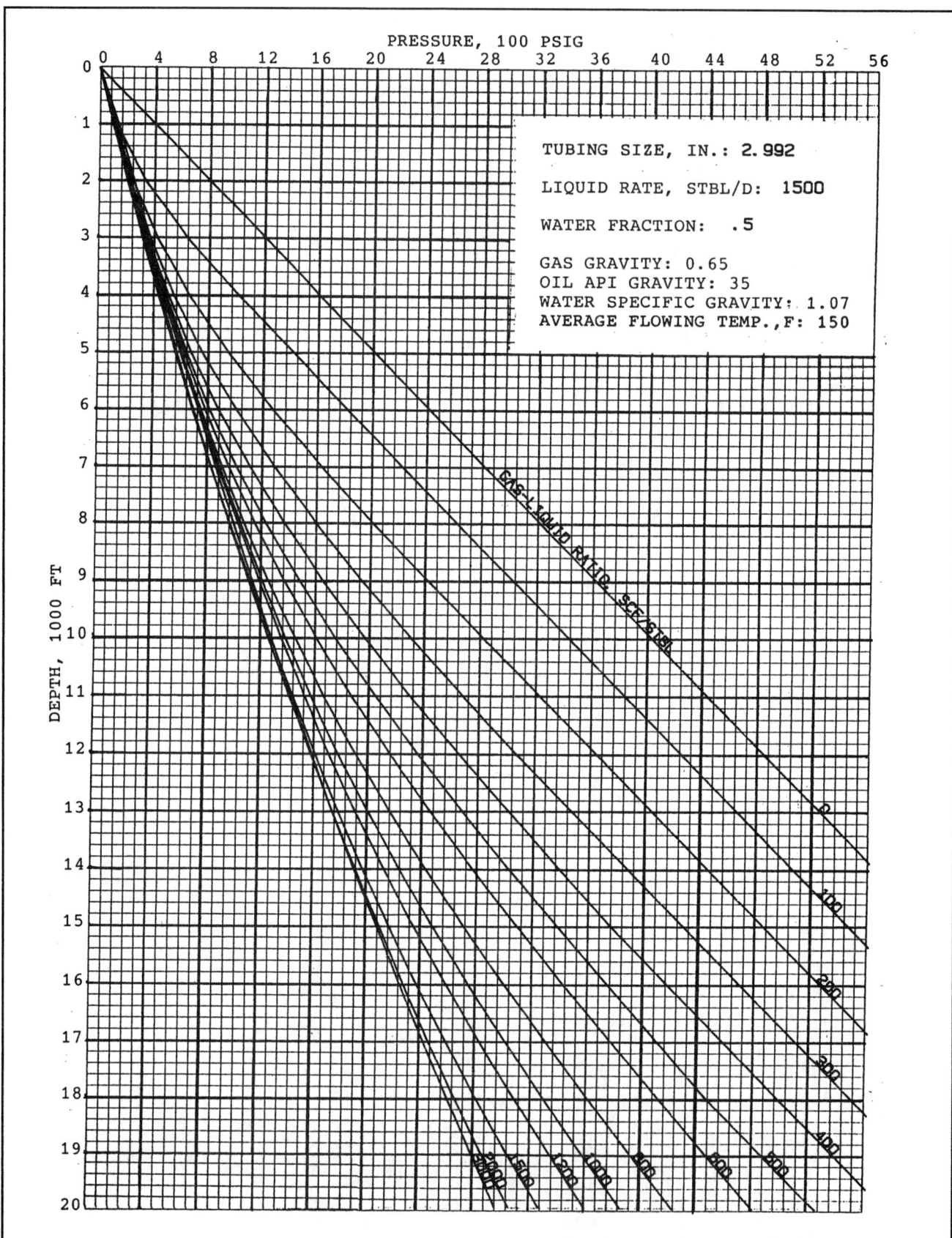

PRESSURE, 100 PSIG

DEPTH, 1000 FT

TUBING SIZE, IN.: **2.992**

LIQUID RATE, STBL/D: **1500**

WATER FRACTION: **.5**

GAS GRAVITY: 0.65
OIL API GRAVITY: 35
WATER SPECIFIC GRAVITY: 1.07
AVERAGE FLOWING TEMP.,F: 150

GAS-LIQUID RATIO, SCF/STBL

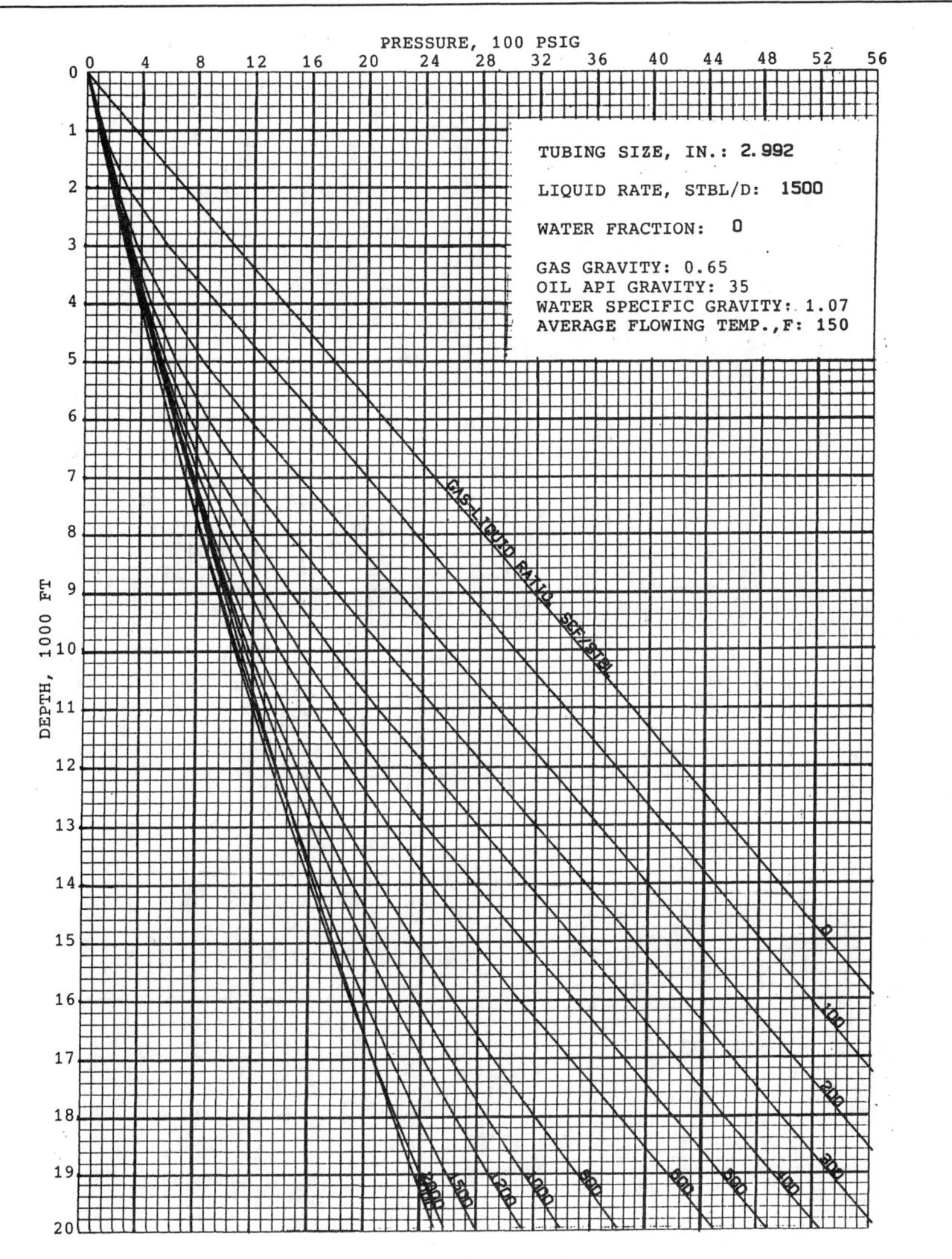

PRESSURE, 100 PSIG

DEPTH, 1000 FT

TUBING SIZE, IN.: **2.992**

LIQUID RATE, STBL/D: **1500**

WATER FRACTION: **0**

GAS GRAVITY: 0.65
OIL API GRAVITY: 35
WATER SPECIFIC GRAVITY: 1.07
AVERAGE FLOWING TEMP.,F: 150

GAS/LIQUID RATIO SCF/STBL

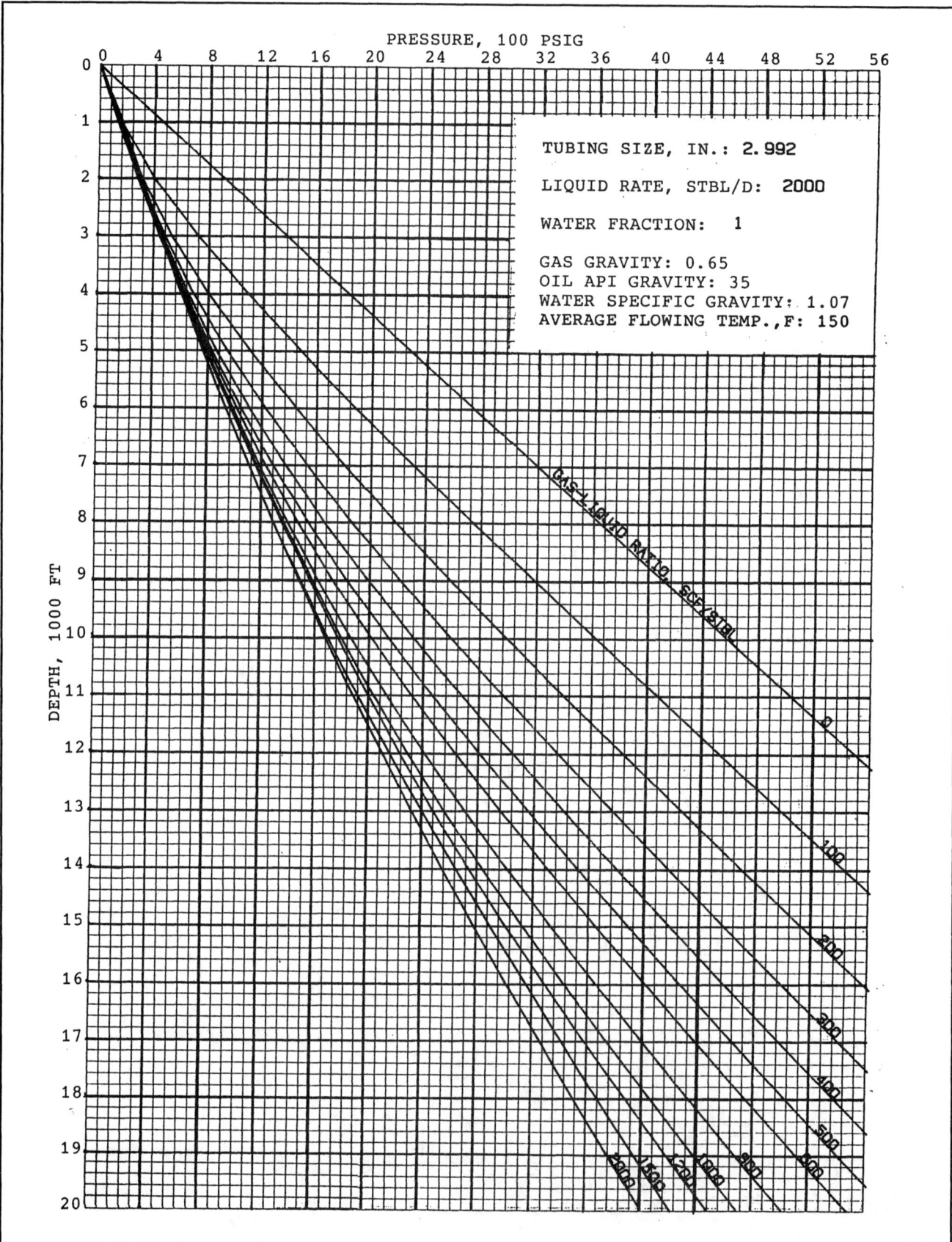

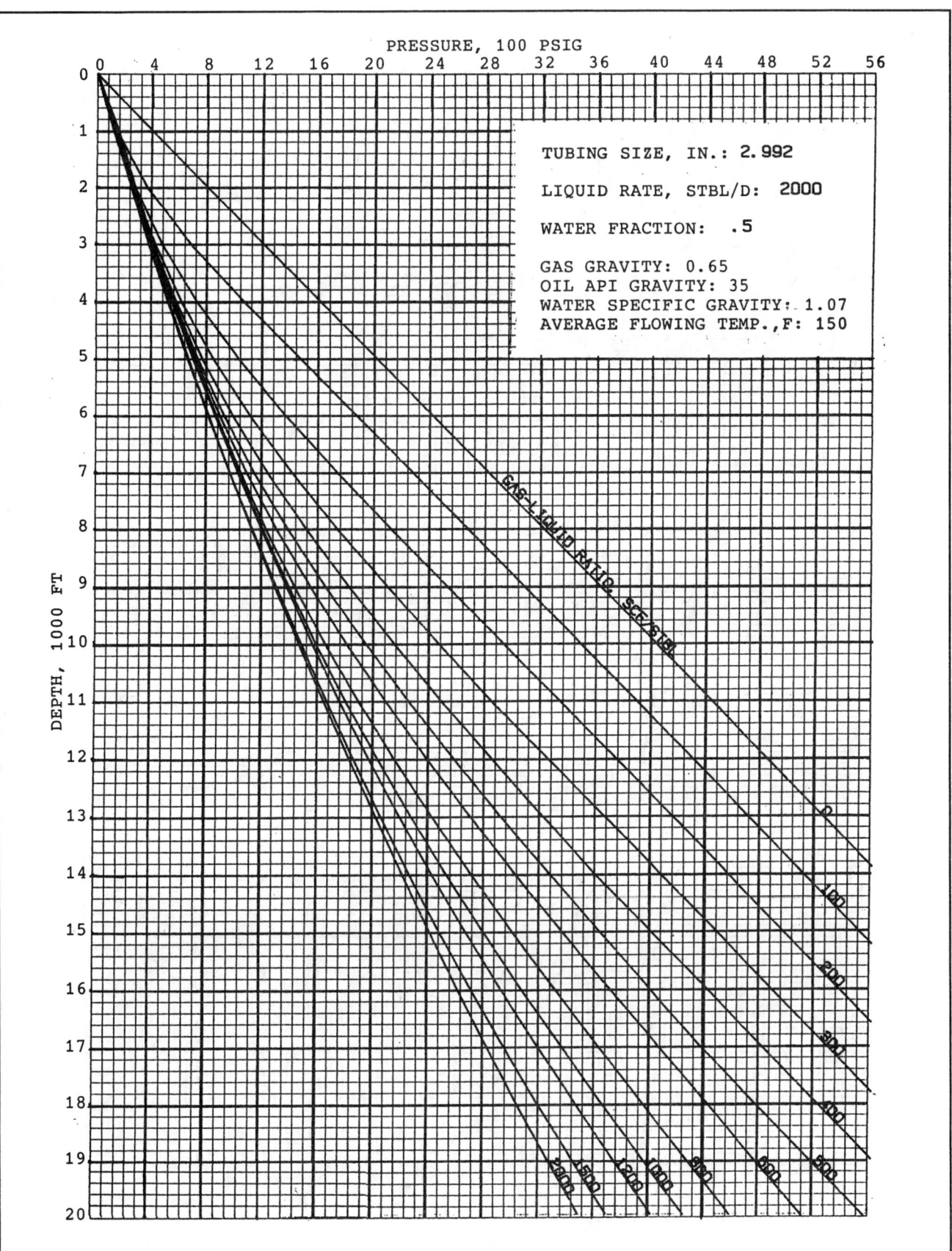

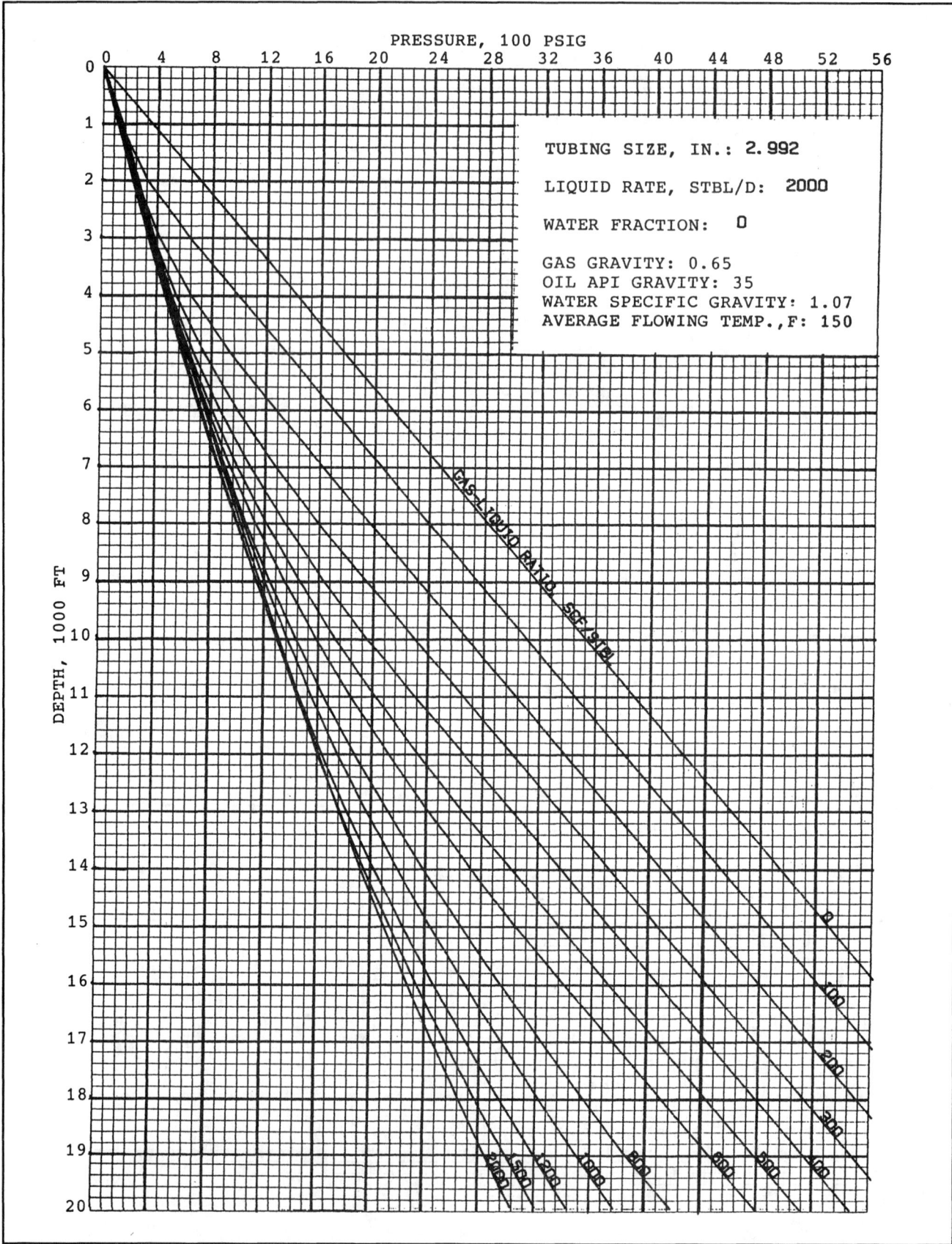

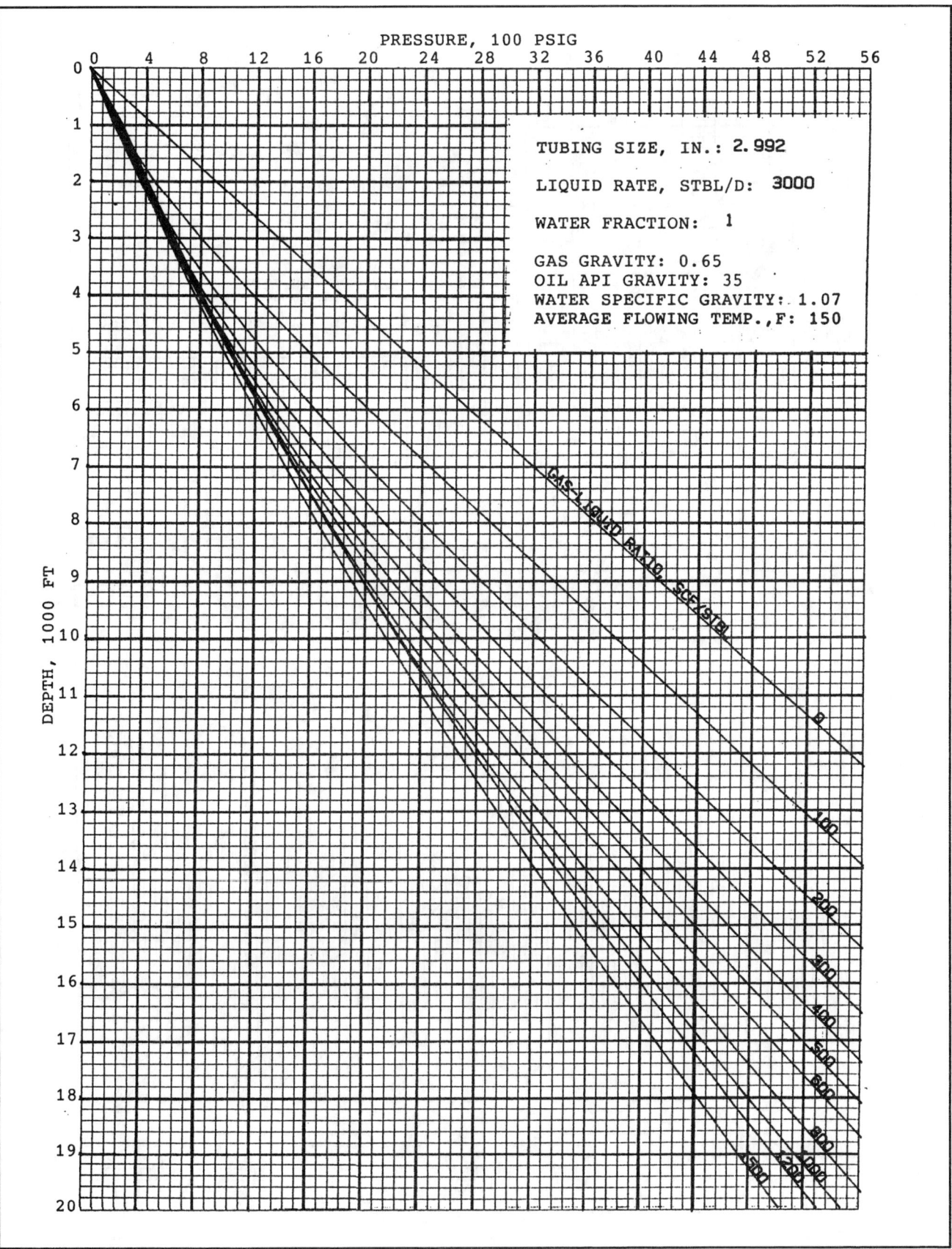

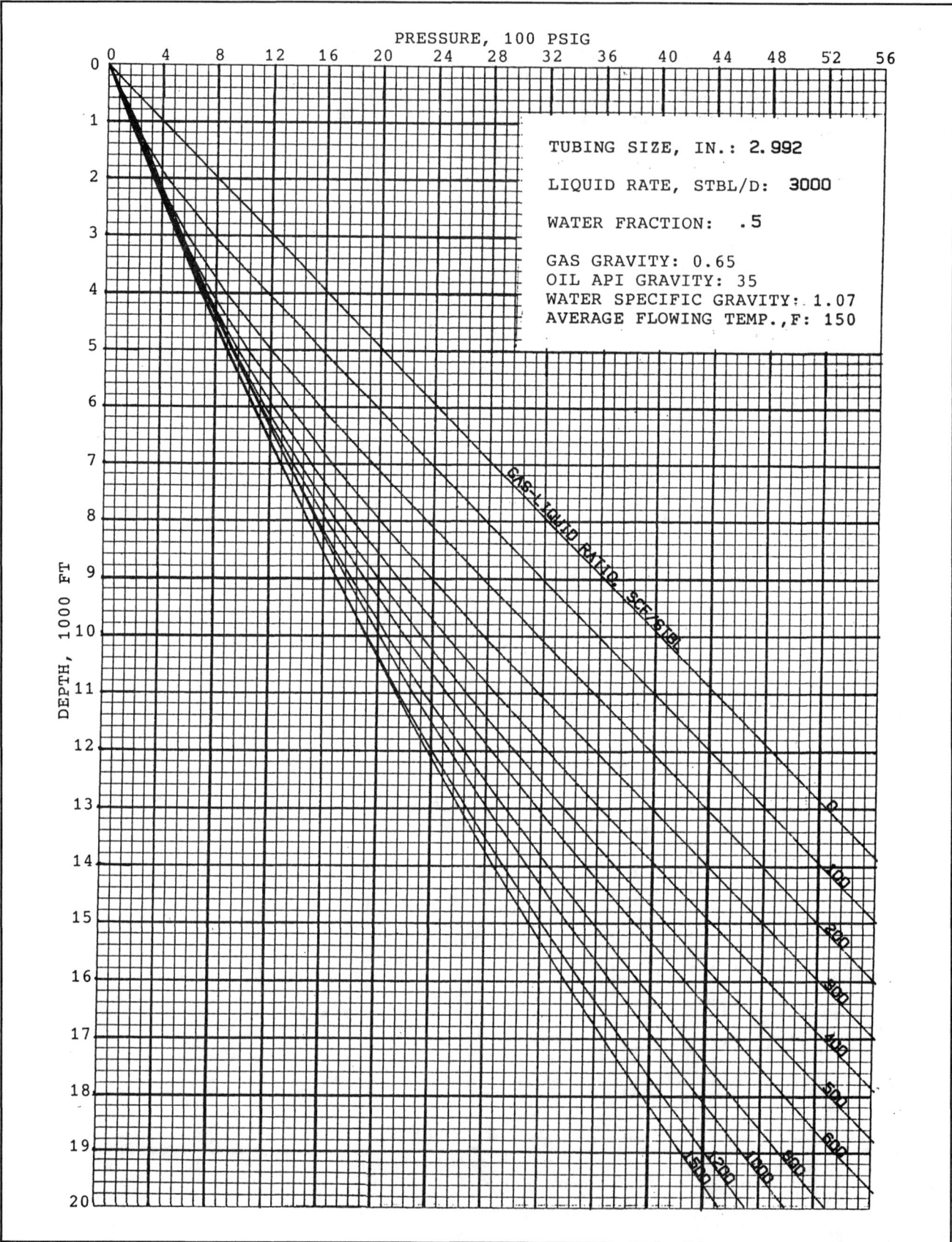

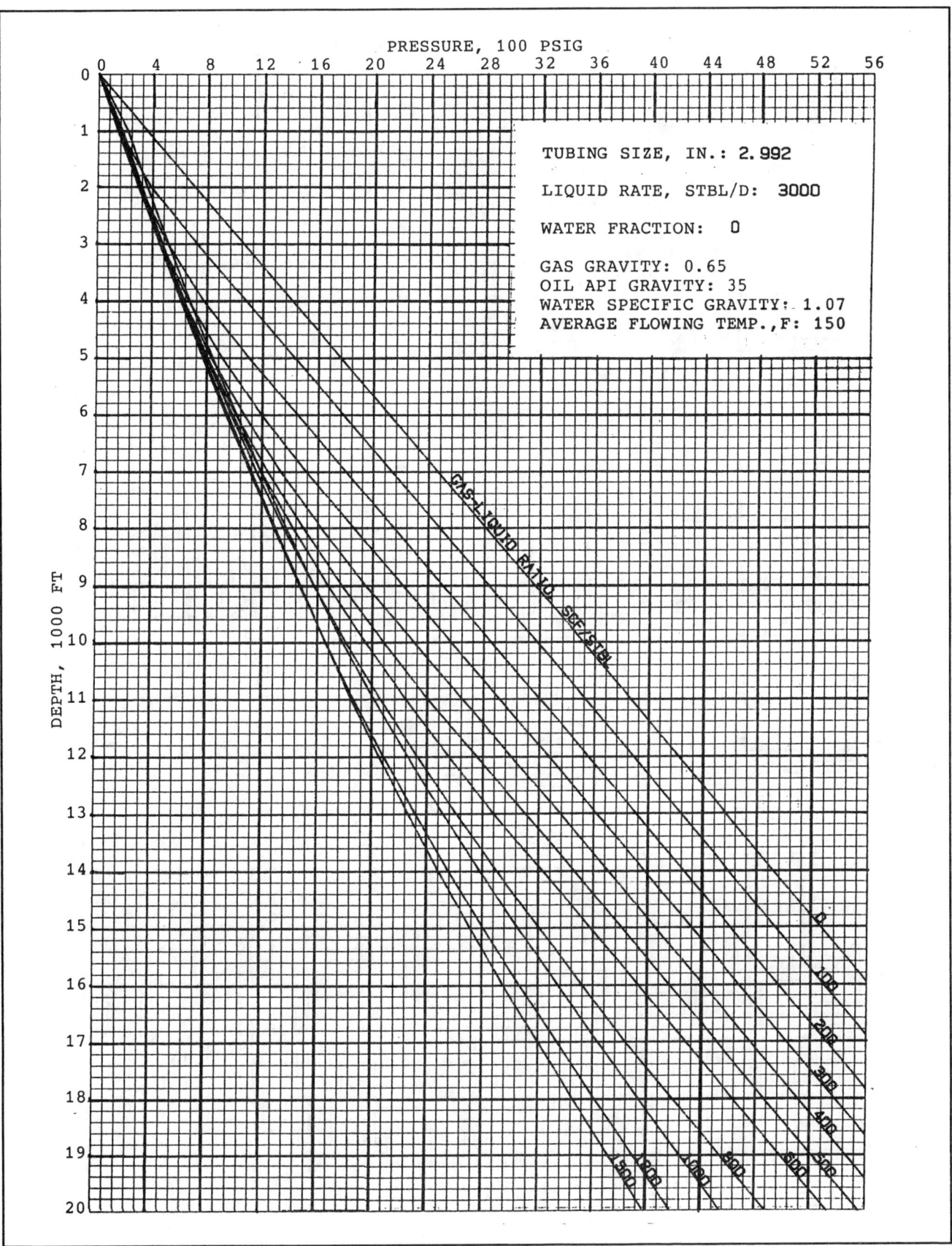

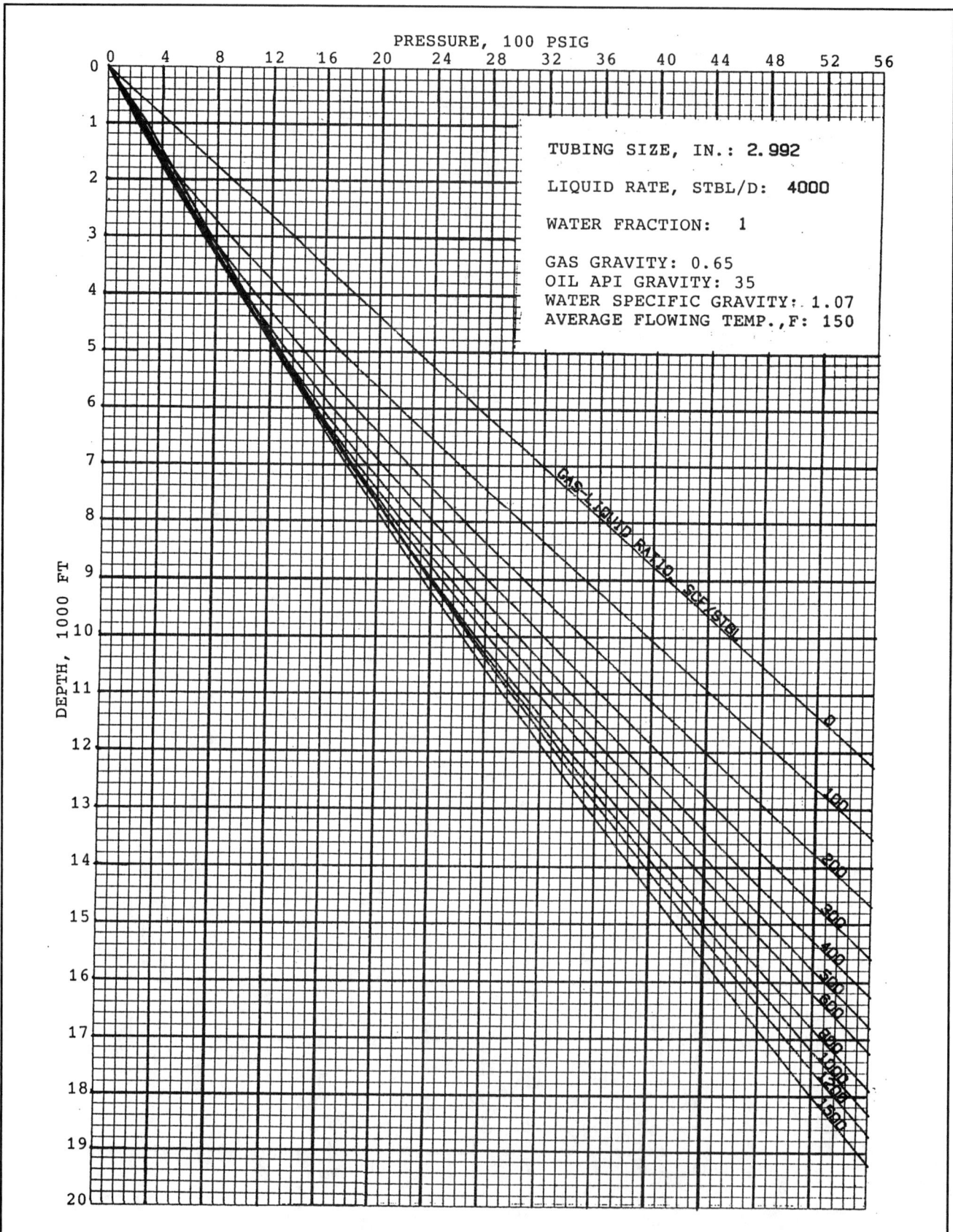

PRESSURE, 100 PSIG

DEPTH, 1000 FT

TUBING SIZE, IN.: **2.992**

LIQUID RATE, STBL/D: **4000**

WATER FRACTION: **1**

GAS GRAVITY: 0.65
OIL API GRAVITY: 35
WATER SPECIFIC GRAVITY: 1.07
AVERAGE FLOWING TEMP.,F: 150

GAS-LIQUID RATIO SCF/STBL

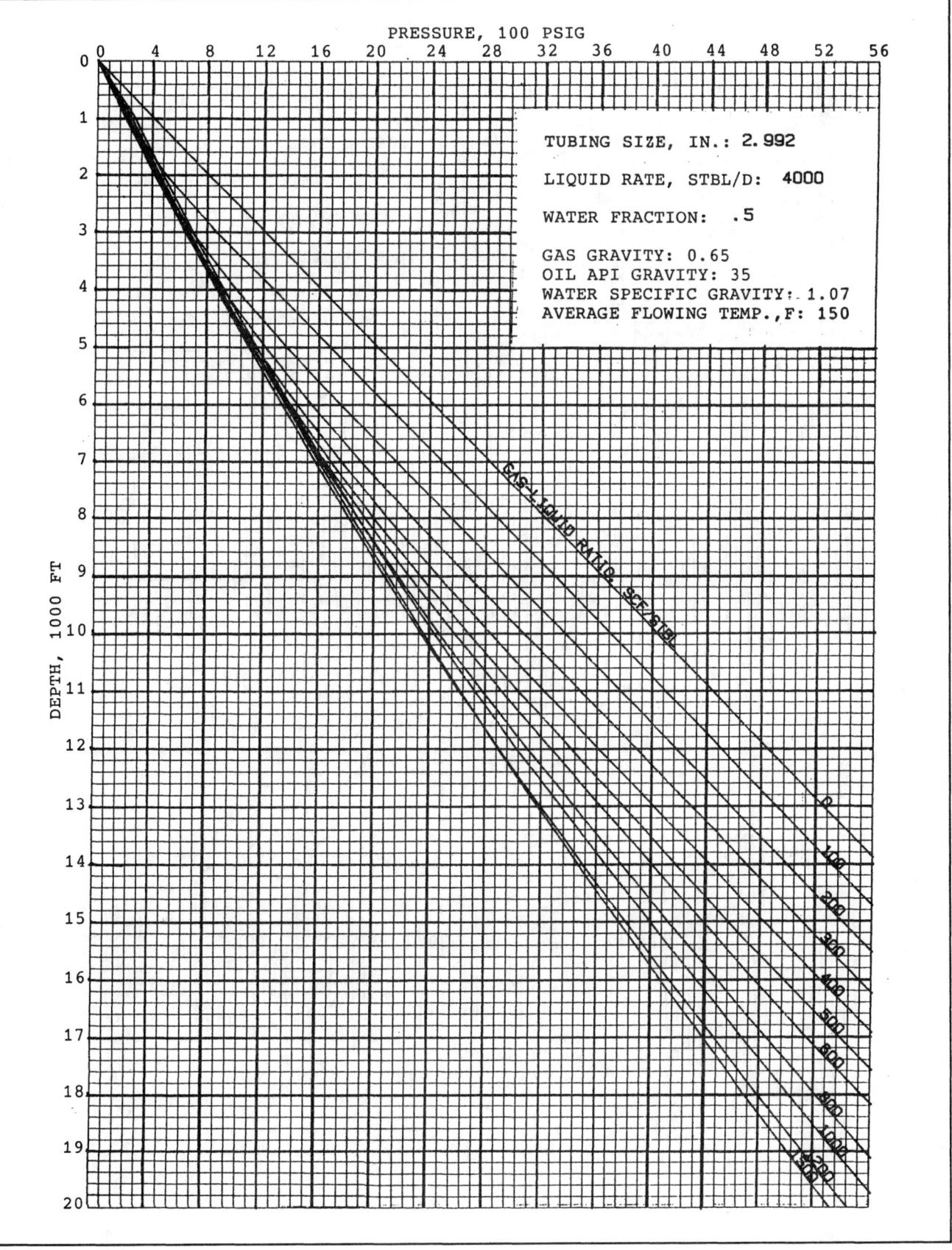

PRESSURE, 100 PSIG

DEPTH, 1000 FT

TUBING SIZE, IN.: 2.992

LIQUID RATE, STBL/D: 4000

WATER FRACTION: .5

GAS GRAVITY: 0.65
OIL API GRAVITY: 35
WATER SPECIFIC GRAVITY: 1.07
AVERAGE FLOWING TEMP.,F: 150

GAS-LIQUID RATIO, SCF/STBL

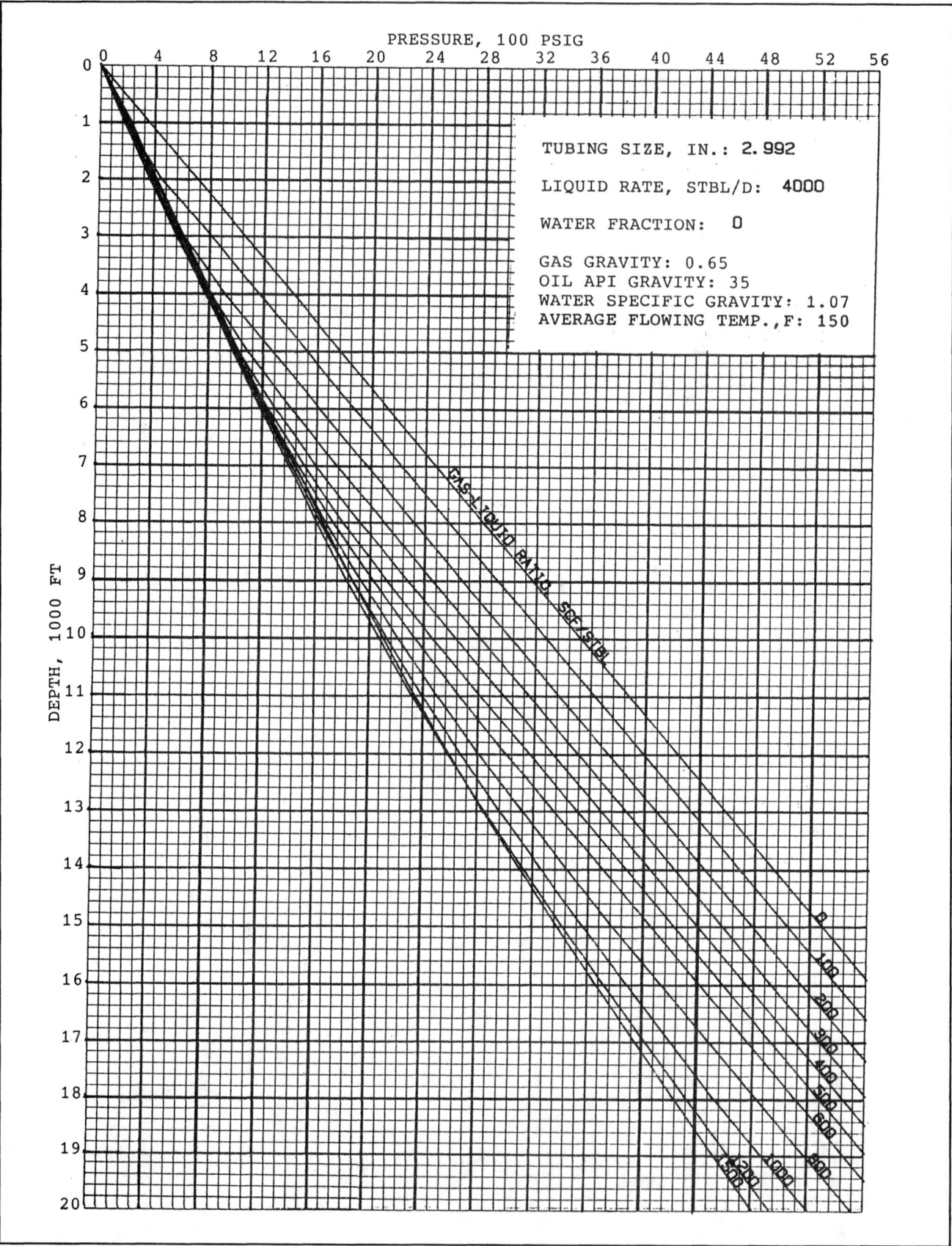

PRESSURE, 100 PSIG

DEPTH, 1000 FT

TUBING SIZE, IN.: **2.992**

LIQUID RATE, STBL/D: **4000**

WATER FRACTION: **0**

GAS GRAVITY: 0.65
OIL API GRAVITY: 35
WATER SPECIFIC GRAVITY: 1.07
AVERAGE FLOWING TEMP.,F: 150

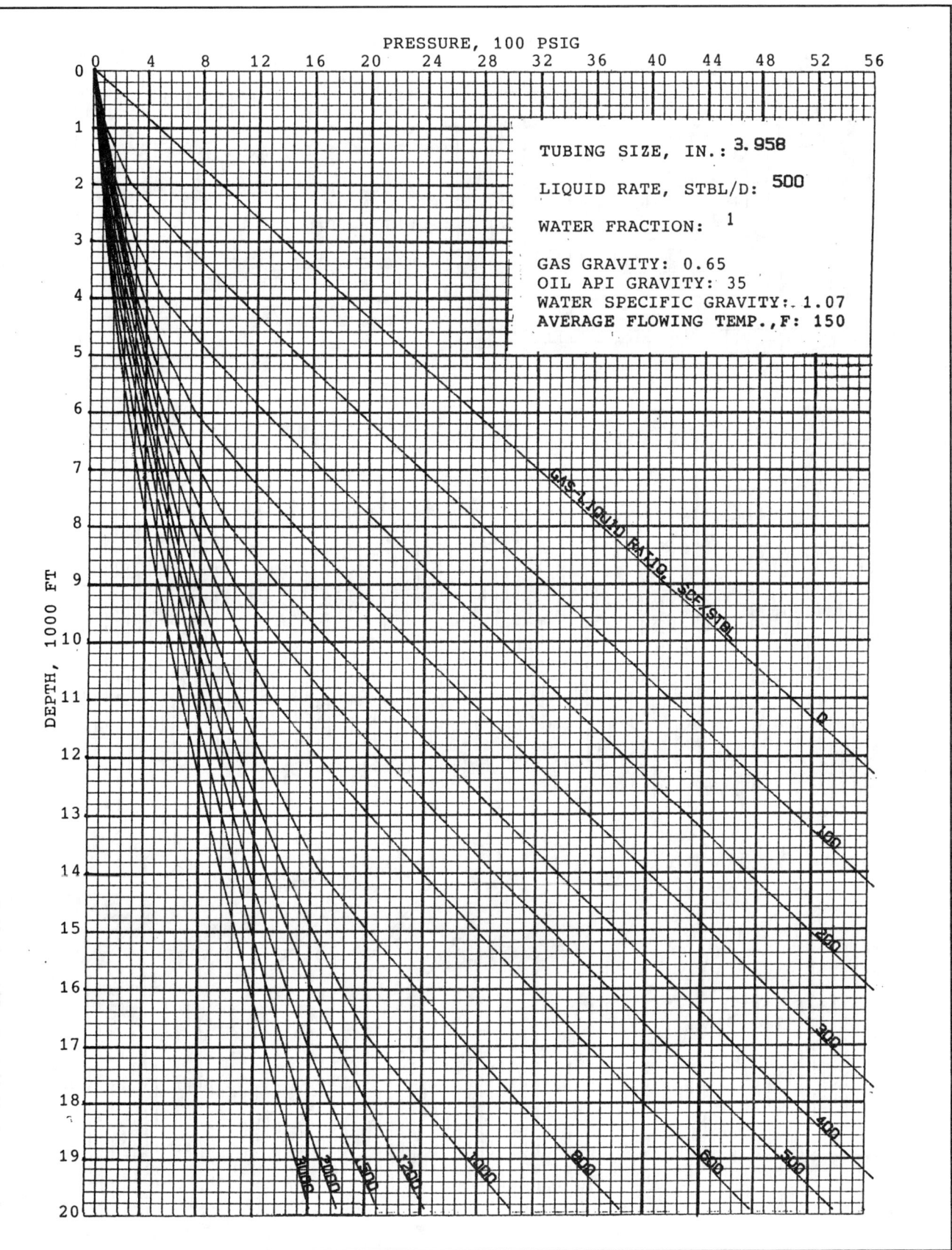

PRESSURE, 100 PSIG

DEPTH, 1000 FT

TUBING SIZE, IN.: 3.958

LIQUID RATE, STBL/D: 500

WATER FRACTION: 1

GAS GRAVITY: 0.65
OIL API GRAVITY: 35
WATER SPECIFIC GRAVITY: 1.07
AVERAGE FLOWING TEMP., F: 150

GAS-LIQUID RATIO, SCF/STBL

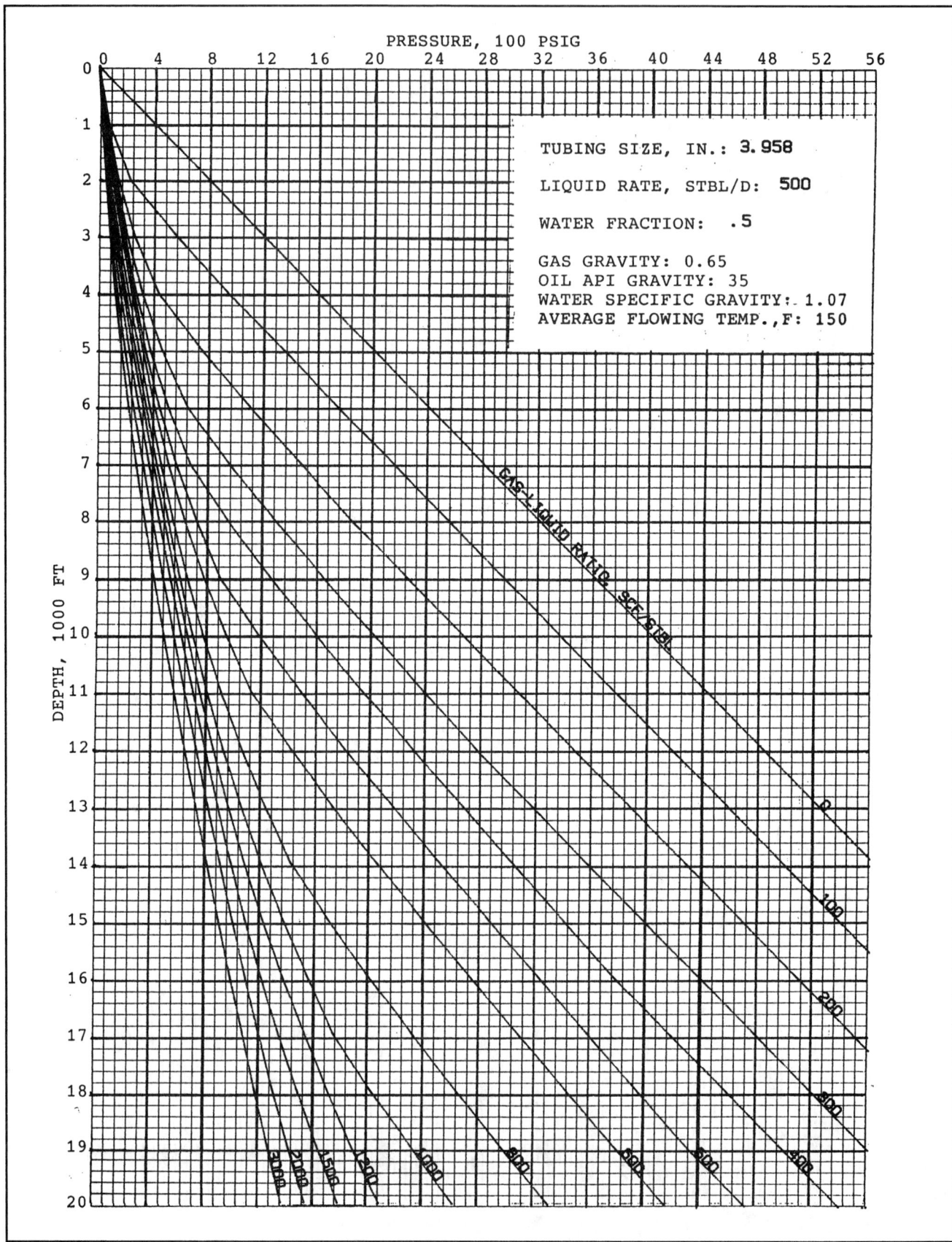

PRESSURE, 100 PSIG

TUBING SIZE, IN.: **3.958**

LIQUID RATE, STBL/D: **500**

WATER FRACTION: **.5**

GAS GRAVITY: 0.65
OIL API GRAVITY: 35
WATER SPECIFIC GRAVITY: 1.07
AVERAGE FLOWING TEMP.,F: 150

DEPTH, 1000 FT

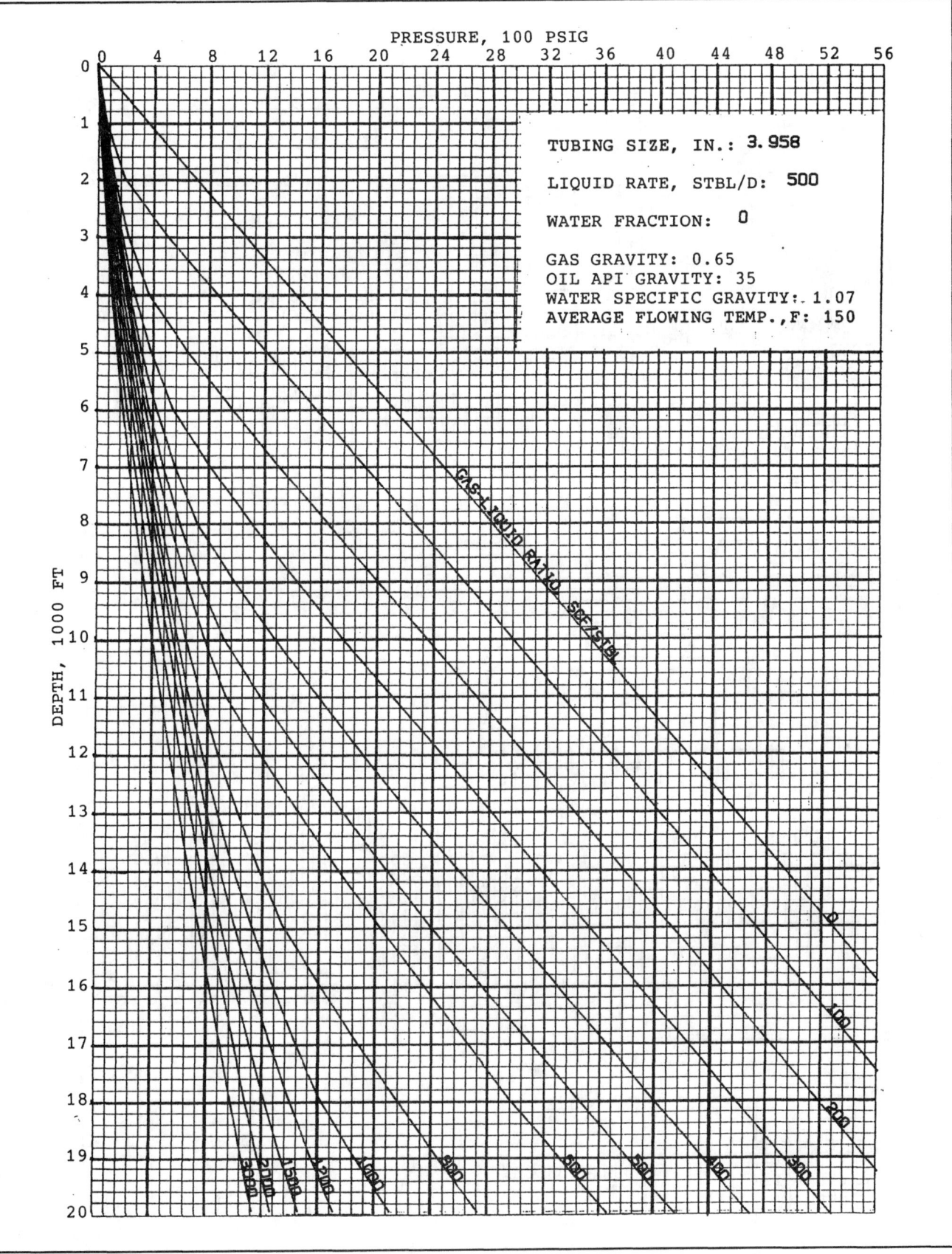

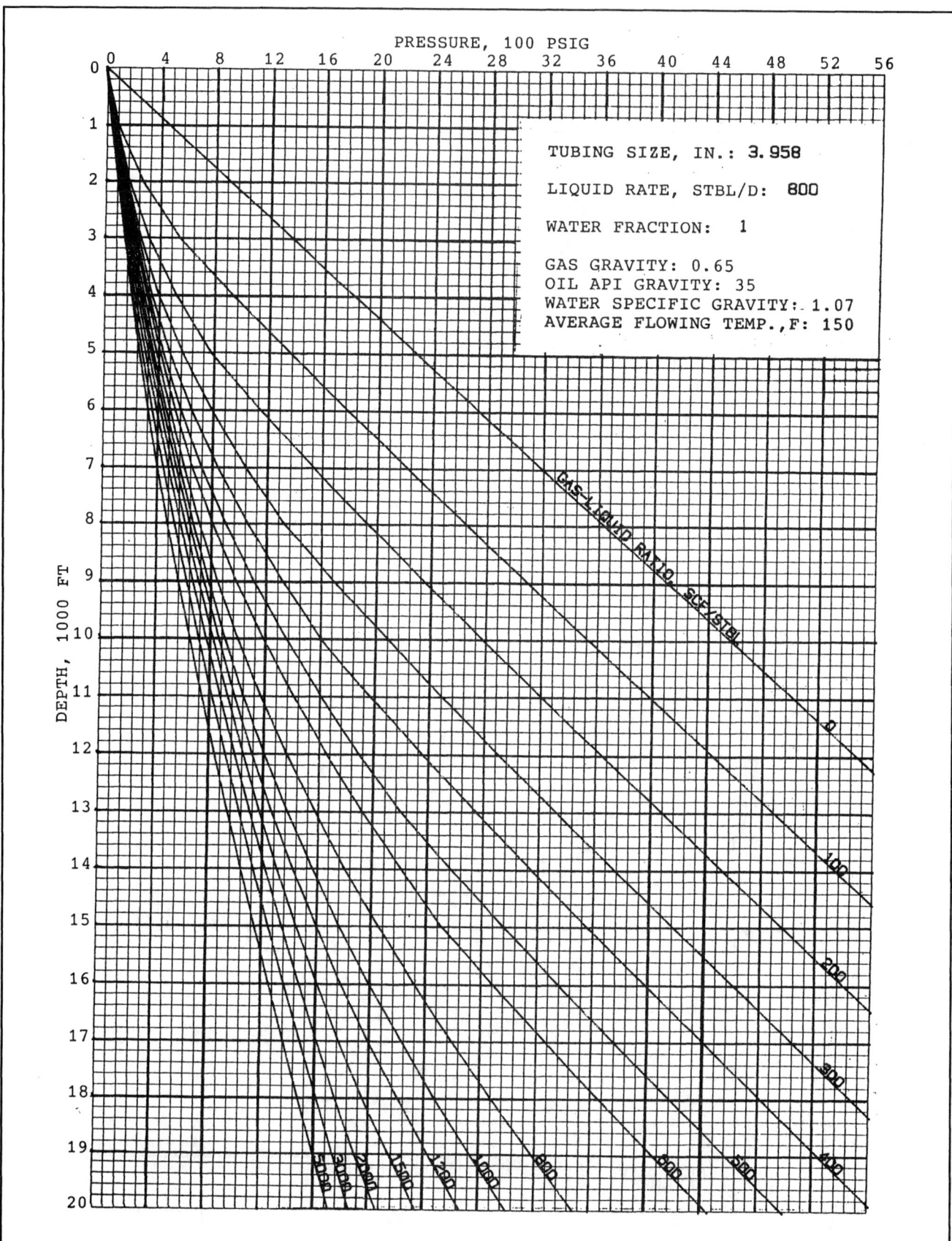

PRESSURE, 100 PSIG

DEPTH, 1000 FT

TUBING SIZE, IN.: **3.958**

LIQUID RATE, STBL/D: **800**

WATER FRACTION: 1

GAS GRAVITY: 0.65
OIL API GRAVITY: 35
WATER SPECIFIC GRAVITY: 1.07
AVERAGE FLOWING TEMP., F: 150

GAS-LIQUID RATIO, SCF/STBL

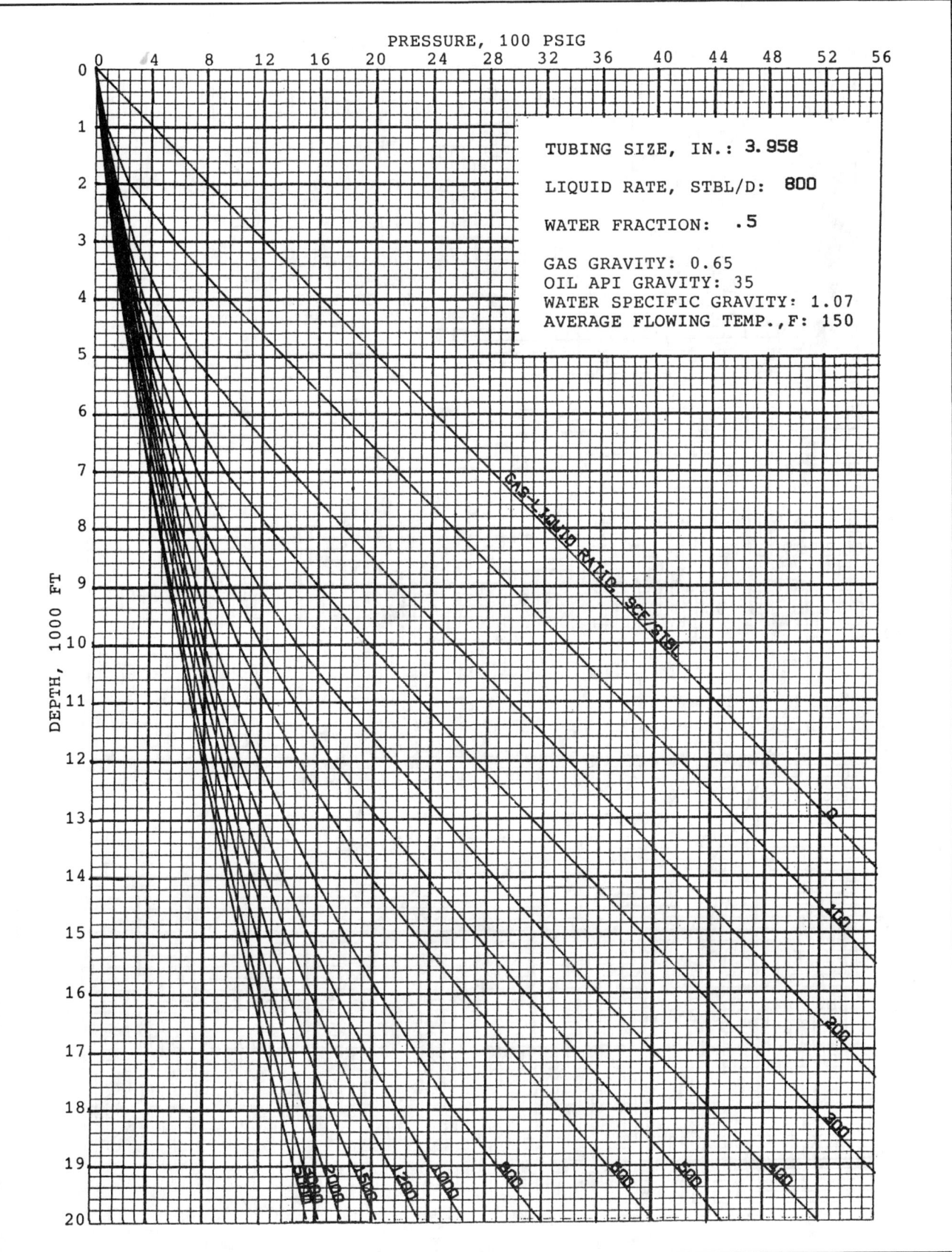

PRESSURE, 100 PSIG

DEPTH, 1000 FT

TUBING SIZE, IN.: **3.958**

LIQUID RATE, STBL/D: **800**

WATER FRACTION: **.5**

GAS GRAVITY: 0.65
OIL API GRAVITY: 35
WATER SPECIFIC GRAVITY: 1.07
AVERAGE FLOWING TEMP.,F: 150

GAS-LIQUID RATIO, SCF/STBL

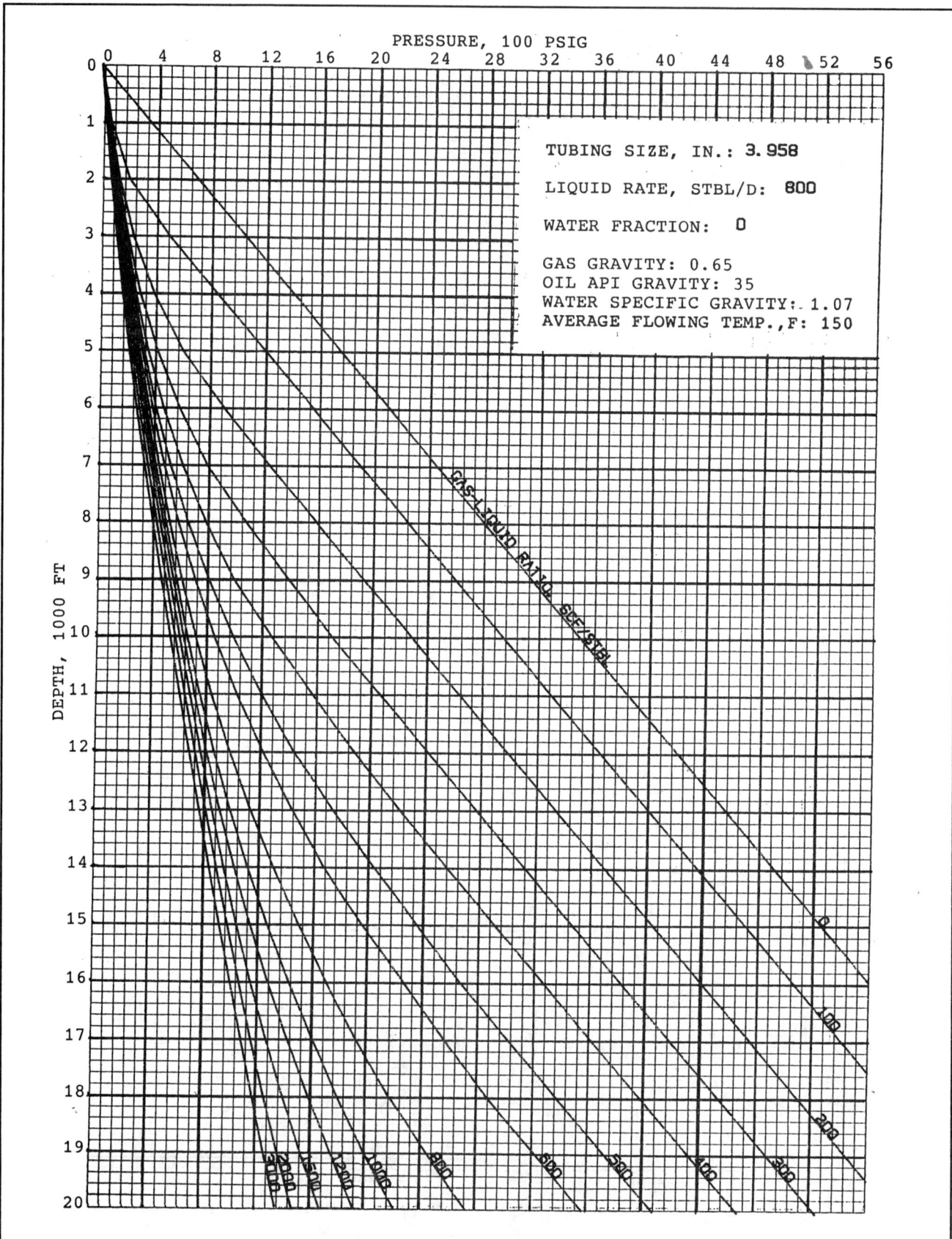

PRESSURE, 100 PSIG

TUBING SIZE, IN.: 3.958

LIQUID RATE, STBL/D: 800

WATER FRACTION: 0

GAS GRAVITY: 0.65
OIL API GRAVITY: 35
WATER SPECIFIC GRAVITY: 1.07
AVERAGE FLOWING TEMP.,F: 150

DEPTH, 1000 FT

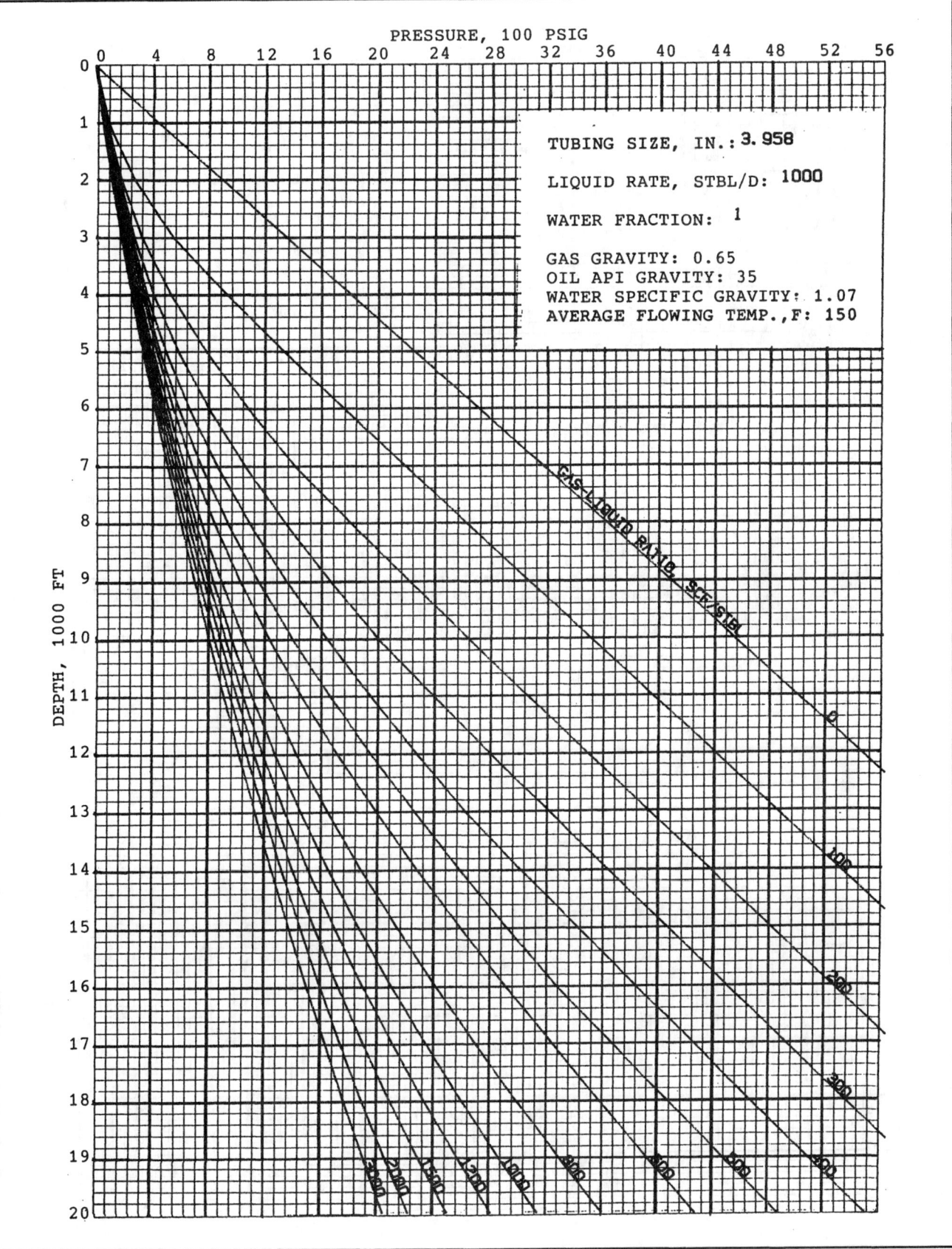

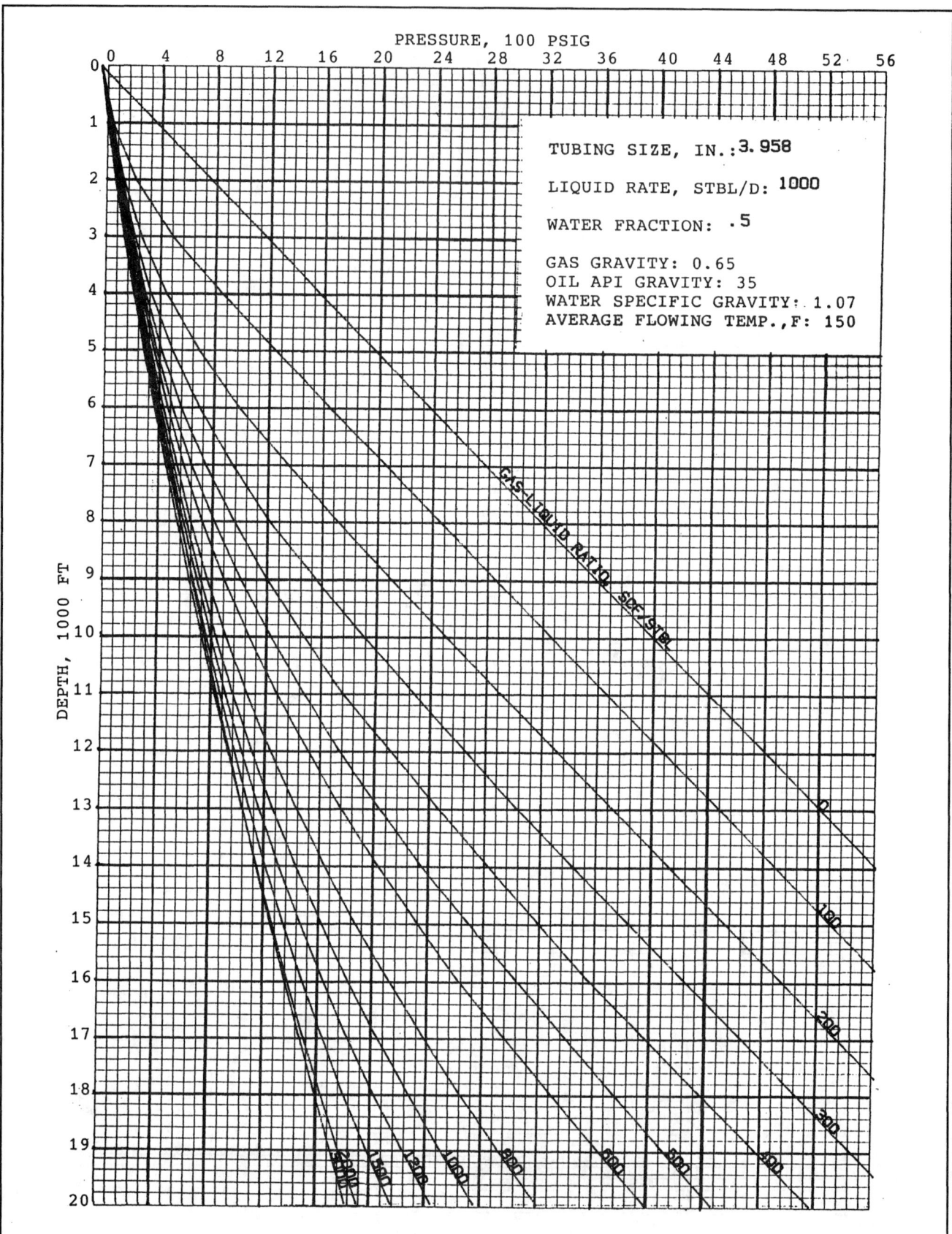

PRESSURE, 100 PSIG

TUBING SIZE, IN.: **3.958**

LIQUID RATE, STBL/D: **1000**

WATER FRACTION: **.5**

GAS GRAVITY: 0.65
OIL API GRAVITY: 35
WATER SPECIFIC GRAVITY: 1.07
AVERAGE FLOWING TEMP., F: 150

DEPTH, 1000 FT

GAS-LIQUID RATIO, SCF/STBL

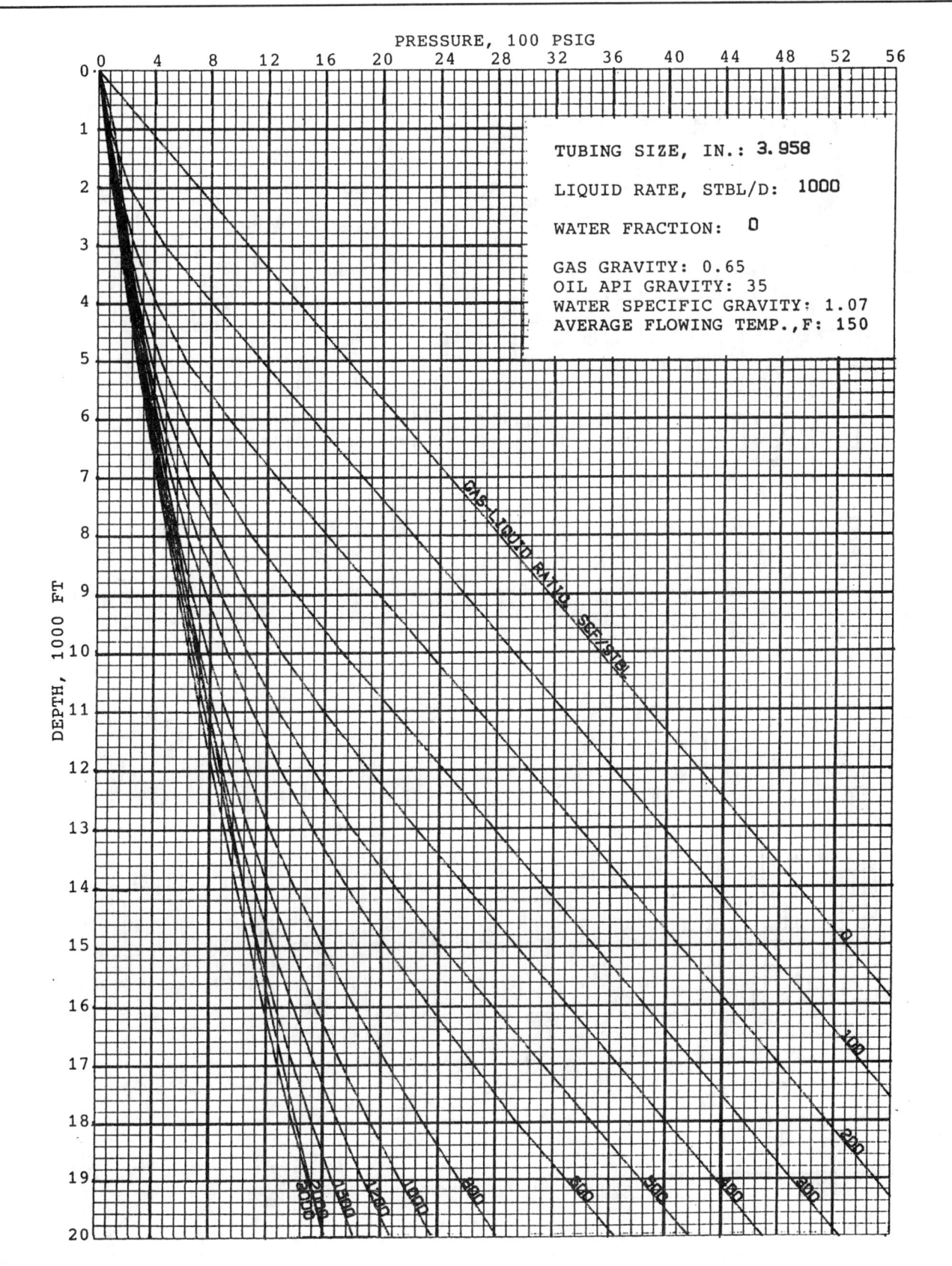

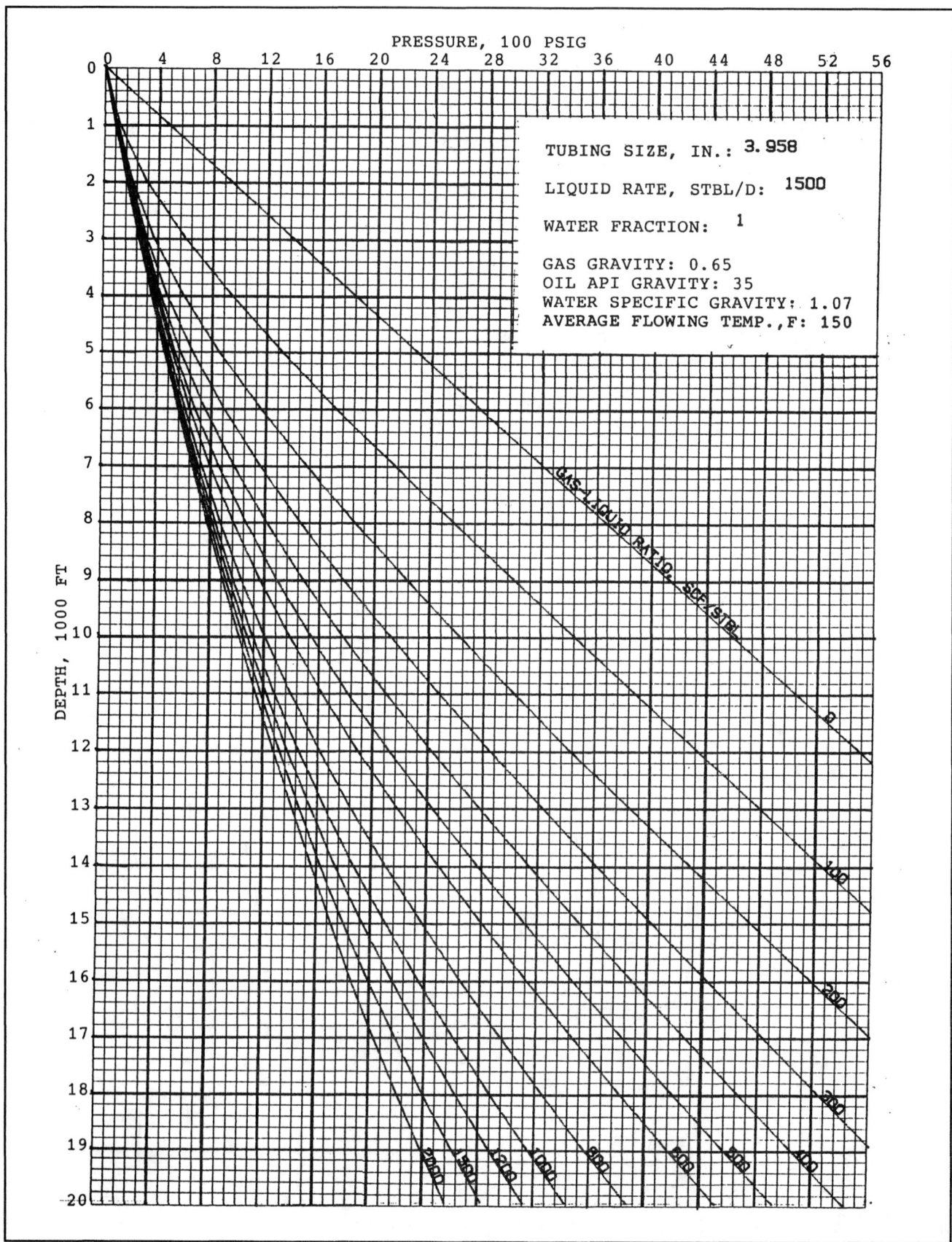

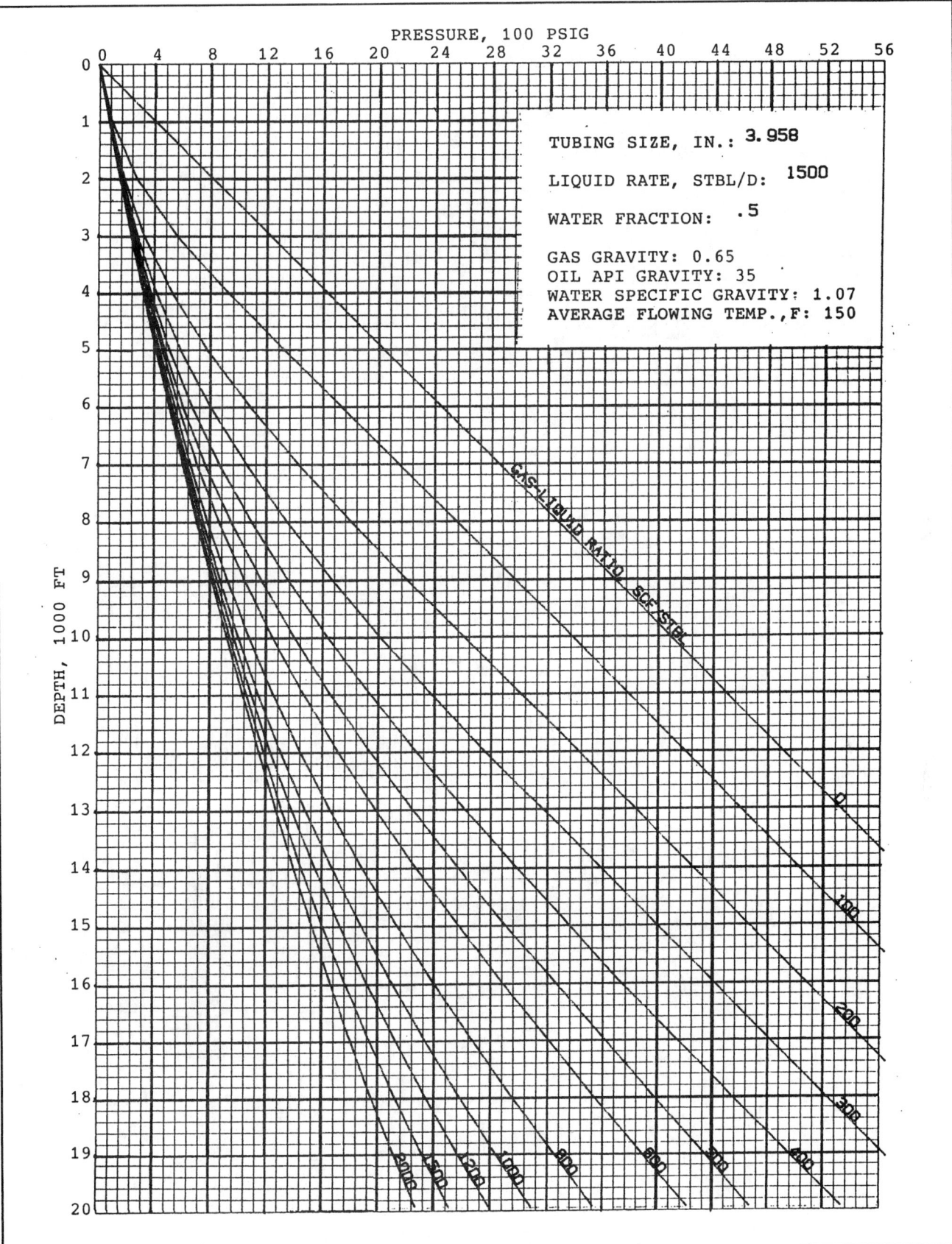

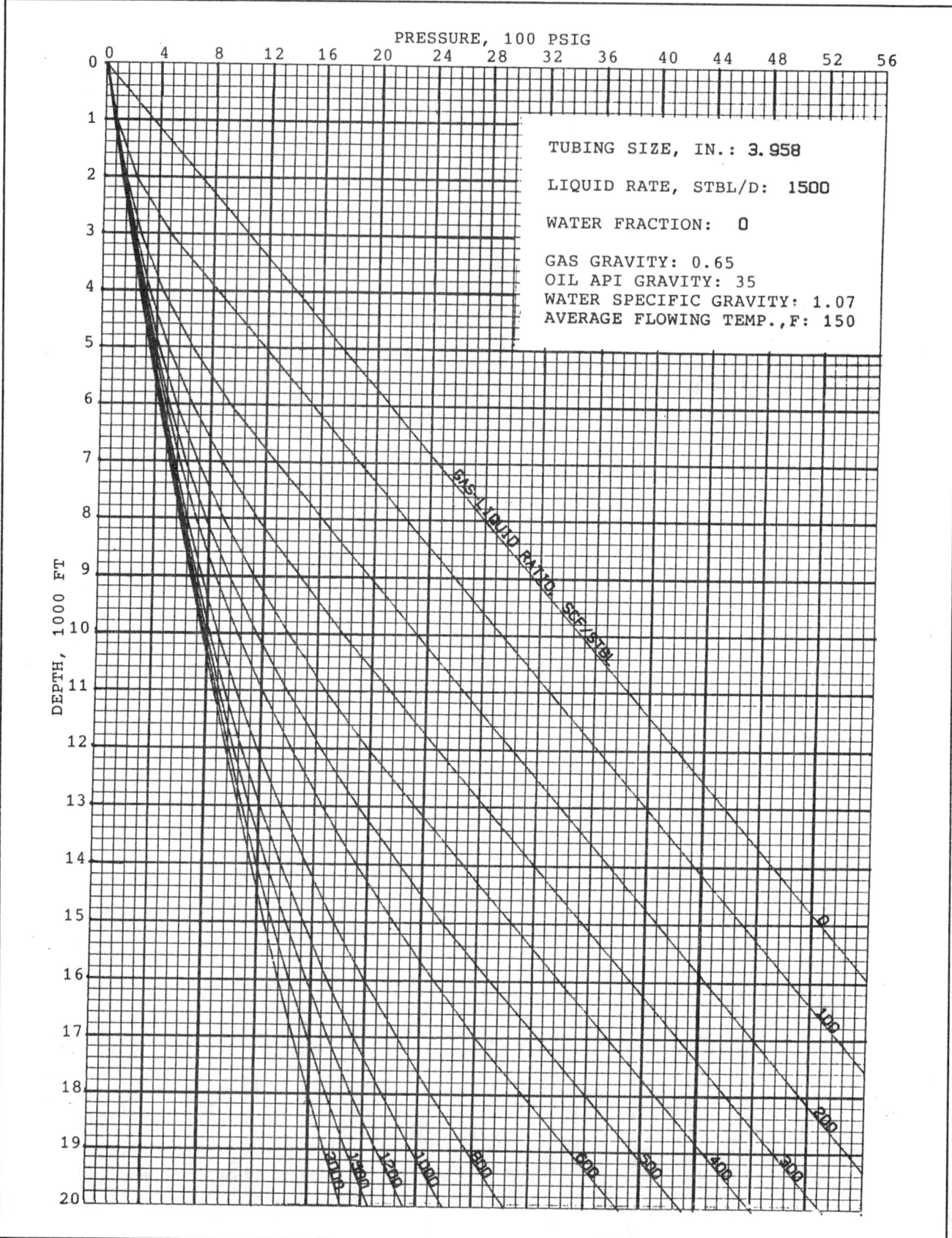

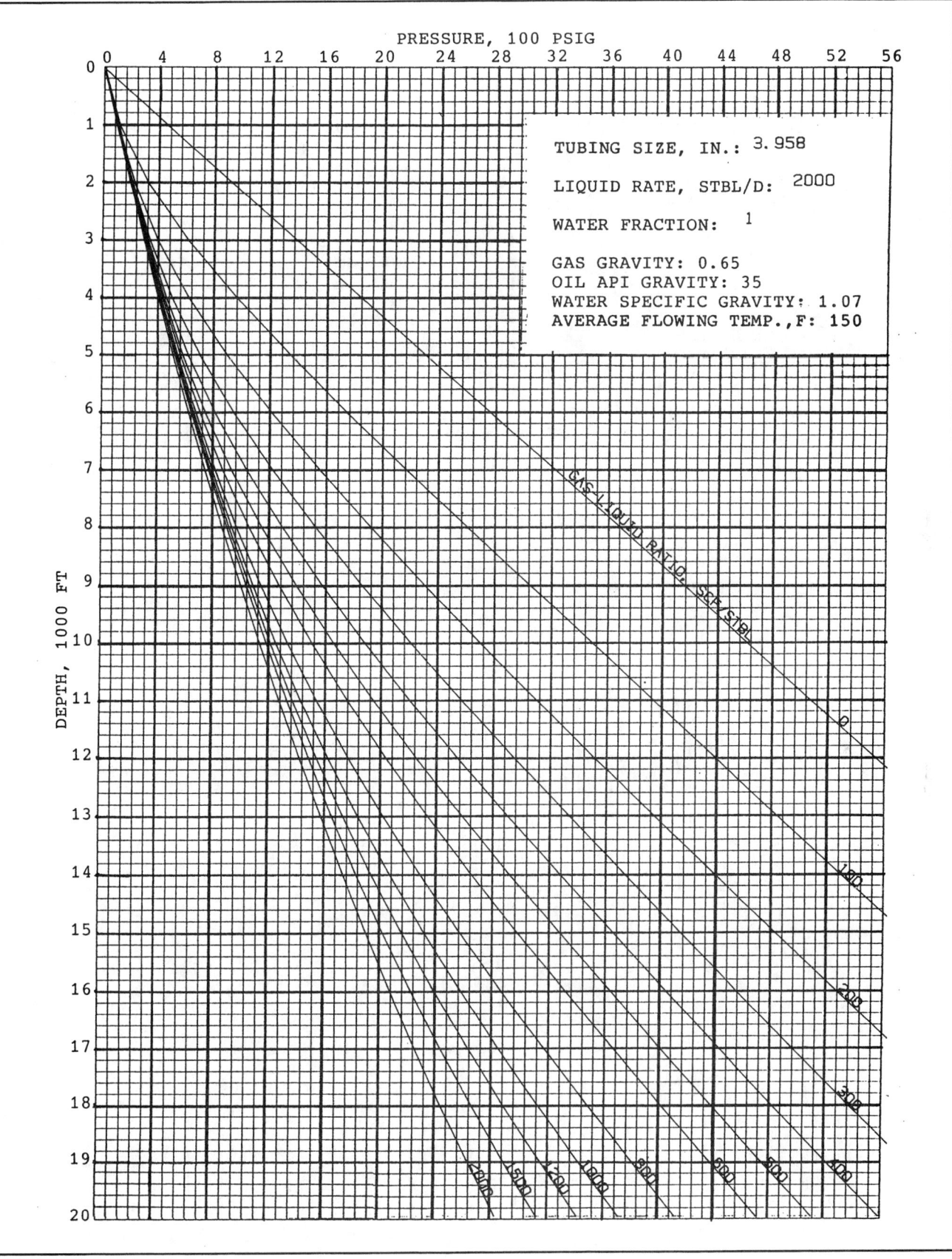

TUBING SIZE, IN.: 3.958

LIQUID RATE, STBL/D: 2000

WATER FRACTION: 1

GAS GRAVITY: 0.65
OIL API GRAVITY: 35
WATER SPECIFIC GRAVITY: 1.07
AVERAGE FLOWING TEMP.,F: 150

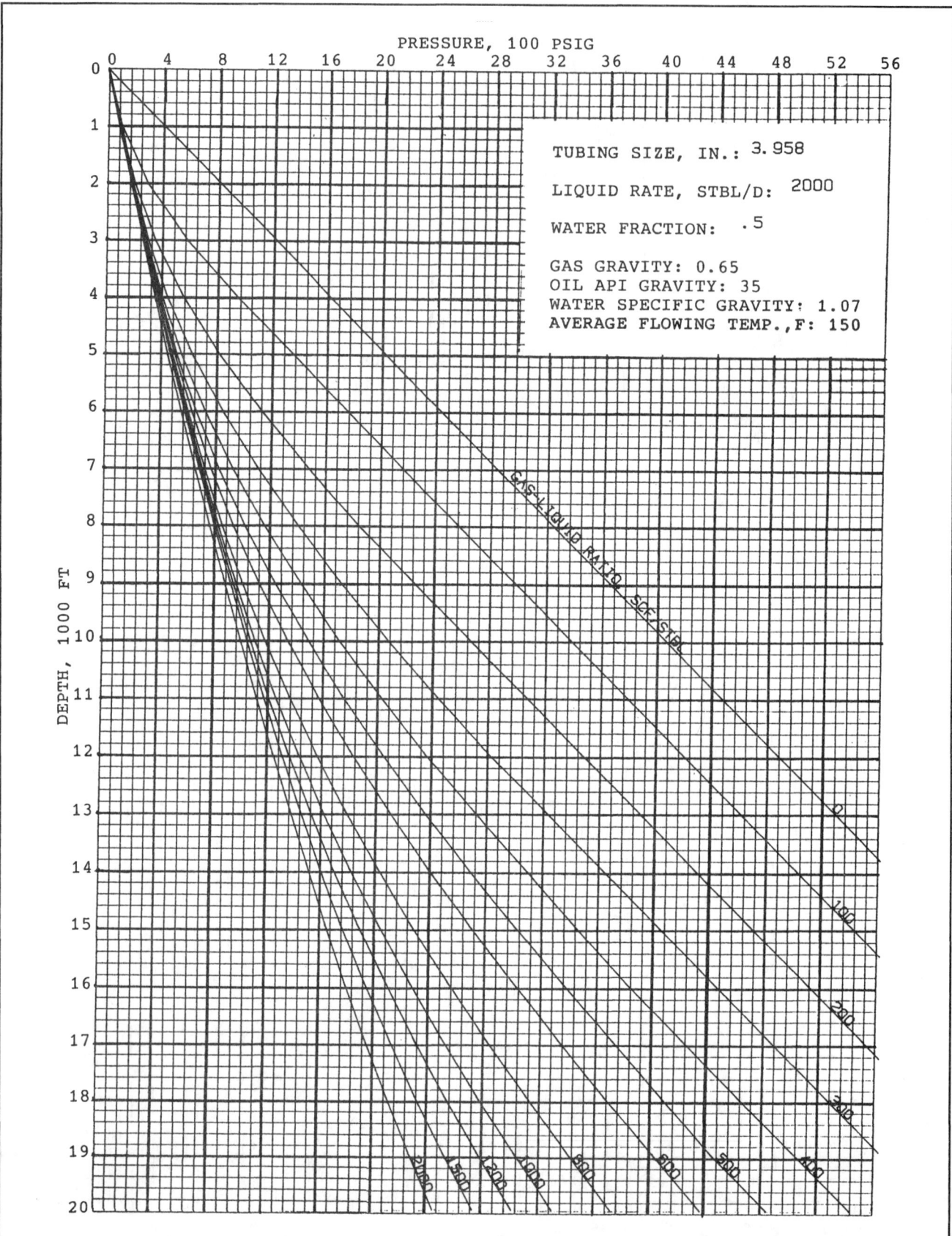

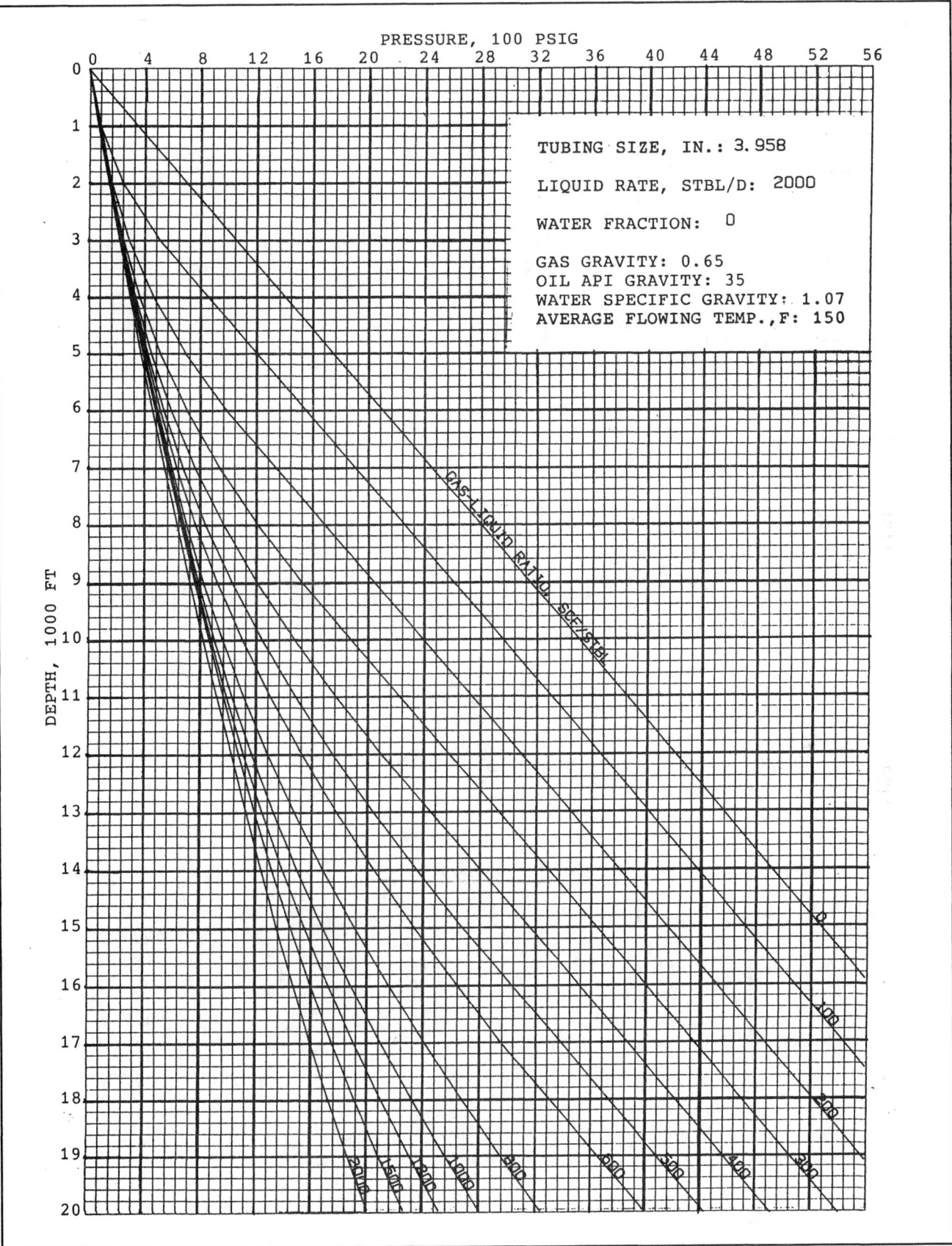

PRESSURE, 100 PSIG

DEPTH, 1000 FT

TUBING SIZE, IN.: 3.958

LIQUID RATE, STBL/D: 2000

WATER FRACTION: 0

GAS GRAVITY: 0.65
OIL API GRAVITY: 35
WATER SPECIFIC GRAVITY: 1.07
AVERAGE FLOWING TEMP.,F: 150

GAS-LIQUID RATIO, SCF/STBL

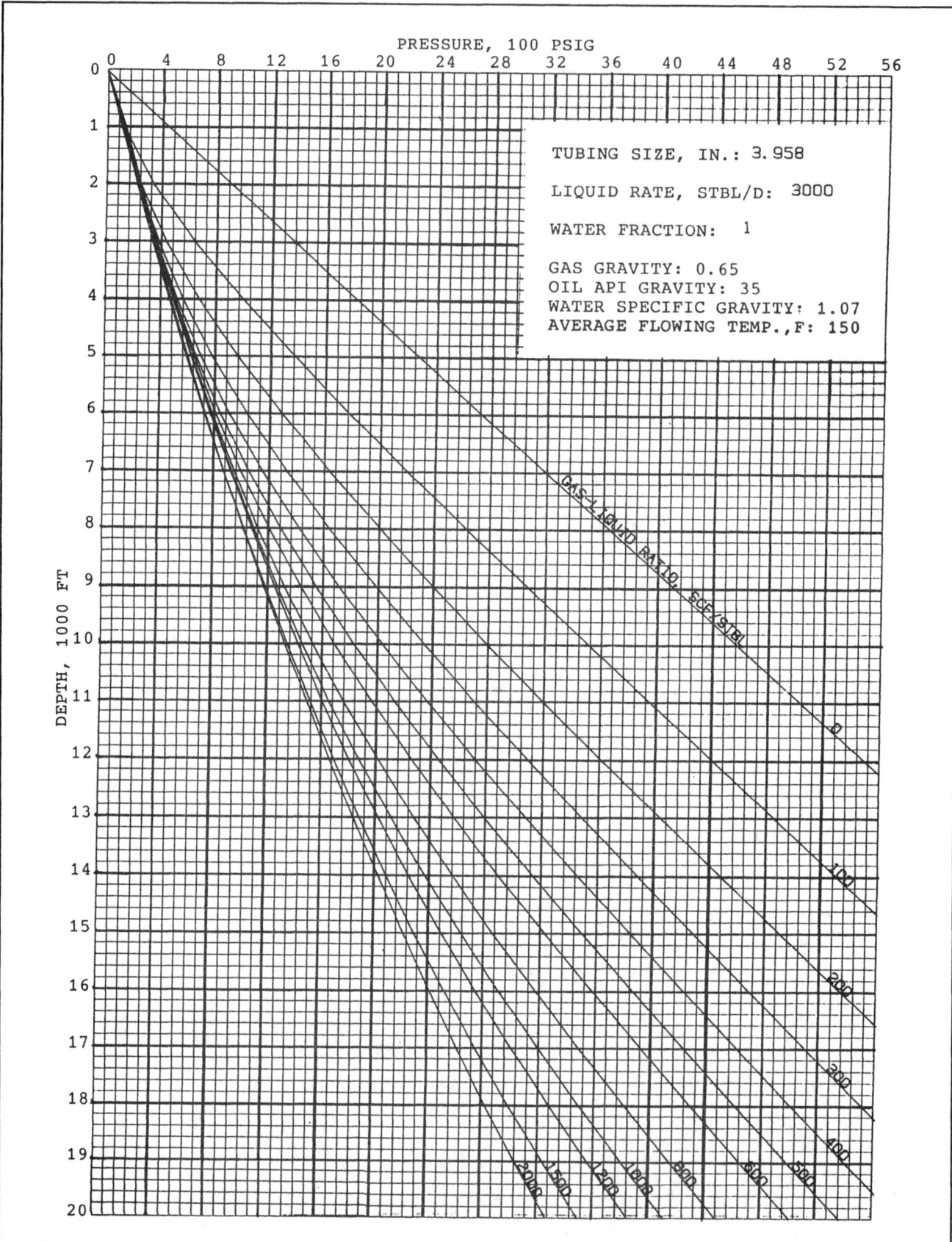

PRESSURE, 100 PSIG

TUBING SIZE, IN.: 3.958

LIQUID RATE, STBL/D: 3000

WATER FRACTION: 1

GAS GRAVITY: 0.65
OIL API GRAVITY: 35
WATER SPECIFIC GRAVITY: 1.07
AVERAGE FLOWING TEMP.,F: 150

DEPTH, 1000 FT

GAS-LIQUID RATIO, scf/STBL

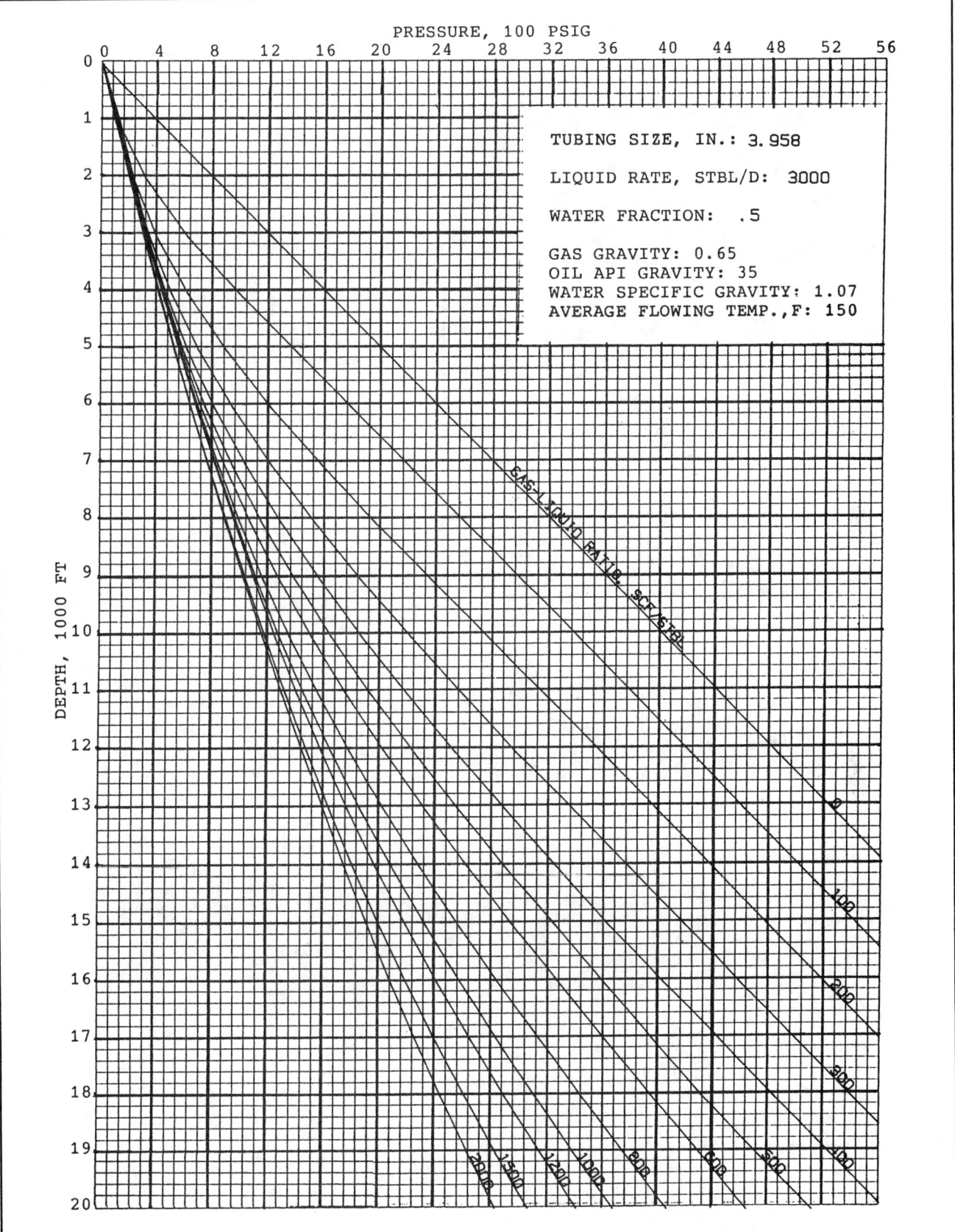

PRESSURE, 100 PSIG

DEPTH, 1000 FT

TUBING SIZE, IN.: 3.958

LIQUID RATE, STBL/D: 3000

WATER FRACTION: .5

GAS GRAVITY: 0.65
OIL API GRAVITY: 35
WATER SPECIFIC GRAVITY: 1.07
AVERAGE FLOWING TEMP.,F: 150

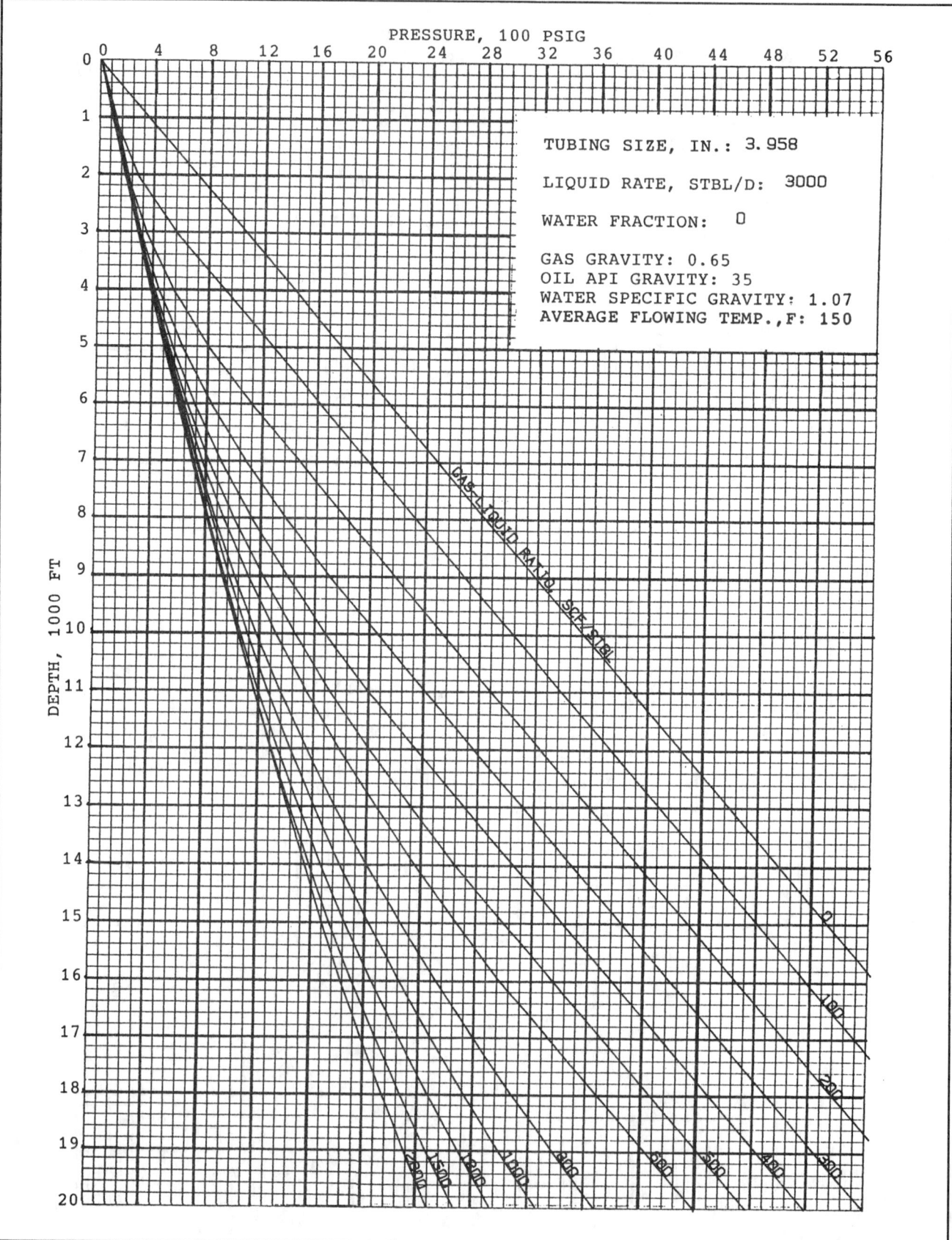

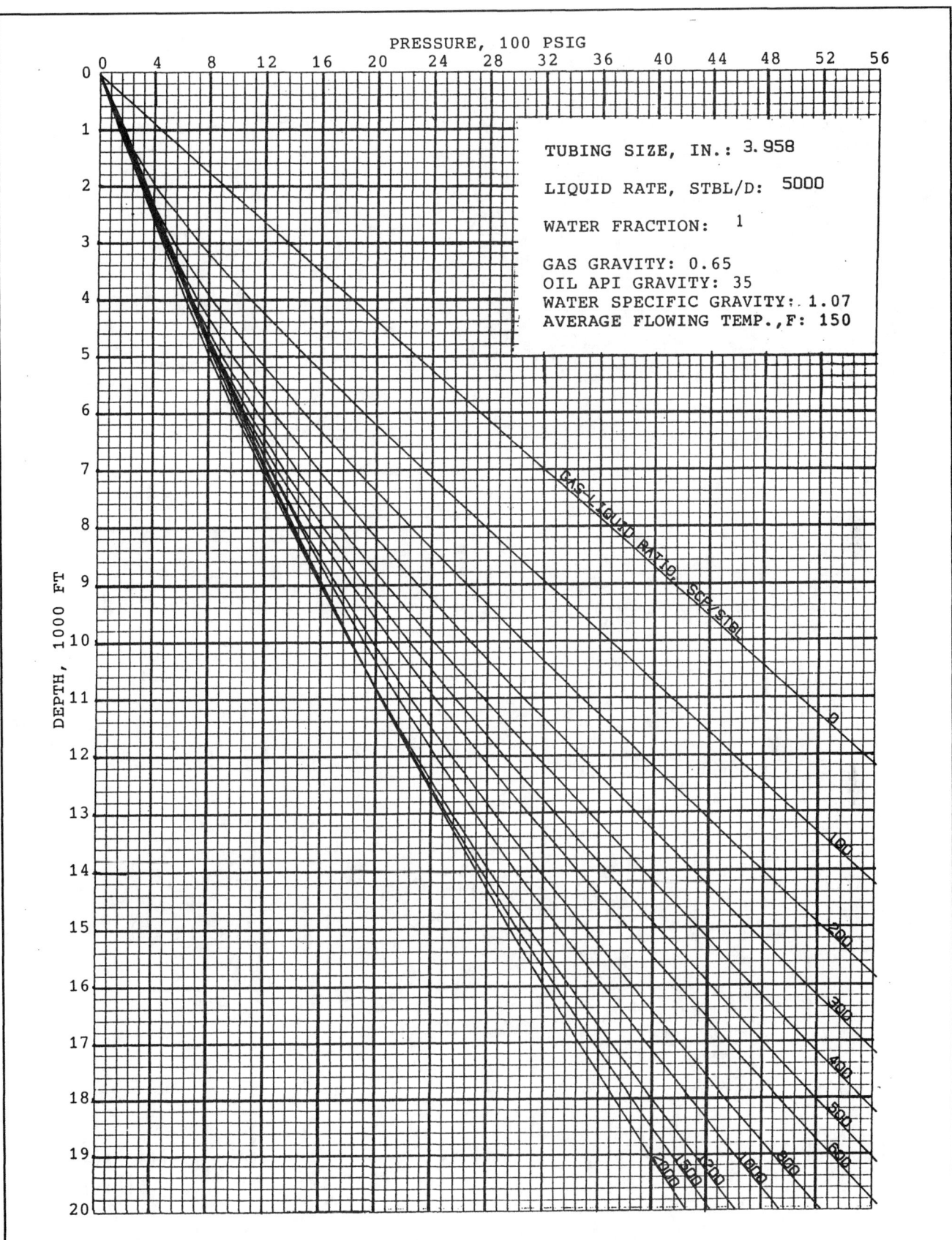

PRESSURE, 100 PSIG

DEPTH, 1000 FT

TUBING SIZE, IN.: 3.958

LIQUID RATE, STBL/D: 5000

WATER FRACTION: 1

GAS GRAVITY: 0.65
OIL API GRAVITY: 35
WATER SPECIFIC GRAVITY: 1.07
AVERAGE FLOWING TEMP.,F: 150

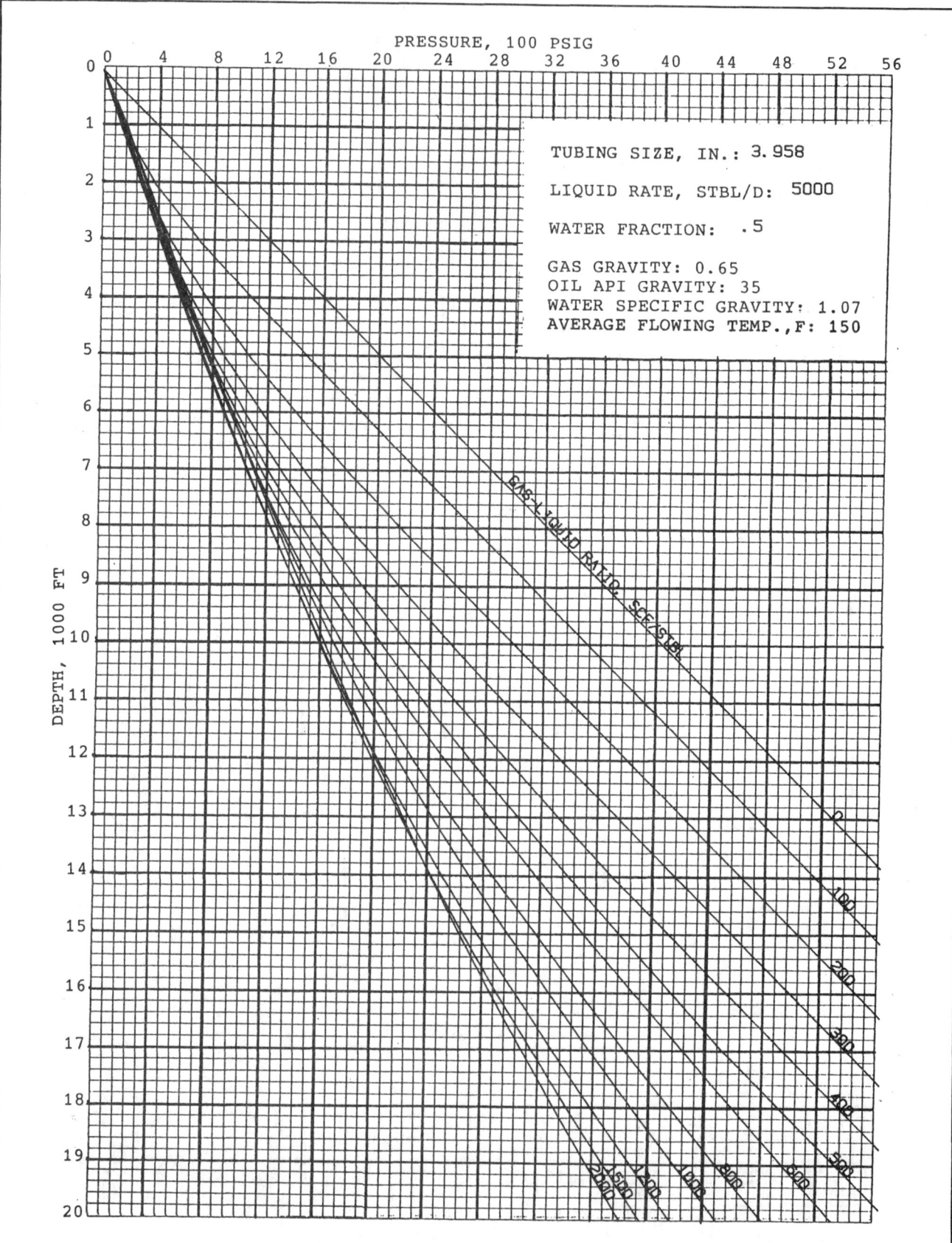

PRESSURE, 100 PSIG

DEPTH, 1000 FT

TUBING SIZE, IN.: 3.958

LIQUID RATE, STBL/D: 5000

WATER FRACTION: .5

GAS GRAVITY: 0.65
OIL API GRAVITY: 35
WATER SPECIFIC GRAVITY: 1.07
AVERAGE FLOWING TEMP.,F: 150

GAS-LIQUID RATIO, SCF/STBL

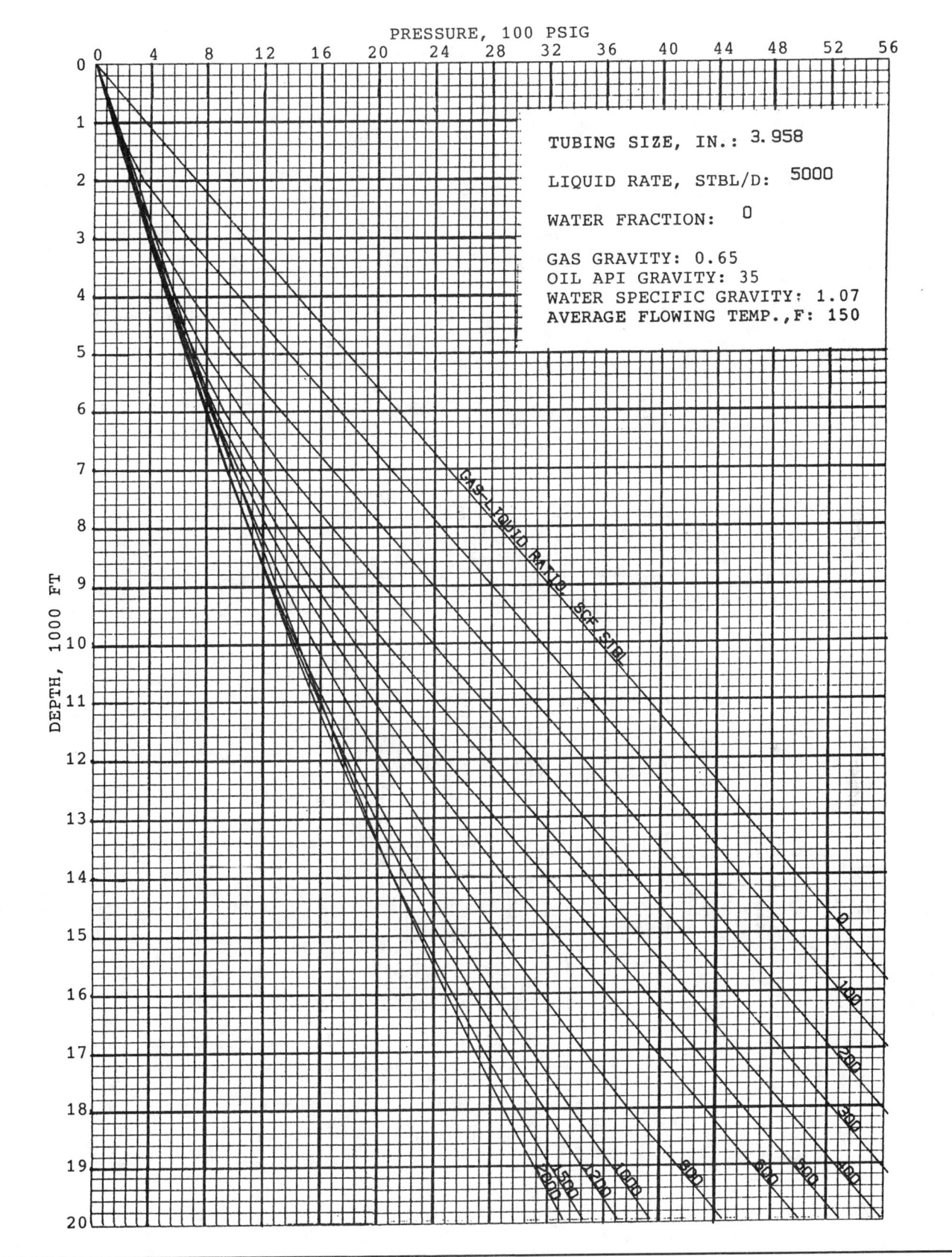

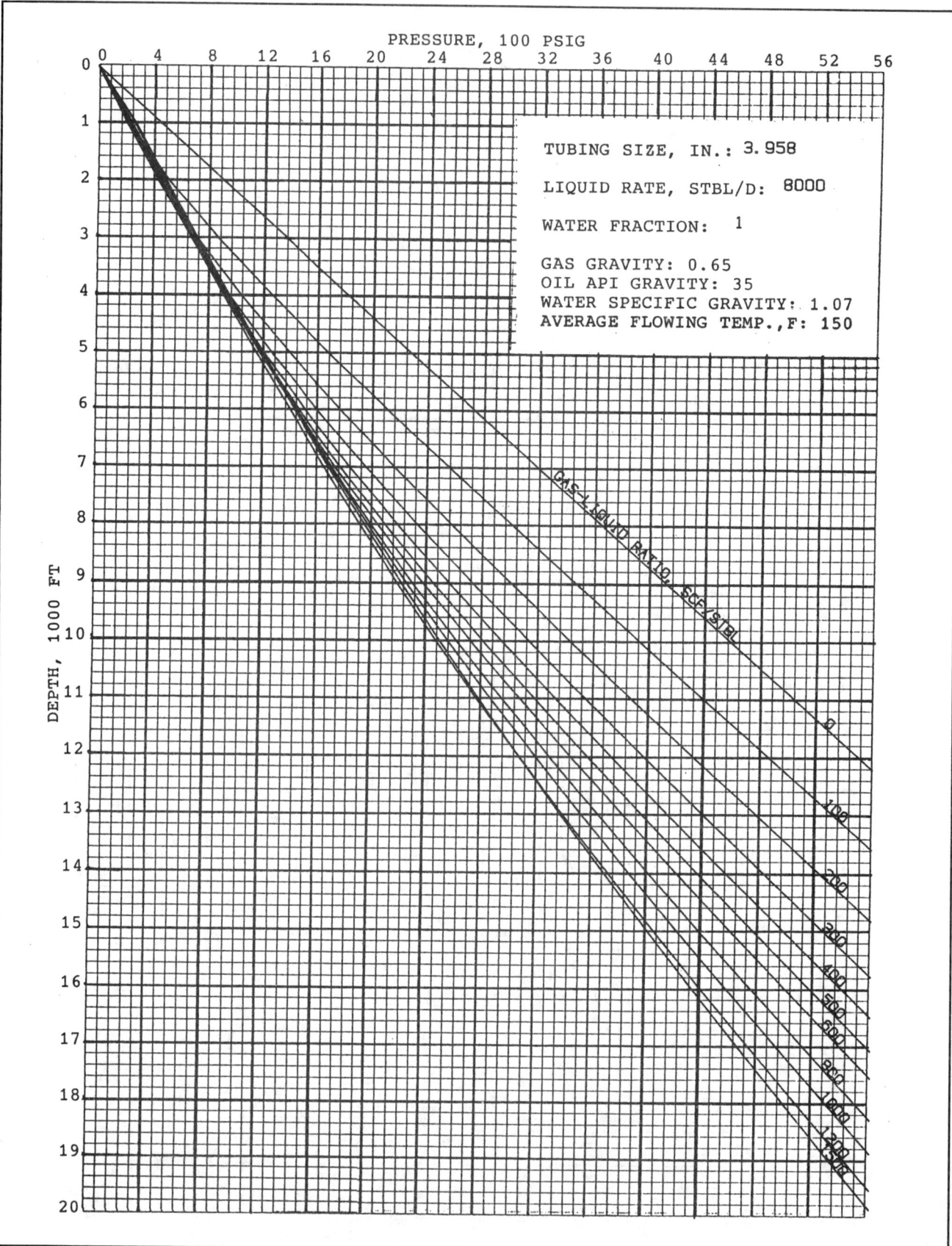

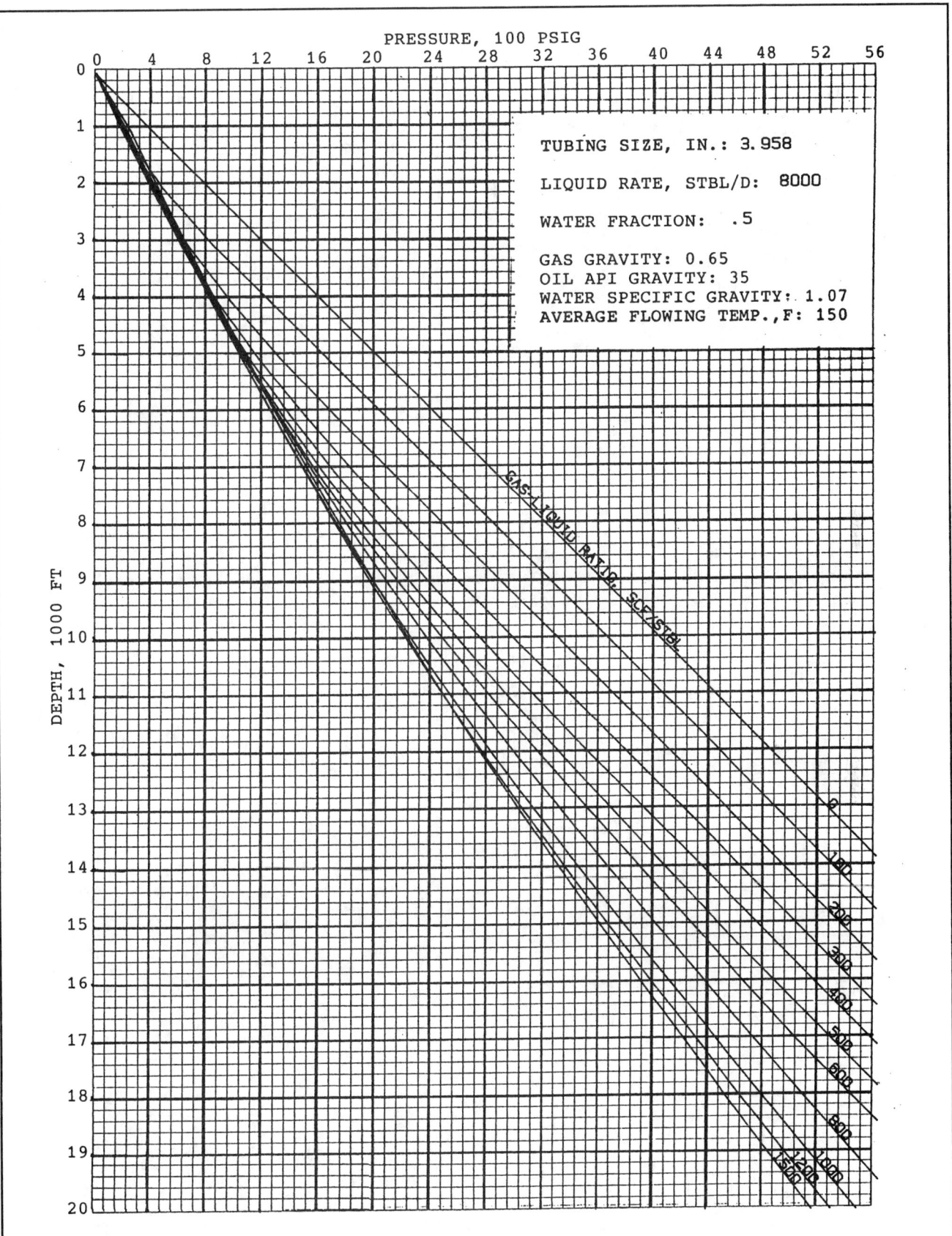

PRESSURE, 100 PSIG

DEPTH, 1000 FT

TUBING SIZE, IN.: 3.958

LIQUID RATE, STBL/D: 8000

WATER FRACTION: .5

GAS GRAVITY: 0.65
OIL API GRAVITY: 35
WATER SPECIFIC GRAVITY: 1.07
AVERAGE FLOWING TEMP.,F: 150

GAS-LIQUID RATIO, SCF/STBL

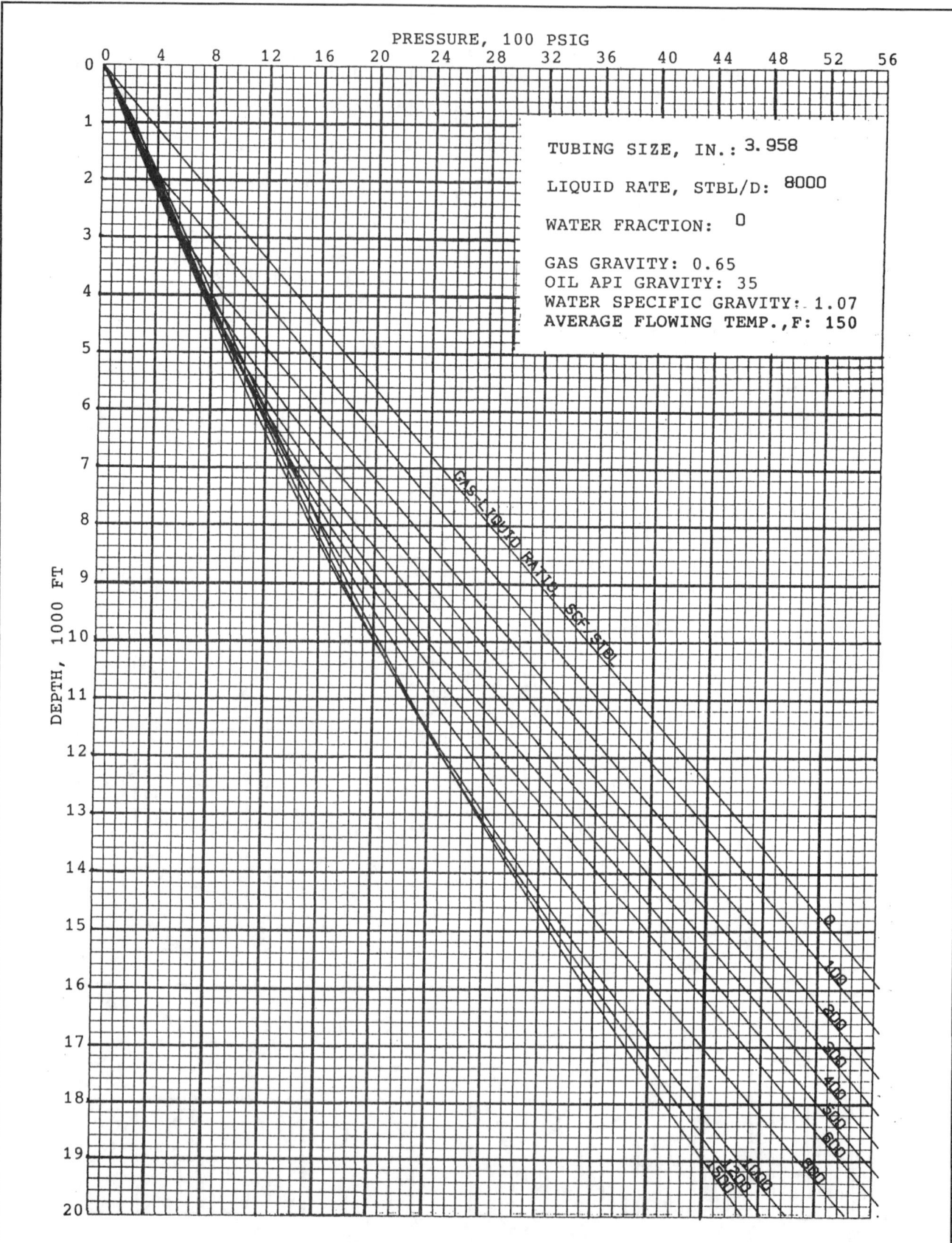

PRESSURE, 100 PSIG

TUBING SIZE, IN.: 3.958

LIQUID RATE, STBL/D: 8000

WATER FRACTION: 0

GAS GRAVITY: 0.65
OIL API GRAVITY: 35
WATER SPECIFIC GRAVITY: 1.07
AVERAGE FLOWING TEMP., F: 150

DEPTH, 1000 FT

GAS-LIQUID RATIO, SCF/STBL

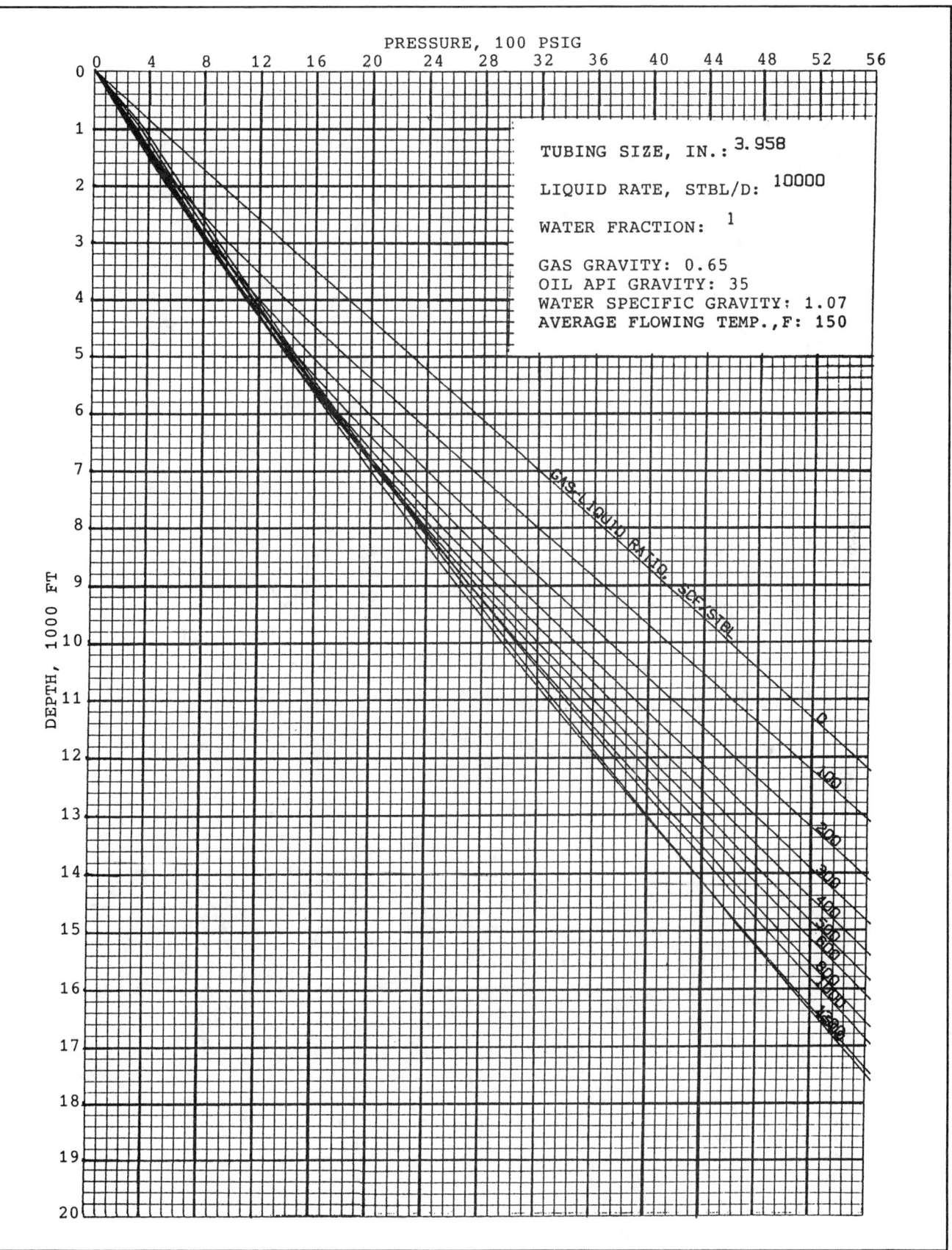

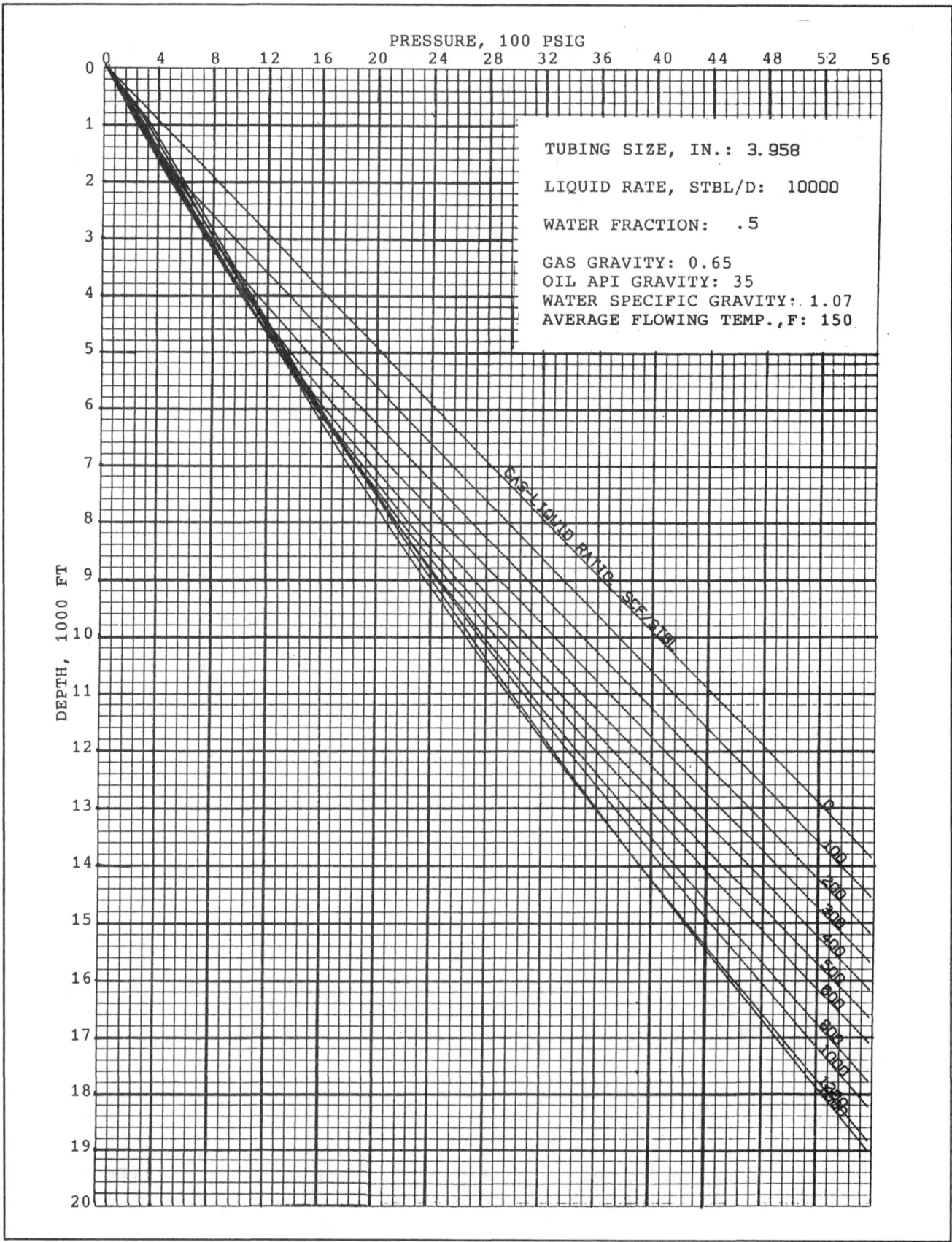

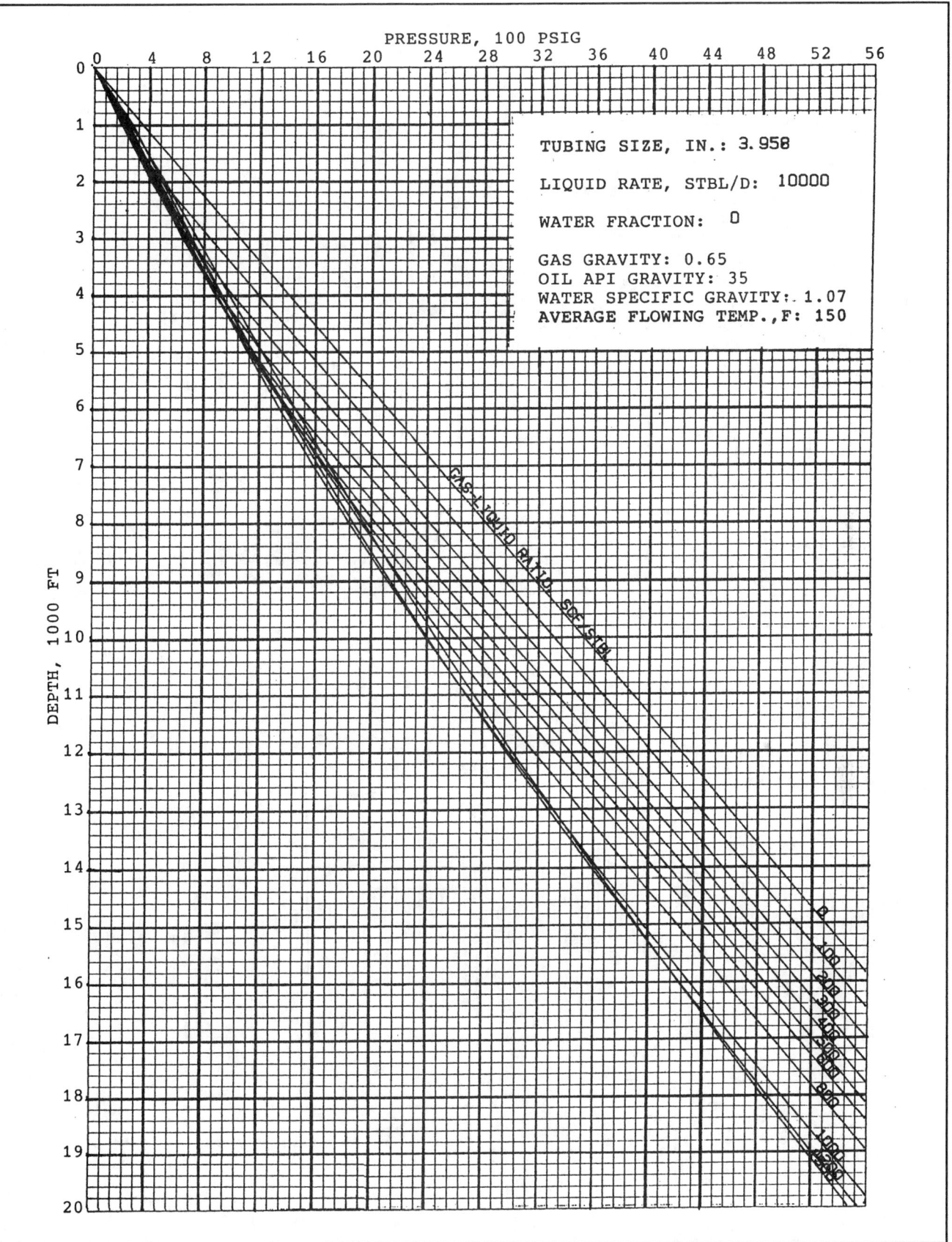

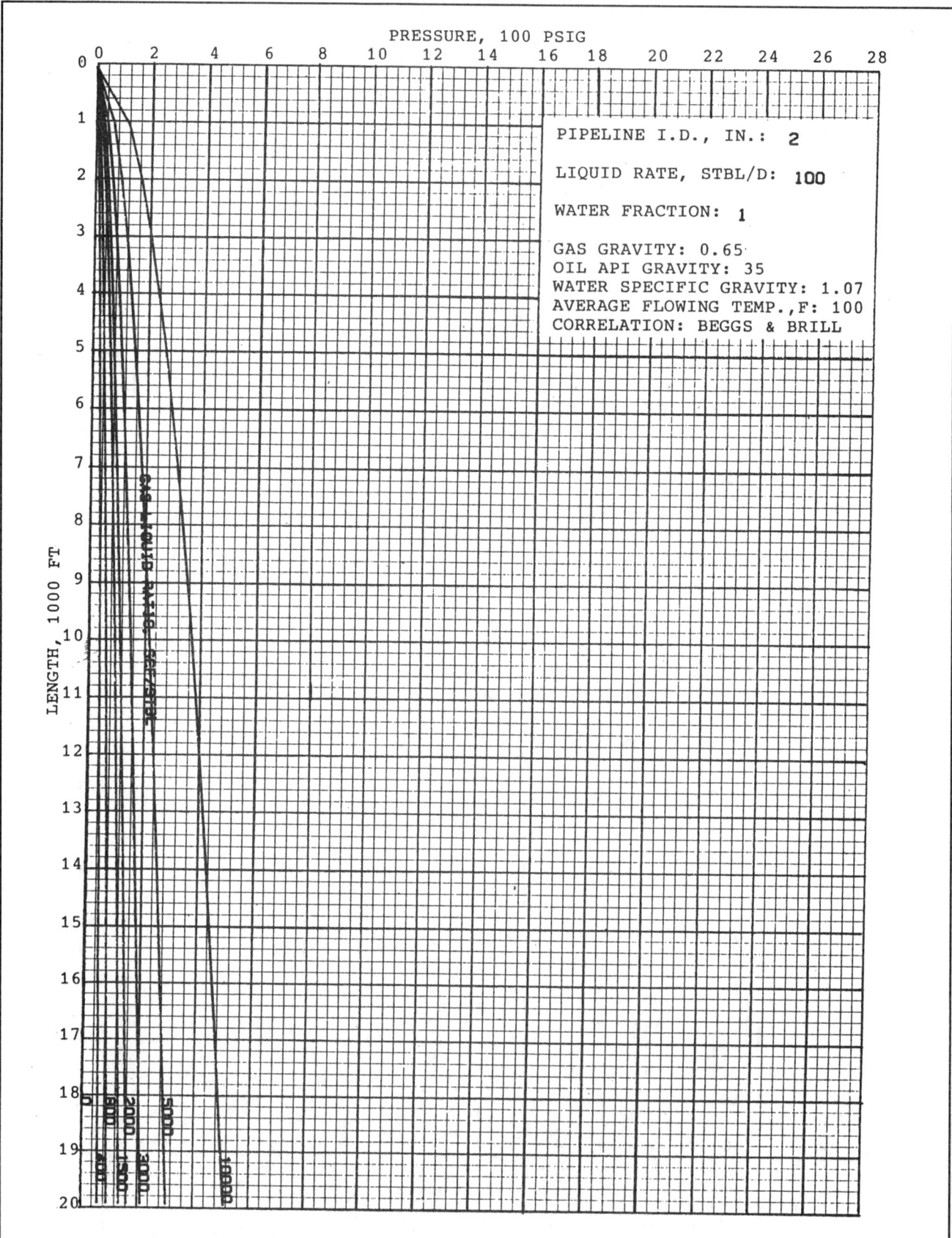

PRESSURE, 100 PSIG

LENGTH, 1000 FT

PIPELINE I.D., IN.: **2**

LIQUID RATE, STBL/D: **100**

WATER FRACTION: **1**

GAS GRAVITY: 0.65
OIL API GRAVITY: 35
WATER SPECIFIC GRAVITY: 1.07
AVERAGE FLOWING TEMP.,F: 100
CORRELATION: BEGGS & BRILL

GAS LIQUID RATIO, SCF/STBL

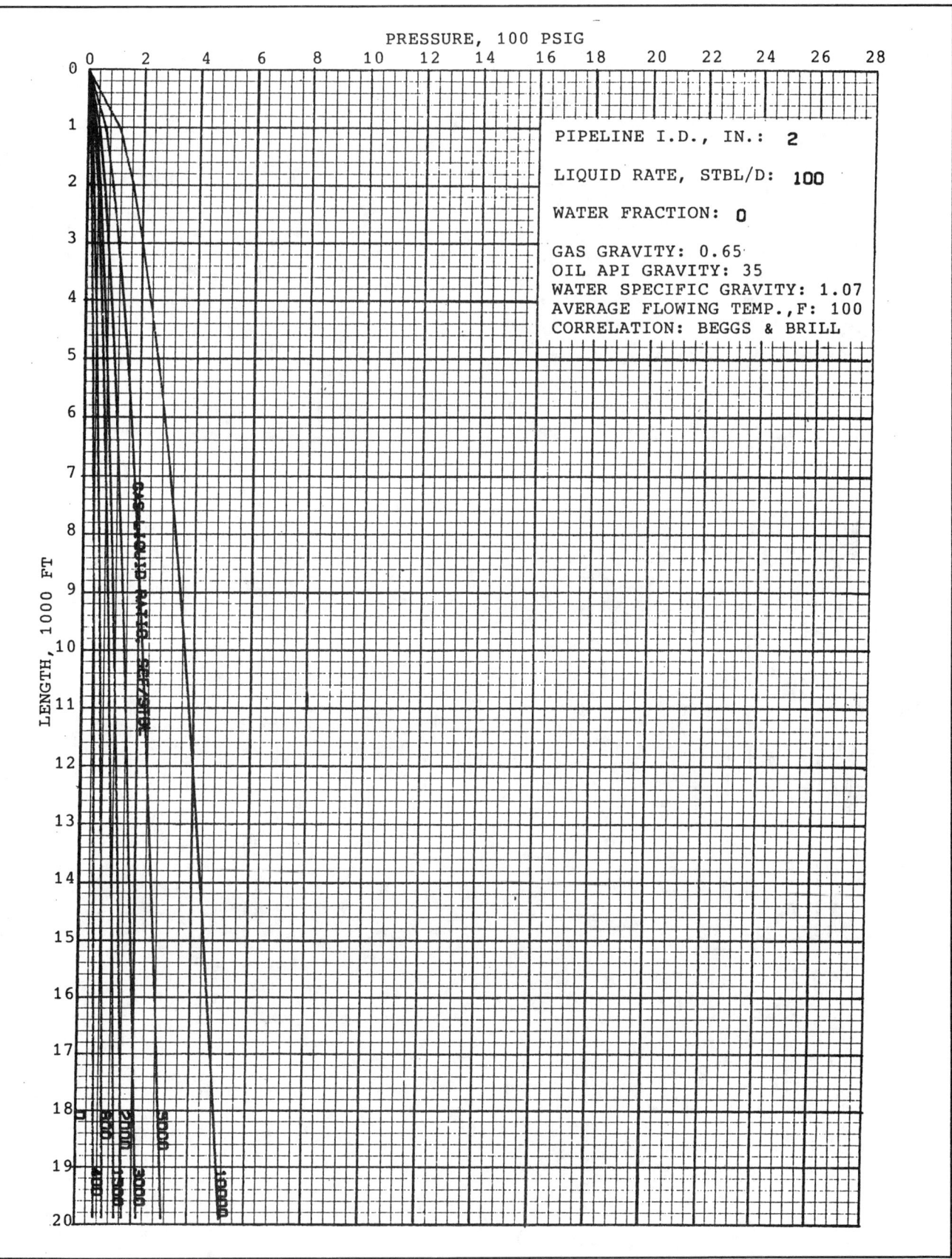

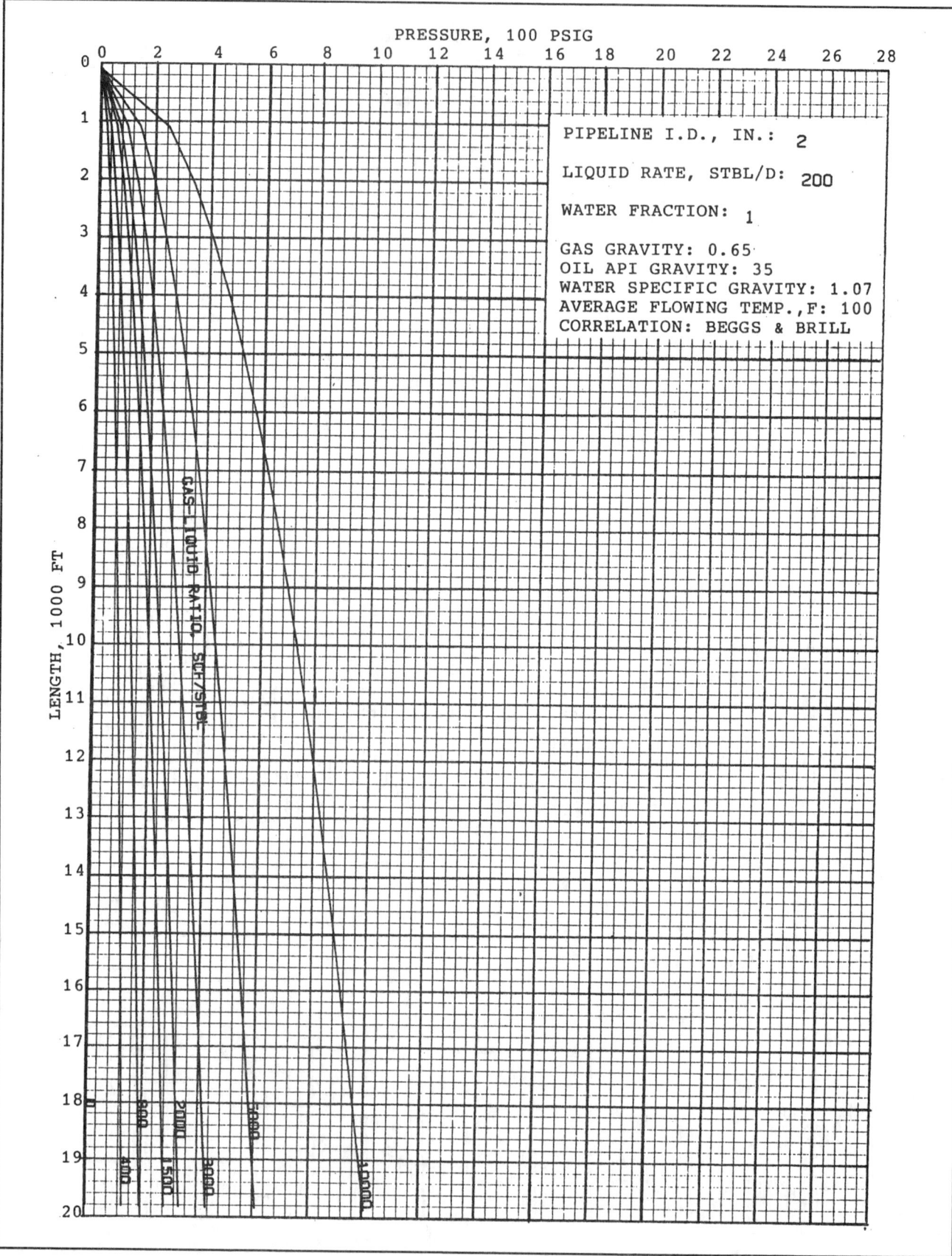

PRESSURE, 100 PSIG

LENGTH, 1000 FT

GAS-LIQUID RATIO, SCF/STBL

PIPELINE I.D., IN.: 2

LIQUID RATE, STBL/D: 200

WATER FRACTION: 1

GAS GRAVITY: 0.65
OIL API GRAVITY: 35
WATER SPECIFIC GRAVITY: 1.07
AVERAGE FLOWING TEMP.,F: 100
CORRELATION: BEGGS & BRILL

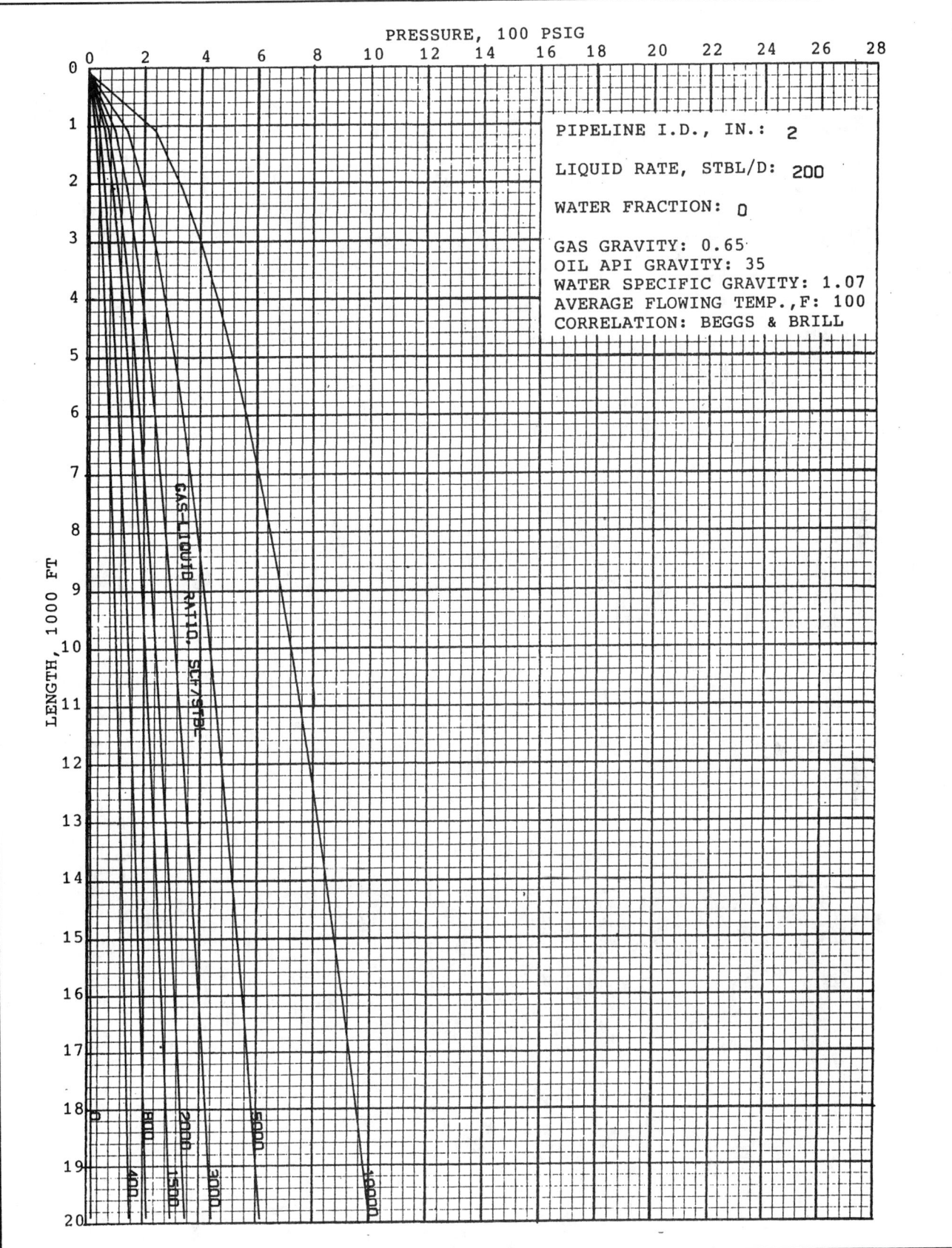

PRESSURE, 100 PSIG

LENGTH, 1000 FT

GAS-LIQUID RATIO, SCF/STBL

PIPELINE I.D., IN.: 2

LIQUID RATE, STBL/D: 200

WATER FRACTION: 0

GAS GRAVITY: 0.65
OIL API GRAVITY: 35
WATER SPECIFIC GRAVITY: 1.07
AVERAGE FLOWING TEMP.,F: 100
CORRELATION: BEGGS & BRILL

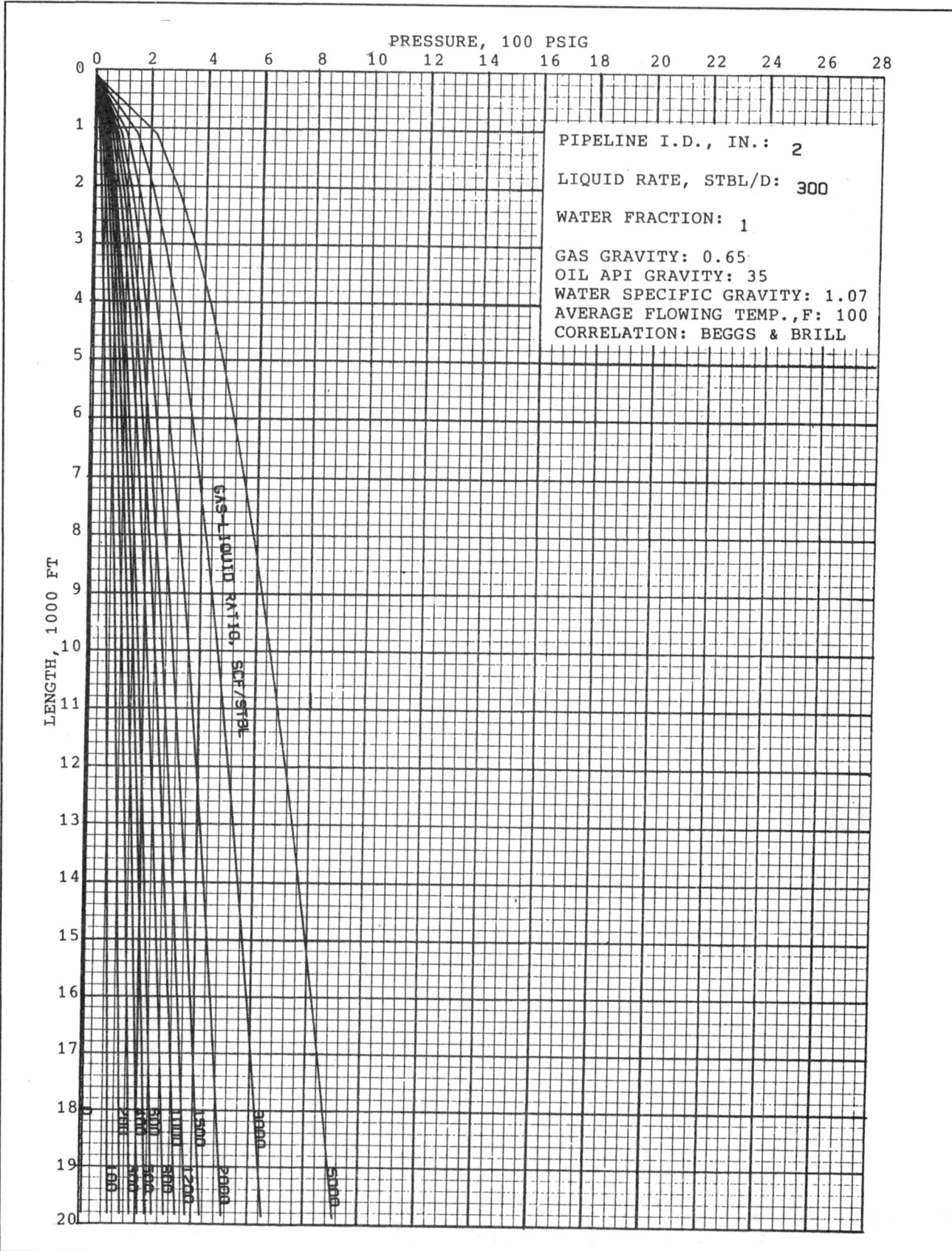

PRESSURE, 100 PSIG

LENGTH, 1000 FT

GAS-LIQUID RATIO, SCF/STBL

PIPELINE I.D., IN.: 2

LIQUID RATE, STBL/D: 300

WATER FRACTION: 1

GAS GRAVITY: 0.65
OIL API GRAVITY: 35
WATER SPECIFIC GRAVITY: 1.07
AVERAGE FLOWING TEMP.,F: 100
CORRELATION: BEGGS & BRILL

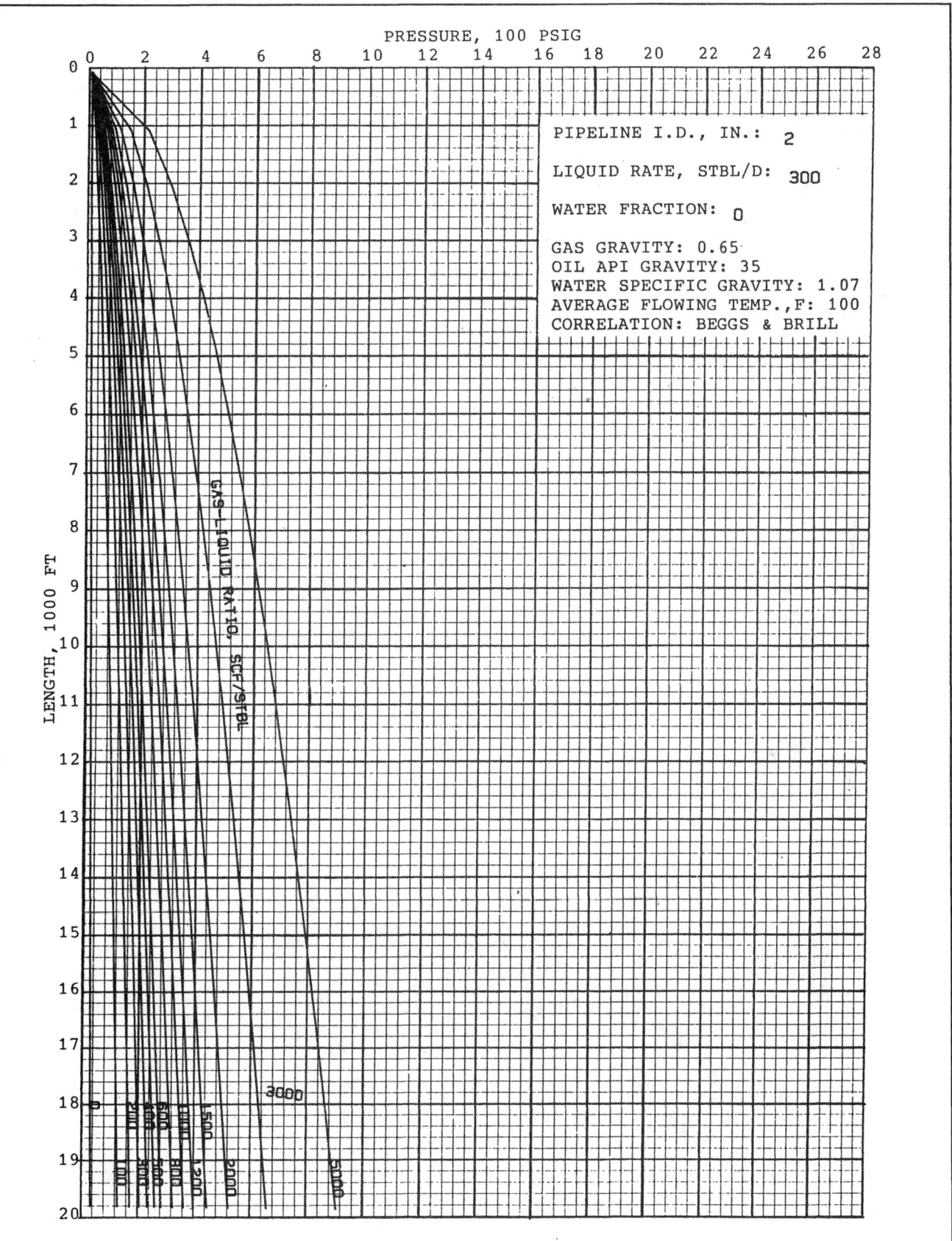

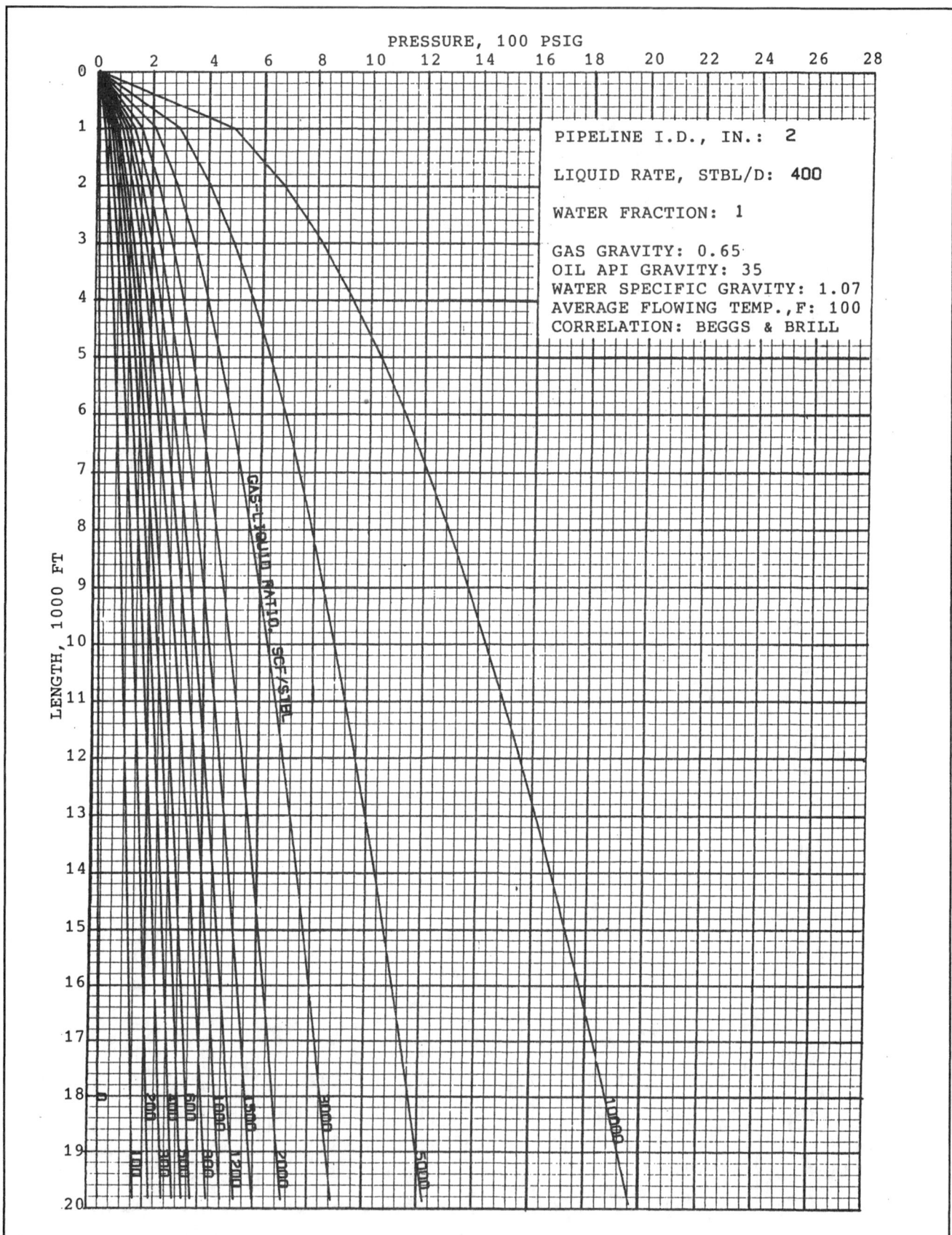

PRESSURE, 100 PSIG

PIPELINE I.D., IN.: 2

LIQUID RATE, STBL/D: 400

WATER FRACTION: 1

GAS GRAVITY: 0.65
OIL API GRAVITY: 35
WATER SPECIFIC GRAVITY: 1.07
AVERAGE FLOWING TEMP.,F: 100
CORRELATION: BEGGS & BRILL

LENGTH, 1000 FT

GAS-LIQUID RATIO, SCF/STBL

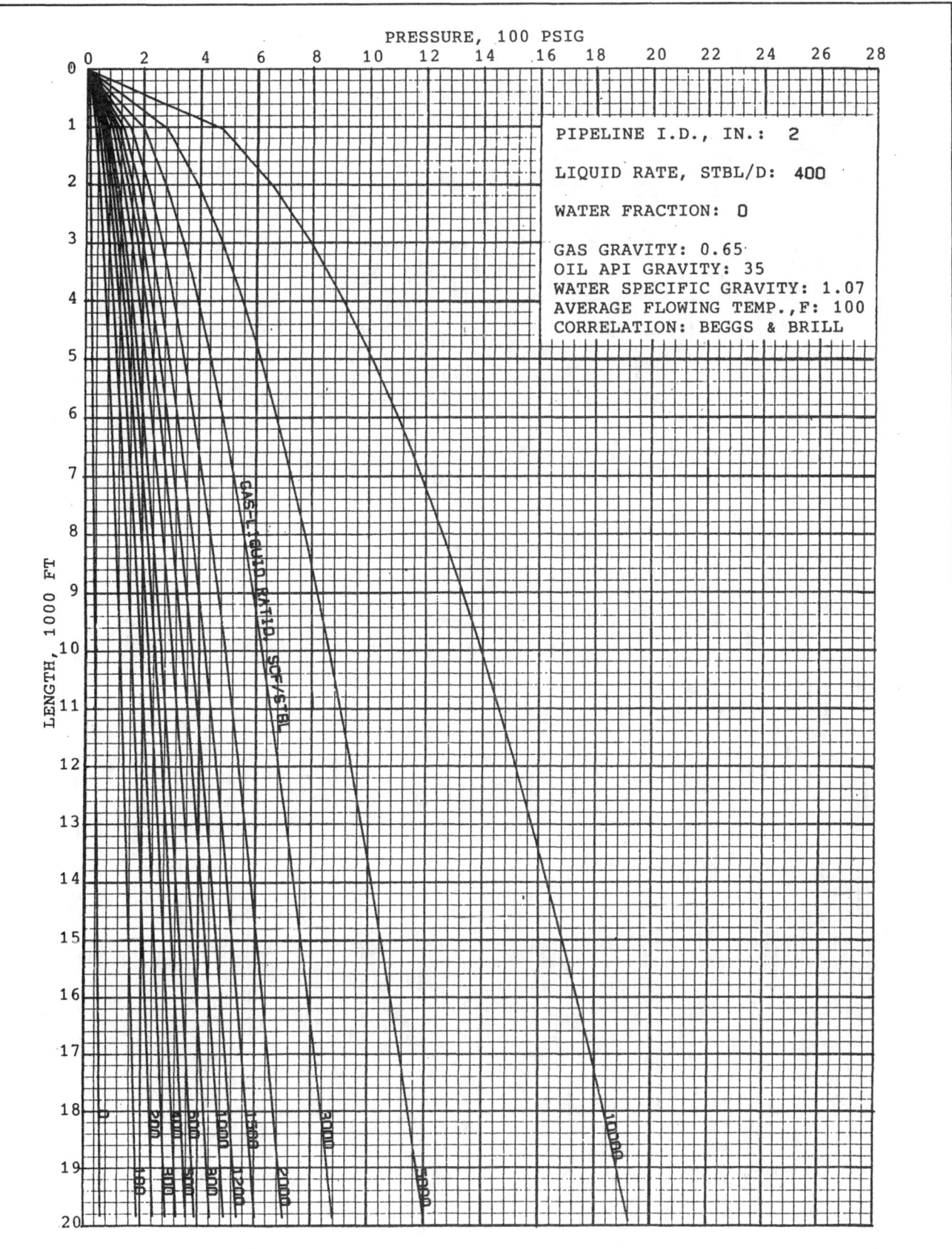

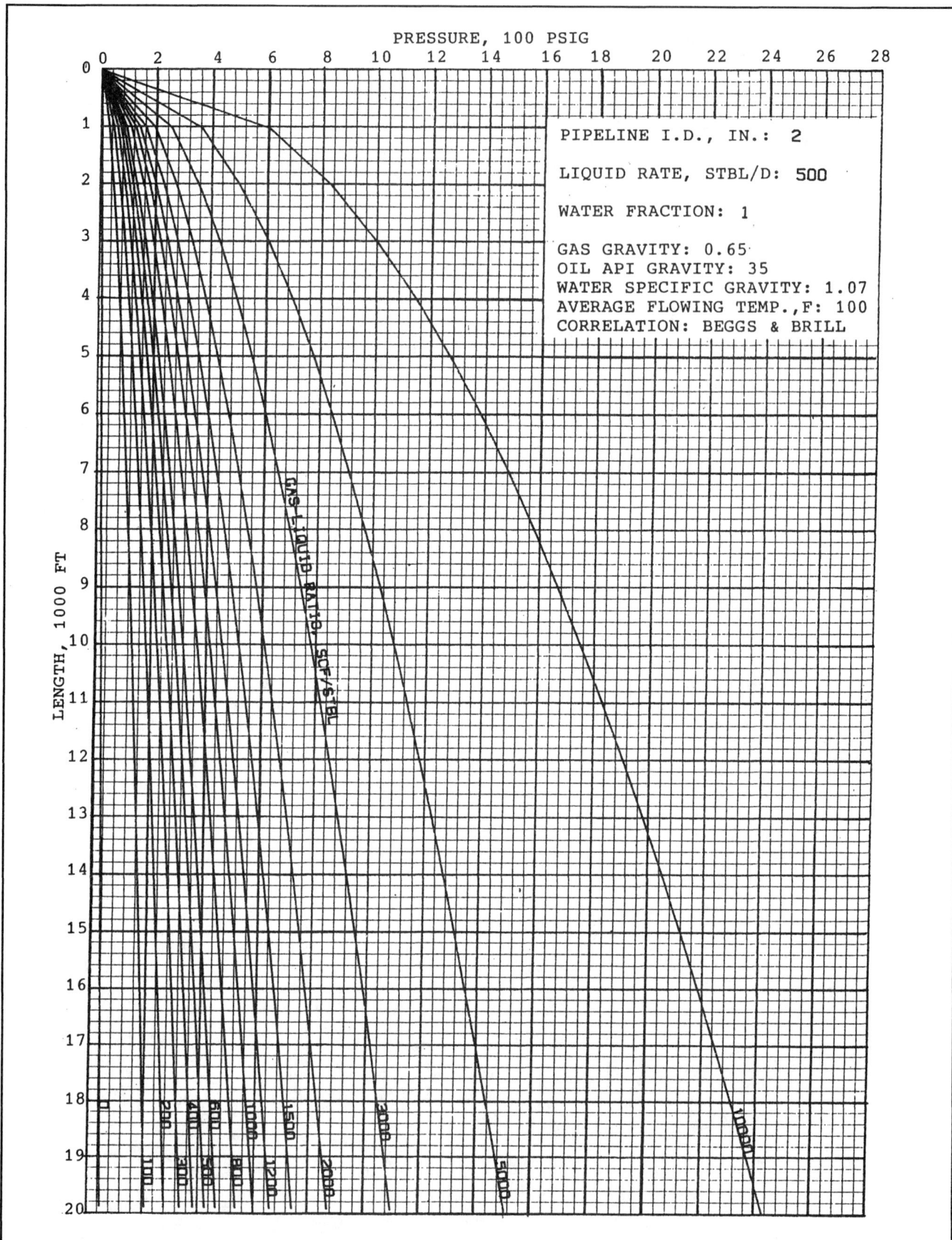

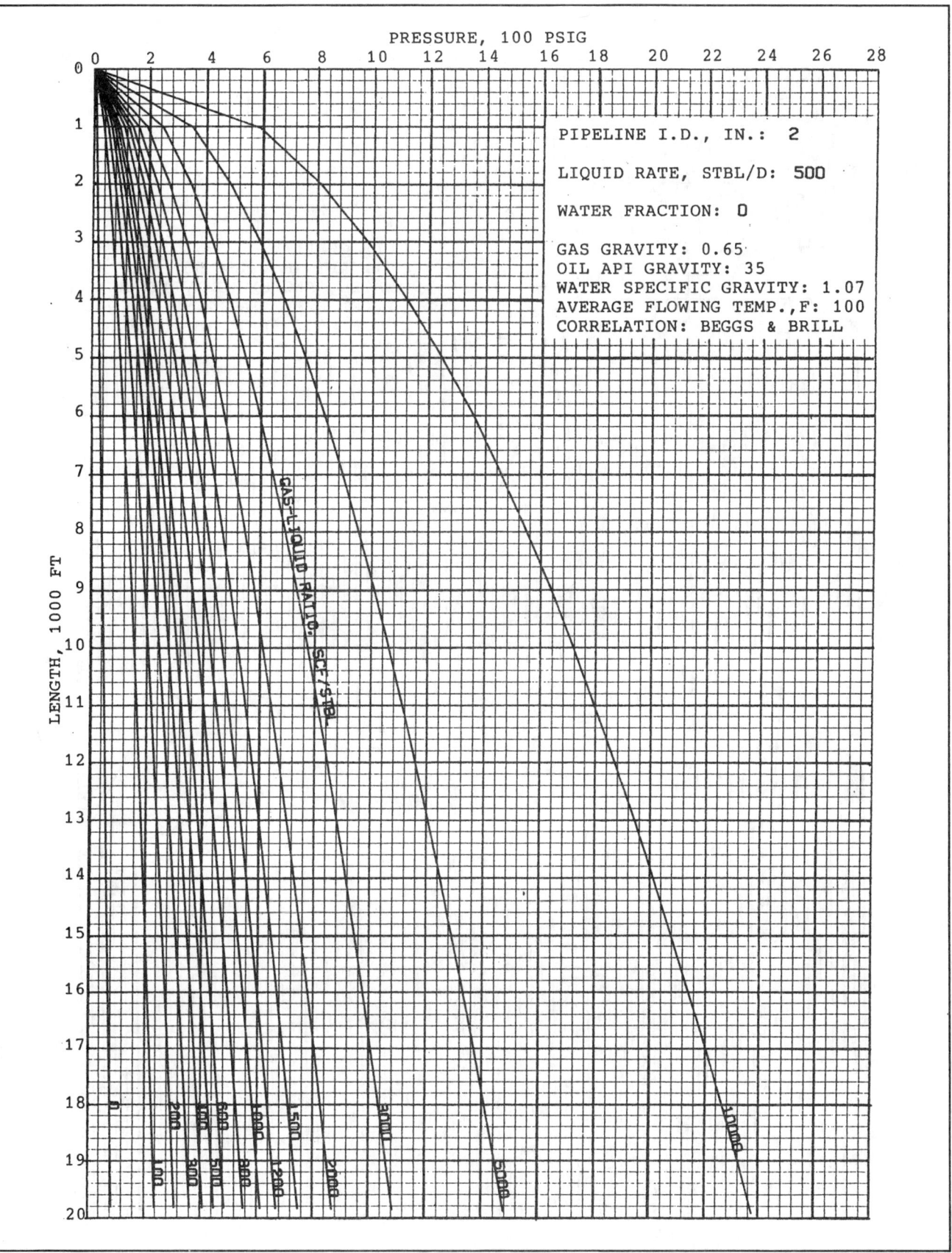

PRESSURE, 100 PSIG

LENGTH, 1000 FT

PIPELINE I.D., IN.: 2

LIQUID RATE, STBL/D: 500

WATER FRACTION: 0

GAS GRAVITY: 0.65
OIL API GRAVITY: 35
WATER SPECIFIC GRAVITY: 1.07
AVERAGE FLOWING TEMP.,F: 100
CORRELATION: BEGGS & BRILL

GAS-LIQUID RATIO, SCF/STBL

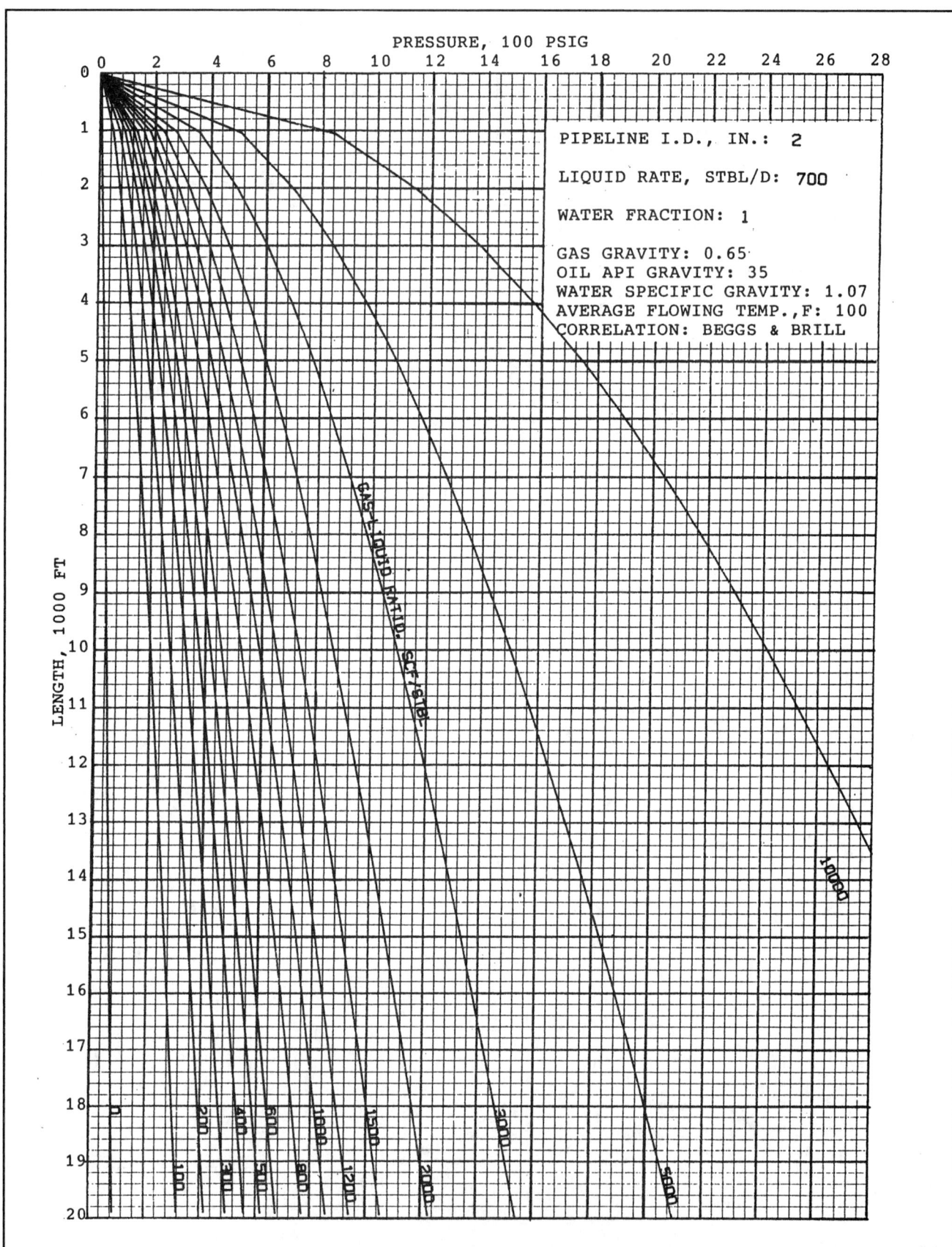

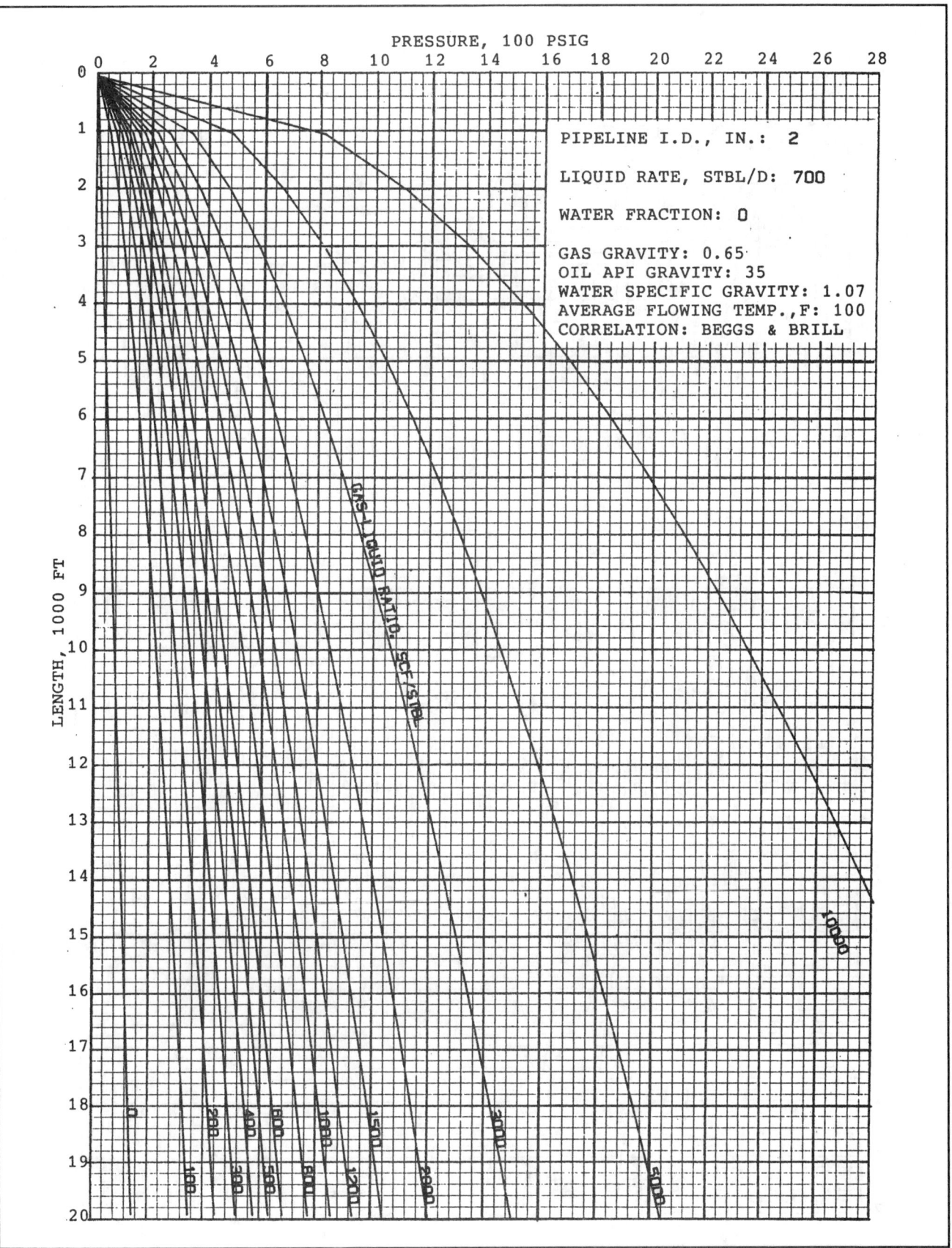

PRESSURE, 100 PSIG

LENGTH, 1000 FT

PIPELINE I.D., IN.: 2

LIQUID RATE, STBL/D: 700

WATER FRACTION: 0

GAS GRAVITY: 0.65
OIL API GRAVITY: 35
WATER SPECIFIC GRAVITY: 1.07
AVERAGE FLOWING TEMP.,F: 100
CORRELATION: BEGGS & BRILL

GAS-LIQUID RATIO, SCF/STBL

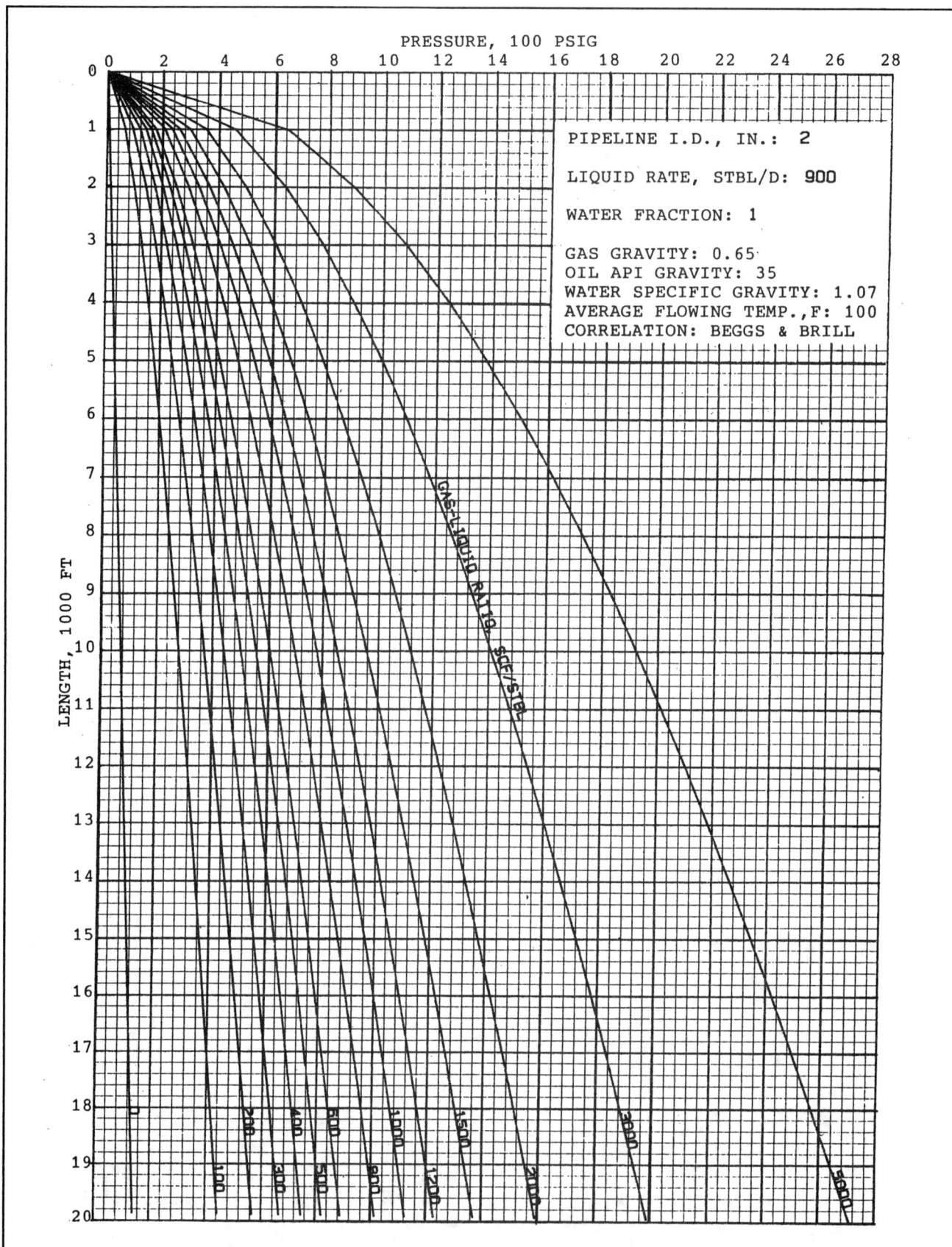

PRESSURE, 100 PSIG

LENGTH, 1000 FT

PIPELINE I.D., IN.: **2**

LIQUID RATE, STBL/D: **900**

WATER FRACTION: **1**

GAS GRAVITY: 0.65
OIL API GRAVITY: 35
WATER SPECIFIC GRAVITY: 1.07
AVERAGE FLOWING TEMP.,F: 100
CORRELATION: BEGGS & BRILL

GAS-LIQUID RATIO, SCF/STBL

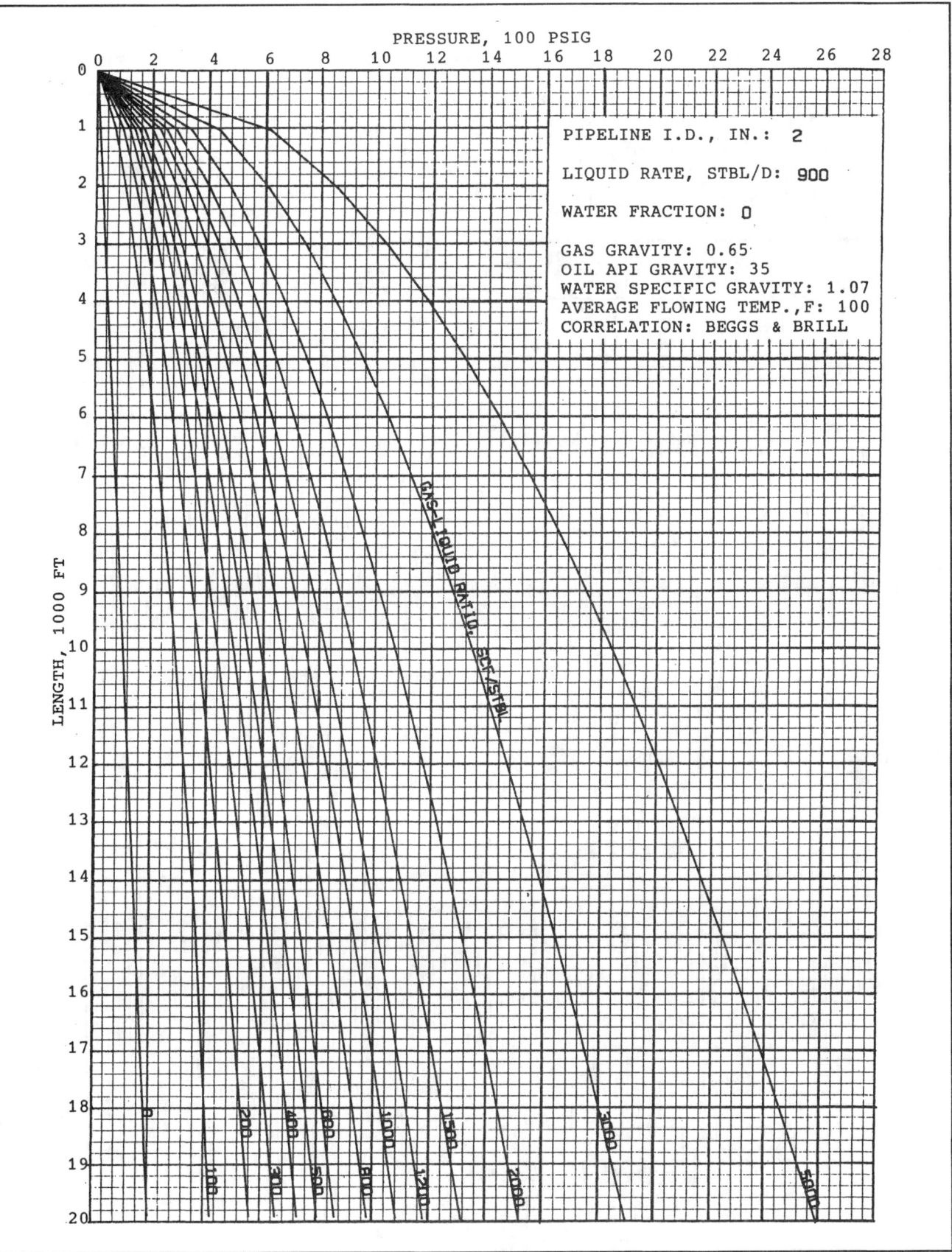

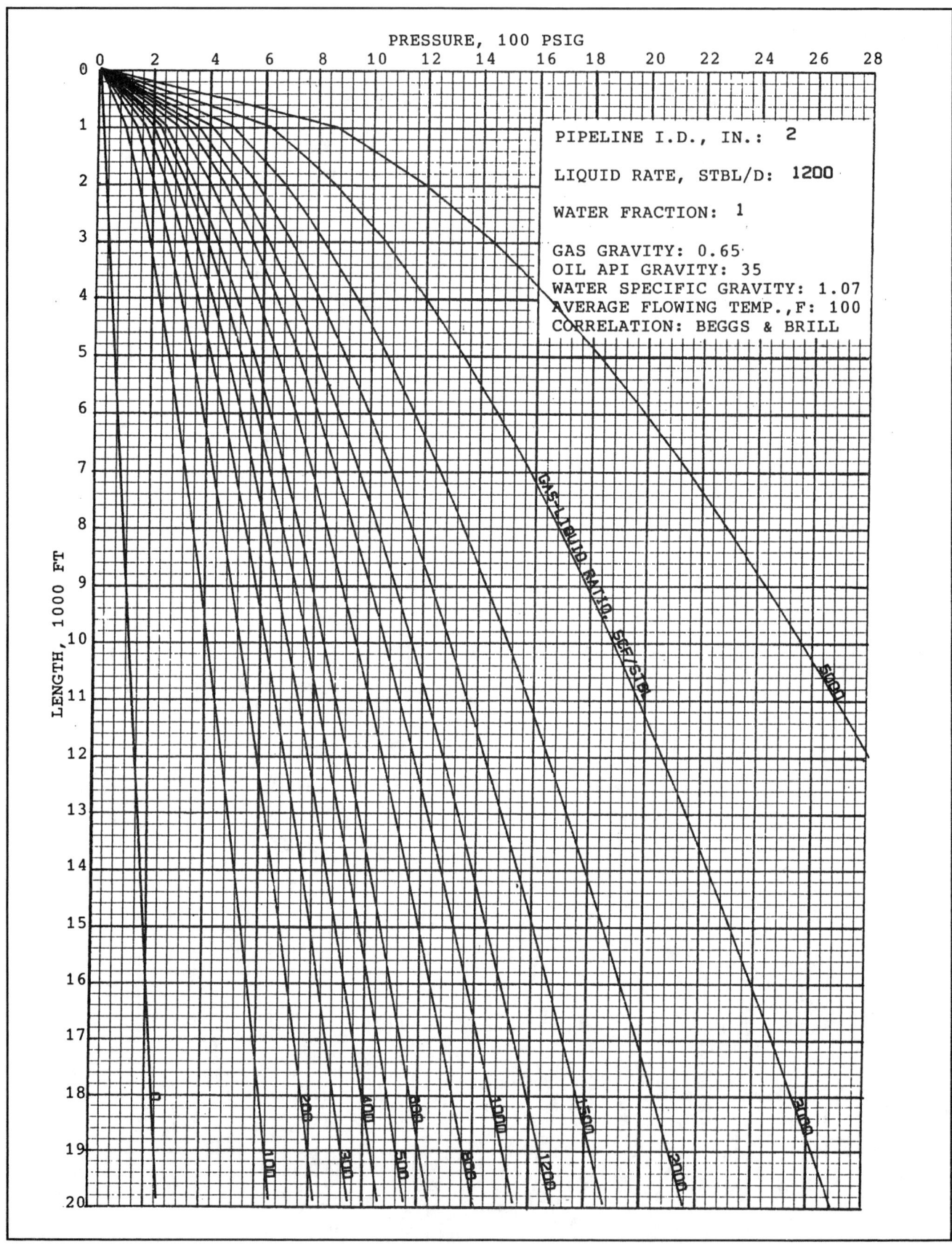

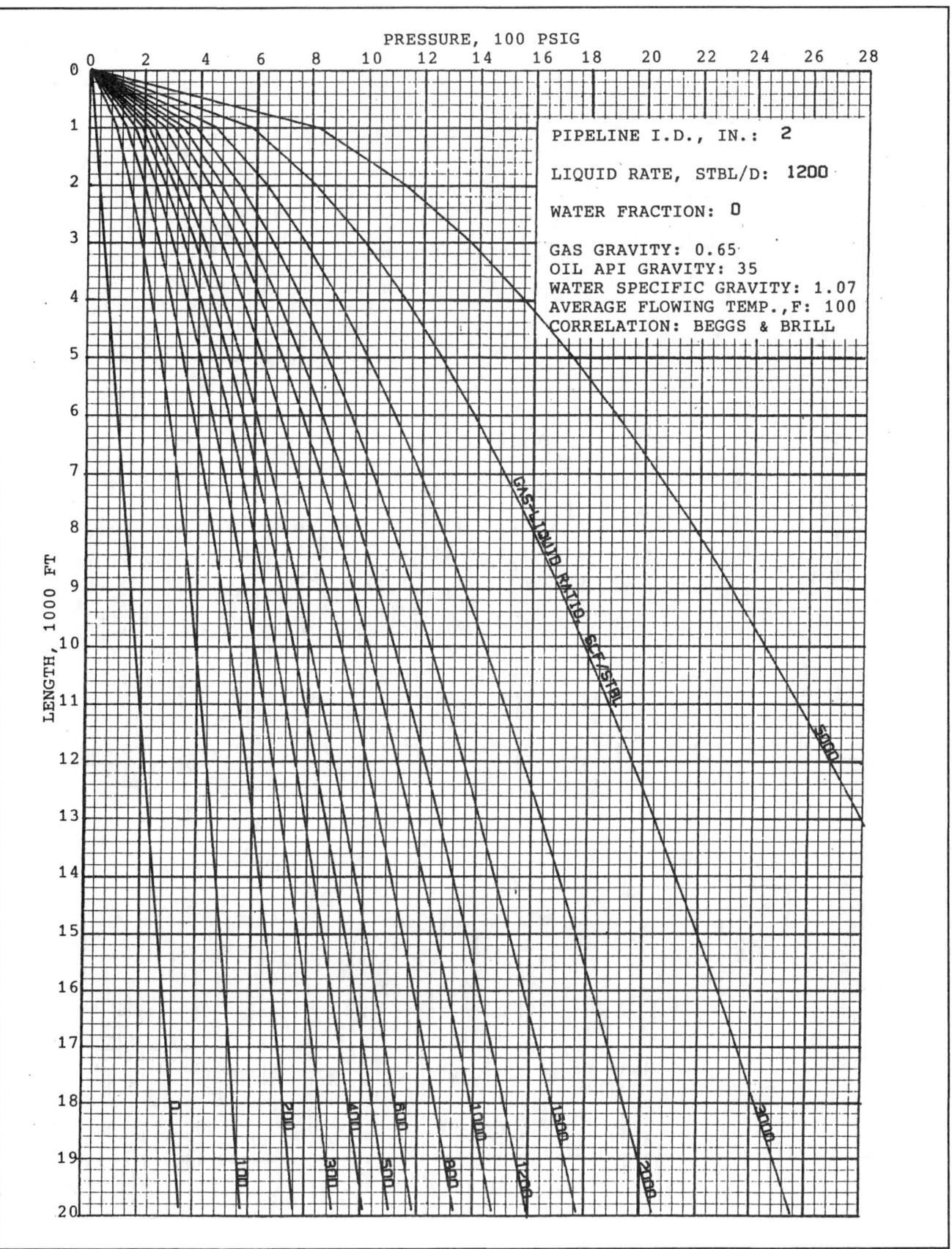

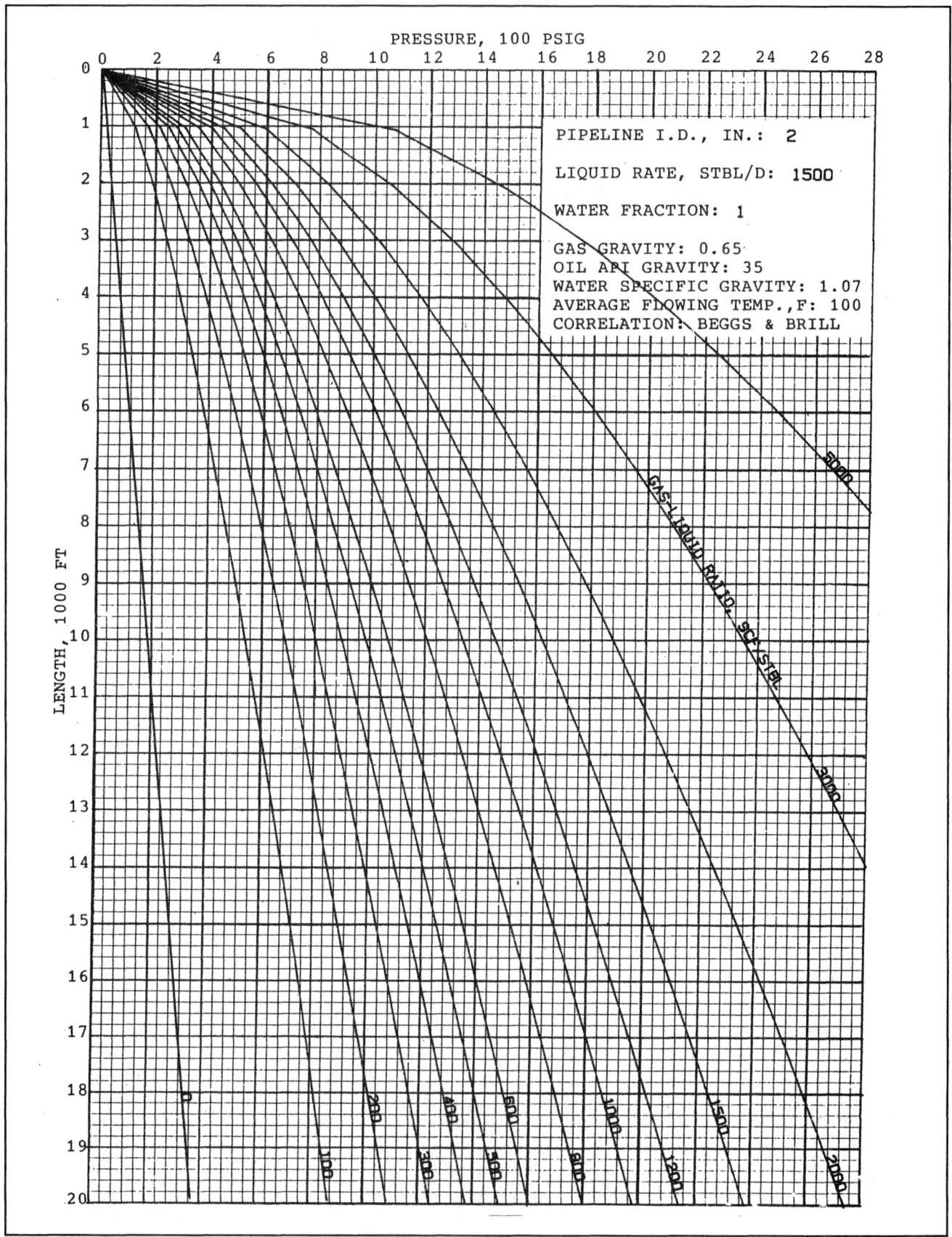

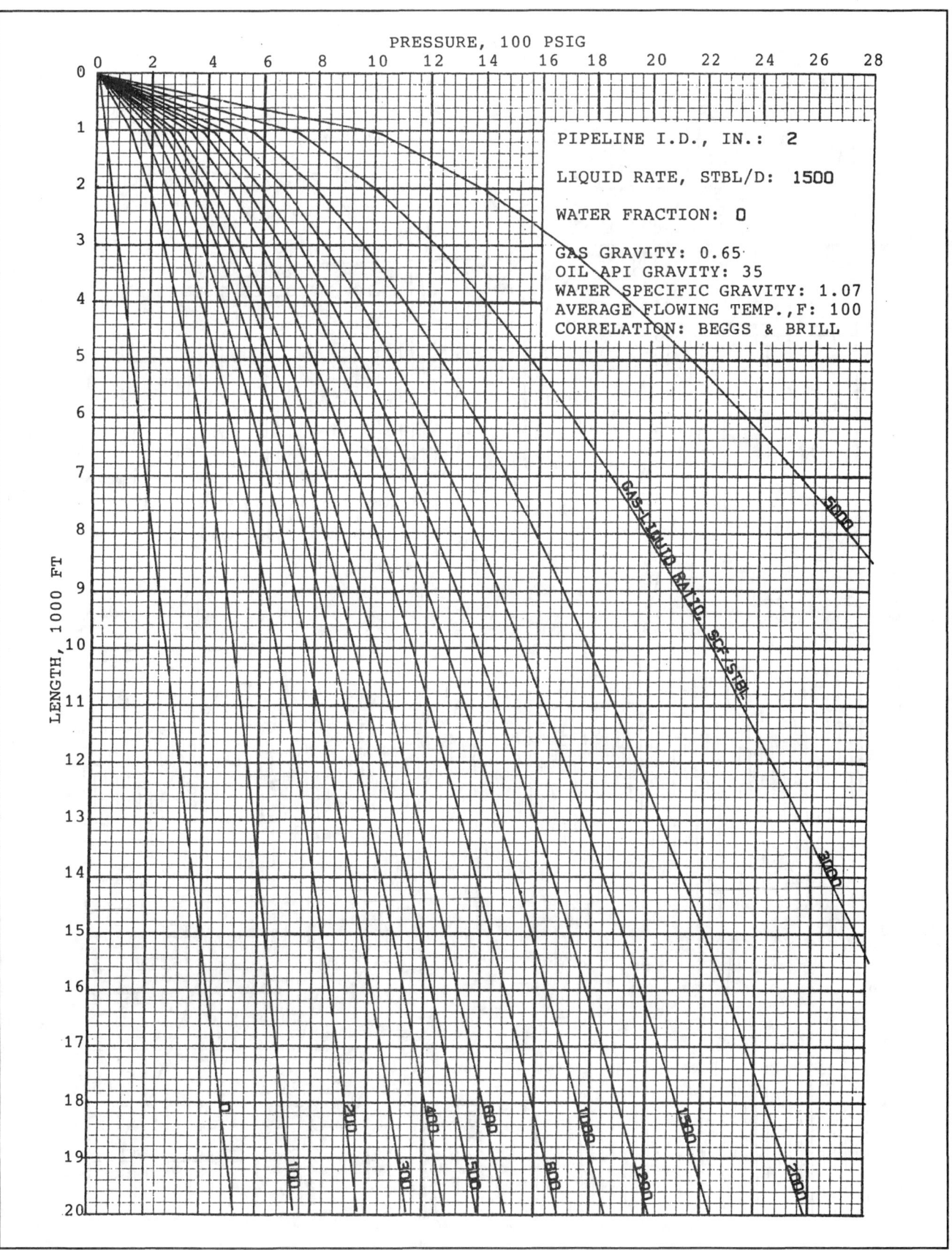

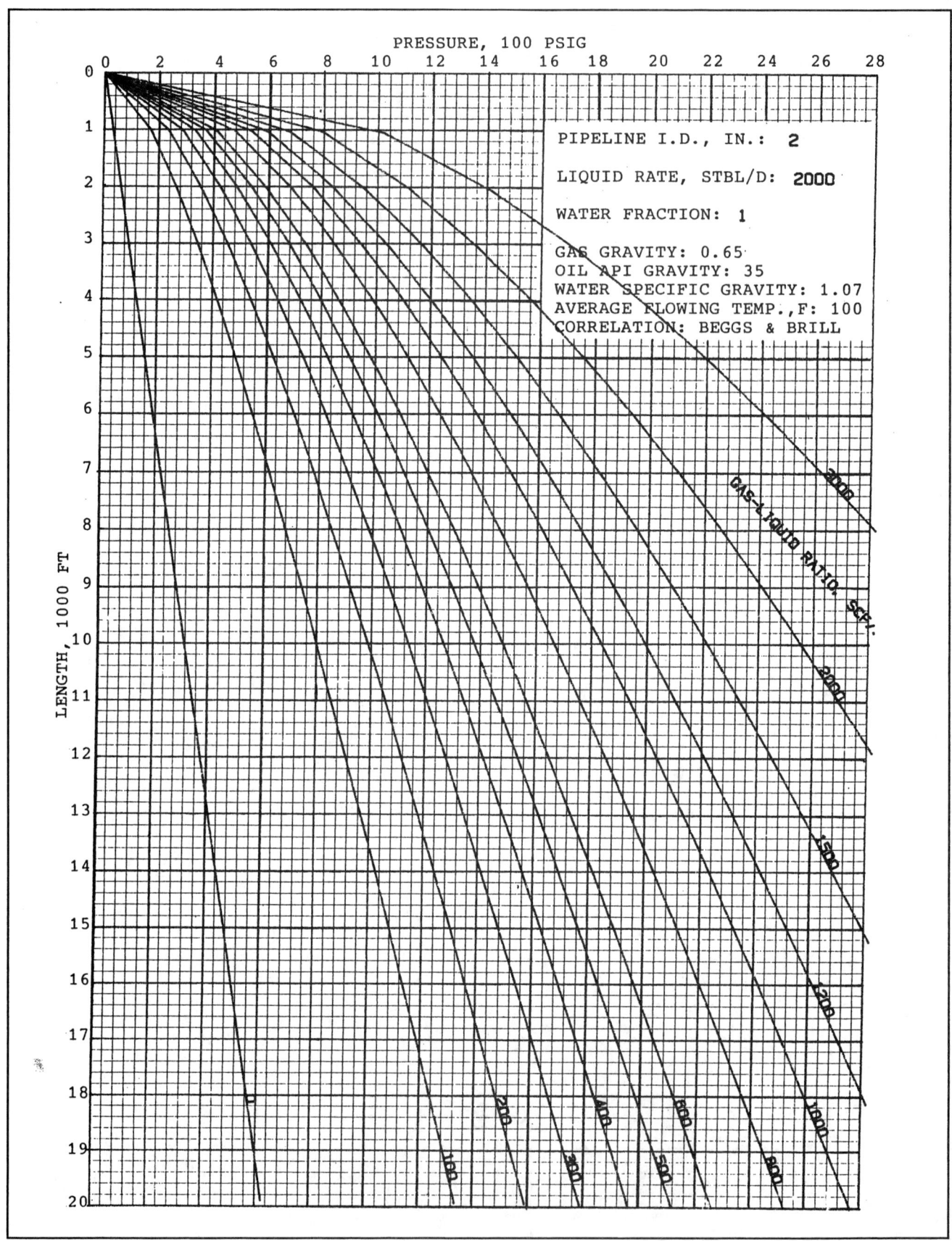

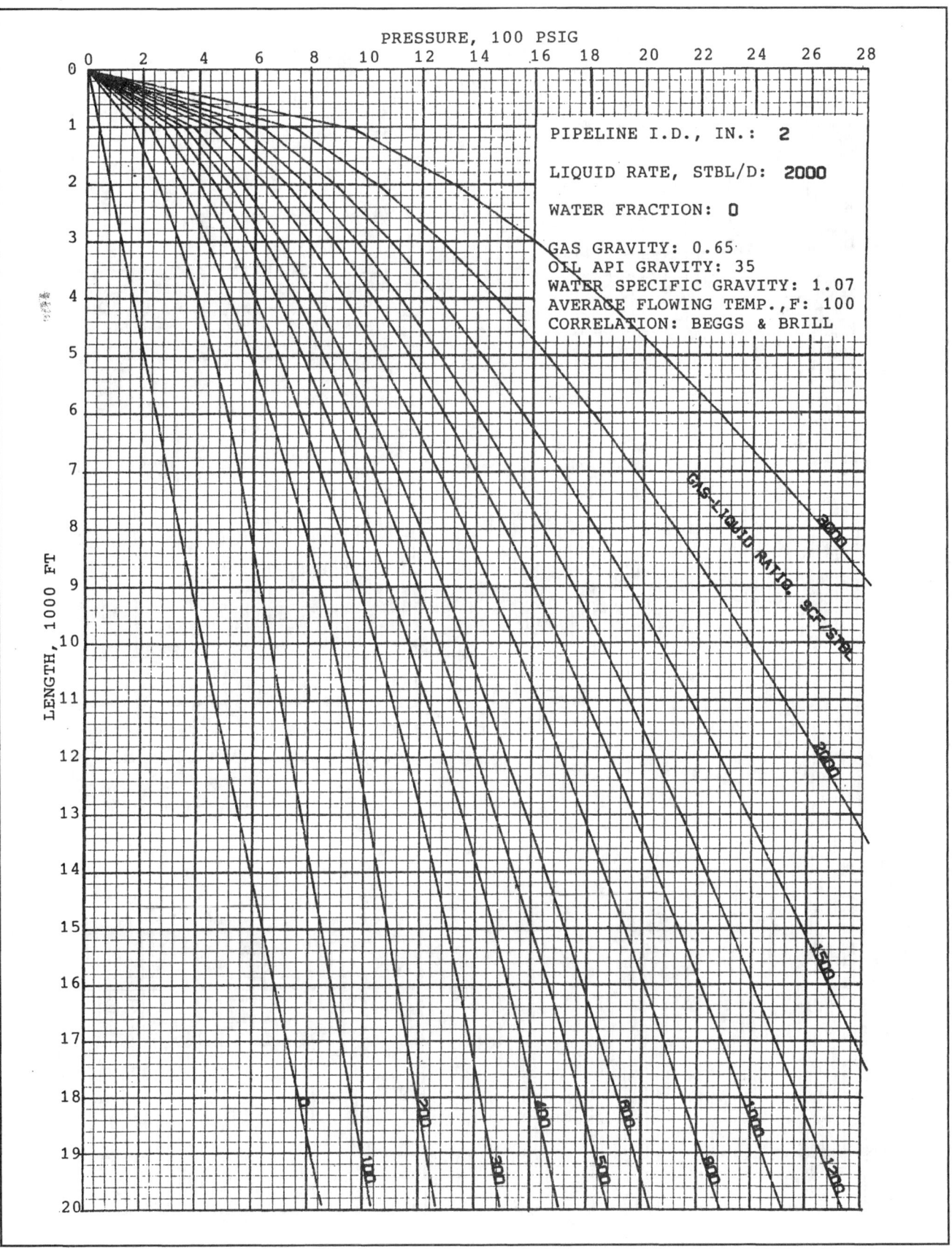

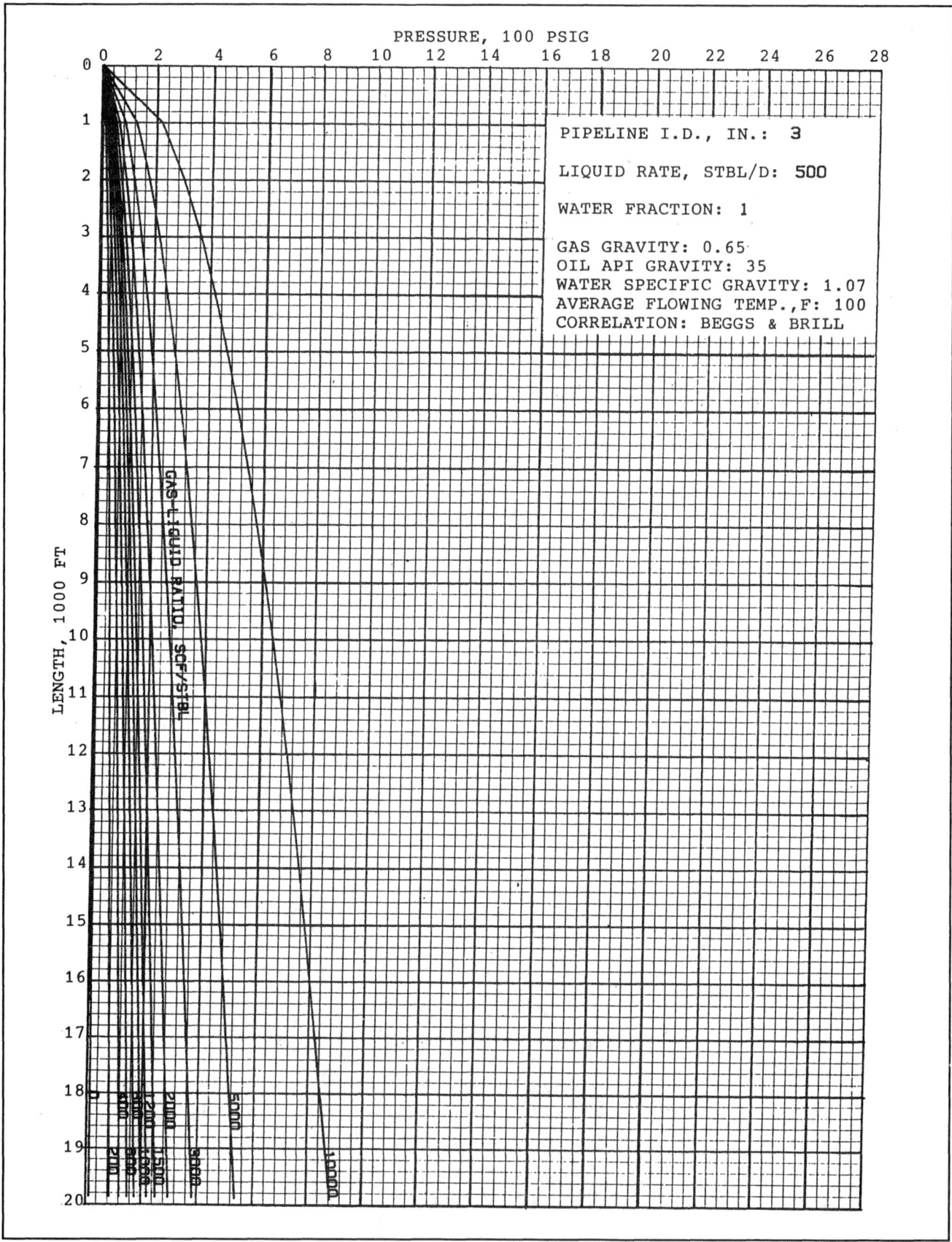

PRESSURE, 100 PSIG

LENGTH, 1000 FT

GAS-LIQUID RATIO, SCF/STBL

PIPELINE I.D., IN.: **3**

LIQUID RATE, STBL/D: **500**

WATER FRACTION: **1**

GAS GRAVITY: 0.65
OIL API GRAVITY: 35
WATER SPECIFIC GRAVITY: 1.07
AVERAGE FLOWING TEMP.,F: 100
CORRELATION: BEGGS & BRILL

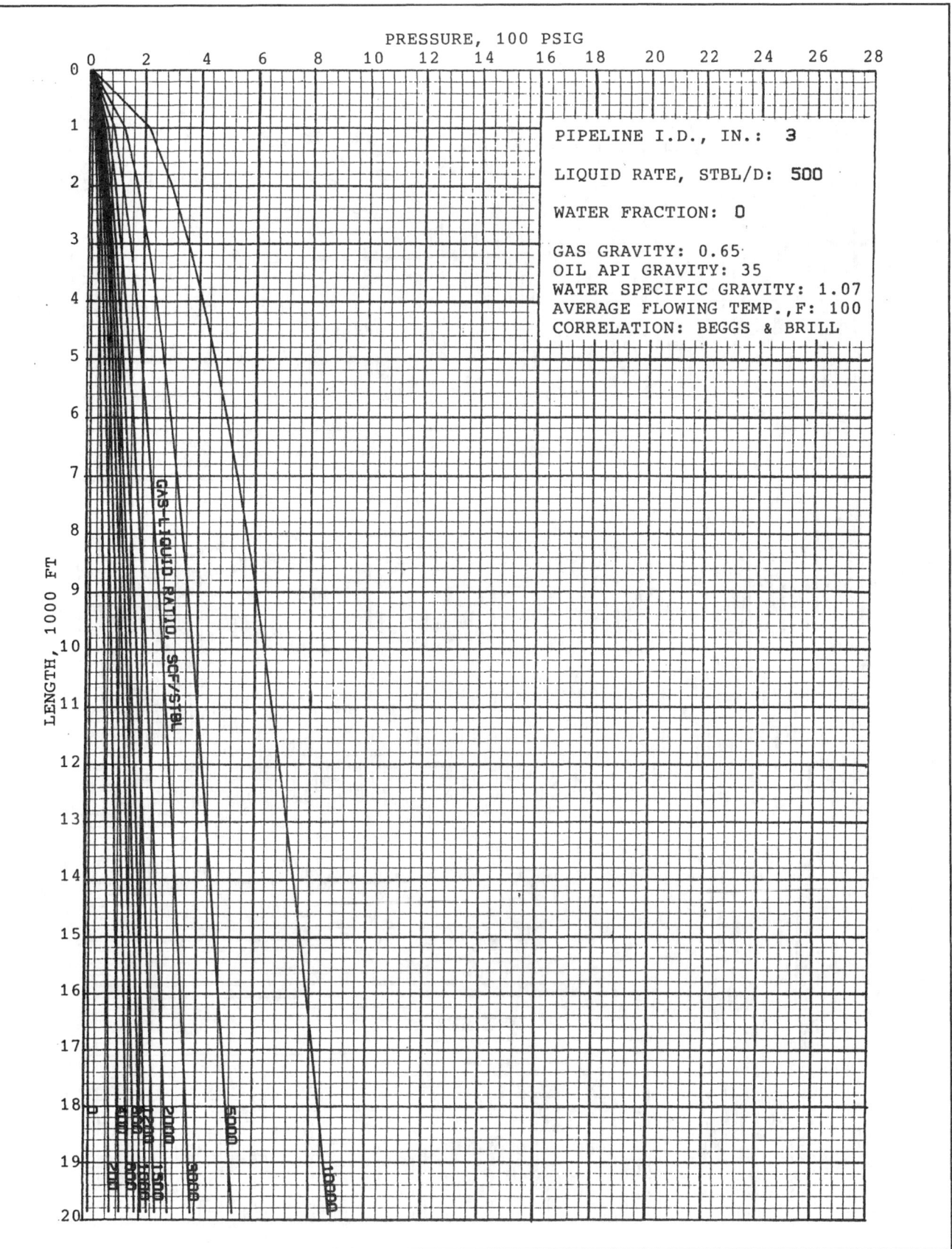

PRESSURE, 100 PSIG

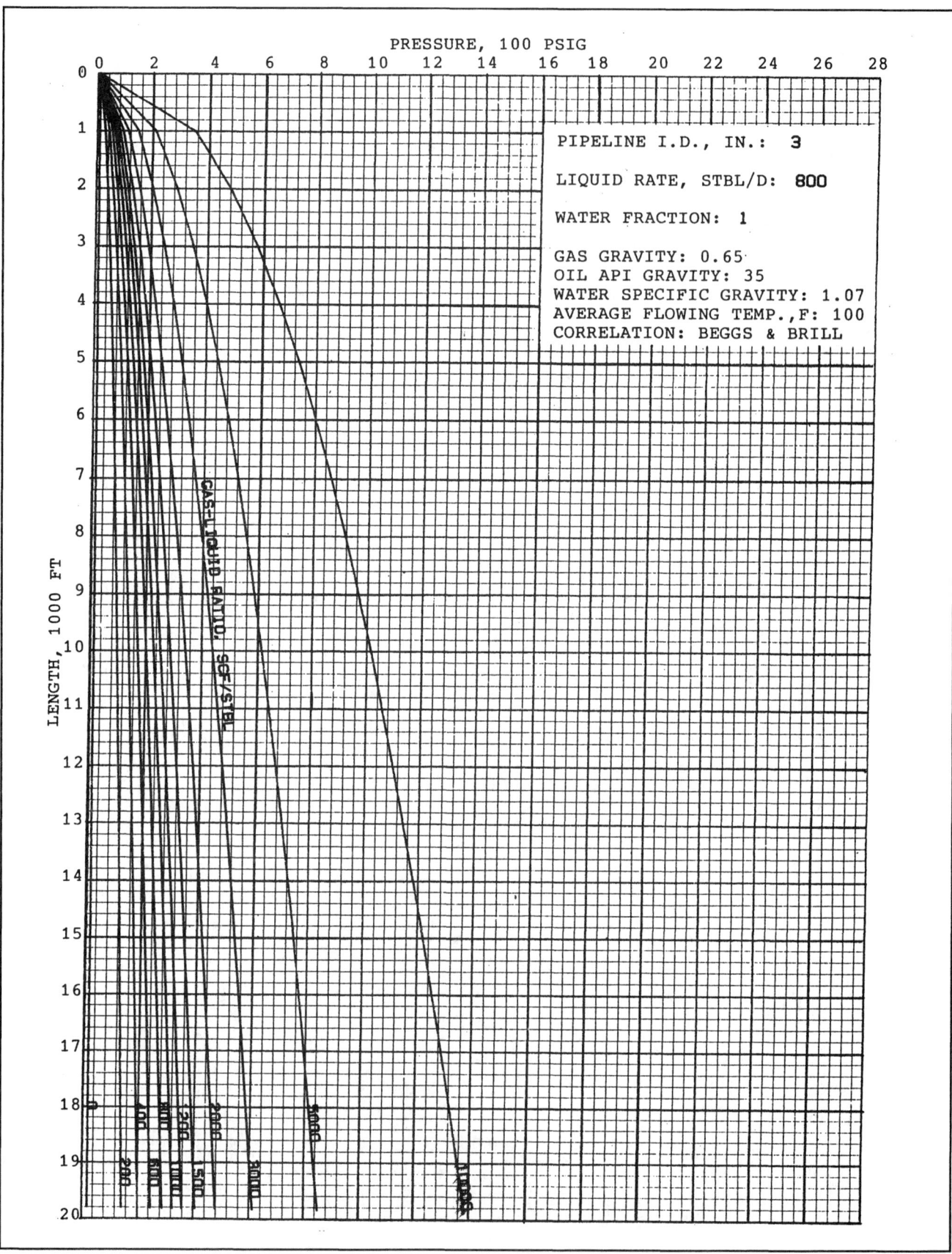

PIPELINE I.D., IN.: **3**

LIQUID RATE, STBL/D: **800**

WATER FRACTION: **1**

GAS GRAVITY: 0.65
OIL API GRAVITY: 35
WATER SPECIFIC GRAVITY: 1.07
AVERAGE FLOWING TEMP.,F: 100
CORRELATION: BEGGS & BRILL

LENGTH, 1000 FT

GAS-LIQUID RATIO, SCF/STBL

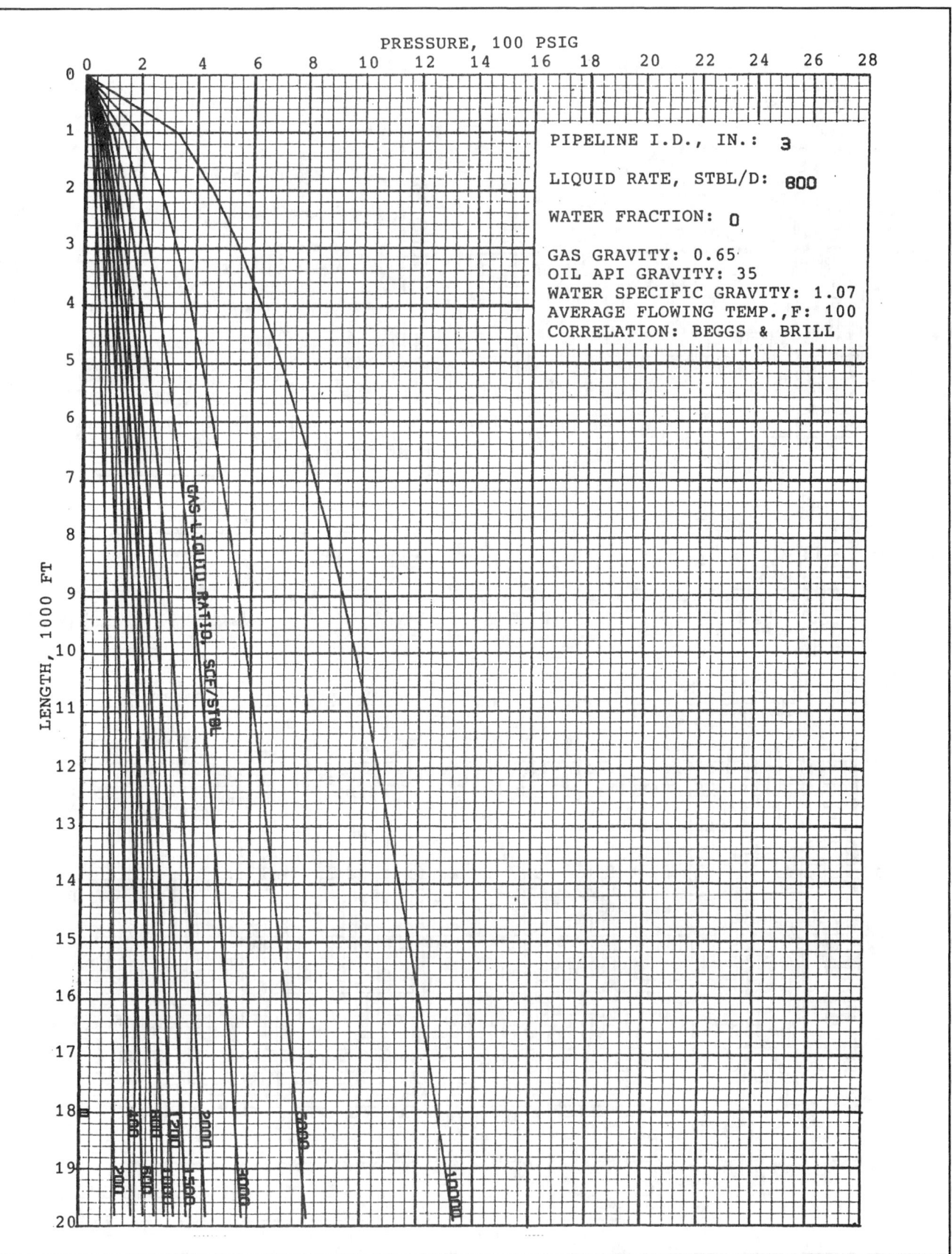

PRESSURE, 100 PSIG

LENGTH, 1000 FT

GAS-LIQUID RATIO, SCF/STBL

PIPELINE I.D., IN.: 3

LIQUID RATE, STBL/D: 800

WATER FRACTION: 0

GAS GRAVITY: 0.65
OIL API GRAVITY: 35
WATER SPECIFIC GRAVITY: 1.07
AVERAGE FLOWING TEMP.,F: 100
CORRELATION: BEGGS & BRILL

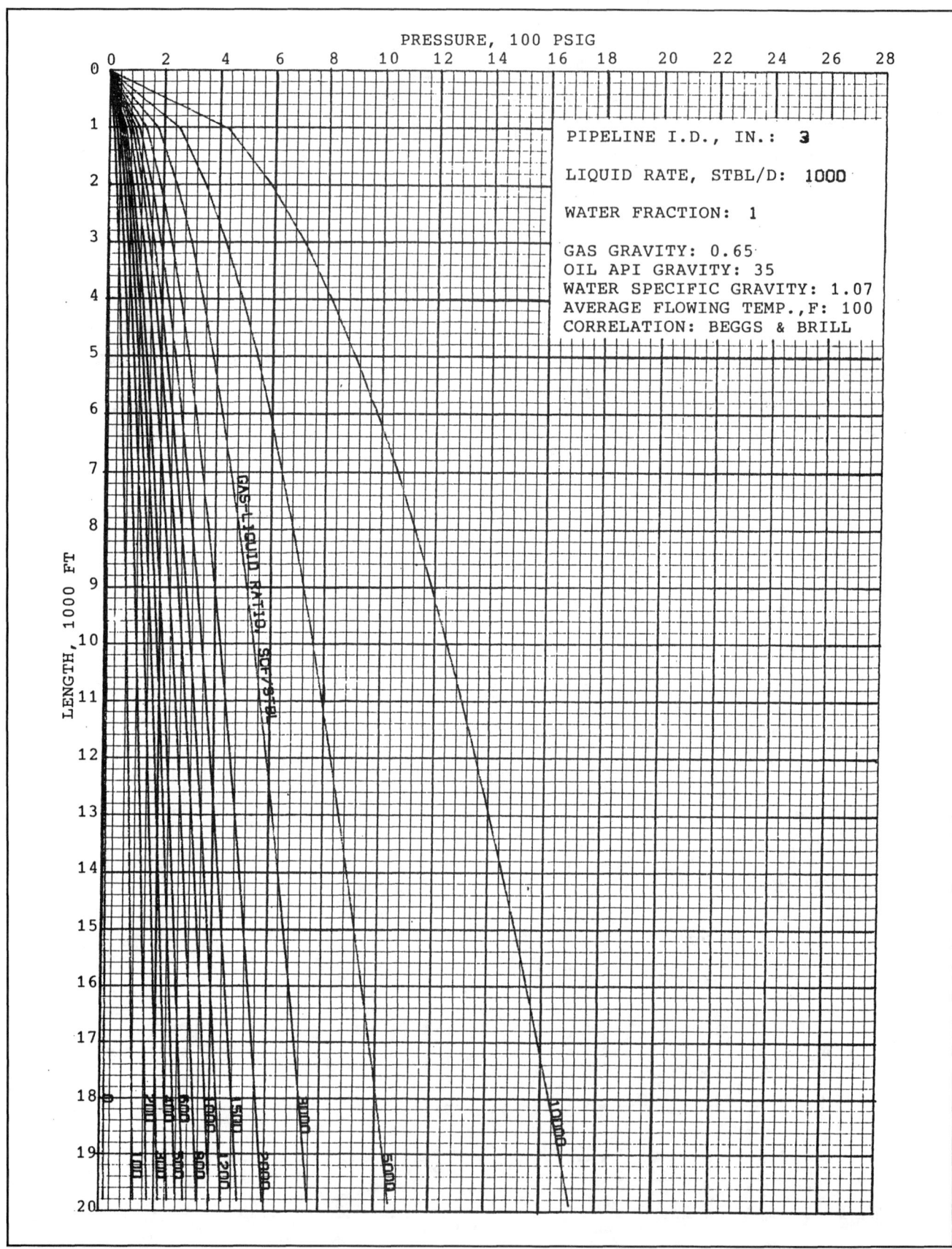

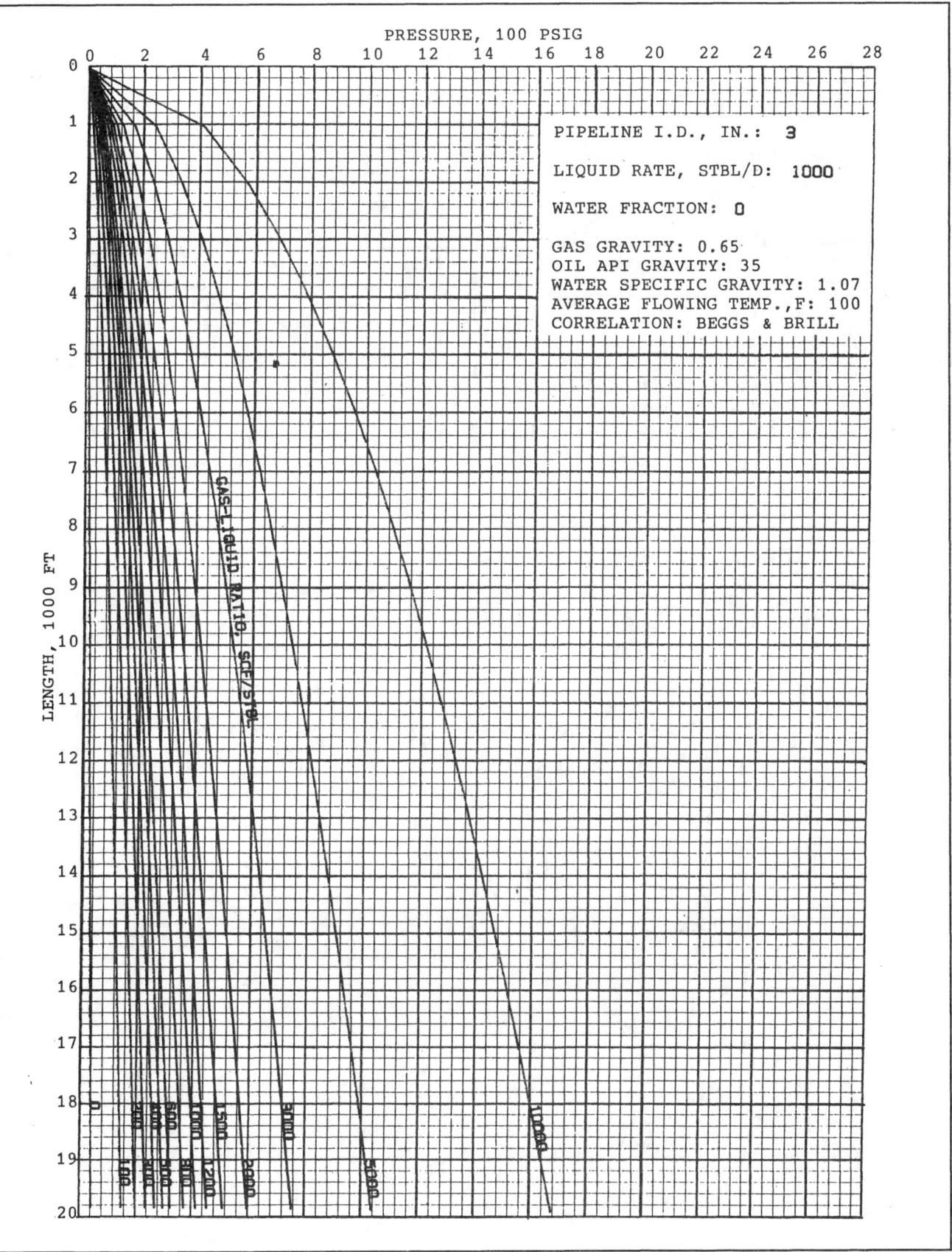

PRESSURE, 100 PSIG

LENGTH, 1000 FT

PIPELINE I.D., IN.: **3**

LIQUID RATE, STBL/D: **1000**

WATER FRACTION: **0**

GAS GRAVITY: 0.65
OIL API GRAVITY: 35
WATER SPECIFIC GRAVITY: 1.07
AVERAGE FLOWING TEMP.,F: 100
CORRELATION: BEGGS & BRILL

GAS-LIQUID RATIO, SCF/STBL

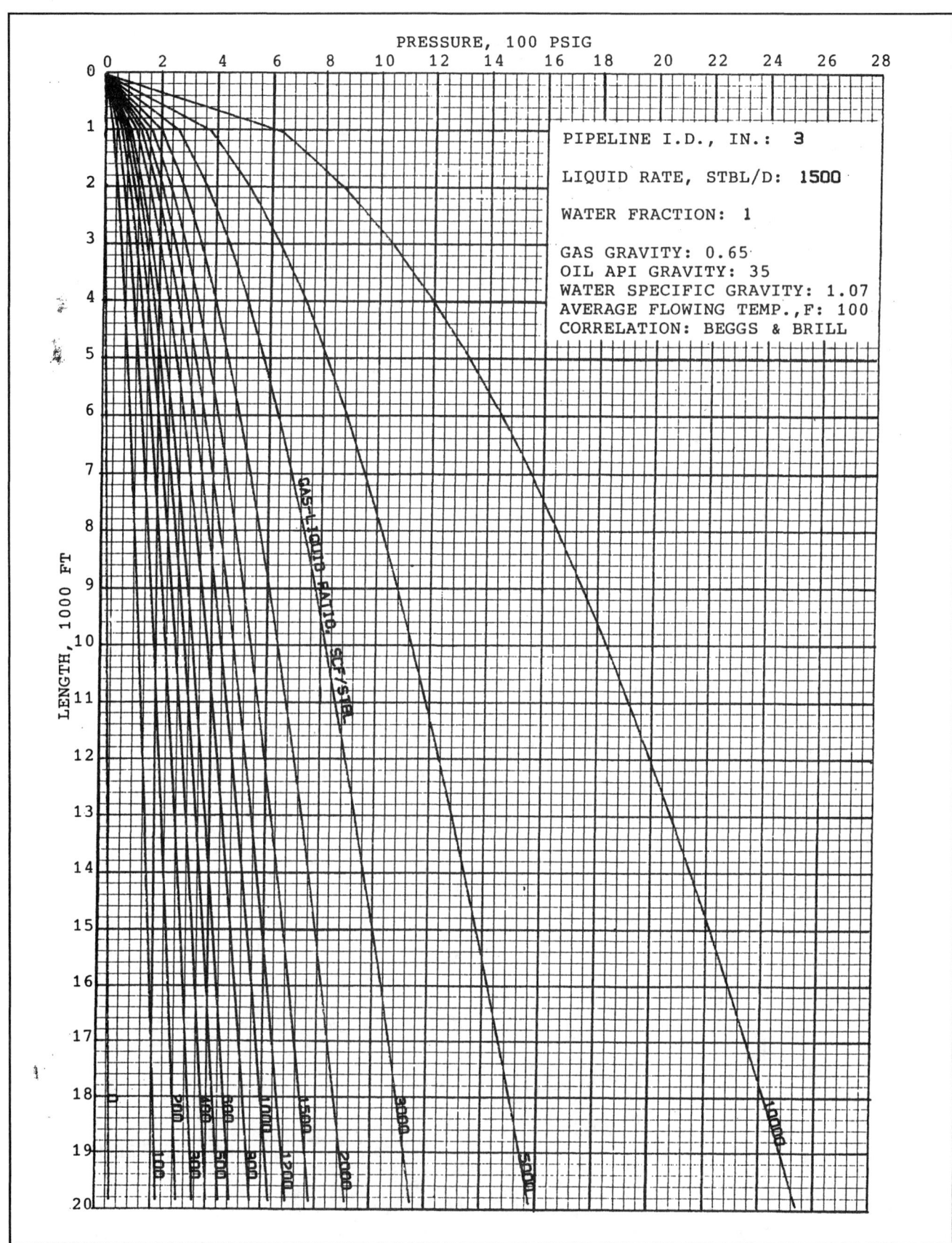

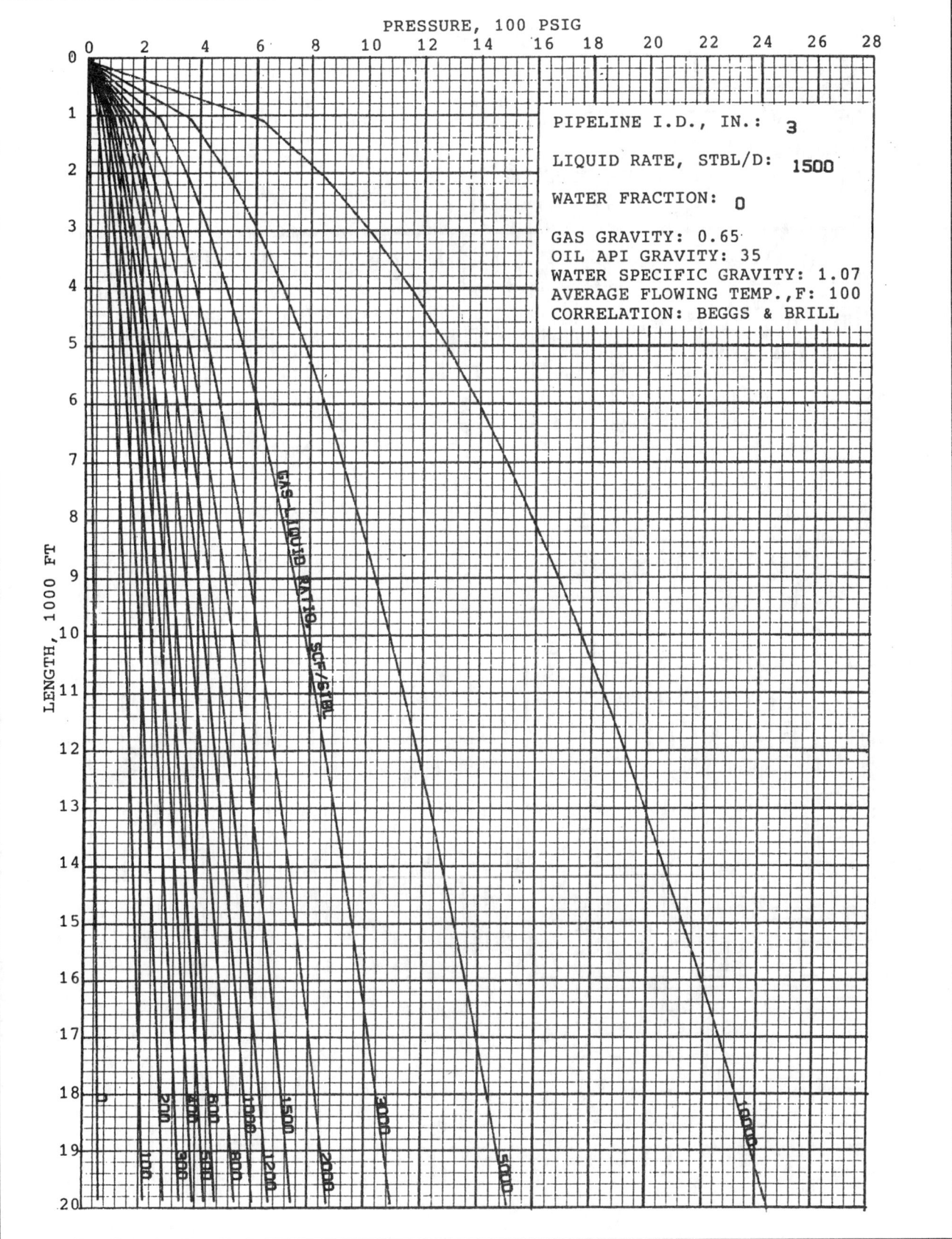

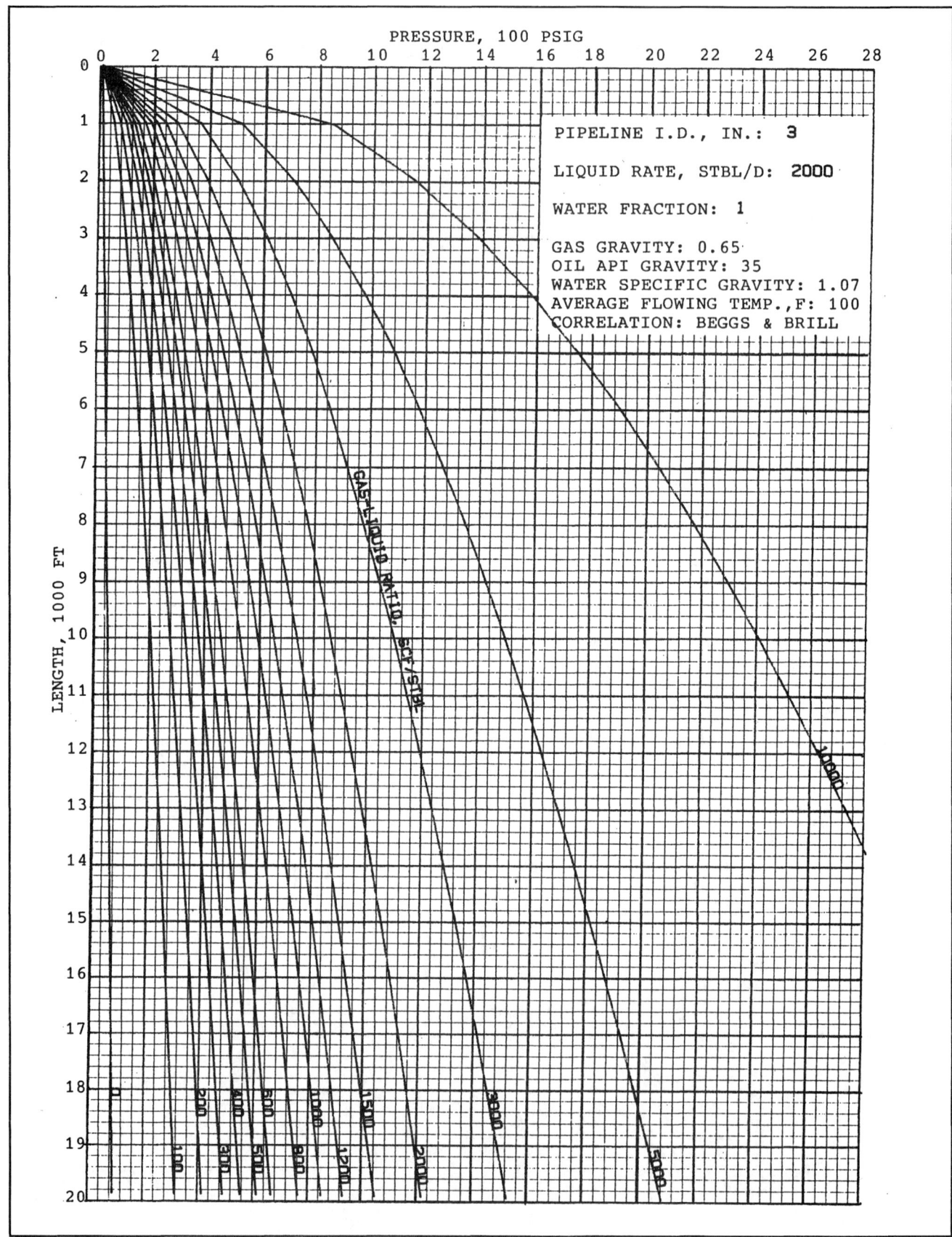

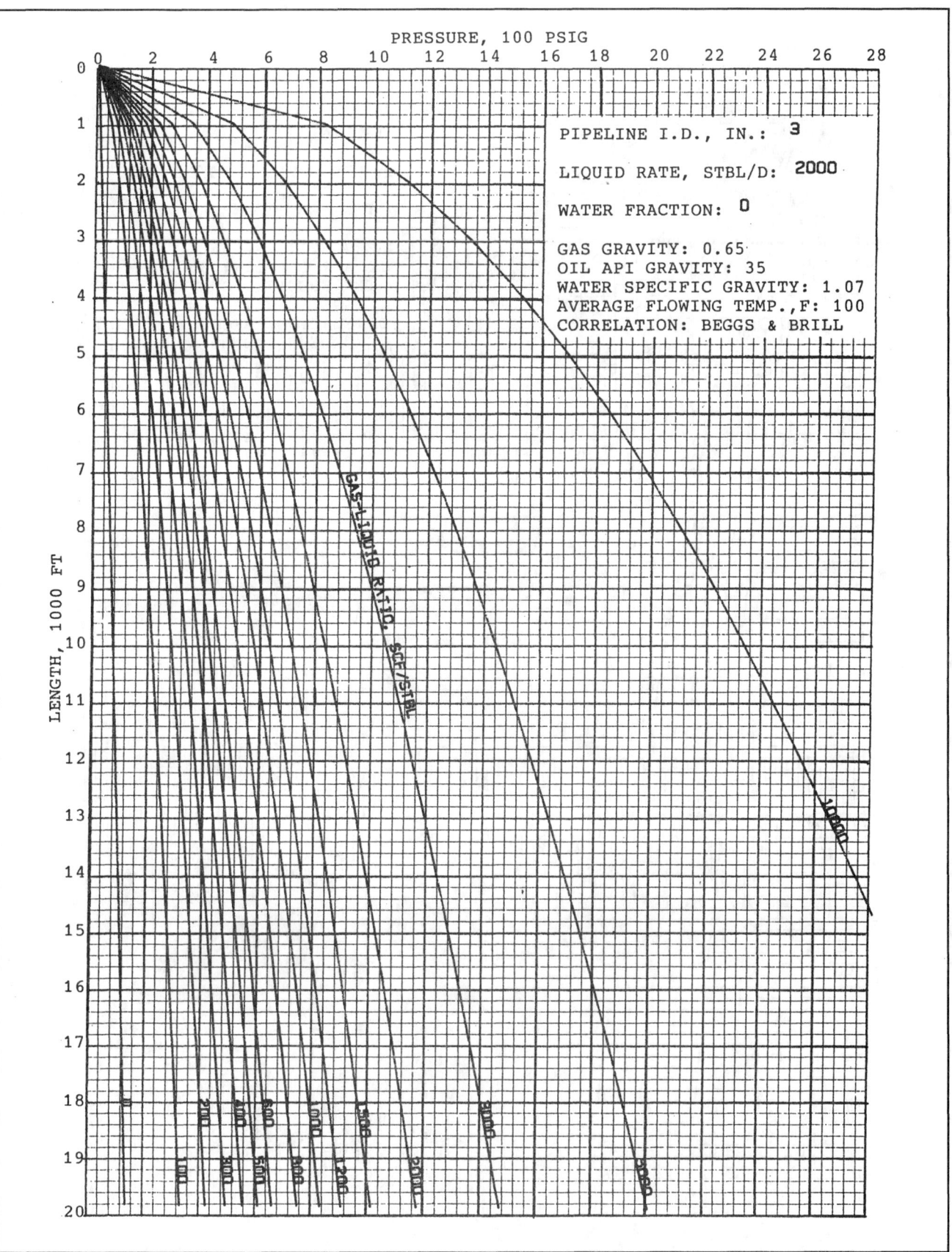

PRESSURE, 100 PSIG

LENGTH, 1000 FT

PIPELINE I.D., IN.: **3**

LIQUID RATE, STBL/D: **2000**

WATER FRACTION: **0**

GAS GRAVITY: 0.65
OIL API GRAVITY: 35
WATER SPECIFIC GRAVITY: 1.07
AVERAGE FLOWING TEMP.,F: 100
CORRELATION: BEGGS & BRILL

GAS-LIQUID RATIO, SCF/STBL

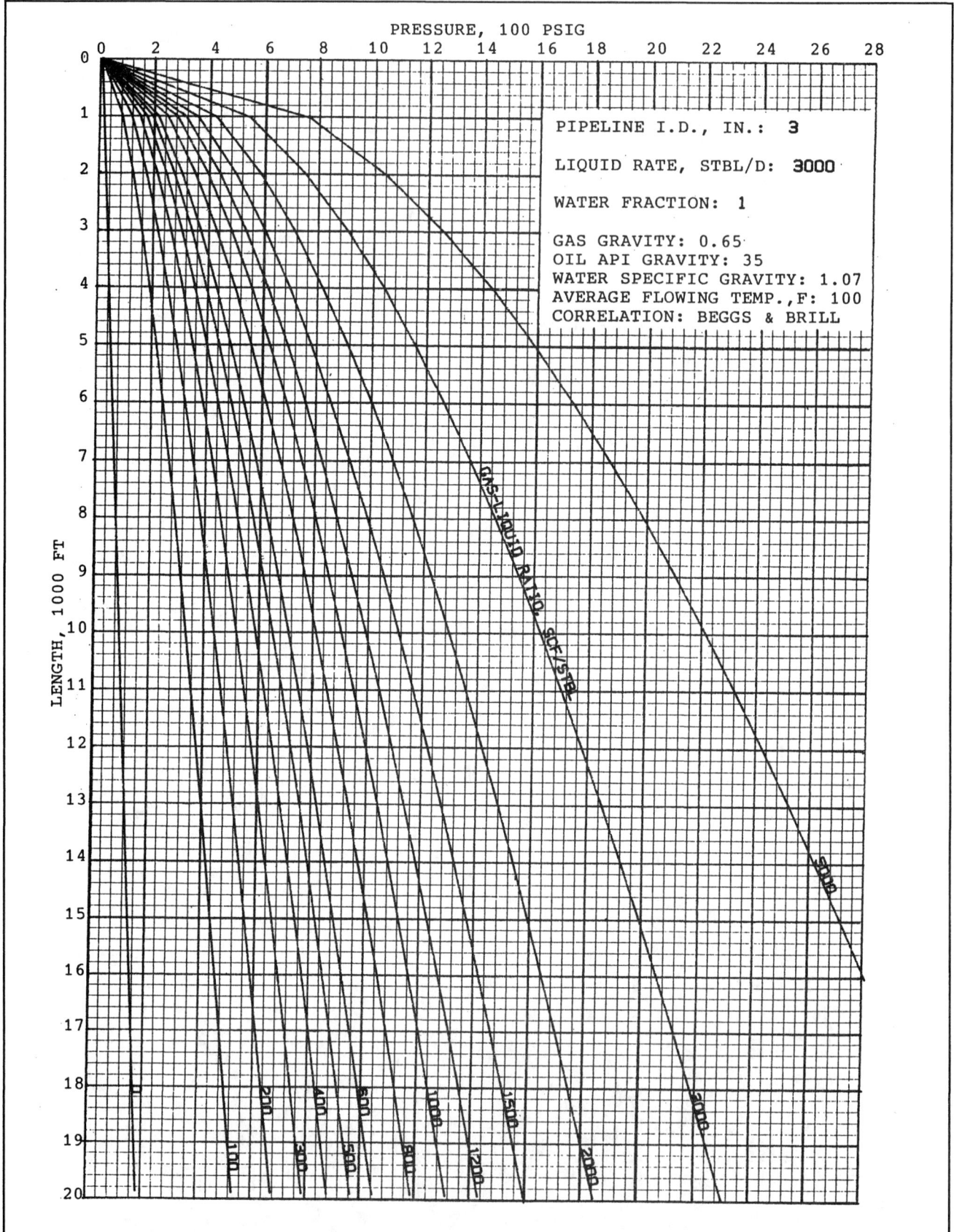

PRESSURE, 100 PSIG

PIPELINE I.D., IN.: **3**

LIQUID RATE, STBL/D: **3000**

WATER FRACTION: **1**

GAS GRAVITY: 0.65
OIL API GRAVITY: 35
WATER SPECIFIC GRAVITY: 1.07
AVERAGE FLOWING TEMP.,F: 100
CORRELATION: BEGGS & BRILL

LENGTH, 1000 FT

GAS-LIQUID RATIO, SCF/STBL

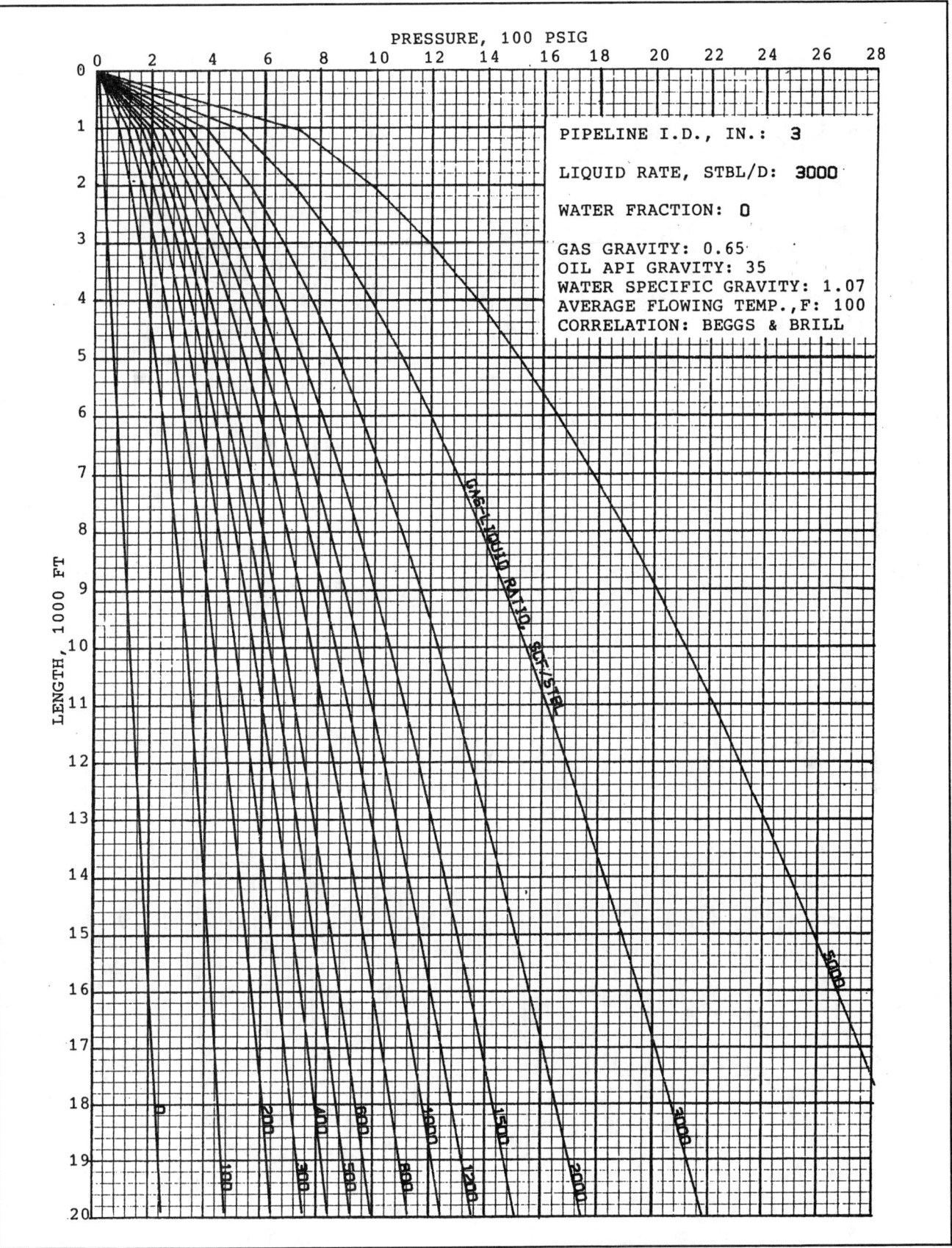

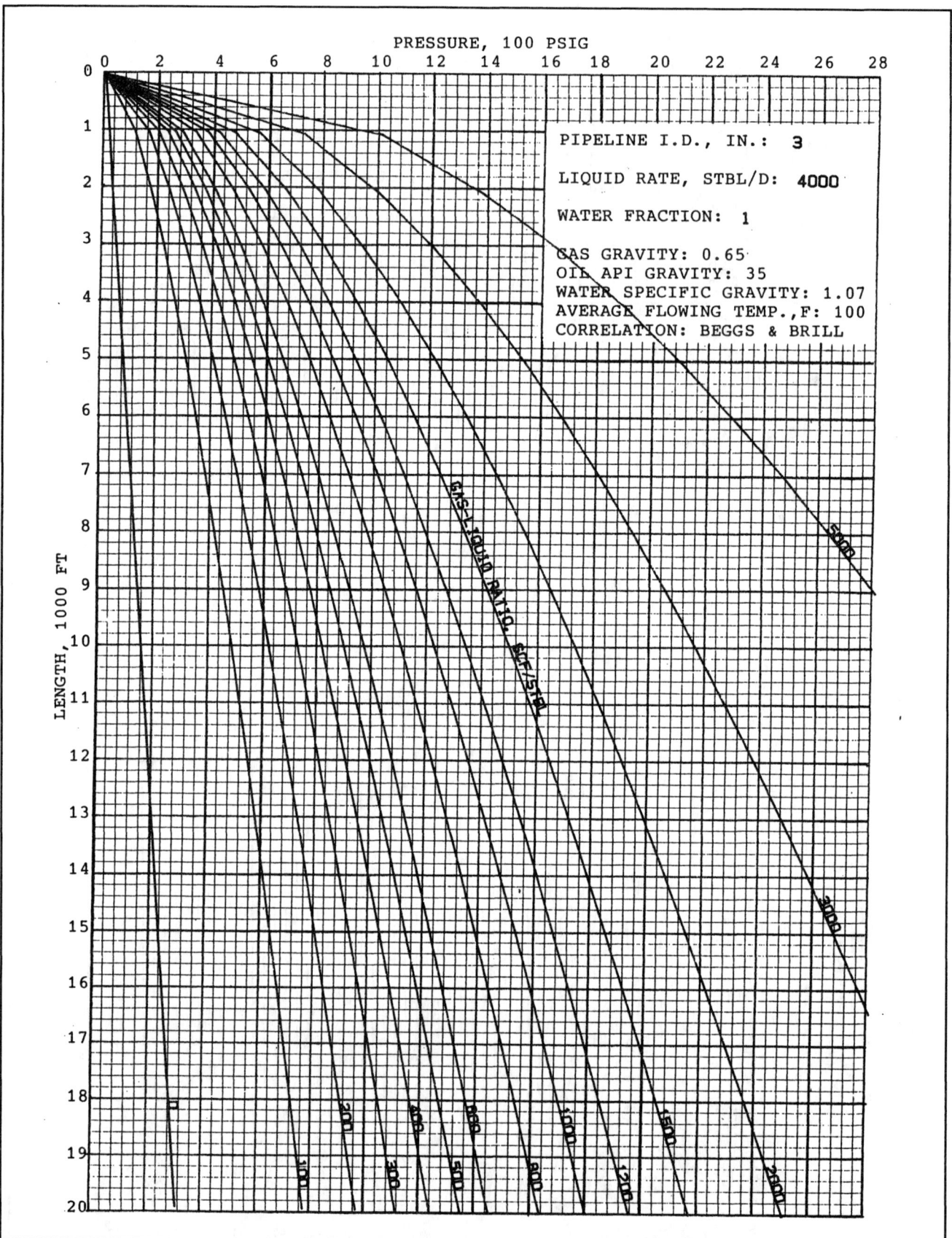

PRESSURE, 100 PSIG

PIPELINE I.D., IN.: **3**

LIQUID RATE, STBL/D: **4000**

WATER FRACTION: **1**

GAS GRAVITY: 0.65
OIL API GRAVITY: 35
WATER SPECIFIC GRAVITY: 1.07
AVERAGE FLOWING TEMP.,F: 100
CORRELATION: BEGGS & BRILL

LENGTH, 1000 FT

GAS-LIQUID RATIO, SCF/STBL

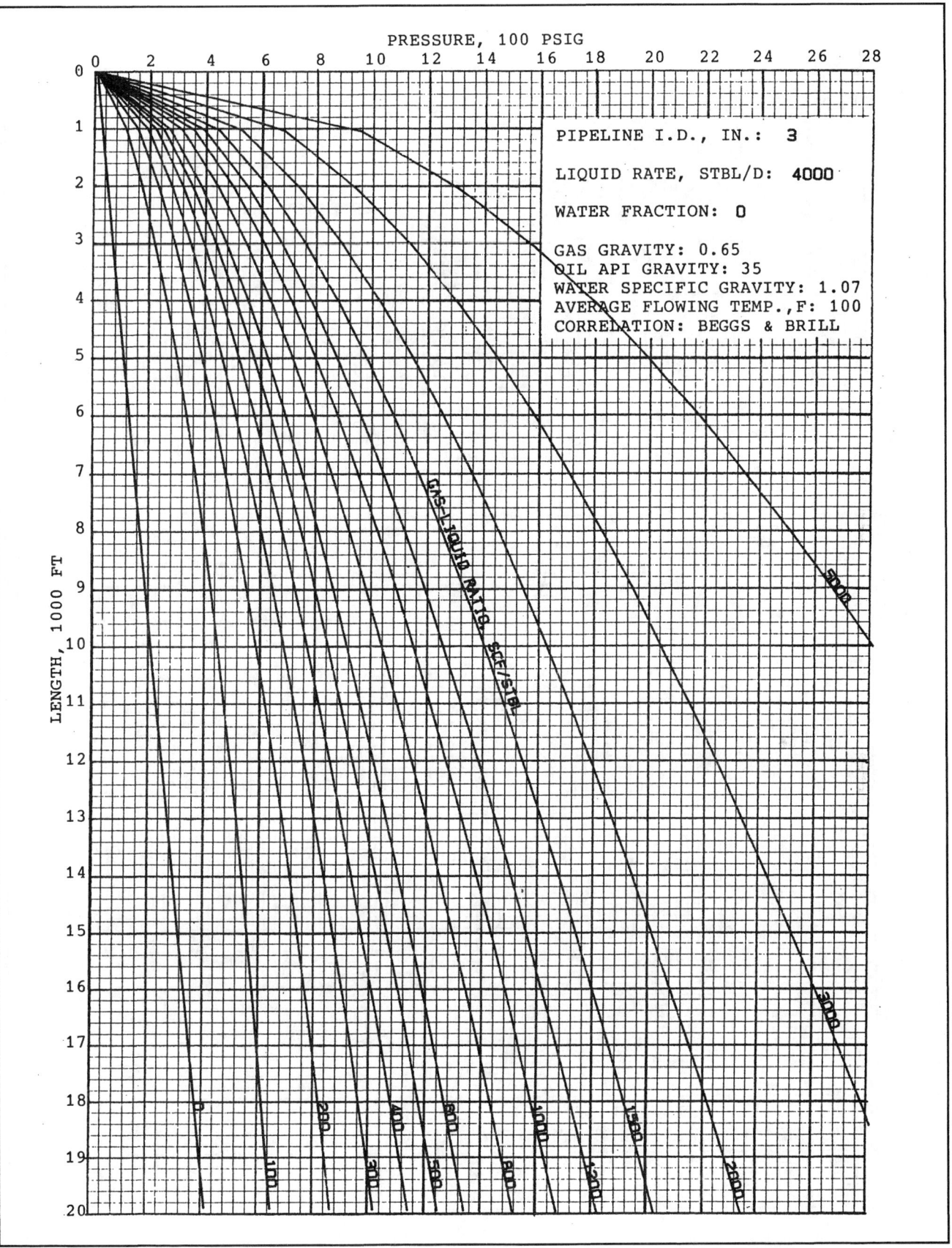

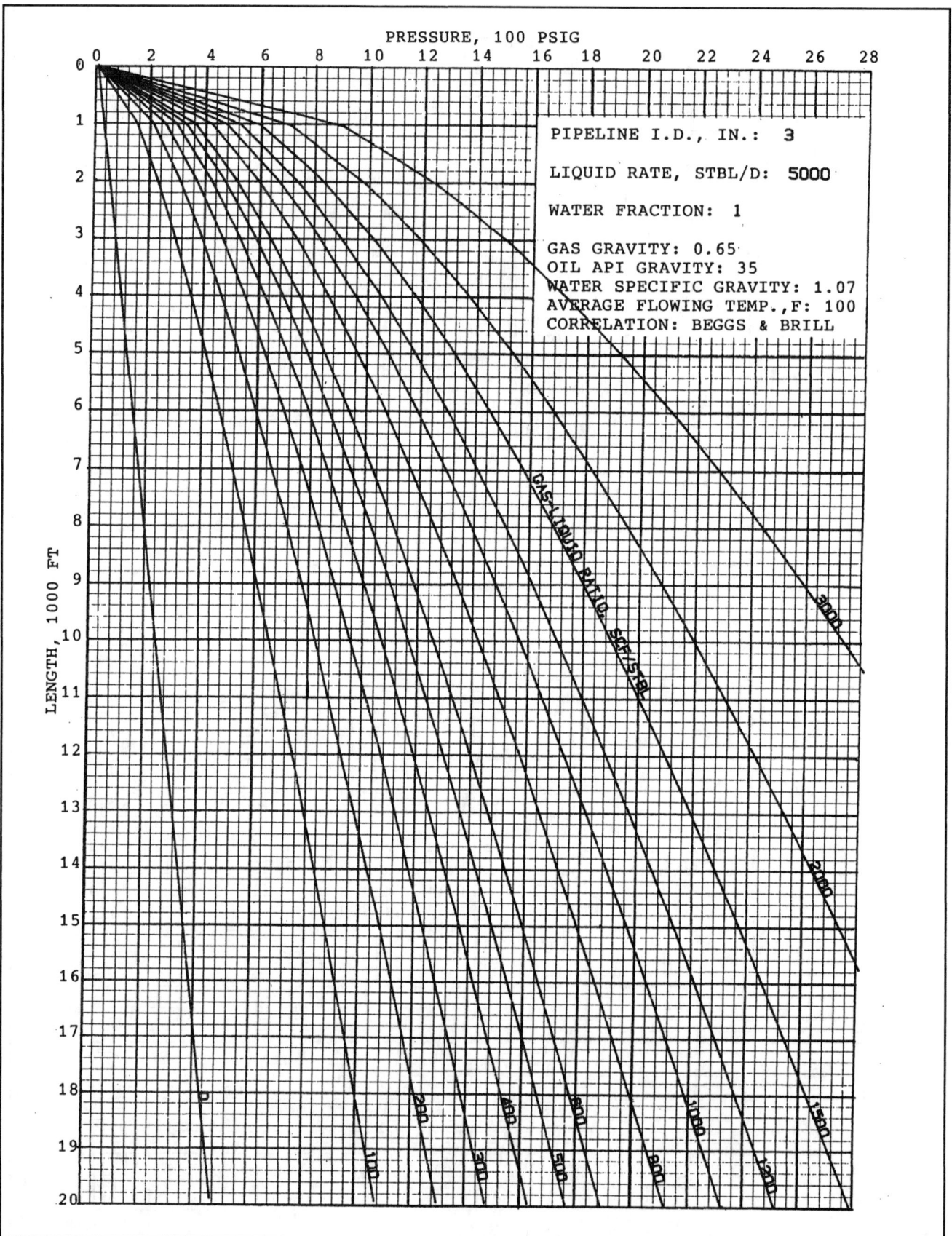

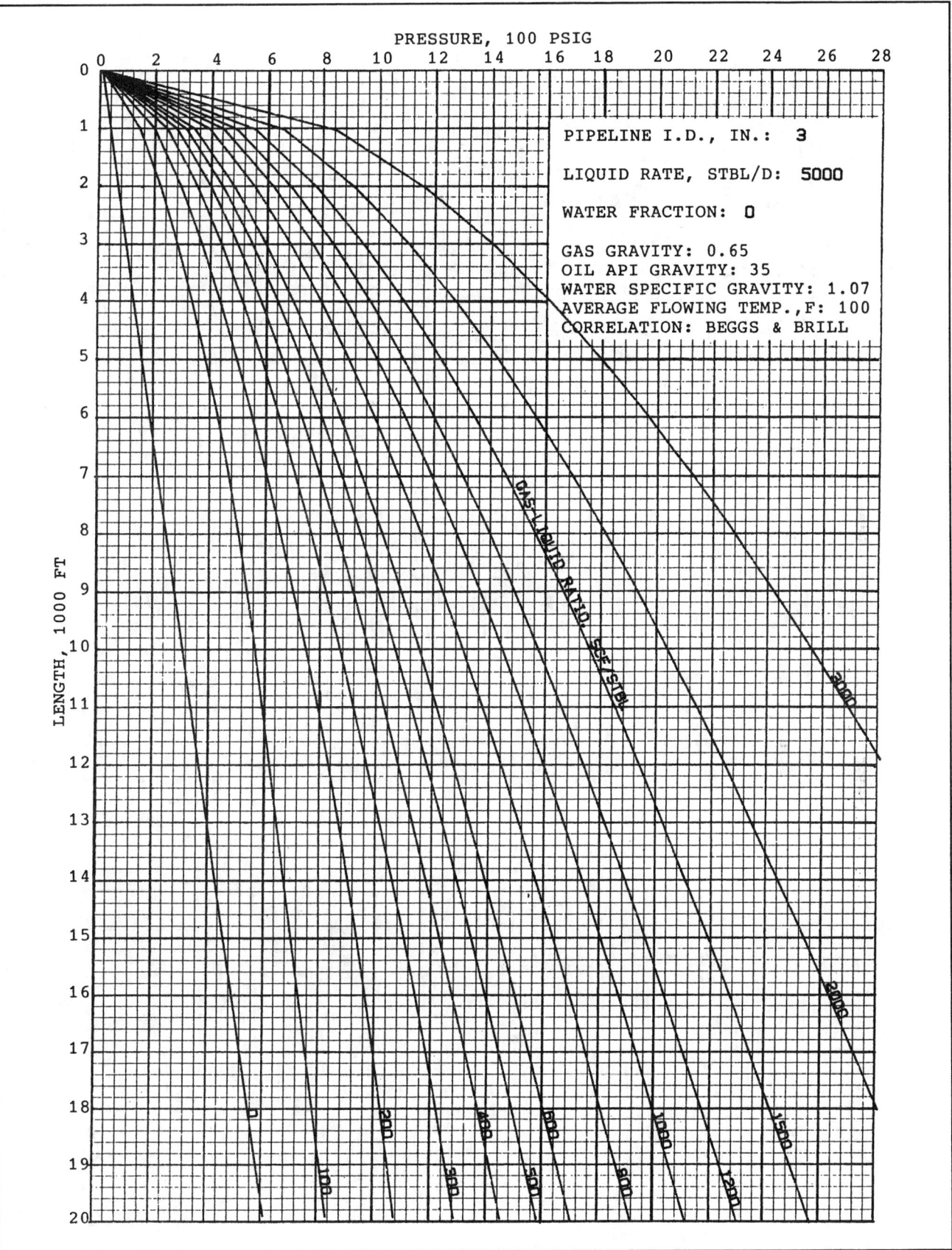

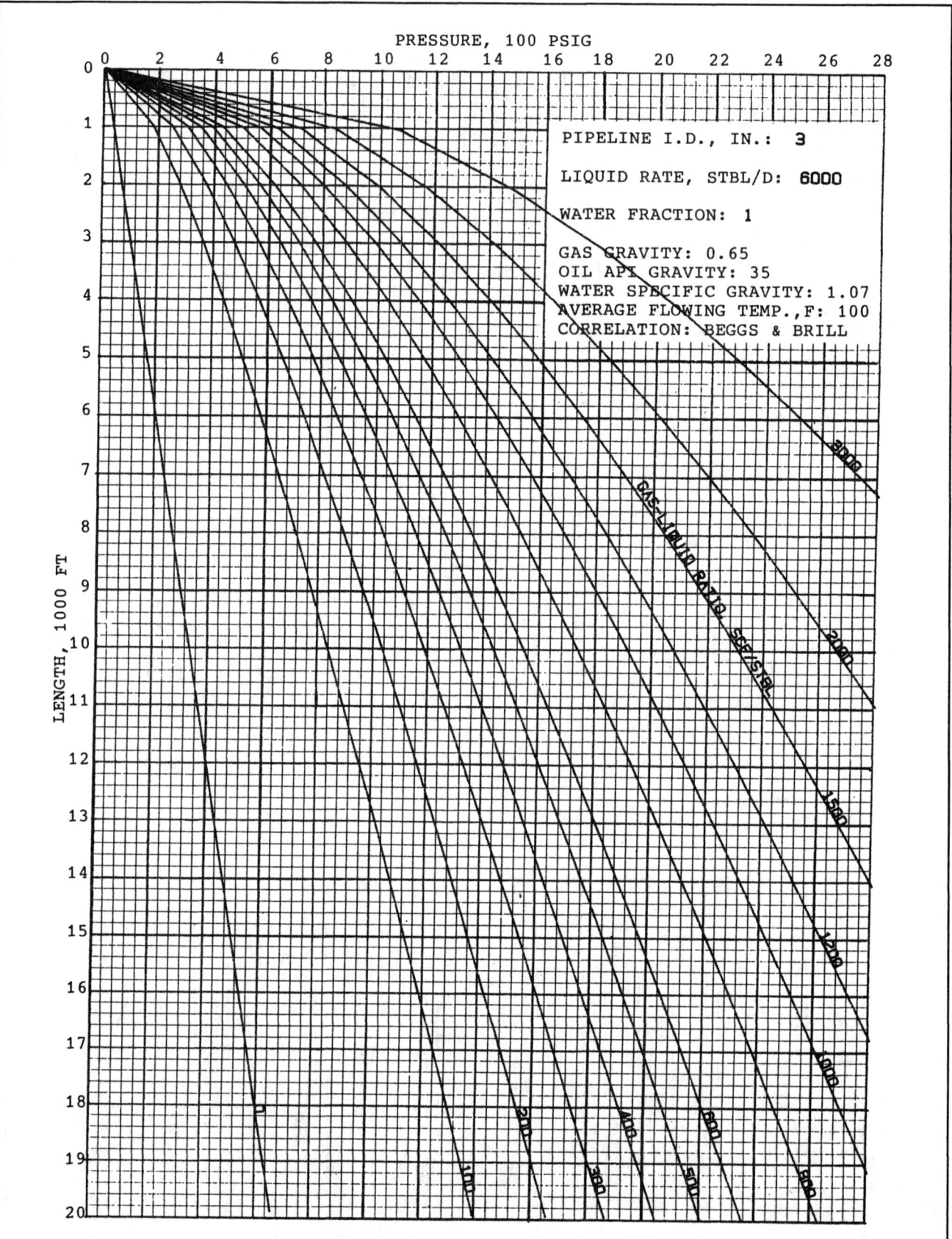

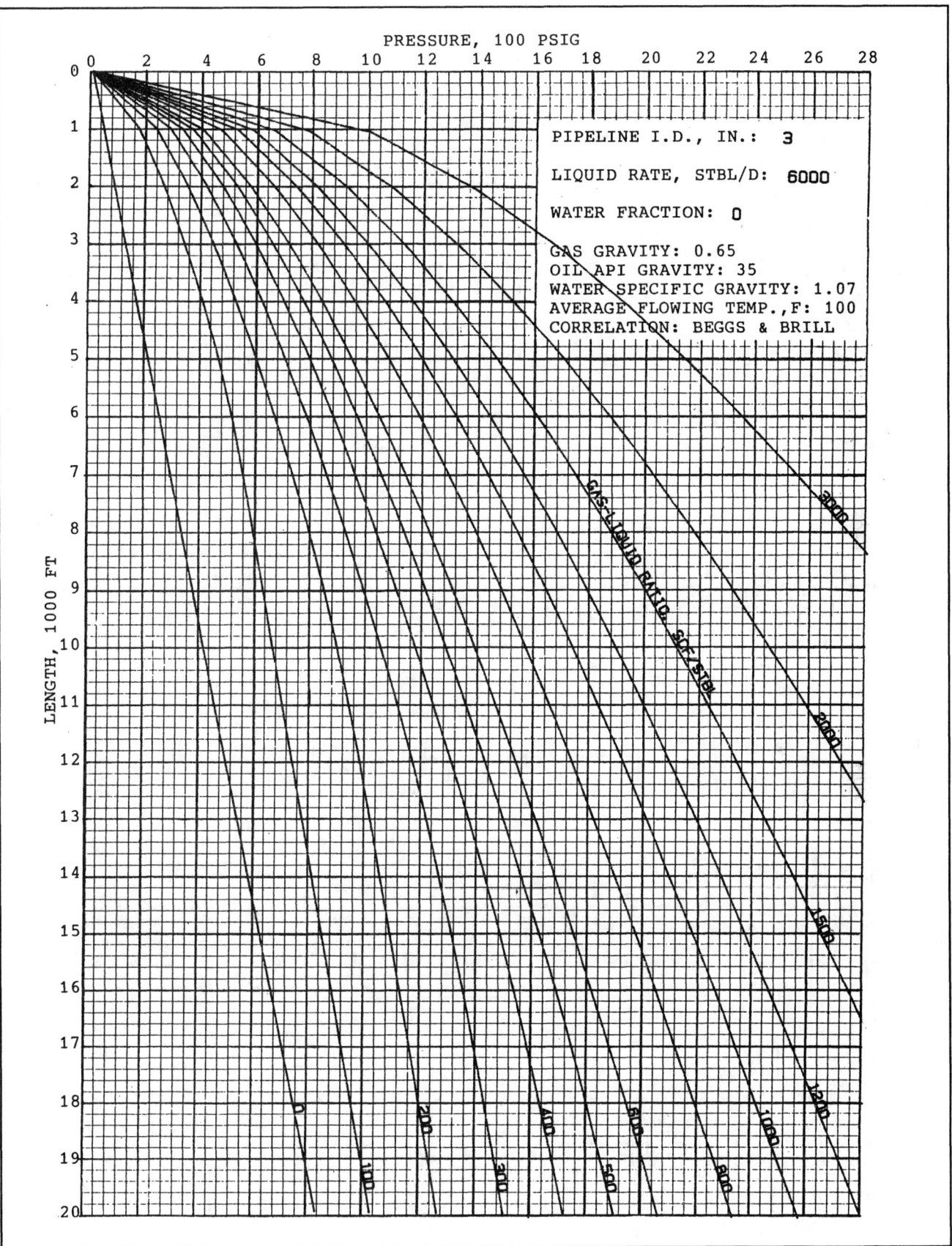

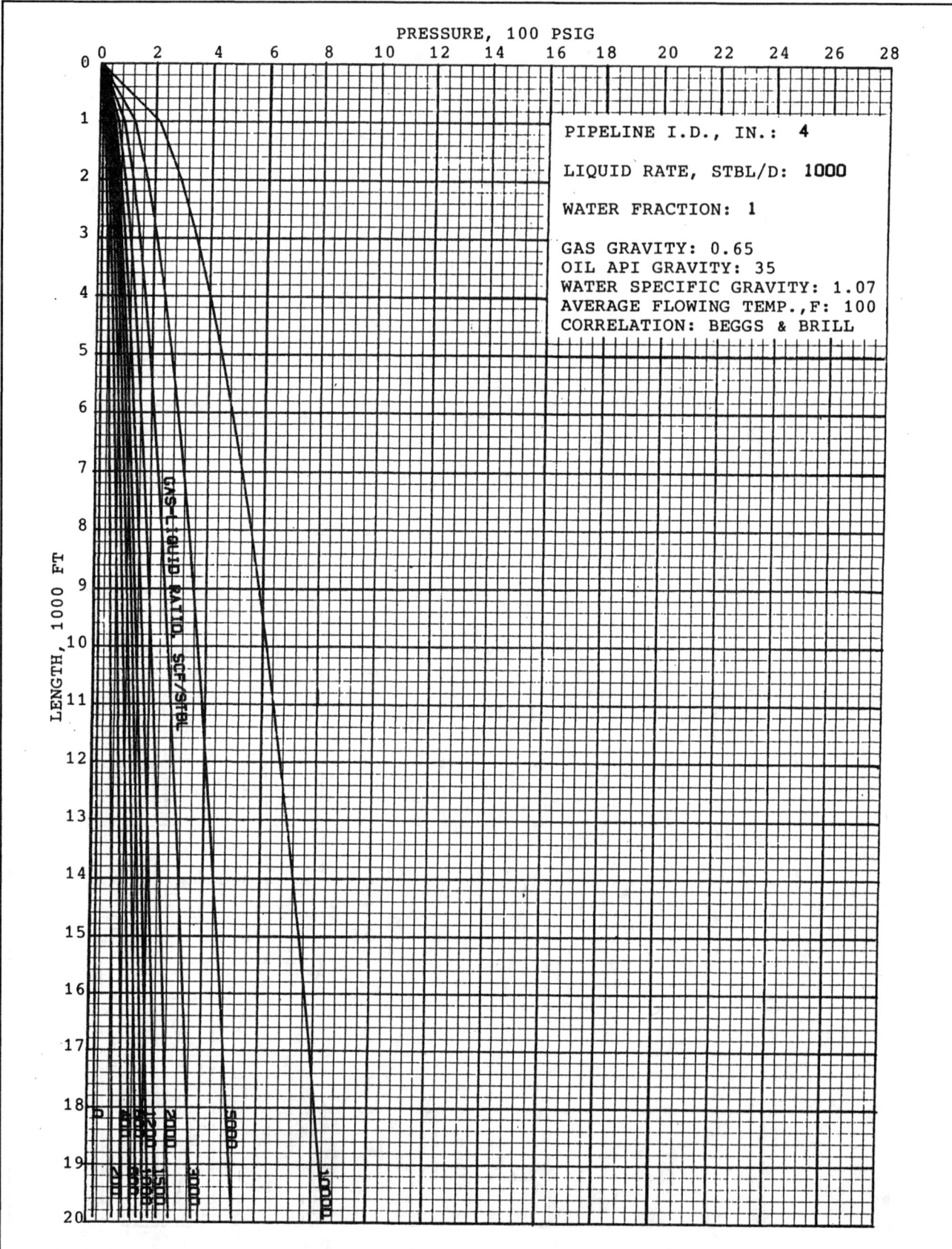

PRESSURE, 100 PSIG

LENGTH, 1000 FT

GAS-LIQUID RATIO, SCF/STBL

PIPELINE I.D., IN.: **4**

LIQUID RATE, STBL/D: **1000**

WATER FRACTION: **1**

GAS GRAVITY: 0.65
OIL API GRAVITY: 35
WATER SPECIFIC GRAVITY: 1.07
AVERAGE FLOWING TEMP.,F: 100
CORRELATION: BEGGS & BRILL

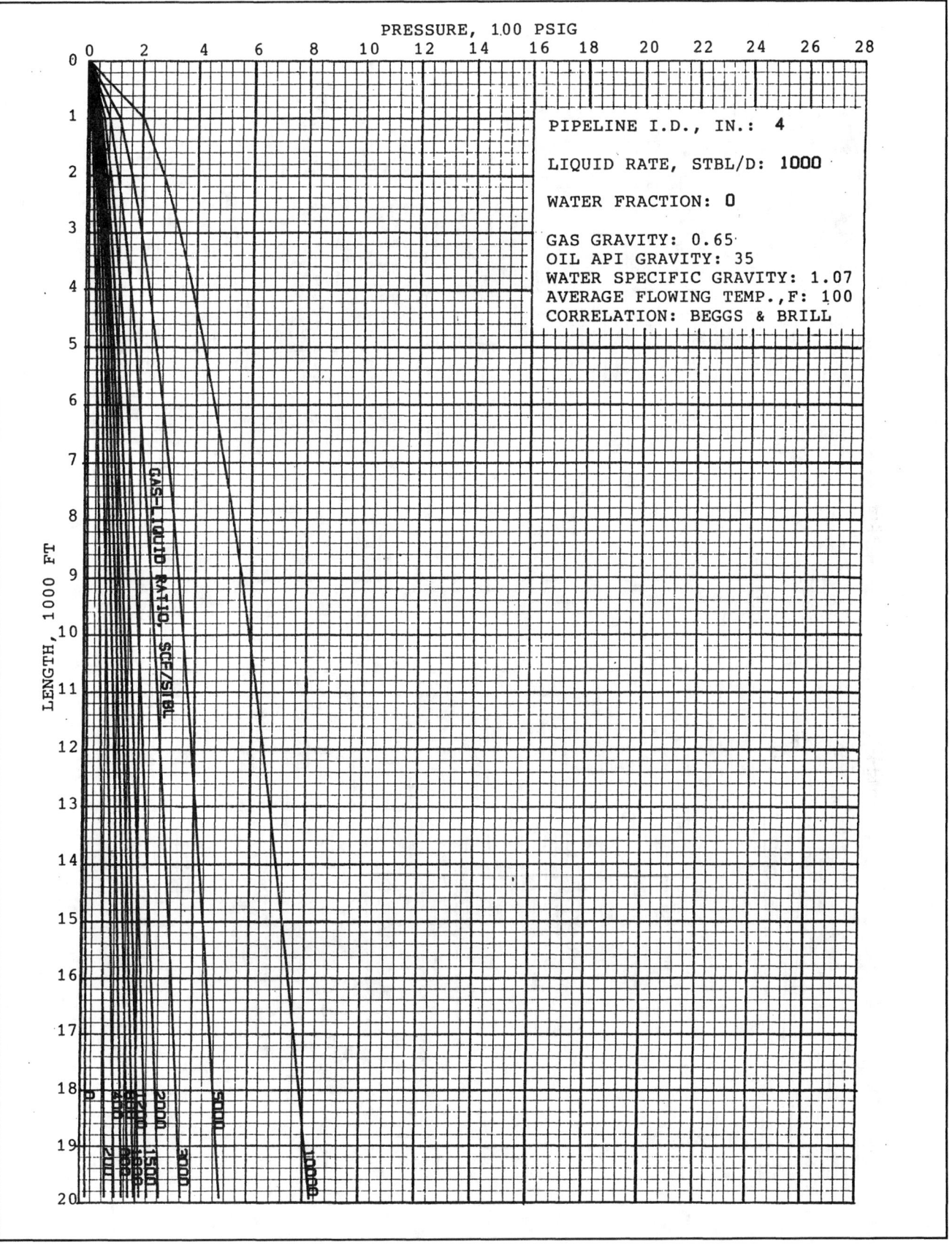

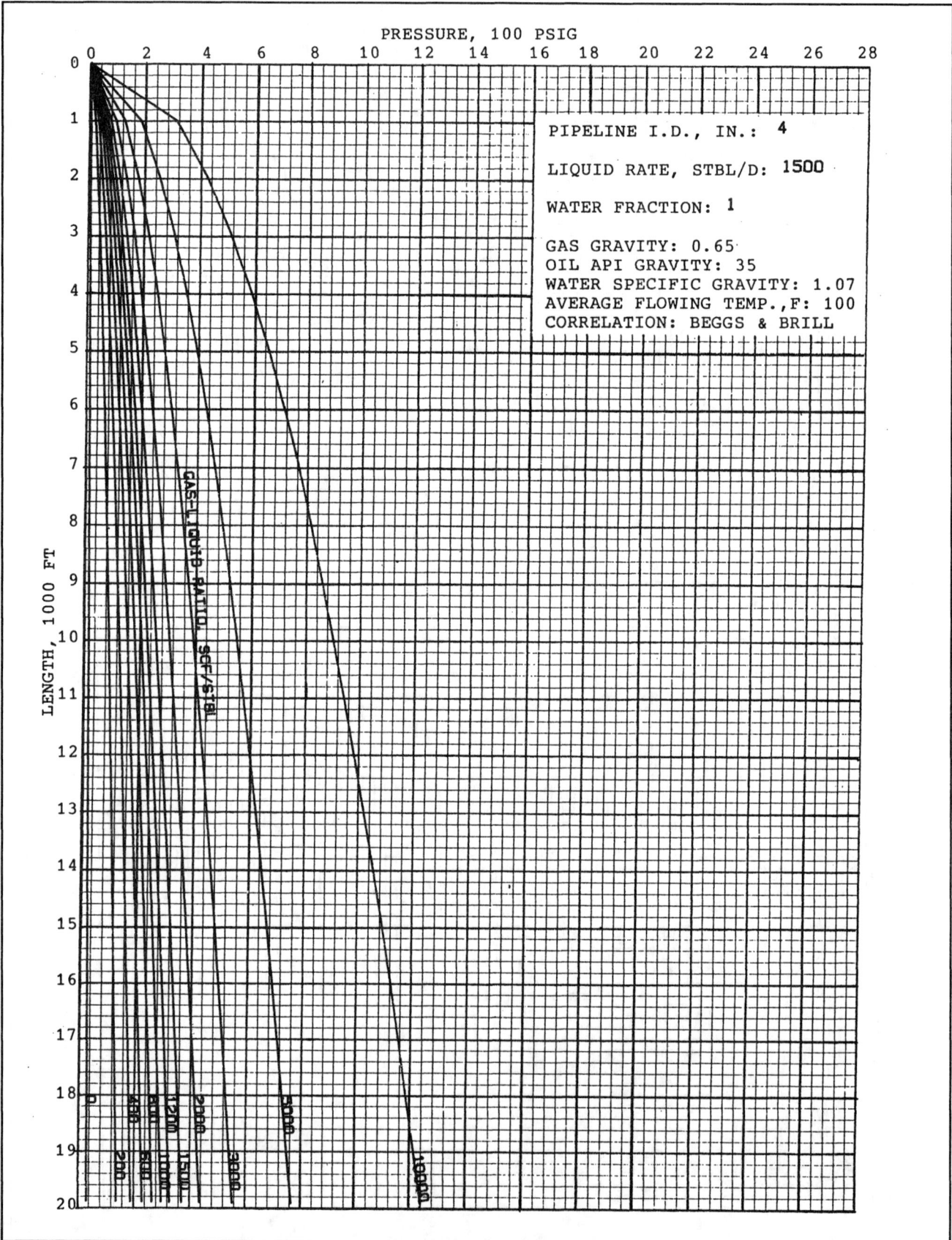

PRESSURE, 100 PSIG

LENGTH, 1000 FT

GAS-LIQUID RATIO, SCF/STBL

PIPELINE I.D., IN.: **4**

LIQUID RATE, STBL/D: **1500**

WATER FRACTION: **1**

GAS GRAVITY: 0.65
OIL API GRAVITY: 35
WATER SPECIFIC GRAVITY: 1.07
AVERAGE FLOWING TEMP.,F: 100
CORRELATION: BEGGS & BRILL

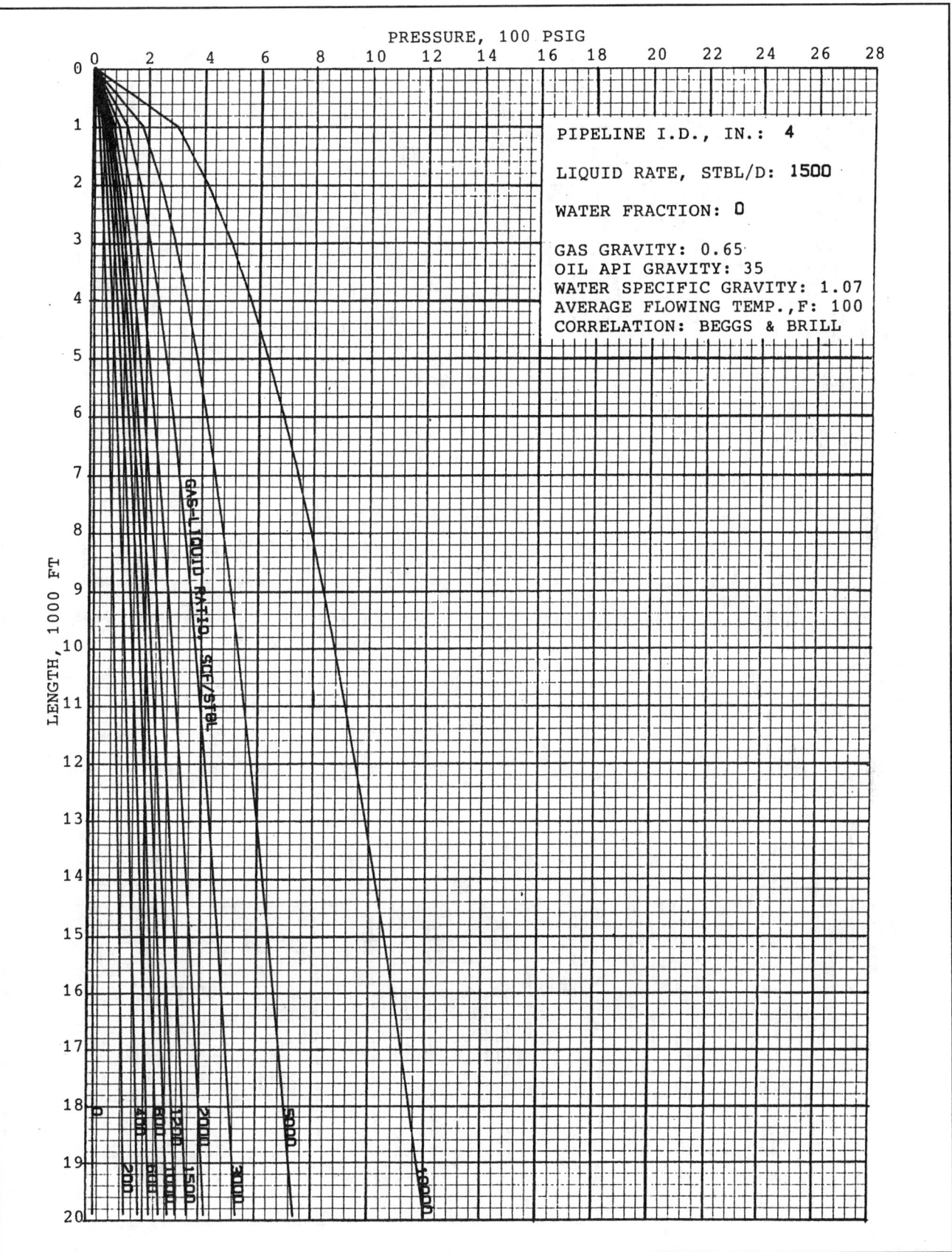

PRESSURE, 100 PSIG

LENGTH, 1000 FT

GAS-LIQUID RATIO, SCF/STBL

PIPELINE I.D., IN.: **4**

LIQUID RATE, STBL/D: **1500**

WATER FRACTION: **0**

GAS GRAVITY: 0.65
OIL API GRAVITY: 35
WATER SPECIFIC GRAVITY: 1.07
AVERAGE FLOWING TEMP., F: 100
CORRELATION: BEGGS & BRILL

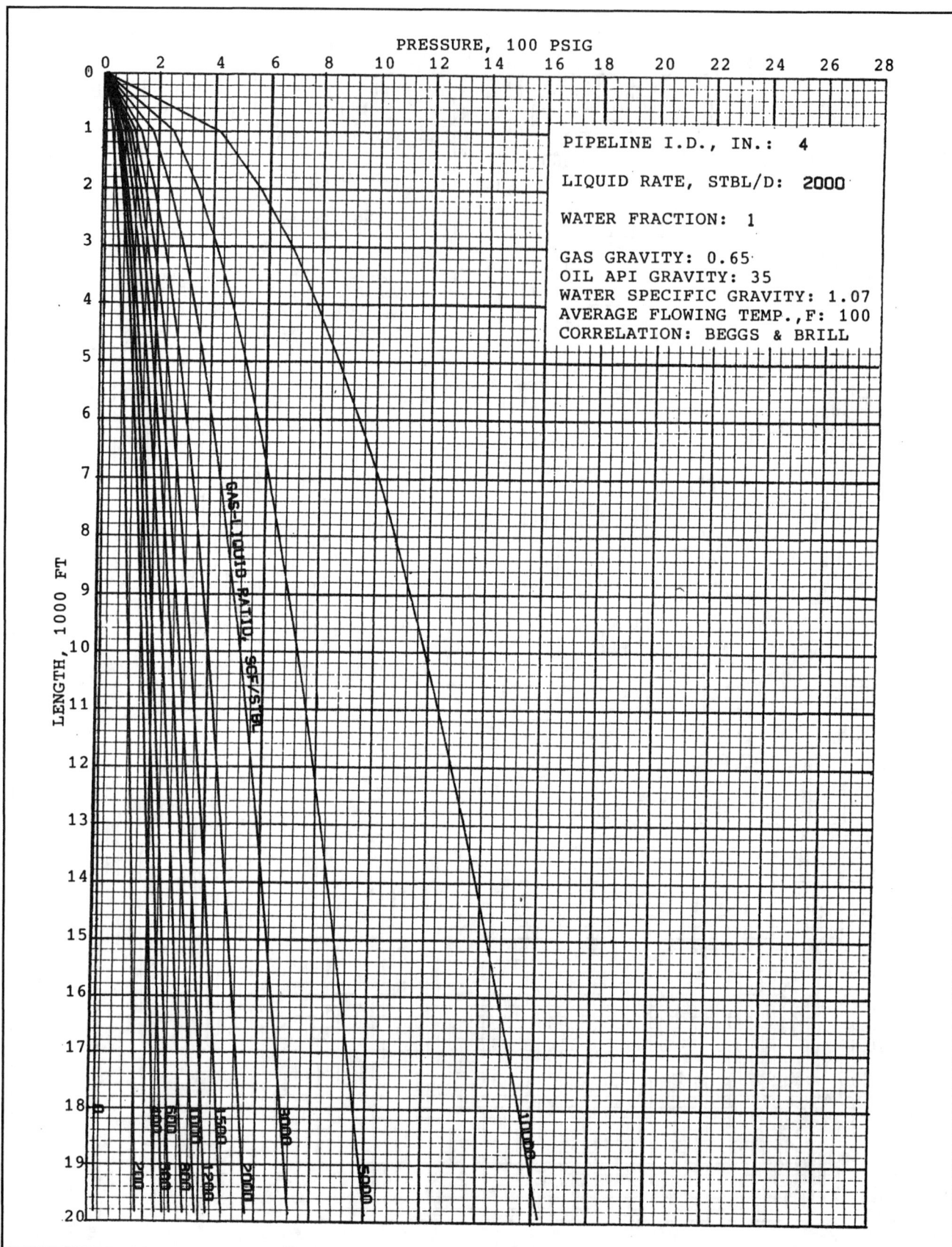

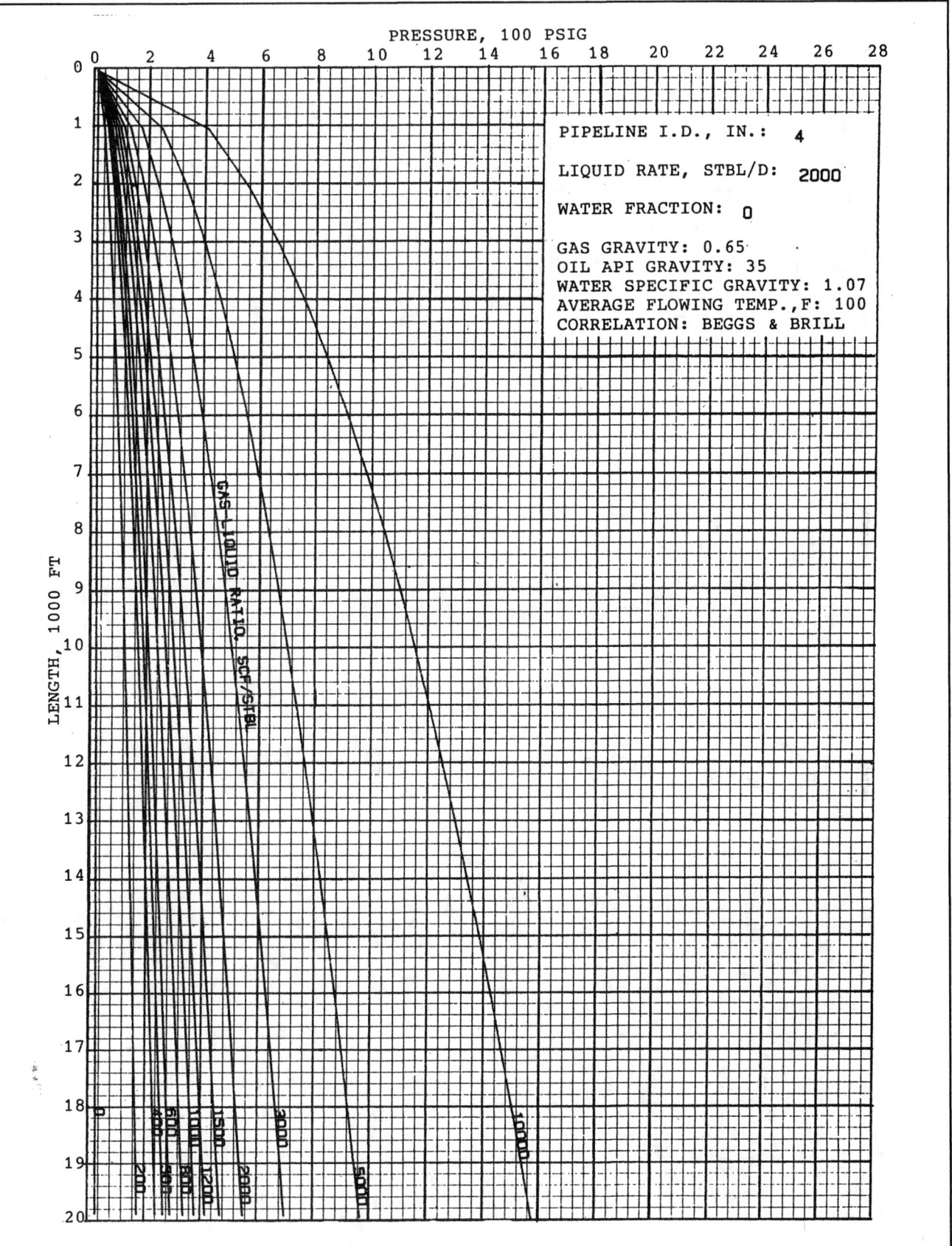

PRESSURE, 100 PSIG

LENGTH, 1000 FT

GAS-LIQUID RATIO, SCF/STBL

PIPELINE I.D., IN.: 4

LIQUID RATE, STBL/D: 2000

WATER FRACTION: 0

GAS GRAVITY: 0.65
OIL API GRAVITY: 35
WATER SPECIFIC GRAVITY: 1.07
AVERAGE FLOWING TEMP.,F: 100
CORRELATION: BEGGS & BRILL

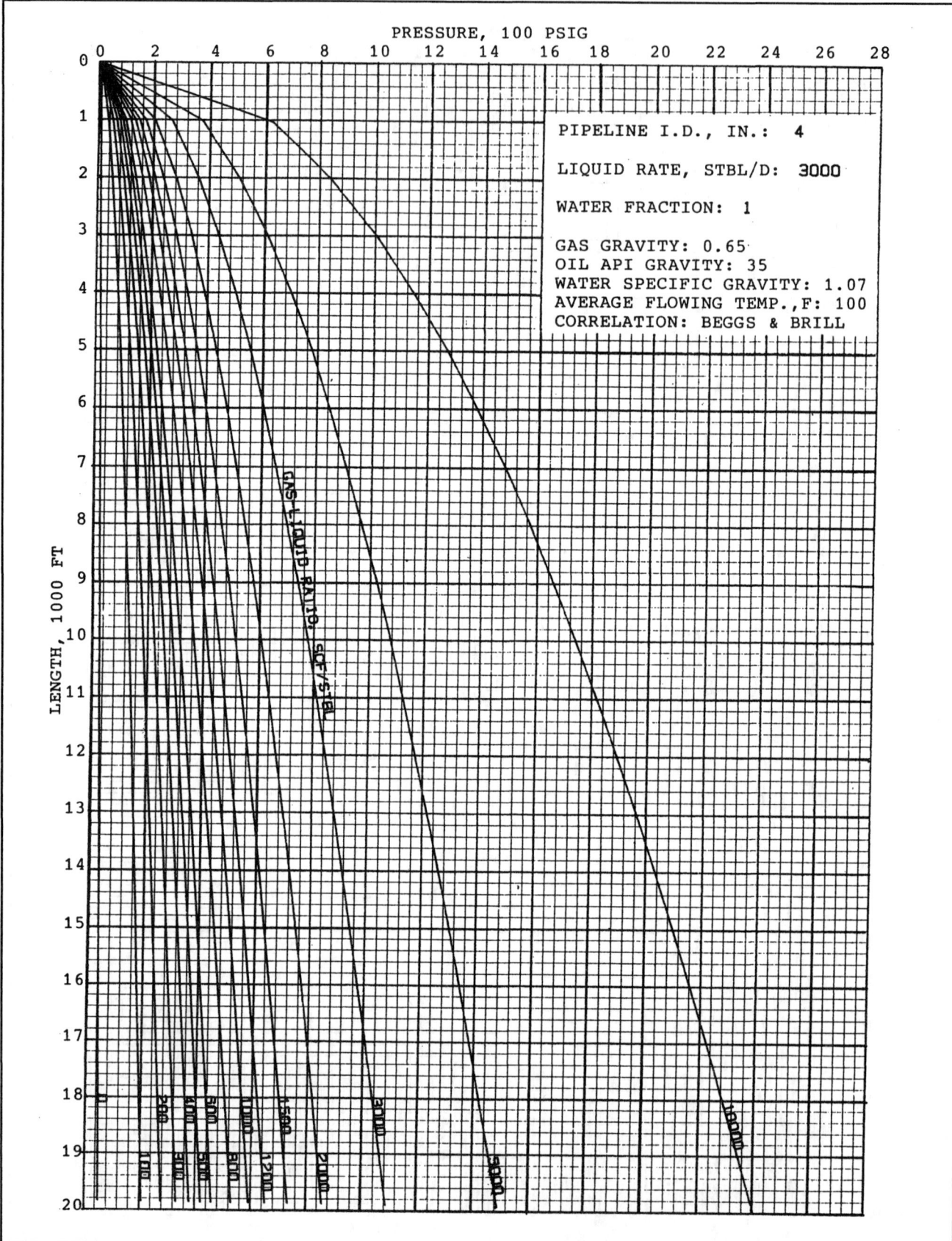

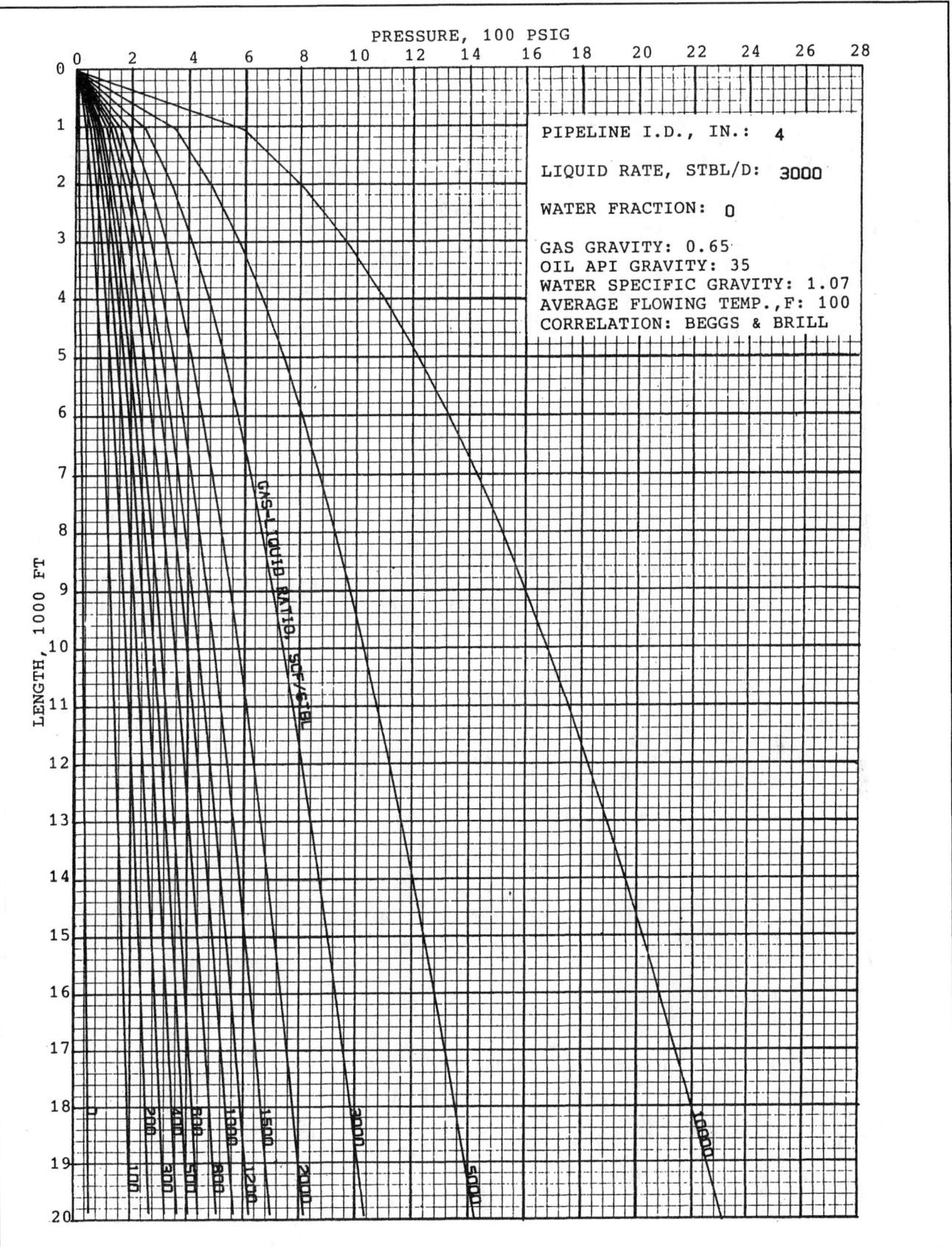

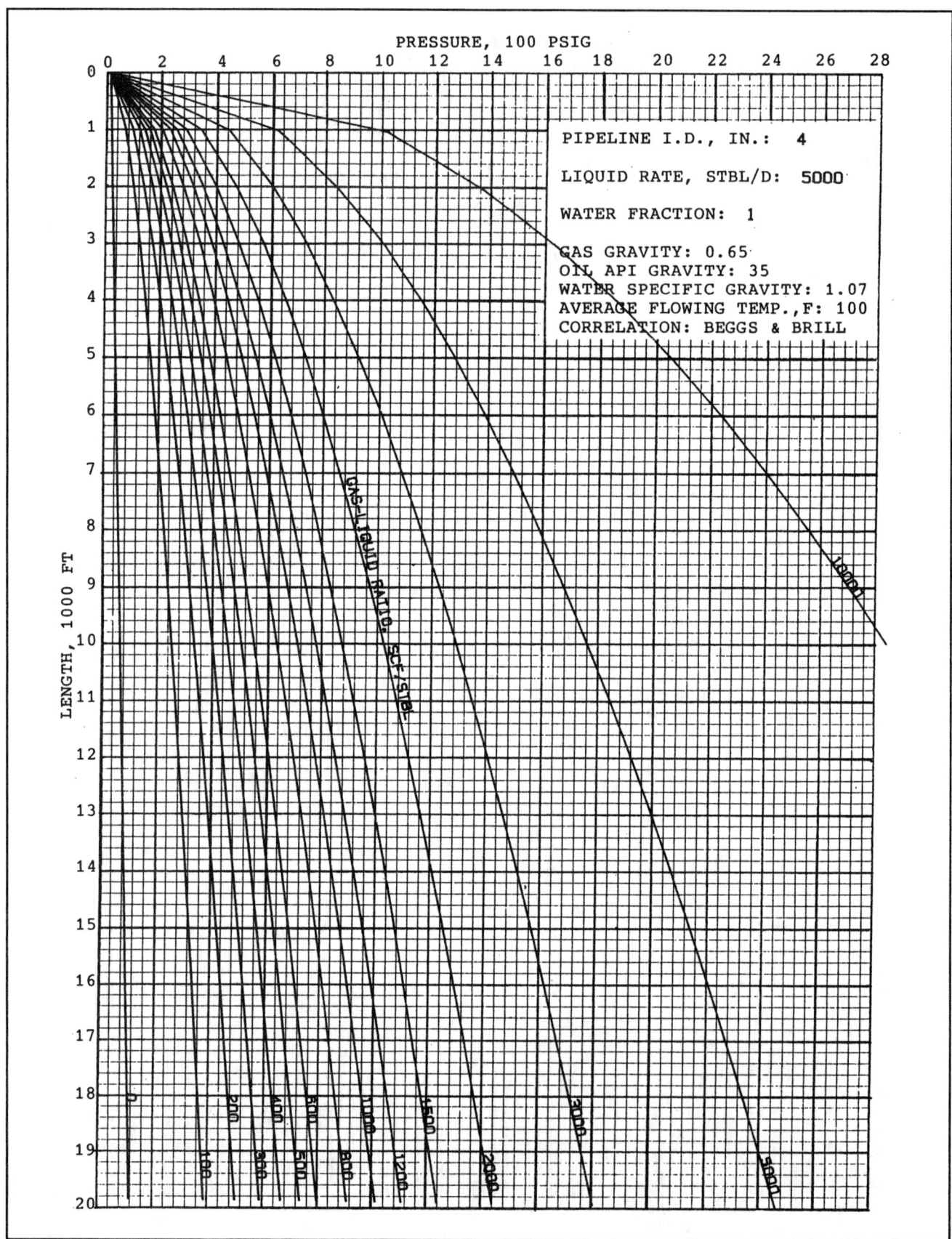

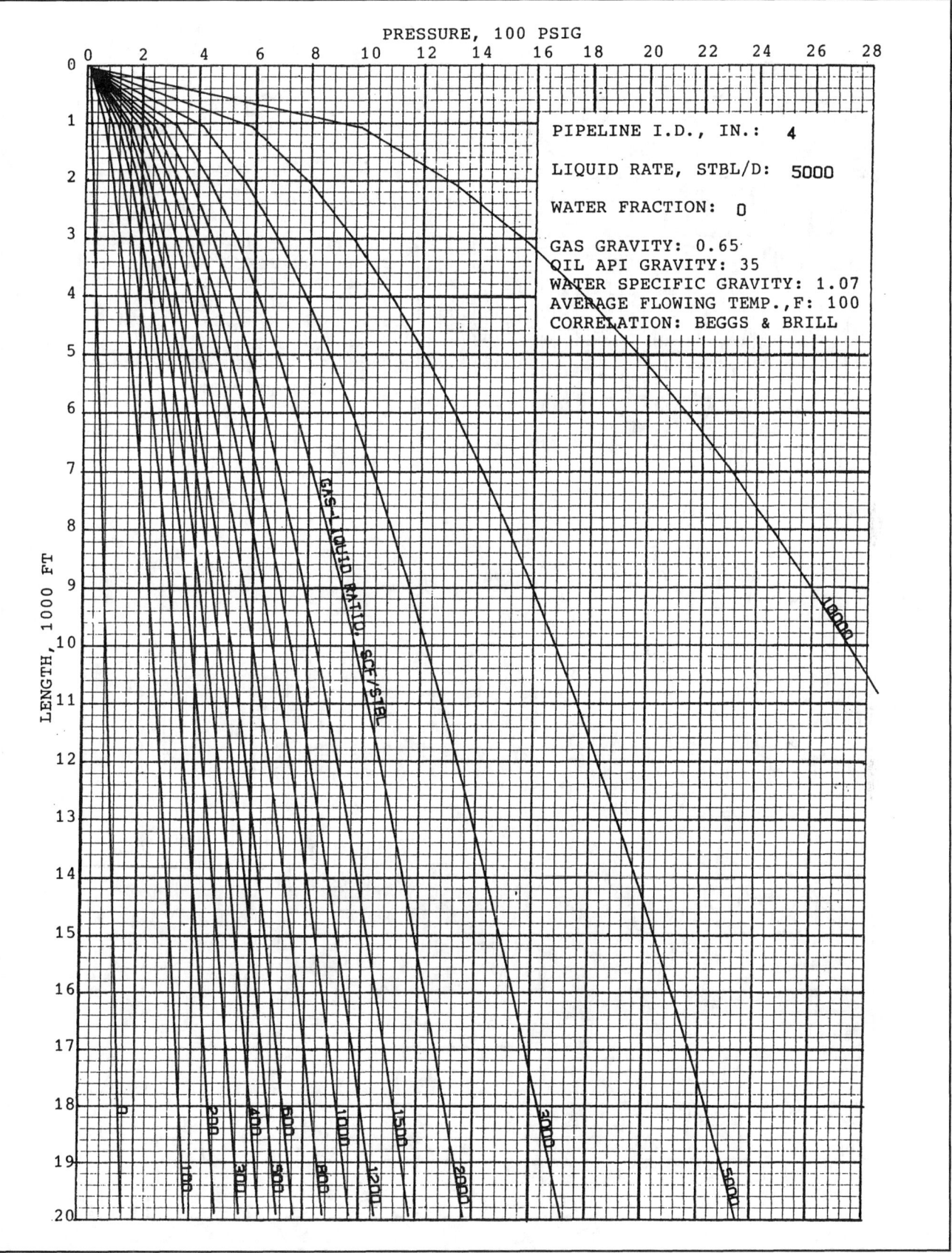

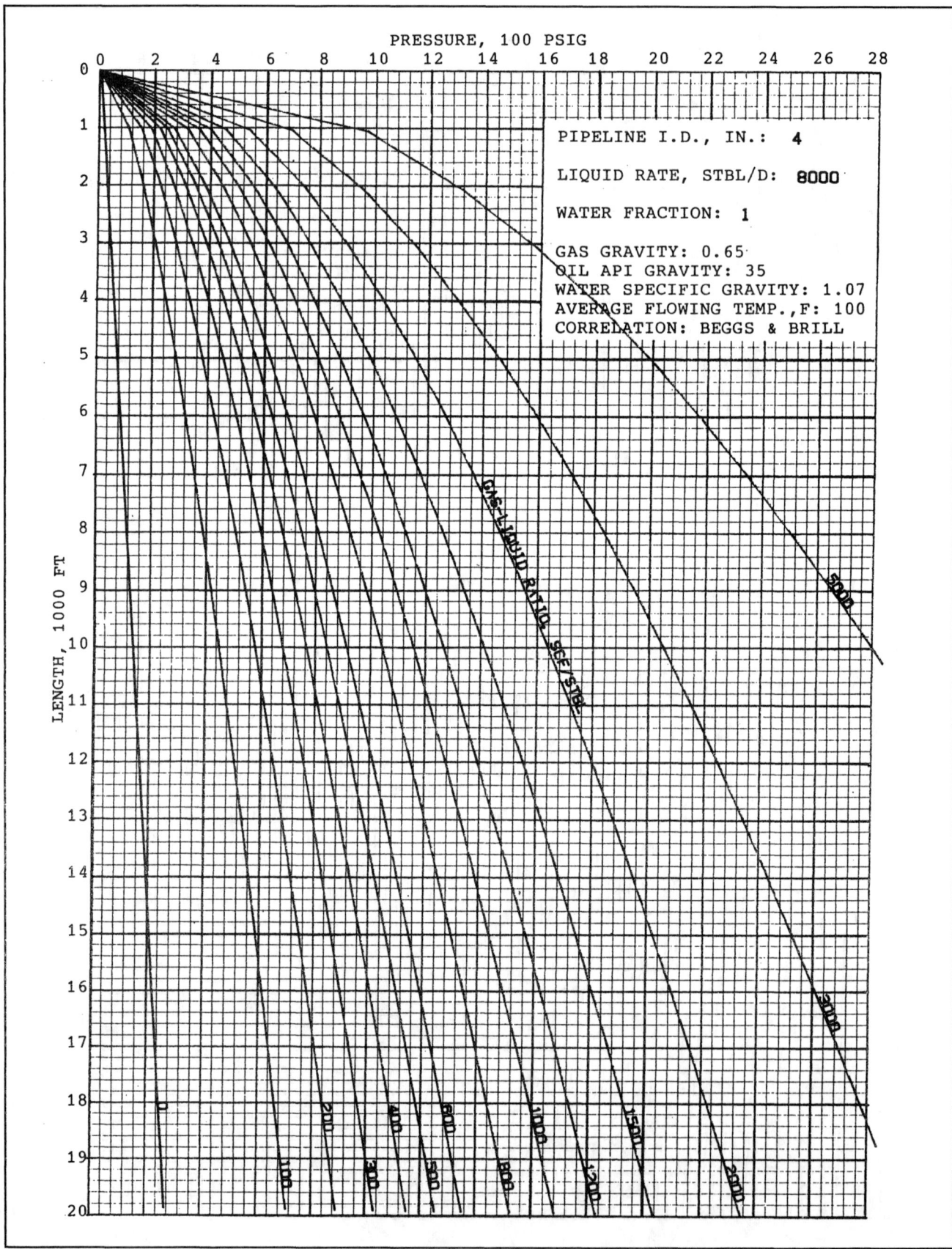

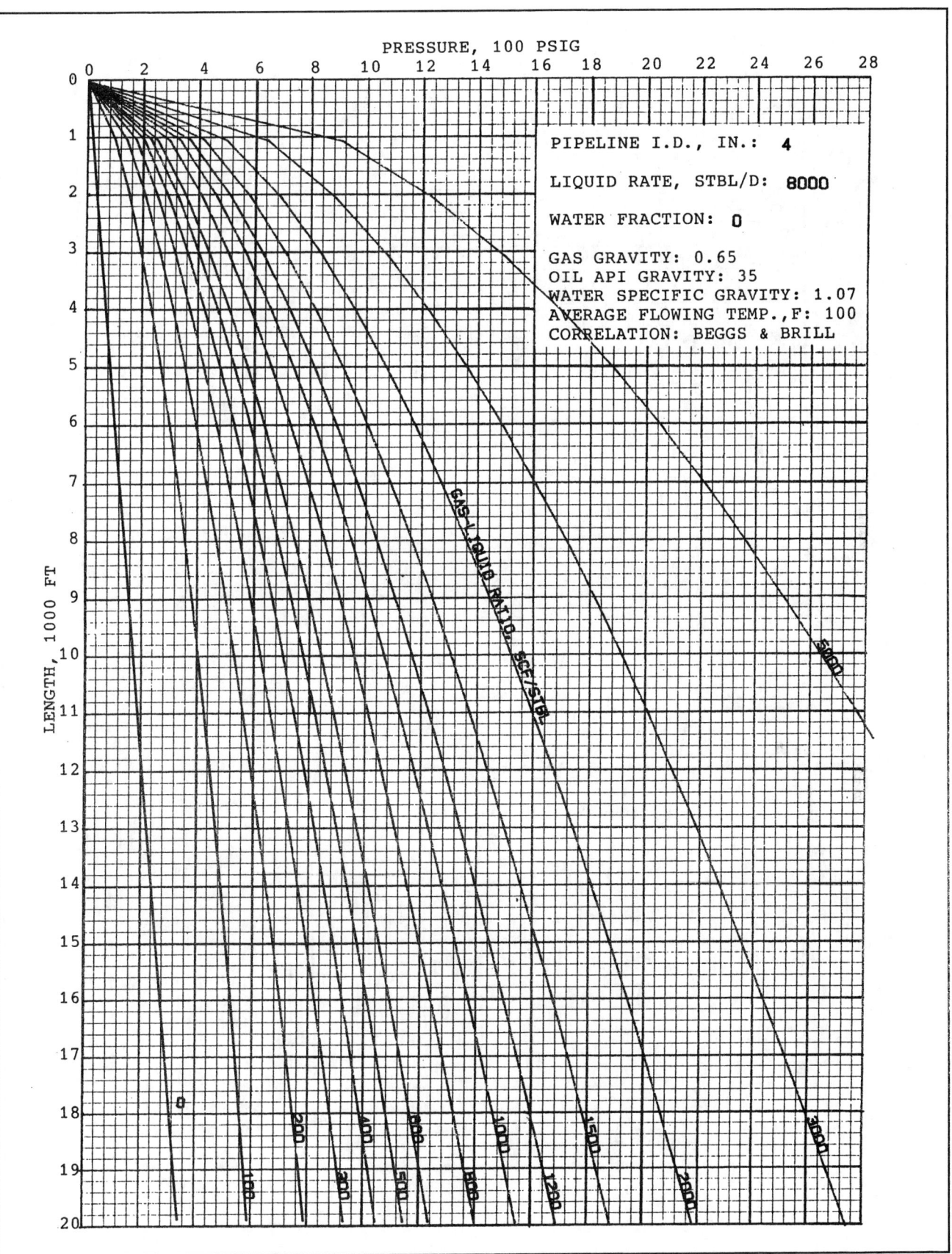

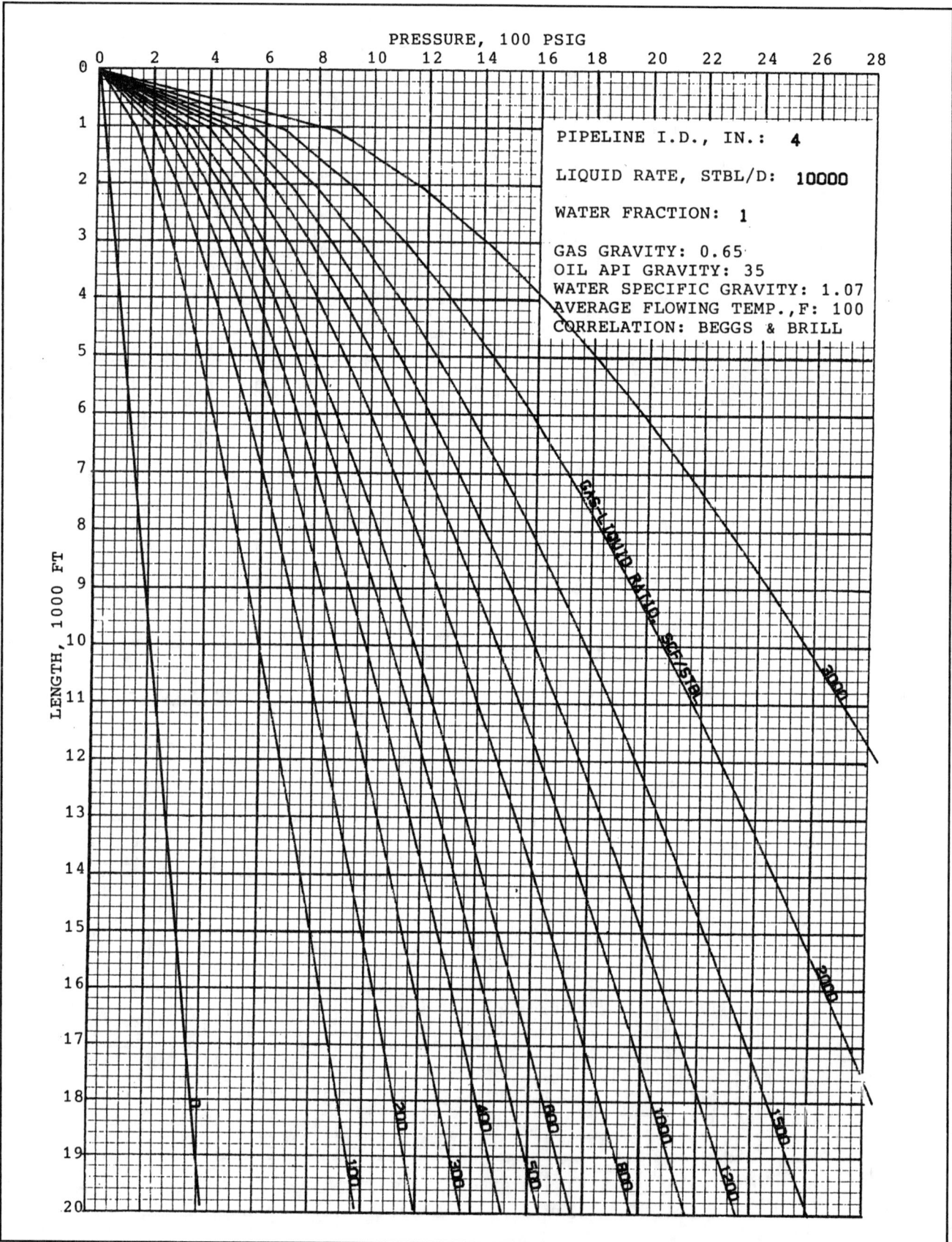

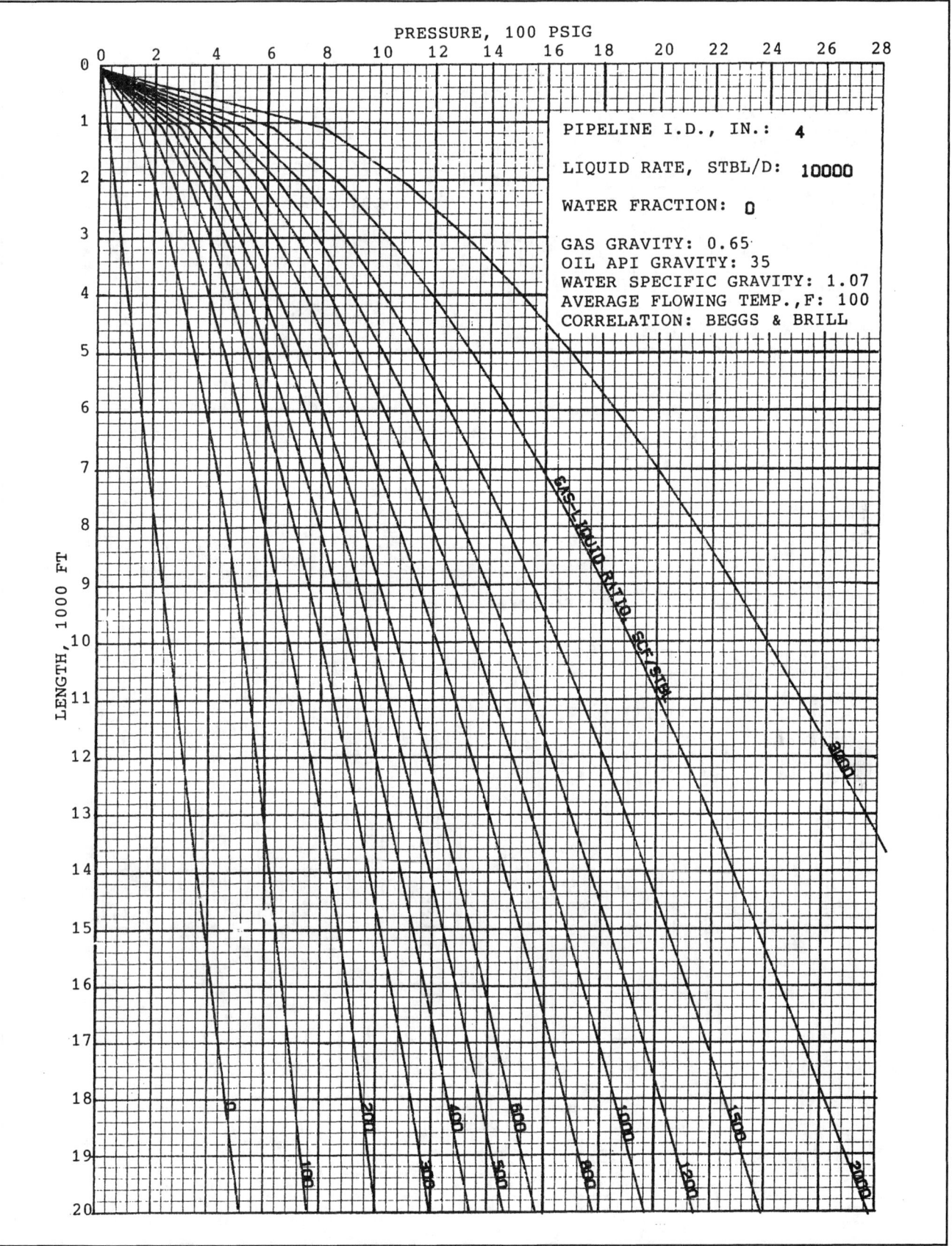

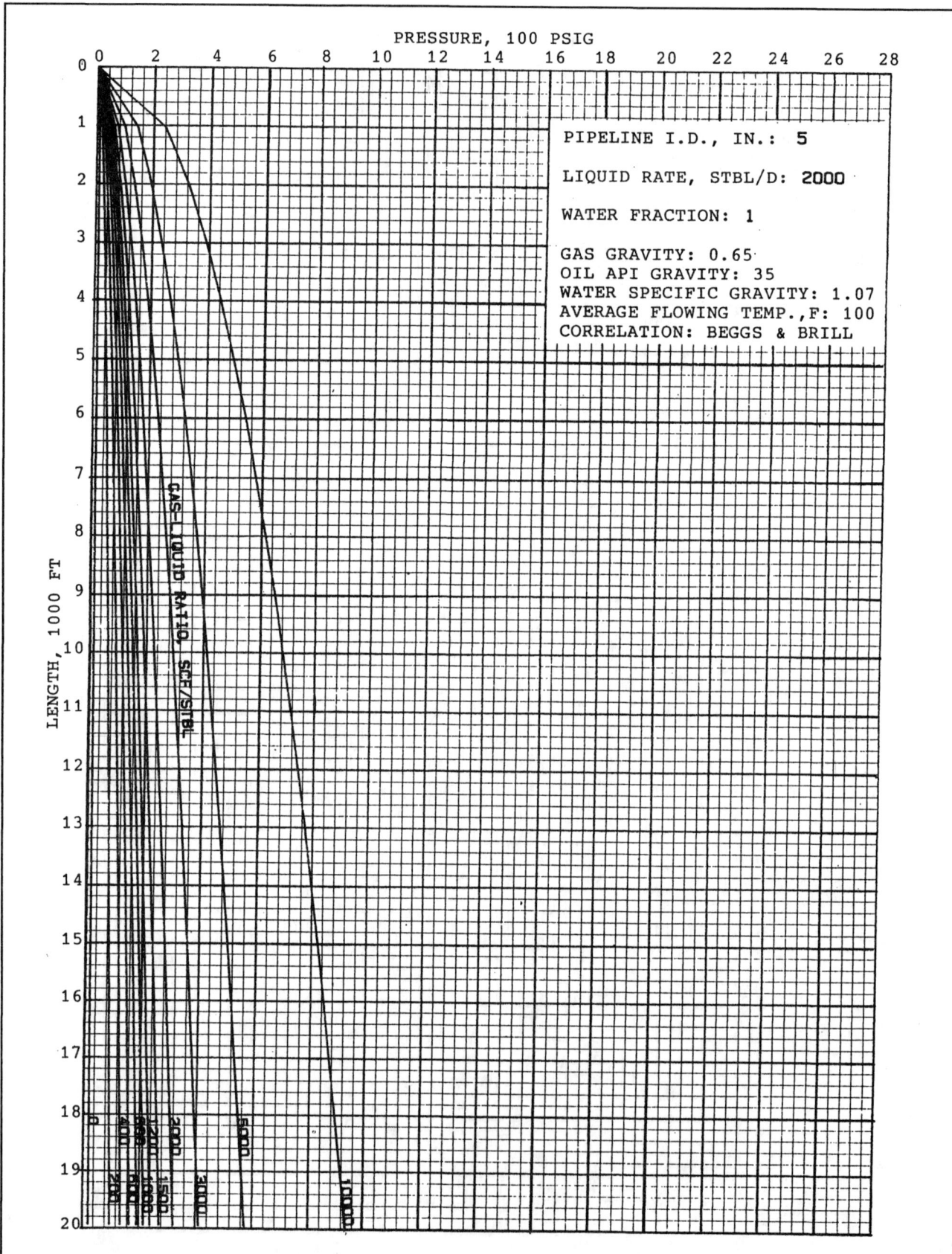

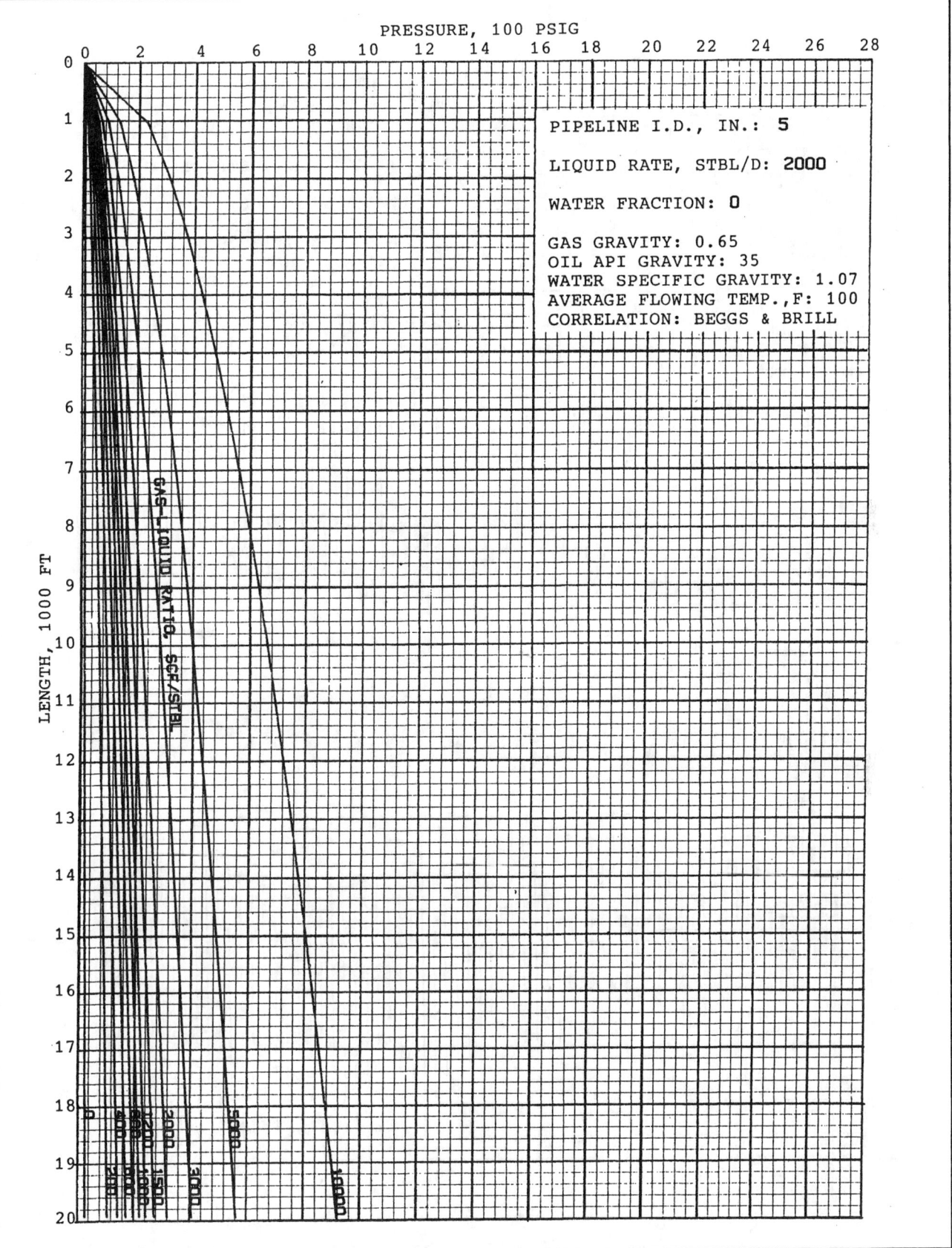

PRESSURE, 100 PSIG

LENGTH, 1000 FT

GAS-LIQUID RATIO, SCF/STBL

PIPELINE I.D., IN.: **5**

LIQUID RATE, STBL/D: **2000**

WATER FRACTION: **0**

GAS GRAVITY: 0.65
OIL API GRAVITY: 35
WATER SPECIFIC GRAVITY: 1.07
AVERAGE FLOWING TEMP.,F: 100
CORRELATION: BEGGS & BRILL

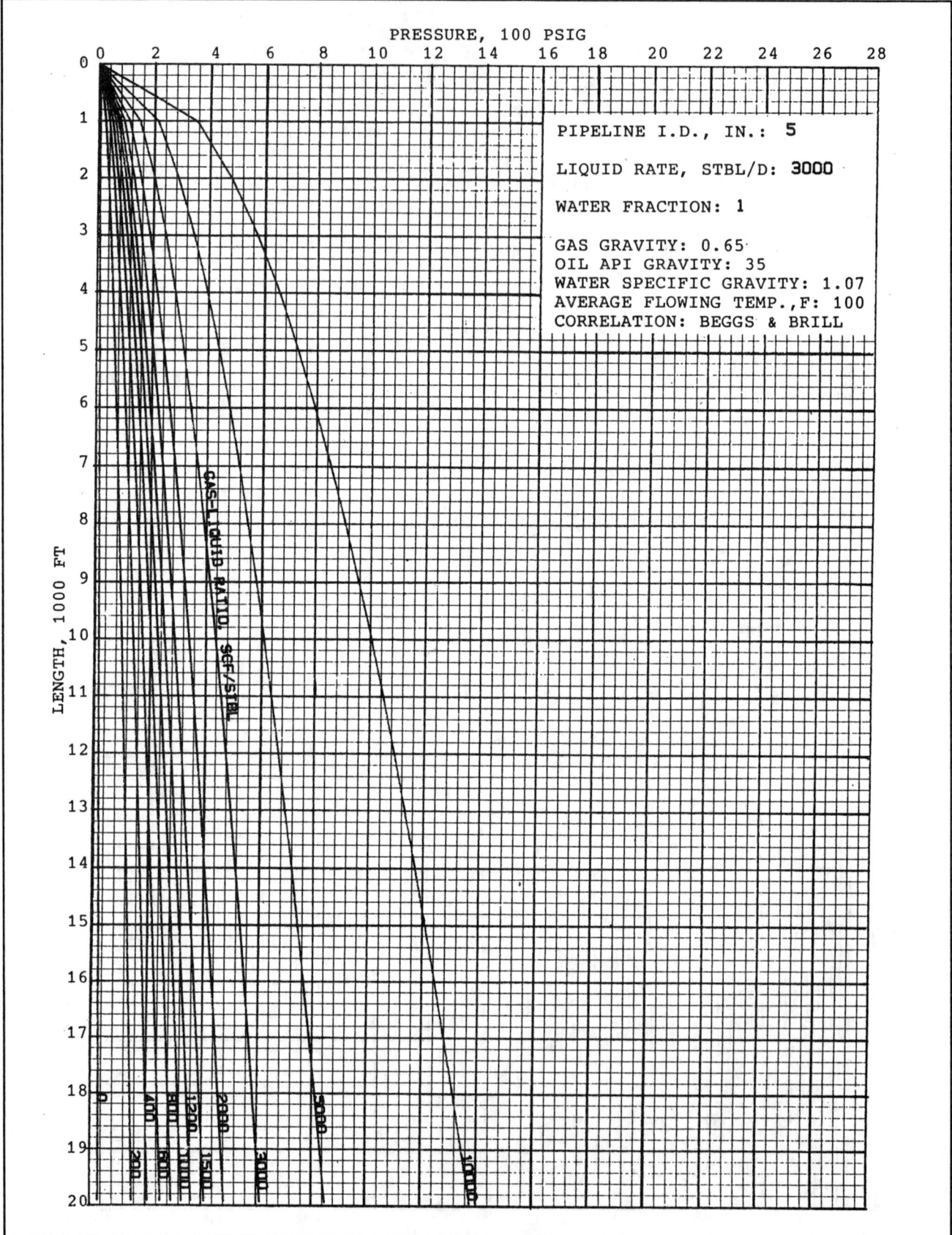

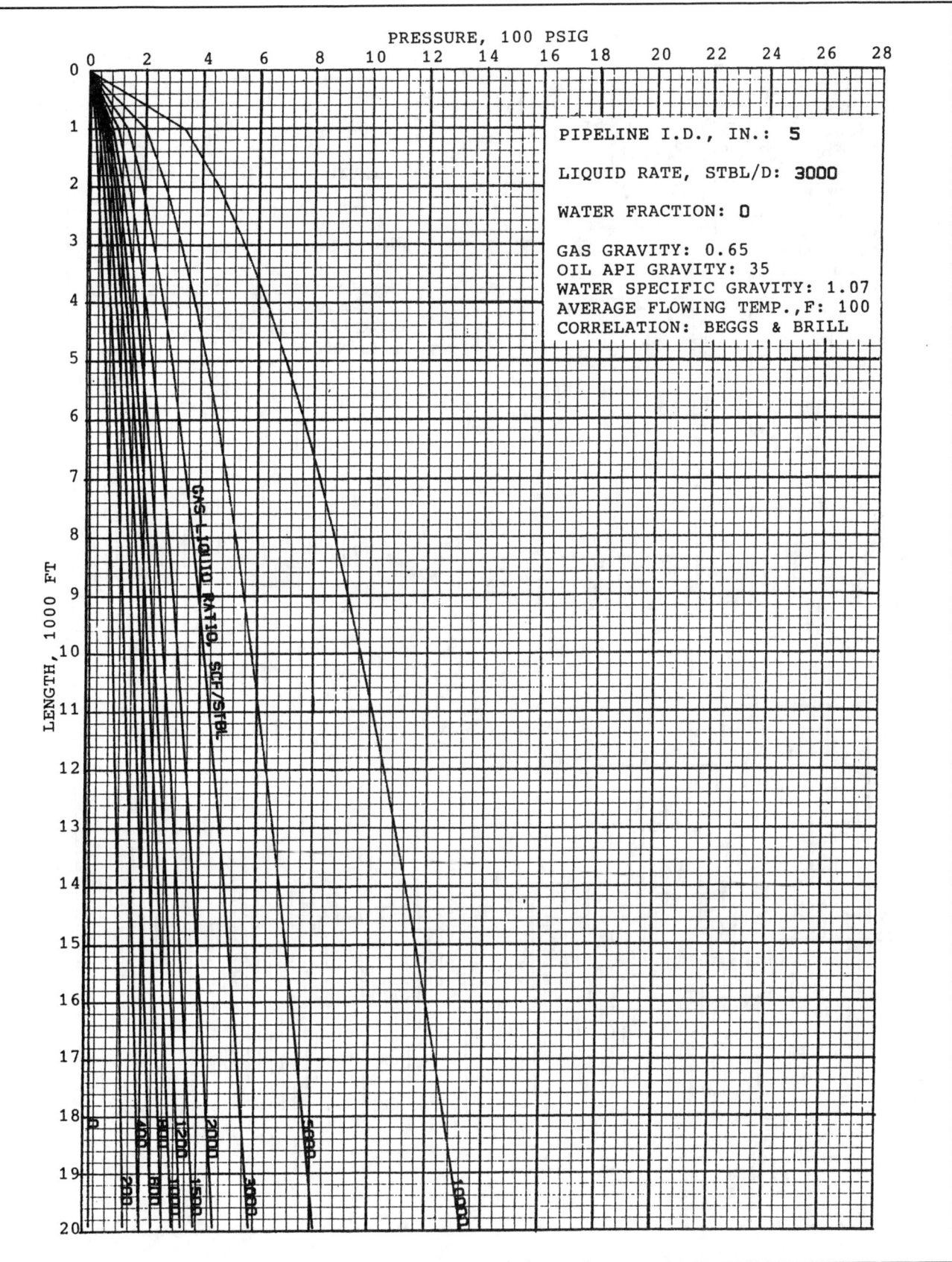

PRESSURE, 100 PSIG

LENGTH, 1000 FT

GAS-LIQUID RATIO, SCF/STBL

PIPELINE I.D., IN.: **5**

LIQUID RATE, STBL/D: **3000**

WATER FRACTION: **0**

GAS GRAVITY: 0.65
OIL API GRAVITY: 35
WATER SPECIFIC GRAVITY: 1.07
AVERAGE FLOWING TEMP.,F: 100
CORRELATION: BEGGS & BRILL

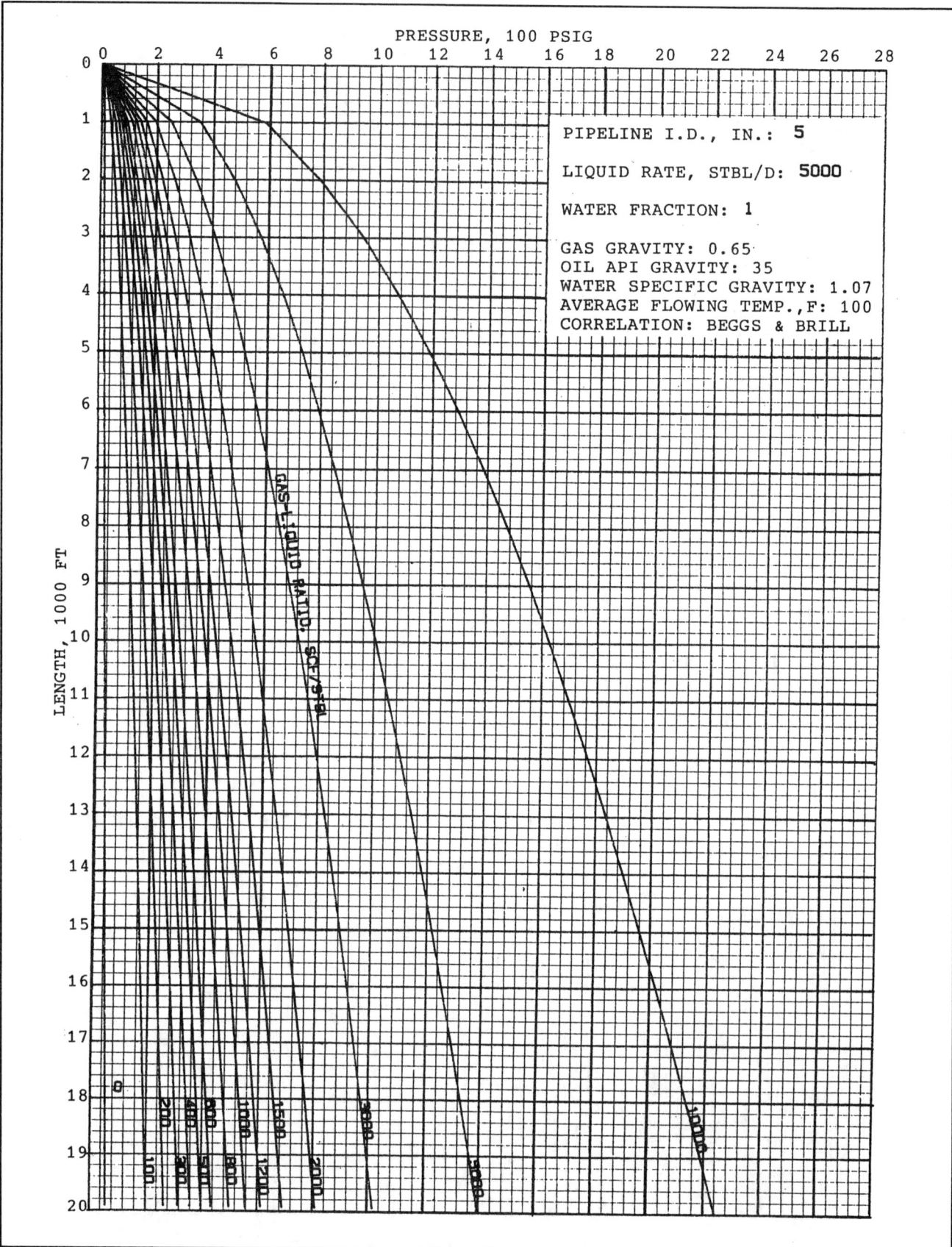

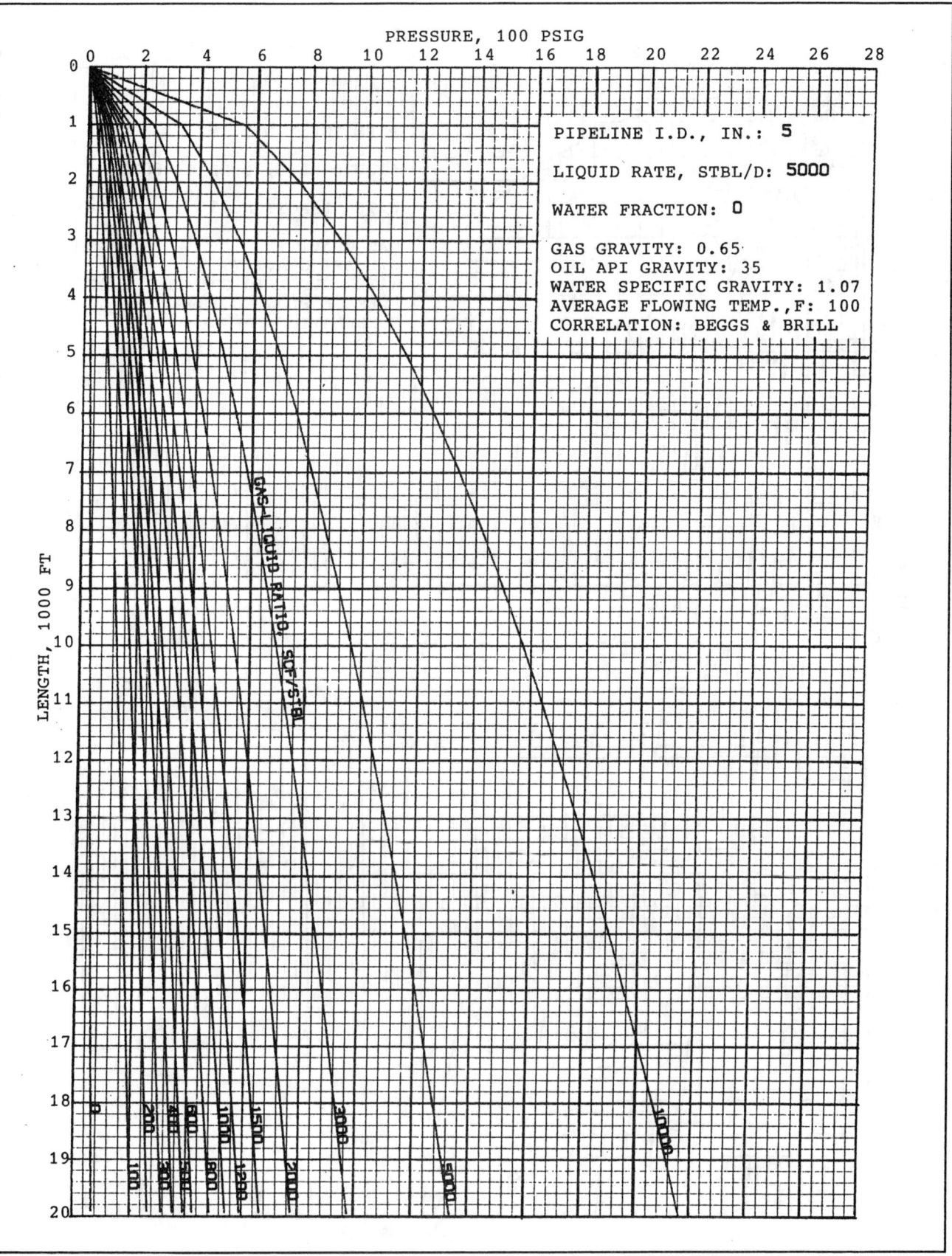

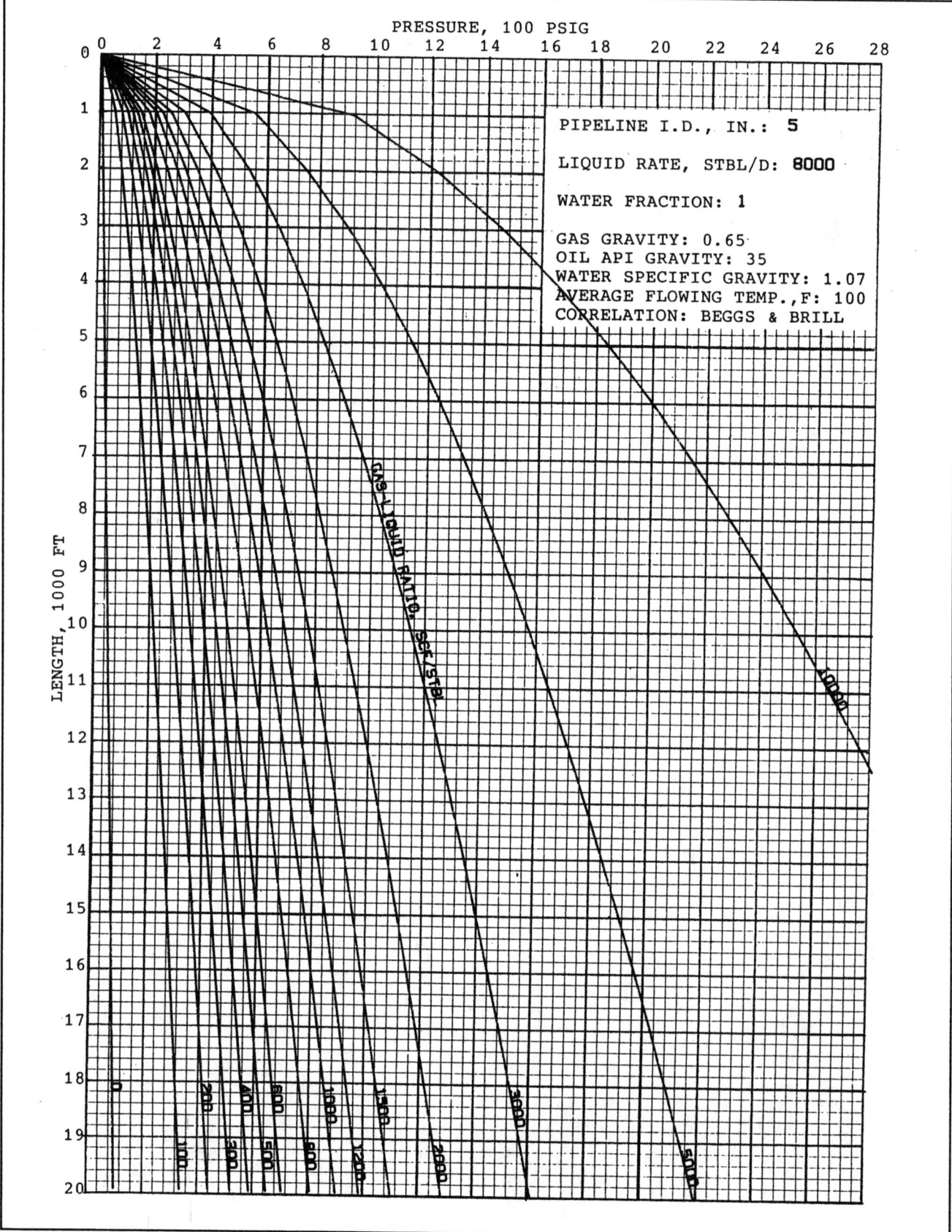

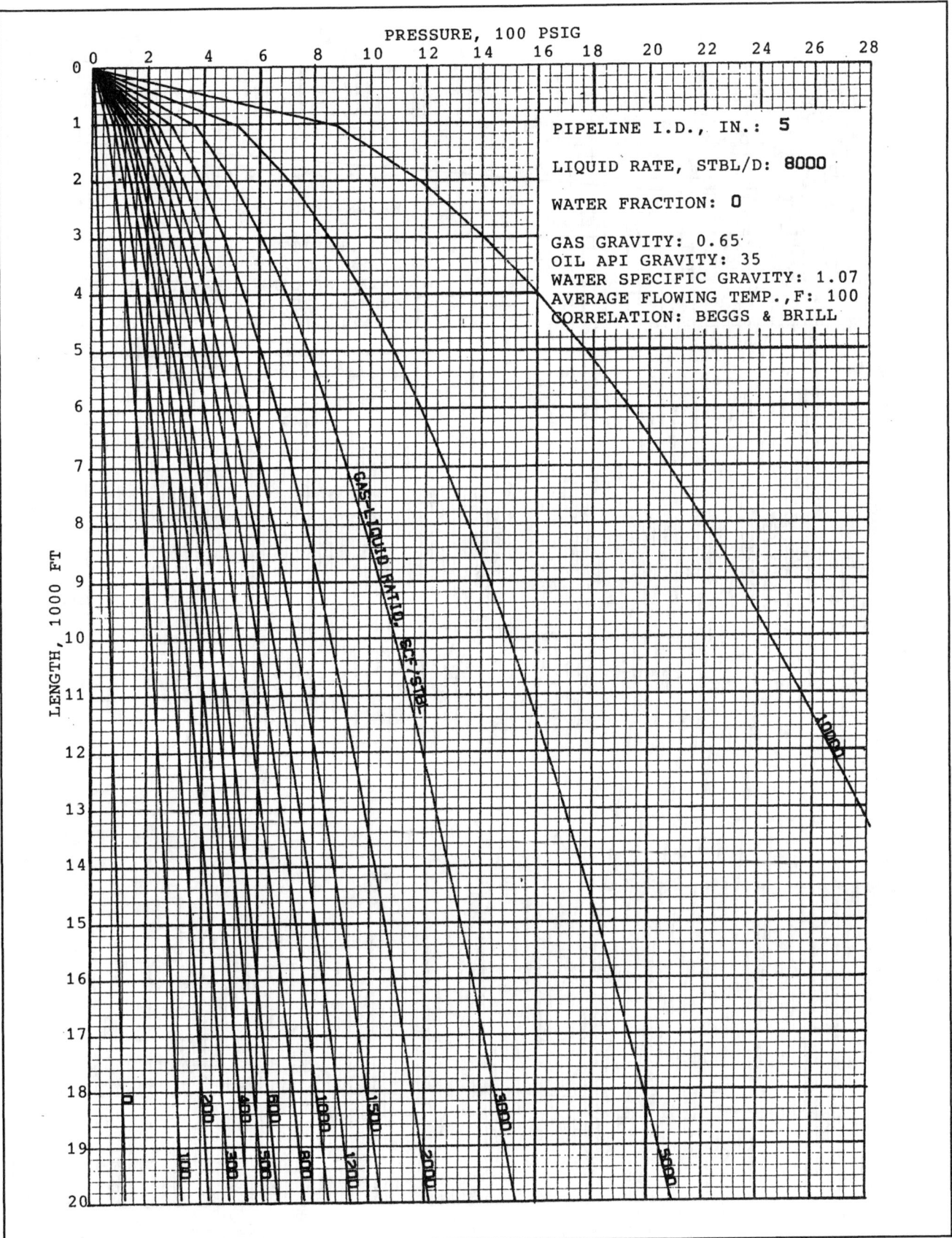

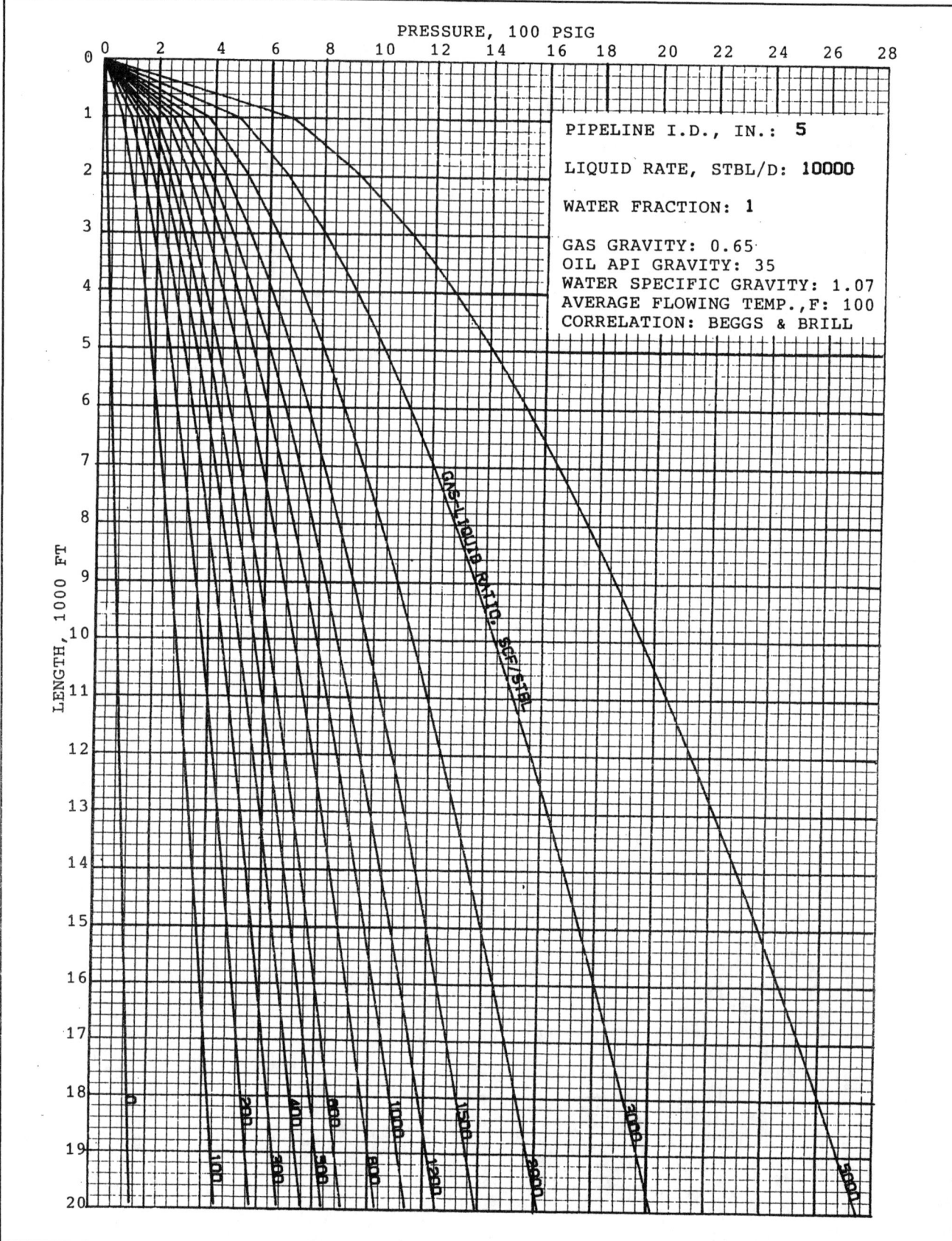

PRESSURE, 100 PSIG

PIPELINE I.D., IN.: **5**

LIQUID RATE, STBL/D: **10000**

WATER FRACTION: **1**

GAS GRAVITY: 0.65
OIL API GRAVITY: 35
WATER SPECIFIC GRAVITY: 1.07
AVERAGE FLOWING TEMP.,F: 100
CORRELATION: BEGGS & BRILL

LENGTH, 1000 FT

GAS-LIQUID RATIO, SCF/STBL

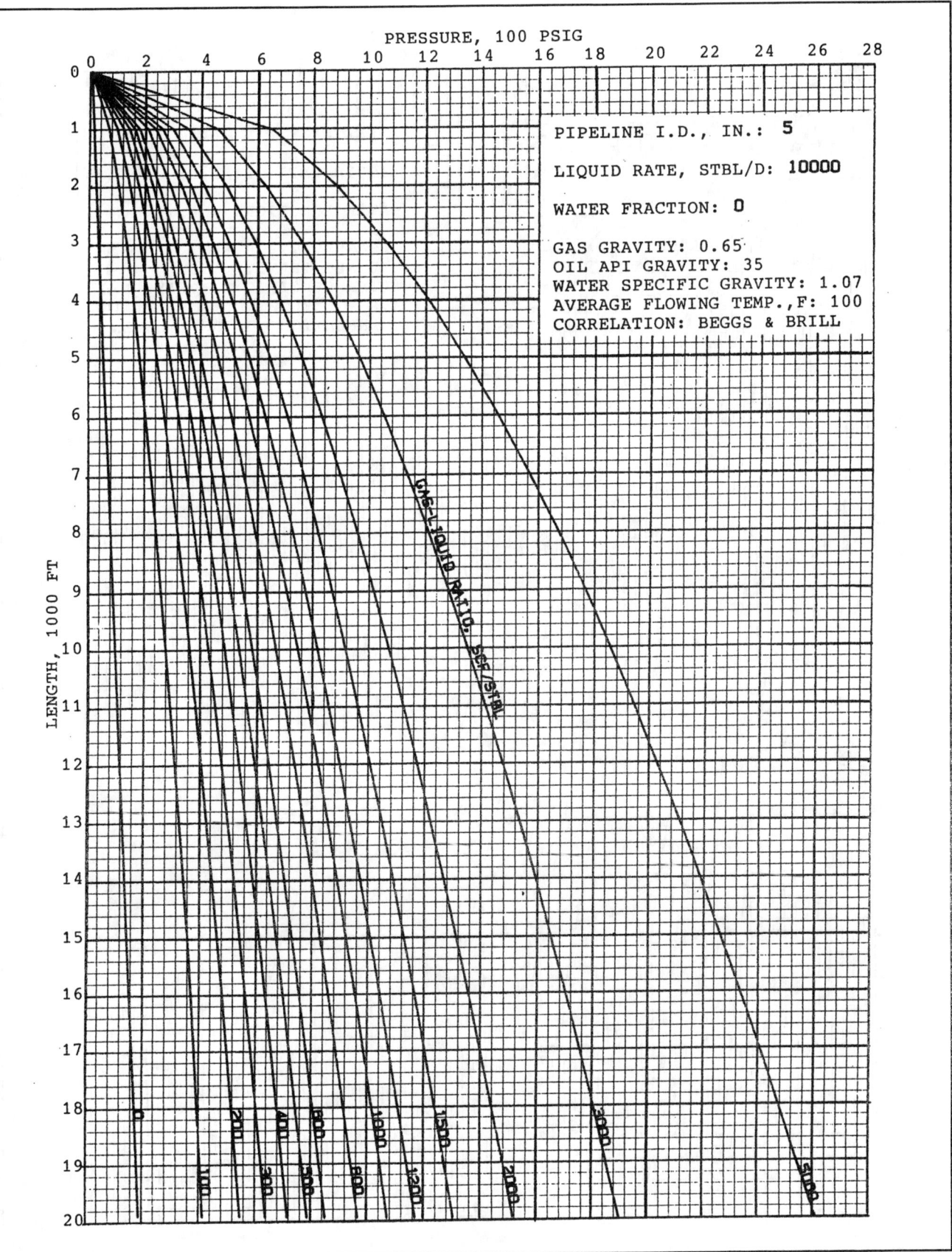

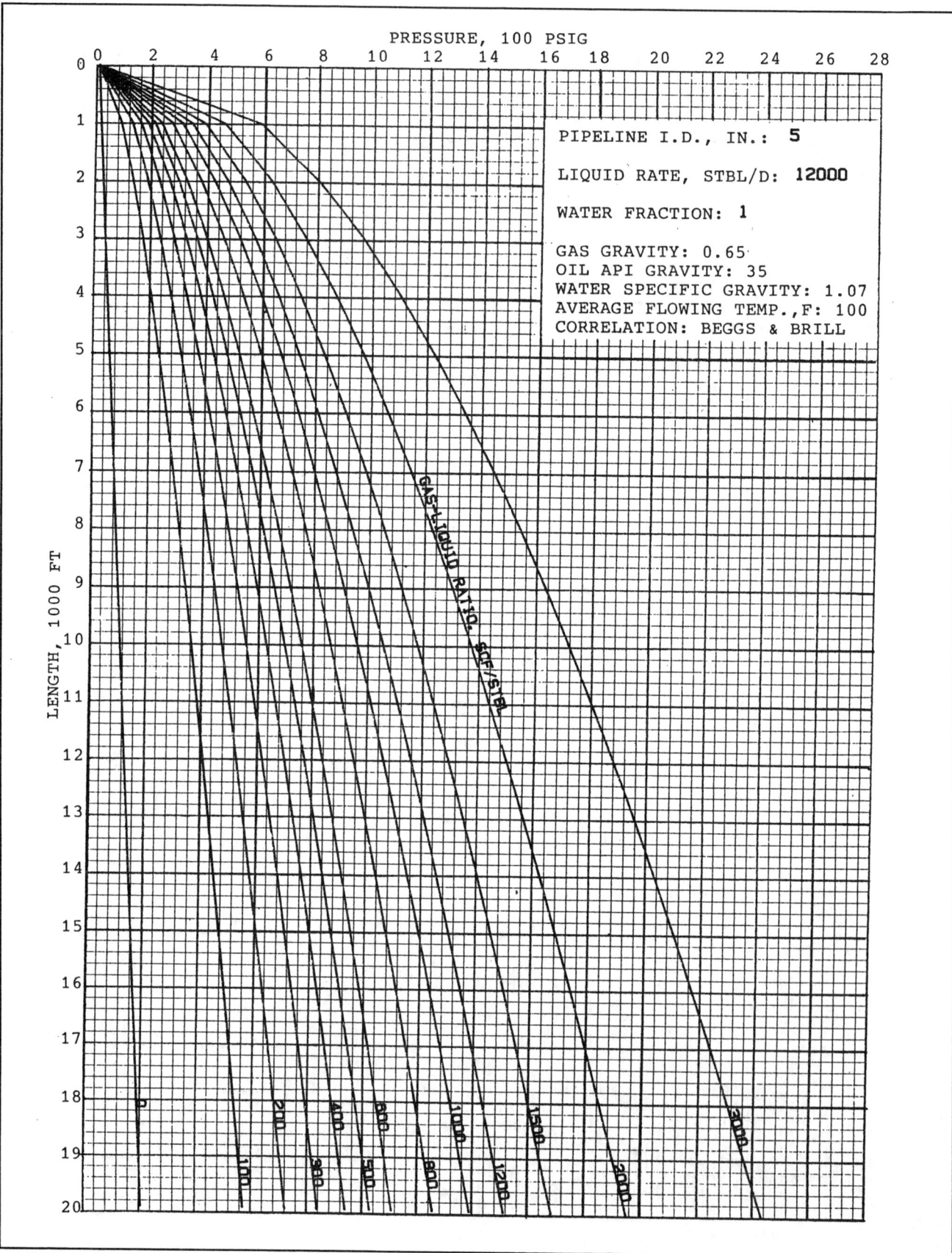

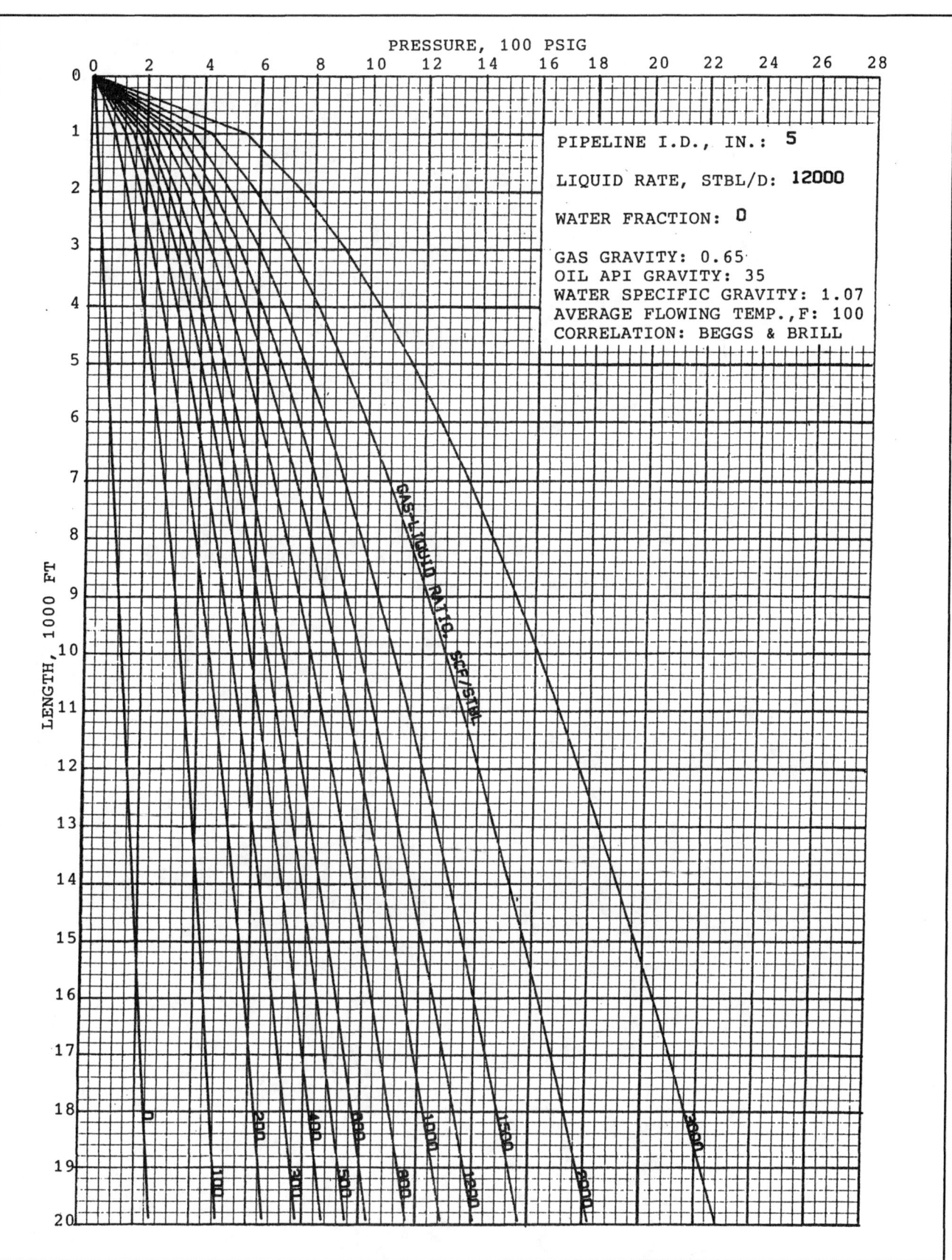

PRESSURE, 100 PSIG

LENGTH, 1000 FT

PIPELINE I.D., IN.: 5

LIQUID RATE, STBL/D: 12000

WATER FRACTION: 0

GAS GRAVITY: 0.65
OIL API GRAVITY: 35
WATER SPECIFIC GRAVITY: 1.07
AVERAGE FLOWING TEMP.,F: 100
CORRELATION: BEGGS & BRILL

GAS-LIQUID RATIO, SCF/STBL

Index

well, 182
well calculation, 182
SI Metric system of units, 123
Side pocket mandrels, 160, 165
Similarity analysis, 111
Single-phase,
   compressible transient flow, 64
   flow, 57, 62, 63, 64, 67, 91, 96, 122, 124,
      128
   fluid, 65, 108
   gas, 123
      flow, 65, 117
   liquid, 126
      flow, 65, 126
   parameters, 71
Sizing lines, 108
Skin, 15
   effect, 14, 15, 45
   factor, 17, 19, 29, 30, 47, 48, 51, 54
      negative, 19
      non-zero, 20
      positive, 19
Slip velocity, 65, 86
Slippage, 65, 95
Slug, 67,87, 109
   catchers, 108
   flow, 65, 88, 108
   pattern, 86,8 7
Small diameter pipelines, 111
Smooth wall pipe, 61
Solution gas, 37, 75, 76, 79
   oil ratio, 37, 76,80, 125
Sonic velocity, 124
Specifications
   for Camco pressure operated gas lift
      valves, 169
   for Otis spreadmaster pressure operated
      gas lift valves, 170
Specified opening pressure, 166
SSSV, 128
Stabilized
   flow, 12, 37
      conditions, 83
   production tests, 39
   test, 32, 43, 44
Standard flow rates, 76
Standing and Katz correlation, 77
Standing's
   graph, 26,28
   method, 40, 48
   modification, 21, 26, 29
Static
   BHP, 163
   fluid levels, 182
   gradient, 160
   liquid level, 172
   pressure, 39, 43
   reservoir pressure, 37, 39,45, 147, 150,
      158, 160
   wellbore pressure, 32
Steady state
   energy balance, 58
   flow, 63, 64
   heat transfer, 71
   laminar flow, 12
Stimulation, 17, 20, 26, 27, 28, 36, 48, 54, 55,
   135, 139, 140

gas wells, 55
Stratification, 67
Stratified
   flow, 67, 109
   formations, 37, 38
   reservoir, 38
Subcritical (subsonic) flow, 123, 124, 125,
   126, 127, 141, 143
Submersible pumps, 174
   assembly, 178
   installation, 178
   schematic, 178
   selection, 179
Submersible pumping, 155, 174
Subsurface
   hydralic motor, 183
   safety valves (SSSV), 123,127, 141, 143
Sucker rod pumping, 153, 155, 177, 183
Suction conditions, 176
Suction pressures, 175
Superficial
   gas velocity, 65, 76, 109, 113
   liquid velocity, 65,76, 109
   velocities, 65, 90
      of liquid phases, 86
      of gas phases, 86
Surface
   casing pressure, 162, 178, 180
   choke, 123, 137, 141
   conditions, 141, 169
   equipment, 36
   flowing temperature, 81, 97
   flowline calculations, 81
   flow rates, 76
   gas injection pressure, 156, 163
   lines, 122
   line sizes, 152
   operating pressure, 157
   pipelines, 90
   pressure, 84, 114, 157
   producing gas/oil ratio, 18
   temperature, 81
   tension, 64, 66
   tubing pressure, 162
Surging fluid, 49,50
System
   analysis, 2, 57, 140, 143, 176, 183
   for wells with restrictions, 141
   plot, 143
   procedures, 133
   capacities, 153
   components, 133, 151
   design, 57
   nodal analysis, 156
   pressure, 109

Tapered tubing string, 136
Temperature, 10, 59, 68, 69, 73, 75, 77, 78,
   79, 80, 82, 96, 97, 129, 165, 175
   change, 57,65
      effects, 72
   Correction Factors, 171
   distribution, 69
   inclination, 68
   loss, 57
   profile, 82
      in pipeline, 82

sensitive valves, 163
Tension,
   interfacial, 66
   oil surface, 66
   water surface, 66
Test rack, 166, 167, 168
   opening pressure, 166, 167, 169
Thermal
   conductivity, 71
   radiation loss, 71
   radiation transfer, 70
Tight wells, 31
Time, 150
   dependent function, 71
   increment, 150
Transient
   flow, 64, 71
   test, 30, 43, 47, 48
   well tests, 36
Transition zone, 86, 88
Traverse curves, 98, 118, 138
   application, 98
True vertical depth (TVD), 97, 98, 157
Tubing, 83, 97, 117, 123, 143, 144, 155, 160,
   162, 174, 178, 183
   design, 111
   diameter, 95, 140
      on minimum rate, 96
   Effect (TB), 166
   Effect Factor (rEP), 166
   flowing pressure traverse, 162
   flowing wellhead pressure, 162
   gradient, 163
   increments, 147
   inside diameter, 100, 128
   length, 104
   outside diameter, 182
   pressure, 160, 163, 165, 166, 168
      at gas lift valve, 157
      drop, 136
      drop calculations, 174
      traverse, 162
   requirements, 158
   size, 3, 4, 5, 36, 85, 93, 95, 100, 141, 146,
      147, 149, 150, 156, 158, 167
      flow capacities, 135
      optimum, 135
      selection, 5, 135
      velocity, 95
Tubing string, 94, 97, 98, 135
Tubing-conveyed
   method, 48
   perforating techniques, 49, 50
Tunnel length, 53
Turbulence, 9, 10, 14, 15, 17, 20, 21, 37, 43,
   45, 53
   coefficient, 14,45,47
   effects, 35, 43, 45, 55
   flow, 20, 52, 60, 61, 64, 71
      calculations, 64
   fluid flow, 60
   pressure drop, 45, 46
   term, 35
Turner, et al., equation, 97
TVD, 104
Twenty percent design gradient method, 163,
   167